$$\begin{array}{cccccc} F & -I & O & N & -N \\ O & O & I & O & O \\ O & & O & O & O \end{array}$$

Calcu... ...

$(A \quad \rightarrow \quad)$ ... calculate λ

Calculate ... ...

$x = a \gamma$

$x_1 = \ldots \gamma + 0.720 \ldots$

$x_2 = \ldots$

... calculate ...

$\rho(a) \rightarrow \rho(\gamma)$

... $\rho(\gamma)$ calculate ...

$D \left[\begin{array}{c} \sqrt{1/q_{11}} \\ \sqrt{1/q_{22}} \end{array} \right]$ ... $y = Dz$

$y_1 = \sqrt{1.399 \cdot z_1}$ ... $q_2 = \sqrt{\ldots}$

Calculate ... z_2

$\rho(\ldots) \rightarrow \rho(\gamma)$ calculate ...

$q = \ldots \frac{}{2 m v^2 \ldots} + x \cdot q \qquad r = QDz$

INTRODUCTION TO OPTIMUM DESIGN

McGraw-Hill Series in Mechanical Engineering

Jack P. Holman, Southern Methodist University
Consulting Editor

Anderson: *Modern Compressible Flow: With Historical Perspective*
Arora: *Introduction to Optimum Design*
Dieter: *Engineering Design: A Materials and Processing Approach*
Eckert and Drake: *Analysis of Heat and Mass Transfer*
Heywood: *Internal Combustion Engine Fundamentals*
Hinze: *Turbulence*
Hutton: *Applied Mechanical Vibrations*
Juvinall: *Engineering Considerations of Stress, Strain, and Strength*
Kays and Crawford: *Convective Heat and Mass Transfer*
Kane and Levinson: *Dynamics: Theory and Applications*
Martin: *Kinematics and Dynamics of Machines*
Phelan: *Dynamics of Machinery*
Phelan: *Fundamentals of Mechanical Design*
Pierce: *Acoustics: An Introduction to Its Physical Principles and Applications*
Raven: *Automatic Control Engineering*
Rosenberg and Karnopp: *Introduction to Physics*
Schlichting: *Boundary-Layer Theory*
Shames: *Mechanics of Fluids*
Shigley: *Kinematic Analysis of Mechanisms*
Shigley and Mitchell: *Mechanical Engineering Design*
Shigley and Uicker: *Theory of Machines and Mechanisms*
Stoecker and Jones: *Refrigeration and Air Conditioning*
Vanderplaats: *Numerical Optimization: Techniques for Engineering Design,*
  with Applications

Also available from McGraw-Hill

Schaum's Outline Series in Mechanical Engineering

Each outline includes basic theory, definitions, and hundreds of solved problems and supplementary problems with answers.

Current List Includes:

Acoustics
Basic Equations of Engineering
Continuum Mechanics
Engineering Economics
Engineering Mechanics, 4th edition
Fluid Dynamics
Fluid Mechanics and Hydraulics
Heat Transfer
Introduction to Engineering Calculations
Lagrangian Dynamics
Machine Design
Mechanical Vibrations
Operations Research
Strength of Materials, 2d edition
Theoretical Mechanics
Thermodynamics

Available at Your College Bookstore

INTRODUCTION TO OPTIMUM DESIGN

Jasbir S. Arora

Professor and Director
Optimal Design Laboratory
College of Engineering
University of Iowa

Boston, Massachusetts   Burr Ridge, Illinois
Dubuque, Iowa   Madison, Wisconsin   New York, New York
San Francisco, California   St. Louis, Missouri

McGraw-Hill

A Division of The **McGraw·Hill** Companies

INTRODUCTION TO OPTIMUM DESIGN

This book was set in Times Roman.
The editors were John Corrigan and John M. Morriss;
the cover was designed by John Hite;
the production supervisor was Louise Karam.
Project supervision was done by The Universities Press.

Printed and bound by Book-mart Press, Inc.

  7 8 9 10 BKM BKM 9 9 8

ISBN 0-07-002460-X

Library of Congress Cataloging-in-Publication Data

Arora, Jasbir S.
    Introduction to optimum design/Jasbir S. Arora.
        p.   cm.
    Bibliography: p.
    Includes index.
    ISBN 0-07-002460-X
    1. Engineering design—Mathematical models.
I. Title. II. Title: Optimum design.
TA174.A76 1989
620'.00425'0724—dc19   88-6526

ABOUT THE AUTHOR

Dr Jasbir S. Arora is a Professor in the departments of Civil and Environmental Engineering and Mechanical Engineering at the University of Iowa. He obtained his Ph.D. degree in Structural Mechanics from the same university in 1971 and has been on the faculty there since 1972. He has taught a variety of undergraduate and graduate courses in his area of specialization. Among them are the courses on engineering design and optimization which he enjoys teaching to undergraduate and graduate students.

Dr Arora is one of the leading authorities in the world on the subject of design optimization, and has developed several new algorithms to solve complex problems. Many of the procedures are currently in use in the industry. He has published over 100 articles in leading national and international journals and conference proceedings. He is a co-author of one of the earliest graduate level books on the subject of applied optimal design, and has also authored or co-authored several design manuals and chapters in books.

Dr Arora is a member of the American Society of Civil Engineers, American Society of Mechanical Engineers, American Institute of Aeronautics and Astronautics, and American Academy of Mechanics. He is also a member of the Electronics Computations Committee of the ASCE where he chairs the Subcommittee on Optimal Design. He is on the Editorial Advisory Board of *Computational Mechanics* and *Structural Optimization*, both international journals. He has organized several sessions and seminars on the subject, and presented invited lectures at national and international conferences. He is a reviewer for several journals and Federal funding agencies, and is consultant to several industrial organizations and laboratories.

TO

Ruhee
Rita
Balwant Kaur
Wazir Singh

CONTENTS

Appendices   521

A   Economic Analysis   521

B   Vector and Matrix Algebra   536

PREFACE

This book is intended for use in a first course on engineering design and optimization. Material for the text has evolved over a period of several years and is based on classroom presentations for an undergraduate core course called *Principles of Design I*. This course covers deterministic design models and a follow-on course called *Principles of Design II* covers probabilistic models and system reliability. In the first course, no prior exposure to design or optimization is assumed. The basic background needed to use the text is vector and matrix algebra, and fundamental vector calculus. The material of the course can be also called mathematical foundations for engineering design. It has been taught to first semester junior students from most engineering disciplines. The students complete all the required mathematics, basic science and engineering science courses by the end of their sophomore year (such as statics, dynamics, deformable bodies, engineering computations, engineering physics, computer graphics, electrical science, engineering mathematics, thermodynamics and fluid mechanics). By introducing the material at the beginning of a student's study of chosen specialization we are emphasizing the use of systematic approaches to the creative process of engineering design. The students then have the opportunity to use optimization techniques in their senior level and graduate design courses as well as independent study projects.

Optimization is really a branch of applied mathematics where it has been researched for some time and a reasonable theory has been developed in the recent past. The subject has been taught at the graduate level in almost all disciplines of engineering. Considerable research has been conducted and continues to be conducted at the present to fully develop the subject for large-scale engineering applications. The *potential of the optimum design methodology has been realized and many industrial organizations have adopted it* as a tool in their design practice. The methodology can be viewed as a means of systematizing the engineering design process. With the use of proper hardware and software, it is of substantial aid in the creative process of conceptual as well as detailed design. With the proper use of these modern tools, the time needed from conception to putting the system into operation can be reduced considerably. It is my sincere belief that methods of optimum design will form a core for the engineering design process. To fully realize this potential of the new methodology, we need to integrate it completely into our undergraduate and graduate curricula and prepare our students to use it effectively in the design process.

The primary purpose of the text is to describe an organized approach to engineering design and its optimization in a rigorous and yet simplified manner. Major emphasis is on the correct formulation of the problem and basic concepts of optimum design. Only a few robust numerical procedures that bring out basic concepts and ideas are presented. It is not the purpose of this text to describe derivations of all the methods and discuss their advantages and disadvantages. References are cited for this purpose. A few modern methods that work well are described and illustrated. The necessary results from optimization theory are stated and their implications are studied through applications to engineering design problems. Theory and concepts of optimum design are explained only through examples and simple engineering applications. Proofs of most of the theorems are omitted. To fully explain the concepts, large numbers of solved problems (over 160) and figures (over 150) are used. *Sections or paragraphs marked with an "*" contain slightly advanced or specialized material and need not be covered in a first course on the subject or in first reading.* They may be covered in senior or graduate level courses.

Many *"what if"* type questions on the subject are fully addressed. Throughout the text, simple design problems involving two to three design variables and three to four constraints are solved in detail to illustrate fundamental concepts and basic ideas. Several simple exercises and challenging project type problems are given for each section. Several of the numerical procedures and concepts described in the text are useful in many other engineering courses and applications.

Virtually any problem for which certain parameters need to be determined to satisfy constraints can be formulated as a design optimization problem. The concepts and methods described in the text are quite general and applicable to all such formulations. Therefore, the range of application of the optimum design methodology is almost limitless, constrained only by the imagination and ingenuity of the user. We describe and discuss the basic concepts and techniques with only a few simple applications. Once they are clearly understood, they can be applied to many other advanced applications that are not discussed in the text.

The material of the text can be also adapted for senior and first-graduate level courses on design optimization. In that case, the pace of material coverage should be accelerated and more advanced and open-ended design problems from the application area should be discussed to meet the needs of the course. The advanced material identified with an "*" has been used, along with other material, in a graduate level course called *Applied Optimal Design*.

The text is a *self-contained* exposition to the subject of design optimization. It contains a large number of examples to illustrate the concepts and can be easily used for *self-learning*. Explanation of the key concepts is kept brief and simple. Professional engineers involved in the design process can adapt optimum design concepts in their work using the material of the text.

Most of the theoretical results given in the text are not new. They are derived from several existing sources. What is new are the emphasis on the designer's viewpoint of the subject, *interactive use of optimization methods* and the use of computer graphics in the engineering design process. Optimization is viewed as a tool for the systematic design of engineering systems. Many of the examples are simple and not related to any particular engineering application. They are used to illustrate important concepts, or to highlight procedures. Therefore, the book can be used for teaching optimum design concepts and procedures in all branches of engineering. To make it suitable for use in the classroom environment, more than 50% of the text is devoted to solved problems and exercises, and the rest of it contains review material, optimization concepts and numerical methods.

The text is organized into eight chapters and four appendices. Synopsis of each section is given at the beginning of each chapter to give the reader an overview of the material. Also, in a few instances, there is repetition of the material to make the text more convenient to read.

The material of the book can be divided roughly into three parts. The first part, consisting of Chapters 1–3, concentrates on design problem formulation, basic concepts and definitions, and optimality conditions that characterize the optimum solution. The second part, consisting of Chapters 4–6, describes numerical search methods for linear, unconstrained and constrained optimization problems. The concept of iterative numerical algorithms is introduced and basic ideas used to construct many algorithms are described. The third part, consisting of Chapters 7 and 8, discusses practical aspects of optimization. Interactive algorithms and their use are described. Several typical engineering design examples are formulated and solved.

Economic considerations have substantial impact on the design of engineering systems. Therefore, an introduction to the subject of *economic analysis* is given in Appendix A. This subject has been generally presented at the beginning of the course, followed by a group project where students learn the value of economic considerations in engineering design decision making.

Vector and matrix algebra – used throughout the text – is a key to the understanding of modern computational methods in engineering design. These topics are thoroughly reviewed in Appendix B, including the solution of a system of equations and eigenvalue problems. Many students get bogged down with the vector and matrix operations and their notation. Therefore, it is critically important to understand thoroughly this material and be comfortable with it. This will help in concentrating on understanding of the optimum design concepts.

In Chapter 1, a basic concept of systems engineering is discussed. The *process of system design,* from the conception to completion, is described. The preliminary design phase is discussed and a clear distinction between analysis and design of a system is given. The role of computers in the design process is explained. Additional notations used throughout the text are defined.

In Chapter 2, formulation of an optimum design problem is discussed in detail. A proper *mathematical statement of the problem* is critical for designing workable systems. In most cases, once the problem has been formulated, user-friendly software is available to solve the problem. Pitfalls in optimum design problem formulation are discussed. The concepts of design variables, cost function, constraint functions, feasible and infeasible designs are described. Graphical solutions and concepts for optimum design problems are presented and discussed.

In Chapter 3, various *optimum design concepts* are described. Global and local minima are defined. Required concepts from vector calculus are reviewed and illustrated. The necessary conditions for unconstrained and constrained problems involving only first derivatives are explained and illustrated. The Kuhn–Tucker optimality conditions for general constrained problems are illustrated with simple design problems. Often, these conditions lead to a nonlinear system of equations that must be solved numerically. *Newton– Raphson* is a method for solving such systems which is described in Appendix C. Also, a sensitivity theorem is explained and used to study the effect of constraint variations on optimum cost function value.

Chapter 4 contains a thorough treatment of *linear programming* methods in optimum design. The two-phase *Simplex method* is developed using a simple example. The notion of duality in linear programming is discussed. Post-optimal analysis for LP problems is described and illustrated.

Chapter 5 contains concepts and numerical methods for *unconstrained optimum design problems.* The concept of iterative numerical procedures is introduced and explained. The one-dimensional minimization problem is addressed. Classical methods, such as the steepest descent and Newton, are described first, then modified Newton, conjugate directions and quasi-Newton methods are given. The methods are illustrated with example problems. Computer programs for some of the methods are given in Appendix D. The chapter also contains an introduction to the techniques of using unconstrained optimization methods to solve constrained problems. These cover the so-called Sequential Unconstrained Minimization Techniques (SUMT) and the modern multiplier methods.

Chapter 6 describes numerical methods for constrained optimization problems. Various fundamental concepts and terms are discussed. The concept of linearization of a nonlinear programming problem is developed. This concept is used in most constrained numerical methods. A sequential linear programming method is described. Then a modern constrained steepest descent method is given and illustrated with simple examples. The method is extended to include approximate Hessian of the Lagrange function for the problem. A brief introduction to three other methods is also given.

The idea of *computer-aided optimum design* is introduced in Chapter 7 and an interactive design optimization environment is described. Interactive design optimization techniques and the role of the designer in the interactive environment are discussed in detail. The use of interactive graphics in the optimum design process is illustrated with examples.

Chapter 8 describes advanced use of optimization techniques. Practical design problems are usually quite complex, requiring considerable analyses and evaluations to come up with a safe and cost-effective final design. Different software components are needed to complete the design task. Integration of software components for optimum design is discussed and a few examples showing advanced application of optimization methodology are discussed.

My attempt has been to keep most examples and exercises simple to put across the fundamental ideas in a straightforward manner. Examples and exercises are taken from several fields in order to make the subject appealing to all engineering disciplines. This is of course a monumental task and I believe that depending on the level and type of the course some augmentation for advanced exercises may be necessary to meet the needs of a particular course.

A project of this magnitude would not be possible without the help and encouragement of many colleagues, associates and students. I learnt considerably from their association and offer my sincere thanks to all. Special thanks go to the following reviewers: Charles Beadle, University of California, Davis; Gary A. Gabriele, Renssaleaer Polytechnic Institute; Raphael Haftka, Virginia Polytechnic Institute; Gary Kinzel, Ohio State University; Charles Mischke, Iowa State University; and Fred Moses, Case Western Reserve University.

The material of the text has improved over a period of several years thanks to the suggestions provided by many students and instructors who endured through several rough drafts. I offer my sincere thanks to all, especially to Professors Goel, Kane, Lance, Liittschwager, Osburn and Trummel who also contributed some homework problems. I welcome additional suggestions and criticism from them and others to further improve the material of the text, for a better understanding of the role of optimization methods in the engineering design process.

I would like to acknowledge special contributions by Drs R. L. Benedict and P. B. Thanedar for their useful discussions on the subject of the text. They gave suggestions for the organization of the text and provided some draft material for it. Invaluable contributions by Gerry Jackson, T. P. Lin, Y. S. Ryu, and C. H. Tseng in formulating and solving several design problems, and assistance in proofreading the material are acknowledged. Also, useful discussions on the subject of optimum design with Professors Ed Haug, B. M.

Kwak, Kyung Choi, Y. S. Ryu, and O. K. Lim, and Drs. S. V. Belsare, J. B. Cardoso, M. Haririan, J. K. Paeng, G. J. Park, T. SreekantaMurthy, J. J. Tsay, and C. C. Wu are acknowledged.

I would like to thank Professors Charles Beadle of the University of California – Davis, Ashok Belegundu of Pennsylvania State University, Gary Gabriele of Renssalaer Polytechnic Institute, Raphael Haftka of Virginia Polytechnic Institute, Gary Kinzel of Ohio State University, Charles Mischke of Iowa State University, Fred Moses of Case Western Reserve University, D. T. Nguyen of Old Dominion University, and Dr. C. C. Hsieh of G.M. Research Laboratories for their careful review of the manuscript and suggestions to improve it. Assistance provided by Professor Jack Holman of Southern Methodist University, and John Corrigan and John Morriss at McGraw-Hill was invaluable in putting the text together.

My special thanks to Anthea Craven and Jane Frank whose help was invaluable in putting the text together. They sweated through endless revisions and additions, keeping their cool without giving up.

Jasbir S. Arora

TO THE TEACHER

The material of the text has been used for a one semester course on *Principles of Design I* at the first semester junior level. However, the text can also be used in the senior and first year graduate level courses with accelerated pace for the material coverage and inclusion of advanced material and examples from the specific area of application. At the junior level students have all the background in *linear algebra and differential calculus* to understand the material of the text. It is assumed that the students have *no prior exposure to the engineering design process,* but they have taken engineering physics, statics and dynamics. Thus they have some idea about analysing given systems. The idea of introducing the Principles of Design I course at the junior level is that the students get exposed to systematic techniques for design and optimization of systems before they go into senior design courses or independent design projects. There, they have an opportunity to use the optimum design techniques.

The *basic idea of the course* is to introduce the design of engineering systems as a systematic and well-organized activity. The methods of optimization are considered as tools that are at the disposal of the designer. Their use as design aids is emphasized. The optimization techniques can be discussed using advanced mathematical analyses. However, use of such mathematics is kept to a minimum and derivations of the theory are avoided. The procedure followed is to discuss the optimization theory with examples only. Various theorems are explained as results without proofs. Use of the theorems, assumptions made in them and their consequences are discussed with examples. Roughly half of the semester is devoted to the topics of economic considerations, optimal design processes, problem formulation, graphical concepts, the review of calculus and matrix algebra, and necessary and sufficiency conditions for unconstrained and constrained optimization problems. The remaining half concentrates on the numerical methods for optimum design, such as the Simplex method for linear problems, one-dimensional minimization, and basic concepts of unconstrained and constrained optimization methods. Most sections of the text contain conceptual exercises having true/false answers. They are carefully designed to force the students to read the appropriate text to answer the questions correctly. They also help to reinforce fundamental concepts and ideas.

Usually three to five group projects from different application areas have been assigned during the semester. They may also be selected from a particular application area, as dictated by the needs of the course. The projects require students to formulate a design problem and solve it using the optimum design methods. These usually involve economic analyses discussed in Appendix A, graphical optimization, use of Kuhn–Tucker necessary conditions for constrained optimization, linear programming, and iterative methods for nonlinear optimization. The projects expose students to the advanced use of optimization methods and computer programs. They have been found to be extremely beneficial. *Many of the projects identified with an "*" have been included as exercises at the end of each section.* Using the given examples (over 160) and exercises (nearly 700), many variations of them can be generated as homework problems.

The course begins with a brief *introduction to the economic considerations* in the design process. *Time value of money and methods of economic comparison given in Appendix A are presented.* The material in Chapter 1, containing an introduction to the process of systematic design of engineering systems, is covered next. The notation used throughout the text is explained. The process of formulating a design problem given in Chapter 2 is described next. This involves transcribing a given problem statement into a mathematical form. This is quite important, so considerable time is spent in discussing proper formulation of the problem. After discussing several sample problems, a *general model for design optimization* is defined and explained. The model has equality as well as inequality constraints. Several *geometrical concepts,* such as feasible/infeasible regions, active/inactive/violated constraints, isocost curves and others are discussed and illustrated. Simple design problems are graphically solved to bring out the fundamental concepts of optimum design.

The *basic vector and matrix algebra* given in Appendix B is reviewed briefly before discussing the optimum design concepts given in Chapter 3. Basic concepts from *vector calculus* contained in Chapter 3 are also reviewed briefly. These include gradient of a function, Hessian matrix, Taylor's expansion, and quadratic forms. The local and global minima of a function are defined and illustrated. The *necessary and sufficient conditions* for a local minimum of a function of several variables are presented and discussed. They are illustrated using simple examples. The necessary conditions for unconstrained or constrained problems can lead to a nonlinear system of equations.

These can be solved using the *Newton–Raphson method* which is discussed in Appendix C. A computer program is used to compute solutions of nonlinear equations. The basic ideas for constrained optimization problems are discussed next. First, the Lagrange theorem for equality constraints is described and illustrated. It is then extended to include inequality constraints. The *Kuhn–Tucker necessary conditions* for a general constrained problem are then described and illustrated with several examples. The question of global optimality is discussed briefly and the idea of convex programming is introduced. Use of the Lagrange multipliers as sensitivity coefficients for the cost function with respect to the constraint variations is described and illustrated. Second-order necessary and sufficient conditions for general constrained problems are usually not presented in the first course.

The basic ideas of *linear programming* are discussed next. After introducing the fundamental concepts and definitions, the Simplex method of linear programming is developed. The method is described in the context of the Gauss–Jordan elimination procedure (described in Appendix B) for solving linear systems of equations. The idea of pivot step is explained and a two-phase Simplex procedure is developed to solve linear programming problems. A computer program is used to solve several problems. The question of *post-optimality analysis* is discussed briefly. Use of the final Simplex tableau to perform *sensitivity analysis* and variation of parameters is explained and illustrated with examples.

Many optimization problems must be solved using *numerical algorithms* because they are difficult to solve by analytical procedures. Chapters 5 and 6 contain such methods for solving unconstrained and constrained optimization problems. The approach followed in presenting this material is to describe the concepts of algorithms without deriving them. The algorithms are then illustrated with examples. Chapter 5 describes *general concepts of iterative algorithms*. These methods start with a trial design and continue to improve it until convergence is obtained. Each improvement involves calculation of a search direction in the design space and a step size along it. Several algorithms, such as *steepest descent, conjugate gradient* and modified *Newton's methods* are discussed to determine the direction vector. The one-dimensional search methods – equal interval and Golden Section – are described and illustrated. Several simple programs based on these algorithms are given in Appendix D. Students use these programs to solve homework problems.

Numerical methods for constrained optimization problems are described in Chapter 6. The approach followed here is to discuss some key concepts and steps used in most algorithms, and illustrate them with examples. Most gradient-based methods *linearize the nonlinear problem* at the current design point using linear Taylor series expansions. The process of linearization is illustrated with examples. The linearized problem can be solved for a change in design using the Simplex method of linear programming. The resulting procedure is usually called *Sequential Linear Programming* or, in short, SLP. The method is illustrated with simple examples and its limitations are pointed out. The other algorithm presented is an extension of the steepest descent

TABLE
Suggested schedule for a junior level undergraduate course

Lec. no.	Topic	Sections
1	Introduction; Time value of money	1.1, 1.2, A.1
2	Economic comparison	A.2, A.3
3	Design process; Notation; Formulation	1.3–1.7
4	Design problem formulation	2.1–2.6
5	Examples (Design Project #1)	2.6, 2.7
6	Graphical solutions	2.7, 2.8
7	Definitions; Math review	3.1, 3.2, B.1–B.3
8	Unconstrained problems; Nec. and suff. conds	B.7, 3.1–3.3
9	Examples; Sol. of linear eqs; Gaussian elim.	3.3, B.4
10	Examples; Newton–Raphson method	3.3, C.1, C.2
11	Constrained problems: Lagrange theorem	3.4
12	Kuhn–Tucker conditions (Design Project #2)	3.5
13	Examples; Constrained nec. and suff. conds	3.4, 3.5
14	Global optimality; Examples	3.6
15	Sensitivity theorem; Examples	3.7
16	Examples; Constrained problems	3.4–3.8
17	Solution of system of equations; Standard LP	B.5, 4.1–4.3
18	LP; Gaussian and Gauss–Jordan procedures	B.4, B.5
19	Simplex method	4.4
20	Simplex method (Design Project #3)	4.4
21	Artificial variables	4.5
22	Significance of the information contained within the final LP tableau	4.6
23	LP examples	4.4–4.6
24	Nonlinear problems; 1D search	5.1–5.3, D.2
25	Golden Section search	5.3, D.3
26	1D examples	5.1–5.3
27	Steepest descent; Conjugate directions	5.4, 5.5, D.4, D.5
28	Conjugate directions, Newton's method	5.5, 5.6
29	Examples; Numerical aspects	5.4–5.6
30	Constrained optimization; Normalization	6.1, 6.2
31	Basic concepts; Linearization (Design Project #4)	6.2, 6.3
32	Linearization, Sequential LP	6.3, 6.4
33	Quadratic Programming (QP) problem	6.5
34	QP; Constrained steepest descent concepts	6.5, 6.6
35	Constrained steepest descent concepts	6.6
36	Constained steepest descent; Examples	6.6
37	Constrained examples	6.1–6.6
38	Interactive design optimization	7.1–7.5
39	Examples; Numerical aspects	7.6, 8.1–8.4
40	Examples; Practical applications	8.5–8.9

method of unconstrained optimization. It is called the *constrained steepest descent* (CSD) method. The method is explained and illustrated with examples. The search direction in the method is determined by solving a quadratic programming (QP) subproblem. The Simplex method of linear programming is extended to solve QP problems. A descent function for the method is defined and a step size determination procedure is illustrated with examples.

Optimum design of realistic systems requires sophisticated computer programs. Interactive graphics and queries can be of tremendous aid in the design optimization process. The role of such a capability is explained in Chapter 7. An *Interactive DESIGN* optimization system called *IDESIGN* is introduced. Capabilities of the system are explained and demonstrated on simple examples. Students solve homework problems as well as a project using the system. Other such systems may also be used.

Chapter 8 contains topics of interest for advanced practical applications. Criteria for selection of an algorithm and a software are discussed. Several optimum design problems are formulated and solved. This chapter is usually not covered in the junior level course.

The Table presented here contains a suggested sequence of material coverage in a first semester junior level course. Each class session is 50 minutes long. There are usually 45 class sessions in a semester. However, the plan shown in the table contains only 40 sessions. The remaining sessions can be devoted to design projects, discussions, laboratories, exams and other material to meet local needs. After each class session, two or three homework problems are usually assigned. The homework is due at the beginning of the next class session. Note that the course is designed to give the students the basic knowledge of design problem formulation, optimization concepts and theory, numerical methods, and their applications. If desired, the plan shown in the Table can be modified to give more emphasis to numerical methods of optimization and their applications, and less emphasis to optimization theory.

After teaching this course since the early 1970s, I have concluded that it is best to present design examples to illustrate optimization theory, especially at the undergraduate level. Each method or theorem should be illustrated only with examples. Also, implications and assumptions of the theorems should be highlighted and explained with examples. It has been determined that with this approach, the students understand the optimum design techniques, and grasp the fundamental concepts quite easily.

Using the material of the text, senior level design courses in specific area of application, technical electives and graduate courses can also be constructed. For senior level courses only a few more applications from the specific area are needed to augment the material of the text. For a graduate level course, the pace of material coverage needs to be accelerated. Also the order of presentation of the material may be reversed, i.e. theory followed by examples. Advanced applications from the field of interest should also be included.

TO THE STUDENT

The goal of many engineers is to design systems for automotive, aerospace, mechanical, civil, chemical, industrial, electrical, biomedical, agricultural, naval and nuclear engineering applications. In the highly competitive world of today, it is no longer sufficient to design a system that performs the required task satisfactorily. It is essential to design the *best system*. "Best" means an efficient, versatile, unique and cost-effective system. To design such systems, proper *analytical, experimental and numerical tools* are needed. Optimum design concepts and methods provide some of the needed tools. When they are properly implemented into a software, they give powerful numerical tools for designing best systems. This is what the present text is all about, i.e. *describes tools for designing the best systems*.

Surprisingly, methods of optimum design are built upon a few very *simple mathematical concepts*. In particular, vector and matrix algebra, and calculus of functions of several variables is all that is needed to develop and understand the methods. *The notations used in the development can be sometimes complex making the subject appear complex. This is, of course, not true.* One needs simply to understand the notation and memorize it. Some general notations and terminologies used throughout the text are explained in Section 1.7. Others are introduced at appropriate places. They must be thoroughly reviewed and understood.

Vector and matrix algebra is reviewed and illustrated in Appendix B. Relevant concepts from vector calculus are reviewed and illustrated in Section 3.2. Students who are not comfortable with this material should review it from time to time. Also, for some students, this may be their first exposure to the subject of the design of systems. Chapters 1 and 2 contain a good introduction to the subject.

The book is a self-contained exposition to the subject of design and optimization. Anyone having a background in vector and matrix algebra and vector calculus should be able to understand the material. Most "*what if*" type questions are addressed. Conceptual exercises having true/false answers are given at the end of most sections. These questions are designed to reinforce basic concepts and must be answered.

It must be realized that the overall process of designing systems in different fields of engineering is roughly the same. Analytical and numerical methods for analyzing the system can vary somewhat. Statement of the design problem can involve terminology that is specific to the particular domain of application. For example, in the fields of structural, mechanical, automotive and aerospace engineering, we are concerned with the integrity of the structure and its components. The performance requirements involve constraints on member stresses, strains, deflections at key points, frequencies of vibration, buckling failure, etc. Thus, we use the terminology that is specific to these fields in describing and formulating the problem. Designers working in the area understand the meaning of the terms and the constraints. Similarly, other fields of engineering have their own terminology to describe the problem. However, *once design problems from different fields have been transcribed into mathematical statements using some standard notation, they all look alike. Methods of optimization* described in the text are quite *general* and can be *used to solve the problems*. The methods can be described and illustrated without reference to any design application. The students must keep this *key point* in mind while studying the text and to get a proper *perspective* of the subject of optimum design. The domain-specific terminology used in some examples *should not* be allowed to *hinder* in understanding the importance of optimization concepts and methods. Simple engineering design applications given throughout the text to motivate the development and use of the concepts and methods should be clearly understood.

CHAPTER
1

INTRODUCTION
TO DESIGN

1.1 INTRODUCTION

Engineering consists of a number of well-established activities, including analysis, design, fabrication, sales, research and the development of systems. The subject of this text – the design of systems – is a major field of the engineering profession. The process of designing and fabricating systems has been developed and used for centuries, and the existence of fine buildings, bridges, highways, automobiles, airplanes, space vehicles and other complex systems is an excellent testimonial. However, evolution of these systems has been slow. The entire process is both time-consuming and costly, requiring substantial human and material resources. Therefore, the procedure has been to design, fabricate and use a system regardless of whether it was the *best one*. Improved systems were designed only after a substantial investment had been recovered. These new systems performed the same or even more tasks, cost less, and were more efficient.

The preceding discussion indicates that several systems can usually accomplish the same task, and that some are better than others. For example, the purpose of a bridge is to provide continuity in traffic from one side to the other. Several types of bridges can serve this purpose. However, to analyze and design all possibilities can be a time-consuming and costly affair. Usually one type is selected and designed in detail.

1

The design of complex systems requires large calculations and data processing. During the last three decades, a revolution in computer technology and numerical computations has taken place. Today's computers can perform complex calculations and process large amounts of data efficiently. The engineering design process benefits greatly from this revolution. Better systems can now be designed by analyzing various options in a short time. This is highly desirable because better designed systems cost less, have more capability, and are easy to maintain and operate.

The design of systems can be *formulated as problems of optimization* where a measure of performance is to be optimized while satisfying all the constraints. In recent years, numerical methods of optimization have been developed extensively. Many of the methods have been used to design better systems. This text describes optimization methods and their applications to the design of engineering systems. Design process is emphasized rather than the optimization theory. Various theorems are stated as results without rigorous proofs. However, their implications from an engineering point of view are studied and discussed in detail. Optimization theory, numerical methods, advanced computer hardware and software can be used as tools to design better engineering systems. The text emphasizes this theme throughout.

Any problem in which certain parameters need to be determined to satisfy constraints can be formulated as an optimum design problem. Once this has been done, the concepts and the methods described in this text can be used to solve the problem. Therefore, the optimization techniques are quite general, having a wide range of applications in diverse fields. The range of applications is limited only by the imagination or ingenuity of design engineers. It is impossible to discuss all the applications of optimization concepts and techniques in this introductory text. However, using simple applications we shall discuss concepts, fundamental principles and basic techniques that can be used in numerous applications. The student should understand them without getting bogged down with the notation, terminology and details of the particular area of application.

The design of a system begins by analyzing different options. Subsystems and their components are identified. They are designed and tested to complete the process. Section 1.2 describes the design process from a systems engineering point of view. The distinction between design and analysis is discussed in Section 1.3. Section 1.4 describes the distinction between conventional and optimum design processes. The role of computers in the design process is described in Section 1.5. In Section 1.6, the distinction between optimum design and optimal control is briefly discussed. Finally, terminologies and notations used throughout the text are defined in Section 1.7.

Chapter 2 describes the *process of formulating a design optimization problem*. A verbal statement of the problem is transcribed into a mathematical one which can be then subjected to further analysis. There are three distinct steps in the process of problem formulation. A simple design example illustrates that. Several other optimum design problems are formulated and a general model for design optimization is described. This model is treated throughout the text. Also, two variable problems are used to explain certain geometrical concepts.

Chapter 3 contains *concepts from optimization theory* for the design of engineering systems. Linear algebra (vectors and matrices), and calculus concepts are needed in various discussions. The relevant material from these fields is reviewed. Unconstrained and constrained optimization problems are discussed. Lagrange theorem for equality constrained problems is discussed and illustrated. It is then extended to problems with inequality constraints. Next, Kuhn–Tucker necessary conditions for constrained optimization problems are developed and illustrated. A constraint variation sensitivity theorem is described which has considerable practical value in the design process. Several design examples are used to illustrate various concepts. Solutions of many optimization problems are the roots of nonlinear equations. Such equations arise when necessary conditions of optimization are written. These equations cannot always be solved analytically, i.e. a closed form solution cannot be obtained. A numerical method for finding the *roots of nonlinear equations* is needed. The Newton-Raphson is such a method which is described in its most rudimentary form in Appendix C. It is illustrated with examples.

Chapter 4 contains a special class of problems having linear functions of design variables. These are called *linear programming problems*. The advantage of this special structure can be realized in the solution process. Basic concepts and theorems of linear programming are given. The Simplex method is developed and illustrated in many examples. Chapter 4 also contains topics of duality and post-optimality analysis.

Methods for solving *unconstrained nonlinear optimum design problems* are described in Chapter 5. The basic concept of an iterative numerical procedure is described. The concept has broad applications and is, therefore, quite important. Steepest descent, Newton's and other numerical procedures are illustrated. Chapter 5 also contains procedures for converting a constrained problem to a series of unconstrained problems. These are called transformation methods. With these procedures, unconstrained optimization methods can be used to solve constrained problems.

Chapter 6 contains various concepts related to the primal methods used in directly solving a *general constrained optimization problem*. In primal methods a sequence of constrained subproblems is solved. Thus, the constrained form of the original problem is retained in each iteration. This philosophy is quite different from the transformation methods discussed in Chapter 5. Some numerical algorithms based on primal philosophy are illustrated.

Chapter 7 introduces the topic of *interactive design optimization*, and interactive algorithms are described and illustrated. The role of interactive graphics and designer interaction during the optimal design process are also discussed and illustrated with examples.

Chapter 8 contains a discussion of *practical considerations in design optimization.* The process of formulating complex practical problems, and the problem of gradient evaluation are discussed. Several topics such as selection of an algorithm, definition of a good algorithm, selection of general optimization software, and interfacing an application to a general purpose software are also discussed. Several interesting design optimization problems are formulated and solved.

There are four appendices at the end of the book. Appendix A contains an introduction to methods of *economic analysis.* These play a direct role in design decision-making. Annual cost and present worth methods of comparison are illustrated. Appendix B describes *vector and matrix operations.* Topics such as determinants, rank of a matrix, solution of linear equations, eigenvalue problems, linear independence of vectors and condition number of a matrix are discussed. Appendix C describes the *Newton-Raphson* method for finding roots of nonlinear equations. These appendices should be reviewed at appropriate times during a course on design optimization. Appendix D contains listings of some simple *computer programs* which can be used to solve unconstrained optimization problems.

1.2   THE DESIGN PROCESS

The process of designing systems results in a set of drawings, calculations and reports, and the system can be fabricated based on these. We shall use a systems engineering model to describe the *design process.* Although a complete discussion on this subject is beyond the scope of the text, some basic concepts will be discussed using a simple block diagram.

Design is an *iterative process.* The designer's experience, intuition and ingenuity are required in the design of systems in most fields of engineering (aerospace, automotive, civil, chemical, industrial, electrical, mechanical, hydraulic, and transportation). *Iterative* implies analyzing several trial systems in a sequence before an acceptable design is obtained. Engineers strive to design the best systems and, depending on the specifications, best can have different connotations for different systems. In general, it implies cost-effective, efficient, reliable and durable systems. The process can involve teams of specialists from different disciplines requiring considerable interaction. The basic concepts are described in the text to aid the engineer in designing systems at the minimum cost and in the shortest time.

The design process should be a well-organized activity. To discuss it, we consider a *system evolution model* shown in Fig. 1.1. The process begins with the identification of a need which may be conceived by engineers or nonengineers. The first step in the evolutionary process is to define precisely specifications for the system. Considerable interaction between the engineer and the sponsor is usually necessary to quantify the *system specifications.* Once these are identified, the task of designing the system can begin. The second important step in the process is to come up with a *preliminary design* of the system. Various concepts for the system are studied. Since this must be done in

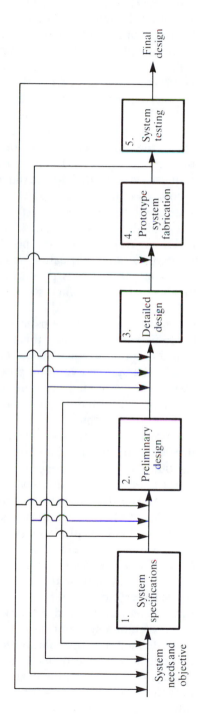

FIGURE 1.1
A system evolution model.

a relatively short time, highly *idealized models* are used. Various subsystems are identified and their preliminary designs estimated. Decisions made at this stage generally effect the final appearance and performance of the system. At the end of the preliminary design phase, a few promising concepts needing further analysis are identified.

The third step in the process is to carry out a *detailed design* for all subsystems. To evaluate various possibilities, this must be done for all the promising concepts identified in the previous step. The design parameters for subsystems must be identified. The parameters must be such that once their numerical values are specified, the subsystem can be fabricated. The design parameters must also satisfy technological and system performance require- ments. Various subsystems must be designed to maximize system worth, or minimize a measure of the cost. Systematic optimization methods can aid the designer in accelerating the detailed design process. At the end of the process, a description of the system is available in the form of reports and drawings.

The last two blocks of Fig. 1.1 may or may not be necessary for all systems. These involve a prototype system fabrication and its testing. The steps are necessary when the system has to be mass produced or human lives are involved. These blocks may appear to be the final steps in the design process. However, they are not, because during tests the system may not perform according to specifications. Therefore, specifications may have to be modified or, other concepts may have to be studied. In fact, this re-examination may be necessary at any step of the design process. This is the reason for having feedback loops at every stage of the system evolution process as shown in Fig. 1.1. The iterative process has to be continued until an acceptable system has evolved. Depending on the complexity of the system, the process may take anywhere from a few days to several months.

The model described above is a simplified block diagram for system evolution. In actual practice, each block may have to be broken down into several sub-blocks to carry out the studies properly and arrive at rational decisions. *The important point is that optimization concepts and methods can help at each stage of the process.* The use of such methods along with proper software can be extremely useful in studying various design possibilities in a short time. The techniques can be of aid for preliminary and detailed design as well as for fabrication and testing. Therefore, in this text, we discuss optimization methods and their use in the design process.

At some stages in the text, it may appear that the design process can be completely automated, the designer can be eliminated from the loop, and optimization methods and programs can be used as black boxes. This may be true in some cases. However, design of a system is a creative process which can be quite complex. It can be ill-defined and the solution to the design problem may not exist. Problem functions may not be defined in certain regions of the design space. Thus, in most practical problems, the designers play a key role in guiding the process to acceptable regions. They must be an integral part of the process and use their intuition and judgement in obtaining the final design.

More details of the interactive design optimization process and the role of the designer are discussed in Chapter 7.

1.3 ENGINEERING DESIGN VERSUS ANALYSIS

It is important to realize differences between *engineering analysis and design activities*. The analysis problem is concerned with determining behavior of an existing system, or a trial system being designed for the given task. Determination of the behavior of the system implies calculation of its response under the specified inputs. Therefore, sizes of various parts and their configurations are given for the analysis problem, i.e. the design of the system is known. On the other hand, the design problem is to calculate sizes and shapes of various parts of the system to meet performance requirements. The design of systems is a trial and error procedure. We estimate a design and analyze it to see if it performs according to the specifications. If it does, we have an acceptable (feasible) design. We may still want to change it to improve its performance. If the trial design does not work, we need to change it to come up with an acceptable system. In both these cases, we must be able to *analyze designs* to make further decisions. Thus analysis capability must be available in the design process.

This book is intended for use in all branches of engineering. It is assumed throughout that students understand analysis methods covered in undergraduate statics, dynamics and physics courses. However, *we will not let the lack of analysis capability hinder the understanding of the systematic process of optimum design* – equations for analysis of the system will be given wherever they are needed.

Considerable progress has been made since 1940 in analyzing engineering systems operating in different environments. It is possible to analyze efficiently complex systems under static and dynamic inputs. Linear and nonlinear systems can be treated. The availability of high-speed digital computers has played a central role in the development of analysis capability. It is now possible to develop similar capabilities for the design of complex systems. Methods of optimization will play a central role in the design process. Therefore, it is important to understand them and the implication of their use in the engineering design process.

1.4 CONVENTIONAL VERSUS OPTIMUM DESIGN PROCESS

It is a challenge for engineers to design efficient and cost-effective systems without compromising their integrity. The conventional design process depends on the designer's intuition, experience and skill. This overwhelming presence of a human element can sometimes lead to dangerous and erroneous

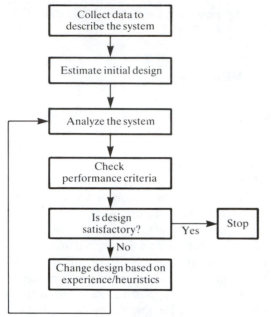

FIGURE 1.2
Conventional design process.

results in the synthesis of complex systems. Figure 1.2 shows the self-explanatory flow chart for a conventional design process that involves the use of information gathered from one or more trial designs together with the designer's experience and intuition.

Scarcity and need for efficiency in today's competitive world has forced engineers to evince greater interest in economical and better designs. With recent advances in computer technology affecting various disciplines of engineering, the design process can hardly remain untouched. Recently, the term computer-aided design optimization (CADO) has been used for summarizing all computer aids in design. Figure 1.3 shows the optimum design process. Design is not only the more or less intuitively guided creation of new information but is comprised of analysis, presentation of results, simulation and optimization. These are essential constituents of an iterative process leading to a feasible and finally optimum design.

Both the conventional and optimum design processes can be used at different stages of system evolution. The main advantage in the conventional design process is that the designer's experience and intuition can go into making conceptual changes in the system or to make additional specifications in the procedure. For example, the designer can choose either a suspension bridge or an arched bridge, add or delete certain components of the structure, and so on. When it comes to detailed design, however, the conventional design process has some disadvantages and difficulties. These difficulties include the treatment of complex constraints (such as limits on vibration frequencies), as well as inputs (for example, when the structure is subjected to a variety of loading conditions). In these cases, the designer would find it difficult to decide

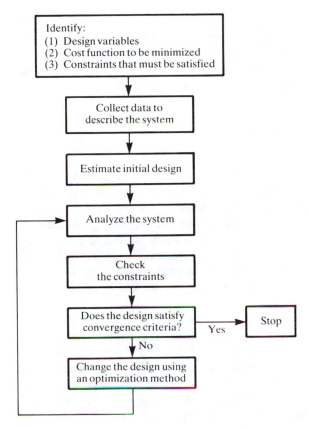

FIGURE 1.3
Optimum design process.

whether to increase or decrease the size of a particular structural element to satisfy the constraints. Furthermore, the conventional design process can lead to uneconomical designs and can involve a lot of calendar time. *The optimum design process forces the designer to identify explicitly a set of design variables, a cost function to be minimized, and the constraint functions for the system.* This rigorous formulation of the design problem helps the designer to gain a better understanding of the problem. Proper mathematical formulation of the design problem is a key to good solutions. The topic is discussed in more detail in Chapter 2.

The foregoing distinction between the two approaches is that the conventional design process is less formal. An objective function that measures the performance of the system is not identified. Trend information is not calculated to make design decisions for improvement of the system. Most of the decisions are made based on the designer's experience and intuition. In contrast, the optimization process is more organized using trend information to make decisions. However, the optimization process can benefit substantially from the designer's experience and intuition. Thus, the best approach would be to have an optimum design process that is aided by the designer's interaction.

It is evident from the foregoing discussion that while the design process can be automated to a certain degree using optimization techniques, it also needs human interaction. In other words, an efficient design process should allow for the designer's creativity to go hand-in-hand with the optimization technique. Interactive design optimization is discussed in more detail in Chapter 7.

1.5 ROLE OF COMPUTERS IN OPTIMUM DESIGN

Engineering systems can be analyzed more accurately using computers. This allows us to understand the behavior of systems more precisely and thus design them more accurately and efficiently. The design process – conventional or optimum – is iterative, requiring the use of the same set of calculations repeatedly. Such repetitive calculations are ideally suited for computer implementation. In addition, each design cycle may require substantial calculations. Therefore, computers play an important role in the design process. They facilitate each step of the design process which will be evident throughout the text.

It can be seen that the amount of data generated in the iterative process can be enormous, which must be presented in a comprehensible form. Graphical representation of data is well suited for this purpose. Computer-generated movies or animations combined with color-coded displays are highly desirable for visualization of complex results. For example, stress concentration in a body can be well depicted by shades of different colors representing various stress levels. In vehicle dynamics, it is possible to animate the vehicle motion on the screen, and thus performance of a design can be simulated before fabrication. On many occasions a number of conceptual and detailed designs can be eliminated using this procedure before they are ever built and tested.

As noted before, the design process can benefit greatly from designer interaction. However, proper interactive aids and graphics must be provided. This means that optimization methods must be implemented in a user-friendly interactive software. Such a software with properly designed user facilities and decision-making capabilities is absolutely necessary for the optimum design of engineering systems. It is evident that proper computer hardware plays a prominent role in the design process. The subject of interactive design optimization and its advantages is discussed in detail in Chapter 7 with examples.

1.6 OPTIMUM DESIGN VERSUS OPTIMAL CONTROL

The optimum design and optimal control of systems are two separate activities. There are numerous applications where methods of optimum design are useful in designing and fabricating systems. There are many other applications where optimal control concepts are needed. In addition, there are some applications where both optimum design and optimal control concepts must be used. Sample applications include *robotics* and *aerospace structures*. In this text, optimal control problems and methods are not described in any detail. However, fundamental differences between the two activities are briefly explained. It turns out that optimal control problems can be transformed to optimum design problems and treated by the methods described in the text. Thus, methods of optimum design are very powerful and should be clearly understood. A simple optimal control problem is described in Chapter 8 and is solved by the methods of optimum design.

The optimal control problem is to find feedback controllers for a system to produce the desired output. The system has active elements that sense fluctuations in the output. System controls get automatically adjusted to correct the situation and optimize a measure of performance. Thus control problems are usually dynamic in nature. In optimum design, on the other hand, we design the system and its elements to optimize an objective function. The system then remains fixed for the entire life. This is the major difference of the two applications.

As an example, consider the cruise control mechanism in passenger cars. The idea of this feedback system is to control fuel injection to maintain constant speed of the car. Thus the system's output is known, i.e. the cruise speed. The job of the control mechanism is to sense fluctuations in the speed and adjust fuel injection accordingly. The amount of fuel injected depends upon road conditions. When the car is going uphill, the fuel injection is greater than when going downhill.

1.7 BASIC TERMINOLOGY AND NOTATION

To understand and be comfortable with the methods of optimum design or modern analysis, familiarity with linear algebra (vector and matrix operations) and basic calculus is essential. Operations of *linear algebra* are described in Appendix B. Students who are not comfortable with that material must review it thoroughly. Calculus of functions of single and multiple variables must also be understood. These concepts will be reviewed wherever they are needed.

In this section, the *standard terminology* and *notations* used throughout the text are defined. It is extremely important to understand these, because without them it will be difficult to follow the rest of the text. The notation defined here is not at all difficult; in fact, it is extremely simple and straight-forward. Anyone with the knowledge of basic calculus and linear algebra will have no difficulty in comprehending it.

1.7.1 US–British Versus SI Units

Concepts and procedures for design problem formulation, and methods of optimization do not depend on the units of measure used. Thus, it does not matter which units are used in defining the problem. However, the final form of some of the analytical expressions for the problem does depend on the units used. In the text, we shall use both US–British and SI units in examples and exercises. Readers unfamiliar with either system of units should not feel handicapped when reading and understanding the text. It is a simple matter to switch from one system of units to the other. To facilitate the conversion from US–British to SI units or vice versa, Table 1.1 gives conversion factors for the most commonly used quantities. For a complete list of the conversion factors, the ASTM [1980] publication can be consulted.

1.7.2   Sets and Points

Since realistic systems generally involve several variables, it is necessary to define and utilize some convenient and compact notation. *Set and vector notations serve this purpose quite well and will be utilized throughout the text. The terms vector and point will be used interchangeably and lower-case letters in boldface will be used to denote them. Upper-case letters in boldface will represent matrices.*

A *point* means an ordered list of numbers. Thus, (x_1, x_2) is a point consisting of the two numbers; $(x_1, x_2, \ldots, x_n)$ is a point consisting of the n numbers. Such a point is often called an n-tuple. Each of the numbers is called a component of the (point) vector. Thus, x_1 is the first component, x_2 is the second, and so forth. The n components $x_1, x_2, \ldots, x_n$ can be collected into a

TABLE 1.1

Conversion factors between US–British and SI units

To convert from US–British	To SI units	Multiply by
Acceleration		
foot/second^2 (ft/s^2)	metre/second^2 (m/s^2)	3.048 000 E – 01*
inch/second^2 (in/s^2)	metre/second^2 (m/s^2)	2.540 000 E – 02*
Area		
foot^2 (ft^2)	metre^2 (m^2)	9.290 304 E – 02*
inch^2 (in^2)	metre^2 (m^2)	6.451 600 E – 04*
Bending moment or torque		
pound force inch (lbf · in)	Newton metre (N · m)	1.129 848 E – 01
pound force foot (lbf · ft)	Newton metre (N · m)	1.355 818
Density		
pound mass/inch^3 (lbm/in^3)	kilogram/metre^3 (kg/m^3)	2.767 990 E + 04
pound mass/foot^3 (lbm/ft^3)	kilogram/metre^3 (kg/m^3)	1.601 846 E + 01
Energy or Work		
British thermal unit (BTU)	Joule (J)	1.055 056 E + 03
foot-pound force (ft · lbf)	Joule (J)	1.355 818
kilowatt-hour (kW h)	Joule (J)	3.600 000 E + 06*
Force		
kip (1000 lbf)	Newton (N)	4.448 222 E + 03
pound force (lbf)	Newton (N)	4.448 222

TABLE 1.1 (*continued*)

To convert from US–British	To SI units	Multiply by
Length		
foot (ft)	metre (m)	3.048 000 E − 01*
inch (in)	metre (m)	2.540 000 E − 02*
mile (mi), US statute	metre (m)	1.609 347 E + 03
mile (mi), international		
nautical	metre (m)	1.852 000 E + 03*
Mass		
pound mass (lbm)	kilogram (kg)	4.535 924 E − 01
slug (lbf · s^2/ft)	kilogram (kg)	1.459 390 E + 01
ton (short, 2000 lbm)	kilogram (kg)	9.071 847 E + 02
ton (long, 2240 lbm)	kilogram (kg)	1.016 047 E + 03
tonne (t, metric ton)	kilogram (kg)	1.000 000 E + 03*
Power		
foot-pound/minute		
(ft · lbf/min)	Watt (W)	2.259 697 E − 02
horsepower (550 ft · lbf/s)	Watt (W)	7.456 999 E + 02
Pressure or stress		
atmosphere (std) (14.7 lbf/in^2)	Newton/metre^2 (N/m^2 or Pa)	1.013 250 E + 05*
one bar (b)	Newton/metre^2 (N/m^2 or Pa)	1.000 000 E + 05*
pound/foot^2 (lbf/ft^2)	Newton/metre^2 (N/m^2 or Pa)	4.788 026 E + 01
pound/inch^2 (lbf/in^2 or psi)	Newton/metre^2 (N/m^2 or Pa)	6.894 757 E + 03
Velocity		
foot/minute (ft/min)	metre/second (m/s)	5.080 000 E − 03*
foot/second (ft/s)	metre/second (m/s)	3.048 000 E − 01*
knot (nautical mi/h),		
international	metre/second (m/s)	5.144 444 E − 01
mile/hour (mi/h),		
international	metre/second (m/s)	4.470 400 E − 01*
mile/hour (mi/h),		
international	kilometre/hour (km/h)	1.609 344*
mile/second (mi/s),		
international	kilometre/second (km/s)	1.609 344*
Volume		
foot^3 (ft^3)	metre^3 (m^3)	2.831 685 E − 02
inch^3 (in^3)	metre^3 (m^3)	1.638 706 E − 05
gallon (Canadian liquid)	metre^3 (m^3)	4.546 090 E − 03
gallon (UK liquid)	metre^3 (m^3)	4.546 092 E − 03
gallon (US dry)	metre^3 (m^3)	4.404 884 E − 03
gallon (US liquid)	metre^3 (m^3)	3.785 412 E − 03
one litre	metre^3 (m^3)	1.000 000 E − 03*
ounce (UK fluid)	metre^3 (m^3)	2.841 307 E − 05
ounce (US fluid)	metre^3 (m^3)	2.957 353 E − 05
pint (US dry)	metre^3 (m^3)	5.506 105 E − 04
pint (US liquid)	metre^3 (m^3)	4.731 765 E − 04
quart (US dry)	metre^3 (m^3)	1.101 221 E − 03
quart (US liquid)	metre^3 (m^3)	9.463 529 E − 04

* An asterisk indicates the exact conversion factor. Also note that scientific notation is used throughout to write the numbers in the exponential form.

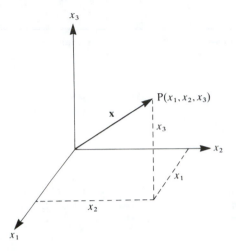

FIGURE 1.4
Vector representation of a point P in 3-dimensional space.

column vector as

$$\mathbf{x} = \begin{bmatrix} x_1 \\ x_2 \\ \vdots \\ x_n \end{bmatrix} = [x_1\, x_2\, \ldots\, x_n]^T \tag{1.1}$$

where the superscript T denotes transpose of a vector or a matrix, a notation that will be used throughout the text (refer to Appendix B for a detailed discussion of vector and matrix algebra). *We shall also use the notation*

$$\mathbf{x} = (x_1, x_2, \ldots, x_n)$$

to represent a point or vector in the n-dimensional space.

In the 3-dimensional space the vector $\mathbf{x} = [x_1\, x_2\, x_3]^T$ represents a point P as shown in Fig. 1.4. Similarly, when there are n components in a vector as in Eq. (1.1), $\mathbf{x}$ is interpreted as a point in the n-dimensional real space denoted as R^n. The space R^n is simply the collection of all n-vectors (points) of real numbers. For example, the real line is R^1, and the plane is R^2, etc.

Often we deal with *sets* of points satisfying certain conditions. For example, we may consider a set of all points having three components with the last component being zero. Denoting the set by S, we can write

$$S = \{\mathbf{x} = (x_1, x_2, x_3) \mid x_3 = 0\} \tag{1.2}$$

Information about the set is contained in braces. Equation (1.2) reads as "S equals the set of all points (x_1, x_2, x_3) with $x_3 = 0$". The vertical bar divides information about the set S into two parts: to the left of the bar is the dimension of points in the set; to the right are special characteristics that distinguish points from those not in the set (e.g. characteristics a point must possess to be in the set S).

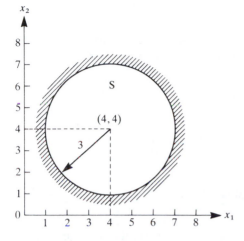

FIGURE 1.5
Geometrical representation for the set $S = \{\mathbf{x} \mid (x_1 - 4)^2 + (x_2 - 4)^2 \leqq 9\}$.

Members of a set are sometimes called *elements*. If a point $\mathbf{x}$ is an element of the set S, then we write $\mathbf{x} \in S$. The expression "$\mathbf{x} \in S$" is read, "$\mathbf{x}$ is an element of (belongs to) S". Conversely, the expression "$\mathbf{y} \notin S$" is read, "$\mathbf{y}$ is not an element of (does not belong to) S".

If all the elements of a set S are also elements of another set T, then S is said to be a *subset* of T. Symbolically, we write $S \subset T$ which is read as, "S is a subset of T", or "S is contained in T". Alternatively, we say T is a superset of S, written as $T \supset S$.

As an example of a set S, we consider a domain of the x_1–x_2 plane enclosed by a circle of radius 3 with the center at the point $(4, 4)$. This is shown in Fig. 1.5. Mathematically, all points within and on the circle can be expressed as

$$S = \{\mathbf{x} = (x_1, x_2) \mid (x_1 - 4)^2 + (x_2 - 4)^2 \leqq 9\} \tag{1.3}$$

Thus the center of the circle $(4, 4)$ is in the set S because it satisfies the inequality in Eq. (1.3). We write this as $(4, 4) \in S$. The origin of coordinates $(0, 0)$ does not belong to the set since it does not satisfy the inequality in Eq. (1.3). We write this as $(0, 0) \notin S$. It can be verified that the following points belong to the set:

$$(3, 3), (2, 2), (3, 2), (6, 6)$$

In fact, set S has an infinite number of points. Many other points are not in the set. It can be verified that the following points are not in the set:

$$(1, 1), (8, 8), (-1, 2)$$

1.7.3 Notation for Constraints

Constraints arise naturally in optimum design problems. For example, material of the system must not fail, demand must be met, resources must not exceed, etc. We shall discuss the constraints in more detail in Chapter 2. Here we discuss the terminology and notations for them.

We have already encountered a constraint in Fig. 1.5. There the set S defines points within and on the circle of radius 3. This is denoted by the following constraint:

$$(x_1 - 4)^2 + (x_2 - 4)^2 \leqq 9$$

A constraint of this form will be called a *less than or equal to type*. It shall be abbreviated as "$\leqq$ type". Similarly, there can be *greater than or equal to type* constraints, abbreviated as "$\geqq$ type".

1.7.4 Superscripts/Subscripts and Summation Notation

Later in the text, we will discuss a set of vectors, components of vectors, and multiplication of matrices and vectors. To write such quantities in a convenient form, consistent and compact notations must be used. We define such a notation here. *Superscripts are used to represent different vectors and matrices.* For example, $\mathbf{x}^{(i)}$ represents the ith vector of a set, and $\mathbf{A}^{(k)}$ represents the kth matrix. *Subscripts are used to represent components of vectors and matrices.* For example, x_j is the jth component of $\mathbf{x}$ and a_{ij} is the i–j element of matrix $\mathbf{A}$. Double subscripts are used to denote elements of a matrix.

To indicate the *range of a subscript* or superscript we use the notation

$$x_i; \qquad i = 1 \text{ to } n \tag{1.4}$$

This represents the numbers $x_1, x_2, \ldots, x_n$. Note that "$i = 1$ to n" represents the range for the index i and is read, "i goes from 1 to n". Similarly, a set of k vectors each having n components will be represented as

$$\mathbf{x}^{(j)}; \qquad j = 1 \text{ to } k \tag{1.5}$$

This represents the k vectors $\mathbf{x}^{(1)}, \mathbf{x}^{(2)}, \ldots, \mathbf{x}^{(k)}$. It is important to note that subscript i in Eq. (1.4) and superscript j in Eq. (1.5) are *free indices*, i.e. they can be replaced by any other variable. For example, Eq. (1.4) can also be written as $x_j; j = 1$ to n and Eq. (1.5) can be written as $\mathbf{x}^{(i)}; i = 1$ to k. Note that the superscript j in Eq. (1.5) does not represent power of $\mathbf{x}$. It is an index that represents the jth vector of a set of vectors.

We shall also use the *summation notation* quite frequently. For example,

$$c = x_1 y_1 + x_2 y_2 + \cdots + x_n y_n \tag{1.6}$$

will be written as

$$c = \sum_{i=1}^{n} x_i y_i \tag{1.7}$$

Also, multiplication of an n-dimensional vector $\mathbf{x}$ by an $m \times n$ matrix $\mathbf{A}$ to obtain an m-dimensional vector **y** is written as

$$\mathbf{y} = \mathbf{Ax} \tag{1.8}$$

Or, in the summation notation, the ith component ($i = 1$ to m) of **y** is

$$y_i = \sum_{j=1}^{n} a_{ij} x_j$$

$$= a_{i1} x_1 + a_{i2} x_2 + \ldots + a_{in} x_n \tag{1.9}$$

There is another way of writing the matrix multiplication of Eq. (1.8). Let m-dimensional vectors $\mathbf{a}^{(i)}$; $i = 1$ to n represent columns of the matrix **A**. Then $\mathbf{y} = \mathbf{Ax}$ is also given as

$$\mathbf{y} = \sum_{j=1}^{n} \mathbf{a}^{(j)} x_j = \mathbf{a}^{(1)} x_1 + \mathbf{a}^{(2)} x_2 + \ldots + \mathbf{a}^{(n)} x_n \tag{1.10}$$

The sum on the right-hand side of Eq. (1.10) is said to be a *linear combination* of columns of the matrix **A** with x_j, $j = 1$ to n as multipliers of the linear combination. Or, **y** is given as a linear combination of columns of **A** (refer to Appendix B for further discussion on the linear combination of vectors).

Occasionally, we will have to use the double summation notation. For example, assuming $m = n$ and substituting for y_i from Eq. (1.9) into Eq. (1.7), we obtain the double sum as

$$c = \sum_{i=1}^{n} x_i \left(\sum_{j=1}^{n} a_{ij} x_j \right) = \sum_{i=1}^{n} \sum_{j=1}^{n} a_{ij} x_i x_j \tag{1.11}$$

The indices of summation i and j in Eq. (1.11) can be interchanged. This is possible because c is a *scalar quantity*, so its value is not affected whether we first sum on i or j. Equation (1.11) can also be written in the matrix form as shown below.

1.7.5 Norm/Length of a Vector

Let **x** and **y** be two n-dimensional vectors. Then their *dot product* is defined as

$$(\mathbf{x} \cdot \mathbf{y}) = \mathbf{x}^T \mathbf{y} = \sum_{i=1}^{n} x_i y_i \tag{1.12}$$

Thus, the dot product is a sum of the product of corresponding elements of the vectors **x** and **y**. Two vectors are said to be *orthogonal* (*normal*) if their dot product is zero, i.e. **x** and **y** are orthogonal if $\mathbf{x} \cdot \mathbf{y} = 0$. If the vectors are not orthogonal, the angle between them can be calculated from the definition of the dot product:

$$\mathbf{x} \cdot \mathbf{y} = \|\mathbf{x}\| \, \|\mathbf{y}\| \cos \theta \tag{1.13}$$

where θ is the angle between vectors **x** and **y**, and $\|\mathbf{x}\|$ represents the length of the vector defined in the following paragraph.

The double sum of Eq. (1.11) can be written in the matrix form as follows:

$$c = \sum_{i=1}^{n} \sum_{j=1}^{n} a_{ij} x_i x_j = \sum_{i=1}^{n} x_i \left(\sum_{j=1}^{n} a_{ij} x_j \right) = \mathbf{x}^T \mathbf{A} \mathbf{x} \tag{1.14}$$

Since $\mathbf{A}\mathbf{x}$ represents a vector, the triple product of Eq. (1.14) will be also written as a dot product:

$$c = \mathbf{x}^T \mathbf{A} \mathbf{x} = (\mathbf{x} \cdot \mathbf{A}\mathbf{x}) \tag{1.15}$$

The *length of a vector* $\mathbf{x}$ will be represented as $\|\mathbf{x}\|$. This is also called *norm of the vector* (for a more general definition of the norm, refer to Appendix B). The length of a vector $\mathbf{x}$ is defined as the square root of the sum of squares of the components, i.e.

$$\|\mathbf{x}\| = \sqrt{\sum_{i=1}^{n} x_i^2} = \sqrt{\mathbf{x}^T \mathbf{x}} = \sqrt{\mathbf{x} \cdot \mathbf{x}} \tag{1.16}$$

1.7.6   Functions

Just as a function of a single variable is represented as $f(x)$, a function of n independent variables $x_1, x_2, \ldots, x_n$ is written as

$$f(\mathbf{x}) = f(x_1, x_2, \ldots, x_n) \tag{1.17}$$

We will deal with many functions of vector variables. To distinguish between functions, subscripts will be used. Thus the ith function will be written as

$$g_i(\mathbf{x}) = g_i(x_1, x_2, \ldots, x_n) \tag{1.18}$$

If there are m functions $g_i(\mathbf{x})$; $i = 1$ to m, these will be represented in the vector form

$$\mathbf{g}(\mathbf{x}) = [g_1(\mathbf{x}) \quad g_2(\mathbf{x}) \ldots g_m(\mathbf{x})]^T \tag{1.19}$$

Throughout the text it is assumed that all functions are continuous and at least twice continuously differentiable. A function $f(\mathbf{x})$ of n variables is called *continuous* at a point $\mathbf{x}^*$ if for any $\varepsilon > 0$, there is a $\delta > 0$ such that

$$|f(\mathbf{x}) - f(\mathbf{x}^*)| < \varepsilon, \tag{1.20}$$

whenever $\|\mathbf{x} - \mathbf{x}^*\| < \delta$. Thus, for all points $\mathbf{x}$ in a small neighborhood of the point $\mathbf{x}^*$, a change in the function value from $\mathbf{x}^*$ to $\mathbf{x}$ is small when the function is continuous. A continuous function need not be differentiable. *Twice-continuous differentiability* of a function implies that it is not only differentiable two times but also its second derivative is continuous. Figures 1.6(a) and (b) show continuous functions. The function shown in Fig. 1.6(a) is differentiable everywhere whereas the function of Fig. 1.6(b) is not differentiable at points x_1, x_2 and x_3. Figures 1.6(c) and (d) show examples of discontinuous functions. As examples, $f(x) = x^3$ and $f(x) = \sin x$ are continuous functions everywhere

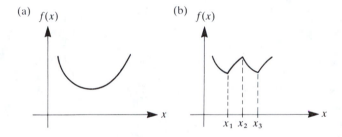

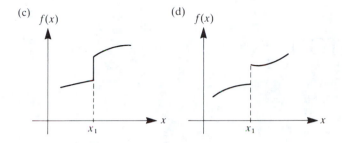

FIGURE 1.6
Continuous and discontinuous functions. (a) Continuous function; (b) continuous function; (c) discontinuous function; and (d) discontinuous function.

and they are also continuously differentiable. However, the function $f(x) = |x|$ is continuous everywhere but not differentiable at $x = 0$.

CHAPTER
2

OPTIMUM DESIGN PROBLEM FORMULATION

2.1 INTRODUCTION

It is a generally accepted fact that the correct formulation of a problem takes roughly 50% of the total effort needed to solve it. Therefore, it is critically important to follow well-defined procedures for formulating design optimization problems. This chapter describes these procedures by considering several design examples.

The design of most engineering systems is a fairly complex process. Many assumptions must be made to develop models that can be subjected to analysis by the available methods and the models must be verified by experiments. Many possibilities and factors must be considered during the problem formulation phase. *Economic considerations* play an important role in designing cost-effective systems. Methods of economic analysis described in Appendix A are useful in this regard. To complete the design of an engineering system, designers from different fields of engineering must usually cooperate. For example, design of a high-rise building involves designers from architectural, structural, mechanical, electrical and environmental engineering as well as construction management experts. Design of a passenger car needs cooperation between structural, mechanical, automotive, electrical, human factors, chemical and hydraulics design engineers. Thus in an *interdisciplinary environment* considerable interaction is needed between various design teams to complete the project. For most applications the entire design project must be broken down into several subproblems which are then treated independently. Each of the subproblems can be posed as a problem of optimum design.

For the most part in this text, *it is assumed that various preliminary analyses have been completed and a detailed design of a concept or a subproblem needs to be carried out.* We shall describe the process of transcribing the subproblem into an optimum design problem. This is consistent with our objective to present concepts and methods of optimum design in the text in a simple and clear manner. Students should bear in mind that considerable analyses usually have to be performed before reaching the final design stage of optimizing the system. This chapter considers several simple and moderately complex applications and formulates them as design optimization problems. More complex and advanced applications are discussed in Chapters 7 and 8.

Formulation of an optimum design problem involves transcribing a verbal description of the problem into a well-defined mathematical statement. The formulation process begins by identifying a set of variables to describe the system, called the *design variables*. Once the variables are given numerical values, we have a design of the system. Whether or not this design works is another question. We shall introduce various concepts to investigate such questions.

All systems are designed to perform within a given set of constraints which include limitations on resources, material failure, response of the system, member sizes, etc. *The constraints must be influenced by the design variables of the system,* because only then can they be imposed. If a design *satisfies all constraints,* we have a *feasible (workable) system.*

A criterion is needed to judge whether or not a given design is better than another. This criterion is called the objective function or cost function. A valid objective function must be influenced by the variables of the design problem, i.e. it must be a function of the design variables. These concepts are further elaborated in the following sections with examples.

The *importance of proper formulation* of a design optimization problem must be clearly understood because the optimum solution will only be as good as the formulation is. For example, if we forget to include a critical constraint in the formulation, the optimum solution will most likely violate it because optimization methods tend to exploit errors or uncertainties in the design models. This is due to the fact that we are trying to optimize the system and, if constraints are not properly formulated, the optimization techniques will take designs into the portion of the design space where either the design is absurd or dangerous. Note also that if we have too many constraints on the system or if they are inconsistent, there may not be any solution to the design problem. Therefore, a careful formulation of the design problem is of paramount importance and proper care should always be exercised in defining and developing expressions for the constraints. In practice, once the problem is properly formulated, good software is usually available to solve it.

The following is an outline of this chapter.

Section 2.2 Design of a Two-bar Truss. The design of a simple two-bar structure is described. It is used as an example in later sections to define design variables, cost function and constraints.

Section 2.3 Design Variables. The first step in the formulation of the design problem is to identify a set of variables that describe the system. These are called design variables. We must be able to change the variables independently to obtain alternate designs; in other words, design variables must be independent of each other as far as possible. For some problems, different sets of variables can be identified to describe the system. The problem formulation depends on the selected set. A simple example is used to illustrate these ideas.

Section 2.4 Cost Function. We must be able to compare different designs of a system. This is needed to label one design as being better than another. The criterion distinguishing alternate designs is called the objective function or the cost function. A valid cost function must be influenced by the design variables of the problem; otherwise, it is not meaningful. A function that is to be minimized is called the cost function.

Section 2.5 Design Constraints. All realistic systems must be designed and fabricated within given resources and performance requirements, i.e. the system must be designed to satisfy all the constraints. For example, structural members should not fail under normal operating loads. Vibration frequencies of a structure must be different from the operating frequency of the machine it supports; otherwise, resonance can occur causing catastrophic failure. Members must fit into available space. All constraints must be expressed in terms of the design variables, because only then do their values change with different designs, i.e. a meaningful constraint must be a function of at least one design variable. Most constraints for a design problem are expressed as inequalities. For example, the actual member stress must not exceed the material allowable stress, maximum deflection must not exceed the specified limit, etc. However, there are some constraints that must be expressed as equalities. For example, we may want deflection at certain points in the system to be exactly Δ, no more or no less. Both types of constraints are illustrated.

Section 2.6 Examples of Optimum Design Problem Formulation. Using the concepts introduced in previous sections, several design problems are considered in this section. A verbal statement of the problem is systematically transcribed into an equivalent mathematical statement for optimization. The formulation process uses three well-defined steps: (i) identification of design variables, (ii) identification of an objective (cost) function and expressing it as a function of the design variables, and (iii) identification of all design constraints and their transcription into mathematical expressions. Many of the examples are used in the subsequent chapters to illustrate various optimization methods.

Section 2.7 A General Mathematical Model for Optimum Design. To describe optimization concepts and methods we need a general mathematical statement for the optimum design problem. Such a mathematical model is defined as minimization of a cost function subject to equality and inequality constraints. The inequality constraints are always transformed as "≦ types". This design problem is treated in the remaining text. All design problems can easily be transcribed into the standard form. The important concept of a feasible region is also introduced.

Section 2.8 Graphical Optimization. Any optimization problem having two design variables can be solved using the graphical method. In this method we plot all the constraint functions on a graph sheet and identify the feasible region. The idea of feasible and infeasible regions is introduced. Any point in the feasible region represents a workable design. To identify the best design, cost function contours must be plotted through the feasible region. It is then possible to visually identify a feasible design having the least cost, and read the optimum design directly from the graph sheet. Examples are used to illustrate these ideas. Most of the geometrical ideas carry over to problems with more variables.

2.2 DESIGN OF A TWO-BAR STRUCTURE

A simple structural design problem is used in the next three sections to illustrate various concepts and considerations that are needed in formulating an optimum design problem. The problem is to design a two-member bracket shown in Fig. 2.1 to support a force W without structural failure. The force is

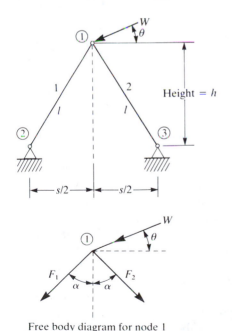

Free body diagram for node 1

FIGURE 2.1
A two-bar structure.

applied at an angle θ which is between 0 and 90°, h is the height and s is the base width for the bracket. Since the brackets will be produced in large quantity, the design objective is to minimize its mass while also satisfying certain fabrication and space limitations

Several formulations for the optimum design problem are possible. In subsequent sections, a material for the bracket with known properties is assumed, although the structure can be optimized with different materials and associated fabrication costs. They can then be compared to select the best possible material for the structure.

In formulating the design problem, *we need to define structural failure more precisely*. Member forces F_1 and F_2 can be used to define failure conditions. To compute member forces, we use the principle of *static equilibrium*. Using the *free-body diagram* for Node 1 (shown in Fig. 2.1), equilibrium of forces in horizontal and vertical directions gives:

$$-F_1 \sin \alpha + F_2 \sin \alpha = W \cos \theta \tag{2.1}$$

$$-F_1 \cos \alpha - F_2 \cos \alpha = W \sin \theta \tag{2.2}$$

From the geometry of Fig. 2.1, $\sin \alpha = s/2l$ and $\cos \alpha = h/l$. Note that forces F_1 and F_2 are shown as tensile in the free-body diagram. Also, *tensile force will be taken as positive*. Thus, the member will be in compression if the force is negative after analysis.

The above two equations are in terms of the unknowns F_1 and F_2. Solving them simultaneously, we obtain

$$F_1 = -0.5Wl \left[\frac{\sin \theta}{h} + \frac{2 \cos \theta}{s} \right] \tag{2.3}$$

$$F_2 = -0.5Wl \left[\frac{\sin \theta}{h} - \frac{2 \cos \theta}{s} \right] \tag{2.4}$$

where l is the length of members given as

$$l = \sqrt{h^2 + (0.5s)^2}$$

2.3   DESIGN VARIABLES

Parameters chosen to describe the design of a system are called the design variables. Once these variables are assigned numerical values, a design of the system is known. These variables are regarded as free because the designer can assign any value to them. If the specified values do not satisfy all constraints of the problem, the design is *not feasible*, which shall be called an *infeasible design*. If the constraints are satisfied, we have a *feasible (workable or usable), design*, i.e. the system fabricated using the design will perform all the stated tasks. A feasible design may not be the best, but it is usable.

An important first step in the proper formulation of the problem is to identify design variables for the system. If proper variables are not selected, the formulation will be either incorrect, or not possible at all. At the initial stage

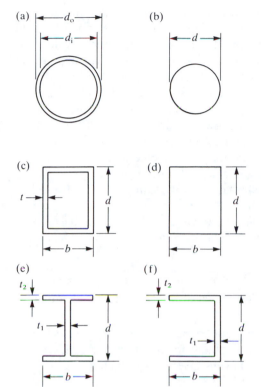

FIGURE 2.2
Cross-sectional shapes for two-bar truss members. (a) Circular tube; (b) solid circular; (c) rectangular tube; (d) solid rectangular; (e) I – section; and (f) channel section.

of problem formulation, all options of identifying design variables should be investigated. Sometimes it is desirable to designate more design variables than may be apparent from the statement of the problem. This gives an added flexibility in the problem formulation. Later, it is possible to assign fixed numerical value to any variable and thus eliminate it from the problem formulation. Another important point should be kept in mind – all design variables should be independent of each other as far as possible. One should be able to assign numerical value to any variable independent of any other variable. It is sometimes possible to have dependent design variables; however, the problem formulation will be unnecessarily complicated. As an example, consider the design of a hollow circular tube shown in Fig. 2.2(a). The inner and outer diameters d_i and d_o and wall thickness t may be identified as design variables, although they are not independent of each other. We cannot specify $d_i = 10$, $d_o = 12$ and $t = 2$ because it violates the physical condition, $t = 0.5(d_o - d_i)$. Therefore, if we formulate the problem with d_i, d_o and t as design variables, we must also impose the constraint, $t = 0.5(d_o - d_i)$. This type of formulation is unnecessary most of the time. We could substitute for t in all equations to eliminate it from the problem, thus reducing the number of design variables as well as constraints.

For the two-bar structure of Section 2.2, several sets of design variables may be identified. The height h and span s can be treated as design variables in the initial formulation. Later, they may be assigned numerical values to eliminate them from the formulation. Other design variables will depend on the cross-sectional shape for members 1 and 2. Several cross-sectional shapes are possible as shown in Fig. 2.2, where design variables for each shape are identified. Note that selection of design variables for many cross-sectional shapes is not unique. For example, in the case of a circular tube of Fig. 2.2(a), the outer diameter d_o and the ratio of inner to outer diameter $r = d_i/d_o$ may be selected as design variables. Or, d_o and d_i may be selected as design variables. However, it is not desirable to designate d_o, d_i and r as design variables because they are not independent. Similar remarks can be made for other cross-sections shown in Fig. 2.2.

We shall represent all the design variables for a problem in the vector **x**.

To summarize, *the following considerations should be given in identifying design variables for a problem:*

1. Design variables should be independent of each other as far as possible.
2. There is a minimum number of design variables required to formulate a design problem properly.
3. It is good to designate as many independent parameters as possible as design variables at the initial design problem formulation phase. Later on, some of the design variables can be always given a fixed value.

2.4 COST FUNCTION

There can be many feasible designs for a system and some are better than others. To make such a claim we must have some criterion to compare various designs. The criterion must be a scalar function whose numerical value can be obtained once a design is specified i.e. it must be a function of the design variables. *Such a criterion is called an objective function for the optimum design problem,* and it is represented by f, or $f(\mathbf{x})$ to emphasize its dependence on design variable vector **x**. The objective function will always be minimized in this text. There is no loss of generality with this treatment, because maximization of $f(\mathbf{x})$ can be transformed into minimization of $-f(\mathbf{x})$. *A function that is to be minimized will be called the cost function.*

Selection of a proper objective function is an important decision in the design process. Several objective functions have been used in the literature – minimize cost, maximize profit, minimize weight, minimize energy expenditure, maximize ride quality of a vehicle, etc. In many situations an obvious function can be identified, e.g. we always want to minimize the cost of manufacturing goods, or maximize return on an investment. In other situations, there may appear to be two or more cost functions. For example, we may want to minimize the weight of a structure and at the same time minimize the deflection or stress at a certain point. These are called *multiobjective design optimization problems.* There is no general and reliable method for solving such optimum design problems. However, several treatments for such situations are possible. For example, a *composite cost function* for the problem can be defined as a weighted sum of all the cost functions. Proper weighting coefficients must be used for various cost functions because no particular cost function should dominate the composite function in locating the optimum design. A second way is to select the most important criterion as the cost function and treat the remaining ones as constraints. By varying the limit values for the constraints, new optimum designs can be obtained. Thus various trade-off curves for all cost functions can be generated and used in the design process.

For many design problems, it is not obvious what the cost function should be and how it should be related to design variables. Considerable insight and experience are needed to identify a proper cost function. For example, consider optimization of a passenger car. What are the design variables for the car? What is the cost function and what is its functional form in terms of design variables? This is a very practical problem; however, it is quite complex. Usually, such problems are divided into several smaller problems and each one is formulated as an optimum design problem. For example, design of the passenger car for a given capacity and performance specifications can be divided into several subproblems – optimization of trunk lid, doors, side panels, roof, seats, suspension system, transmission system, chassis, hood, power plant, bumpers, etc. Each of the problems is now manageable and may be formulated as an optimum design problem. Complex systems from other fields can be also broken down into smaller, more manageable pieces so that each one can be designed more or less independently. The pieces can then be put together to optimize the system.

For the case of the two-bar structure in Section 2.2., mass is identified as the cost function, whose expression is determined by the cross-sectional shape and its design variables. For the case of a circular tube, the design variables are defined as:

x_1 = height h of the truss,

x_2 = span s of the truss,

x_3 = outer diameter of member 1,

x_4 = inner diameter of member 1,

x_5 = outer diameter of member 2, and

x_6 = inner diameter of member 2.

Total mass of the truss is (material volume × density):

$$\text{Mass} = \frac{\pi\rho}{8}(4x_1^2 + x_2^2)^{1/2}(x_3^2 + x_5^2 - x_4^2 - x_6^2) \tag{2.5}$$

where ρ is the mass density. Note that if the outer diameter and ratio of the inner to outer diameter are selected as design variables, the form of the cost function changes. Thus the final form depends on design variables identified for the problem.

Cost function expressions for other cross-sectional shapes in Fig. 2.2 can also easily be written.

2.5   DESIGN CONSTRAINTS

2.5.1   Feasible Design

The design of a system is a set of numerical values for the design variables (i.e. a particular design variable vector **x**). Even if this design is absurd (e.g. negative radius or thickness) or inadequate in terms of its function, it can still be called a design. Clearly, some designs are useful and others are not. A design meeting all the requirements is called a *feasible* (*acceptable* or *workable*) *design*. An *infeasible* (*unacceptable*) *design* does not meet one or more requirements.

2.5.2   Implicit Constraints

All restrictions placed on a design are collectively called *constraints*. Each constraint must be influenced by one or more design variables. Only then is it meaningful and does it have influence on optimum design. Some constraints are quite simple, such as minimum and maximum value of design variables, while more complex ones may be indirectly influenced by design variables. For example, deflection at a point in large structure depends obviously on its design. However, it is impossible to express deflection as an explicit function of the design variables except for very simple structures. These are called *implicit constraints* which we shall treat in more detail in Chapter 8.

2.5.3   Linear and Nonlinear Constraints

Many constraint functions have only first-order terms in design variables. These are called *linear constraints*. *Linear programming problems have only linear constraints,* whereas more general problems have nonlinear constraint functions as well. Therefore, methods to treat both linear and nonlinear constraints must be developed.

2.5.4   Equality and Inequality Constraints

Design problems may have equality as well as inequality constraints. For example, to perform the desired operation, a machine component must move precisely by Δ, so we must treat this as an equality constraint. A feasible design must satisfy precisely all the equality constraints. Also, there are inequality constraints in most design problems. Examples of such constraints are that calculated stresses must not exceed allowable stress of the material,

(a)

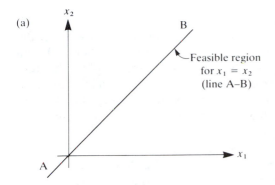

(b)

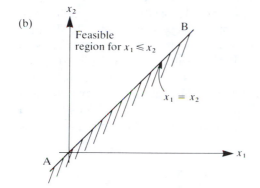

FIGURE 2.3
Distinction between equality and inequality constraints. (a) Feasible region for constraint $x_1 = x_2$ (line A–B); (b) feasible region for constraint $x_1 \leqq x_2$ (line A–B and region above it).

fundamental vibration frequency (an intrinsic dynamic property of a structural or mechanical system) must be higher than the operating frequency, deflections must not exceed specified limits, resources must not be exceeded, demand must be met, load for structure must not exceed the buckling load, etc. Note that there are many feasible designs with respect to an inequality constraint. For example, any design having calculated stress less than or equal to the allowable stress is feasible with respect to that constraint. A large number of designs satisfy this constraint. A feasible design with respect to an equality constraint, however, must lie on its surface. Thus, the *feasible region* for the inequality constraints is much larger than the one for the same constraint expressed as an equality. It is easier to find feasible designs for a system having only inequality constraints.

To further illustrate the difference between equality and inequality constraints, we consider a constraint written in both equality and inequality forms. Figure 2.3(a) shows the equality constraint $x_1 = x_2$. Feasible designs with respect to the constraint must lie on the straight line A–B. However, if the constraint is written as an inequality $x_1 \leqq x_2$, the feasible region is much larger as shown in Fig. 2.3(b). Any point on the line A–B or above it gives a feasible design.

2.5.5 Two-bar Structure

Note that it is extremely important to state and include all constraints in the problem statement, because the final solution depends on them. As an example of formulating design constraints, we consider the two-bar structure of Section 2.2. Constraints for the problem are member stress shall not exceed the allowable stress, and various limitations on design variables shall be met. These constraints will be formulated for hollow circular tubes using the previously identified design variables.

Stress σ in a member is defined as force divided by cross-sectional area (stress = force/area). Its SI units are Newtons/m^2, also called Pascals (Pa). US–British units are pounds/in^2 (written as psi). To avoid over-stressing of a member, the calculated stress σ must be less than or equal to material allowable stress σ_a; i.e. $|\sigma| \leq \sigma_a$, where $\sigma_a > 0$. *Allowable stress is defined as* the material failure stress divided by a factor of safety greater than 1. We may also call it the design stress. Note that to treat positive and negative stresses (tension and compression), we must use absolute value of the calculated stress.

We are now ready to express the stress constraints in terms of design variables. From Eq. (2.3), F_1 (force in member 1) is always negative, so it is compressive. If F_2 is also a compressive force (i.e. $(\sin \theta)/x_1 \geq 2 (\cos \theta)/x_2$ in Eq. (2.4) where $x_1 = h$ and $x_2 = s$ have been used), then member stress constraints are given as ($l^2 = x_1^2 + 0.25x_2^2$):

$$\frac{2Wl}{\pi(x_3^2 - x_4^2)} \left[\frac{\sin \theta}{x_1} + \frac{2 \cos \theta}{x_2} \right] \leq \sigma_a \tag{2.6}$$

$$\frac{2Wl}{\pi(x_5^2 - x_6^2)} \left[\frac{\sin \theta}{x_1} - \frac{2 \cos \theta}{x_2} \right] \leq \sigma_a \tag{2.7}$$

where the following expressions for the cross-sectional areas A_1 and A_2 of members 1 and 2 have been used:

$$A_1 = \frac{\pi}{4}(x_3^2 - x_4^2), \qquad A_2 = \frac{\pi}{4}(x_5^2 - x_6^2)$$

If $(\sin \theta)/x_1 < 2 (\cos \theta)/x_2$ in Eq. (2.4), then F_2 is a tensile force and the stress constraint for member 2 becomes

$$\frac{-2Wl}{\pi(x_5^2 - x_6^2)} \left[\frac{\sin \theta}{x_1} - \frac{2 \cos \theta}{x_2} \right] \leq \sigma_a \tag{2.8}$$

Finally, constraints on design variables are written as

$$x_{il} \leq x_i \leq x_{iu}; \qquad i = 1 \text{ to } 6 \tag{2.9}$$

where x_{il} and x_{iu} are the minimum and maximum values for the ith design variable. These constraints are necessary to impose fabrication and physical space limitations.

Note that the form of member stress constraints changes if different design variables are chosen for circular tubes. For example, inner and outer radii, mean radius and wall thickness, or outside diameter and the ratio of inside to outside diameter as design variables will give different expressions for Eqs (2.6) to (2.8). This shows that the choice of design variables greatly influences the problem formulation. It should also be obvious that if another cross-section is selected for the members, expressions for the cross-sectional area as well as the constraints change.

Note also that we had to first *analyze* the structure (calculate its response to given inputs) to write the constraints properly. We had to calculate forces in the members; only then were we able to write the constraints. This is an important step in any engineering design problem formulation. The designer must be able to analyze the system before he can formulate the problem. We shall further illustrate this point in the next section by formulating several design problems.

Finally, the problem of optimizing two-bar structures can be summarized as follows:

Find design variables x_1, x_2, x_3, x_4, x_5 and x_6 to minimize the cost function of Eq. (2.5) subject to the constraints of Eqs (2.6) to (2.9).

Note that there are 15 constraints for this simple problem.

2.6 EXAMPLES OF OPTIMUM DESIGN PROBLEM FORMULATION

2.6.1 Introduction

In this section several problems are stated and formulated as optimum design problems. Some problems can be formulated in several ways, so different formulations are given for them. The *problem formulation procedure begins by identifying the design variables* which is sometimes the most difficult part of the entire process. One should identify independent variables to describe the design. *A cost function for the problem must then be identified* which measures the performance of the system. The cost function must depend on some or all of the design variables. Different designs for the system can be compared using values of the associated cost function. *The last step in the process is to identify design constraints* which must also depend on some or all of the design variables; only then can they be enforced in the design process.

It should be clear that formulation of an optimum design problem hinges on proper identification of design variables. In addition, the form of all the problem functions depends on them, so the meaning of all the design variables must be clearly given. Careful adherence to the foregoing simple rules can facilitate in the correct formulation of a design problem.

For some of the problems, all functions – cost and constraint – are linear in design variables. These are called *linear programming problems.* Numerical methods for solving such problems are well developed. They are described in Chapter 4.

In summary, the following three steps shall be followed to transcribe a verbal statement of the design problem to mathematical formulation:

1. Identify and define design variables.
2. Identify the cost function and develop an expression for it in terms of design variables.
3. Identify constraints and develop expressions for them in terms of design variables.

2.6.2 Design of a Beer Can

As a first example, we define and formulate a simple problem of designing a can to hold at least the specified amount of beer and meet other design requirements. The cans will be produced in billions, so it is desirable to minimize the cost of manufacturing them. Since the cost can be related directly to the surface area of the sheet metal used, it is reasonable to minimize the sheet metal required to fabricate the can. Fabrication, handling, asthetic, and shipping considerations impose the following restrictions on the size of the can:

1. The diameter of the can should be no more than 8 cm. Also, it should not be less than 3.5 cm.
2. The height of the can should be no more than 18 cm and no less than 8 cm.

The can is required to hold at least 400 ml of fluid (400 ml = 400 cm^3).

Following the three-step procedure to transcribe the problem into a mathematical form, the *two design variables* are defined as D = diameter of the can (cm) and H = height of the can (cm).

The design objective is to minimize the total surface area of the sheet metal which consists of two parts:

1. Surface area of the cylinder of diameter D and height H is circumference × height = πDH (cm^2).
2. Surface area of the two ends: $2(\pi D^2/4) = (\pi/2)D^2$ (cm^2).

Therefore, the *cost function* (total sheet metal) is

$$f(D, H) = \pi DH + \frac{\pi}{2} D^2, \qquad \text{cm}^2$$

The *constraints* must be formulated in terms of design variables. The first constraint is that the can must hold at least 400 cm^3 of fluid. Since volume of the can is given as $\pi D^2 H/4$, this constraint is expressed as

$$\frac{\pi}{4} D^2 H \geqq 400, \qquad \text{cm}^3$$

The other constraints on the size of the can are:

$$3.5 \leqq D \leqq 8; \qquad 8 \leqq H \leqq 18, \qquad \text{cm}$$

The explicit constraints on the design variables have many different names in the literature, such as the *side constraints, technological constraints, simple bounds, sizing constraints,* and *upper and lower limits on design variables.* Note that these are really four constraints for the present problem: $3.5 \leqq D$, $D \leqq 8$, $8 \leqq H$, $H \leqq 18$. Thus the problem has two design variables and five inequality constraints. Note that the cost function and first constraint are nonlinear in design variable whereas remaining constraints are linear.

2.6.3 Saw Mill Operation

A company owns two saw mills and two forests. Table 2.1 shows the capacity of each mill in logs/day and the distances between forests and mills. Each forest can yield up to 200 logs/day for the duration of the project and the cost to transport the logs is estimated at 15 cents/km/log. At least 300 logs are needed each day. Formulate the problem to minimize the cost of transportation of logs each day.

The design variables for the problem are defined as

$x_1 = $ number of logs shipped from Forest 1 to Mill A,
$x_2 = $ number of logs shipped from Forest 2 to Mill A,
$x_3 = $ number of logs shipped from Forest 1 to Mill B, and
$x_4 = $ number of logs shipped from Forest 2 to Mill B.

The next step is to express cost of transportation (which is to be minimized) in terms of the design variables x_1, x_2, x_3 and x_4. The cost depends

TABLE 2.1
Data for saw mill operation

Mill	Distance (km)		Mill capacity/day
	Forest 1	Forest 2	
A	24.0	20.5	240 logs
B	17.2	18.0	300 logs

on the distance of the forests from the mills and is given as

$$\text{cost} = 24(0.15)x_1 + 20.5(0.15)x_2 + 17.2(0.15)x_3 + 18(0.15)x_4$$
$$= 3.6x_1 + 3.075x_2 + 2.58x_3 + 2.7x_4$$

The constraints for the problem are on the capacity of the mills and the yield of the forests. The constraints for mill capacities are expressed as

$$x_1 + x_2 \leq 240 \qquad \text{(Mill A)}$$
$$x_3 + x_4 \leq 300 \qquad \text{(Mill B)}$$

Constraints on the yield of forests are expressed as

$$x_1 + x_3 \leq 200 \qquad \text{(Forest 1)}$$
$$x_2 + x_4 \leq 200 \qquad \text{(Forest 2)}$$

Constraint on the number of logs needed for each day is expressed as

$$x_1 + x_2 + x_3 + x_4 \geq 300$$

For a realistic problem formulation, all the design variables must be nonnegative, i.e.

$$x_i \geq 0; \qquad i = 1 \text{ to } 4$$

The problem has four design variables, five inequality constraints, and four nonnegativity constraints on the variables. Note that all functions of the problem are linear in design variables, so it is a *linear programming problem*. Note also that for a meaningful solution, all the design variables must have integer values. Such problems are called *integer programming problems* which need special methods for their solution. A simple but approximate method for solving the problems is discussed later in Section 2.7.

2.6.4 Design of a Cabinet

A cabinet is assembled from components C_1, C_2 and C_3. Each cabinet requires eight C_1, five C_2 and fifteen C_3 components. Assembly of C_1 needs either five bolts or five rivets; C_2 six bolts or six rivets; and C_3 three bolts or three rivets. The cost of putting a bolt, including the cost of the bolt, is $0.70 for C_1, $1.00 for C_2 and $0.60 for C_3. Similarly, riveting costs are $0.60 for C_1, $0.80 for C_2 and $1.00 for C_3. A total of 100 cabinets must be assembled daily. Bolting and riveting capacities per day are 6000 and 8000, respectively. We wish to determine the number of components to be bolted and riveted to minimize the cost [after Siddall, 1972].

This interesting problem has several formulations. For each formulation proper design variables are identified and expressions for the cost and constraint functions are derived.

2.6.4.1 FORMULATION 1 FOR CABINET DESIGN. For the first formulation the following design variables are identified: For 100 cabinets, let

x_1 = number of C_1 to be bolted,
x_2 = number of C_1 to be riveted,
x_3 = number of C_2 to be bolted,
x_4 = number of C_2 to be riveted,
x_5 = number of C_3 to be bolted, and
x_6 = number of C_3 to be riveted.

The design objective is to minimize the cost of fabricating cabinets which is obtained from the specified costs for bolting and riveting each device:

$$\text{cost} = 0.70(5)x_1 + 0.60(5)x_2 + 1.00(6)x_3 + 0.80(6)x_4 + 0.60(3)x_5 + 1.00(3)x_6$$
$$= 3.5x_1 + 3.0x_2 + 6.0x_3 + 4.8x_4 + 1.8x_5 + 3.0x_6$$

The constraints for the problem consist of the riveting and bolting capacities, and the number of cabinets fabricated every day. Since 100 cabinets must be fabricated every day, the required numbers of C_1, C_2 and C_3 are given in the following constraints:

$$x_1 + x_2 = 800 \quad \text{(number of } C_1\text{'s)}$$
$$x_3 + x_4 = 500 \quad \text{(number of } C_2\text{'s)}$$
$$x_5 + x_6 = 1500 \quad \text{(number of } C_3\text{'s)}$$

The capacity for bolting and for riveting must not be exceeded. Thus

$$5x_1 + 6x_3 + 3x_5 \leqq 6000 \quad \text{(bolting capacity)}$$
$$5x_2 + 6x_4 + 3x_6 \leqq 8000 \quad \text{(riveting capacity)}$$

Finally, all design variables must be nonnegative for a meaningful solution to the problem:

$$x_i \geqq 0; \quad i = 1 \text{ to } 6$$

2.6.4.2 FORMULATION 2 FOR CABINET DESIGN. If we relax the constraint that each component must be either bolted or riveted, then the following design variables can be defined:

x_1 = total number of bolts required for all C_1,
x_2 = total number of bolts required for all C_2,
x_3 = total number of bolts required for all C_3,
x_4 = total number of rivets required for all C_1,
x_5 = total number of rivets required for all C_2, and
x_6 = total number of rivets required for all C_3.

The objective is still to minimize the total cost of fabricating 100 cabinets which is given as

$$\text{cost} = 0.70x_1 + 1.00x_2 + 0.60x_3 + 0.60x_4 + 0.80x_5 + 1.00x_6$$

Since 100 cabinets must be built every day, we require $800\,C_1$, $500\,C_2$ and $1500\,C_3$ components. The total numbers of bolts and rivets needed for all C_1, C_2 and C_3 components are described by the following equality constraints:

$$x_1 + x_4 = 4000 \qquad \text{(for } C_1\text{)}$$
$$x_2 + x_5 = 3000 \qquad \text{(for } C_2\text{)}$$
$$x_3 + x_6 = 4500 \qquad \text{(for } C_3\text{)}$$

Constraints on capacity for bolting and riveting are

$$x_1 + x_2 + x_3 \leqq 6000$$
$$x_4 + x_5 + x_6 \leqq 8000$$

Finally, all design variables must be nonnegative:

$$x_i \geqq 0; \qquad i = 1 \text{ to } 6$$

Thus, this formulation also has six design variables, three equality and two inequality constraints. After an optimum solution for the problem has been obtained, we can decide the number of components to bolt and the number to rivet.

2.6.4.3 FORMULATION 3 FOR CABINET DESIGN. Another formulation of the problem is possible if we require all cabinets to be identical. The following design variables can be identified:

$x_1 =$ number of C_1's to be bolted on one cabinet,
$x_2 =$ number of C_1's to be riveted on one cabinet,
$x_3 =$ number of C_2's to be bolted on one cabinet,
$x_4 =$ number of C_2's to be riveted on one cabinet,
$x_5 =$ number of C_3's to be bolted on one cabinet, and
$x_6 =$ number of C_3's to be riveted on one cabinet.

With the above design variable definitions, the cost of fabricating 100 cabinets each day is given as

$$\text{cost} = 100[5(0.7)x_1 + 5(0.6)x_2 + 6(1.0)x_3 + 6(0.8)x_4$$
$$+ 3(0.6)x_5 + 3(1.0)x_6]$$
$$= 350x_1 + 300x_2 + 600x_3 + 480x_4 + 180x_5 + 300x_6$$

Since each cabinet needs $8\,C_1$, $5\,C_2$ and $15\,C_3$ components, the following equality constraints can be identified:

$$x_1 + x_2 = 8 \qquad \text{(for } C_1)$$
$$x_3 + x_4 = 5 \qquad \text{(for } C_2)$$
$$x_5 + x_6 = 15 \qquad \text{(for } C_3)$$

Constraints on capacity to rivet and bolt are expressed as the following inequalities

$$100(5x_1 + 6x_3 + 3x_5) \leqq 6000$$
$$100(5x_2 + 6x_4 + 3x_6) \leqq 8000$$

Finally, all the design variables must be nonnegative:

$$x_i \geqq 0; \qquad i = 1 \text{ to } 6$$

Note that the cost and constraint functions are *linear* in all the three formulations. Therefore, these are linear programming problems. It is conceivable that the three formulations will yield three different optimum solutions. After solving the problems the designer can select the best strategy for fabricating the cabinets.

Note also that all the formulations have *three equality constraints*, each involving two design variables. We can use these constraints to express three variables in terms of the remaining three and thus reduce the dimension of the problem. This is desirable from a computational standpoint because the numbers of variables and constraints are reduced. However, elimination of variables is not possible for many complex problems and, therefore, we must develop methods to treat both equality and inequality constraints.

Note that for a meaningful solution with these formulations, all design variables must have integer values. These are called *integer programming problems* which occur quite often in the real world. Whereas some numerical procedures have been developed to treat this class of problems, most of them are inexact and based on heuristics. We will not discuss details of such methods in the text. A simple procedure to obtain good feasible solutions will be described in Section 2.7.

2.6.5 Insulated Spherical Tank Design

The goal is to choose insulation thickness t to minimize the cooling cost for a spherical tank. The cooling costs include both the cost of installing and running the refrigeration equipment, and the cost of installing the insulation. Assume a 10-year life, 10% annual interest rate and no salvage value.

The *design variable* for this problem is the insulation thickness, $t(m)$. The surface area of the spherical tank is

$$A = 4\pi r^2, \qquad m^2$$

where $r(m)$ is the radius of the sphere. The insulation costs c_1 dollars per cubic meter. If $t \ll r$, the insulation cost is

$$c_1 At = c_1 4\pi r^2 t$$

The annual heat gain is

$$G = \frac{(365)(24)(\Delta T)A}{c_2 t}, \qquad \text{Watt-hours}$$

where ΔT is the average difference between the internal and external temperatures in Kelvin and c_2 is the thermal resistivity per unit thickness in Kelvin-meter per Watt. The cost of purchasing the refrigeration equipment depends on its capacity and is $c_3 G$, where c_3 is the dollar cost per Watt-hour of capacity. The annual running cost is $c_4 G$, where c_4 is the annual dollar cost per Watt-hour. Therefore, the total cost of cooling the tank over 10 years is

$$f(t) = c_1 4\pi r^2 t + \frac{A}{c_2 t}(c_3 + \text{uspwf}(0.1, 10)c_4)(365)(24)(\Delta T)$$

$$= at + b/t$$

where $a = c_1 4\pi r^2$, $b = (c_3 + \text{uspwf}(0.1, 10)c_4)(365)(24)(\Delta T)A/(c_2)$, and uspwf is the uniform series present worth factor (see Appendix A).

Thus, the design problem is to minimize the total cooling cost with $t \geqq 0$ as the only constraint. Note that in reality t cannot be zero, so the constraint should be expressed as $t > 0$. However, strict inequalities cannot be mathematically or numerically treated. We must allow the possibility of satisfying inequalities as equalities. A more realistic constraint is $t \geqq t_{min}$, where t_{min} is selected to satisfy manufacturing, material failure and other considerations.

2.6.6. Minimum Cost Cylindrical Tank Design

Design a minimum cost cylindrical tank closed at both ends to contain a fixed volume of fluid V. The conceptual design stage indicates a welded cylindrical steel tank. The cost is found to depend directly on the area of sheet metal used.

One possible set of design variables for the problem is the tank radius R and height H. The cost function for the problem is the dollar cost of the sheet metal for the tank. Total surface area of the sheet metal consisting of the end plates and cylinder is given as

$$A = 2\pi R^2 + 2\pi RH$$

Therefore, if c is the dollar cost per unit area of sheet metal, then the cost function for the problem is given as

$$f = c(2\pi R^2 + 2\pi RH)$$

The volume of the tank ($\pi R^2 H$) is required to be V. Therefore,

$$\pi R^2 H = V$$

Also, both design variables must be within some minimum and maximum values:

$$R_{\min} \leqq R \leqq R_{\max}; \qquad H_{\min} \leqq H \leqq H_{\max}$$

This problem is quite similar to the beer can problem in Section 2.6.2. The only difference is in the volume constraint. There the constraint is an inequality and here it is an equality.

2.6.7 Minimum Weight Tubular Column Design

Straight columns as structural elements are used in many civil, mechanical, aerospace, agricultural and automotive structures. Many applications of columns can be observed in daily life, e.g. street light pole, traffic light post, flag pole, water tower support, highway sign post, power transmission poles, etc. It is important to design them as well as possible.

The problem is to design a minimum weight tubular column of length l supporting a load P without buckling or overstressing. The column is fixed at the base and free at the top. This type of structure is called a cantilever column. Buckling load for such a column is given as $\pi^2 EI/4l^2$ (buckling load for a column with other support conditions will be different from this formula [Crandall, Dahl and Lardner, 1978]). Here I is the moment of inertia for the cross-section of the column and E is the material property called modulus of elasticity (Young's modulus). The material stress σ for the column is defined as P/A, where A is the cross-sectional area of the column material. The material allowable stress under axial load is σ_a, and material mass density is ρ (mass per unit volume). Formulate the design problem.

The tubular column and its cross-section are shown in Fig. 2.4. Many

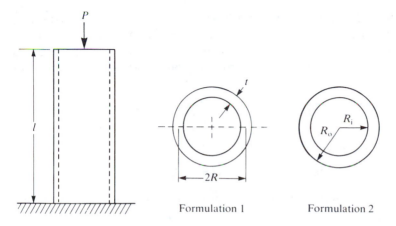

Formulation 1   Formulation 2

FIGURE 2.4
Tubular column.

formulations for the design problem are possible depending on the definition of the design variables. Two formulations are described.

2.6.7.1 FORMULATION 1 FOR COLUMN DESIGN. For the first formulation, the following design variables are defined: R = mean radius of the column and t = wall thickness. Assuming that the column wall is thin $(R \gg t)$, the material cross-sectional area and moment of inertia are

$$A = 2\pi R t; \qquad I = \pi R^3 t$$

The cost function is total mass of the column:

$$\text{mass} = \rho(lA) = 2\rho l\pi R t$$

The first constraint is that the stress (P/A) should not exceed σ_a to avoid material crushing. This is expressed as the inequality $\sigma \leq \sigma_a$. Replacing σ by P/A and then substituting for A, we obtain

$$\frac{P}{2\pi R t} \leq \sigma_a$$

The column should not buckle under the applied load P. Using the given expression for the buckling load, this constraint is expressed as $P \leq \pi^2 EI/4l^2$. Substituting for I, we obtain

$$P \leq \frac{\pi^3 E R^3 t}{4l^2}$$

Finally, both R and t must be within the specified minimum and maximum values:

$$R_{min} \leq R \leq R_{max}; \qquad t_{min} \leq t \leq t_{max}$$

2.6.7.2 FORMULATION 2 FOR COLUMN DESIGN. Another formulation of the design problem is possible, if the following design variables are defined: R_o = outer radius of the column and R_i = inner radius of the column. In terms of these design variables, the material cross-sectional area A and moment of inertia I are

$$A = \pi(R_o^2 - R_i^2); \qquad I = \frac{\pi}{4}(R_o^4 - R_i^4)$$

The total mass of the column is given as

$$\text{mass} = \rho(lA) = \pi \rho l(R_o^2 - R_i^2)$$

As before, the material crushing constraint is $P/A \leq \sigma_a$. Or substituting for A,

$$\frac{P}{[\pi(R_o^2 - R_i^2)]} \leq \sigma_a$$

Using the foregoing expression for I, the buckling load constraint is

$$P \leq \frac{\pi^3 E}{16l^2}(R_o^4 - R_i^4)$$

Finally, the design variables must be within the specified limits:

$$R_{o_{min}} \leq R_o \leq R_{o_{max}} \quad \text{and} \quad R_{i_{min}} \leq R_i \leq R_{i_{max}}$$

When this problem is solved using a numerical method, a constraint $R_o > R_i$ must also be imposed. Otherwise, some methods may take the design to a point where $R_o < R_i$. This situation is physically not possible and must be excluded explicitly for numerical solution of the design problem.

Note that in the second formulation, the assumption of thin-walled sections is not imposed. Thus, optimum solutions with the two formulations can differ. If required, the assumption of thin-walled section, must be explicitly imposed by requiring mean radius to wall thickness ratio to be larger than a constant k:

$$\frac{(R_o + R_i)}{2(R_o - R_i)} \geq k$$

Usually $k \geq 20$ gives reasonable approximation for thin-walled sections.

2.6.8 Minimum Weight Design of a Symmetric Three-bar Truss

As an example of a slightly more complex design problem, consider the three-bar structure shown in Fig. 2.5 [Sun, Arora and Haug, 1975; Haug and Arora, 1979], The structure has been used in numerous investigations since its introduction by Schmit [1960]. The structure is to be designed for minimum volume (or, equivalently, minimum mass) to support a force P. It must satisfy various performance and technological constraints, such as member crushing, member buckling, failure by excessive deflection of Node 4, and failure by

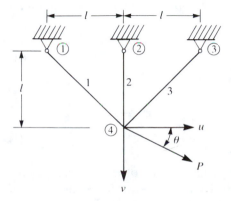

FIGURE 2.5
Three-bar truss.

resonance when natural frequency of the structure is below a given threshold. This is a slightly more complicated design problem than the previous ones. The structure is statically indeterminate. We need to use advanced analysis procedures to determine member forces, nodal displacements and the natural frequency.

The structure must be symmetric. Therefore, the following design variables are defined:

A_1 = cross sectional area of material for members 1 and 3, and

A_2 = cross sectional area of material for member 2.

Other design variables for the problem are possible depending on the cross-sectional shape of members, as shown in Fig. 2.2. The relative merit of any design for the problem is measured in its material volume. Therefore, total material volume for the structure (volume of a member = area × length) serves as a cost function,

$$\text{volume} = l(2\sqrt{2}\,A_1 + A_2) \tag{a}$$

where l is defined in Fig. 2.5.

To define constraint functions for the problem, we must calculate stresses, deflections and fundamental natural frequency for the structure. Using analysis procedures for statically indeterminate structures, horizontal and vertical displacements u and v of the Node 4 of the truss are

$$u = \frac{\sqrt{2}\,l P_u}{A_1 E} \tag{b}$$

$$v = \frac{\sqrt{2}\,l P_v}{(A_1 + \sqrt{2}\,A_2)E} \tag{c}$$

where E is the modulus of elasticity for the material, and P_u and P_v are horizontal and vertical components of the load P:

$$P_u = P \cos \theta; \qquad P_v = P \sin \theta$$

Stresses σ_1, σ_2 and σ_3 in members 1, 2 and 3 under the load P are computed from member forces as

$$\sigma_1 = \frac{1}{\sqrt{2}} \left[\frac{P_u}{A_1} + \frac{P_v}{(A_1 + \sqrt{2}\,A_2)} \right] \tag{d}$$

$$\sigma_2 = \frac{\sqrt{2}\,P_v}{(A_1 + \sqrt{2}\,A_2)} \tag{e}$$

$$\sigma_3 = \frac{1}{\sqrt{2}} \left[\frac{P_v}{(A_1 + \sqrt{2}\,A_2)} - \frac{P_u}{A_1} \right] \tag{f}$$

Many structures support moving machinery and other dynamic loads. These structures vibrate with a certain frequency known as the *natural frequency*. This is an intrinsic dynamic property of a structural system. There can be several modes of vibration each having its own frequency. *Resonance* causes catastrophic failure of the structure which occurs when any vibration frequency coincides with the frequency of the operating machinery. Therefore, it is reasonable to demand that no structural frequency be close to the frequency of the operating machinery. The mode of vibration corresponding to the lowest natrual frequency is important because that mode is excited first. It is important to make the lowest (fundamental) natural frequency of the structure as high as possible to avoid any possibility of resonance. This also makes the structure stiffer. Frequencies of a structure are obtained by solving an eigenvalue problem involving stiffness and mass properties of the structure. The lowest eigenvalue ζ related to the lowest natural frequency of the symmetric three-bar truss is computed using a consistent mass model as (Sun *et al.*, 1975):

$$\zeta = \frac{3EA_1}{[\rho l^2 (4A_1 + \sqrt{2}\,A_2)]} \tag{g}$$

where ρ is the material mass density/volume. This completes analysis of the structure. We can now develop various constraint expressions.

The structure is designed for use in two applications. In each application, it supports different loads. These are called loading conditions for the structure. In the present application, a symmetric structure would be obtained if the following two loading conditions are considered for the structure. First load is applied at an angle θ and the second one at an angle $(\pi - \theta)$, where angle θ is shown in Fig. 2.5. If we let member one be the same as member three, then the second loading condition can be ignored. Therefore, we consider only one load applied at an angle θ ($0 \leq \theta \leq 90$).

Note from Eqs (d) and (f) that σ_1 is always larger than σ_3. Therefore, we need to impose constraints on only σ_1 and σ_2. If σ_a is an allowable stress for the material, then the stress constraints ($\sigma_1 \leq \sigma_a$ and $\sigma_2 \leq \sigma_a$) are

$$\frac{1}{\sqrt{2}} \left[\frac{P_u}{A_1} + \frac{P_v}{(A_1 + \sqrt{2}\,A_2)} \right] \leq \sigma_a \tag{h}$$

and

$$\frac{\sqrt{2}\,P_v}{(A_1 + \sqrt{2}\,A_2)} \leq \sigma_a \tag{i}$$

Horizontal and vertical deflections of node 4 must be within the specified limits Δ_u and Δ_v, respectively ($u \leq \Delta_u$, $v \leq \Delta_v$). From Eqs (b) and (c), the deflection constraints are

$$\frac{\sqrt{2}\,l P_u}{A_1 E} \leq \Delta_u \tag{j}$$

and

$$\frac{\sqrt{2}\,l P_v}{E(A_1 + \sqrt{2}\,A_2)} \leq \Delta_v \tag{k}$$

As discussed above, the fundamental natural frequency of the structure should be higher than a specified frequency ω_0 Hertz (Hz). This constraint can be written in terms of the lowest eigenvalue for the structure. The eigenvalue corresponding to a frequency of ω_0 (Hz) is given as $(2\pi\omega_0)^2$. The lowest eigenvalue ζ for the structure should be higher than $(2\pi\omega_0)^2$. Therefore, from Eq. (g), the frequency constraint becomes

$$\frac{3EA_1}{[\rho l^2(4A_1 + \sqrt{2}\,A_2)]} \geqq (2\pi\omega_0)^2 \tag{l}$$

To impose buckling constraints for members under compression, the dependence of moment of inertia I on the cross-sectional area of the members must be specified. A form with quite general applicability is $I = \beta A^2$, where A is the cross-sectional area and β is a nondimensional constant. This relation follows if the shape of the cross-section is fixed and all its dimensions are varied in the same proportion. The axial force for the ith member is given as $F_i = A_i\sigma_i$, where $i = 1, 2, 3$ with tensile force taken as positive. Members of the truss are considered columns with pin ends. Therefore, buckling load for the ith member is given as $\pi^2 EI_i/l_i^2$, where l_i is the length of the ith member [Crandall, Dahl and Lardner, 1978]. Buckling constraints are expressed as $-F_i \leqq \pi^2 EI_i/l_i^2$, where $i = 1, 2, 3$. Negative sign for F_i is used to make the left-hand side of the constraints positive when the member is in compression. Also, there is no need to impose buckling constraints for members in tension. With the foregoing formulation, the buckling constraint for tensile members is automatically satisfied. Substituting various quantities, member buckling constraints are

$$-\frac{1}{\sqrt{2}}\left[\frac{P_u}{A_1} + \frac{P_v}{(A_1 + \sqrt{2}\,A_2)}\right] \leqq \frac{\pi^2 E\beta A_1}{2l^2} \tag{m}$$

$$-\frac{\sqrt{2}\,P_v}{(A_1 + \sqrt{2}\,A_2)} \leqq \frac{\pi^2 E\beta A_2}{l^2} \tag{n}$$

$$-\frac{1}{\sqrt{2}}\left[\frac{P_v}{(A_1 + \sqrt{2}\,A_2)} - \frac{P_u}{A_1}\right] \leqq \frac{\pi^2 E\beta A_1}{2l^2} \tag{o}$$

Note that the buckling load on the right-hand side has been divided by the member area in the foregoing expressions. Also, constraints of Eqs (m) and (n) are automatically satisfied, since both the members are always in tension (forces in them are always positive for the direction of load shown in Fig. 2.5).

Finally, A_1 and A_2 must be both non-negative, i.e. $A_1, A_2 \geqq 0$. Most practical design problems would require each member to have a certain minimum area, A_{min}. The minimum area constraints can be written as

$$A_1, A_2 \geqq A_{min} \tag{p}$$

The optimum design problem then is to find cross-sectional areas $A_1, A_2 \geqq A_{min}$ to minimize volume of Eq. (a) subject to the constraints of Eqs

(h) to (p). This small-scale problem has 10 inequality constraints and 2 design variables.

2.7 A GENERAL MATHEMATICAL MODEL FOR OPTIMUM DESIGN

2.7.1 Design Optimization Model

In the previous section, several design problems were formulated. All problems have a cost function which can be used to compare various designs of the system. Most design problems must also satisfy certain constraints. Some design problems have only inequality constraints, others have only equality constraints, and some have both inequalities and equalities. We can define a general mathematical model for optimum design to encompass all the possibilities. A *standard form of the model* that is treated throughout the text is first stated. Then transformation of various problems into the standard form is explained.

2.7.1.1 STANDARD DESIGN OPTIMIZATION MODEL. The standard design optimization model is defined as follows: Find an n-vector $\mathbf{x} = (x_1, x_2, \ldots, x_n)$ of design variables to minimize a cost function

$$f(\mathbf{x}) = f(x_1, x_2, \ldots, x_n) \tag{2.10}$$

subject to the p equality constraints

$$h_j(\mathbf{x}) \equiv h_j(x_1, x_2, \ldots, x_n) = 0; \qquad j = 1 \text{ to } p \tag{2.11}$$

and the m inequality constraints

$$g_i(\mathbf{x}) \equiv g_i(x_1, x_2, \ldots, x_n) \leq 0; \qquad i = 1 \text{ to } m \tag{2.12}$$

where p is the total number of equality constraints and m is the total number of inequality constraints.

Note that the simple bounds on design variables such as $x_i \geq 0$, $i = 1$ to n, or $x_{il} \leq x_i \leq x_{iu}$, $i = 1$ to n where x_{il} and x_{iu} are the smallest and largest allowed value for x_i, are included in the inequalities of Eq. (2.12). In numerical methods, these constraints can be treated more efficiently in the original form without converting them to the form of Eq. (2.12). However, in discussing the basic concepts and theory, we shall assume that they have been included in the inequalities of Eq. (2.12).

Design optimization problems from different fields of engineering can be transcribed into the standard model. For example, using the standard notation, all the problems formulated in Section 2.6 can be transformed into the form of Eqs (2.10) to (2.12). Therefore, the standard model is quite general. *It is important to note that once design problems from different fields are transcribed into the standard model, they all look alike. Therefore, same solution strategies, as described in this text can be used. Thus the concepts and methods contained in the text are applicable in many fields.*

2.7.1.2 OBSERVATIONS ON THE STANDARD MODEL. Several points must clearly be understood about the standard model:

1. First of all, it is obvious that the functions $f(\mathbf{x})$, $h_j(\mathbf{x})$ and $g_i(\mathbf{x})$ must depend on some or all of the design variables. Only then are they valid for the design problem. Functions that do not depend on any variable have no relation to the problem and can be safely ignored.

2. *The number of independent equality constraints must be less than or at the most equal to the number of design variables,* i.e. $p \leq n$. When $p > n$, we have an *overdetermined system* of equations. In that case, either there are some *redundant equality constraints* (linearly dependent on other constraints), or the *formulation is inconsistent.* In the former case, redundant constraints should be deleted and, if $p < n$, the optimum solution for the problem is possible. In the latter case, no solution for the design problem is possible and the designer should closely re-examine the formulation. When $p = n$, no optimization of the system is necessary because solutions of the equality constraints are the only candidates for optimum design. These solutions can be obtained using an appropriate method of solving the equations.

3. *Note that all inequality constraints in Eq. (2.12) are written as "≤ 0".* This is standard practice throughout the text. In the example problems of Section 2.6, we encountered "$\leq$ type" as well as "$\geq$ type" constraints. "$\leq$ type" constraints can be converted to the standard form of Eq. (2.12) by transferring the right-hand side to the left-hand side. "$\geq$ type" constraints can also be transformed to the "$\leq$ form" quite easily by multiplying them by -1 as explained later in Section 2.7.3. Note, however, that while there is a restriction on the number of independent equality constraints, *there is no restriction on the number of inequality constraints.* Some inequalities may be strictly satisfied at the optimum design. The total number of active constraints (satisfied at equality) at the optimum is usually less than or at the most equal to the number of design variables.

4. *Some design problems may not have any constraints. These are called unconstrained optimization problems,* and others are called constrained optimization problems. The theory of unconstrained optimization has been known for some time and is described in Chapter 3. The theory of constrained optimization is more recent and is also described there.

5. If all the functions $f(\mathbf{x})$, $h_j(\mathbf{x})$ and $g_i(\mathbf{x})$ are linear in design variables $\mathbf{x}$, then *the problem is called a linear programming problem.* If any of these functions is nonlinear the problem is called a *nonlinear programming problem.* Linear programming problems are inherently easier to solve

compared to nonlinear programming problems. Linear programming methods are well-developed and one of the methods is described in Chapter 4.

6. *It is important to note that if the cost function is scaled by multiplying it with a positive constant, the optimum design does not change.* The optimum cost function value, however, changes. Also, any constant can be added to the cost function without affecting the optimum design. Similarly, the inequality constraints can be scaled by any positive constant, and equalities by any constant. This will not affect the feasible region and hence the optimum solution. All the foregoing transformations, however, affect the values of the Lagrange multipliers (defined in Chapter 3) as we shall see later in Section 3.7.

2.7.2 Maximization Problem Treatment

The general design model treats only minimization problems. This is no restriction as maximization of a function $F(\mathbf{x})$ is the same as minimization of a transformed function $f(\mathbf{x}) = -F(\mathbf{x})$. To see this graphically, consider a plot of the function $F(\mathbf{x})$ shown in Fig. 2.6(a). The function $F(\mathbf{x})$ takes its maximum value at the point $\mathbf{x}^*$. Next consider a graph of the function $f(\mathbf{x}) = -F(\mathbf{x})$ shown in Fig. 2.6(b). It is clear that $f(\mathbf{x})$ is a reflection of $F(\mathbf{x})$ about the x axis. It is also clear from the graph that $f(\mathbf{x})$ takes on a minimum value at the same

(a) $F(x)$

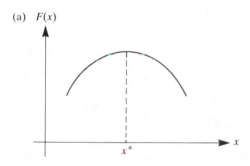

(b) $f(x)$

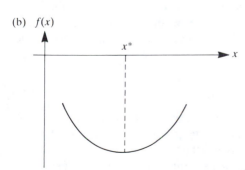

FIGURE 2.6
Point maximizing $F(x) =$ point minimizing $-F(x)$.

point $\mathbf{x}^*$ where the maximum of $F(\mathbf{x})$ occurs. Therefore, minimization of $f(\mathbf{x})$ is the equivalent to maximization of $F(\mathbf{x})$.

2.7.3 Treatment of "Greater Than Type" Constraints

Note that the general design model treats only "$\leqq$ type" inequality constraints. Many design problems may also have "$\geqq$ type" inequalities. Such constraints can be converted to the standard form without much difficulty. A "$\geqq$ type" constraint of the form

$$G_j(\mathbf{x}) \geqq 0$$

is equivalent to the following "$\leqq$ type" inequality

$$g_j(\mathbf{x}) \equiv -G_j(\mathbf{x}) \leqq 0$$

Therefore, we can multiply any "$\geqq$ type" constraint by -1 to convert it to the "$\leqq$ type".

Thus, a large class of design problems can be represented by the general model.

2.7.4 Constraint Set

The term "constraint set" will be used throughout the text. *A constraint set for the design problem is a collection of all feasible designs.* The letter S will be used to represent a constraint set. Mathematically, the set S is a collection of design points satisfying all the constraints:

$$S = \{\mathbf{x} \mid h_j(\mathbf{x}) = 0; \quad j = 1 \text{ to } p; \quad g_i(\mathbf{x}) \leqq 0; \quad i = 1 \text{ to } m\} \qquad (2.13)$$

Note that S represents *a set of feasible designs* and is sometimes referred to as the *feasible region*.

It is important to note that the feasible region usually shrinks when more constraints are added in the design model and expands when some constraints are deleted. When the feasible region shrinks, the number of possible designs that can optimize the cost function reduces, i.e. there are fewer feasible designs. In this event, the minimum value of the cost function is likely to increase. The effect is completely the opposite when some constraints are dropped. This observation is significant in practical design and should be clearly understood.

2.7.5 Active/Inactive/Violated Constraints

We will quite frequently refer to a constraint as active, tight, inactive or violated. We define these terms precisely. An inequality constraint $g_i(\mathbf{x}) \leqq 0$ is said to be *active* at a design point x^* if it is satisfied at equality, i.e. $g_i(\mathbf{x}^*) = 0$. This will be also called a *tight* or *binding* constraint. For a feasible design, an inequality constraint may or may not be active. However, all equality constraints are active for all feasible designs.

An inequality constraint $g_i(\mathbf{x}) \leqq 0$ is said to be *inactive* at a design point $\mathbf{x}^*$ if it is strictly satisfied, i.e. $g_i(\mathbf{x}^*) < 0$. An inequality constraint $g_i(\mathbf{x}) \leqq 0$ is said to be *violated* at a design point $\mathbf{x}^*$ if its value is positive, i.e. $g_i(\mathbf{x}^*) > 0$. An *equality constraint* $h_i(\mathbf{x}) = 0$ is violated at a design point $\mathbf{x}^*$ if $h_i(\mathbf{x}^*)$ is not identically zero. Note that by these definitions, an equality constraint is either active or violated at any given design point.

2.7.6 Discrete and Integer Design Variables

So far, we have assumed in the general model that variables x_i can have any numerical value within the feasible region. Many times, however, some variables are required to have discrete or integer values. Such variables appear quite often in engineering design problems. We have already come across problems in Sections 2.6.3 and 2.6.4 having integer design variables. Before describing how to treat them, let us define what we mean by discrete and integer variables.

A *design variable* is called *discrete* if its value must be selected from a given finite set of values. For example, the plate thickness must be the one that is available commercially, i.e. $\frac{1}{8}, \frac{1}{4}, \frac{3}{8}, \frac{1}{2}, \frac{5}{8}, \frac{3}{4}, 1, \ldots$ etc. Similarly, structural members must be selected off-the-shelf to reduce the fabrication cost.

An *integer variable,* as the name implies, must have integer value, e.g. the number of logs to be shipped, number of bolts used, number of items to be shipped, etc.

Some methods to treat discrete and integer variables have been investigated. However, most of them are quite complex requiring enormous computational effort. They are also not reliable. Therefore, a few simple and practical procedures are suggested to treat *discrete and integer programming problems*.

In some sense, discrete and integer variables impose additional constraints on the design problem. Therefore, as noted before, the optimum value of the cost function is likely to increase with their presence compared to that with continuous variables. If we treat all the design variables as continuous, the minimum value of the cost function represents a lower bound on the true minimum value with the discrete or integer variables. This gives some idea about the "best" optimum solution if all design variables were continuous. The optimum cost function value is likely to increase when discrete values are assigned to variables. Thus, the first suggested procedure is to solve the problem assuming continuous design variables. Then nearest discrete/integer values are assigned to the variables and the design is checked for feasibility. With a few trials and errors, the best feasible design close to the continuous optimum can be obtained. Note that there can be numerous combinations of variables that can give feasible designs.

The second approach is to use an *adaptive numerical optimization procedure.* Optimum solution with continuous variables is first obtained. Then, only the variables that are close to their discrete or integer value are assigned that value. They are then held fixed and the problem is optimized again. The procedure is continued until all the variables have proper values. The final design thus obtained is feasible. A few further trials can be made to improve the optimum cost function value. This procedure has been demonstrated by Arora and Tseng [1988].

Both the procedures require additional computational effort and do not guarantee true minimum value. However, they are quite straightforward and do not require any additional methods or software.

2.8 GRAPHICAL OPTIMIZATION

Some optimum design problems can be solved by visually inspecting their graphical representation. All constraint functions can be plotted and the constraint set (a set of feasible designs) for the problem is identified. Then cost function contours can be drawn and optimum design located by visual inspection. Since the functions must be plotted on a graph paper, we can solve problems with two, or at the most three design variables. With three variables, however, surfaces must be plotted, which can be quite tedious. We can introduce all the important geometric concepts with two variable problems. Therefore, only such problems are considered in this section. All the geometric terms and concepts introduced are applicable to more general problems.

2.8.1 Profit Maximization Problem

A company manufactures two machines, A and B. Using available resources either 28 A or 14 B machines can be manufactured each day. The sales department can sell up to 14 A machines or 24 B machines. The shipping facility can handle no more than 16 machines per day. The company makes a profit of $400 on each A machine and $600 on each B machine. How many A and B machines should the company manufacture every day to maximize profit?

The design variables for this problem are identified as $x_1 = $ number of A machines manufactured each day and $x_2 = $ number of B machines manufactured each day. The objective is to maximize profit. This can be expressed as a function of x_1 and x_2:

$$\text{profit} = 400x_1 + 600x_2 \tag{a}$$

Writing in the standard minimization form, the cost function for the problem is

$$f(x_1, x_2) = -(400x_1 + 600x_2) \tag{b}$$

Design constraints are on manufacturing capacity, limitations on sales personnel and restrictions on the shipping and handling facility. The constraint on shipping and handling facility is quite straightforward, and can be expressed as

$$x_1 + x_2 \leqq 16 \qquad \text{(shipping and handling constraint)} \qquad (c)$$

Constraints on the manufacturing and sales facilities are a bit tricky. Consider first the manufacturing limitation. It is assumed that if the company is manufacturing x_1 of A machines per day, then the remaining resources and equipment can be utilized to manufacture proportionately the B machines. Therefore, this constraint can be expressed as

$$\frac{x_1}{28} + \frac{x_2}{14} \leqq 1 \qquad \text{(manufacturing constraint)} \qquad (d)$$

Similarly, the constraint on resources of the sales department is given as

$$\frac{x_1}{14} + \frac{x_2}{24} \leqq 1 \qquad \text{(limitation on sales department)} \qquad (e)$$

Finally, the design variables must be nonnegative as $x_1, x_2 \geqq 0$, or in the standard form

$$-x_1 \leqq 0; \qquad -x_2 \leqq 0 \qquad (f)$$

Note that for this problem, the formulation remains valid even when a design variable has zero value.

This problem has two design variables and five inequality constraints. All functions of the problem are linear in variables x_1 and x_2. Therefore, this is a *linear programming problem*. Constraints for the problem are plotted in Fig. 2.7. Since $x_1, x_2 \geqq 0$, the optimum design must lie in the first quadrant. Lines

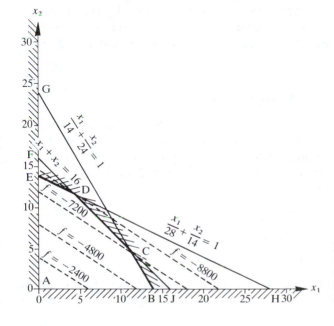

FIGURE 2.7

Graphical solution for the profit maximization problem. Optimum point = (4, 12); optimum cost = −8800.

J–F, H–E and B–G represent the constraint equations (c), (d) and (e), respectively. All the points on and within the polygon ABCDE give *feasible designs* (the constraint set S). The area ABCDE is also called the *feasible region*. Its complement – the set of points outside the region ABCDE – is called the infeasible region. In Fig. 2.7, *the infeasible region is hatched out. This notation will be used throughout the text.* Once the feasible region has been identified, we can proceed to locate the best feasible (optimum) design. We must plot *cost function contours (iso-cost lines)* through the feasible region. A visual inspection of the figure then determines the minimum point. In general, we must continue to draw iso-cost lines through the feasible region as long as the cost function can be improved (reduced). Iso-cost lines -2400, -4800, -7200 and -8800 are drawn in Fig. 2.7. It can be seen that point D is a feasible design having least value for the cost function. We simply read coordinates of point D to obtain the optimum design. Thus the best strategy for the company is to manufacture 4 A and 12 B machines to maximize its profit. The maximum profit is \$8800. Constraints of Eqs (c) and (d) are active at the optimum. These represent limitations on shipping and handling facilities, and manufacturing. The company can think about relaxing these constraints to improve its profit.

Note that in this example, the design variables must be integers. Fortunately, the optimum solution has integer values for the variables. If this were not the case, we would have used the procedures suggested in Section 2.7.6.

For this example, all functions are linear in design variables. Therefore, all curves in Fig. 2.7 are *straight lines*. In general, the functions of a design problem may not be linear. In such a case, curves must be plotted to identify the feasible region. Iso-cost curves must be drawn to identify the optimum design. To *plot nonlinear functions* numerical tables for x_1 versus x_2 must be generated. These points are then plotted on a graph and connected by a smooth curve.

2.8.2 Design Problem with Multiple Solutions

Some problems have many optimum designs. This situation can arise when a constraint is parallel to the cost function. If the constraint is active at the optimum, then there are multiple solutions for the problem. To illustrate the situation, let us consider the following design problem. Minimize $f = -x_1 - 0.5x_2$ subject to the constraints

$$2x_1 + 3x_2 \leqq 12$$
$$2x_1 + x_2 \leqq 8$$
$$-x_1 \leqq 0; \qquad -x_2 \leqq 0$$

For the problem second constraint is parallel to the cost function. Therefore, there is a *possibility of multiple optimum designs*. Figure 2.8 shows the

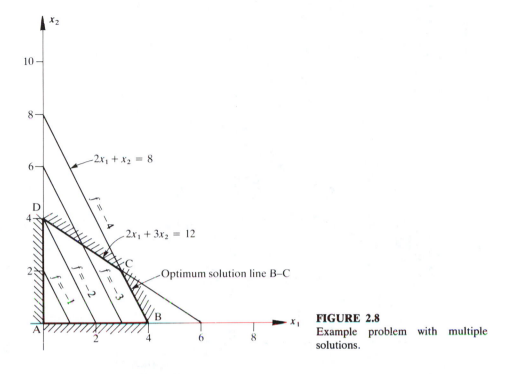

FIGURE 2.8
Example problem with multiple solutions.

graphical solution for the problem. It can be seen that any point on the line B–C gives an optimum design.

2.8.3   Problem with Unbounded Solution

Some design problems may not have a bounded solution. This situation can arise if we forget a constraint or incorrectly formulate the problem. To illustrate such a situation, we consider the following design problem. Maximize $x_1 - 2x_2$ subject to the constraints

$$2x_1 - x_2 \geqq 0$$
$$-2x_1 + 3x_2 \leqq 6$$
$$x_1, x_2 \geqq 0$$

We transform the problem to the standard form as minimize $f_1 = -x_1 + 2x_2$ subject to the constraints

$$-2x_1 + x_2 \leqq 0$$
$$-2x_1 + 3x_2 \leqq 6$$
$$-x_1 \leqq 0; \qquad -x_2 \leqq 0$$

The constraint set for the problem is shown in Fig. 2.9. Several isocost lines

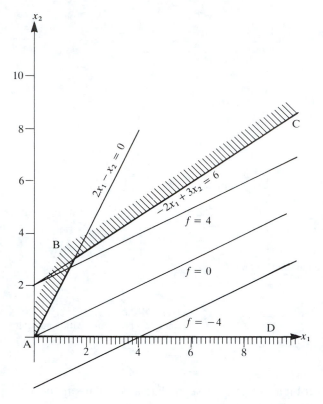

FIGURE 2.9
Example problem with unbounded solution.

are shown. It can be seen that the constraint set (feasible region) for the problem is unbounded. Therefore, there is no finite solution. We must re-examine the problem formulation to correct the situation. It can be seen in Fig. 2.9, that the problem is underconstrained.

2.8.4   Graphical Solution for Minimum Weight Tubular Column

The design problem has been formulated in Section 2.6.7. The following data will be used to solve the problem by the graphical method: $P = 10$ MN, $E = 207$ GPa, $\rho = 7833$ kg/m^3, $l = 5.0$ m and $\sigma_a = 248$ MPa. Using this data, Formulation 1 for the problem is defined as: find mean radius R and thickness t to minimize

$$f(R, t) = 2\pi\rho lRt$$
$$= 2\pi(7833)(5)Rt = (2.4608\text{E}+05)Rt, \qquad \text{kg}$$

subject to the constraints

$$g_1(R, t) \equiv \frac{P}{2\pi Rt} - \sigma_a \leqq 0$$

$$\equiv \frac{10(1.0E+06)}{2\pi Rt} - 248(1.0E+06) \leqq 0$$

$$g_2(R, t) \equiv P - \frac{\pi^3 ER^3 t}{4l^2} \leqq 0$$

$$\equiv 10(1.0E+06) - \frac{\pi^3(207.0E+09)R^3 t}{4(5)(5)} \leqq 0$$

$$g_3(R, t) \equiv -R \leqq 0$$

$$g_4(R, t) \equiv -t \leqq 0$$

The constraints for the problem are plotted in Fig. 2.10 and the feasible region is indicated. Cost function contours for $f = 1000$, 1500, 1579 kg are also

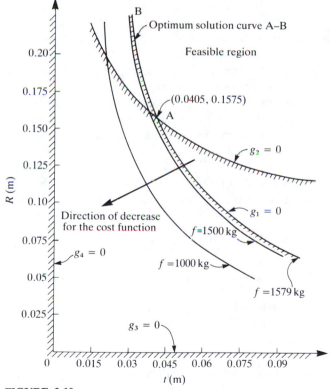

FIGURE 2.10
Graphical solution for a minimum weight tubular column.

shown. Note that for this example the cost function contours are parallel to the constraint g_1. Since g_1 is active at the optimum, the problem has an infinite number of optimum designs, i.e. the entire curve A–B in Fig. 2.10. We can read coordinates of any point on the curve A–B as an optimum solution. In particular, point A, where constraints g_1 and g_2 intersect, is also an optimum point where $R^* = 0.1575$ m and $t^* = 0.0405\ m$.

Note that this problem has nonlinear functions. To plot them, we generate tables of data points t versus R and connect them with smooth curves on the graph. For example, to plot the constraint boundary for g_2 ($R^3 t = 1.558\mathrm{E}{-}04$), we select values for t as 0.015, 0.03, 0.06, 0.075, 0.09 and calculate the values for R from $g_2 = 0$ as 0.218, 0.173, 0.1374, 0.1275 and 0.12. This procedure can be used in general to plot any nonlinear function of two variables.

2.8.5 Graphical Solution for a Beam Design Problem

A beam of rectangular cross-section is subjected to a bending moment of M ($\mathrm{N \cdot m}$) and a maximum shear force of V (N). The bending stress in the beam is calculated as $\sigma = 6M/bd^2$ (Pa) and average shear stress is calculated as $\tau = 3V/2bd$ (Pa), where b is the width and d is the depth of the beam. The allowable stresses in bending and shear are 10 MPa and 2 MPa respectively. It is also desired that the depth of the beam shall not exceed twice its width. It is desired to minimize cross-sectional area of the beam.

Let us first formulate the problem using a consistent set of units. Let d = depth of the beam (mm) and b = width of the beam (mm). The cost function for the problem is the cross-sectional area which is expressed as

$$f(b, d) = bd \tag{a}$$

Constraints for the problem consist of bending stress, shear stress and depth to width ratio. For the numerical example, let $M = 40\ \mathrm{kN \cdot m}$ and $V = 150\ \mathrm{kN}$. Bending stress is given as

$$\sigma = \frac{6(40)(1000)(1000)}{bd^2}, \qquad \mathrm{N/mm}^2 \tag{b}$$

The shear stress is given as

$$\tau = \frac{3(150)(1000)}{2bd}, \qquad \mathrm{N/mm}^2 \tag{c}$$

The allowable bending stress is

$$10\ \mathrm{MPa} = 10(1.0\mathrm{E}{+}06)\ \mathrm{N/m}^2 = 10\ \mathrm{N/mm}^2 \tag{d}$$

The allowable shear stress is

$$2\ \mathrm{MPa} = 2(1.0\mathrm{E}{+}06)\ \mathrm{N/m}^2 = 2\ \mathrm{N/mm}^2 \tag{e}$$

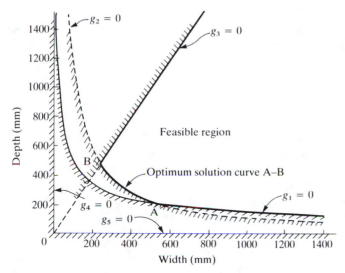

FIGURE 2.11
Graphical solution of the minimum area beam design problem.

Using Eqs (b) to (e), we get the bending and shear stress constraints as

bending stress: $g_1 \equiv \dfrac{(2.40E+08)}{bd^2} - 10 \leqq 0$ (f)

shear stress: $g_2 \equiv \dfrac{(2.25E+05)}{bd} - 2 \leqq 0$ (g)

The constraint requiring depth to be no more than twice the width can be expressed as

$$g_3 \equiv d - 2b \leqq 0 \qquad (h)$$

Finally, both the design variables should be nonnegative:

$$g_4 \equiv -b \leqq 0; \qquad g_5 \equiv -d \leqq 0 \qquad (i)$$

In reality, both b and d cannot have zero value, so we should use some minimum value as lower bounds on them, i.e. $b \geqq b_{min}$ and $d \geqq d_{min}$.

The constraints for the problem are plotted in Fig. 2.11 and the feasible region is identified. Note that the cost function is parallel to constraint g_2 (both functions have same form, $bd = $ constant). Therefore any point along the curve A–B represents an optimum solution. There are an infinite number of optimum designs and the designer can choose any one of them to meet his needs. From a practical standpoint, this is a highly desirable situation, because it offers a wide choice of optimum solutions to the designer.

The optimum cross-sectional area is $112\,500\ mm^2$. Point B corresponds to an optimum design of $b = 237$ mm and $d = 474$ mm. Point A corresponds to $b = 527.3$ mm and $d = 213.3$ mm. These points represent the two extreme optimum solutions; all other solutions lie in between this range on the curve A–B.

2.8.6 Infeasible Problem

If we are not careful in formulating a design problem, it may not have any solution. This happens when there are conflicting requirements or inconsistent constraint equations. Another situation where no solution may exist is when we put *too many constraints* on the system, i.e. constraints are such that there is no feasible solution. These are called *infeasible problems*. To illustrate the situation, we consider the following problem. Minimize $x_1 + 2x_2$ subject to

$$3x_1 + 2x_2 \leqq 6$$
$$2x_1 + 3x_2 \geqq 12$$
$$x_1 \leqq 5$$
$$x_2 \leqq 5$$
$$x_1, x_2 \geqq 0$$

Constraints for the problem are plotted in Fig. 2.12. It can be seen that there is no region of the design space that satisfies all the constraints. Thus, the problem is infeasible. Basically, the first two constraints impose conflicting requirements on the design problem. The first constraint requires the feasible designs to be below the line A–G and the second one above the line C–F. Since the two lines do not intersect in the first quadrant, there is no feasible region for the problem.

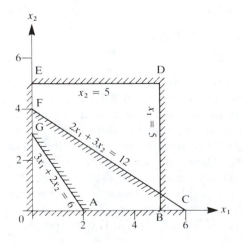

FIGURE 2.12
Example of infeasible design optimiation problem.

EXERCISES FOR CHAPTER 2

Section 2.6   Examples of Optimum Design Problem Formulation

2.1 A 100×100 m lot is available to construct a multistory office building. At least 20 000 m² total floor space is needed. According to a zoning ordinance, the maximum height of the building can be only 21 m, and the area for parking outside the building must be at least 25% of the total floor area. It has been decided to fix the height of each story at 3.5 m. The cost of the building in millions of dollars is estimated at $(0.6h + 0.001A)$ where A is the cross-sectional area of the building per floor and h is the height of the building. Formulate the minimum cost design problem.

2.2 A refinery has two crude oils:

1. crude A costs \$30/barrel (bbl) and 20 000 bbl are available, and
2. crude B costs \$36/bbl and 30 000 bbl are available.

The company manufactures gasoline and lube oil from the crudes. Yield and sale price per barrel of the product and markets are shown in Table 2.2. How much crude oils should the company use to maximize its profit? Formulate the optimum design problem.

TABLE 2.2
Data for refinery operations

Product	Yield/bbl		Sale price per bbl	Market (bbl)
	Crude A	Crude B		
Gasoline	0.6	0.8	\$50	20 000
Lube oil	0.4	0.2	\$120	10 000

2.3 Design a beer mug shown in Fig. E2.3 to hold as much beer as possible. The height and radius of the mug should be not more than 20 cm. The mug must be at least 5 cm in radius. The surface area of the sides must not be greater than 900 cm² (ignore the area of the bottom of the mug and ignore the mug handle – see Fig.). Formulate the optimum design problem.

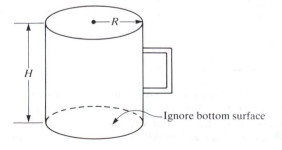

Ignore bottom surface

FIGURE E2.3
Beer mug.

2.4 A Company is redesigning its parallel flow heat exchanger of length l to increase its heat transfer. An end view of the unit is shown in the Fig. E2.4. There are certain limitations on the design problem. The smallest available conducting tube has a radius of 0.5 cm and all tubes must be of the same size. Further, the total cross-sectional area of all the tubes cannot exceed 2000 cm^2 to ensure adequate space inside the outer shell. Formulate the problem to determine the number of tubes, and the radius of each tube to maximize the surface area of the tubes in the exchanger.

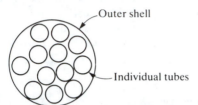

FIGURE E2.4
Cross-section of heat exchanger.

2.5 Proposals for a parking ramp having been defeated, we plan to build a parking lot in the downtown urban renewal section. The cost of land is $200W + 100D$, where W is the width along the street, and D the depth of the lot in meters. The available width along the street is 100 m, while the maximum depth available is 200 m. We want to have at least 10 000 m^2 in the lot. To avoid unsightly lots, the city requires the longer dimension of any lot be no more than twice the shorter dimension. Formulate the minimum cost design problem.

2.6 A manufacturer sells products A and B. Profit from A is $10/kg and from B $8/kg. Available raw materials for the products are: 100 kg of C and 80 kg of D. To produce 1 kg of A, 0.4 kg of C and 0.6 kg of D are needed. To produce 1 kg of B, 0.5 kg of C and 0.5 kg of D are needed. The markets for the products are 70 kg for A and 110 kg for B. How much of A and B should be produced to maximize profit? Formulate the design optimization problem.

2.7 Design a diet of bread and milk to get at least 5 units of vitamin A and 4 units of vitamin B each day. The amount of vitamins A and B in 1 kg of each food and the cost per kg of food are given in Table 2.3. Formulate the design optimization problem so that we get at least the basic requirements of vitamins at the minimum cost.

TABLE 2.3
Data for the diet problem

Vitamin	Bread	Milk
A	1	2
B	3	2
Cost/kg	2	1

2.8 Enterprising chemical engineering students have set up a still in a bathtub. They can produce 225 bottles of pure alcohol each week. They bottle two products from alcohol: (i) wine, 20 proof; and (ii) whiskey, 80 proof. Recall that pure alcohol is 200 proof. They have an unlimited supply of water, but can only obtain 800 empty bottles per week because of stiff competition. The weekly supply of sugar is enough for either 600 bottles of wine or 1200 bottles of whiskey. They make $1.0 profit on each bottle of wine and $2.00 profit on each bottle of whiskey. They can sell whatever they produce. How many bottles of wine and whiskey should they produce each week to maximize profit. Formulate the design optimization problem. (Created by D. Levy.)

2.9 Design a can closed at one end using the smallest area of sheet metal for a specified interior volume of 600 cm^3. The can is a right circular cylinder with interior height h and radius r. The ratio of height to diameter must not be less than 1.0 and not greater than 1.5. The height cannot be more than 20 cm. Formulate the design optimization problem.

2.10 Design a shipping container closed at both ends with dimensions $b \times b \times h$ to minimize the ratio: (round-trip cost of shipping the container only)/(one-way cost of shipping the contents only). Use the following data:

> Mass of the container/area: 80 kg/m^2
>
> Maximum b: 10 m
>
> Maximum h: 18 m
>
> One-way shipping cost,
>   full or empty: $18/kg gross mass
>
> Mass of the contents: 150 kg/m^3

Formulate the design optimization problem.

2.11 Certain mining operations require an open top rectangular container to transport materials. The data for the problem are:

> Construction costs:
>   sides: $50/m^2$
>   ends: $60/m^2$
>   bottom: $90/m^2$
> Salvage value: 25% of the construction cost
> Useful life: 20 years
> Yearly maintenance: $12/m^2$ of outside surface area
> Minimum volume needed: 150 m^3
> Interest rate: 12% per annum

Formulate the problem of determining the container dimensions for minimum cost.

2.12 Design a circular tank closed at both ends to have a volume of 150 m^3. The fabrication cost is proportional to the surface area of the sheet metal and is $400/m^2$. The tank is to be housed in a shed with sloping roof. Therefore, height H of the tank is limited by the relation $H \le 10 - D/2$, where D is the diameter of the tank. Formulate the minimum cost design problem.

2.13 Design the steel framework shown in Fig. E2.13 at a minimum cost. The cost of all horizontal members in one direction is $20w$ and in the other direction it is

$30d$. The cost of a vertical column is $50h$. The frame must enclose a total volume of at least 600 m³. Formulate the design optimization problem.

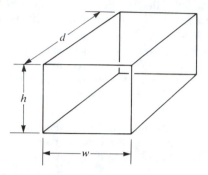

FIGURE E2.13
Steel frame.

2.14 Two electrical generators are interconnected to provide total power to meet the load. Each generator's cost is a function of the power output, as shown in Fig. E2.14. All costs and power are expressed on a per unit basis. The total power needed is at least 60 units. Formulate a minimum cost design problem.

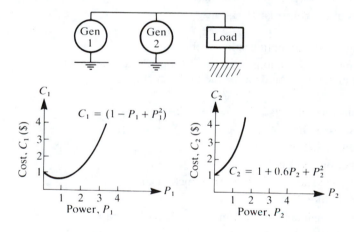

In the left plot: $C_1 = (1 - P_1 + P_1^2)$

In the right plot: $C_2 = 1 + 0.6P_2 + P_2^2$

FIGURE E2.14
Power generator.

2.15 Transportation problem. A company has m manufacturing facilities. The facility at the ith location has capacity to produce b_i units of an item. The product should be shipped to n distribution centers. The distribution center at the jth location requires at least a_j units of the item to satisfy demand. The cost of shipping an item from the ith plant to the jth distribution center is c_{ij}. Formulate a minimum cost transportation system to meet each distribution center's demand without exceeding the capacity of any manufacturing facility.

2.16 Design of a two-bar truss. Design a symmetric two-bar truss (both members have the same cross-section) shown in Fig. E2.16 to support a load W. The truss consists of two steel tubes pinned together at one end and supported on the ground at the other. The span of the truss is fixed at s. Formulate the minimum mass truss design problem using height and the cross-sectional dimensions as

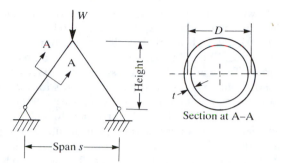

FIGURE E2.16
Two-bar structure.

design variables. The design should satisfy the following constraints:

1. Because of space limitations, the height of the truss must not exceed b_1, and must not be less than b_2.
2. The ratio of the mean diameter to thickness of tube must not exceed b_3.
3. The compressive stress in the tubes must not exceed the allowable stress σ_a for steel.
4. The height, diameter and thickness must be chosen to safeguard against member buckling.

Use the following data: $W = 10\,\text{kN}$, span $s = 2\,\text{m}$, $b_1 = 5\,\text{m}$, $b_2 = 2\,\text{m}$, $b_3 = 90$, allowable stress, $\sigma_a = 250\,\text{MPa}$, modulus of elasticity, $E = 210\,\text{GPa}$, mass density, $\rho = 7850\,\text{kg/m}^3$, factor of safety against buckling, $\text{FS} = 2$, $0.1 \leqq D \leqq 2$ (m), and $0.01 \leqq t \leqq 0.1$ (m).

2.17 A beam of rectangular cross-section (Fig. E2.17) is subjected to a maximum bending moment of M and a maximum shear of V. The allowable bending and shearing stresses are σ_a and τ_a, respectively. The bending stress in the beam is calculated as

$$\sigma = \frac{6M}{bd^2}$$

and average shear stress in the beam is calculated as

$$\tau = \frac{3V}{2bd}$$

where d is the depth and b is the width of the beam. It is also desired that the depth of the beam shall not exceed twice its width. Formulate the design problem for minimum cross-sectional area using the following data: $M = 140\,\text{kN} \cdot \text{m}$, $V = 24\,\text{kN}$, $\sigma_a = 165\,\text{MPa}$, $\tau_a = 50\,\text{MPa}$.

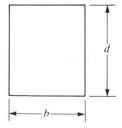

FIGURE E2.17
Cross-section of a rectangular beam.

2.18 A vegetable oil processor wishes to determine how much shortening, salad oil and margarine to produce to optimize the use of his current oil stocks. At the current time, he has 250 000 kg of soybean oil, 110 000 kg of cottonseed oil and 2000 kg of milk base substances. The milk base substances are required only in the production of margarine. There are certain processing losses associated with each product; 10% for shortening, 5% for salad oil, and no loss for margarine. The producer's back orders require him to produce at least 100 000 kg of shortening, 50 000 kg of salad oil and 10 000 kg of margarine. In addition, sales forecasts indicate a strong demand for all products in the near future. The profit per kg and the base stock required per kg of each product are given in Table 2.4. Formulate the problem to maximize profit over the next production scheduling period. (Created by J. Liittschwager.)

TABLE 2.4
Data for the vegetable oil processing problem

Product	Profit per kg	Parts per kg of base stock requirements		
		Soybean	Cottonseed	Milk base
Shortening	0.10	2	1	0
Salad oil	0.08	0	1	0
Margarine	0.05	3	1	1

Section 2.7   A General Mathematical Model for Optimum Design

2.19 *Answer True of False*

1. Design of a system implies specification for the design variable values.
2. All design problems have only linear inequality constraints.
3. All design variables should be independent of each other as far as possible.
4. If there is an equality constraint in the design problem, the optimum solution must satisfy it.
5. Each optimization problem must have certain parameters called the design variables.
6. A feasible design may violate equality constraints.
7. A feasible design may violate "$\geq$ type" constraints.
8. A "$\leq$ type" constraint expressed in the standard form is active at a design point if it has zero value there.
9. The constraint set for a design problem consists of all the feasible points.
10. The number of independent equality constraints can be larger than the number of design variables for the problem.
11. The number of "$\leq$ type" constraints must be less than the number of design variables for a valid problem formulation.

12. The feasible region for an equality constraint is a subset of that for the same constraint expressed as an inequality.
13. Maximization of $f(x)$ is equivalent to minimization of $1/f(x)$.
14. A lower minimum value for the cost function is obtained if more constraints are added to the problem formulation
15. Let f_n be the minimum value for the cost function with n design variables for a problem. If the number of design variables for the same problem is increased to, say $m = 2n$, then $f_m > f_n$ where f_m is the minimum value for the cost function with m design variables.

2.20* A trucking firm is considering to purchase several new trucks. It has $2 million to spend. The investment should yield a maximum of capacity in tonnes × kilometers per day. Three available models are given in Table 2.5. Some additional restrictions should be considered.

   The company employs 150 drivers and the labor market does not allow hiring of additional drivers. Garage and maintenance facilities can handle 30 trucks at the most.

   How many trucks of each type should the company purchase? Formulate the design optimization problem. Transcribe it into the standard optimization model.

TABLE 2.5
Data for the available trucks

Truck model	Load capacity (tonnes)	Average speed (km/h)	Crew req'd	No. of hours of operation per day (3 shifts)	Initial investment per truck ($)
A	10	55	1	18	40 000
B	20	50	2	18	60 000
C	18	50	2	21	70 000

2.21* A large steel corporation has two iron ore reduction plants. Each plant processes iron ore into two different ingot stocks. They are shipped to any of the three fabricating plants where they are made into either of the two finished products. In total, there are two reduction plants, two ingot stocks, three fabricating plants, and two finished products.

   For the coming season, the company wants to minimize total tonnage of iron ore processed in its reduction plants, subject to production and demand constraints. Formulate the design optimization problem and transcribe it into the standard model.

Nomenclature

$a(r, s)$ tonnage yield of ignot stock s from 1 ton of iron ore processed at reduction plant r,

$b(s, f, p)$ total yield from 1 ton of ingot stock s shipped to fabricating plant f and manufactured into product p,

$c(r)$ iron ore processing capacity in tonnage at reduction plant r,

$k(f)$ capacity of the fabricating plant f in tonnage for all stocks, and

$D(p)$ tonnage demand requirement for product p.

Production and demand constraints

1. The total tonnage of iron ore processed by both reduction plants must equal the total tonnage processed into ingot stocks for shipment to the fabricating plants.
2. The total tonnage of iron ore processed by each reduction plant cannot exceed its capacity.
3. The total tonnage of ingot stock manufactured into products at each fabricating plant must equal the tonnage of ingot stock shipped to it by the reduction plants.
4. The total tonnage of ingot stock manufactured into products at each fabricating plant cannot exceed its available capacity.
5. The total tonnage of each product must equal its demand.

Constants for the problem

$$a(1, 1) = 0.39 \quad c(1) = 1\,200\,000 \quad k(1) = 190\,000 \quad D(1) = 330\,000$$
$$a(1, 2) = 0.46 \quad c(2) = 1\,000\,000 \quad k(2) = 240\,000 \quad D(2) = 125\,000$$
$$a(2, 1) = 0.44 \quad\quad\quad\quad\quad\quad\quad\quad k(3) = 290\,000$$
$$a(2, 2) = 0.48$$

$$b(1, 1, 1) = 0.79 \quad b(1, 1, 2) = 0.84$$
$$b(2, 1, 1) = 0.68 \quad b(2, 1, 2) = 0.81$$
$$b(1, 2, 1) = 0.73 \quad b(1, 2, 2) = 0.85$$
$$b(2, 2, 1) = 0.67 \quad b(2, 2, 2) = 0.77$$
$$b(1, 3, 1) = 0.74 \quad b(1, 3, 2) = 0.72$$
$$b(2, 3, 1) = 0.62 \quad b(2, 3, 2) = 0.78$$

2.22 Optimization of a water canal (created by V. K. Goel). Design a water canal having a cross-sectional area of 150 m². Least construction costs occur when the volume of the excavated material equals the amount of material required for the dyke, as shown in Fig. E2.22. Formulate the problem to minimize the dugout material A_1. Transcribe the problem into the standard design optimization model.

2.23 A cantilever beam is subjected to the point load P (kN), as shown in Fig. E2.23. The maximum bending moment in the beam is Pl (kN · m) and the maximum shear is P (kN). Formulate the minimum mass design problem using hollow circular cross-section. The material should not fail under bending stress or shear

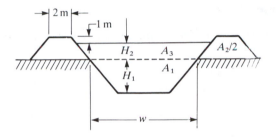

FIGURE E2.22
Cross-section of a canal.

stress. The maximum bending stress is calculated as

$$\sigma = \frac{Pl}{I} R_o$$

where $I =$ moment of inertia of the cross-section. The maximum shearing stress is calculated as

$$\tau = \frac{P}{3I}(R_o^2 + R_o R_i + R_i^2)$$

Transcribe the problem into the standard design optimization model (also use $R_0 \leqq 40.0$ cm, $R_i \leqq 40.0$ cm). Use the following data: $P = 14$ kN, $l = 10$ m, mass density, $\rho = 7850$ kg/m^3, allowable bending stress, $\sigma_b = 165$ MPa, allowable shear stress, $\tau_a = 50$ MPa.

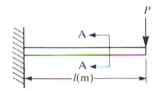

FIGURE E2.23
Section A–A   Cantilever beam.

2.24 Design a hollow circular beam-column shown in Fig. E2.24 for two conditions: when $P = 50$ (kN), the axial stress σ should be less than σ_a, and when $P = 0$, deflection δ due to self-weight should satisfy $\delta \leqq 0.001l$. The limits for dimensions are $t = 0.10$ to 1.0 cm, $R = 2.0$ to 20.0 cm and $R/t \geqq 20$. Formulate the minimum weight design problem and transcribe it into the standard form. Use the following data: $\delta = 5wl^4/384EI$, $w =$ self weight force/length (N/m), $\sigma_a = 250$ MPa, modulus of elasticity, $E = 210$ GPa, mass density, $\rho = 7800$ kg/m^3, $\sigma = P/A$, gravitational constant, $g = 9 \cdot 80$ m/s^2.

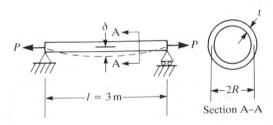

FIGURE E2.24
Section A–A   Hollow circular beam.

Section 2.8   Graphical Optimization

Transcribe the following problems into the standard form, and then solve them using the graphical method:

2.25 Minimize $(x_1 - 3)^2 + (x_2 - 3)^2$
subject to $x_1 + x_2 \leq 4$
$x_1, x_2 \geq 0.$

2.26 Maximize $x_1 + 2x_2$
subject to $2x_1 + x_2 \leq 4$
$x_1, x_2 \geq 0$

2.27 Minimize $x_1 + 3x_2$
subject to $x_1 + 4x_2 \geq 48$
$5x_1 + x_2 \geq 50$
$x_1, x_2 \geq 0$

2.28 Maximize $x_1 + x_2 + 2x_3$
subject to $1 \leq x_1 \leq 4$
$3x_2 - 2x_3 = 6$
$-1 \leq x_3 \leq 2$
$x_2 \geq 0$

2.29 Maximize $4x_1 x_2$
subject to $x_1 + x_2 \leq 20$
$x_2 - x_1 \leq 10$
$x_1, x_2 \geq 0$

2.30 Minimize $5x_1 + 10x_2$
subject to $10x_1 + 5x_2 \leq 50$
$5x_1 - 5x_2 \geq -20$
$x_1, x_2 \geq 0$

2.31 Minimize $3x_1 + x_2$
subject to $2x_1 + 4x_2 \leq 21$
$5x_1 + 3x_2 \leq 18$
$x_1, x_2 \geq 0$

2.32 Minimize $x_1^2 - 2x_2^2 - 4x_1$
subject to $x_1 + x_2 \leq 6$
$x_2 \leq 3$
$x_1, x_2 \geq 0$

2.33 Minimize $x_1 x_2$
subject to $x_1 + x_2^2 \leq 0$
$x_1^2 + x_2^2 \leq 9$

2.34 Minimize $3x_1 + 6x_2$
subject to $-3x_1 + 3x_2 \leq 2$
$4x_1 + 2x_2 \leq 4$
$-x_1 + 3x_2 \geq 1$

2.35 Solve the rectangular beam problem of Exercise 2.17 graphically for the following data: $M = 80$ kN · m, $V = 150$ kN, $\sigma_a = 8$ MPa and $\tau_a = 3$ MPa.

2.36 Solve the cantilever beam problem of Exercise 2.23 graphically for the following data: $P = 10$ kN, $l = 5.0$ m, modulus of elasticity, $E = 210$ GPa, allowable bending stress, $\sigma_a = 250$ MPa, allowable shear stress, $\tau_a = 90$ MPa, mass density, $\rho = 7850$ kg/m^3, $R_o \leq 20.0$ cm, $R_i \leq 20.0$ cm.

2.37 For the minimum mass tubular column design problem formulated in Section 2.6.7, consider the following data: $P = 50$ kN, $l = 5.0$ m, modulus of elasticity, $E = 210$ GPa, allowable stress, $\sigma_a = 250$ MPa, mass density, $\rho = 7850$ kg/m^3.

Treating mean radius R and wall thickness t as design variables, solve the design problem graphically imposing an additional constraint $R/t \leq 50$. This constraint is needed to avoid local crippling of the column. Also impose the member size constraints as

$$0.01 \leq R \leq 1.0 \text{ m}; \qquad 5 \leq t \leq 200 \text{ mm}$$

2.38 For Exercise 2.37, treat outer radius R_o and inner radius R_i as design variables, and solve the design problem graphically. Impose the same constraints as in Exercise 2.37.

2.39 Formulate the minimum mass column design problem of Section 2.6.7 using hollow square cross-section with outside dimension w and thickness t as design variables. Solve the problem graphically using the constraints and the data given in Exercise 2.37.

2.40 Consider the symmetric (members are identical) case of the two-bar truss problem discussed in Sections 2.2–2.5 with the following data: $W = 10$ kN, $\theta = 30°$, height $h = 1.0$ m, span $s = 1.5$ m, allowable stress, $\sigma_a = 250$ MPa, modulus of elasticity, $E = 210$ GPa.

Formulate the minimum mass design problem with constraints on member stresses and bounds on design variables. Solve the problem graphically using circular tubes as members.

2.41 Formulate and solve the problem of Exercise 2.1 graphically.

2.42 In design of a closed-end, thin-walled cylindrical pressure vessel shown in Fig. E2.42, the design objective is to select the mean radius R and wall thickness t to minimize the total mass. The vessel should contain at least 25.0 m^3 of gas at a pressure of 3.5 MPa. It is required that the circumferential stress in the pressure vessel not exceed 210 MPa and the circumferential strain not exceed (1.0E−03). The circumferential stress and strain are calculated from the equations

$$\sigma_c = \frac{PR}{t}, \qquad \varepsilon_c = \frac{PR(2 - v)}{2Et}$$

where $\rho =$ mass density (7850 kg/m^3), $\sigma_c =$ circumferential stress (Pa), $\varepsilon_c =$ circumferential strain, $P =$ internal pressure (Pa), $E =$ Young's modulus (210 GPa), and $v =$ Poisson's ratio (0.3).

(i) Formulate the optimum design problem and (ii) solve the problem graphically.

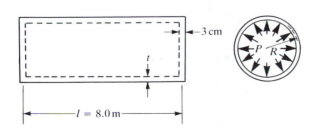

FIGURE E2.42
Cylindrical pressure vessel.

2.43 Consider the symmetric three-bar truss design problem formulated in Section 2.6.8. Formulate and solve the problem graphically for the following data: $l = 1.0$ m, $P = 100$ kN, $\theta = 30°$, mass density, $\rho = 2800$ kg/m^3 modulus of elasticity, $E = 70$ GPa, allowable stress, $\sigma_a = 140$ MPa, $\Delta_u = 0.5$ cm, $\Delta_v = 0.5$ cm, $\omega_0 = 50$ Hz, $\beta = 1.0$, and $A_1, A_2 \geqq 2$ cm^2.

2.44 Consider the cabinet design problem given in Section 2.6.4. Use the equality constraints to eliminate three design variables from the problem. Restate the problem in terms of the remaining three variables, transcribing it into the standard form.

2.45 Solve the insulated spherical tank design problem formulated in Section 2.6.5 graphically for the following data: $r = 3.0$ m, $c_1 = \$100$, $c_2 = 500$, $c_3 = \$10$, $c_4 = \$5$, $\Delta T = 10$.

2.46 Solve the cylindrical tank design problem given in Section 2.6.6 graphically for the following data: $c = \$1500/$m^2, $V = 3000$ m^3.

2.47 Consider the minimum mass tubular column problem formulated in Section 2.6.7. Find the optimum solution for both formulations of the problem using the graphical method for the data: load, $P = 100$ kN, length, $l = 5.0$ m, Young's modulus, $E = 210$ GPa, allowable stress, $\sigma_a = 250$ MPa, mass density, $\rho = 7850$ kg/m^3, $R \leqq 0.4$ m, $t \leqq 0.1$ m, and $R, t \geqq 0$.

2.48* Design a hollow torsion rod shown in Fig. E2.48 to satisfy the following requirements (created by J. M. Trummel):

1. The calculated shear stress, τ, shall not exceed the allowable shear stress τ_a under the normal operating torque T_o (N · m).
2. The calculated angle of twist, θ, shall not exceed the allowable twist, θ_a (radians).
3. The member shall not buckle under a short duration torque of T_{max} (N · m).

Requirements for the rod and material properties are given in Tables 2.6 and 2.7 (select a material for one rod). Use the following design variables: $x_1 =$ outside diameter of the rod and $x_2 =$ ratio of inside/outside diameter, d_i/d_o.

Using graphical optimization, determine the inside and outside diameters for a minimum mass rod to meet the above design requirements. Compare the hollow rod with an equivalent solid rod ($d_i/d_o = 0$). Use consistent set of units (e.g. Newtons and millimeters) and let the minimum and maximum values for design variables be given as

$$0.02 \leqq d_o \leqq 0.5 \text{ m}$$

$$0.60 \leqq \frac{d_i}{d_o} \leqq 0.999$$

Useful expressions for the rod are:

Mass of rod: $M = \dfrac{\pi}{4} \rho l (d_o^2 - d_i^2),$ kg

Calculated shear stress: $\tau = \dfrac{c}{J} T_o,$ Pa

Calculated angle of twist: $\theta = \dfrac{l}{GJ} T_{\text{o}}$,   radians

Critical buckling torque: $T_{\text{cr}} = \dfrac{\pi d_{\text{o}}^3 E}{12\sqrt{2}\,(1-v^2)^{0.75}} \left(1 - \dfrac{d_{\text{i}}}{d_{\text{o}}}\right)^{2.5}$,   N · m

Notation

M = mass of the rod (kg),
d_0 = outside diameter of the rod (m),
d_{i} = inside diameter of the rod (m),
ρ = mass density of material (kg/m³),
l = length of the rod (m),
T_{o} = normal operating torque (N · m),
c = distance from rod axis to extreme fiber (m),
J = polar moment of inertia (m⁴),
θ = angle of twist (radians),
G = modulus of rigidity (Pa),
T_{cr} = critical buckling torque (N · m),
E = modulus of elasticity (Pa), and
v = Poisson's ratio.

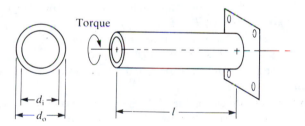

FIGURE E2.48
Hollow torsion rod.

TABLE 2.6
Rod requirements

Torsion rod	Length, l (m)	Normal torque, T_{o} (kN · m)	Maximum, T_{max} (kN · m)	Allowable twist, θ_{a} (degrees)
1	0.50	10.0	20.0	2
2	0.75	15.0	25.0	2
3	1.00	20.0	30.0	2

TABLE 2.7
Materials and properties for the torsion rod

Material	Density, ρ (kg/m^3)	Allowable shear stress, τ_a (MPa)	Elastic modulus, E (GPa)	Shear modulus, G (GPa)	Poisson ratio (v)
1. 4140 alloy steel	7850	275	210	80	0.30
2. Aluminum alloy 24 ST4	2750	165	75	28	0.32
3. Magnesium alloy A261	1800	90	45	16	0.35
4. Beryllium	1850	110	300	147	0.02
5. Titanium	4500	165	110	42	0.30

2.49* Formulate and solve Exercise 2.48 using the outside diameter d_o and the inside diameter d_i as design variables.

2.50* Formulate and solve Exercise 2.48 using the mean radius R and wall thickness t as design variables. Let the bounds on design variables be given as $5 \leq R \leq 20$ cm and $0.2 \leq t \leq 4$ cm.

2.51 Formulate the problem of Exercise 2.3 and solve it.

2.52 Formulate the problem of Exercise 2.4 and solve it.

2.53 Solve Exercises 2.37 and 2.38 for a column pinned at both ends. The buckling load for such a column is given as $\pi^2 EI/l^2$.

2.54 Solve Exercises 2.37 and 2.38 for a column fixed at both ends. The buckling load for such a column is given as $4\pi^2 EI/l^2$.

2.55 Solve Exercises 2.37 and 2.38 for a column fixed at one end and pinned at the other. The buckling load for such a column is given as $2\pi^2 EI/l^2$.

2.56 Solve the beer can design problem formulated in Section 2.6.2 using the graphical approach.

2.57 Consider the two-bar truss shown in Fig. 2.1. Using the given data, design a minimum mass structure where $W = 100$ kN, $\theta = 30°$, $h = 1$ m, $s = 1.5$ m, modulus of elasticity, $E = 210$ GPa, allowable stress, $\sigma_a = 250$ MPa, mass density, $\rho = 7850$ kg/m^3. Use Newtons and millimeters as units. The members should not fail in stress and their buckling should be avoided. Deflection at the top in either direction should not be more than 5 cm.

Use cross-sectional areas A_1 and A_2 of the two members as design variables and let the moment of inertia of the members be given as $I = A^2$. Areas must also satisfy the constraint $1 \leq A_i \leq 50$ cm^2.

2.58 For Exercise 2.57, use hollow circular tubes as members with mean radius R and wall thickness t as design variables. Make sure that $R/t \leq 50$. Design the structure to be symmetric with member 1 the same as member 2. The radius and thickness must also satisfy the constraints $2 \leq t \leq 40$ mm and $2 \leq R \leq 40$ cm.

2.59 Design a symmetric structure defined in Exercise 2.57 treating cross-sectional area A and height h as design variables. The design variables must also satisfy the constraints $1 \leq A \leq 50$ cm^2 and $0.5 \leq h \leq 3$ m.

2.60 Design a symmetric structure defined in Exercise 2.57 treating cross-sectional area A and the span s as design variables. The design variables must also satisfy the constraints $1 \leqq A \leqq 50$ cm^2 and $0.5 \leqq s \leqq 4$ m.

2.61 A minimum mass symmetric (area of member 1 is the same as member 3) three-bar truss is to be designed to support a load P as shown in Fig. 2.5. The following notation may be used: $P_u = P \cos \theta$, $P_v = P \sin \theta$, $A_1 =$ cross-sectional area of members 1 and 3, $A_2 =$ cross-sectional area of member 2.

    The members must not fail under the stress, and deflection at node 4 must not exceed 2 cm in either direction. Use Newtons and millimeters as units. The data is given as $P = 50$ kN, $\theta = 30°$, mass density, $\rho = 7850$ kg/m^3, modulus of elasticity, $E = 210$ GPa and allowable stress, $\sigma_a = 150$ MPa. The design variables must also satisfy the constraints $50 \leqq A_i \leqq 5000$ mm^2.

2.62* **Design of a water tower support column** (created by G. Baenziger). As a member of the ABC Consulting Engineers you have been asked to design a cantilever cylindrical support column of minimum mass for a new water tank. The tank itself has already been designed in the tear-drop shape shown in Fig. E2.62. The height of the base of the tank (H), the diameter of the tank (D), and the wind pressure on the tank (w) are given as $H = 30$ m, $D = 10$ m and $w = 700$ N/m^2. Formulate the design optimization problem and solve it graphically.

    In addition to designing for combined axial and bending stresses and buckling, several limitations have been placed on the design. The support column must have an inside diameter of at least 0.70 m (d_i) to allow for piping and ladder access to the interior of the tank. To prevent local buckling of the column walls the diameter/thickness ratio (d_o/t) shall not be greater than 92. The large mass of water and steel makes deflections critical as they add to the bending moment. The deflection effects as well as an assumed construction eccentricity (e) of 10 cm must be accounted for in the design process. Deflection at C.G. of the tank should not be greater than Δ.

    Limits on the inner radius and wall thickness are $0.35 \leqq R \leqq 2.0$ m and $1.0 \leqq t \leqq 20$ cm.

Pertinent constants and formulae

Height of water tank,	$h = 10$ m
Allowable deflection,	$\Delta = 20$ cm
Unit weight of water,	$\gamma_w = 10$ kN/m^3
Unit weight of steel,	$\gamma_s = 80$ kN/m^3
Modulus of elasticity,	$E = 210$ GPa

Moment of inertia of the column, $\quad I = \dfrac{\pi}{64}[d_o^4 - (d_o - 2t)^4]$

Cross-sectional area of column material, $\quad A = \pi t(d_o - t)$

Allowable bending stress, $\sigma_b = 165$ MPA

Allowable axial stress, $\qquad\qquad\qquad\qquad \sigma_a = \dfrac{12\pi^2 E}{92(H/r)^2}$ (calculated using

the critical buckling load with a factor of safety of $\frac{23}{12}$)

Radius of gyration, $\qquad\qquad\qquad\qquad r = \sqrt{I/A}$

Average thickness of tank wall, $t_t = 1.5 \text{ cm}$

Volume of tank, $V = 1.2\pi D^2 h$

Surface area of tank, $A_s = 1.25\pi D^2$

Projected area of tank, for wind loading, $A_p = \dfrac{2Dh}{3}$

Load on the column due to weight of water and steel tank,

$$P = V\gamma_w + A_s t_t \gamma_s$$

Lateral load at the tank C.G. due to wind pressure, $W = wA_p$.

Deflection at C.G. of tank, $\delta = \delta_1 + \delta_2$, where

$$\delta_1 = \frac{WH^2}{12EI}(4H + 3h)$$

$$\delta_2 = \frac{H}{2EI}(0.5Wh + Pe)(H + h)$$

Moment at base, $M = W(H + 0.5h) + (\delta + e)P$

Bending stress, $f_b = \dfrac{M}{2I}d_o$

Axial stress, $f_a(= P/A) = \dfrac{V\gamma_w + A_s\gamma_s t_t}{\pi t(d_o - t)}$

Combined stress constraint, $\dfrac{f_a}{\sigma_a} + \dfrac{f_b}{\sigma_b} \leq 1$

Gravitational acceleration, $g = 9.81 \text{ m/s}^2$

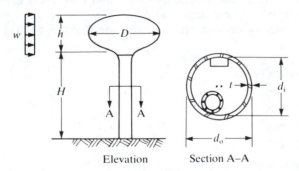

| Elevation | Section A–A | **FIGURE E2.62**
Water tower support column. |

2.63* Design of a flag pole. Your consulting firm has been asked to design a minimum mass flag pole of height H. The pole will be made of uniform hollow circular tube with d_o and d_i as outer and inner diameters, respectively. The pole must not fail under the action of high winds.

For design purposes, the pole will be treated as a cantilever that is subjected to a uniform lateral wind load of w (kN/m). In addition to the uniform load, the wind induces a concentrated load of P (kN) at the top of the pole, as shown in Fig. E2.63. The flag pole must not fail in bending or shear.

The deflection at the top should not exceed 10 cm. The ratio of mean diameter to thickness must not exceed 60. The pertinent data are given below. Assume any other data if needed. The minimum and maximum values of design variables are $5 \leq d_o \leq 50$ cm and $4 \leq d_i \leq 45$ cm.

Pertinent constants and equations

Cross-sectional area, $\qquad A = \dfrac{\pi}{4}(d_o^2 - d_i^2)$

Moment of inertia, $\qquad I = \dfrac{\pi}{64}(d_o^4 - d_i^4)$

Modulus of elasticity, $\qquad E = 210$ GPa

Allowable bending stress, $\sigma_b = 165$ MPa

Allowable shear stress, $\qquad \tau_s = 50$ MPa

Mass density, $\qquad \rho = 7800$ kg/m^3

Wind load, $\qquad w = 2.0$ kN/m

Height of flag pole, $\qquad H = 10$ m

Concentrated load at top, $\quad P = 4.0\,kN$

Moment at the base, $\qquad M = (PH + 0.5wH^2)$, kN $\cdot$ m

Bending stress, $\qquad \sigma = \dfrac{M}{2I}d_o$, kPa

Shear at the base, $\qquad S = (P + wH)$, kN

Shear stress, $\qquad \tau = \dfrac{S}{12I}(d_o^2 + d_o d_i + d_i^2)$, kPa

Deflection at the top, $\qquad \delta = \dfrac{PH^3}{3EI} + \dfrac{wH^4}{8EI}$

Minimum and maximum thickness, 0.5 and 2 cm

Formulate the design problem and solve it using the graphical optimization technique.

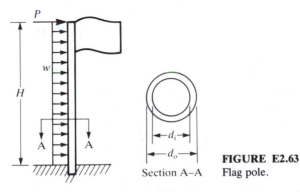

FIGURE E2.63
Section A–A   Flag pole.

2.64* Design of a sign support column (created by H. Kane). The design department of a company has been asked to design a support column of minimum weight for

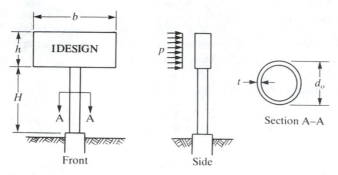

FIGURE E2.64
Sign support column.

the sign shown. The height to the bottom of the sign H, the width of the sign b, and the wind pressure p on the sign are as follows: $H = 20$ m, $b = 8$ m, $p = 800$ N/m² (Fig. E2.64).

The sign itself weighs 2.5 kN/m²(w). The column must be safe with respect to combined axial and bending stresses. The allowable axial stress includes a factor of safety with respect to buckling. To prevent local buckling of the plate the diameter/thickness ratio d_o/t must not exceed 92. Note that the bending stress in the column will increase as a result of the deflection of the sign under the wind load. The maximum deflection at the center of gravity of the sign should not exceed 0.1 m. The minimum and maximum values of design variables are $25 \leq d_o \leq 150$ cm and $0.5 \leq t \leq 10$ cm.

Pertinent constants and equations

Height of the sign, $\qquad h = 4.0$ m

For column section

Area, $\qquad A = \dfrac{\pi}{4}[d_o^2 - (d_o - 2t)^2]$

Moment of inertia, $\qquad I = \dfrac{\pi}{64}(d_o^4 - (d_o - 2t)^4)$

Radius of gyration, $\qquad r = \sqrt{I/A}$

Young's modulus (aluminum alloy), $\qquad E = 75$ GPa

Unit weight of steel, $\qquad \gamma = 80$ kN/m³

Allowable bending stress, $\qquad \sigma_b = 140$ MPa

Allowable axial stress, $\qquad \sigma_a = \dfrac{12\pi^2 E}{92\,(H/r)^2}$

Wind force, $\qquad F = pbh$

Weight of sign, $\qquad W = wbh$

Deflection at center of gravity of sign, $\quad \delta = \dfrac{F}{EI}\left(\dfrac{H^3}{3} + \dfrac{H^2 h}{2} + \dfrac{Hh^2}{4}\right)$

Moment in column bending stress, $f_b = \dfrac{M}{2I} d_o$

Axial stress, $\qquad\qquad\qquad\qquad f_a = \dfrac{W}{A}$

Moment at the base, $\qquad\qquad M = F\left(H + \dfrac{h}{2}\right) + W\delta$

Combined stress requirement, $\qquad \dfrac{f_a}{\sigma_a} + \dfrac{f_b}{\sigma_b} \leqq 1$

2.65* Design of a tripod. Design a minimum mass tripod of height H to support a vertical load $W = 60\,\text{kN}$. The tripod base is an equilateral triangle with sides $B = 1200\,\text{mm}$. The struts have a solid circular cross-section of diameter D (Fig. E2.65).

The axial stress in the struts must not exceed the allowable stress in compression, and the axial load in the strut P must not exceed the critical buckling load P_{cr} divided by a safety factor $\text{FS} = 2$. Use consistent units of Newtons and centimeters. The minimum and maximum values for design variables are $0.5 \leqq H \leqq 5\,\text{m}$ and $0.5 \leqq D \leqq 50\,\text{cm}$. Material properties and other relationships are given below:

Material: aluminum alloy 2014-T6

Allowable compressive stress, $\sigma_a = 150\,\text{MPa}$

Young's modulus, $\qquad\qquad E = 75\,\text{GPa}$

Mass density, $\qquad\qquad\quad \rho = 2800\,\text{kg/m}^3$

Strut length, $\qquad\qquad\qquad l = (H^2 + \tfrac{1}{3}B^2)^{0.5}$

Critical buckling load, $\qquad P_{cr} = \dfrac{\pi^2 EI}{l^2}$

Moment of inertia, $\qquad\qquad I = \dfrac{\pi}{64} D^4$

Strut load, $\qquad\qquad\qquad\quad P = \dfrac{Wl}{3H}$

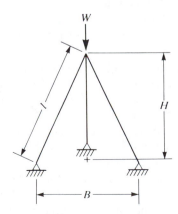

FIGURE E2.65
A tripod.

CHAPTER
3

OPTIMUM DESIGN CONCEPTS

3.1 INTRODUCTION

In Chapter 2, a proper mathematical formulation of a design optimization problem is described. The problem is always converted to minimization of a cost function with proper constraints on design variables. In this chapter, we discuss *basic ideas, concepts and theories used for design optimization* (the minimization problem). Several theorems on the subject are stated without proofs. Their implications and use in the optimization process are discussed. The student is reminded to review the basic terminology and notation explained in Section 1.7 as these are used throughout the present chapter and the remaining text.

It must be realized that the overall process of designing systems in different fields of engineering is roughly the same. Analytical and numerical methods for analyzing various systems can differ somewhat. Statement of the design problem can contain terminology that is specific to the particular domain of application. For example, in the fields of structural, mechanical and aerospace engineering, we are concerned with the integrity of the structure and its components. The performance requirements involve constraints on member stresses, strains, deflections at key points, frequencies of vibration, buckling failure, etc. These terms are specific to the fields and designers working in the area understand their meaning and the constraints. Similarly, other fields of engineering have

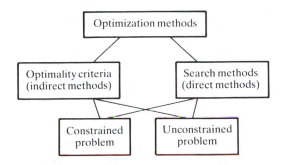

FIGURE 3.1
Classification of optimization methods.

their own terminology to describe *design optimization problems. However, once the problems from different fields have been transcribed into mathematical statements using a standard notation, they all look alike.* They are contained in the standard design optimization model defined in Eqs (2.10) to (2.12). For example, all the problems formulated in Section 2.6 can be transformed into the form of Eqs (2.10) to (2.12). *Methods of optimization described in the text are quite general and can be used to solve the problems. They can be developed without reference to any design application.* This *key point* must be kept in mind while studying the optimization concepts and the methods.

As an overview of the material of the present and the remaining chapters, we show a broad classification of the optimization techniques in Fig. 3.1. Two philosophically different viewpoints are shown. It is important to understand the features – limitations and advantages – of the two approaches to gain insights for practical applications of optimization. The two categories are indirect or optimality criteria methods and direct or search methods. *Optimality criteria* are the conditions a function must satisfy at its minimum point. Minimization techniques seeking solutions to optimality conditions are often called *indirect methods*. The *direct (search) techniques* are based on a different philosophy. There we start with an estimate of the optimum design for the problem. Usually the starting design will not satisfy optimality criteria. Therefore, it is improved iteratively until they are satisfied. Thus, in this approach we search the design space for optimum points. We shall address unconstrained and constrained optimization problems under both categories.

A thorough knowledge of optimality conditions is important to understand the performance of various numerical methods discussed later in the text. This chapter discusses these conditions and methods based on them, and uses simple examples to explain the underlying concepts and ideas. The examples will also show practical limitations of the methods based on optimality criteria. The search methods are discussed in the remaining chapters. Those methods refer to the results discussed in this chapter. Therefore, the *material in the present chapter should be understood thoroughly*. We will first discuss the concept of local optimum of a function and the conditions that characterize it. The problem of global optimality of a function is discussed later in this chapter. It is assumed throughout the text that all the problem functions are *twice continuously differentiable*.

The following is an outline of this chapter.

Section 3.2 Fundamental Concepts. To discuss optimal design concepts we need basic ideas from vector and matrix algebra, and vector calculus. A thorough review of this basic material is given. The concepts of *local and global minima* are defined and illustrated. The *differentiation notation* for functions of several variables is also introduced. The *gradient vector* for a function of several variables is defined. It requires first partial derivatives of the function. The *Hessian matrix* for the function is then defined. It requires second partial derivatives of the function. *Taylor series expansions* for functions of single and several variables are discussed. The idea of Taylor series expansion is fundamental to the development of optimal design concepts and numerical methods, so it should be thoroughly understood. The concept of *quadratic forms* is needed to develop sufficiency conditions for optimality. Therefore, notation and analyses related to quadratic forms are described. The concepts of necessary and sufficient conditions are explained.

Section 3.3 Unconstrained Optimum Design Problems. Here, we consider the problem of minimizing a function with no constraints on design variables. Necessary and sufficient conditions for a local minimum are discussed and illustrated in several examples.

Section 3.4 Constrained Optimum Design Problems. This section describes theory for constrained optimization problems. The necessary conditions for an equality constrained problem are first discussed. These conditions are contained in the Lagrange Multiplier Theorem generally discussed in textbooks on calculus. The necessary conditions for the general constrained problem are obtained as an extension of the Lagrange Multiplier Theorem. Kuhn–Tucker necessary conditions for a general optimum design problem are then discussed. They are illustrated in several examples.

Section 3.5 Global Optimality. A question about global optimality of an optimum design always arises. In general, it is difficult to answer the question satisfactorily. However, for a class of problems, known as the convex programming problems, it is possible to determine the global optimum solution. To study them, we need the concept of convex functions. Therefore, basic ideas of convex sets and functions are developed. Checks for convexity of a function and the feasible region are described. A theorem is given to identify global optimum designs.

Section 3.6 Second-order Conditions for Constrained Optimization. These are first discussed for convex programming problems and then for the general problem. Second-order necessary and sufficiency conditions are discussed and illustrated with examples. They require calculation of Hessians of cost and constraint functions.

Section 3.7 Post-optimality Analysis: Physical Meaning of Lagrange Multipliers. The physical interpretation allows a designer to study the effect of loosening or tightening constraints of the design problem. The Lagrange multipliers at optimum point provide the necessary information to study this effect. Thus the multipliers can be used to great advantage in practical design. This is explained in detail. Effects of scaling the cost and constraint functions on the Lagrange multipliers are also explained.

Section 3.8 Engineering Design Examples. Methods described in the previous sections are used to solve some engineering design problems. Convexity of the problems is checked, necessary and sufficient conditions are illustrated and sensitivity analysis is performed.

3.2 FUNDAMENTAL CONCEPTS

Optimality conditions for a minimum point are discussed in later sections. These conditions use ideas from fundamental calculus. Therefore, in this section, *we review basic concepts from calculus using the vector and matrix notations.* It is extremely important to understand and be comfortable with these fundamental concepts and linear algebra which is described in Appendix B.

The concepts of *gradient and Hessian* of a function are explained. *Taylor's* expansion of a function plays a key role in optimization methods and theory. Therefore it is discussed for functions of single and several variables. Quadratic forms appear in second-order conditions of optimality and various concepts related to them are discussed.

3.2.1 Minimum

In Section 2.7, we defined the *constraint set S* (also called the *feasible region*) for a design problem as simply a collection of feasible designs. Since there are no constraints in unconstrained problems, the entire design space is feasible for them.

The *optimization problem* is to find a design in the feasible region which gives a minimum value to the cost function. Methods to locate optimum designs are discussed throughout the text. We must first carefully define what is meant by an optimum. In the following discussion, $\mathbf{x}^*$ is used to designate a particular point of the constraint set.

3.2.1.1 GLOBAL (ABSOLUTE) MINIMUM. A function $f(\mathbf{x})$ of n variables has global (absolute) minimum at $\mathbf{x}^*$ if

$$f(\mathbf{x}^*) \leqq f(\mathbf{x}) \tag{3.1}$$

for all $\mathbf{x}$ in the feasible region (the set S). If strict inequality holds for all $\mathbf{x}$ other than $\mathbf{x}^*$ in Eq. (3.1), then $\mathbf{x}^*$ is called a *strict global minimum*.

3.2.1.2 LOCAL (RELATIVE) MINIMUM. A function $f(\mathbf{x})$ of n variables has a local (relative) minimum at $\mathbf{x}^*$ if Inequality (3.1) holds for all $\mathbf{x}$ in a small *neighborhood N* of $\mathbf{x}^*$ in the feasible region (the set S). If strict inequality holds, then $\mathbf{x}^*$ is called a *strict local minimum*. Neighborhood N of the point $\mathbf{x}^*$ is mathematically defined as the set of points

$$N = \{\mathbf{x} \mid \mathbf{x} \in S \quad \text{with} \quad \|\mathbf{x} - \mathbf{x}^*\| < \delta\}$$

for some small $\delta > 0$. Geometrically, it is a small feasible region containing the point $\mathbf{x}^*$.

Note that a function $f(\mathbf{x})$ can have *strict global minimum* at only one point. It may, however, have global minimum at several points if it has the same value at each of those points. Similarly, a function $f(\mathbf{x})$ can have *strict local minimum* at only one point in the neighborhood N of $\mathbf{x}^*$. It may, however, have local minimum at several points in N if the function value is the same at each of those points. We note here that these definitions do not provide a method for locating minimum points. Based on them, however, we can develop analyses and computational procedures to locate them. Note that we can define *global* and *local maxima* in a similar manner.

To understand *graphical significance* of global and local minima, consider graphs of a function $f(x)$ shown in Fig. 3.2. In Part (a) of the figure where x is between $-\infty$ and ∞ ($-\infty \leqq x \leqq \infty$), points B and D are local minima since the function has smallest value in their neighborhood. Similarly, both A and C are points of local maxima for the function. There is, however, no global minimum or maximum for the function since the domain and the function $f(x)$ are

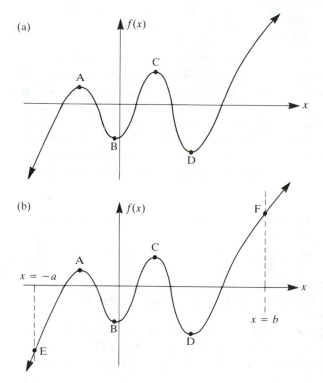

FIGURE 3.2
Graphical representation of optimum points. (a) Unbounded domain and function (no global optimum). (b) Bounded domain and function (global minimum and maximum exist).

unbounded, i.e. x and $f(x)$ are allowed to have any value between $-\infty$ and $+\infty$. If we restrict x to lie between $-a$ and b as in Part (b) of Fig. 3.2, then point E gives the global minimum and F the global maximum for the function.

In general we do not know before attempting to solve a problem if a minimum even exists. In certain cases we can *ensure existence of a minimum* even though we may not know how to find it. The Weierstrass theorem guarantees this when certain conditions are satisfied.

> **Theorem 3.1 Weierstrass Theorem – existence of global minimum.** If $f(\mathbf{x})$ is continuous on a nonempty feasible set S which is closed and bounded, then $f(\mathbf{x})$ has a global minimum in S. ‖

To use the theorem we must understand the meaning of a *closed and bounded set*. A set S is *closed* if it includes all its boundary points and every sequence of points has a subsequence that converges to a point in the set. A set is *bounded* if for any point $\mathbf{x} \in S$, $\mathbf{x}^T\mathbf{x} < c$, where c is a finite number. Since domain of the function in Fig. 3.2(a) is not closed and the function is also unbounded, a global minimum or maximum for the function is not assured. Actually, there is no global minimum or maximum for the function. However, in Fig. 3.2(b), since the feasible region is closed and bounded with $-a \leqq x \leqq b$ and the function is continuous, it has global minimum as well as maximum points. The following example further illustrates these concepts.

> **Example 3.1 Existence of global minimum using Weierstrass Theorem.** Consider a function $f(x) = -1/x$ defined on the set $S = \{x \mid 0 < x \leqq 1\}$. The feasible set S is not closed since it does not include the boundary point $x = 0$. Conditions of the Weierstrass theorem are not satisfied, although f is continuous on S. Existence of a global minimum is not guaranteed and indeed there is no point x^* satisfying $f(x^*) \leqq f(x)$ for all $x \in S$. If we define $S = \{x \mid 0 \leqq x \leqq 1\}$, then the feasible region is closed and bounded. However, f is not defined at $x = 0$ (hence not continuous), so the conditions of the theorem are still not satisfied and there is no guarantee of a global minimum for f in the set S. ‖

Note that when conditions of the Weierstrass theorem are satisfied, existence of global optimum is guaranteed. It is important, however, to realize that when they are not satisfied, global solution may still exist. The theorem does not rule out this possibility. The difference is that we cannot guarantee its existence. Note also that the theorem does not give a method of finding global solution even if its conditions are satisfied. It is only the so-called existence theorem. Later we shall discuss methods for determining optimum solutions.

3.2.2 Gradient Vector

Since gradient of a function is used while discussing methods of optimum design, we define and discuss its geometrical significance. Also, the *differentiation notation* defined here is used throughout the text. Therefore, it should be clearly understood.

Consider a function $f(\mathbf{x})$ of n variables $x_1, x_2, \ldots, x_n$. The partial derivative of the function with respect to x_1 at a given point $\mathbf{x}^*$ is defined as $\partial f(\mathbf{x}^*)/\partial x_1$, with respect to x_2 as $\partial f(\mathbf{x}^*)/\partial x_2$, and so on. Let c_i represent the partial derivative of $f(\mathbf{x})$ with respect to x_i at the point $\mathbf{x}^*$. Then using the index notation of Section 1.7, we can represent all partial derivatives of $f(\mathbf{x})$ as follows:

$$c_i = \frac{\partial f(\mathbf{x}^*)}{\partial x_i}; \qquad i = 1 \text{ to } n \tag{3.2}$$

For convenience and compactness of notation, we arrange the partial derivatives $\partial f(\mathbf{x}^*)/\partial x_1$, $\partial f(\mathbf{x}^*)/\partial x_2, \ldots, \partial f(\mathbf{x}^*)/\partial x_n$ into a column vector, called the *gradient vector* and represent it by any of the following symbols: $\mathbf{c}$, ∇f, $\partial f/\partial \mathbf{x}$, grad f. Thus gradient of a function $f(\mathbf{x})$ of n variables $x_1, x_2, \ldots, x_n$ at a point $\mathbf{x}^*$ is defined as a column vector:

$$\nabla f(\mathbf{x}^*) = \begin{bmatrix} \dfrac{\partial f(\mathbf{x}^*)}{\partial x_1} \\[2mm] \dfrac{\partial f(\mathbf{x}^*)}{\partial x_2} \\[1mm] \vdots \\[1mm] \dfrac{\partial f(\mathbf{x}^*)}{\partial x_n} \end{bmatrix} = \left[\dfrac{\partial f(\mathbf{x}^*)}{\partial x_1} \ \ \dfrac{\partial f(\mathbf{x}^*)}{\partial x_2} \cdots \dfrac{\partial f(\mathbf{x}^*)}{\partial x_n} \right]^T \tag{3.3}$$

where superscript T denotes transpose of the row vector. Note that all partial derivatives are calculated at the given point $\mathbf{x}^*$. That is, each component of the gradient vector is a function in itself which must be evaluated at the given point $\mathbf{x}^*$.

Geometrically, the gradient vector is normal to the tangent plane at the point $\mathbf{x}^*$ as shown in Fig. 3.3 for a function of three variables. Also, it points in

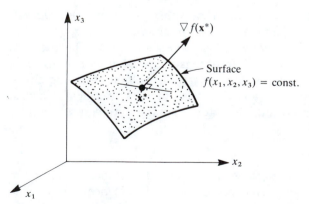

FIGURE 3.3
Gradient vector for $f(x_1, x_2, x_3)$ at the point $\mathbf{x}^*$.

the direction of *maximum increase* in the function. These properties are quite important, and will be proved and discussed in Chapter 5. They will be used in developing optimality conditions and numerical methods for optimum design.

Example 3.2 **Calculation of gradient vector.** Calculate the gradient vector for the function

$$f(\mathbf{x}) = (x_1 - 1)^2 + (x_2 - 1)^2$$

at the point $\mathbf{x}^* = (1.8, 1.6)$.

Solution. The given function is the equation for a circle with center at the point $(1, 1)$. The value of the function at $\mathbf{x}^*$ is given as

$$f(1.8, 1.6) = (1.8 - 1)^2 + (1.6 - 1)^2 = 1$$

So, point $(1.8, 1.6)$ lies on a circle of radius 1, shown as point A in Fig. 3.4. The partial derivatives for the function at point $(1.8, 1.6)$ are calculated as

$$\frac{\partial f}{\partial x_1}(1.8, 1.6) = 2(x_1 - 1) = 2(1.8 - 1) = 1.6$$

$$\frac{\partial f}{\partial x_2}(1.8, 1.6) = 2(x_2 - 1) = 2(1.6 - 1) = 1.2$$

Thus, the gradient vector for $f(\mathbf{x})$ at point $(1.8, 1.6)$ is given as

$$\mathbf{c} = \begin{bmatrix} 1.6 \\ 1.2 \end{bmatrix}$$

This is shown in Fig. 3.4. It can be seen that vector $\mathbf{c}$ is normal to the circle at point $(1.8, 1.6)$. This is consistent with the observation that gradient is normal to the surface. ‖

3.2.3 Hessian Matrix

Differentiating the gradient vector once again (each component of the gradient vector is differentiated with respect to each x_i), we obtain a matrix of second

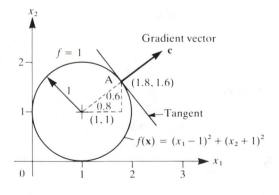

FIGURE 3.4
Gradient vector for the function $f(\mathbf{x})$ of Example 3.2 at the point $(1.8, 1.6)$.

partial derivatives for the function $f(\mathbf{x})$ called the *Hessian matrix,* or simply the Hessian. That is, differentiating each component of the gradient vector given in Eq. (3.3) with respect to $x_1, x_2, \ldots, x_n$, we obtain

$$\frac{\partial^2 f}{\partial \mathbf{x}\, \partial \mathbf{x}} = \begin{bmatrix} \dfrac{\partial^2 f}{\partial x_1^2} & \dfrac{\partial^2 f}{\partial x_1\, \partial x_2} & \cdots & \dfrac{\partial^2 f}{\partial x_1\, \partial x_n} \\[2mm] \dfrac{\partial^2 f}{\partial x_2\, \partial x_1} & \dfrac{\partial^2 f}{\partial x_2^2} & \cdots & \dfrac{\partial^2 f}{\partial x_2\, \partial x_n} \\[2mm] \vdots & \vdots & & \vdots \\[2mm] \dfrac{\partial^2 f}{\partial x_n\, \partial x_1} & \dfrac{\partial^2 f}{\partial x_n\, \partial x_2} & \cdots & \dfrac{\partial^2 f}{\partial x_n^2} \end{bmatrix} \tag{3.4}$$

where all derivatives are calculated at the given point $\mathbf{x}^*$. Hessian is an $n \times n$ matrix usually denoted as $\mathbf{H}$ or $\nabla^2 f$. *It is important to note that each element of the Hessian is a function in itself which is evaluated at the given point* $\mathbf{x}^*$. Also, since $f(\mathbf{x})$ is assumed to be twice continuously differentiable, the cross partial derivatives are equal, i.e.

$$\frac{\partial^2 f}{\partial x_i\, \partial x_j} = \frac{\partial^2 f}{\partial x_j\, \partial x_i}; \qquad i = 1 \text{ to } n,\ j = 1 \text{ to } n$$

Therefore, *the Hessian is always a symmetric matrix.* It plays a prominent role in the sufficiency conditions for optimality discussed later in this chapter. It will be written as

$$\mathbf{H} = \left[\frac{\partial^2 f}{\partial x_i\, \partial x_j} \right]; \qquad i = 1 \text{ to } n,\ j = 1 \text{ to } n \tag{3.5}$$

Example 3.3 Evaluation of gradient and Hessian of a function. For the following function, calculate the gradient vector and the Hessian matrix at the point $(1, 2)$:

$$f(\mathbf{x}) = x_1^3 + x_2^3 + 2x_1^2 + 3x_2^2 - x_1 x_2 + 2x_1 + 4x_2$$

Solution. The first partial derivatives of the function are given as

$$\frac{\partial f}{\partial x_1} = 3x_1^2 + 4x_1 - x_2 + 2$$

$$\frac{\partial f}{\partial x_2} = 3x_2^2 + 6x_2 - x_1 + 4$$

Substituting the point $x_1 = 1$, $x_2 = 2$, the gradient vector is given as

$$\nabla f(1, 2) = \begin{bmatrix} 7 \\ 27 \end{bmatrix}$$

The second partial derivatives of the function are calculated as

$$\frac{\partial^2 f}{\partial x_1^2} = 6x_1 + 4; \qquad \frac{\partial^2 f}{\partial x_1 \partial x_2} = -1$$

$$\frac{\partial^2 f}{\partial x_2 \partial x_1} = -1; \qquad \frac{\partial^2 f}{\partial x_2^2} = 6x_2 + 6.$$

Therefore, Hessian matrix at the point $(1, 2)$ is given as

$$\mathbf{H}(1, 2) = \begin{bmatrix} 10 & -1 \\ -1 & 18 \end{bmatrix} \qquad\qquad \|$$

3.2.4 Taylor Series Expansion

A function can be approximated by polynomials in a neighborhood of any point in terms of its value and derivatives using Taylor series expansion. Consider first a function $f(x)$ of a single variable. Taylor's expansion about the point x^* is

$$f(x) = f(x^*) + \frac{df(x^*)}{dx}(x - x^*) + \frac{1}{2}\frac{d^2 f(x^*)}{dx^2}(x - x^*)^2 + R \qquad (3.6)$$

where R is the *remainder* term that is smaller in magnitude than the previous terms if x is sufficiently close to x^*. If we let $x - x^* = d$ (a small change in the point x^*), Taylor's expansion of Eq. (3.6) becomes

$$f(x^* + d) = f(x^*) + \frac{df(x^*)}{dx}d + \frac{1}{2}\frac{d^2 f(x^*)}{dx^2}d^2 + R \qquad (3.7)$$

For a *function of two variables* $f(x_1, x_2)$, Taylor's expansion at the point (x_1^*, x_2^*) is

$$f(x_1, x_2) = f(x_1^*, x_2^*) + \frac{\partial f}{\partial x_1}(x_1 - x_1^*) + \frac{\partial f}{\partial x_2}(x_2 - x_2^*)$$

$$+ \frac{1}{2}\left[\frac{\partial^2 f}{\partial x_1^2}(x_1 - x_1^*)^2 + 2\frac{\partial^2 f}{\partial x_1 \partial x_2}(x_1 - x_1^*)(x_2 - x_2^*)\right.$$

$$\left. + \frac{\partial^2 f}{\partial x_2^2}(x_2 - x_2^*)^2\right] + R \qquad (3.8)$$

where all partial derivatives are calculated at the given point (x_1^*, x_2^*). The arguments of these partial derivatives are omitted in Eq. (3.8) and in all subsequent discussions for notational compactness. Taylor's expansion in Eq. (3.8) can be written using the *summation notation* defined in Chapter 1 as

$$f(x_1, x_2) = f(x_1^*, x_2^*) + \sum_{i=1}^{2} \frac{\partial f}{\partial x_i}(x_i - x_i^*)$$

$$+ \frac{1}{2}\sum_{i=1}^{2}\sum_{j=1}^{2} \frac{\partial^2 f}{\partial x_i \partial x_j}(x_i - x_i^*)(x_j - x_j^*) + R \qquad (3.9)$$

It can be seen that by expanding the summations in Eq. (3.9), Eq. (3.8) is obtained. Recognizing the quantities $\partial f / \partial x_i$ as components of the gradient of the function given in Eq. (3.3) and $\partial^2 f / \partial x_i \, \partial x_j$ as the Hessian of Eq. (3.5) evaluated at the given point $\mathbf{x}^*$, Taylor's expansion can also be written in *matrix notation* as

$$f(\mathbf{x}) = f(\mathbf{x}^*) + \nabla f^T (\mathbf{x} - \mathbf{x}^*) + \tfrac{1}{2}(\mathbf{x} - \mathbf{x}^*)^T \mathbf{H}(\mathbf{x} - \mathbf{x}^*) + R \qquad (3.10)$$

where $\mathbf{x} = (x_1, x_2)$, $\mathbf{x}^* = (x_1^*, x_2^*)$, and $\mathbf{H}$ is the 2×2 Hessian matrix. Note that with matrix notation, Taylor's expansion in Eq. (3.10) can be easily generalized to functions of n variables. In that case, $\mathbf{x}$, $\mathbf{x}^*$ and ∇f are n dimensional vectors and $\mathbf{H}$ is the $n \times n$ Hessian matrix. Defining $\mathbf{x} - \mathbf{x}^* = \mathbf{d}$, Eq. (3.10) becomes

$$f(\mathbf{x}^* + \mathbf{d}) = f(\mathbf{x}^*) + \nabla f^T \mathbf{d} + \tfrac{1}{2}\mathbf{d}^T \mathbf{H} \mathbf{d} + R \qquad (3.11)$$

Often a change in the function is desired when $\mathbf{x}^*$ moves to a neighboring point $\mathbf{x}$. Defining the change as $\Delta f = f(\mathbf{x}^* + \mathbf{d}) - f(\mathbf{x}^*)$, Eq. (3.11) gives

$$\Delta f = \nabla f^T \mathbf{d} + \tfrac{1}{2}\mathbf{d}^T \mathbf{H} \mathbf{d} + R \qquad (3.12)$$

A *first-order change* in $f(\mathbf{x})$ at $\mathbf{x}^*$ (denoted as δf) is obtained by retaining only the first term in Eq. (3.12),

$$\delta f = \nabla f^T \delta \mathbf{x} \qquad (3.13)$$

where $\delta \mathbf{x}$ is a small change in $\mathbf{x}^*(\delta \mathbf{x} = \mathbf{x} - \mathbf{x}^*)$. Note that the first-order change of the function given in Eq. (3.13) is simply a dot product of the vectors ∇f and $\delta \mathbf{x}$. A first order change is an acceptable approximation for change in the original function when $\mathbf{x}$ is near $\mathbf{x}^*$.

We now consider some functions and approximate them at the given point $\mathbf{x}^*$ using Taylor's expansion. The remainder R will be dropped while using Eq. (3.11).

Example 3.4 Taylor's expansion of a function of one variable. Approximate $f(x) = \cos x$ around the point $x^* = 0$.

Solution. Derivatives of the function $f(x)$ are given as

$$\frac{df}{dx} = -\sin x, \qquad \frac{d^2 f}{dx^2} = -\cos x$$

Therefore, using Eq. (3.6), second-order Taylor's expansion for $\cos x$ at the point $x^* = 0$ is given as

$$\cos x \approx \cos 0 - \sin 0 (x - 0) + \tfrac{1}{2}(-\cos 0)(x - 0)^2$$
$$\approx 1 - \tfrac{1}{2}x^2 \qquad \qquad \|$$

Example 3.5 Taylor's expansion of a function of two variables. Obtain second-order Taylor's expansion for the function

$$f(\mathbf{x}) = 3x_1^3 x_2$$

at the point $\mathbf{x}^* = (1, 1)$.

Solution. Gradient and Hessian of the function $f(\mathbf{x})$ at the point $\mathbf{x}^* = (1, 1)$ using Eqs (3.3) and (3.5) are

$$\nabla f = \begin{bmatrix} \dfrac{\partial f}{\partial x_1} \\ \dfrac{\partial f}{\partial x_2} \end{bmatrix} = \begin{bmatrix} 9x_1^2 x_2 \\ 3x_1^3 \end{bmatrix} = \begin{bmatrix} 9 \\ 3 \end{bmatrix}$$

$$\mathbf{H} = \begin{bmatrix} 18x_1 x_2 & 9x_1^2 \\ 9x_1^2 & 0 \end{bmatrix} = \begin{bmatrix} 18 & 9 \\ 9 & 0 \end{bmatrix}$$

Substituting these in the matrix form of Taylor's expression given in Eq. (3.10), and using $\mathbf{d} = \mathbf{x} - \mathbf{x}^*$, we obtain an approximation $\bar{f}(\mathbf{x})$ for $f(\mathbf{x})$ as

$$\bar{f}(\mathbf{x}) = 3 + \begin{bmatrix} 9 \\ 3 \end{bmatrix}^T \begin{bmatrix} (x_1 - 1) \\ (x_2 - 1) \end{bmatrix} + \frac{1}{2} \begin{bmatrix} (x_1 - 1) \\ (x_2 - 1) \end{bmatrix}^T \begin{bmatrix} 18 & 9 \\ 9 & 0 \end{bmatrix} \begin{bmatrix} (x_1 - 1) \\ (x_2 - 1) \end{bmatrix}$$

where $f(\mathbf{x}^*) = 3$ has been used. Simplifying the expression by expanding vector and matrix products, we obtain Taylor's expansion for $f(\mathbf{x})$ about the point $(1, 1)$ as

$$\bar{f}(\mathbf{x}) = 9x_1^2 + 9x_1 x_2 - 18x_1 - 6x_2 + 9$$

This expression is a second-order approximation of the function $3x_1^3 x_2$ about the point $\mathbf{x}^* = (1, 1)$. That is, in a small neighborhood of $\mathbf{x}^*$, the expression will give almost the same value as the original function $f(\mathbf{x})$. To see how accurately $\bar{f}(\mathbf{x})$ approximates $f(\mathbf{x})$, Table 3.1 compares values of the approximate function with the original function for some points in the neighborhood of $(1, 1)$. It can be seen that even up to 30% change in the point, the approximate function has an error of only 4%. After that point, the error starts to increase. Thus, the approximate function $\bar{f}(\mathbf{x})$ represents the original function quite accurately in a small neighborhood of the point $(1, 1)$. ‖

TABLE 3.1

Accuracy of the second-order Taylor series expansion about $(1, 1)$ for the function $f(\mathbf{x}) = 3x_1^3 x_2$ of Example 3.5

x_1	x_2	$f(\mathbf{x})$	$\bar{f}(\mathbf{x})$	% Error
1.00	1.00	3.0000	3.0000	0.000
1.05	1.05	3.6465	3.6450	0.041
1.10	1.10	4.3923	4.3800	0.280
1.15	1.15	5.2470	5.2050	0.800
1.20	1.20	6.2208	6.1200	1.620
1.25	1.25	7.3242	7.1250	2.720
1.30	1.30	8.5683	8.2200	4.065
1.35	1.35	9.9645	9.4050	5.615
1.40	1.40	11.5248	10.6800	7.330
1.45	1.45	13.2615	12.0450	9.173
1.50	1.50	15.1875	13.5000	11.111

Example 3.6 Linear Taylor series expansion of a function. Obtain linear Taylor series expansion for the function

$$f(\mathbf{x}) = x_1^2 + x_2^2 - 4x_1 - 2x_2 + 4$$

at the point $\mathbf{x}^* = (1, 2)$. Compare the approximate function with the original function in a neighborhood of the point $(1, 2)$.

Solution. The gradient of the function at the point $(1, 2)$ is given as

$$\nabla f = \begin{bmatrix} \dfrac{\partial f}{\partial x_1} \\[2mm] \dfrac{\partial f}{\partial x_2} \end{bmatrix} = \begin{bmatrix} (2x_1 - 4) \\ (2x_2 - 2) \end{bmatrix} = \begin{bmatrix} -2 \\ 2 \end{bmatrix}$$

Since $f(1, 2) = 1$, Eq. (3.10) gives linear Taylor series approximation for $f(\mathbf{x})$ as

$$\bar{f}(\mathbf{x}) = 1 + [-2 \quad 2]\begin{bmatrix} (x_1 - 1) \\ (x_2 - 2) \end{bmatrix}$$

$$= -2x_1 + 2x_2 - 1$$

To see how accurately $\bar{f}(\mathbf{x})$ approximates the original $f(\mathbf{x})$ in the neighborhood of $(1, 2)$, we calculate the function values at several points as shown in Table 3.2. We see that even up to a change of 20% in one of the variables, the maximum error between $f(\mathbf{x})$ and $\bar{f}(\mathbf{x})$ is only about 8%. When both the variables are changed by 20% the maximum error is about 12.5%. These errors are quite acceptable in many applications. Note, however, that the errors will be different for different functions and can be larger for highly nonlinear functions.   ‖

3.2.5 Quadratic Forms and Definite Matrices

3.2.5.1 QUADRATIC FORM. *Quadratic form is a special nonlinear function having only second-order terms,* e.g. the function

$$F(\mathbf{x}) = x_1^2 + 2x_2^2 + 3x_3^2 + 2x_1x_2 + 2x_2x_3 + 2x_3x_1$$

Quadratic forms play a prominent role in optimization theory and methods. Therefore, in this subsection, we discuss some results related to them.

Consider a special function of n variables $F(\mathbf{x}) = F(x_1, x_2, \ldots, x_n)$ written in the double summation notation as (refer to Chapter 1 for discussion on the summation notation):

$$F(\mathbf{x}) = \tfrac{1}{2}\sum_{i=1}^{n}\sum_{j=1}^{n} p_{ij}x_ix_j \tag{3.14}$$

where p_{ij} are known constants and the factor $\frac{1}{2}$ is used for convenience only to match $F(\mathbf{x})$ with the second-order term in Taylor series expansion of Eq. (3.11). The results of this section will not be affected if the factor $\frac{1}{2}$ is not used, as is done in many other texts. The value of the quadratic form is affected by the factor. Expanding Eq. (3.14) by setting $i = 1$ and letting j vary

TABLE 3.2

Accuracy of linear approximation for the function $f(\mathbf{x}) = x_1^2 + x_2^2 - 4x_1 - 2x_2 + 4$ **of Example 3.6 at point (1, 2)**

x_1	x_2	$f(\mathbf{x})$	$\bar{f}(\mathbf{x})$	% Error
1.00	2.00	1.0000	1.0000	0.000
1.02	2.00	0.9604	0.9600	0.042
1.04	2.00	0.9216	0.9200	0.174
1.06	2.00	0.8836	0.8800	0.407
1.08	2.00	0.8464	0.8400	0.756
1.10	2.00	0.8100	0.8000	1.235
1.12	2.00	0.7744	0.7600	1.860
1.14	2.00	0.7396	0.7200	2.650
1.16	2.00	0.7056	0.6800	3.628
1.18	2.00	0.6724	0.6400	4.818
1.20	2.00	0.6400	0.6000	6.250
1.00	2.04	1.0816	1.0800	0.148
1.00	2.08	1.1632	1.1600	0.275
1.00	2.12	1.2544	1.2400	1.148
1.00	2.16	1.3456	1.3200	1.903
1.00	2.20	1.4400	1.4000	2.778
1.00	2.24	1.5376	1.4800	3.746
1.00	2.28	1.6384	1.5600	4.785
1.00	2.32	1.7424	1.6400	5.877
1.00	2.36	1.8496	1.7200	7.007
1.00	2.40	1.9600	1.8000	8.163
1.02	2.04	1.0420	1.0400	0.192
1.04	2.08	1.0880	1.0800	0.735
1.06	2.12	1.1380	1.1200	1.582
1.08	2.16	1.1920	1.1600	2.684
1.10	2.20	1.2500	1.2000	4.000
1.12	2.24	1.3120	1.2400	5.488
1.14	2.28	1.3780	1.2800	7.112
1.16	2.32	1.4480	1.3200	8.840
1.18	2.36	1.5220	1.3600	10.640
1.20	2.40	1.6000	1.4000	12.500

from 1 to n, and then setting $i = 2$ and letting j vary again from 1 to n, and so on, we get

$$
\begin{aligned}
F(\mathbf{x}) = \tfrac{1}{2}[&(p_{11}x_1^2 + p_{12}x_1x_2 + \ldots + p_{1n}x_1x_n) \\
&+ (p_{21}x_2x_1 + p_{22}x_2^2 + \ldots + p_{2n}x_2x_n) \\
&+ \ldots + (p_{n1}x_nx_1 + \ldots + p_{nn}x_n^2)]
\end{aligned}
\tag{3.15}
$$

Note that with coefficients p_{ij} specified and the variables x_i given, $F(\mathbf{x})$ in

Eq. (3.15) *is just a number (scalar).* The function is called a *quadratic form* because each of its terms is either the square of a variable or product of two different variables.

3.2.5.2 MATRIX OF THE QUADRATIC FORM. The quadratic form can be written in the matrix notation. Let $\mathbf{P} = [p_{ij}]$ be an $n \times n$ matrix, $\mathbf{x} = (x_1, x_2, \ldots, x_n)$ an n dimensional vector, and $\mathbf{y} = (y_1, y_2, \ldots, y_n)$ another n dimensional vector obtained by multiplying $\mathbf{P}$ by $\mathbf{x}$. Writing $\mathbf{y} = \mathbf{Px}$ in the summation notation we get

$$y_i = \sum_{j=1}^{n} p_{ij} x_j; \qquad i = 1 \text{ to } n \tag{3.16}$$

Also, we can rewrite Eq. (3.14) as

$$F(\mathbf{x}) = \tfrac{1}{2} \sum_{i=1}^{n} x_i \left(\sum_{j=1}^{n} p_{ij} x_j \right) \tag{3.17}$$

Substituting Eq. (3.16) into Eq. (3.17),

$$F(\mathbf{x}) = \tfrac{1}{2} \sum_{i=1}^{n} x_i y_i \tag{3.18}$$

But, the summation on the right-hand side of Eq. (3.18) represents the scalar product of vectors $\mathbf{x}$ and $\mathbf{y}$ as $\mathbf{x}^T \mathbf{y}$. Substituting $\mathbf{y} = \mathbf{Px}$ in this, we obtain the matrix representation for $F(\mathbf{x})$,

$$F(\mathbf{x}) = \tfrac{1}{2} \mathbf{x}^T \mathbf{y} = \tfrac{1}{2} \mathbf{x}^T \mathbf{Px} \tag{3.19}$$

$\mathbf{P}$ is called the *matrix of the quadratic form $F(\mathbf{x})$.* Elements of $\mathbf{P}$ are identified as coefficients of the terms in the function $F(\mathbf{x})$. For example, in Eq. (3.15) element p_{ij} is twice the coefficient of the term $x_i x_j$ in $F(\mathbf{x})$. We see that except for the squared terms, each product $x_i x_j (i \neq j)$ appears twice. Therefore, Eq. (3.15) can be rewritten as

$$\begin{aligned} F(\mathbf{x}) = \tfrac{1}{2} \{ &[p_{11} x_1^2 + p_{22} x_2^2 + \ldots + p_{nn} x_n^2] \\ &+ [(p_{12} + p_{21}) x_1 x_2 + (p_{13} + p_{31}) x_1 x_3 + \ldots + (p_{1n} + p_{n1}) x_1 x_n] \\ &+ [(p_{23} + p_{32}) x_2 x_3 + (p_{24} + p_{42}) x_2 x_4 + \ldots + (p_{2n} + p_{n2}) x_2 x_n] \\ &+ \ldots + [(p_{n-1,n} + p_{n,n-1}) x_{n-1} x_n] \} \end{aligned} \tag{3.20}$$

Thus, for $i \neq j$, the coefficient of $x_i x_j$ is $(p_{ij} + p_{ji})$. Define coefficients of an $n \times n$ matrix $\mathbf{A}$ as

$$a_{ij} = \tfrac{1}{2}(p_{ij} + p_{ji}), \qquad \text{for all } i \text{ and } j \tag{3.21}$$

Using this definition, it can be easily seen that

$$a_{ij} + a_{ji} = p_{ij} + p_{ji}$$

Therefore, $(p_{ij} + p_{ji})$ in Eq. (3.20) can be replaced with $(a_{ij} + a_{ji})$ and the quadratic form of Eq. (3.19) becomes

$$F(\mathbf{x}) = \tfrac{1}{2}\mathbf{x}^T \mathbf{P}\mathbf{x} = \tfrac{1}{2}\mathbf{x}^T \mathbf{A}\mathbf{x} \tag{3.22}$$

The value of the quadratic form does not change with $\mathbf{P}$ replaced by $\mathbf{A}$. The matrix $\mathbf{A}$, however, is always *symmetric* $(a_{ij} = a_{ji})$ whereas $\mathbf{P}$ may not be. Symmetry of $\mathbf{A}$ can be easily seen from the definition of a_{ij} given in Eq. (3.21), i.e. interchanging the indices i and j, we get $a_{ji} = a_{ij}$. Thus, given any quadratic form $\tfrac{1}{2}\mathbf{x}^T \mathbf{P}\mathbf{x}$ we can always replace $\mathbf{P}$ with a symmetric matrix. The preceding discussion also shows that many matrices can be associated with the same quadratic form. All of them are asymmetric except one. The symmetric matrix associated with it is always unique. Asymmetric matrices are not very useful. The symmetric matrix, however, determines the nature of the quadratic form which will be discussed later.

Comparing Eq. (3.11) with Eq. (3.22), we observe that the third term of Taylor's expansion is a quadratic form in the variables $\mathbf{d}$. Therefore, the Hessian $\mathbf{H}$ is a matrix associated with that quadratic form.

Example 3.7 Matrix of the Quadratic form. Identify a matrix associated with the quadratic form

$$F(x_1, x_2, x_3) = \tfrac{1}{2}(2x_1^2 + 2x_1x_2 + 4x_1x_3 - 6x_2^2 - 4x_2x_3 + 5x_3^2)$$

Solution. Writing F in the matrix form $(F(\mathbf{x}) = \tfrac{1}{2}\mathbf{x}^T \mathbf{P}\mathbf{x})$, we obtain

$$F(\mathbf{x}) = \tfrac{1}{2}[x_1 \quad x_2 \quad x_3] \begin{bmatrix} 2 & 2 & 4 \\ 0 & -6 & -4 \\ 0 & 0 & 5 \end{bmatrix} \begin{bmatrix} x_1 \\ x_2 \\ x_3 \end{bmatrix}$$

The matrix $\mathbf{P}$ of the quadratic form can be easily identified by comparing the expression with Eq. (3.20). The ith diagonal element p_{ii} is the coefficient of x_i^2. Therefore, $p_{11} = 2$, the coefficient of x_1^2; $p_{22} = -6$, the coefficient of x_2^2; and $p_{33} = 5$, the coefficient of x_3^2. The coefficient of $x_i x_j$ can be divided between the elements p_{ij} and p_{ji} of the matrix $\mathbf{P}$. However, the sum $p_{ij} + p_{ji}$ must be equal to the coefficient of $x_i x_j$. In the above matrix $p_{12} = 2$ and $p_{21} = 0$, giving $p_{12} + p_{21} = 2$ which is the coefficient of $x_1 x_2$. Similarly, we can calculate the elements p_{13}, p_{31}, p_{23} and p_{32}.

Since the coefficient of $x_i x_j$ can be divided between p_{ij} and p_{ji} in any proportion there are many matrices associated with a quadratic form. For example, the following matrix is also associated with the same quadratic form:

$$\begin{bmatrix} 2 & 0.5 & 1 \\ 1.5 & -6 & -6 \\ 3 & 2 & 5 \end{bmatrix}$$

Dividing the coefficients equally between p_{ij} and p_{ji}, we obtain

$$F(\mathbf{x}) = \tfrac{1}{2}[x_1 \quad x_2 \quad x_3] \begin{bmatrix} 2 & 1 & 2 \\ 1 & -6 & -2 \\ 2 & -2 & 5 \end{bmatrix} \begin{bmatrix} x_1 \\ x_2 \\ x_3 \end{bmatrix}$$

Any of the preceding expressions gives a matrix associated with the quadratic form. However, the first two matrices are asymmetric and the third one is symmetric. The diagonal elements of the symmetric matrix are obtained from the coefficient of x_i^2 as before. The off-diagonal elements are obtained by dividing the coefficient of the term $x_i x_j$ equally between a_{ij} and a_{ji}. This satisfies Eq. (3.21).

‖

3.2.5.3 FORM OF A MATRIX. Quadratic forms $F(\mathbf{x}) = \frac{1}{2}\mathbf{x}^T \mathbf{A}\mathbf{x}$ may be either positive, negative or zero for any fixed $\mathbf{x}$. It may also have the property of being always positive [except for $F(\mathbf{0})$]. Such a form is called positive definite. Similarly, it is called negative definite if $\mathbf{x}^T \mathbf{A}\mathbf{x} < 0$ for all $\mathbf{x}$ except $\mathbf{x} = \mathbf{0}$. If a quadratic form has the property $\mathbf{x}^T \mathbf{A}\mathbf{x} \geq 0$ for all $\mathbf{x}$ and there exists at least one $\mathbf{x} \neq \mathbf{0}$ (nonzero $\mathbf{x}$) with $\mathbf{x}^T \mathbf{A}\mathbf{x} = 0$, then it is called positive semidefinite. A similar definition for negative semidefinite is obtained by reversing the sense of the inequality. A quadratic form which is positive for some vectors $\mathbf{x}$ and negative for others is called indefinite. A symmetric matrix $\mathbf{A}$ is often referred to as a *positive definite, positive semidefinite, negative definite, negative semidefinite,* or *indefinite* if the quadratic form associated with $\mathbf{A}$ is positive definite, positive semidefinite, negative definite, negative semidefinite, or indefinite, respectively.

Example 3.8 Determination of the form of a matrix. Determine the form of the following matrices:

$$\text{(a) } \mathbf{A} = \begin{bmatrix} 2 & 0 & 0 \\ 0 & 4 & 0 \\ 0 & 0 & 3 \end{bmatrix} \qquad \text{(b) } \mathbf{A} = \begin{bmatrix} -1 & 1 & 0 \\ 1 & -1 & 0 \\ 0 & 0 & -1 \end{bmatrix}$$

Solution. The quadratic form associated with the matrix (a) is always positive, i.e.

$$\mathbf{x}^T \mathbf{A}\mathbf{x} = (2x_1^2 + 4x_2^2 + 3x_3^2) > 0$$

unless $x_1 = x_2 = x_3 = 0$ ($\mathbf{x} = \mathbf{0}$). Thus the matrix is positive definite.

The quadratic form associated with the matrix (b) is negative semidefinite, since

$$\mathbf{x}^T \mathbf{A}\mathbf{x} = (-x_1^2 - x_2^2 + 2x_1 x_2 - x_3^2) = \{-x_3^2 - (x_1 - x_2)^2\} \leq 0$$

for all $\mathbf{x}$, and $\mathbf{x}^T \mathbf{A}\mathbf{x} = 0$ when $x_3 = 0$, and $x_1 = x_2$ [e.g. $\mathbf{x} = (1, 1, 0)$]. The quadratic form is not negative definite but is negative semidefinite since it can have zero value for nonzero $\mathbf{x}$. Therefore, the matrix associated with it is also negative semidefinite.

‖

We will now discuss *methods for checking* positive definiteness or semidefiniteness (form) of a quadratic form or a matrix. Since this involves calculation of eigenvalues of a matrix, Section B.7 in Appendix B should be reviewed at this point.

Theorem 3.2   Eigenvalue check for the form of a matrix. Let λ_i, $i = 1$ to n be n eigenvalues of a symmetric $n \times n$ matrix $\mathbf{A}$ associated with the quadratic form $F(\mathbf{x}) = \frac{1}{2}\mathbf{x}^T\mathbf{A}\mathbf{x}$. The following results can be stated regarding the quadratic form $F(\mathbf{x})$ or the matrix $\mathbf{A}$:

1. $F(\mathbf{x})$ is a *positive definite if and only if* all eigenvalues of $\mathbf{A}$ are strictly positive, i.e. $\lambda_i > 0$, $i = 1$ to n.
2. $F(\mathbf{x})$ is *positive semidefinite if and only if* all eigenvalues of $\mathbf{A}$ are non-negative, i.e. $\lambda_i \geq 0$, $i = 1$ to n (note that at least one eigenvalue must be zero for it to be called positive semidefinite).
3. $F(\mathbf{x})$ is *negative definite if and only if* all eigenvalues of $\mathbf{A}$ are strictly negative, i.e. $\lambda_i < 0$, $i = 1$ to n.
4. $F(\mathbf{x})$ is *negative semidefinite if and only if* all eigenvalues of $\mathbf{A}$ are nonpositive, i.e. $\lambda_i \leq 0$, $i = 1$ to n (note that at least one eigenvalue must be zero for it to be called negative semidefinite).
5. $F(\mathbf{x})$ is *indefinite if* some $\lambda_i < 0$ and some other $\lambda_i > 0$.   ‖

Another way of checking the form of a matrix is provided by the following theorem:

Theorem 3.3   Check for the form of a matrix using principal minors. Let M_k be the kth principal minor of the $n \times n$ symmetric matrix $\mathbf{A}$ defined as the determinant of a $k \times k$ submatrix obtained by deleting last $(n - k)$ rows and columns of $\mathbf{A}$ (Appendix B, Section B.4). Assume that *no two consecutive principal minors* are zero. Then

1. A is *positive definite if and only if* all the principal minors are positive, i.e. $M_k > 0$, $k = 1$ to n.
2. A is *positive semidefinite if and only if* $M_k \geq 0$, $k = 1$ to n (note that at least one principal minor must be zero for it to be called positive semidefinite).
3. A is *negative definite if and only if* $M_k < 0$ for odd k and $M_k > 0$ for k even.
4. A is *negative semidefinite if and only if* $M_k \leq 0$ for k odd and $M_k \geq 0$ for k even (note that at least one principal minor must be zero for it to be called negative semidefinite).
5. A is *indefinite if* it does not satisfy any of the preceding criteria.   ‖

This theorem is applicable only if the assumption of no two consecutive principal minors being zero is satisfied. When there are consecutive zero principal minors, we may resort to the eigenvalue check of Theorem 3.2. Note also that a *positive definite matrix cannot have negative or zero diagonal elements.*

The theory of quadratic forms is used in second-order conditions for a local optimum point. Also, it is used to determine convexity of functions of the optimization problem. Convex functions play a role in determining the global optimum point. These topics are discussed in later sections.

Example 3.9 **Determination of the form of a matrix.** Determine the form of the matrices given in Example 3.8.

Solution. For a given matrix $\mathbf{A}$, the eigenvalue problem (refer to Section B.7 in Appendix B for more details) is defined as $\mathbf{A}\mathbf{x} = \lambda\mathbf{x}$, where λ is an eigenvalue and $\mathbf{x}$ is the corresponding eigenvector. To determine the eigenvalues, we set $|(\mathbf{A} - \lambda\mathbf{I})| = 0$. For the matrix (a), this gives

$$\begin{vmatrix} 2-\lambda & 0 & 0 \\ 0 & 4-\lambda & 0 \\ 0 & 0 & 3-\lambda \end{vmatrix} = 0$$

The three eigenvalues are $\lambda_1 = 2$, $\lambda_2 = 3$, and $\lambda_3 = 4$. Since all eigenvalues are strictly positive, the matrix is positive definite. The principal minor check of Theorem 3.3 also gives the same conclusion.

For the matrix (b), the characteristic determinant of the eigenvalue problem is

$$\begin{vmatrix} -1-\lambda & 1 & 0 \\ 1 & -1-\lambda & 0 \\ 0 & 0 & -1-\lambda \end{vmatrix} = 0$$

Expanding the determinant by the third row, we obtain

$$(-1-\lambda)[(-1-\lambda)^2 - 1] = 0$$

Therefore, the three roots give the eigenvalues as $\lambda_1 = -2$, $\lambda_2 = -1$, and $\lambda_3 = 0$. Since all eigenvalues are nonpositive, the matrix is negative semidefinite. To use Theorem 3.3, we calculate the three principal minors as

$$M_1 = -1, \quad M_2 = \begin{vmatrix} -1 & 1 \\ 1 & -1 \end{vmatrix} = 0, \quad M_3 = \begin{vmatrix} -1 & 1 & 0 \\ 1 & -1 & 0 \\ 0 & 0 & -1 \end{vmatrix} = 0$$

Since there are two consecutive zero principal minors, we cannot use Theorem 3.3. ‖

3.2.5.4 DIFFERENTIATION OF A QUADRATIC FORM. On several occasions we would like to find gradient and Hessian matrix for the quadratic form. We consider the symmetric quadratic form of Eq. (3.22) and write it in summation notation as

$$F(\mathbf{x}) = \tfrac{1}{2}\sum_{i=1}^{n}\sum_{j=1}^{n} a_{ij}x_i x_j \tag{3.23}$$

To calculate derivatives of $F(\mathbf{x})$, we first expand the summations and then differentiate the expression with respect to x_i to obtain

$$\frac{\partial F(\mathbf{x})}{\partial x_i} = \sum_{j=1}^{n} a_{ij}x_j \tag{3.24}$$

Writing the partial derivatives of Eq. (3.24) in a column vector, we get the *gradient of the quadratic form* as

$$\nabla F(\mathbf{x}) = \mathbf{A}\mathbf{x} \tag{3.25}$$

Differentiating Eq. (3.24) once again with respect to x_j, we get

$$\frac{\partial^2 F(\mathbf{x})}{\partial x_j \, \partial x_i} = a_{ij} \tag{3.26}$$

Equation (3.26) shows that the components a_{ij} of the matrix $\mathbf{A}$ are the components of the *Hessian matrix for the quadratic form*. Note that $\mathbf{A}$ must be the symmetric matrix associated with the quadratic form. This can be easily seen by differentiating $F(\mathbf{x})$ in Eq. (3.23) with respect to x_j first and then with respect to x_i.

> **Example 3.10 Calculations for the gradient and Hessian of the quadratic form.** Calculate gradient and Hessian of the following quadratic form:
>
> $$F(\mathbf{x}) = \tfrac{1}{2}(2x_1^2 + 2x_1x_2 + 4x_1x_3 - 6x_2^2 - 4x_2x_3 + 5x_3^2)$$
>
> ***Solution.*** Differentiating $F(\mathbf{x})$ with respect to x_1, x_2 and x_3, we get gradient components as
>
> $$\frac{\partial F}{\partial x_1} = 2x_1 + x_2 + 2x_3$$
>
> $$\frac{\partial F}{\partial x_2} = x_1 - 6x_2 - 2x_3$$
>
> $$\frac{\partial F}{\partial x_3} = 2x_1 - 2x_2 + 5x_3$$
>
> Differentiating the gradient components once again, we get the Hessian components as
>
> $$\frac{\partial^2 F}{\partial x_1^2} = 2, \qquad \frac{\partial^2 F}{\partial x_1 \, \partial x_2} = 1, \qquad \frac{\partial^2 F}{\partial x_1 \, \partial x_3} = 2$$
>
> $$\frac{\partial^2 F}{\partial x_2 \, \partial x_1} = 1, \qquad \frac{\partial^2 F}{\partial x_2^2} = -6, \qquad \frac{\partial^2 F}{\partial x_2 \, \partial x_3} = -2$$
>
> $$\frac{\partial^2 F}{\partial x_3 \, \partial x_1} = 2, \qquad \frac{\partial^2 F}{\partial x_3 \, \partial x_2} = -2, \qquad \frac{\partial^2 F}{\partial x_3^2} = 5$$
>
> Writing the given quadratic form in a matrix form, we identify matrix $\mathbf{A}$ as
>
> $$\mathbf{A} = \begin{bmatrix} 2 & 1 & 2 \\ 1 & -6 & -2 \\ 2 & -2 & 5 \end{bmatrix}$$
>
> Comparing elements of the matrix $\mathbf{A}$ with second partial derivatives of F, we observe that the Hessian $\mathbf{H} = \mathbf{A}$. Using Eq. (3.25), the gradient of the quadratic

form is also given as

$$
\nabla F(\mathbf{x}) = \begin{bmatrix} 2 & 1 & 2 \\ 1 & -6 & -2 \\ 2 & -2 & 5 \end{bmatrix} \begin{bmatrix} x_1 \\ x_2 \\ x_3 \end{bmatrix}
$$

$$
= \begin{bmatrix} (2x_1 + x_2 + 2x_3) \\ (x_1 - 6x_2 - 2x_3) \\ (2x_1 - 2x_2 + 5x_3) \end{bmatrix}
$$

which is the same as before. ‖

3.2.6 Concept of Necessary and Sufficient Conditions

In the remainder of this chapter, we shall describe necessary and sufficient conditions for optimality of unconstrained and constrained optimization problems. It is important to understand the meaning of the terms *necessary* and *sufficient*. These terms have general meaning in mathematical analyses. We shall, however, discuss them for the optimization problem only. The optimality conditions are derived by assuming that we are at an optimum point and then studying the behavior of the functions and their derivatives at the point. *The conditions that must be satisfied at the optimum point are called necessary. Stated differently, if any point does not satisfy the necessary conditions, it cannot be optimum.* Note, however, that satisfaction of necessary conditions does not guarantee an optimum point, i.e. there can be non-optimum points that also satisfy the same conditions. This indicates that the number of points satisfying necessary conditions can be more than the number of optima. Points satisfying the necessary conditions are called *candidate optimum points*. We must, therefore, perform further tests to distinguish between optimum and nonoptimum points, both satisfying the necessary conditions.

The *sufficient conditions* provide tests to distinguish between optimum and nonoptimum points. *If a candidate optimum point satisfies the sufficient conditions, then it is indeed optimum.* We do not need any further tests. If the sufficient conditions, however, are not satisfied or cannot be used, we may not be able to conclude that the candidate design is not optimum. Our conclusion will depend on the assumptions and restrictions used in defining the sufficient conditions. Further analysis of the problem or other conditions are needed to make a definite statement about optimality of the candidate point. *In summary*,

1. Optimum points must satisfy the necessary conditions. Points that do not satisfy them cannot be optimum.
2. A point satisfying the necessary conditions need not be optimum, i.e. nonoptimum points may also satisfy the necessary conditions.

3. A candidate point satisfying a sufficient condition is indeed optimum.

4. If sufficiency conditions cannot be used or they are not satisfied, we may not be able to draw any conclusions about optimality of the candidate point.

We shall further explain these conditions and illustrate them with examples in later sections when proper theorems are stated.

3.3 UNCONSTRAINED OPTIMUM DESIGN PROBLEMS

3.3.1 Introduction

We are now ready to discuss the theory and concepts of optimum design. In this section, we illustrate the concepts for unconstrained problems defined as follows: minimize $f(\mathbf{x})$ without any constraints on $\mathbf{x}$. Such problems arise infrequently in practical engineering applications. However, we consider them here because optimality conditions for constrained problems are a logical extension of them. In addition, one numerical strategy for solving a constrained problem (discussed in Chapter 6) is to convert it into a sequence of unconstrained problems. Thus, it is important to completely understand unconstrained optimization concepts.

The optimality conditions for unconstrained or constrained problems can be used in two ways:

1. If a design point is given, the optimality conditions can be used to check whether or not that point is a candidate optimum.

2. The optimality conditions can be solved for candidate optimum points.

3.3.2 Procedure for Derivation of Optimality Conditions

Optimality conditions can be used to determine candidate minimum points for a function $f(\mathbf{x})$. We will discuss only the *local optimality conditions* for unconstrained problems. Global optimality will be discussed in Section 3.5. First the necessary and then the sufficient conditions will be discussed. *The necessary conditions must be satisfied at the minimum point, otherwise it cannot be a minimum. These conditions, however, may also be satisfied by a point that is not minimum. A point satisfying the conditions is simply a candidate local minimum.* The sufficient conditions distinguish minimum points from others. We shall elaborate these concepts further with some examples.

The procedure for deriving local optimality conditions is to assume that we are at a minimum point $\mathbf{x}^*$ and then examine a small neighborhood to study properties of the function and its derivatives. Since we examine only a small neighborhood, the conditions we obtain are called local.

Let $\mathbf{x}^*$ be a *local minimum* point for $f(\mathbf{x})$. To investigate its neighborhood, let $\mathbf{x}$ be any point near $\mathbf{x}^*$. Define increments $\mathbf{d}$ and Δf in $\mathbf{x}^*$ and $f(\mathbf{x}^*)$, respectively, as

$$\mathbf{d} = \mathbf{x} - \mathbf{x}^* \quad \text{and} \quad \Delta f = f(\mathbf{x}) - f(\mathbf{x}^*)$$

Since $f(\mathbf{x})$ has a local minimum at $\mathbf{x}^*$, it cannot reduce any further if we move a small distance away. Therefore, a change in the function for any move in a small neighborhood of $\mathbf{x}^*$ must be nonnegative, i.e. the function value must either remain constant or increase. This condition can be expressed as the following inequality:

$$\Delta f = f(\mathbf{x}) - f(\mathbf{x}^*) \geqq 0 \tag{3.27}$$

for all small changes $\mathbf{d}$. The inequality can be used to derive necessary and sufficient conditions for a local minimum point. Since $\mathbf{d}$ is small, we can approximate Δf by Taylor's expansion at $\mathbf{x}^*$ and derive optimality conditions using it.

3.3.3 Optimality Conditions for Functions of Single Variable

3.3.3.1 FIRST-ORDER NECESSARY CONDITIONS. Let us first consider a *function of one variable only*. The Taylor series of $f(x)$ at the point x^* gives

$$f(x) = f(x^*) + f'(x^*)d + \tfrac{1}{2}f''(x^*)d^2 + R$$

where R is the remainder containing higher order terms in d and "primes" indicate order of the derivatives. From this equation, the change in the function at x^*, i.e. $\Delta f = f(x) - f(x^*)$, is given as

$$\Delta f = f'(x^*)d + \tfrac{1}{2}f''(x^*)d^2 + R \tag{3.28}$$

The Inequality (3.27) shows that expression for Δf must be nonnegative ($\geqq 0$) as x^* is a local minimum. Since d is small, the first-order term $f'(x^*)d$ dominates other terms. Focusing on this term we observe that Δf in Eq. (3.28) can be positive or negative depending on the sign of the term $f'(x^*)d$. Since d is an arbitrary small increment in x^*, it may be positive or negative. Therefore, if $f'(x^*) \neq 0$, the term $f'(x^*)d$ (and hence Δf) can be negative. To see this more clearly, let the term be positive for some increment d_1 which satisfies the Inequality (3.27), i.e. $\Delta f = f'(x^*)d_1 > 0$. Since the increment d is arbitrary, it is reversible, so $d_2 = -d_1$ is another possible increment. For d_2, Δf in Eq. (3.28) becomes negative which violates the Inequality (3.27). Thus, the quantity $f'(x^*)d$ can be negative regardless of the sign of $f'(x^*)$, unless it is zero. The only way it can be nonnegative for all d in a neighborhood of x^* is when

$$f'(x^*) = 0 \tag{3.29}$$

Equation (3.29) is a *first-order necessary condition* for local minimum of

$f(x)$ at x^*. It is called "first-order" because it only involves the first derivative of the function. Note that preceding arguments can be used to show that the condition of Eq. (3.29) are also necessary for local maximum points. Therefore, since the points satisfying Eq. (3.29) can be local minima, maxima, or neither minimum nor maximum (*inflection or saddle points*), they are called *stationary points*.

3.3.3.2 SUFFICIENT CONDITIONS. Now we need a *sufficient condition* to determine which of the stationary points are actually minimum for the function. Since stationary points satisfy the necessary condition $f'(x^*) = 0$, the change in function Δf of Eq. (3.28) becomes

$$\Delta f = \tfrac{1}{2} f''(x^*) d^2 + R \tag{3.30}$$

Since the second-order term dominates all other higher-order terms, we need to focus on it. Note that the term can be positive for all $d \neq 0$, if

$$f''(x^*) > 0 \tag{3.31}$$

Stationary points satisfying Inequality (3.31) must be at least local minima because they satisfy Inequality (3.27) ($\Delta f > 0$). That is, the function has positive curvature at the minimum points. Inequality (3.31) is then *sufficient* for x^* to be a local minimum. Thus, if we have a point x^* satisfying both conditions in Eqs (3.29) and (3.31), then any small move away from it will either increase the function value or keep it constant. This indicates that $f(x^*)$ has the smallest value in a small neighborhood (local minimum) of the point x^*.

3.3.3.3 SECOND-ORDER NECESSARY CONDITION. If Inequality (3.31) is not satisfied (e.g. $f''(x^*) = 0$), we cannot conclude that x^* is not a minimum point. Note, however, from Eqs (3.27) and (3.28) that $f(x^*)$ cannot be a minimum unless

$$f''(x^*) \geqq 0 \tag{3.32}$$

That is, if f'' evaluated at the candidate point x^* is less than zero, then x^* is not a local minimum point. Inequality (3.32) is known as a *second-order necessary condition*, so any point violating it (i.e. $f''(x^*) < 0$) cannot be a local minimum.

If $f''(x^*) = 0$, we need to evaluate higher-order derivatives to determine if the point is a local minimum or not. By the arguments used to derive Eq. (3.29), $f'''(x^*)$ must be zero for the stationary point (necessary condition) and $f^{IV}(x^*) > 0$ for x^* to be a local minimum. In general, *the lowest nonzero derivative must be even ordered for stationary points (necessary conditions), and it must be positive for local minimum points (sufficiency condition). All odd ordered derivatives lower than the nonzero even ordered derivative must be zero as the necessary condition.*

Example 3.11 Determination of local minimum points using necessary conditions. Find local minima for the function $f(x) = \sin x$.

Solution. Differentiating the function twice,

$$f' = \cos x; \qquad f'' = -\sin x$$

Stationary points are obtained as roots of $f'(x) = 0$ ($\cos x = 0$). These are

$$x = \pm\pi/2, \ \pm3\pi/2, \ \pm5\pi/2, \ \pm7\pi/2, \ldots$$

Local minima are identified as

$$x^* = 3\pi/2, \ 7\pi/2, \ldots; \qquad -\pi/2, \ -5\pi/2, \ldots$$

since these points satisfy the sufficiency condition of Eq. (3.31) ($f'' = -\sin x > 0$). The minimum value of $\sin x$ at the points x^* is -1. This is true from the graph of the function $\sin x$. There are infinite minimum points and they are all actually global minima. The points $\pi/2, \ 5\pi/2, \ldots$, and $-3\pi/2, \ -7\pi/2, \ldots$ are global maximum points where $\sin x$ has a value of 1.   ‖

Example 3.12 Determination of local minimum points using necessary conditions. Find local minima for the function

$$f(x) = x^2 - 4x + 4$$

Solution. Figure 3.5 shows a graph for the function $f(x) = x^2 - 4x + 4$. It can be

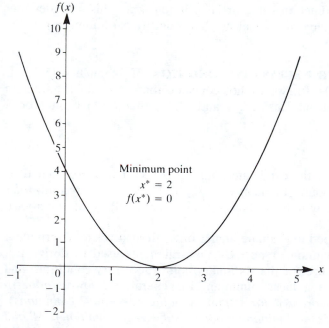

FIGURE 3.5
Graph of $f(x) = x^2 - 4x + 4$ of Example 3.12.

seen that the function always has positive value except at $x = 2$ where it is zero. Therefore, this is the minimum point for the function. Let us see how this point will be determined using the necessary and sufficient conditions.

Differentiating the function twice,

$$f' = 2x - 4; \qquad f'' = 2$$

The necessary condition $f' = 0$ implies that $x^* = 2$ is a stationary point. Since $f'' > 0$ at $x^* = 2$ (actually for all x), the sufficiency condition of Eq. (3.31) is satisfied. Therefore $x^* = 2$ is a local minimum for $f(x)$. Minimum value of f is 0 at $x^* = 2$. ‖

Example 3.13 Determination of local minimum points using necessary conditions. Find local minima for the function

$$f(x) = x^3 - x^2 - 4x + 4$$

Solution. Figure 3.6 shows the graph of the function. It can be seen that point A is a local minimum point and point B is a local maximum point. We shall use the necessary and sufficient conditions to prove that this is indeed true.

Differentiating the function,

$$f' = 3x^2 - 2x - 4; \qquad f'' = 6x - 2$$

For this example there are two points satisfying the necessary condition of Eq. (3.29). These are obtained as roots of the equation $f'(x) = 0$,

$$x_1^* = \tfrac{1}{6}(2 + 7.211) = 1.535 \qquad \text{(Point A)}$$
$$x_2^* = \tfrac{1}{6}(2 - 7.211) = -0.8685 \qquad \text{(Point B)}$$

Evaluating f'' at these points,

$$f''(1.535) = 7.211 > 0$$
$$f''(-0.8685) = -7.211 < 0$$

We see that only x_1^* satisfies the sufficiency condition $(f'' > 0)$ of Eq. (3.31).

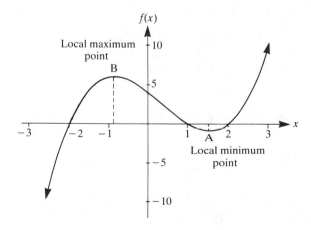

FIGURE 3.6
Graph of $f(x) = x^3 - x^2 - 4x + 4$ of Example 3.13.

Therefore, it is a local minimum point. From the graph (Fig. 3.6) of $f(x)$ we can see that the local minimum $f(x_1^*)$ is not the global minimum. *A global minimum for $f(x)$ does not exist since domain as well as the function are not bounded.* The value of the function at the local minimum is obtained as -0.88 by substituting $x_1^* = 1.535$ in $f(x)$. Note that $x_2^* = -0.8685$ is a local maximum point since $f''(x_2^*) < 0$. The value of the function at the maximum point is 6.065. There is no global maximum point for the function. ‖

Example 3.14 Determination of local minimum points using necessary conditions. Find the minimum for the function $f(x) = x^4$.

Solution. Differentiating the function twice,

$$f' = 4x^3; \qquad f'' = 12x^2$$

The necessary condition gives $x^* = 0$ as a stationary point. Since $f''(x^*) = 0$, we cannot conclude from the sufficiency condition of Eq. (3.31) that x^* is a minimum point. However, the second-order necessary condition of Eq. (3.32) is satisfied, so we cannot rule out the possibility of x^* being a minimum point. In fact, a graph of $f(x)$ versus x will show that x^* is indeed the global minimum point. $f''' = 24x$ which is zero at $x^* = 0$. $f^{IV}(x^*) = 24$, which is strictly greater than zero. Therefore, the fourth-order sufficiency condition is satisfied, and $x^* = 0$ is indeed a minimum point. It is actually a global minimum point with $f(0) = 0$. ‖

3.3.4 Optimality Conditions for Functions of Several Variables

For the general case of a function of several variables $f(\mathbf{x})$ where $\mathbf{x}$ is an n-vector, we can repeat the derivation of *necessary and sufficient* conditions using the multidimensional form of Taylor's expansion:

$$f(\mathbf{x}) = f(\mathbf{x}^*) + \nabla f(\mathbf{x}^*)^T \mathbf{d} + \tfrac{1}{2}\mathbf{d}^T \mathbf{H}(\mathbf{x}^*)\mathbf{d} + R$$

Or, change in the function is given as

$$\Delta f = \nabla f(\mathbf{x}^*)^T \mathbf{d} + \tfrac{1}{2}\mathbf{d}^T \mathbf{H}(\mathbf{x}^*)\mathbf{d} + R \tag{3.33}$$

If we assume a local minimum at $\mathbf{x}^*$ then Δf must be nonnegative, i.e. $\Delta f \geqq 0$. Concentrating only on the first-order term in Eq. (3.33), we observe (as before) that Δf can be nonnegative for all possible $\mathbf{d}$ when

$$\nabla f(\mathbf{x}^*) = \mathbf{0} \tag{3.34}$$

That is, the gradient of the function at $\mathbf{x}^*$ be zero. In the component form this necessary condition becomes

$$\frac{\partial f(\mathbf{x}^*)}{\partial x_i} = 0; \qquad i = 1 \text{ to } n \tag{3.35}$$

Points satisfying Eq. (3.35) are called *stationary points*. Considering the second term in Eq. (3.33) evaluated at a stationary point the positivity of Δf is assured if

$$\mathbf{d}^T\mathbf{H}(\mathbf{x}^*)\mathbf{d} > 0 \tag{3.36}$$

for all $\mathbf{d} \neq 0$. This will be true if the Hessian $\mathbf{H}(\mathbf{x}^*)$ is a positive definite matrix (see Section 3.2) which is then the sufficient condition for a local minimum of $f(\mathbf{x})$ at $\mathbf{x}^*$.

Conditions (3.35) and (3.36) are the multidimensional equivalent of Conditions (3.29) and (3.31), respectively. We summarize the development of this section in the following theorem:

Theorem 3.4   Necessary and sufficient conditions for local minimum.
Necessary condition. If $f(\mathbf{x})$ has a local minimum at $\mathbf{x}^*$ then

$$\frac{\partial f(\mathbf{x}^*)}{\partial x_i} = 0; \qquad i = 1 \text{ to } n$$

Second-order necessary condition. If $f(\mathbf{x})$ has a local minimum at $\mathbf{x}^*$, then the Hessian matrix of Eq. (3.5)

$$\mathbf{H}(\mathbf{x}^*) = \left[\frac{\partial^2 f}{\partial x_i \, \partial x_j} \right]_{(n \times n)}$$

is positive semidefinite or positive definite at the point $\mathbf{x}^*$.

Second-order sufficiency condition. If the matrix $\mathbf{H}(\mathbf{x}^*)$ is positive definite at the stationary point $\mathbf{x}^*$, then $\mathbf{x}^*$ is a local minimum point for the function $f(\mathbf{x})$.   ‖

Note that these conditions involve derivatives of $f(\mathbf{x})$ and not value of the function. If we *add a constant* to $f(\mathbf{x})$, the solution $\mathbf{x}^*$ of the minimization problem remains unchanged, although the value of the cost function is altered. In a graph of $f(\mathbf{x})$ versus $\mathbf{x}$, adding a constant to $f(\mathbf{x})$ changes the origin of the coordinate system but leaves the shape of the curve unchanged. Similarly, if we multiply $f(\mathbf{x})$ by any positive constant the minimum point $\mathbf{x}^*$ is unchanged but the value $f(\mathbf{x}^*)$ is altered. In a graph of $f(\mathbf{x})$ versus $\mathbf{x}$ this is equivalent to a uniform change of scale of the graph along the $f(\mathbf{x})$ axis which again leaves the shape of the curve unaltered. Multiplying $f(\mathbf{x})$ by a negative constant changes the minimum at $\mathbf{x}^*$ to a maximum. We may use this property to convert maximization problems to minimization problems by multiplying $f(\mathbf{x})$ by -1.

Example 3.15   Effects of scaling or adding constant to a function. Discuss the effect of preceding variations for the function

$$f(x) = x^2 - 2x + 2$$

Solution. Consider the graphs of Fig. 3.7. Figure 3.7(a) represents the function $f(x) = x^2 - 2x + 2$ that has a minimum at $x^* = 1$. Figures 3.7(b), (c) and (d) show the effect of adding a constant to the function $(f(x) + 1)$, multiplying $f(x)$ by positive number $(2f(x))$, and multiplying it by a negative number $(-f(x))$. In all cases, the stationary point remains unchanged.   ‖

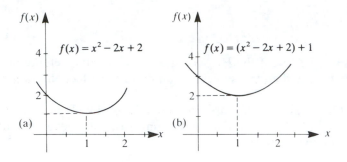

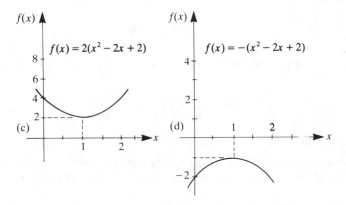

FIGURE 3.7
Graphs for Example 3.15. Effects of scaling or adding constant to a function. (a) Graph of $f(x) = x^2 - 2x + 2$. (b) Effect of addition of a constant to $f(x)$. (c) Effect of multiplying $f(x)$ by a positive constant. (d) Effect of multiplying $f(x)$ by -1.

Example 3.16 Minimum cost spherical tank using necessary conditions. The result of a problem formulation in Section 2.6.5 is a cost function that represents the life-time cooling related cost of an insulated spherical tank as

$$f(x) = ax + b/x$$

where x is the thickness of insulation, and a and b are positive constants. To minimize f, we solve the equation (necessary condition)

$$f' = a - b/x^2 = 0.$$

The solution is $x^* = \sqrt{b/a}$. To check if the stationary point is a local minimum, evaluate

$$f''(x^*) = 2b/x^{*3}$$

Since b and x^* are positive, $f''(x^*)$ is positive and x^* is a local minimum point. The value of the function at x^* is $2\sqrt{ab}$. Note that since the function cannot have negative value, x^* represents a global minimum for the problem. ∥

Example 3.17 Local minima for a function of two variables using necessary conditions. Find local minimum points for the function

$$f(\mathbf{x}) = x_1^2 + 2x_1x_2 + 2x_2^2 - 2x_1 + x_2 + 8$$

Solution. The necessary conditions for the problem give

$$\frac{\partial f}{\partial \mathbf{x}} \equiv \begin{bmatrix} (2x_1 + 2x_2 - 2) \\ (2x_1 + 4x_2 + 1) \end{bmatrix} = \begin{bmatrix} 0 \\ 0 \end{bmatrix}$$

These equations are linear in variables x_1 and x_2. If a solution exists for the system then it is unique (Appendix B). Solving the equations simultaneously, we get the stationary point as $\mathbf{x}^* = (\frac{5}{2}, -\frac{3}{2})$. To check if the stationary point is a local minimum, we evaluate **H** at $\mathbf{x}^*$.

$$\mathbf{H} = \begin{bmatrix} \dfrac{\partial^2 f}{\partial x_1^2} & \dfrac{\partial^2 f}{\partial x_1\,\partial x_2} \\ \dfrac{\partial^2 f}{\partial x_2\,\partial x_1} & \dfrac{\partial^2 f}{\partial x_2^2} \end{bmatrix} = \begin{bmatrix} 2 & 2 \\ 2 & 4 \end{bmatrix}$$

By either of the tests of Theorems 3.2 and 3.3, **H** is positive definite at the stationary point $\mathbf{x}^*$. Thus it is a local minimum with $f(\mathbf{x}^*) = 4.75$. Figure 3.8 shows a few iso-cost lines for the function of this problem. It can be seen that the point (2.5, −1.5) is the minimum for the function. ‖

Example 3.18 Cylindrical tank design using necessary conditions. In Section 2.6.6, a minimum cost cylindrical storage tank problem is formulated. The tank is closed at both ends and is required to have volume V. The radius R and length l

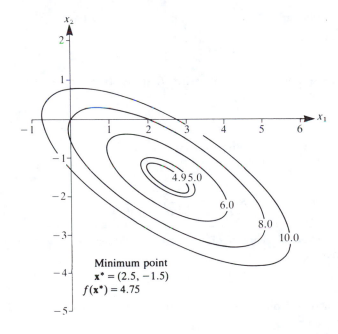

Minimum point
$\mathbf{x}^* = (2.5, -1.5)$
$f(\mathbf{x}^*) = 4.75$

FIGURE 3.8
Iso-cost lines for the function of Example 3.17.

are selected as design variables. It is desired to design a tank having minimum surface area. For the solution we may simplify the cost function as

$$\bar{f} = R^2 + Rl \tag{a}$$

The volume constraint is an equality,

$$h \equiv \pi R^2 l - V = 0 \tag{b}$$

This constraint cannot be satisfied if either R or l is zero. We may then neglect the non-negativity constraints on R and l if we agree to choose only the positive root. We may further use the equality constraint (b) to eliminate l from the cost function,

$$l = \frac{V}{\pi R^2} \tag{c}$$

Therefore, the cost function of Eq. (a) becomes

$$\bar{\bar{f}} = R^2 + \frac{V}{\pi R} \tag{d}$$

This is an unconstrained problem in terms of R for which the necessary condition gives

$$\frac{d\bar{\bar{f}}}{dR} \equiv 2R - \frac{V}{\pi R^2} = 0 \tag{e}$$

The solution is

$$R^* = \left(\frac{V}{2\pi}\right)^{1/3} \tag{f}$$

Using Eq. (c), we obtain

$$l^* = \left(\frac{4V}{\pi}\right)^{1/3} \tag{g}$$

Using Eq. (e), the second derivative of $\bar{\bar{f}}$ w.r.t. R at the stationary point is

$$\frac{d^2\bar{\bar{f}}}{dR^2} = \frac{2V}{\pi R^3} + 2 = 6 \tag{h}$$

Since the second derivative is positive for all positive R the solution in Eqs (f) and (g) is a local minimum. Using Eqs (a) or (d) the cost function at the optimum is given as

$$\bar{\bar{f}}(R^*, l^*) = 3\left(\frac{V}{2\pi}\right)^{2/3} \qquad \|$$

Example 3.19   Numerical solution of necessary conditions. Find a stationary point for the function

$$f(x) = \tfrac{1}{3}x^2 + \cos x \tag{a}$$

Solution. The function is plotted in Fig. 3.9. It can be seen that there are three

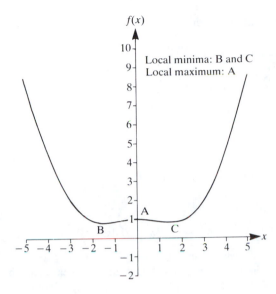

FIGURE 3.9
Graph of $f(x) = x^2/3 + \cos x$ of Example 3.19.

stationary points: $x = 0$ (Point A), x between 1 and 2 (Point C), and x between -1 and -2 (Point B). The point $x = 0$ is a local maximum for the function and the other two are minima.

The necessary condition is

$$f'(x) \equiv \frac{2x}{3} - \sin x = 0 \tag{b}$$

It can be seen that $x = 0$ satisfies Eq. (b), so it is a stationary point. We must find other roots of Eq. (b). Analytical solution for the equation is difficult, so we must use numerical methods. We can either plot $f'(x)$ versus x on a graph sheet and locate the point where $f'(x) = 0$, or use a numerical method for solving nonlinear equations. A numerical method for solving such an equation known as the *Newton-Raphson method* is given in Appendix C. By either of the two methods, we find that $x^* = 1.496$ and -1.496 satisfy $f'(x) = 0$ in Eq. (b). Therefore, these are additional stationary points. To determine whether they are local minimum, maximum or inflection points, we must determine f'' at the stationary points and use sufficient conditions of Theorem 3.4. Since $f'' = \frac{2}{3} - \cos x$, we have

1. $x^* = 0$; $f'' = -\frac{1}{3} < 0$, so x^* is a local maximum with $f(0) = 1$.
2. $x^* = 1.496$; $f'' = 0.592 > 0$, so x^* is a local minimum with $f(1.496) = 0.821$.
3. $x^* = -1.496$; $f'' = 0.592 > 0$, so x^* is a local minimum with $f(1.496) = 0.821$.

These results agree with the graphical solutions observed in Fig. 3.9. Note that $x^* = 1.496$ and -1.496 are actually global minimum points for the function.   ‖

Example 3.20 Local minima for a function of two variables using necessary conditions. Find a local minimum point for the function

$$f(\mathbf{x}) = x_1 + \frac{(4.0E + 06)}{x_1 x_2} + 250x_2$$

Solution. The necessary conditions for optimality are

$$\frac{\partial f}{\partial x_1} = 0; \qquad 1 - \frac{(4.0\mathrm{E}+06)}{x_1^2 x_2} = 0 \tag{a}$$

$$\frac{\partial f}{\partial x_2} = 0; \qquad 250 - \frac{(4.0\mathrm{E}+06)}{x_1 x_2^2} = 0 \tag{b}$$

Equations (a) and (b) give

$$x_1^2 x_2 - (4.0\mathrm{E}+06) = 0$$

$$250 x_1 x_2^2 - (4.0\mathrm{E}+06) = 0$$

These equations give

$$x_1^2 x_2 = 250 x_1 x_2^2, \qquad \text{or} \qquad x_1 x_2 (x_1 - 250 x_2) = 0$$

Since neither x_1 nor x_2 can be zero (the function has singularity at $x_1 = 0$, or $x_2 = 0$), the preceding equation gives $x_1 = 250 x_2$. Substituting this into Eq. (b), we obtain $x_2 = 4$. Therefore, $x_1^* = 1000$, and $x_2^* = 4$ is a stationary point for the function $f(\mathbf{x})$. Using Eqs (a) and (b), the Hessian matrix for $f(\mathbf{x})$ is

$$\mathbf{H} = \frac{(4.0\mathrm{E}+06)}{x_1^2 x_2^2} \begin{bmatrix} \dfrac{2x_2}{x_1} & 1 \\[2ex] 1 & \dfrac{2x_1}{x_2} \end{bmatrix}$$

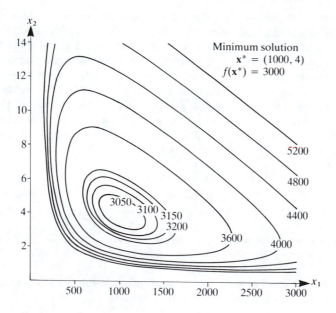

FIGURE 3.10
Iso-cost lines for the function of Example 3.20.

Hessian at the point $\mathbf{x}^*$ is given as

$$\mathbf{H}(1000, 4) = \frac{(4.0\text{E}+06)}{(4000)^2} \begin{bmatrix} 0.008 & 1 \\ 1 & 500 \end{bmatrix}$$

Eigenvalues of the above Hessian (without the constant of $\frac{1}{4}$) are $\lambda_1 = 0.006$ and $\lambda_2 = 500.002$. Since both the eigenvalues are positive, the Hessian of $f(\mathbf{x})$ at the point $\mathbf{x}^*$ is positive definite. Therefore, $\mathbf{x}^* = (1000, 4)$ is a local minimum point with $f(\mathbf{x}^*) = 3000$. Figure 3.10 shows some iso-cost lines for the function of this problem. It can be seen that $x_1 = 1000$ and $x_2 = 4$ is the minimum point. (Note that the horizontal and vertical scales are quite different in Fig. 3.10; this is done to obtain reasonable iso-cost lines.)   ∥

3.4 CONSTRAINED OPTIMUM DESIGN PROBLEMS

3.4.1 Introduction

We have seen in Chapter 2 that most design problems include constraints on variables and on the performance of the system. This section describes concepts related to constrained optimization problems. The necessary conditions for optimality are explained and illustrated with examples. All optimum designs must satisfy these conditions.

The general design optimization model (defined in Chapter 2) is to find a design variable vector $\mathbf{x} = (x_1, x_2, \ldots, x_n)$ to minimize a cost function

$$f(\mathbf{x}) = f(x_1, x_2, \ldots, x_n) \tag{3.37}$$

subject to the equality constraints

$$h_i(\mathbf{x}) = 0; \quad i = 1 \text{ to } p \tag{3.38}$$

and the inequality constraints

$$g_i(\mathbf{x}) \leqq 0; \quad i = 1 \text{ to } m \tag{3.39}$$

The inequality constraints of Eq. (3.39) will be initially ignored to discuss the Lagrange theorem given in calculus books. The theorem will be then extended for inequality constraints to obtain the Kuhn–Tucker necessary conditions for the general model defined in Eqs (3.37) to (3.39).

Based on the discussion of unconstrained optimization problems, one might conclude that only the nature of the cost function $f(\mathbf{x})$ for the constrained problems will determine the location of the minimum point. This, however, is not true. The constraint functions can play a prominent role in determining the optimum solution. The following examples illustrate these situations.

Example 3.21 Constrained optimum point. Minimize

$$f(\mathbf{x}) = (x_1 - 1.5)^2 + (x_2 - 1.5)^2$$

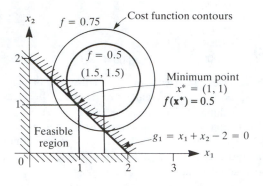

FIGURE 3.11
Graphical representation for Example 3.21. Constrained optimum point.

subject to

$$g_1(\mathbf{x}) \equiv x_1 + x_2 - 2 \leqq 0$$
$$g_2(\mathbf{x}) \equiv -x_1 \leqq 0$$
$$g_3(\mathbf{x}) \equiv -x_2 \leqq 0$$

Solution. The constraint set for the problem is a triangular region shown in Fig. 3.11. If constraints are ignored, $f(\mathbf{x})$ has a minimum at the point $(1.5, 1.5)$ which violates the constraint g_1. Note that contours of $f(\mathbf{x})$ are circles. They increase in diameter as $f(\mathbf{x})$ increases. It is clear that the minimum value for $f(\mathbf{x})$ corresponds to a circle with the smallest radius intersecting the feasible region (constraint set). This is the point $(1, 1)$ at which $f(\mathbf{x}) = 0.5$. The point is on the boundary of the feasible region. Thus location of the optimum point is governed by the constraints for this problem. ‖

Example 3.22 Unconstrained optimum point for a constrained problem. Minimize

$$f(\mathbf{x}) = (x_1 - 0.5)^2 + (x_2 - 0.5)^2$$

subject to the same constraints as in Example 3.21.

Solution. The constraint set is the same as in Example 3.21. The cost function, however, has been modified. If constraints are ignored, $f(\mathbf{x})$ has a minimum at $(0.5, 0.5)$. Since the point also satisfies all the constraints, it is the optimum solution. The solution for this problem therefore occurs in the interior of the feasible region and the constraints play no role in its location. ‖

Note that a *solution to a constrained optimization problem may not exist.* This can happen if we over-constrain the system. The requirements can be conflicting such that it is impossible to build a system to satisfy them. In such a case we must re-examine the problem formulation and relax constraints. The following example illustrates the situation.

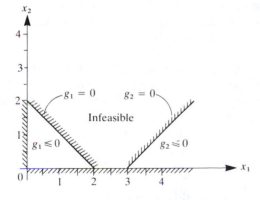

FIGURE 3.12
Plot of constraints for Example 3.23. Infeasible problem.

Example 3.23   Infeasible problem. Minimize

$$f(\mathbf{x}) = (x_1 - 2)^2 + (x_2 - 2)^2$$

subject to

$$g_1(\mathbf{x}) \equiv x_1 + x_2 - 2 \leqq 0$$
$$g_2(\mathbf{x}) \equiv -x_1 + x_2 + 3 \leqq 0$$
$$g_3(\mathbf{x}) \equiv -x_1 \leqq 0$$
$$g_4(\mathbf{x}) \equiv -x_2 \leqq 0$$

Solution. Figure 3.12 shows a plot of the constraints for the problem. It can be seen that there is no design satisfying all the constraints. The constraint set for the problem is empty and there is no solution (i.e. no feasible design).   ‖

3.4.2 Necessary Conditions: Equality Constraints

We first discuss the necessary conditions for constrained minimization problems with only equality constraints. Just as for the unconstrained case, solutions of these conditions give candidate minimum points. The sufficient conditions – discussed later in the chapter – can be used to determine if a candidate point is indeed a local minimum. Extensions to include inequalities are described in the next subsection.

3.4.2.1 REGULAR POINT. Before discussing the necessary conditions, we define a *regular point* of the design space (feasible region). Consider the constrained optimization problem of minimizing $f(\mathbf{x})$ subject to the constraints $h_i(\mathbf{x}) = 0$, $i = 1$ to p. A point $\mathbf{x}^*$ satisfying the constraints $\mathbf{h}(\mathbf{x}^*) = \mathbf{0}$ is said to be a *regular point* of the design space if gradient vectors of all constraints at the point $\mathbf{x}^*$ are linearly independent. *Linear independence* (Appendix B) means that no two gradients are parallel to each other, and no gradient can be expressed as a linear combination of the others. When inequality constraints are also included in the problem definition, then for a point to be regular, gradients of active inequalities must be also linearly independent.

3.4.2.2 LAGRANGE MULTIPLIERS AND NECESSARY CONDITIONS. It turns out that each constraint has a scalar multiplier associated with it, called the *Lagrange multiplier*. These multipliers play a prominent role in optimization theory as well as numerical methods discussed in Chapters 4 to 8. The multipliers have a geometrical as well as physical meaning. Their values depend on the form of the cost and constraint functions. If these functions change, the value of the Lagrange multipliers also change. We shall discuss this aspect in Section 3.7.

To introduce the idea of Lagrange multipliers, we consider the following example of minimizing a cost function of two variables with one equality constraint.

Example 3.24 Introduction of Lagrange multipliers and their geometrical meaning. Find x_1 and x_2 to minimize

$$f(x_1, x_2) = (x_1 - 1.5)^2 + (x_2 - 1.5)^2 \qquad \text{(a)}$$

subject to

$$h(x_1, x_2) \equiv x_1 + x_2 - 2 = 0 \qquad \text{(b)}$$

Solution. The problem has two variables and can be solved easily by the graphical procedure. Figure 3.13 shows a graphical representation for the problem. The straight line A–B represents the equality constraint and the feasible region for the problem. Therefore, the optimum solution must lie on the line A–B. The cost function is an equation of a circle with its center at point $(1.5, 1.5)$. The iso-cost lines, having values of 0.5 and 0.75, are shown in the figure. It can be seen that point C, having coordinates $(1, 1)$, gives the optimum solution for the problem. The cost function contour of value 0.5 just touches the line A–B, so this is the minimum value for the cost function.

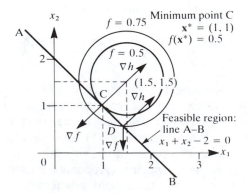

FIGURE 3.13
Graphical solution for Example 3.24. Geometrical interpretation of necessary conditions.

Introduction of Lagrange multipliers. Now let us see what mathematical conditions are satisfied at the minimum point C. Let the optimum point be represented as (x_1^*, x_2^*). To derive the conditions and to introduce the Lagrange multiplier, we first assume that the equality constraint can be used to solve for one variable in terms of the other (at least symbolically), i.e. assume that we can write

$$x_2 = \phi(x_1) \tag{c}$$

where ϕ is an appropriate function of x_1. In many problems, it may not be possible to write explicitly the function $\phi(x_1)$, but for derivation purposes, we assume its existence. It will be seen later that the explicit form of the function is not needed. For the present example, $\phi(x_1)$ from Eq. (b) is given as

$$\phi(x_1) = -x_1 + 2 \tag{d}$$

Substituting Eq. (c) into Eq. (a), we eliminate x_2 from the cost function and obtain the unconstrained minimization problem in terms of x_1 only:

$$\text{minimize } f(x_1, \phi(x_1)) \tag{e}$$

For the present example, substituting Eq. (d) into Eq. (a), we eliminate x_2 and obtain the minimization problem in terms of x_1 alone:

$$f(x_1) = (x_1 - 1.5)^2 + (-x_1 + 2 - 1.5)^2$$

The necessary condition $df/dx_1 = 0$ gives $x_1^* = 1$. Then Eq. (d) gives $x_2^* = 1$ and the cost function at the point $(1, 1)$ is 0.5. It can be checked that the sufficiency condition $d^2f/dx_1^2 > 0$ is satisfied, so the point is indeed a local minimum as seen in Fig. 3.13.

If we assume that the explicit form of the function $\phi(x_1)$ cannot be obtained (which is generally the case), then some alternate procedure must be developed to obtain the optimum solution. We shall derive such a procedure and see that the Lagrange multiplier for the constraint gets defined naturally in the process. Using the chain rule of differentiation, we write the necessary condition $df/dx_1 = 0$ for the problem defined in Eq. (e) as

$$\frac{df(x_1, x_2)}{dx_1} = \frac{\partial f(x_1, x_2)}{\partial x_1} + \frac{\partial f(x_1, x_2)}{\partial x_2}\frac{dx_2}{dx_1} = 0$$

Substituting Eq. (c), the preceding equation can be written at the optimum point (x_1^*, x_2^*) as

$$\frac{\partial f(x_1^*, x_2^*)}{\partial x_1} + \frac{\partial f(x_1^*, x_2^*)}{\partial x_2}\frac{d\phi}{dx_1} = 0 \tag{f}$$

Since ϕ is not known, we need to eliminate $d\phi/dx_1$ from Eq. (f). To accomplish this, we differentiate the constraint equation $h(x_1, x_2) = 0$ at the point (x_1^*, x_2^*) as

$$\frac{dh(x_1^*, x_2^*)}{dx_1} = \frac{\partial h(x_1^*, x_2^*)}{\partial x_1} + \frac{\partial h(x_1^*, x_2^*)}{\partial x_2}\frac{d\phi}{dx_1} = 0$$

or solving for $d\phi/dx_1$, we obtain

$$\frac{d\phi}{dx_1} = -\frac{\partial h(x_1^*, x_2^*)/\partial x_1}{\partial h(x_1^*, x_2^*)/\partial x_2} \tag{g}$$

Now substituting for $d\phi/dx_1$ from Eq. (g) into Eq. (f), we obtain

$$\frac{\partial f(x_1^*, x_2^*)}{\partial x_1} - \frac{\partial f(x_1^*, x_2^*)}{\partial x_2}\left(\frac{\partial h(x_1^*, x_2^*)/\partial x_1}{\partial h(x_1^*, x_2^*)/\partial x_2}\right) = 0 \tag{h}$$

If we define a quantity v as

$$v = -\frac{\partial f(x_1^*, x_2^*)/\partial x_2}{\partial h(x_1^*, x_2^*)/\partial x_2} \tag{i}$$

and substitute it into Eq. (h), we obtain

$$\frac{\partial f(x_1^*, x_2^*)}{\partial x_1} + v\frac{\partial h(x_1^*, x_2^*)}{\partial x_1} = 0 \tag{j}$$

Also, rearranging Eq. (i) that defines v, we obtain

$$\frac{\partial f(x_1^*, x_2^*)}{\partial x_2} + v\frac{\partial h(x_1^*, x_2^*)}{\partial x_2} = 0 \tag{k}$$

Equations (j) and (k) along with the equality constraint $h(x_1, x_2) = 0$ are the *necessary conditions of optimality. Any point that violates these conditions cannot be a minimum point for the problem.* The scalar quantity v defined in Eq. (i) is called the Lagrange multiplier. If the minimum point is known, Eq. (i) can be used to calculate its value. For the present example, $\partial f(1, 1)/\partial x_2 = -1$ and $\partial h(1, 1)/\partial x_2 = 1$; therefore, Eq. (i) gives $v^* = 1$ as the Lagrange multiplier at the optimum point.

Recall that the necessary conditions can be used to solve for the candidate minimum points, i.e. Eqs (j), (k) and $h(x_1, x_2) = 0$ can be used to solve for x_1, x_2 and v. For the present example, these equations give

$$2(x_1 - 1.5) + v = 0$$
$$2(x_2 - 1.5) + v = 0$$
$$x_1 + x_2 - 2 = 0$$

Solution of the preceding equations is indeed $x_1^* = 1$, $x_2^* = 1$ and $v^* = 1$.

Geometrical meaning of Lagrange multipliers. It is customary to use what is known as the *Lagrange function* in writing the necessary conditions. The Lagrange function is denoted as L and defined using cost and constraint functions as

$$L(x_1, x_2, v) = f(x_1, x_2) + vh(x_1, x_2) \tag{l}$$

It is seen that the necessary conditions of Eqs (j) and (k) are given in terms of L as

$$\frac{\partial L(x_1^*, x_2^*)}{\partial x_1} = 0, \qquad \frac{\partial L(x_1^*, x_2^*)}{\partial x_2} = 0 \tag{m}$$

Or, in the vector notation we see that the gradient of L is zero at the candidate minimum point, i.e. $\nabla L(x_1^*, x_2^*) = \mathbf{0}$. Writing this condition, using Eq. (l), or writing Eqs (j) and (k) in the vector form, we obtain

$$\nabla f(\mathbf{x}^*) + v\nabla h(\mathbf{x}^*) = \mathbf{0} \tag{n}$$

where gradients of the cost and constraint functions are given as

$$\nabla f(\mathbf{x}^*) = \begin{bmatrix} \dfrac{\partial f(x_1^*, x_2^*)}{\partial x_1} \\[2ex] \dfrac{\partial f(x_1^*, x_2^*)}{\partial x_2} \end{bmatrix}, \qquad \nabla h = \begin{bmatrix} \dfrac{\partial h(x_1^*, x_2^*)}{\partial x_1} \\[2ex] \dfrac{\partial h(x_1^*, x_2^*)}{\partial x_2} \end{bmatrix}$$

Equation (n) can be rearranged as

$$\nabla f(\mathbf{x}^*) = -v\nabla h(\mathbf{x}^*) \tag{o}$$

The preceding equation brings out the geometrical meaning of the necessary conditions. It shows that *at the candidate minimum point, gradients of the cost and constraint functions are along the same line and proportional to each other*; *the Lagrange multiplier v is the proportionality constant.*

For the present example, the gradients of cost and constraint functions at the candidate optimum point are given as

$$\nabla f(1, 1) = \begin{bmatrix} -1 \\ -1 \end{bmatrix}, \qquad \nabla h(1, 1) = \begin{bmatrix} 1 \\ 1 \end{bmatrix}$$

These vectors are shown at point C in Fig. 3.13. It can be seen that they are along the same line. For any feasible point on the line A–B other than the candidate minimum, say (0.4, 1.6), the gradients of cost and constraint functions will not be along the same line, as seen in the following:

$$\nabla f(0.4, 1.6) = \begin{bmatrix} -2.2 \\ 0.2 \end{bmatrix}, \qquad \nabla h(0.4, 1.6) = \begin{bmatrix} 1 \\ 1 \end{bmatrix}$$

As another example, point D in Fig. 3.13 is not a candidate minimum, since gradients of cost and constraint functions are not along the same line. Also, the cost function has higher value at these points compared to the one at the minimum point, i.e. we can move away from point D towards point C and reduce the cost function.

It is interesting to note that the equality constraint can be multiplied by -1 without affecting the minimum point, i.e. the constraint can be written as $-x_1 - x_2 + 2 = 0$. The minimum solution is still the same $x_1^* = 1$, $x_2^* = 1$ and $f(\mathbf{x}^*) = 0.5$; however, the sign of the Lagrange multiplier is reversed, i.e. $v^* = -1$. *This shows that the Lagrange multiplier for the equality constraint is free in sign*, i.e. the sign is determined by the form of the constraint function.

It is also interesting to note that any small move from point C in the feasible region (i.e. along the line A–B) increases the cost function value, and *any further reduction to the cost function is accompanied by violation of the constraint.* Thus point C satisfies the sufficiency condition for a local minimum point because it has the smallest value in a neighborhood of point C. Thus, it is indeed a local minimum point.   ‖

The concept of Lagrange multipliers is quite general. It is encountered in many engineering applications other than the optimum design. *The Lagrange multiplier for a constraint can be interpreted as a force required to impose the constraint.* We shall discuss a physical meaning of the Lagrange multipliers in Section 3.7.

The idea of a Lagrange multiplier for an equality constraint can be generalized to many equality constraints. It can also be extended for inequality constraints. We shall first discuss the necessary conditions with multiple equality constraints, and then describe their extensions in the next subsection to include the inequality constraints.

Theorem 3.5 Lagrange Multiplier Theorem. Consider the problem of minimizing $f(\mathbf{x})$ subject to the equality constraints $h_i(\mathbf{x}) = 0$, $i = 1$ to p. Let $\mathbf{x}^*$ be a regular point that is a local minimum for the problem. Then there exist Lagrange multipliers v_j^*, $j = 1$ to p such that

$$\frac{\partial f(\mathbf{x}^*)}{\partial x_i} + \sum_{j=1}^{p} v_j^* \frac{\partial h_j(\mathbf{x}^*)}{\partial x_i} = 0; \qquad i = 1 \text{ to } n \tag{3.40}$$

$$h_j(\mathbf{x}^*) = 0; \qquad j = 1 \text{ to } p$$

It is convenient to write these conditions in terms of a Lagrange function defined as

$$L(\mathbf{x}, \mathbf{v}) = f(\mathbf{x}) + \sum_{j=1}^{p} v_j h_j(\mathbf{x}) = f(\mathbf{x}) + \mathbf{v}^T \mathbf{h}(\mathbf{x}) \tag{3.41}$$

Then Eq. (3.40) becomes

$$\nabla L(\mathbf{x}^*, \mathbf{v}^*) = \mathbf{0}, \quad \text{or} \quad \frac{\partial L(\mathbf{x}^*, \mathbf{v}^*)}{\partial x_i} = 0; \qquad i = 1 \text{ to } n \tag{3.42}$$

Differentiating $L(\mathbf{x}, \mathbf{v})$ with respect to v_j, we can recover the equality constraints as

$$\frac{\partial L(\mathbf{x}^*, \mathbf{v}^*)}{\partial v_j} \equiv h_j(\mathbf{x}^*) = 0; \qquad j = 1 \text{ to } p \tag{3.43} \quad \|$$

The gradient conditions of Eqs (3.42) and (3.43) show that the Lagrange function is stationary with respect to both $\mathbf{x}$ and $\mathbf{v}$. Therefore, it may be treated as an unconstrained function in the variables $\mathbf{x}$ and $\mathbf{v}$ to determine the stationary points. *Note that any point that does not satisfy conditions of the theorem cannot be a local minimum point.* However, a point satisfying the conditions need not be a minimum point either. It is simply a candidate minimum point which can actually be an inflection or maximum point. The second-order necessary and sufficient conditions given in Section 3.6 can distinguish between the minimum, maximum and inflection points.

The n variables $\mathbf{x}$ and the p multipliers $\mathbf{v}$ are the unknowns and the necessary conditions of Eqs (3.42) and (3.43) provide enough equations to solve for them. Note also that the Lagrange multipliers v_i are free in sign, i.e. they can be positive, negative or zero. This is in contrast to the Lagrange multipliers for the inequality constraints, which we shall see later are required to be nonnegative.

The gradient condition of Eq. (3.40) can be rearranged as

$$\frac{\partial f(\mathbf{x}^*)}{\partial x_i} = -\sum_{j=1}^{p} v_j^* \frac{\partial h_j(\mathbf{x}^*)}{\partial x_i}; \qquad i = 1 \text{ to } n$$

This form shows that the *gradient of the cost function is a linear combination of the gradients of the constraints at the candidate minimum point.* The Lagrange multipliers v_j^* act as the scalars of the linear combination. This linear combination interpretation of the necessary conditions is a generalization of the concept discussed in Example 3.24 for one constraint: "at the candidate minimum point gradients of the cost and constraint functions are along the same line".

> **Example 3.25   Cylindrical tank design – use of Lagrange multipliers for equality constrained problem.** We will re-solve the cylindrical storage tank problem (Example 3.18) using the Lagrange multiplier approach. The problem is to find radius R and length l of the cylinder to minimize
>
> $$\bar{f} = R^2 + Rl$$
>
> subject to
>
> $$h \equiv \pi R^2 l - V = 0$$
>
> **Solution.** The Lagrange function L for the problem is given as
>
> $$L = R^2 + Rl + v(\pi R^2 l - V)$$
>
> The necessary conditions of the Lagrange Multiplier Theorem 3.5 give
>
> $$\frac{\partial L}{\partial R} \equiv 2R + l + 2\pi v Rl = 0$$
>
> $$\frac{\partial L}{\partial l} \equiv R + v\pi R^2 = 0$$
>
> $$\frac{\partial L}{\partial v} \equiv \pi R^2 l - V = 0$$
>
> These are three equations in three unknown v, R and l. Note that they are nonlinear. However, they can be easily solved by the elimination process giving the solution as
>
> $$R^* = \left(\frac{V}{2\pi}\right)^{1/3}$$
>
> $$l^* = \left(\frac{4V}{\pi}\right)^{1/3}$$
>
> $$v^* = -\frac{1}{\pi R} = -\left(\frac{2}{\pi^2 V}\right)^{1/3}$$
>
> This is the same solution as obtained for Example 3.18, treating it as an unconstrained problem. It can be also verified that the gradients of the cost and constraint functions are along the same line at the optimum point.   ‖

Often, *the necessary conditions of the Lagrange Multiplier Theorem lead to a nonlinear set of equations that cannot be solved analytically. In such cases, we must use a numerical algorithm such as the Newton-Raphson method (Appendix C) to solve for their roots and the candidate minimum points.*

3.4.3   Necessary Conditions: Inequality Constraints

The design problem formulations in Chapter 2 often included inequality constraints of the form

$$g_i(\mathbf{x}) \leqq 0, \qquad i = 1 \text{ to } m$$

We can transform an inequality constraint to an equality by adding a new variable to it, called the *slack variable*. Since the constraint is of the form "$\leqq$", its value is either negative or zero. Thus the slack variable must always be nonnegative (i.e. positive or zero) to make the inequality an equality. An inequality constraint $g_i(\mathbf{x}) \leqq 0$ is equivalent to the equality constraint $g_i(\mathbf{x}) + s_i = 0$, where $s_i \geqq 0$ is a slack variable. The variables s_i are treated as unknowns of the design problem along with the original variables. Their values are determined as a part of the solution. When the variable s_i has zero value, the corresponding inequality constraint is satisfied at equality. Such inequality is called an *active (tight) constraint,* i.e. there is no "slack" in the constraint. For any $s_i > 0$, the corresponding constraint is a strict inequality. It is called an *inactive constraint,* and has slack given by s_i.

Note that with the preceding procedure, we must introduce one additional design variable s_i and an additional constraint $s_i \geqq 0$ to treat each inequality constraint. This increases the dimension of the design problem. The constraint $s_i \geqq 0$ can be avoided if we use s_i^2 as the slack variable instead of just s_i. Therefore, the inequality $g_i \leqq 0$ is converted to equality as

$$g_i + s_i^2 = 0 \tag{3.44}$$

where s_i can have any real value. This form can be used in the Lagrange Multiplier Theorem to treat inequality constraints and to derive the corresponding necessary conditions. The m new equations needed for determining the slack variables are obtained by requiring the Lagrangian L to be stationary with respect to the slack variables ($\partial L / \partial \mathbf{s} = \mathbf{0}$).

Note that once a design point is specified, Eq. (3.44) can be used to calculate the slack variable s_i^2. If the constraint is satisfied at the point (i.e. $g_i \leqq 0$), then $s_i^2 \geqq 0$. If it is violated, then s_i^2 is negative which is not acceptable, i.e. the point is not a candidate minimum point.

There is an *additional necessary condition* for the Lagrange multipliers of "$\leqq$ type" constraints given as

$$u_j^* \geqq 0; \qquad j = 1 \text{ to } m \tag{3.45}$$

where u_j^* is the Lagrange multiplier for the jth inequality constraint. Thus, *the Lagrange multiplier for each "$\leqq$" inequality constraint must be nonnegative.* If the constraint is inactive at the optimum, its associated Lagrange multiplier is zero. If it is active ($g_i = 0$), then the associated multiplier must be nonnegative. We will explain the condition of Eq. (3.45) from a physical point of view in Section 3.7.

Example 3.26 Inequality constrained problem – use of necessary conditions. We will re-solve Example 3.24 by treating the constraint as an inequality. The problem is to minimize

$$f(x_1, x_2) = (x_1 - 1.5)^2 + (x_2 - 1.5)^2$$

subject to

$$g(\mathbf{x}) \equiv x_1 + x_2 - 2 \leqq 0$$

Solution. The graphical representation for the problem remains the same as in Fig. 3.13 for Example 3.24, except that the feasible region is enlarged; it is line A–B and the region below it. The minimum point for the problem is same as before, i.e. $x_1^* = 1$, $x_2^* = 1$, $f(\mathbf{x}^*) = 0.5$.

Introducing a slack variable s^2 for the inequality, the Lagrangian for the problem is defined as

$$L = (x_1 - 1.5)^2 + (x_2 - 1.5)^2 + u(x_1 + x_2 - 2 + s^2)$$

The necessary conditions of the Lagrange Theorem give (treating x_1, x_2, u and s as unknowns)

$$\frac{\partial L}{\partial x_1} \equiv 2(x_1 - 1.5) + u = 0 \tag{a}$$

$$\frac{\partial L}{\partial x_2} \equiv 2(x_2 - 1.5) + u = 0 \tag{b}$$

$$\frac{\partial L}{\partial u} \equiv x_1 + x_2 - 2 + s^2 = 0 \tag{c}$$

$$\frac{\partial L}{\partial s} \equiv 2us = 0 \tag{d}$$

These are four equations in four unknowns x_1, x_2, u and s. The equations must be solved simultaneously for all the unknowns. Note that the equations are nonlinear. Therefore they can have many roots.

One solution can be obtained by setting s to zero to satisfy the condition $2us = 0$ in Eq. (d). Equations (a)–(c) are solved to obtain

$$x_1^* = x_2^* = 1, \qquad u^* = 1, \qquad s = 0$$

When $s = 0$, the inequality constraint is active. x_1, x_2 and u are solved from the remaining three equations (a) – (c), which are linear in the variables. This is a stationary point of L, so it is a candidate minimum point. Note from Fig. 3.13 that it is actually a minimum point, since any move away from $\mathbf{x}^*$ either violates the constraint or increases the cost function.

The second stationary point is obtained by setting u = 0 to satisfy the condition of Eq. (d) and solving the remaining equations for x_1, x_2, and s. This gives

$$x_1^* = x_2^* = 1.5, \qquad u^* = 0, \qquad s^2 = -1$$

This is not a valid solution as the constraint is violated at the point $\mathbf{x}^*$, since $g = -s^2 = 1 > 0$.

It is interesting to observe the geometrical representation of the necessary conditions for inequality constrained problems. The gradients of the cost and constraint functions at the candidate point $(1, 1)$ are calculated as

$$\nabla f = \begin{bmatrix} 2(x_1 - 1.5) \\ 2(x_2 - 1.5) \end{bmatrix} = \begin{bmatrix} -1 \\ -1 \end{bmatrix}; \quad \nabla g = \begin{bmatrix} 1 \\ 1 \end{bmatrix}$$

These gradients are along the same line but in opposite directions as shown in Fig. 3.13. Observe also that any small move from point C either increases the cost function or takes the design into the infeasible region to reduce the cost function any further. Thus, point $(1, 1)$ is indeed a local minimum point. This geometrical condition is called the *sufficient condition* for a local minimum point.

It turns out that the necessary condition $u \geqq 0$ *insures that the gradients of the cost and the constraint functions point in opposite directions*. This way f cannot be reduced any further by stepping in the negative gradient direction without violating the constraint. That is, any further reduction in the cost function leads to leaving the feasible region at the candidate minimum point. This can be observed in Fig. 3.13.
‖

The necessary conditions for the equality and inequality constraints can be summed up in what are commonly known as the *Kuhn–Tucker (K–T) necessary conditions*. Although the conditions can be expressed in several forms, we shall discuss only two forms, one with slack variables and the other without them. The conditions with and without the slack variables are completely equivalent. We shall explain this equivalence later.

Theorem 3.6   Kuhn–Tucker (K–T) necessary conditions. Let $\mathbf{x}^*$ be a *regular point* of the constraint set that is a local minimum for $f(\mathbf{x})$ subject to the constraints

$$h_i(\mathbf{x}) = 0; \quad i = 1 \text{ to } p$$

$$g_i(\mathbf{x}) \leqq 0; \quad i = 1 \text{ to } m$$

Define the Lagrange function for the problem as

$$L(\mathbf{x}, \mathbf{v}, \mathbf{u}, \mathbf{s}) = f(\mathbf{x}) + \sum_{i=1}^{p} v_i h_i(\mathbf{x}) + \sum_{i=1}^{m} u_i(g_i(\mathbf{x}) + s_i^2)$$

$$= f(\mathbf{x}) + \mathbf{v}^T \mathbf{h}(\mathbf{x}) + \mathbf{u}^T (\mathbf{g}(\mathbf{x}) + \mathbf{s}^2) \tag{3.46}$$

Then there exist Lagrange multipliers $\mathbf{v}^*$(a p-vector) and $\mathbf{u}^*$ (an m-vector) such that the Lagrangian is stationary with respect to x_j, v_i, u_i and s_i, i.e.

$$\frac{\partial L}{\partial x_j} \equiv \frac{\partial f}{\partial x_j} + \sum_{i=1}^{p} v_i^* \frac{\partial h_i}{\partial x_j} + \sum_{i=1}^{m} u_i^* \frac{\partial g_i}{\partial x_j} = 0; \quad j = 1 \text{ to } n \tag{3.47}$$

$$h_i(\mathbf{x}^*) = 0; \quad i = 1 \text{ to } p \tag{3.48}$$

$$g_i(\mathbf{x}^*) + s_i^2 = 0; \quad i = 1 \text{ to } m \tag{3.49}$$

$$u_i^* s_i = 0; \quad i = 1 \text{ to } m \tag{3.50}$$

and

$$u_i^* \geqq 0; \quad i = 1 \text{ to } m \tag{3.51}$$

where all derivatives are evaluated at point $\mathbf{x}^*$.
‖

The foregoing conditions are sometimes called *first-order necessary conditions*. It is important to understand their use to (i) check possible optimality of a given point, and (ii) determine the candidate local minimum points. Note first from Eqs (3.48) and (3.49) that *the candidate minimum point must be feasible,* so we must check all the constraints to ensure their satisfaction. The gradient conditions of Eq. (3.47) must also be satisfied simultaneously. These conditions have a *geometrical meaning*. To see this rewrite Eq. (3.47) as

$$-\frac{\partial f}{\partial x_j} = \sum_{i=1}^{p} v_i^* \frac{\partial h}{\partial x_j} + \sum_{i=1}^{m} u_i^* \frac{\partial g_i}{\partial x_j}; \qquad j = 1 \text{ to } n \qquad (3.52)$$

which shows that at the stationary point, negative gradient direction (steepest descent direction) for the cost function is a linear combination of the gradients of the constraints with Lagrange multipliers as the scalar parameters of the linear combination.

The *m* conditions in Eq. (3.50) *are known as the switching conditions or complementary slackness conditions*. They can be satisfied by setting either $s_i = 0$ (zero slack implies active inequality, i.e. $g_i = 0$), or $u_i = 0$ (in this case g_i must be $\leqq 0$ to satisfy feasibility). These conditions determine several cases in actual calculations and their use must be clearly understood. In Example 3.26, there was only one switching condition which gave two possible cases; case 1 where the slack variable was zero and case 2 where the Lagrange multiplier u for the inequality constraint was zero. Each of the two cases was solved for the unknowns. For general problems, there are more than one switching condition in Eq. (3.50); the number of switching conditions is equal to the number of inequality constraints for the problem. Various combinations of these conditions can give many solution cases. In general, with *m* inequality constraints, the switching conditions lead to 2^m distinct *normal* solution cases (abnormal case is the one where both $u_i = 0$ and $s_i = 0$). For each case, we need to solve the remaining necessary conditions for candidate local minimum points. Depending on functions of the problem, it may or may not be possible to solve analytically the necessary conditions of each case. If the functions are nonlinear, we will have to use numerical methods to find their roots. In addition, each case may give several candidate minimum points.

We shall illustrate the use of the K–T conditions in several example problems. In Example 3.26 there were only two variables, one Lagrange multiplier and one slack variable. For general problems, the unknowns are **x, u, s** and **v**. These are *n, m, m* and *p* dimensional vectors. There are thus $(n + 2m + p)$ unknown variables and we need $(n + 2m + p)$ equations to determine them. The equations needed for their solution are available in the Kuhn-Tucker necessary conditions. If we count the number of equations in Eqs (3.47) to (3.51), we find that there are indeed $(n + 2m + p)$ equations. These equations then must be solved simultaneously for the candidate local minimum points. After the solutions are found, the remaining necessary condition of Eq. (3.51) must be checked. Conditions of Eq. (3.49) ensure feasibility of candidate local minimum points with respect to the inequality constraints $g_i(\mathbf{x}) \leqq 0$; $i = 1$ to m. And, conditions of Eq. (3.51) say that the Lagrange multipliers of the "$\leqq$ type" inequality constraints must be nonnegative.

Note that evaluation of s_i essentially implies evaluation of the constraint function $g_i(\mathbf{x})$, since $s_i^2 = -g_i(\mathbf{x})$. This way, evaluation of s_i allows us to check feasibility of the candidate points with respect to the constraint $g_i(\mathbf{x}) \leqq 0$.

It is important to note that if an inequality constraint $g_i(\mathbf{x}) \leqq 0$ is inactive at the candidate minimum point $\mathbf{x}^*$ (i.e., $g_i(\mathbf{x}^*) < 0$, or $s_i^2 > 0$), then the corresponding *Lagrange multiplier* $u_i^* = 0$ to satisfy the switching condition of Eq. (3.50). If, however, it is active (i.e., $g_i(\mathbf{x}^*) = 0$), then the Lagrange multiplier must be nonnegative, $u_i^* \geqq 0$. This condition ensures that there are no feasible directions with respect to the ith constraint $g_i(\mathbf{x}) \leqq 0$ at the candidate point $\mathbf{x}^*$ along which the cost function can reduce any further. Stated differently, the condition ensures that any reduction in the cost function at $\mathbf{x}^*$ can occur only by stepping into the infeasible region for the constraint $g_i(\mathbf{x}) \leqq 0$.

Note further that the necessary conditions of Eqs (3.47) to (3.51) are generally a nonlinear system of equations in the variables $\mathbf{x}$, $\mathbf{u}$, $\mathbf{s}$ *and* $\mathbf{v}$. It may not be easy to solve the system analytically. Therefore, we may have to use numerical methods such as the Newton-Rhapson method of Appendix C to find roots of the system. Fortunately, programs are available in most computer center libraries to solve a nonlinear set of equations. Such programs are of great help in solving for candidate local minimum points.

The following important points should be noted relative to the Kuhn-Tucker first-order necessary conditions:

1. K–T conditions are *not applicable* at the points that are not *regular*.
2. Any point that *does not satisfy* K–T conditions *cannot be a local minimum* unless it is an irregular point (in that case K–T conditions are not applicable). Points satisfying the conditions are called Kuhn–Tucker Points.
3. *The points satisfying K–T conditions can be constrained or unconstrained.* They are unconstrained when there are no equalities and all inequalities are inactive. If the candidate point is unconstrained, it can be a local minimum, maximum or inflection point depending on the form of the Hessian matrix of the cost function (refer to Section 3.3 for the necessary and sufficient conditions for unconstrained problems).
4. *If there are equality constraints and no inequalities are active (i.e.* $\mathbf{u} = \mathbf{0}$), *then the points satisfying K–T conditions are only stationary.* They can be minimum, maximum or inflection points.
5. *If some inequality constraints are active and their multipliers are positive, then the points satisfying K–T conditions cannot be local maxima for the cost function* (they may be local maximum points if active inequalities have zero multipliers). They may not be local minima either; this will depend on the second-order necessary and sufficient conditions discussed in Section 3.6.

6. It is important to note that value of the *Lagrange multiplier* for each constraint depends on the functional form for the constraint. For example, Lagrange multiplier for the constraint $x/y - 10 \leq 0$ ($y > 0$) is different for the same constraint expressed as $x - 10y \leq 0$, or $0.1x/y - 1 \leq 0$. The optimum solution for the problem does not change by changing the form of the constraint, but its Lagrange multiplier is changed. This is further explained in Section 3.7.

These points will be illustrated with the following simple example.

> **Example 3.27 Various solutions of Kuhn–Tucker necessary conditions.** For the following problem, write K–T necessary conditions and solve them. Minimize
>
> $$f(\mathbf{x}) = \tfrac{1}{3}x^3 - \tfrac{1}{2}(b + c)x^2 + bcx + f_0$$
>
> subject to
>
> $$a \leq x \leq d$$
>
> where $0 < a < b < c < d$ and f_0 are specified constants (created by Y. S. Ryu).

> ***Solution.*** A graph for the function is shown in Fig. 3.14. It can be seen that Point A is a constrained minimum, Point B is an unconstrained maximum, Point C is an unconstrained minimum, and Point D is a constrained maximum. We shall show how the K–T conditions distinguish between these points.
>
> There are two inequality constraints,
>
> $$g_1 \equiv a - x \leq 0$$
> $$g_2 \equiv x - d \leq 0$$
>
> so the Lagrangian for the problem is given as
>
> $$L = \tfrac{1}{3}x^3 - \tfrac{1}{2}(b + c)x^2 + bcx + f_0 + u_1(a - x + s_1^2) + u_2(x - d + s_2^2)$$
>
> where u_1 and u_2 are the Lagrange multipliers and s_1 and s_2 are the slack variables for $g_1 \equiv a - x \leq 0$ and $g_2 \equiv x - d \leq 0$, respectively. The K–T conditions give
>
> $$\frac{\partial L}{\partial x} = x^2 - (b + c)x + bc - u_1 + u_2 = 0 \tag{a}$$
>
> $$(a - x) + s_1^2 = 0; \qquad (x - d) + s_2^2 = 0 \tag{b}$$
>
> $$u_1 s_1 = 0; \qquad u_2 s_2 = 0 \tag{c}$$
>
> $$u_1 \geq 0; \qquad u_2 \geq 0 \tag{d}$$

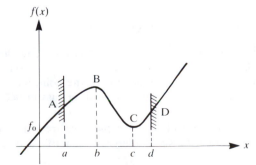

FIGURE 3.14
Graphical representation for Example 3.27. Point A, constrained local minimum; B, unconstrained local maximum; C, unconstrained local minimum; D, constrained local maximum.

The switching conditions in Eqs (c) give four cases for the solution of K–T conditions. Each case will be considered separately and solved.

Case 1: $u_1 = 0$, $u_2 = 0$. For this case, Eq. (a) gives two solutions as $x = b$ and $x = c$. For these points both the inequalities are strictly satisfied because slack variables calculated from Eqs (b) are

$$\text{for } x = b: \qquad s_1^2 = b - a > 0; \qquad s_2^2 = d - b > 0$$
$$\text{for } x = c: \qquad s_1^2 = c - a > 0; \qquad s_2^2 = d - c > 0$$

Thus, all the K–T conditions are satisfied, and these are candidate minimum points. Since the points are unconstrained, so they are actually stationary points. We can check the sufficient condition by calculating *curvature* of the cost function at the two candidate points:

$$x = b; \qquad \frac{d^2f}{dx^2} = 2x - (b + c) = b - c < 0$$

Since $b < c$, d^2f/dx^2 is negative. Therefore, the sufficient condition for local minimum is violated. Actually, the second-order necessary condition of Eq. (3.32) is also violated, so the point cannot be a local minimum for the function. It is actually a local maximum point because it satisfies sufficient condition for that, as also seen in Fig. 3.14.

$$x = c; \qquad \frac{d^2f}{dx^2} = c - b > 0$$

Since $b < c$, d^2f/dx^2 is positive. Therefore, the second-order sufficient condition of Eq. (3.31) is satisfied, and this is a local minimum point, as also seen in Fig. 3.14.

Case 2: $u_1 = 0$, $s_2 = 0$. $s_2 = 0$ implies that g_2 is active and, therefore, $x = d$. Equation (a) gives

$$u_2 = -[d^2 - (b + c)d + bc] = -(d - c)(d - b)$$

Since $d > c > b$, u_2 is < 0. Actually the term within the square brackets is also the slope of the function at $x = d$ which is positive (Fig. 3.14), so $u_2 < 0$. The K–T necessary conditions are violated, so there is no solution for this case, i.e. $x = d$ is not a candidate minimum point. This is true as can be observed for the point D in Fig. 3.14.

Case 3: $s_1 = 0$, $u_2 = 0$. $s_1 = 0$ implies that g_1 is active and, therefore, $x = a$. Equation (a) gives

$$u_1 = a^2 - (b + c)a + bc = (a - b)(a - c) > 0$$

Also, since $u_1 = $ slope of the function at $x = a$ (Fig. 3.14), it is positive and all the K–T conditions are satisfied. Thus $x = a$ is a candidate minimum point. Actually $x = a$ is a local minimum point because a feasible move from the point increases the cost function. This is a sufficient condition which we shall discuss in Section 3.6.

Case 4: $s_1 = 0$, $s_2 = 0$. This case for which both the constraints are active does not give any valid solution, since x cannot be simultaneously equal to a and d. ‖

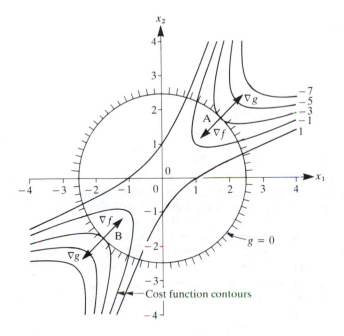

FIGURE 3.15
Graphical solution for Example 3.28. Local minimum points, A and B.

Example 3.28 Solution of Kuhn-Tucker necessary conditions. Consider the optimization problem. Minimize

$$f(\mathbf{x}) = x_1^2 + x_2^2 - 3x_1x_2$$

subject to

$$g \equiv x_1^2 + x_2^2 - 6 \leqq 0$$

Solution. The feasible region for the problem is a circle with its center at $(0,0)$ and radius as $\sqrt{6}$. This is plotted in Fig. 3.15. Several cost function contours are shown there. It can be seen that points A and B give minimum value for the cost function. The gradients of cost and constraint functions at these points are along the same line but in opposite directions, so Kuhn-Tucker necessary conditions are satisfied. We shall verify this by writing these conditions and solving them for candidate minimum points.

The Lagrange function for the problem is

$$L = x_1^2 + x_2^2 - 3x_1x_2 + u(x_1^2 + x_2^2 - 6 + s^2)$$

Since there is only one constraint for the problem, all points of the feasible region are regular, so the Kuhn-Tucker necessary conditions are applicable. They are given as

$$\frac{\partial L}{\partial x_1} \equiv 2x_1 - 3x_2 + 2ux_1 = 0 \tag{a}$$

$$\frac{\partial L}{\partial x_2} \equiv 2x_2 - 3x_1 + 2ux_2 = 0 \tag{b}$$

$$x_1^2 + x_2^2 - 6 + s^2 = 0 \tag{c}$$

$$us = 0 \tag{d}$$

$$u \geqq 0 \tag{e}$$

Equations (a)–(d) are the four equations in four unknowns x_1, x_2, s and u. Thus, in principle, we have enough equations to solve for all the unknowns. The system of equations is nonlinear; however, it is possible to analytically solve for all the roots.

There are three possible ways of satisfying the switching condition of Eq. (d): (i) $u = 0$, (ii) $s = 0$, implying g is active, or (iii) $u = 0$ and $s = 0$. We will consider each case separately and solve for roots of the necessary conditions.

Case 1: $u = 0$. In this case, the inequality constraint may be inactive at the solution point. We shall solve for x_1 and x_2 and then check the constraint. Equations (a) and (b) reduce to

$$2x_1 - 3x_2 = 0$$
$$-3x_1 + 2x_2 = 0$$

This is a 2×2 homogeneous system of linear equations (right-hand side is zero). Such a system has a nontrivial solution only if the determinant of the coefficient matrix is zero. However, since the determinant of the matrix is -5, the system has only a trivial solution, $x_1 = x_2 = 0$. We can also solve the system using Gaussian elimination procedures. This solution gives $s^2 = 6$ from Eq. (c), so the inequality is not active. Thus the candidate minimum point for this case is

$$x_1^* = 0, \ x_2^* = 0, \ u^* = 0, \ f(0, 0) = 0$$

Case 2: $s = 0$. In this case, $s = 0$ implies inequality as active. We must solve Eqs (a) – (c) simultaneously for x_1, x_2 and u. Note that this is a nonlinear set of equations, so there can be multiple roots. Equation (a) gives

$$u = -1 + 3x_2/2x_1$$

Substituting for u in Eq. (b), we obtain

$$x_1^2 = x_2^2$$

Using this in Eq. (c), solving for x_1 and x_2, and then solving for u, we obtain four roots of Eqs (a), (b) and (c) as

$$\begin{aligned}
x_1 = x_2 = \sqrt{3}, && u = \tfrac{1}{2} \\
x_1 = x_2 = -\sqrt{3}, && u = \tfrac{1}{2} \\
x_1 = -x_2 = \sqrt{3}, && u = -\tfrac{5}{2} \\
x_1 = -x_2 = -\sqrt{3}, && u = -\tfrac{5}{2}
\end{aligned}$$

The last two roots do not satisfy Kuhn-Tucker necessary conditions of Eq. (e) as $u < 0$. Therefore, there are two candidate minimum points for this case. The first point corresponds to point A and the second one to B in Fig. 3.15.

Case 3: $u = 0$, $s = 0$. With these conditions, Eqs (a) and (b) give $x_1 = 0$, $x_2 = 0$. Substituting these into Eq. (c), we obtain $s^2 = 6 \neq 0$. Therefore, all K–T conditions cannot be satisfied.

The case where both u and s are zero usually does not occur in most practical problems. This can also be explained using the *physical interpretation* of the Lagrange multipliers discussed later in this chapter. The multiplier u for a constraint $g \leq 0$ actually gives the first derivative of the cost function with respect to variation in the right-hand side of the constraint, i.e. $u = -(\partial f / \partial e)$ where e is a small change in the constraint limit as $g \leq e$. Therefore, $u = 0$ when $g = 0$ implies that, any change in the right-hand side of the constraint $g \leq 0$ has no effect on the optimum cost function value. This usually does not happen in practice. When the right-hand side of a constraint is changed, the feasible region for the problem changes which usually has some effect on the optimum solution.

Finally, the points satisfying Kuhn–Tucker necessary conditions for the problem are summarized as:

	x_1^*	x_2^*	u^*	f	Point in Fig. 3.15
(1)	0	0	0	0	0
(2)	$\sqrt{3}$	$\sqrt{3}$	$\frac{1}{2}$	-3	A
(3)	$-\sqrt{3}$	$-\sqrt{3}$	$\frac{1}{2}$	-3	B

It is interesting to note that points A and B satisfy the sufficient condition for local minima. As can be observed from Fig. 3.15, any feasible move from the points results in an increase in the cost, and any further reduction in the cost results in violation of the constraint. It can also be observed that point 0 does not satisfy the sufficient condition because there are feasible directions that result in a decrease in the cost function. So, point 0 is only a stationary point. We shall check the sufficient conditions for this problem later in Section 3.6. ‖

The foregoing two examples illustrate the procedure of solving Kuhn-Tucker necessary conditions for candidate local minimum points. It is extremely important to understand the procedure clearly. The second example had only one inequality constraint. The switching condition of Eq. (d) gave only two normal cases – either $u = 0$ or $s = 0$ (the abnormal case where $u = 0$ and $s = 0$ rarely gives additional candidate points, so it will be ignored). Each of the cases gave candidate minimum point $\mathbf{x}^*$. For case 1 ($u = 0$), there was only one point $\mathbf{x}^*$ satisfying Eqs (a), (b) and (c). However, for case 2 ($s = 0$), there were four roots for Eqs (a), (b) and (c). Two of the four roots did not satisfy nonnegativity conditions on the Lagrange multipliers. Therefore, the corresponding two roots were not candidate local minimum points.

The preceding procedure is valid for more general nonlinear optimization problems. We illustrate the procedure for an example with two design variables and two inequality constraints.

Example 3.29 Solution of Kuhn-Tucker necessary conditions. Minimize

$$f(x_1, x_2) = x_1^2 + x_2^2 - 2x_1 - 2x_2 + 2$$

subject to

$$g_1 \equiv -2x_1 - x_2 + 4 \leqq 0$$
$$g_2 \equiv -x_1 - 2x_2 + 4 \leqq 0$$

Solution. Figure 3.16 gives a graphical representation for the problem. The two constraint functions are plotted and the feasible region is identified. It can be seen that point A $(\frac{4}{3}, \frac{4}{3})$, where both the inequality constraints are active, is the optimum solution for the problem. Since it is a two-variable problem, only two vectors can be linearly independent. It can be seen in Fig. 3.16 that the constraint gradients ∇g_1 and ∇g_2 are linearly independent (hence the optimum point is regular), so any other vector can be expressed as a linear combination of them. In particular, $-\nabla f$ (the negative gradient of the cost function) can be expressed as linear combination of ∇g_1 and ∇g_2, with positive scalars as the multipliers of the linear combination which is precisely the Kuhn-Tucker necessary condition of Eq. (3.47). In the following, we shall write these conditions and solve them to verify the graphical solution.

The Lagrange function for the problem is given as

$$L = x_1^2 + x_2^2 - 2x_1 - 2x_2 + 2 + u_1(-2x_1 - x_2 + 4 + s_1^2) + u_2(-x_1 - 2x_2 + 4 + s_2^2)$$

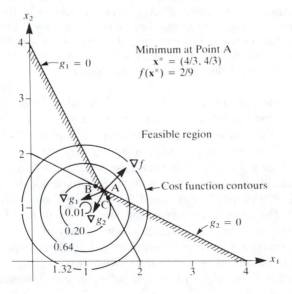

Minimum at Point A
$\mathbf{x}^* = (4/3, 4/3)$
$f(\mathbf{x}^*) = 2/9$

Feasible region

Cost function contours

$g_1 = 0$

$g_2 = 0$

FIGURE 3.16
Graphical solution for Example 3.29.

The Kuhn-Tucker necessary conditions are

$$\frac{\partial L}{\partial x_1} \equiv 2x_1 - 2 - 2u_1 - u_2 = 0 \tag{a}$$

$$\frac{\partial L}{\partial x_2} \equiv 2x_2 - 2 - u_1 - 2u_2 = 0 \tag{b}$$

$$g_1 \equiv -2x_1 - x_2 + 4 + s_1^2 = 0 \tag{c}$$

$$g_2 \equiv -x_1 - 2x_2 + 4 + s_2^2 = 0 \tag{d}$$

$$u_1 s_1 = 0, \qquad u_2 s_2 = 0 \tag{e}$$

$$u_1 \geq 0, \qquad u_2 \geq 0$$

Equations (a) – (e) are the six equations in six unknown x_1, x_2, s_1, s_2, u_1 and u_2. We must solve them simultaneously for candidate local minimum points. One way to satisfy the switching conditions of Eqs (e) is to identify various cases and then solve them for the roots. There are four cases:

1. $u_1 = 0$, $u_2 = 0$
2. $u_1 = 0$, $s_2 = 0$ (or $g_2 = 0$)
3. $s_1 = 0$ (or $g_1 = 0$), $u_2 = 0$
4. $s_1 = 0$ (or $g_1 = 0$), $s_2 = 0$ (or $g_2 = 0$)

We will consider each case separately and solve for all the unknowns.

Case 1. $u_1 = 0$, $u_2 = 0$. Equations (a) and (b) give $x_1 = x_2 = 1$. This is not a valid solution as it gives $s_1^2 = -1$ (or $g_1 = 1$), $s_2^2 = -1$ (or $g_2 = 1$) from Eqs (c) and (d), which implies that the two inequality constraints are violated, so $x_1 = 1$ and $x_2 = 1$ is not a feasible design. Therefore, this case does not give any candidate local minimum point.

Case 2: $u_1 = 0$, $s_2 = 0$. With these conditions, Eqs (a), (b) and (d) become

$$2x_1 - 2 - u_2 = 0 \tag{f}$$

$$2x_2 - 2 - 2u_2 = 0 \tag{g}$$

$$-x_1 - 2x_2 + 4 = 0 \tag{h}$$

These are three linear equations in the three unknowns x_1, x_2 and u_2. Any method of solving a linear system of equations such as Gaussian elimination, or method of determinants (Cramer's rule), can be used to find roots. We use the simple elimination procedure. Multiplying Eq. (f) by 2 and subtracting it from Eq. (g) gives

$$-4x_1 + 2x_2 + 2 = 0 \tag{i}$$

Adding Eqs (h) and (i), we obtain $x_1 = 1.2$. Using Eq. (i), we get $x_2 = 1.4$, and from either Eq. (f) or Eq. (g), $u_2 = 0.4$. Therefore, the solution for this case is

$$x_1 = 1.2, \, x_2 = 1.4; \qquad u_1 = 0, \, u_2 = 0.4; \qquad f = 0.2$$

We need to check for feasibility of the design point with respect to constraint g_1 before it can be claimed as a candidate local minimum point. Substituting $x_1 = 1.2$ and $x_2 = 1.4$ into Eq. (c), we find that $s_1^2 = -0.2 < 0$ (or $g_1 = 0.2$) which is not possible. This is not a feasible point as constraint g_1 is violated, so case 2 does not give any candidate local minimum point. It can be seen in Fig. 3.16 that point $(1.2, 1.4)$ corresponds to point B which is not in the feasible region.

Case 3: $s_1 = 0$, $u_2 = 0$. With these conditions Eqs (a), (b) and (c) give

$$2x_1 - 2 - 2u_1 = 0$$
$$2x_2 - 2 - u_1 = 0$$
$$-2x_1 - x_2 + 4 = 0$$

This is again a linear system of equations for the variables x_1, x_2 and u_1. Solving the system, we get the solution as

$$x_1 = 1.4,\ x_2 = 1.2; \qquad u_1 = 0.4,\ u_2 = 0; \qquad f = 0.2$$

Checking the design for feasibility with respect to constraint g_2, we find $s_2^2 = -0.2 < 0$ (or $g_2 = 0.2$) from Eq. (d). This is not a feasible design. Therefore, Case 3 also does not give any candidate local minimum point. It can be observed in Fig. 3.16 that point $(1.4, 1.2)$ corresponds to point C, which is not in the feasible region.

Case 4: $s_1 = 0$, $s_2 = 0$. For this case Eqs (a) to (d) must be solved for the four unknowns x_1, x_2, u_1 and u_2. This system of equations is again linear and can be solved easily. Using the elimination procedure as before, we obtain $x_1 = \frac{4}{3}$ and $x_2 = \frac{4}{3}$ from Eqs (c) and (d). Solving for u_1 and u_2 from Eqs (a) and (b), we get $u_1 = \frac{2}{9}$ and $u_2 = \frac{2}{9}$. Since both Lagrange multipliers are nonnegative and both constraints are satisfied, the preceding solution is a candidate local minimum point. The solution corresponds to point A in Fig. 3.16. The cost function at the point has a value of $\frac{2}{9}$.

It can be observed in Fig. 3.16 that the vector $-\nabla f$ can be expressed as a linear combination of the vectors ∇g_1 and ∇g_2 at point A. This satisfies the necessary condition of Eq. (3.52). It can also be seen from the figure that point A is indeed a local minimum because any further reduction in the cost function is possible only if we go into the infeasible region. Any feasible move from point A results in an increase in the cost function. ||

In all the examples that have been considered thus far it is implicitly assumed that conditions of the Kuhn-Tucker Theorem 3.6 or the Lagrange Theorem 3.5 are satisfied. In particular, we have assumed that $\mathbf{x}^*$ is a *regular point* of the feasible region (constraint set). That is, gradients of all the active constraints at $\mathbf{x}^*$ are linearly independent (i.e. they are not parallel to each other, nor any gradient can be expressed as a linear combination of others). It must be realized that necessary conditions are *applicable only if the assumption for regularity* of $\mathbf{x}^*$ is satisfied. To show that the necessary conditions are not applicable if $\mathbf{x}^*$ is not a regular point, we consider the following example.

Example 3.30 Check for Kuhn-Tucker conditions at irregular points. Minimize

$$f(x_1, x_2) = x_1^2 + x_2^2 - 4x_1 + 4$$

$$\mathbf{x}^* = (1,0), \, \nabla f(\mathbf{x}^*) = (-2, 0)$$

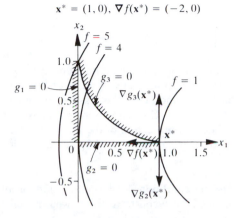

FIGURE 3.17
Graphical solution for Example 3.30. Irregular optimum point.

subject to

$$g_1 \equiv -x_1 \leq 0$$
$$g_2 \equiv -x_2 \leq 0$$
$$g_3 \equiv x_2 - (1-x_1)^3 \leq 0$$

Check if the minimum point $(1, 0)$ satisfies K–T conditions [McCormick, 1967].

Solution. Using the graphical solution procedure (Fig. 3.17) we see that the global minimum occurs at $\mathbf{x}^* = (1, 0)$. Let us see if the solution satisfies the Kuhn-Tucker necessary conditions. The Lagrangian for the problem is given as

$$L = x_1^2 + x_2^2 - 4x_1 + 4 + u_1(-x_1 + s_1^2) + u_2(-x_2 + s_2^2) + u_3[x_2 - (1-x_1)^3 + s_3^2]$$

The Kuhn–Tucker necessary conditions are

$$\frac{\partial L}{\partial x_1} \equiv 2x_1 - 4 - u_1 + u_3(3)(1-x_1)^2 = 0 \tag{a}$$

$$\frac{\partial L}{\partial x_2} \equiv 2x_2 - u_2 + u_3 = 0 \tag{b}$$

$$-x_1 + s_1^2 = 0 \tag{c}$$

$$-x_2 + s_2^2 = 0 \tag{d}$$

$$x_2 - (1-x_1)^3 + s_3^2 = 0 \tag{e}$$

$$u_i \geq 0; \quad u_i s_i = 0; \quad i = 1 \text{ to } 3 \tag{f}$$

At $\mathbf{x}^* = (1, 0)$ the first constraint (g_1) is inactive and the second and third constraints are active. The switching conditions (f) identify the case as

$$u_1 = 0; \quad s_2 = 0; \quad s_3 = 0$$

Substituting the solution into Eq. (a), we find that it is not satisfied. Thus, K–T necessary conditions are not satisfied.

This apparent contradiction can be resolved by noting that at $\mathbf{x}^* = (1, 0)$, gradients of the active constraints g_2 and g_3,

$$\nabla g_2(\mathbf{x}^*) = \begin{bmatrix} 0 \\ -1 \end{bmatrix}; \qquad \nabla g_3(\mathbf{x}^*) = \begin{bmatrix} 0 \\ 1 \end{bmatrix}$$

are not independent vectors. They are along the same line but in opposite directions, as shown in Fig. 3.17. Thus $\mathbf{x}^*$ is not a regular point of the feasible region. Since this is assumed in the K–T conditions, their use is invalid here. Note also that geometrical interpretation of the K–T conditions is violated – that is, ∇f at $(1,0)$ cannot be written as a linear combination of the gradients of active constraints g_2 and g_3. Actually ∇f is normal to both ∇g_2 and ∇g_3 as shown in the figure.   ‖

Example 3.31   Check for Kuhn–Tucker conditions. An optimization problem has one equality constraint h and one inequality constraint g. At what is believed to be the optimum point, the following information is known:

$$h = 0, \ g = 0, \ \nabla f = (2, 3, 2),$$
$$\nabla h = (1, -1, 1), \ \nabla g = (-1, -2, -1)$$

Check if the point satisfies K–T conditions.

Solution. At the candidate minimum point, the gradients of h and g are linearly independent, so the given point is regular. The K–T conditions are

$$\nabla L = \nabla f + v\nabla h + u\nabla g = 0$$
$$us = 0, \ u \geq 0, \ g + s^2 = 0, \text{ and } h = 0$$

Substituting for ∇f, ∇h and ∇g, we get

$$2 + v - u = 0$$
$$3 - v - 2u = 0$$
$$2 + v - u = 0$$

Solving for u and v, we get $u = \frac{5}{3}$, $v = -\frac{1}{3}$. Thus all the K–T conditions are satisfied.   ‖

3.4.3.1 ALTERNATE FORM FOR THE KUHN–TUCKER CONDITIONS. There is an alternate but entirely equivalent form for the Kuhn-Tucker necessary condition. In this form, the slack variables are not added to the inequality constraints and the Kuhn-Tucker conditions of Eqs. (3.47) to (3.51) are written without them. It can be seen that in the necessary conditions of Eqs (3.47) to (3.51), the slack variable s_i^2 appears only in two equations: Eq. (3.49), $g_i(\mathbf{x}^*) + s_i^2 = 0$; and Eq. (3.50), $u_i^* s_i = 0$. Both the equations can be written in an equivalent form without the slack variable s_i^2. We shall show this as follows.

Consider first Eq. (3.49), $g_i(\mathbf{x}^*) + s_i^2 = 0$ for $i = 1$ to m. The purpose of this equation is to ensure that at the candidate minimum point, all the inequalities remain satisfied. The equation can be written as $s_i^2 = -g_i(\mathbf{x}^*)$ and since $s_i^2 \geq 0$ ensures satisfaction of the constraint, we get $-g_i(\mathbf{x}^*) \geq 0$, or

$g_i(\mathbf{x}^*) \leq 0$ for $i = 1$ to m. Thus Eq. (3.49), $g_i(\mathbf{x}^*) + s_i^2 = 0$, can be simply replaced by $g_i(\mathbf{x}^*) \leq 0$.

The second equation (Eq. (3.50)) involving the slack variable is $u_i^* s_i = 0$, $i = 1$ to m. Multiplying the equation by s_i, we get $u_i^* s_i^2 = 0$. Now substituting $s_i^2 = -g_i(\mathbf{x}^*)$, we get $u_i^* g_i(\mathbf{x}^*) = 0$, $i = 1$ to m. This way the slack variable is eliminated from the equation and the switching condition of Eq. (3.50) can be written as $u_i^* g_i(\mathbf{x}^*) = 0$, $i = 1$ to m. These conditions can be used to define various cases as $u_i = 0$ or $g_i = 0$ (instead of $s_i = 0$).

In order to write the alternate form of Kuhn–Tucker conditions, the Lagrange function is defined as follows

$$L(\mathbf{x}, \mathbf{u}, \mathbf{v}) = f(\mathbf{x}) + \sum_{i=1}^{p} v_i h_i(\mathbf{x}) + \sum_{i=1}^{m} u_i g_i(\mathbf{x})$$
$$= f(\mathbf{x}) + \mathbf{v}^T \mathbf{h}(\mathbf{x}) + \mathbf{u}^T \mathbf{g}(\mathbf{x}) \tag{3.53}$$

Requiring stationarity of the Lagrangian with respect to design variables, i.e. $\nabla L = \mathbf{0}$, gives

$$\frac{\partial L}{\partial x_j} = \frac{\partial f}{\partial x_j} + \sum_{i=1}^{p} v_i \frac{\partial h_i}{\partial x_j} + \sum_{i=1}^{m} u_i \frac{\partial g_i}{\partial x_j} = 0; \qquad j = 1 \text{ to } n \tag{3.54}$$

which is same as Eq. (3.47). The constraints $h_i(\mathbf{x}^*) = 0$ of Eq. (3.48) and nonnegativity of the Lagrange multipliers in Eq. (3.51) for the inequality constraints are still required to be satisfied. The remaining necessary conditions are

$$g_i(\mathbf{x}^*) \leq 0, \qquad i = 1 \text{ to } m \tag{3.55}$$
$$u_i^* g_i(\mathbf{x}^*) = 0, \qquad i = 1 \text{ to } m \tag{3.56}$$

Example 3.32 Use of alternate K–T conditions. Consider Example 3.29 and write the alternate form of K–T conditions.

Solution. The Lagrangian of Eq. (3.53) is defined as

$$L = x_1^2 + x_2^2 - 2x_1 - 2x_2 + 2 + u_1(-2x_1 - x_2 + 4) + u_2(-x_1 - 2x + 4)$$

The alternate K–T conditions are

$$\frac{\partial L}{\partial x_1} \equiv 2x_1 - 2 - 2u_1 - u_2 = 0$$

$$\frac{\partial L}{\partial x_2} = 2x_2 - 2 - u_1 - 2u_2 = 0$$

$$g_1 \equiv -2x_1 - x_2 + 4 \leq 0$$
$$g_2 \equiv -x_1 - 2x_2 + 4 \leq 0$$
$$u_1 g_1 = 0, \qquad u_2 g_2 = 0$$
$$u_1 \geq 0, \qquad u_2 \geq 0$$

The switching conditions $u_i g_i = 0$, give the following four cases:

(1) $u_1 = 0$, $u_2 = 0$
(2) $u_1 = 0$, $g_2 = 0$ (or $s_2 = 0$)
(3) $g_1 = 0$ (or $s_1 = 0$), $u_2 = 0$
(4) $g_1 = 0$ (or $s_1 = 0$), $g_2 = 0$ (or $s_2 = 0$)

These four cases are entirely equivalent to the four cases identified in Example 3.29.   ‖

In summary, the following points should be noted regarding Kuhn–Tucker first-order necessary conditions:

1. *The conditions can be used to check* whether or not a given point is a candidate minimum; it must be feasible, gradient of the Lagrangian with respect to the design variables must be zero, and the Lagrange multipliers for inequality constraints must be nonnegative.
2. For a given problem, *the conditions can be used to find* candidate minimum points. Several cases defined by the switching conditions must be considered and solved. Each case can give multiple solutions.
3. The conditions written *with or without the slack variables are completely equivalent* leading to same solution cases.
4. For each solution case, *remember* to
   (i) check all inequality constraints for feasibility (i.e. $g_i \leqq 0$ or $s_i^2 \geqq 0$);
   (ii) calculate all the Lagrange multipliers; and
   (iii) ensure that the Lagrange multipliers for all the inequality constraints are nonnegative.

3.5 GLOBAL OPTIMALITY

In the optimum design of systems, the question of global optimum always arises. The question can be answered in two ways:

1. If the cost function $f(\mathbf{x})$ is continuous on a closed and bounded feasible region, then Weierstrass Theorem 3.1 guarantees the existence of a global minimum. For this situation, if we can calculate all the optimum points, and then select a solution that gives the least value to the cost function.
2. By showing the optimization problem to be convex because in that case any local minimum is also a global minimum.

Both the procedures can involve considerable computations. In this section we discuss topics of convexity and convex programming problems. Such problems are defined in terms of convex sets and convex functions. Therefore, we introduce these concepts and discuss some results regarding global optimum solutions.

(a)

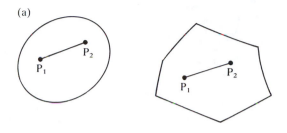

(b)

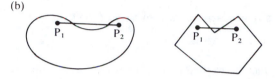

FIGURE 3.18
(a) Convex sets; (b) Nonconvex sets.

3.5.1 Convex Sets

A convex set S is a collection of points (vectors $\mathbf{x}$) having the following property: if P_1 and P_2 are any points in S, then the entire line segment P_1–P_2 is also in S. Figure 3.18 shows some examples of convex and nonconvex sets. *To explain convex sets further, let us consider points on a real line along the x-axis* (Fig. 3.19). Points in any interval on the line represent a convex set. Consider an interval between points a and b as shown in Fig. 3.19. To show that it is a convex set, let x_1 and x_2 be two points in the interval. *The line segment between the points can be written as*

$$x = \alpha x_2 + (1 - \alpha)x_1; \qquad 0 \le \alpha \le 1 \tag{3.57}$$

when $\alpha = 0$, $x = x_1$ and when $\alpha = 1$, $x = x_2$. It is clear that the line defined in Eq. (3.57) is in the interval $[a, b]$.

In general, for the n-dimensional space, the *line segment* between any two points $\mathbf{x}^{(1)}$ and $\mathbf{x}^{(2)}$ can be written as

$$\mathbf{x} = \alpha \mathbf{x}^{(2)} + (1 - \alpha)\mathbf{x}^{(1)}; \qquad 0 \le \alpha \le 1 \tag{3.58}$$

If the entire line segment of Eq. (3.58) is in the set S, then it is a convex set. Equation (3.58) is a generalization of Eq. (3.57) and is called the *parametric representation of a line segment* between the points $\mathbf{x}^{(1)}$ and $\mathbf{x}^{(2)}$.

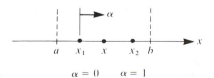

$\alpha = 0 \qquad \alpha = 1$

FIGURE 3.19
Convex interval between a and b on a real line.

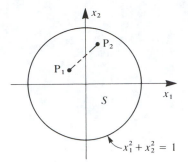

FIGURE 3.20
Convex set S for Example 3.33.

Example 3.33 Check for convexity of a set. Show convexity of the set

$$S = \{\mathbf{x} \mid x_1^2 + x_2^2 - 1.0 \leq 0\}$$

Solution. To show the set S graphically, we first plot the constraint as an equality which represents a circle of radius 1 centered at $(0, 0)$ (Fig. 3.20). Points inside or on the circle are in S. Geometrically we see that for any two points in the circle, the line segment between them is also in the circle. Therefore, S is a convex set. We can also use Eq. (3.58) to show convexity of S. To do this take any two points $\mathbf{x}^{(1)}$ and $\mathbf{x}^{(2)}$ in the set S. Use of Eq. (3.58) to calculate $\mathbf{x}$ and the condition that the distance between $\mathbf{x}^{(1)}$ and $\mathbf{x}^{(2)}$ is nonnegative ($\|\mathbf{x}^{(1)} - \mathbf{x}^{(2)}\| \geq 0$), will show $\mathbf{x} \in S$. This will prove convexity of S and is left as an exercise. ‖

3.5.2 Convex Functions

Consider a function of single variable $f(x) = x^2$. A graph of the function is shown in Fig. 3.21. Note that if a straight line is constructed between any two points $(x_1, f(x_1))$ and $(x_2, f(x_2))$ on the curve, the line lies above the graph of $f(x)$ at all points between x_1 and x_2. This property characterizes convex functions.

 A convex function $f(x)$ is defined on a convex set, i.e. the independent variable x must lie in a convex set. A function $f(x)$ is called convex on the convex set S if it lies below the line joining any two points on the curve $f(x)$. Figure 3.22 shows geometrical representation of a convex function. Using the

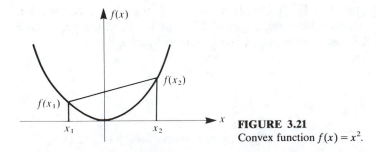

FIGURE 3.21
Convex function $f(x) = x^2$.

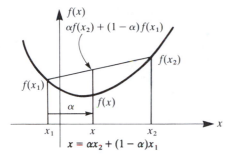

FIGURE 3.22
Characterization of a convex function.

geometry, the foregoing definition of a convex function can be expressed by the inequality $f(x) \leqq \alpha f(x_2) + (1 - \alpha)f(x_1)$. Since $x = \alpha x_2 + (1 - \alpha)x_1$, the inequality becomes

$$f(\alpha x_2 + (1 - \alpha)x_1) \leqq \alpha f(x_2) + (1 - \alpha)f(x_1) \qquad (3.59)$$

for $0 \leqq \alpha \leqq 1$. The definition can be generalized to functions of n variables. A function $f(\mathbf{x})$ defined on a convex set S is convex if it satisfies the inequality

$$f(\alpha \mathbf{x}^{(2)} + (1 - \alpha)\mathbf{x}^{(1)}) \leqq \alpha f(\mathbf{x}^{(2)}) + (1 - \alpha)f(\mathbf{x}^{(1)}) \qquad (3.60)$$

for $0 \leqq \alpha \leqq 1$ and any two points $\mathbf{x}^{(1)}$ and $\mathbf{x}^{(2)}$ in S. Note that convex set S is a region in the n dimensional space satisfying the convexity condition.

Equations (3.59) and (3.60) give *necessary and sufficient conditions for convexity of a function*. However, they are difficult to use in practice because we will have to check infinite number of pairs of points. Fortunately, the following theorem gives an easier way of checking convexity of a function.

> **Theorem 3.7   Check for convexity of a function.** A function of n variables $f(x_1, x_2, \ldots, x_n)$ defined on a *convex set* S is convex if and only if the Hessian matrix of the function is *positive semidefinite* or positive definite at all points in the set S.
>
> If the Hessian matrix is positive definite for all points in the set then f is called a *strictly convex function*. (Note that the converse of this is not true, i.e. a strictly convex function may have only positive semidefinite Hessian at some points, e.g. $f(x) = x^4$ is a strictly convex function but its second derivative is zero, $x = 0$.)   ‖

In one dimension, the convexity check of Theorem 3.7 reduces to the condition that the second derivative of the function be nonnegative. The graph of such a function has nonnegative curvature, as for the functions in Figs 3.21 and 3.22. The theorem can be proved by writing a Taylor's expansion for the function $f(\mathbf{x})$ and then using the definition of Eqs (3.59) and (3.60).

> **Example 3.34   Check for convexity of a function.** Check convexity of the function
>
> $$f(\mathbf{x}) = x_1^2 + x_2^2 - 1$$

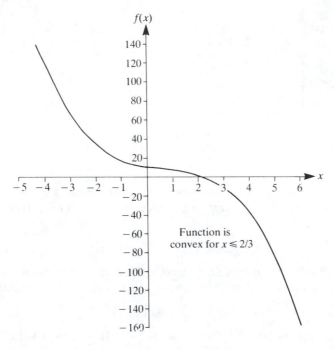

FIGURE 3.23
Graph of the function $f(x) = 10 - 4x + 2x^2 - x^3$ of Example 3.35.

Solution. The domain for the function (which is all values of x_1 and x_2) is convex. The gradient and Hessian of the function are given as

$$\nabla f = \begin{bmatrix} 2x_1 \\ 2x_2 \end{bmatrix}, \qquad \mathbf{H} = \begin{bmatrix} 2 & 0 \\ 0 & 2 \end{bmatrix}$$

By either of the tests given in Theorems 3.2 and 3.3 we see that **H** is positive definite everywhere. Therefore, f is a strictly convex function. ||

Example 3.35   Check for convexity of a function. Check convexity of the function $f(\mathbf{x}) = 10 - 4x + 2x^2 - x^3$.

Solution. The second derivative of the function is

$$\frac{d^2 f}{dx^2} = 4 - 6x$$

For the function to be convex, $d^2 f/dx^2 \geq 0$. Thus, the function is convex if $4 - 6x \geq 0$ or $x \leq \frac{2}{3}$. *The convexity check actually defines a domain for the function over which it is convex.* The function $f(x)$ is plotted in Fig. 3.23. It can be seen that the function is convex for $x \leq \frac{2}{3}$ and concave for $x \geq \frac{2}{3}$ (a function $f(x)$ is called *concave* if $-f(x)$ is convex). ||

3.5.3 Convex Programming Problem

If a function $g_i(\mathbf{x})$ is convex, then the set $g_i(\mathbf{x}) \leq e_i$ is convex, where e_i is any constant. If functions $g_i(\mathbf{x})$ for $i = 1$ to m are convex, then the set defined by $g_i(\mathbf{x}) \leq e_i$ for $i = 1$ to m is also convex. The set $g_i(\mathbf{x}) \leq e_i$ for $i = 1$ to m is called intersection of sets defined by the individual constraints $g_i(\mathbf{x}) \leq e_i$. Therefore, intersection of convex sets is a convex set. We can relate convexity of functions and sets by the following theorem:

> **Theorem 3.8 Convex functions and convex sets.** Let a set S be defined with constraints of the general optimization problem of Section 2.7 as
>
> $$S = \{\mathbf{x} \mid g_i(\mathbf{x}) \leq 0,\ i = 1 \text{ to } m;\ h_j(\mathbf{x}) = 0,\ j = 1 \text{ to } p\} \tag{3.61}$$
>
> Then S is a convex set if functions g_i are convex and h_j are linear. ‖

The set S of Example 3.33 is convex because it is defined by a convex function. *It is important to realize that if we have a nonlinear equality constraint $h_i(\mathbf{x}) = 0$, then the constraint set S is always nonconvex.* This can be easily seen from the definition of a convex set. For an equality constraint, the set S is a collection of points lying on the surface $h_i(\mathbf{x}) = 0$. If we take any two points on the surface, the straight line joining them cannot be on the surface, unless it is a plane (linear equality). Therefore, a set defined by any nonlinear equality constraint is always nonconvex. On the contrary, *a constraint set defined by a linear equality or inequality is always convex.*

If all inequality constraint functions for an optimum design problem are convex, and all equality constraint are linear, then the constraint set S is convex by Theorem 3.8. If the cost function is also convex over S, then we have what is known as a *convex programming problem.* Such problems have a very useful property that Kuhn–Tucker necessary conditions are also sufficient, and any local minimum is also a global minimum. This will be summarized in the Theorem 3.9.

It is important to note that Theorem 3.8 does not say that the feasible set S cannot be convex if a constraint function $g_i(\mathbf{x})$ fails the convexity check, i.e. it is not an "if and only if" theorem. There are some problems having inequality constraint functions that fail the convexity check, but the feasible region is still convex. Thus, the condition that $g_i(\mathbf{x})$ be convex for the region $g_i(\mathbf{x}) \leq 0$ to be convex are only sufficient but not necessary.

> **Theorem 3.9 Global minimum.** If $f(\mathbf{x}^*)$ is a local minimum for a convex function $f(\mathbf{x})$ defined on a convex set S, then it is also a global minimum.
>
> ***Proof.*** We can prove the theorem by contradiction. Let $\mathbf{x}^*$ be a local minimum of $f(\mathbf{x})$ on the set S. Assume that $f(\mathbf{x})$ has another local minimum at $\mathbf{x}'$ in the set S with the condition
>
> $$f(\mathbf{x}') < f(\mathbf{x}^*) \tag{a}$$
>
> Since S is a convex set, the line segment $\mathbf{x}^{(\alpha)} = \alpha\mathbf{x}' + (1-\alpha)\mathbf{x}^*$ for $0 \leq \alpha \leq 1$ lies

in the set S. Also by the convexity of $f(\mathbf{x})$ we can write

$$f(\mathbf{x}^{(\alpha)}) \leq \alpha f(\mathbf{x}') + (1 - \alpha) f(\mathbf{x}^*)$$
$$= f(\mathbf{x}^*) + \alpha[f(\mathbf{x}') - f(\mathbf{x}^*)] \qquad (b)$$

Using Eq. (a) into Eq. (b), we have

$$f(\mathbf{x}^{(\alpha)}) < f(\mathbf{x}^*) \qquad (c)$$

Therefore, for a positive small α, $\mathbf{x}^{(\alpha)}$ is in the neighborhood of $\mathbf{x}^*$ where the inequality (c) holds. That is, function at $\mathbf{x}^{(\alpha)}$ is smaller than that at $\mathbf{x}^*$. This contradicts the hypothesis of $\mathbf{x}^*$ being a local minimum of $f(\mathbf{x})$. Therefore, there cannot be two local minima with different values of the cost function for a convex problem.

∥

It is important to note that the theorem does not say that $\mathbf{x}^*$ cannot be a global minimum point if functions of the problem fail the convexity test. The point may indeed be a global minimum; however, we cannot claim global optimality using Theorem 3.9. We will have to use some other procedure, such as exhaustive search.

Example 3.36   Check for convexity of a problem. Check convexity of the following problem. Minimize

$$f(x_1, x_2) = x_1^3 - x_2^3$$

subject to the constraints

$$x_1 \geq 0, \qquad x_2 \leq 0$$

Solution. The constraints actually define the domain for the function $f(\mathbf{x})$ which is the 4th quadrant of a plane (shown in Fig. 3.24). This domain is convex. The Hessian of f is given as

$$\nabla^2 f = \begin{bmatrix} 6x_1 & 0 \\ 0 & -6x_2 \end{bmatrix}$$

The Hessian is positive semidefinite or positive definite over the domain defined by the constraints ($x_1 \geq 0, x_2 \leq 0$). Therefore, the cost function is convex and the problem is convex. Note that if constraints $x_1 \geq 0$ and $x_2 \leq 0$ are not imposed then the problem will not be convex. This can be observed in Fig. 3.24 where several cost function contours are also shown. Thus, the condition of positive semidefiniteness of the Hessian can define domain for the function over which it is convex.

∥

Example 3.37   Check for convexity of a problem. Check for convexity of the following problem. Minimize

$$f(x_1, x_2) = 2x_1 + 3x_2 - x_1^3 - 2x_2^2$$

subject to

$$x_1 + 3x_2 \leq 6$$
$$5x_1 + 2x_2 \leq 10$$
$$x_1, x_2 \geq 0$$

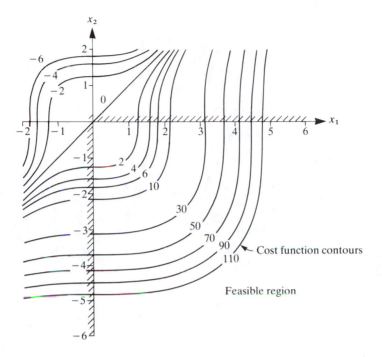

FIGURE 3.24
Graphical representation of Example 3.36.

Solution. Since constraints are linear in the variables x_1 and x_2, the feasible region (constraint set) for the problem is convex. If the cost function $f(x_1, x_2)$ is also convex then the problem is convex. Hessian of the cost function is

$$\mathbf{H} = \begin{bmatrix} -6x_1 & 0 \\ 0 & -4 \end{bmatrix}$$

The eigenvalues of $\mathbf{H}$ are $-6x_1$ and -4. Since the first eigenvalue is nonpositive for $x_1 \geqq 0$, and the second eigenvalue is negative, the function is not convex (Theorem 3.7), so the problem cannot be classified as a convex programming problem. Global optimality of a local minimum is not guaranteed. Figure 3.25 shows the feasible region for the problem along with several iso-cost lines. It can be seen that the feasible region is convex but the cost function is not.   ‖

Example 3.38   Check for convexity of a problem. Check for convexity of the following problem. Minimize

$$f(x_1, x_2) = 9x_1^2 - 18x_1x_2 + 13x_2^2 - 4$$

subject to

$$x_1^2 + x_2^2 + 2x_1 \geqq 16$$

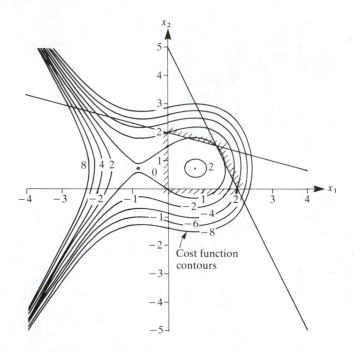

FIGURE 3.25
Graphical representation of Example 3.37.

Solution. To check for convexity of the problem, we need to write the constraint in the standard form as

$$g(\mathbf{x}) \equiv -x_1^2 - x_2^2 - 2x_1 + 16 \leq 0$$

Hessian of $g(\mathbf{x})$ is

$$\mathbf{H} = \begin{bmatrix} -2 & 0 \\ 0 & -2 \end{bmatrix}$$

Eigenvalues of the Hessian are -2 and -2. Since, it is neither positive definite nor positive semidefinite, $g(\mathbf{x})$ is not convex (in fact, Hessian is negative definite, so $g(\mathbf{x})$ is concave). Therefore, the problem cannot be classified as a convex programming problem. Global optimality for the solution cannot be guaranteed by Theorem 3.9. ||

3.5.4 Transformation of a Constraint

A constraint can be transformed to a different form which is equivalent to the original constraint, i.e. the constraint boundary and the feasible region for the problem do not change but the form of the function changes. Transformation of a constraint function, however, may affect its convexity, i.e. *a convex constraint function may become nonconvex and vice versa.* Convexity of the feasible region is, however not affected by the transformation.

To illustrate the effect of transformations, let us consider the following constraint:

$$g_1 \equiv \frac{a}{x_1 x_2} - b \leq 0$$

with $x_1 > 0$, $x_2 > 0$, and a and b as the given positive constants. To check convexity of the constraint, we calculate the Hessian matrix as

$$\nabla^2 g_1 = \frac{2a}{x_1^2 x_2^2} \begin{bmatrix} \dfrac{x_2}{x_1} & 0.5 \\ 0.5 & \dfrac{x_1}{x_2} \end{bmatrix}$$

Both eigenvalues of the preceding matrix are strictly positive, so the matrix is positive definite, and the constraint function g_1 is convex. The feasible region for g_1 is convex.

Now let us transform the constraint by multiplying throughout by $x_1 x_2$ (since $x_1 > 0$, $x_2 > 0$, the sense of the inequality is not changed) to obtain

$$g_2 \equiv a - b x_1 x_2 \leq 0$$

The constraints g_1 and g_2 are equivalent and will give the same optimum solution for the problem. To check convexity of the constraint function, we calculate the Hessian matrix as

$$\nabla^2 g_2 = \begin{bmatrix} 0 & -b \\ -b & 0 \end{bmatrix}$$

The eigenvalues of the preceding matrix are $\lambda_1 = -b$ and $\lambda_2 = b$. Therefore, the matrix is indefinite by Theorem 3.2, and by Theorem 3.7 the constraint function g_2 is not convex. Thus, we lose convexity of the constraint function and we cannot claim convexity of the feasible region by Theorem 3.8. Since the problem cannot be shown to be convex, we cannot use results related to convex programming problems. ‖

We summarize major results of this section as follows:

1. *A function is convex* if and only if its Hessian is at least positive semidefinite or positive definite at all points in the domain of the function; it is called *strictly convex* if the Hessian is positive definite at all points.
2. *A linear equality or inequality constraint* always defines a convex feasible region for the problem.
3. *A nonlinear equality constraint* always defines a nonconvex feasible region for the problem.

4. If all the equality constraint functions are linear and all the inequality constraint functions of a problem written in the standard form (minimization of a function with equality and "$\leqq$ type" inequality constraints) are convex, the feasible region is convex; otherwise it may or may not be convex.

5. If the cost function is convex over the convex feasible region, the problem is called the *convex programming problem.*

6. For a convex programming problem, *Kuhn–Tucker first-order necessary conditions are also sufficient, and any local minimum is also a global minimum.*

7. Nonconvex problems can also have global minimum points.

3.6* SECOND-ORDER CONDITIONS FOR CONSTRAINED OPTIMIZATION

Solutions of the necessary conditions are candidate local minimum designs. The sufficiency conditions determine whether a candidate design is indeed a local minimum or not. In this section, we shall discuss second-order necessary and sufficiency conditions for constrained optimization problems. These conditions also involve Hessians of the functions, as for the unconstrained problems. We shall first discuss sufficiency conditions for convex programming problems and then for general optimization problems.

3.6.1 Sufficient Condition for Convex Problems

For convex programming problems, the first-order Kuhn–Tucker necessary conditions of Theorem 3.6 also become sufficient. Thus, if we can show convexity of a problem, any solution of the necessary conditions will automatically satisfy sufficient conditions. In addition, the solution will be a global minimum due to Theorem 3.9.

> **Theorem 3.10   Sufficient condition for convex problem.** If $f(\mathbf{x})$ is a convex cost function defined on a convex feasible region (constraint set), then first-order Kuhn–Tucker conditions are necessary as well as sufficient for a global minimum.
>
> ‖

To use the theorem we must show that the *constraint set S* for the problem is *convex,* where S is given as

$$S = \{\mathbf{x} \mid h_i(\mathbf{x}) = 0, \; i = 1 \text{ to } p; \; g_i(\mathbf{x}) \leqq 0, \; i = 1 \text{ to } m\}$$

As noted in Section 3.5.3, all equality constraint functions $h_i(\mathbf{x})$ must be linear and the Hessian of all inequality constraint functions $g_i(\mathbf{x})$ must be positive semidefinite or positive definite for convexity of the set S. Therefore, if any constrained problem has a nonlinear equality constraint function, it cannot be convex. If all functions of the problem are linear in design variables, then the problem is convex. Once we have shown *convexity of S,* we need to show that $f(\mathbf{x})$ is also convex on S for the problem to be convex. For such problems, any point satisfying the Kuhn–Tucker necessary conditions will give a global minimum design. Following the procedure of Section 3.4, we consider various cases defined by the switching conditions of Eq. (3.50) until a solution is found. We can stop there as the solution is a *global optimum design.*

Example 3.39   Check for convexity of a problem. Let us consider Example 3.26 again and check for its convexity. Minimize

$$f(\mathbf{x}) = (x_1 - 1.5)^2 + (x_2 - 1.5)^2$$

subject to

$$g(\mathbf{x}) \equiv x_1 + x_2 - 2 \leqq 0$$

Solution. The Kuhn–Tucker necessary conditions give the candidate local minimum as

$$x_1 = 1, \ x_2 = 1; \text{ and } u = 1$$

The constraint function $g(\mathbf{x})$ is linear, so it is convex. Since inequality constraint function is convex and there is no equality constraint, the constraint set S is convex. The Hessian matrix for the cost function is

$$\mathbf{H} = \begin{bmatrix} 2 & 0 \\ 0 & 2 \end{bmatrix}$$

Since $\mathbf{H}$ is positive definite everywhere by Theorem 3.2 or Theorem 3.3, the cost function $f(\mathbf{x})$ is strictly convex by Theorem 3.7. Therefore, the problem is convex and the solution $x_1 = x_2 = 1$ satisfies sufficiency condition of Theorem 3.10. It is a strict global minimum point for the problem.   ‖

3.6.2   Second-order Conditions for General Problems

We are now ready to discuss the second-order conditions for the general optimization problem. As in the unconstrained case, we can use *second-order information* about the functions (i.e. curvature) at the candidate point $\mathbf{x}^*$ to determine if it is indeed a local minimum. Recall for the unconstrained problem that the local sufficiency of Theorem 3.4 requires quadratic part of the Taylor's expansion for the function at $\mathbf{x}^*$ to be positive for all nonzero changes $\mathbf{d}$. *In the constrained case, we must also consider active constraints at $\mathbf{x}^*$ to determine feasible changes* $\mathbf{d}$. We will consider only the points $\mathbf{x} = \mathbf{x}^* + \mathbf{d}$ in the neighborhood of $\mathbf{x}^*$ that satisfy the active constraint equations. *Any $\mathbf{d} \neq \mathbf{0}$ satisfying active constraints to the first-order must be in the constraint tangent plane* (See Fig. 3.26). Such $\mathbf{d}$'s are then orthogonal to the gradients of the active constraints (constraint gradients are normal to the constraint tangent plane). Therefore, dot product of $\mathbf{d}$ with each of the constraint gradients ∇h_i

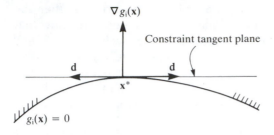

FIGURE 3.26
Directions **d** used in constrained sufficiency conditions.

and ∇g_i must be zero, i.e. $\nabla h_i^T \mathbf{d} = 0$ and $\nabla g_i^T \mathbf{d} = 0$. Thus, directions **d** are determined to define a feasible region around the point $\mathbf{x}^*$. Note that only active inequalities constraints ($g_i = 0$) are used in determining **d**. The situation is depicted in Fig. 3.26 for one inequality constraint.

To derive the second-order conditions, we write Taylor's expansion of the Lagrange function and consider only those **d** that satisfy the preceding conditions. $\mathbf{x}^*$ *is then a local minimum point if the second-order term of Taylor's expansion is positive for all* **d** *in the constraint tangent plane.* This is then the sufficient condition. As a necessary condition the second-order term must be nonnegative. We summarize these results precisely in the following theorems.

Theorem 3.11 Second-order necessary condition for general constrained problems. Let $\mathbf{x}^*$ satisfy the first order K–T necessary conditions for the general optimum design problem. Define Hessian of the Lagrange function L at $\mathbf{x}^*$ as

$$\nabla^2 L = \nabla^2 f + \sum_{i=1}^{p} v_i^* \nabla^2 h_i + \sum_{i=1}^{m} u_i^* \nabla^2 g_i \qquad (3.62)$$

Let there be nonzero feasible directions ($\mathbf{d} \neq \mathbf{0}$) satisfying the following linear systems at the point $\mathbf{x}^*$:

$$\nabla h_i^T \mathbf{d} = 0; \quad i = 1 \text{ to } p \qquad (3.63)$$

$$\nabla g_i^T \mathbf{d} = 0; \quad \text{for all active inequalities (i.e. for those } i \text{ with } g_i(\mathbf{x}^*) = 0) \quad (3.64)$$

Then if $\mathbf{x}^*$ is a local minimum point for the optimum design problem, then it must be true that

$$Q \geqq 0 \quad \text{where} \quad Q = \mathbf{d}^T \nabla^2 L(\mathbf{x}^*) \mathbf{d} \qquad (3.65) \quad \|$$

Note that any point that does not satisfy the second-order necessary conditions cannot be a local minimum point.

Theorem 3.12 Sufficient conditions for general constrained problems. Let $\mathbf{x}^*$ satisfy the first-order K–T necessary conditions for the general optimum design problem. Define Hessian of the Lagrange function L at $\mathbf{x}^*$ as in Eq. (3.62). Define nonzero feasible directions ($\mathbf{d} \neq \mathbf{0}$) as solutions of the linear systems

$$\nabla h_i^T \mathbf{d} = 0; \quad i = 1 \text{ to } p \qquad (3.66)$$

$$\nabla g_i^T \mathbf{d} = 0; \quad i = 1 \text{ to } m \text{ for active inequalities with } u_i > 0 \qquad (3.67)$$

Also let $\nabla g_i^T \mathbf{d} \leqq 0$ for those constraints with $u_i = 0$. If

$$Q > 0, \text{ where } Q = \mathbf{d}^T \nabla^2 L(\mathbf{x}^*)\mathbf{d} \tag{3.68}$$

then $\mathbf{x}^*$ is an *isolated local minimum* point (isolated means that there are no other local minimum points in the neighborhood of $\mathbf{x}^*$). ‖

Note first the difference in the conditions for the directions $\mathbf{d}$ in Eq. (3.64) for the necessary condition and Eq. (3.67) for the sufficient condition. In Eq. (3.64) all active inequalities with nonnegative multipliers are included whereas in Eq. (3.67) only those active inequalities with a positive multiplier are included.

Equations (3.66) and (3.67) simply say that the dot product of vectors ∇h_i and $\mathbf{d}$ and ∇g_i (having $u_i > 0$) and $\mathbf{d}$ should be zero. So, only the $\mathbf{d}$ orthogonal to the gradients of equality and active inequality constraints (with $u_i > 0$) are considered. Or, stated differently, only $\mathbf{d}$ in the tangent hyperplane to the active constraints at the candidate minimum point are considered. Equation (3.68) says that Hessian of the Lagrangian be positive definite for all $\mathbf{d}$ lying in the constraint tangent hyperplane. Note that ∇h_i, ∇g_i and $\nabla^2 L$ are calculated at the candidate local minimum points $\mathbf{x}^*$ satisfying the K–T necessary conditions.

It is important to note that if matrix $\nabla^2 L(\mathbf{x}^*)$ *is positive definite (i.e. Q in Eq. (3.68) is positive for any $\mathbf{d} \neq \mathbf{0}$) then $\mathbf{x}^*$ satisfies the sufficiency condition for an isolated local minimum and no further checks are needed.* The reason is that if $\nabla^2 L$ is positive definite, then it is also positive definite for those $\mathbf{d}$ that satisfy Eqs (3.66) and (3.67). However, *if $\nabla^2 L(\mathbf{x}^*)$ is not positive definite then we cannot conclude that $\mathbf{x}^*$ is not an isolated local minimum.* We must calculate $\mathbf{d}$ to satisfy Eqs (3.66) and (3.67) and carry out the sufficiency test given in the Theorem 3.12. This result is summarized in the following theorem:

Theorem 3.13 Strong sufficient condition. Let $\mathbf{x}^*$ satisfy the first-order K–T necessary conditions for the general optimum design problem. Define Hessian $\nabla^2 L(\mathbf{x}^*)$ for the Lagrange function at $\mathbf{x}^*$ as in Eq. (3.62). Then if $\nabla^2 L(\mathbf{x}^*)$ is positive definite, $\mathbf{x}^*$ is an isolated minimum point. ‖

It should also be emphasized that if Eq. (3.68) is not satisfied, we cannot conclude that $\mathbf{x}^*$ is not a local minimum. It may still be a local minimum but not an isolated one. Note also that the theorem cannot be used for any $\mathbf{x}^*$ if its assumptions are not satisfied. In that case, we cannot draw any conclusions for the point $\mathbf{x}^*$. We shall illustrate these points with examples later in this section and Section 3.8.

One case arising in some applications needs special mention. This occurs when the total number of active constraints (with at least one inequality) at the candidate minimum point $\mathbf{x}^*$ is equal to the number of independent design variables. Since $\mathbf{x}^*$ satisfies K–T conditions, gradients of all the active constraints are linearly independent. Thus the only solution for the system of

Eqs (3.66) and (3.67) is $\mathbf{d} = \mathbf{0}$ and Theorem 3.12 cannot be used. However, since $\mathbf{d} = \mathbf{0}$ is the only solution, there are no feasible directions in the neighborhood that can reduce the cost function any further. So, point $\mathbf{x}^*$ is indeed a local minimum for the cost function.

We consider several examples to illustrate the use of necessary and sufficient conditions of optimality.

Example 3.40 Check for sufficient conditions. Check sufficient condition for Example 3.27. Minimize

$$f(\mathbf{x}) = \tfrac{1}{3}x^3 - \tfrac{1}{2}(b + c)x^2 + bcx + f_0$$

subject to

$$a \leqq x \leqq d$$

where $0 < a < b < c < d$ and f_0 are specified constants.

Solution. There is only one constrained candidate local minimum point, $x = a$. Since there is only one design variable and one active constraint, the condition $\nabla g_1 \bar{d} = 0$ of Eq. (3.67) gives $\bar{d} = 0$ as the only solution (note that $\bar{d}$ is used as a direction for sufficiency check since d is used as a constant in the example). Therefore, Theorem 3.12 cannot be used for a sufficiency check. Also note that at $x = a$, $d^2L/dx^2 = 2a - b - c$ which can be positive, negative or zero depending on the values of a, b, and c. So, we cannot use curvature of Hessian to check the sufficient condition (Strong Sufficient Theorem 3.13). However, from Fig. 3.14 we observe that $x = a$ is indeed an isolated local minimum point. From this example we can conclude that if the number of active inequality constraints is equal to the number of independent design variables and all other K–T conditions are satisfied, then the candidate point is indeed a local minimum design. ‖

Example 3.41 Check for sufficient conditions. Consider the optimization problem of Example 3.28. Minimize

$$f(\mathbf{x}) = x_1^2 + x_2^2 - 3x_1x_2$$

subject to

$$g \equiv x_1^2 + x_2^2 - 6 \leqq 0$$

Check for sufficient conditions for the candidate minimum points.

Solution. The points satisfying the Kuhn–Tucker necessary conditions are

	x_1^*	x_2^*	u^*
(1)	0	0	0
(2)	$\sqrt{3}$	$\sqrt{3}$	$\tfrac{1}{2}$
(3)	$-\sqrt{3}$	$-\sqrt{3}$	$\tfrac{1}{2}$

It was previously observed in Example 3.28 and Fig. 3.15 that the point $(0, 0)$ did not satisfy the sufficient condition, and the other two points did satisfy it. Those geometrical observations shall be mathematically verified using the sufficient theorems of optimality.

The Hessian matrices for the cost and constraint functions are

$$\nabla^2 f = \begin{bmatrix} 2 & -3 \\ -3 & 2 \end{bmatrix}, \quad \nabla^2 g = \begin{bmatrix} 2 & 0 \\ 0 & 2 \end{bmatrix}$$

By the method of Appendix·B, eigenvalues of $\nabla^2 g$ are $\lambda_1 = 2$, and $\lambda_2 = 2$. Since both eigenvalues are positive, the function g is convex. So the constraint set defined by $g(\mathbf{x}) \leqq 0$ is convex by Theorem 3.8. However, since eigenvalues of $\nabla^2 f$ are -1 and 5, f is not convex. Therefore, it cannot be classified as a convex programming problem and sufficiency cannot be shown by Theorem 3.10. We must resort to the general sufficiency Theorem 3.12. Hessian of the Lagrangian is given as

$$\nabla^2 L = \nabla^2 f + u \nabla^2 g$$
$$= \begin{bmatrix} 2 + 2u & -3 \\ -3 & 2 + 2u \end{bmatrix}$$

For the first point $x_1^* = x_2^* = 0$, $\mathbf{u}^* = 0$, $\nabla^2 L$ becomes $\nabla^2 f$ (the constraint $g(\mathbf{x}) \leqq 0$ is inactive). In this case the problem is unconstrained and the local sufficiency requires $\mathbf{d}^T (\nabla^2 f(\mathbf{x}^*)) \mathbf{d} > 0$ for all $\mathbf{d}$. Or, $\nabla^2 f$ should be positive definite at $\mathbf{x}^*$. Since both eigenvalues of $\nabla^2 f$ are not positive, we conclude that the above condition is not satisfied. Therefore, $x_1^* = x_2^* = 0$ does not satisfy the second-order sufficiency condition. Note that since $\lambda_1 = -1$ and $\lambda_2 = 5$, the matrix $\nabla^2 f$ is indefinite at $\mathbf{x}^*$. Therefore the point $x_1 = 0$, $x_2 = 0$ also *violates the second-order necessary condition* of Theorem 3.4 requiring $\nabla^2 f$ to be positive semidefinite or definite at the candidate local minimum point. Thus $x_1^* = 0$, $x_2^* = 0$ cannot be a local minimum point. This agrees with graphical observation made in Example 3.28.

At points $x_1^* = x_2^* = \sqrt{3}$ and $x_1^* = x_2^* = -\sqrt{3}$ with $u^* = \frac{1}{2}$

$$\nabla^2 L(\mathbf{x}^*) = \begin{bmatrix} 3 & -3 \\ -3 & 3 \end{bmatrix}$$
$$\nabla g = \pm(2\sqrt{3}, 2\sqrt{3}) = \pm 2\sqrt{3}(1, 1)$$

It may be checked that $\nabla^2 L$ is not positive definite at either of the two points. Therefore, we cannot use Theorem 3.13 to conclude that $\mathbf{x}^*$ is a minimum point. We must find $\mathbf{d}$ satisfying Eqs (3.66) and (3.67). If we let $\mathbf{d} = (d_1, d_2)$ then $\nabla g^T \mathbf{d} = 0$ gives

$$\pm 2\sqrt{3}[1 \; 1]\begin{bmatrix} d_1 \\ d_2 \end{bmatrix} = 0; \quad \text{or} \quad d_1 + d_2 = 0$$

Thus, $d_1 = -d_2 = c$, where $c \neq 0$ is an arbitrary constant, and a $\mathbf{d} \neq 0$ satisfying $\nabla g^T \mathbf{d} = 0$ is given as

$$\mathbf{d} = c(1, -1).$$

The sufficiency condition of Eq. (3.68) gives

$$Q = \mathbf{d}^T(\nabla^2 L)\mathbf{d} = c[1 \quad -1]\begin{bmatrix} 3 & -3 \\ -3 & 3 \end{bmatrix}c\begin{bmatrix} 1 \\ -1 \end{bmatrix}$$

$$= 12c^2 > 0 \text{ for } c \neq 0$$

The points $x_1^* = x_2^* = \sqrt{3}$ and $x_1^* = x_2^* = -\sqrt{3}$ satisfy the sufficiency conditions. They are therefore isolated local minimum points as was observed graphically in Example 3.28 and Fig. 3.15. We see for this example that $\nabla^2 L(\mathbf{x}^*)$ is not positive definite, but $\mathbf{x}^*$ is still an isolated minimum point.

Note that since f is continuous and the feasible region is closed and bounded we are guaranteed the existence of a global minimum by the Weierstrass Theorem 3.1. Also we have examined every possible point satisfying necessary conditions. Therefore, we must conclude by elimination that $x_1^* = x_2^* = \sqrt{3}$ and $x_1^* = x_2^* = -\sqrt{3}$ are global minimum points. The value of the cost function for both the points is $f(\mathbf{x}^*) = -3$. ‖

Example 3.42 Check for sufficient conditions. Consider Example 3.29. Minimize

$$f(x_1, x_2) = x_1^2 + x_2^2 - 2x_1 - 2x_2 + 2$$

subject to

$$g_1 \equiv -2x_1 - x_2 + 4 \leq 0$$
$$g_2 \equiv -x_1 - 2x_2 + 4 \leq 0$$

Check the sufficient condition for the candidate minimum point.

Solution. The Kuhn–Tucker necessary conditions are satisfied for the point

$$x_1^* = \tfrac{4}{3}, \; x_2^* = \tfrac{4}{3}, \; u_1^* = \tfrac{2}{9}, \; u_2^* = \tfrac{2}{9}$$

Since all the constraint functions are linear the feasible region (constraint set S) is convex. The Hessian of the cost function is positive definite. Therefore, it is also convex and the problem is convex. By Theorem 3.10, $x_1^* = x_2^* = \tfrac{4}{3}$ *satisfies sufficiency conditions for a global minimum* with the cost function as $f(\mathbf{x}^*) = \tfrac{2}{9}$.

Note that local sufficiency cannot be shown by the method of Theorem 3.12. The reason is that the conditions of Eq. (3.67) give

$$-2d_1 - d_2 = 0 \quad (\nabla g_1^T\mathbf{d} = 0)$$
$$-d_1 - 2d_2 = 0 \quad (\nabla g_2^T\mathbf{d} = 0)$$

This is a homogeneous system of equations with a nonsingular coefficient matrix. Therefore, its only solution is $d_1 = d_2 = 0$. Thus, we cannot find a $\mathbf{d} \neq \mathbf{0}$ for use in Condition (3.68), and Theorem 3.12 cannot be used. However, we have seen in the foregoing and in Fig. 3.16 that the point is actually an isolated global minimum point. Since it is a two-variable problem and two constraints are active at the Kuhn–Tucker point, the sufficient condition for local minimum is satisfied. ‖

3.7 POST-OPTIMALITY ANALYSIS: PHYSICAL MEANING OF LAGRANGE MULTIPLIERS

The study of variations in the optimum solution as some of the original problem parameters are changed is known as *post-optimality analysis* or *sensitivity analysis*. This is an important topic in the area of optimum design that is currently under active development [Vanderplaats, 1984; Vanderplaats and Yoshida, 1985]. Variations to the optimum solution (cost function and design variables) due to the variations of many parameters can be studied. However, we shall focus on the question of sensitivity of the cost function to the variatons in the constraint limit values. We shall assume that the minimization problem has been solved with $h_i(\mathbf{x}) = 0$ and $g_j(\mathbf{x}) \leqq 0$, i.e. with the current limit values for the constraints as zero. Thus, we like to know what happens to the optimum cost function when the constraint limits are changed from zero.

It turns out that the Lagrange multipliers $(\mathbf{v}^*, \mathbf{u}^*)$ at the optimum design provide information to answer the foregoing sensitivity question. The investigation of this question leads to a physical interpretation of the Lagrange multipliers that can be very useful in practical applications. The interpretation will also show why the Lagrange multipliers for the "$\leqq$ type" constraints have to be nonnegative. In addition, the multipliers can be used to study the *benefit of relaxing a constraint or the penalty associated with tightening it*; relaxation enlarges the feasible region (constraint set), while tightening contracts it. The sensitivity results are stated in a theorem.

Later in this section we shall also discuss what happens to the Lagrange multipliers if the cost and constraint functions for the problem are scaled.

3.7.1 Effect of Changing Constraint Limits

To discuss changes in the cost function due to changes in the constraint limits, we consider the modified problem of minimizing $f(\mathbf{x})$ subject to the constraints

$$h_i(\mathbf{x}) = b_i; \qquad i = 1 \text{ to } p \tag{3.69}$$

and

$$g_j(\mathbf{x}) \leqq e_j; \qquad j = 1 \text{ to } m \tag{3.70}$$

where b_i and e_j are small variations in the neighborhood of zero. It is clear that the optimum point for the perturbed problem depends on vectors $\mathbf{b}$ and $\mathbf{e}$, i.e. it is a function of $\mathbf{b}$ and $\mathbf{e}$ which can be written as $\mathbf{x}^* = \mathbf{x}^*(\mathbf{b}, \mathbf{e})$. Also, optimum cost function value depends on $\mathbf{b}$ and $\mathbf{e}$, i.e. $f = f(\mathbf{b}, \mathbf{e})$. However, explicit dependence of the cost function on $\mathbf{b}$ and $\mathbf{e}$ is not known, i.e. an expression for f in terms of b_i and e_j is not known. The following theorem gives a way of calculating the partial derivatives $\partial f / \partial b_i$ and $\partial f / \partial e_j$. *These are the implicit derivatives of f with respect to b_i and e_j, i.e. the right-hand side parameters. The* *derivatives can be used to calculate changes in the cost function as b_i and e_j are changed.*

Theorem 3.14   Constraint variation sensitivity theorem. Let $f(\mathbf{x})$, $h_i(\mathbf{x})$, $i = 1$ to p, and $g_j(\mathbf{x})$, $j = 1$ to m, have two continuous derivatives. Let $\mathbf{x}^*$ be a regular point that, together with the multipliers u_j^* and v_i^*, satisfies both the Kuhn–Tucker necessary conditions and the sufficient conditions of Theorem 3.12 for an isolated local minimum point for the problem defined in Eqs (3.37) to (3.39). If for each $g_j(\mathbf{x}^*) = 0$, it is true that $u_j^* > 0$, then the solution $\mathbf{x}^*(\mathbf{b}, \mathbf{e})$ of the above modified optimization problem defined in Eqs (3.69) and (3.70) is a continuously differentiable function of $\mathbf{b}$ and $\mathbf{e}$ in some neighborhood of $\mathbf{b} = \mathbf{0}$ and $\mathbf{e} = \mathbf{0}$. Furthermore,

$$\frac{\partial f(\mathbf{x}^*(\mathbf{0}, \mathbf{0}))}{\partial b_i} = -v_i^*; \qquad i = 1 \text{ to } p \tag{3.71}$$

$$\frac{\partial f(\mathbf{x}^*(\mathbf{0}, \mathbf{0}))}{\partial e_j} = -u_j^*; \qquad j = 1 \text{ to } m \tag{3.72} \,\|$$

The theorem gives values for implicit first-order derivatives of the cost function with respect to the right-hand side parameters of the constraints. Note that the theorem is applicable only when the inequality constraints are written in the "$\leq$" form. Using the theorem we can estimate changes in the cost function if we decide to adjust the right-hand side of constraints in the neighborhood of zero. For this purpose, Taylor's expansion for the cost function in terms of b_i and e_j can be used. Let us assume that we want to vary the right-hand sides b_i and e_j of ith equality and jth inequality constraints. First-order Taylor series expansion for the cost function about the point $b_i = 0$ and $e_j = 0$ is given as

$$f(b_i, e_j) = f(0, 0) + \frac{\partial f(0, 0)}{\partial b_i} b_i + \frac{\partial f(0, 0)}{\partial e_j} e_j$$

Or, substituting from Eqs (3.71) and (3.72), we obtain

$$f(b_i, e_j) = f(0, 0) - v_i^* b_i - u_j^* e_j \tag{3.73}$$

where $f(0, 0)$ is the optimum cost function value obtained with $b_i = 0$, and $e_j = 0$. From Eq. (3.73), change in the cost function Δf due to small changes in b_i and e_j is given as

$$\Delta f = f(b_i, e_j) - f(0, 0) = -v_i^* b_i - u_j^* e_j \tag{3.74}$$

For given values of b_i and e_j, we can estimate the new value of the cost function from Eq. (3.73). If we want to change the right-hand side of more constraints, we simply include them in Eq. (3.74) and obtain the change in cost function as

$$\Delta f = -\sum_i v_i^* b_i - \sum_j u_j^* e_j \tag{3.75}$$

It is useful to note that if conditions of Theorem 3.14 *are not satisfied, existence of implicit derivatives of Eqs* (3.71) *and* (3.72) *is not ruled out by the theorem. That is, the derivatives may still exist but their existence cannot be guaranteed by Theorem* 3.14. This observation shall be verified later in an example problem.

Equation (3.74) can also be used to show that the *Lagrange multiplier corresponding to a "$\leq$ type" constraint must be nonnegative.* To see this, let us assume that we want to relax an inequality constraint $g_j \leq 0$ that is active ($g_j = 0$) at the optimum point, i.e. we select $e_j > 0$. When a constraint is relaxed, the feasible region (constraint set) for the design problem expands. We allow more feasible designs to be candidate minimum points. Therefore, with the expanded feasible region we should be able to improve the optimum value of the cost function, i.e. we should be able to reduce it further or keep it unchanged. We observe from Eq. (3.74) that if $u_j^* < 0$, then relaxation of the constraint ($e_j > 0$) results in an increase in cost ($\Delta f = -u_j^* e_j > 0$). This is a contradiction, as it implies that we have to pay a penalty to relax the constraint. Therefore, the Lagrange multiplier corresponding to a "$\leq$ type" constraint must be nonnegative.

Example 3.43 Effect of variations of constraint limits on optimum cost function. To illustrate the use of constraint variation sensitivity theorem, we consider the following problem solved as Example 3.28 and discuss the effect of changing the limit for the constraint. Minimize

$$f(x_1, x_2) = x_1^2 + x_2^2 - 3x_1 x_2$$

subject to

$$g(x_1, x_2) = x_1^2 + x_2^2 - 6 \leq 0$$

Solution. The graphical solution for the problem is given in Fig. 3.15. A point satisfying both necessary and sufficient conditions is

$$x_1^* = x_2^* = \sqrt{3}, \qquad u^* = \tfrac{1}{2}, \qquad f(\mathbf{x}^*) = -3$$

We like to see what happens if we change the right-hand side of the constraint equation to a value 'e' from zero. Note that the constraint $g(x_1, x_2) \leq 0$ gives a circular feasible region with its center at $(0, 0)$ and its radius as $\sqrt{6}$, as shown in Fig. 3.15. From Theorem 3.14, we have

$$\frac{\partial f}{\partial e} = -u^* = -\tfrac{1}{2}$$

If we set $e = 1$, the new value of cost function will be approximately -3.5 from Eq. (3.73). This is consistent with the constraint set, because with $e = 1$, the radius of the circle becomes $\sqrt{7}$ and the feasible region is expanded (as can be seen in Fig. 3.15). We should expect some reduction in the cost function. If we set $e = -1$, then the effect is opposite. The feasible region becomes smaller and the cost function increases to -2.5. This is again obtained using Eq. (3.73). $\quad \parallel$

From the foregoing discussion and example we see that *optimum Lagrange multipliers give very useful information.* The designer can compare the magnitude of the multipliers for the active constraints. The multipliers with relatively larger values will have a significant effect on optimum cost if the corresponding constraint parameters are changed. *The larger the value of the Lagrange multiplier, the higher is the dividend to relax the constraint, or the higher is the penalty to tighten the constraint.* Knowing this, the designer can select a few critical constraints having the greatest influence on the cost function. He can then analyze to see if these constraints can be relaxed to further reduce the optimum cost function value.

3.7.2 Effect of Scaling Cost Function on the Lagrange Multipliers

On many occasions, a cost function for the problem is multiplied by a positive constant. As noted in Section 3.4, any scaling of the cost function does not alter the optimum point. It does, however, change the optimum value for the cost function. The scaling should also affect the implicit derivatives of Eqs (3.71) and (3.72) for the cost function with respect to the right-hand side parameters of the constraints. *We observe from these equations that all the Lagrange multipliers also get multiplied by the same constant.* Let u_i^* and v_j^* be the Lagrange multipliers for inequality and equality constraints, respectively, and $f(\mathbf{x}^*)$ be the optimum value of the cost function at the solution point $\mathbf{x}^*$. Let the cost function be scaled as $\bar{f}(\mathbf{x}) = Kf(\mathbf{x})$, where $K > 0$ is a given constant, and $\bar{u}_i^*$ and $\bar{v}_j^*$ be the optimum Lagrange multipliers for the inequality and equality constraints, respectively, for the changed problem. Then the optimum solution for the perturbed problem is $\mathbf{x}^*$ and the optimum Lagrange multipliers of the two problems are related as

$$\bar{u}_i^* = Ku_i^*; \qquad i = 1 \text{ to } m$$
$$\bar{v}_j^* = Kv_j^*; \qquad i = 1 \text{ to } p \tag{3.76}$$

Example 3.44 Effect of scaling cost function on the Lagrange Multipliers. Consider Example 3.28. Minimize

$$f(\mathbf{x}) = x_1^2 + x_2^2 - 3x_1x_2$$

subject to

$$g(\mathbf{x}) = x_1^2 + x_2^2 - 6 \leq 0$$

Study the effect on the optimum solution of scaling the cost function by a constant $K > 0$.

Solution. The graphical solution for the problem is given in Fig. 3.15. A point satisfying both the necessary and sufficient condition is

$$x_1^* = x_2^* = \sqrt{3}, \ u^* = \tfrac{1}{2}, \ f(\mathbf{x}^*) = -3$$

Let us solve the scaled problem by writing K–T conditions. The Lagrangian for the problem is given as

$$L = K(x_1^2 + x_2^2 - 3x_1x_2) + \bar{u}(x_1^2 + x_2^2 - 6 + \bar{s}^2)$$

The necessary conditions give

$$\frac{\partial L}{\partial x_1} \equiv 2Kx_1 - 3Kx_2 + 2\bar{u}x_1 = 0 \qquad \text{(a)}$$

$$\frac{\partial L}{\partial x_2} \equiv 2Kx_2 - 3Kx_1 + 2\bar{u}x_2 = 0 \qquad \text{(b)}$$

$$x_1^2 + x_2^2 - 6 + \bar{s}^2 = 0 \qquad \text{(c)}$$

$$\bar{u}\bar{s} = 0, \qquad \bar{u} \geqq 0$$

As in Example 3.28, the case where $\bar{s} = 0$ gives candidate minimum points. Solving Eqs (a) – (c), we get

$$x_1^* = x_2^* = \sqrt{3}, \qquad \bar{u}^* = K/2, \qquad \bar{f}(\mathbf{x}^*) = -3K$$
$$x_1^* = x_2^* = -\sqrt{3}, \qquad \bar{u}^* = K/2, \qquad \bar{f}(\mathbf{x}^*) = -3K$$

Therefore, we observe that $\bar{u}^* = Ku^*$. ‖

3.7.3 Effect of scaling a constraint on its Lagrange Multiplier

Many times, a constraint is scaled by a positive constant. We like to know what is the effect of this scaling on the Lagrange multiplier for the constraint. *It should be noted that scaling of a constraint does not change the constraint boundary, so it has no effect on the optimum solution. Only the Lagrange multiplier for the scaled constraint is affected. Looking at the implicit derivatives of the cost function with respect to the constraint right-hand side parameters, we observe that the Lagrange multiplier for the scaled constraint gets divided by the scaling parameter.* Let $M_j > 0$ and P_i be the scale parameters for the jth inequality and ith equality constraints, and u_j^* and v_i^*, and $\bar{u}_j^*$ and $\bar{v}_i^*$ the corresponding Lagrange multipliers for the original and the scaled constraints, respectively. Then the following relations hold for the Lagrange multipliers:

$$\bar{u}_j^* = u_j^*/M_j$$
$$\bar{v}_i^* = v_i^*/P_i \qquad (3.77)$$

Example 3.45   Effect of scaling a constraint on its Lagrange Multiplier. Consider Example 3.28 and study the effect of multiplying the inequality by a constant $M > 0$.

Solution. The Lagrange function for the problem is given as

$$L = x_1^2 + x_2^2 - 3x_1x_2 + \bar{u}[M(x_1^2 + x_2^2 - 6) + \bar{s}^2]$$

The K–T conditions give

$$\frac{\partial L}{\partial x_1} \equiv 2x_1 - 3x_2 + 2\bar{u}Mx_1 = 0 \tag{a}$$

$$\frac{\partial L}{\partial x_2} = 2x_2 - 3x_1 + 2\bar{u}Mx_2 = 0 \tag{b}$$

$$M(x_1^2 + x_2^2 - 6) + \bar{s}^2 = 0 \tag{c}$$

$$\bar{u}\bar{s} = 0, \quad \bar{u} \geqq 0$$

As in Example 3.28, only the case with $\bar{s} = 0$ gives candidate optimum points. Solving this case, we get

$$x_1^* = x_2^* = \sqrt{3}, \qquad \bar{u}^* = \frac{1}{2M}, \qquad f(\mathbf{x}^*) = -3$$

$$x_1^* = x_2^* = -\sqrt{3}, \qquad \bar{u}^* = \frac{1}{2M}, \qquad f(\mathbf{x}^*) = -3$$

Therefore, we observe that $\bar{u}^* = u^*/M$. ‖

3.8 ENGINEERING DESIGN EXAMPLES

The procedures described in the previous sections are used to solve two engineering design examples. The problems are formulated, convexity is checked, K–T necessary conditions are written and solved, sufficiency conditions are checked, and the constraint variation sensitivity theorem is used to study changes in the constraint limits.

3.8.1 Design of a Wall Bracket

A wall bracket shown in Fig. 3.27 is to be designed to support a load of $W = 1.2$ MN. The material for the bracket should not fail under the action of forces in the bars. These are expressed as the following stress constraints:

$$\text{Bar 1: } \sigma_1 \leqq \sigma_a$$
$$\text{Bar 2: } \sigma_2 \leqq \sigma_a$$

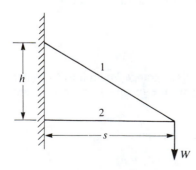

FIGURE 3.27
Wall bracket. $h = 30$ cm, $s = 40$ cm, and $W = 1\cdot2$ MN.

where σ_a = allowable stress for the material $(16\,000\ \text{N/cm}^2)$,
$\quad\quad\sigma_1$ = stress in Bar 1 which is given as F_1/A_1 (N/cm^2),
$\quad\quad\sigma_2$ = stress in Bar 2 which is given as F_2/A_2 (N/cm^2),
$\quad\quad A_1$ = cross-sectional area of Bar 1 (cm^2),
$\quad\quad A_2$ = cross-sectional area of Bar 2 (cm^2),
$\quad\quad F_1$ = force due to load W in Bar 1 (N), and
$\quad\quad F_2$ = force due to load W in Bar 2 (N).

Total volume of the bracket is to be minimized.

3.8.1.1 PROBLEM FORMULATION. The cross-sectional areas A_1 and A_2 are the two design variables and the cost function for the problem is the volume which is given as

$$f(A_1, A_2) = l_1 A_1 + l_2 A_2, \quad \text{cm}^3 \tag{a}$$

where $l_1 = \sqrt{30^2 + 40^2} = 50$ cm is the length of member 1, and $l_2 = 40$ cm is the length of member 2. To write the stress constraints, we need forces in the members which can be obtained using static equilibrium analysis as follows: $F_1 = (2.0\text{E}+06)$ N, $F_2 = (1.6\text{E}+06)$ N. Therefore stress constraints are given as

$$g_1 = \frac{(2.0\text{E}+06)}{A_1} - 16\,000 \leqq 0 \tag{b}$$

$$g_2 = \frac{(1.6\text{E}+06)}{A_2} - 16\,000 \leqq 0 \tag{c}$$

The cross-sectional areas must both be nonnegative:

$$g_3 \equiv -A_1 \leqq 0, \quad\quad g_4 \equiv -A_2 \leqq 0 \tag{d}$$

Constraints for the problem are plotted in Fig. 3.28 and the feasible region is identified. A few cost function contours are also shown. It can be seen that the optimum solution is at point A with $A_1 = 125\ \text{cm}^2$, $A_2 = 100\ \text{cm}^2$, and $f = 10\,250\ \text{cm}^3$.

3.8.1.2 CONVEXITY. Since the cost function of Eq. (a) is linear in terms of design variables, it is convex. The Hessian matrix for the constraint g_1 is

$$\nabla^2 g_1 = \begin{bmatrix} \dfrac{(4.0\text{E}+06)}{A_1^3} & 0 \\ 0 & 0 \end{bmatrix}$$

which is a positive semidefinite matrix for $A_1 > 0$, so g_1 is convex. Similarly, g_2 is convex, and since g_3 and g_4 are linear, they are convex. Thus the problem is *convex* and Kuhn–Tucker necessary conditions are also *sufficient* due to Theorem 3.10. Any design satisfying the K–T conditions is a global optimum.

3.8.1.3 KUHN–TUCKER CONDITIONS. To use the K–T conditions, we introduce slack variables into the constraints and define the Lagrange function

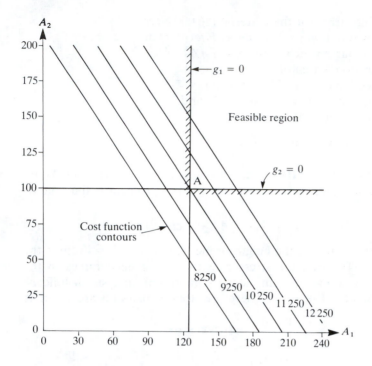

FIGURE 3.28
Graphical Solution for the Wall Bracket Problem.

for the problem as

$$L = (l_1 A_1 + l_2 A_2) + u_1 \left[\frac{(2.0\text{E}+06)}{A_1} - 16\,000 + s_1^2 \right] + u_2 \left[\frac{(1.6\text{E}+06)}{A_2} - 16\,000 + s_2^2 \right]$$

$$+ u_3(-A_1 + s_3^2) + u_4(-A_2 + s_4^2) \tag{e}$$

The necessary conditions become (note that gradients of constraints are linearly independent, so feasible points are regular)

$$\frac{\partial L}{\partial A_1} = l_1 - u_1 \frac{(2.0\text{E}+06)}{A_1^2} - u_3 = 0 \tag{f}$$

$$\frac{\partial L}{\partial A_2} = l_2 - u_2 \frac{(1.6\text{E}+06)}{A_2^2} - u_4 = 0 \tag{g}$$

$$u_i s_i = 0, \; u_i \geqq 0, \; g_i + s_i^2 = 0; \qquad i = 1 \text{ to } 4 \tag{h}$$

The switching conditions in Eqs (h) give the following 16 cases:

1. $u_1 = 0, \; u_2 = 0, \; u_3 = 0, \; u_4 = 0$
2. $u_1 = 0, \; s_2 = 0, \; u_3 = 0, \; u_4 = 0$

3. $s_1 = 0,\ u_2 = 0,\ u_3 = 0,\ u_4 = 0$
4. $s_1 = 0,\ s_2 = 0,\ u_3 = 0,\ u_4 = 0$
5. $u_1 = 0,\ u_2 = 0,\ u_3 = 0,\ s_4 = 0$
6. $u_1 = 0,\ u_2 = 0,\ s_3 = 0,\ u_4 = 0$
7. $u_1 = 0,\ u_2 = 0,\ s_3 = 0,\ s_4 = 0$
8. $u_1 = 0,\ s_2 = 0,\ s_3 = 0,\ u_4 = 0$
9. $s_1 = 0,\ u_2 = 0,\ s_3 = 0,\ u_4 = 0$
10. $s_1 = 0,\ u_2 = 0,\ u_3 = 0,\ s_4 = 0$
11. $u_1 = 0,\ s_2 = 0,\ u_3 = 0,\ s_4 = 0$
12. $u_1 = 0,\ s_2 = 0,\ s_3 = 0,\ s_4 = 0$
13. $s_1 = 0,\ u_2 = 0,\ s_3 = 0,\ s_4 = 0$
14. $s_1 = 0,\ s_2 = 0,\ u_3 = 0,\ s_4 = 0$
15. $s_1 = 0,\ s_2 = 0,\ s_3 = 0,\ u_4 = 0$
16. $s_1 = 0,\ s_2 = 0,\ s_3 = 0,\ s_4 = 0$

Note that any case that requires $s_3 = 0$ (i.e. $g_3 = 0$) makes the area $A_1 = 0$. For such a case the constraint g_1 of Eq. (b) is violated, so it does not give a candidate solution. Similarly, $s_4 = 0$ makes $A_2 = 0$ which violates the constraint of Eq. (c). In addition, A_1 and A_2 cannot be negative because the corresponding solution has no physical meaning. Therefore, all the cases that require either $s_3 = 0$ and/or $s_4 = 0$ do not give any candidate solution, so they need not be considered any further. This leaves only cases 1 to 4 for further consideration, and we solve them as follows (any case giving $A_1 < 0$ or $A_2 < 0$ will also be discarded):

Case 1: $u_1 = 0,\ u_2 = 0,\ u_3 = 0,\ u_4 = 0$. This case gives $l_1 = 0$ and $l_2 = 0$ in Eqs (f) and (g) which is *not acceptable*.

Case 2: $u_1 = 0,\ s_2 = 0,\ u_3 = 0,\ u_4 = 0$. This gives $l_1 = 0$ in Eq. (f) which is *not acceptable*.

Case 3: $s_1 = 0,\ u_2 = 0,\ u_3 = 0,\ u_4 = 0$. This gives $l_2 = 0$ in Eq. (g) which is *not acceptable*.

Case 4: $s_1 = 0,\ s_2 = 0,\ u_3 = 0,\ u_4 = 0$. Equations (b) and (c) give $A_1 = 125$ cm^2, $A_2 = 100$ cm^2. Equations (f) and (g) give the Lagrange multipliers as $u_1 = 0.391$ and $u_2 = 0.25$ and since both are nonnegative, all the K–T conditions are satisfied. Cost function at optimum is obtained as $f = 50(125) + 40(100)$ or $f = 10\,250$ cm^3.

3.8.1.4 SENSITIVITY ANALYSIS. If the allowable stress changes to 16 500 N/cm^2 compared to 16 000 N/cm^2, we need to know how the cost function will change. Using Eq. (3.74) we get change in the cost function as

$$\Delta f = -u_1 e_1 - u_2 e_2$$

where $e_1 = e_2 = 16\,500 - 16\,000 = 500\,\text{N/cm}^2$. Therefore, change in the cost function is

$$\Delta f = -0.391(500) - 0.25(500)$$
$$= -320.5\,\text{cm}^3$$

Thus the volume of the bracket will reduce by $320.5\,\text{cm}^3$.

3.8.2 Design of a Rectangular Beam

In Section 2.8.5, a rectangular beam design problem is formulated and solved graphically. We will solve the same problem using the Kuhn–Tucker necessary conditions. The problem is formulated as follows. Find b and d to minimize

$$f(b, d) = bd \tag{a}$$

subject to the inequality constraints

$$g_1 \equiv \frac{(2.40\text{E}+08)}{bd^2} - 10 \leq 0 \tag{b}$$

$$g_2 \equiv \frac{(2.25\text{E}+05)}{bd} - 2 \leq 0 \tag{c}$$

$$g_3 \equiv d - 2b \leq 0 \tag{d}$$

$$g_4 \equiv -b \leq 0, \qquad g_5 \equiv -d \leq 0 \tag{e}$$

3.8.2.1 CONVEXITY. Constraints g_3, g_4 and g_5 are linear in terms of b and d, and are therefore convex. The Hessian for the constraint g_1 is given as

$$\nabla^2 g_1 = \frac{(4.80\text{E}+08)}{b^3 d^4} \begin{bmatrix} d^2 & bd \\ bd & 3b^2 \end{bmatrix}$$

Since this matrix is positive definite for $b > 0$ and $d > 0$, g_1 is a strictly convex function. Hessian for the constraint g_2 is given as

$$\nabla^2 g_2 = \frac{(2.25\text{E}+05)}{b^3 d^3} \begin{bmatrix} 2d^2 & bd \\ bd & 2b^2 \end{bmatrix}$$

Since this matrix is positive definite, the constraint g_2 is also strictly convex. Since all the constraints of the problem are convex, the constraint set (feasible region) is convex.

It is interesting to note that *constraints* g_1 and g_2 can be *transformed* as (since $b > 0$ and $d > 0$, the sense of inequality is not changed):

$$\bar{g}_1 \equiv (2.40\text{E}+08) - 10bd^2 \leq 0 \tag{f}$$

$$\bar{g}_2 \equiv (2.25\text{E}+05) - 2bd \leq 0 \tag{g}$$

Hessians of the functions $\bar{g}_1$ and $\bar{g}_2$ are given as

$$\nabla^2 \bar{g}_1 = \begin{bmatrix} 0 & -20d \\ -20d & -20b \end{bmatrix}; \qquad \nabla^2 \bar{g}_2 = \begin{bmatrix} 0 & -2 \\ -2 & 0 \end{bmatrix}$$

Note that both of the preceding matrices are not positive semidefinite. Therefore, the constraint functions $\bar{g}_1$ and $\bar{g}_2$ given in Eqs (f) and (g) are not convex. This goes to show that convexity of a function can be lost if it is transformed to another form. This is an important observation and it shows that we should be careful in transformation of constraint functions. Note, however, that transformation of constraints does not change the optimum solution. It does change the values of the Lagrange multipliers for the constraints as discussed in Section 3.7.3.

In order to check convexity of the cost function, we write its Hessian as

$$\nabla^2 f = \begin{bmatrix} 0 & 1 \\ 1 & 0 \end{bmatrix} \tag{h}$$

This matrix is indefinite, so the cost function is nonconvex. The problem fails the convexity check of Theorem 3.8 and we cannot guarantee global optimality of the solution by Theorem 3.9. Note that this does not say that a local solution cannot be a global minimum. It may still be a global minimum, but cannot be guaranteed by Theorem 3.9.

3.8.2.2 KUHN–TUCKER CONDITIONS. To use the K–T conditions, we introduce slack variables into the constraints and define the Lagrange function for the problem as

$$L = bd + u_1\left(\frac{(2.40\text{E}+08)}{bd^2} - 10 + s_1^2\right) + u_2\left(\frac{(2.25\text{E}+05)}{bd} - 2 + s_2^2\right) + u_3(d - 2b + s_3^2)$$
$$+ u_4(-b + s_4^2) + u_5(-d + s_5^2)$$

The necessary conditions give

$$\frac{\partial L}{\partial b} = d + u_1\frac{(-2.40\text{E}+08)}{b^2d^2} + u_2\frac{(-2.25\text{E}+05)}{b^2d} - 2u_3 - u_4 = 0 \tag{i}$$

$$\frac{\partial L}{\partial d} = b + u_1\frac{(-4.80\text{E}+08)}{bd^3} + u_2\frac{(-2.25\text{E}+05)}{bd^2} + u_3 - u_5 = 0 \tag{j}$$

$$u_i s_i = 0, \qquad u_i \geq 0, \qquad g_i + s_i^2 = 0; \qquad i = 1 \text{ to } 5 \tag{k}$$

The switching conditions of Eqs (k) give 32 cases for the necessary conditions. However, note that the cases requiring either $s_4 = 0$ or $s_5 = 0$, or both as zero, do not give any candidate optimum points because they violate constraint of either Eqs (b) and (c) or Eq. (d). Therefore, these cases shall not be considered, which can be done by setting $u_4 = 0$ and $u_5 = 0$ in the remaining cases. This leaves the following eight cases for further consideration:

1. $u_1 = 0$, $u_2 = 0$, $u_3 = 0$, $u_4 = 0$, $u_5 = 0$
2. $u_1 = 0$, $u_2 = 0$, $s_3 = 0$, $u_4 = 0$, $u_5 = 0$
3. $u_1 = 0$, $s_2 = 0$, $u_3 = 0$, $u_4 = 0$, $u_5 = 0$
4. $s_1 = 0$, $u_2 = 0$, $u_3 = 0$, $u_4 = 0$, $u_5 = 0$

5. $u_1 = 0$, $s_2 = 0$, $s_3 = 0$, $u_4 = 0$, $u_5 = 0$
6. $s_1 = 0$, $s_2 = 0$, $u_3 = 0$, $u_4 = 0$, $u_5 = 0$
7. $s_1 = 0$, $u_2 = 0$, $s_3 = 0$, $u_4 = 0$, $u_5 = 0$
8. $s_1 = 0$, $s_2 = 0$, $s_3 = 0$, $u_4 = 0$, $u_5 = 0$

We consider each case at a time and solve for the candidate optimum points. Note that any solution having $b < 0$ or $d < 0$ violates constraints g_4 and g_5 respectively and shall be discarded.

Case 1: $u_1 = 0$, $u_2 = 0$, $u_3 = 0$, $u_4 = 0$, $u_5 = 0$. This case gives $d = 0$, $b = 0$ in Eqs (i) and (j). Therefore, this case does not give a solution.

Case 2: $u_1 = 0$, $u_2 = 0$, $s_3 = 0$, $u_4 = 0$, $u_5 = 0$. Equation (d) gives $d = 2b$. Equations (i) and (j) give $d - 2u_3 = 0$ and $b + u_3 = 0$. These three equations give $b = 0$ and $d = 0$, which is not feasible.

Case 3: $u_1 = 0$, $s_2 = 0$, $u_3 = 0$, $u_4 = 0$, $u_5 = 0$. Equations (i), (j) and (c) give

$$d - u_2 \frac{(2.25\mathrm{E}+05)}{b^2 d} = 0$$

$$b - u_2 \frac{(2.25\mathrm{E}+05)}{bd^2} = 0$$

$$\frac{(2.25\mathrm{E}+05)}{bd} - 2 = 0$$

These equations give a solution as $u_2 = (5.625\mathrm{E}+03)$ and $bd = (1.125\mathrm{E}+05)$. Since $u_2 > 0$, this is a valid solution. Actually, there is a family of solutions given by $bd = (1.125\mathrm{E}+05)$; for a value of d, b can be found from this equation. However, there must be some limits on the values of b and d for which this family of solutions is valid. These ranges are provided by requiring $s_1^2 \geqq 0$ and $s_3^2 \geqq 0$, or $g_1 \leqq 0$ and $g_3 \leqq 0$.
Substituting $b = (1.125\mathrm{E}+05)/d$ into g_1 (Eq. b),

$$\frac{(2.40\mathrm{E}+08)}{(1.125\mathrm{E}+05)d} - 10 \leqq 0, \quad \text{or} \quad d \geqq 213.33 \,\text{mm} \tag{l}$$

Substituting $b = (1.125\mathrm{E}+05)/d$ into g_3 (Eq. d),

$$d - \frac{(2.25\mathrm{E}+05)}{d} \leqq 0; \quad \text{or} \quad d \leqq 474.34 \,\text{mm} \tag{m}$$

This gives limits on the depth d. We can find limits on the width b by substituting Eqs (l) and (m) into $bd = (1.125\mathrm{E}+05)$:

$$d \geqq 213.33, \qquad b \leqq 527.34$$
$$d \leqq 474.33, \qquad b \geqq 237.17$$

Therefore, for this case the possible solutions are

$$237.17 \leqq b \leqq 527.34 \text{ mm}; \qquad 213.33 \leqq d \leqq 474.33 \text{ mm}$$
$$bd = (1.125\text{E}+05) \text{ mm}^2$$

Case 4: $s_1 = 0$, $u_2 = 0$, $u_3 = 0$, $u_4 = 0$, $u_5 = 0$. Equations (i) and (j) reduce to

$$d - \frac{(2.40\text{E}+08)}{b^2 d^2} = 0; \qquad \text{or } b^2 d^3 = (2.40\text{E}+08)$$

$$b - \frac{(4.80\text{E}+08)}{bd^3} = 0; \qquad \text{or } b^2 d^3 = (4.80\text{E}+08)$$

Since the above two equations are inconsistent, there is no solution for this case.

Case 5: $u_1 = 0$, $s_2 = 0$, $s_3 = 0$, $u_4 = 0$, $u_5 = 0$. Equations (c) and (d) can be solved for b and d, e.g. substituting $d = 2b$ from Eq. (d) into Eq. (c), we get $b = 237.17$ mm. Therefore, $d = 2(237.17) = 474.34$ mm. We can calculate u_2 and u_3 from Eqs (i) and (j) as $u_2 = (5.625\text{E}+04)$, $u_3 = 0$. Substituting values of b and d into Eq. (b), we get $g_1 = -5.5 < 0$, so the constraint is satisfied (i.e. $s_1^2 > 0$). Since all the necessary conditions are satisfied, this is a valid solution. The constraint sensitivity Theorem 3.14 and Eq. (3.72) tell us that since $u_3 = 0$, we can move away from that constraint towards the feasible region without affecting the optimum cost function value. This can also be observed from Fig. 2.11 where graphical solution for the problem is given. In the figure, point B represents the solution for this case. We can leave point B towards point A and remain on the constraint $g_2 = 0$ for optimum designs.

Case 6: $s_1 = 0$, $s_2 = 0$, $u_3 = 0$, $u_4 = 0$, $u_5 = 0$. Equations (b) and (c) can be solved for the b and d as $b = 527.34$ mm and $d = 213.33$ mm. We can solve for u_1 and u_2 from Eqs (i) and (j) as $u_1 = 0$ and $u_2 = (5.625\text{E}+04)$. Substituting values of b and d into Eq. (d), we get $g_3 = -841.35 < 0$, so the constraint is satisfied (i.e. $s_3^2 \geqq 0$). Since all the K–T conditions are satisfied, this is a valid solution. This solution is quite similar to the one for case 5. The solution corresponds to point A on Fig. 2.11. If we leave constraint $g_1 = 0$ (point A) and remain on the curve A–B, we obtain other optimum designs near the point A.

Case 7: $s_1 = 0$, $u_2 = 0$, $s_3 = 0$, $u_4 = 0$, $u_5 = 0$. Equations (b) and (d) can be solved as $b = 181.71$ mm and $d = 363.42$ mm. Equations (i) and (j) give the Lagrange multipliers $u_1 = 4402.35$ and $u_3 = -60.57$. Since $u_3 < 0$, this case does not give a valid solution.

Case 8: $s_1 = 0$, $s_2 = 0$, $s_3 = 0$, $u_4 = 0$, $u_5 = 0$. This case gives three equations in two unknowns (overdetermined system) which has no solution.

3.8.2.3 SUFFICIENCY CHECK. Cases 3, 5 and 6 give solutions that satisfy the K–T conditions. Cases 5 and 6 have two active constraints but Eq. (3.67) in sufficiency theorem requires only constraints with $u_i > 0$ to be considered, so only the g_2 constraint needs to be included in the check for sufficient conditions. Thus all three cases have the same sufficiency check.

We need to calculate Hessians of the cost function and the second constraint:

$$\nabla^2 f = \begin{bmatrix} 0 & 1 \\ 1 & 0 \end{bmatrix}, \qquad \nabla^2 g_2 = \frac{(2.25\text{E}+05)}{b^3 d^3} \begin{bmatrix} 2d^2 & bd \\ bd & 2b^2 \end{bmatrix}$$

Since $bd = (1.125\text{E}+05)$, $\nabla^2 g_2$ becomes

$$\nabla^2 g_2 = 2 \begin{bmatrix} \dfrac{2}{b^2} & (1.125\text{E}+05)^{-1} \\ (1.125\text{E}+05)^{-1} & \dfrac{2}{d^2} \end{bmatrix}$$

Hessian of the Lagrangian is given as

$$\nabla^2 L = \nabla^2 f + u_2 \nabla^2 g_2 = \begin{bmatrix} 0 & 1 \\ 1 & 0 \end{bmatrix} + 2 \begin{bmatrix} \dfrac{(1.125\text{E}+05)}{b^2} & 0.5 \\ 0.5 & \dfrac{(1.125\text{E}+05)}{d^2} \end{bmatrix}$$

$$= 2 \begin{bmatrix} \dfrac{(1.125\text{E}+05)}{b^2} & 1.0 \\ 1.0 & \dfrac{1.125\text{E}+05}{d^2} \end{bmatrix}$$

The determinant of $\nabla^2 L$ is 0 for $bd = (1.125\text{E}+05)$, therefore, it is not a positive definite matrix. So, Theorem 3.13 cannot be used to show sufficiency of x^*. We must check the sufficient condition of Eq. (3.68). In order to do that, we must find y satisfying Eq. (3.67). The gradient of g_2 is given as

$$\nabla g_2 = \begin{bmatrix} \dfrac{-(2.25\text{E}+05)}{b^2 d}, & \dfrac{(-2.25\text{E}+05)}{bd^2} \end{bmatrix}$$

and $\nabla g_2^T y = 0$ gives

$$y_1/b + y_2/d = 0, \qquad \text{or } y_2 = -(d/b)y_1$$

Therefore, vector y is given as

$$y = (1, -d/b)c; \qquad c = y_1 \text{ is any constant}$$

Using $\nabla^2 L$ and y, Q of Eq. (3.68) is given as

$$Q = y^T \nabla^2 L y = 0$$

Thus the sufficiency condition of Theorem 3.12 is not satisfied. The points satisfying $bd = (1.125E+05)$ need not be isolated minimum points. This is, of course, true from Fig. 2.11. Note, however, that since $Q = 0$, the second-order necessary condition of Theorem 3.11 is satisfied.

It is important to note that this problem does not satisfy condition for a convex programming problem and all the points satisfying K–T conditions do not satisfy the sufficiency condition for isolated minimum. Yet, all the points are actually global optimum designs. Two conclusions can be drawn from this example:

1. *Global optimum solutions* can be obtained for problems that cannot be classified as convex programming problems. We cannot show global optimality unless we find all the local optimum solutions.
2. *If sufficiency conditions are not satisfied,* the only conclusion we can draw is that the candidate point need not be an isolated minimum. It may have many local optima in the neighborhood and they may all be actually global solutions.

3.8.2.4 SENSITIVITY ANALYSIS. It should be observed that none of the candidate minimum points (Points A and B and curve A–B in Fig. 2.11) satisfies the sufficiency conditions. Therefore, existence of partial derivatives of the cost function with respect to the right-hand side parameters of Eqs (3.71) and (3.72) is not guaranteed by Theorem 3.14. However, since we have a graphical solution for the problem in Fig. 2.11, we can check what happens if we use the sensitivity theorem.

For Point A in Fig. 2.11 (case 6), constraints g_1 and g_2 are active, $b = 527.34$ mm, $d = 213.33$ mm, $u_1 = 0$ and $u_2 = (5.625E+04)$. Since $u_1 = 0$, Eq. (3.72) gives $\partial f / \partial e_1 = 0$. This means any small change in the constraint limit does not change the optimum cost function value. This is true, which can be observed from Fig. 2.11. The optimum point is changed but constraint g_1 remains active; i.e. $bd = (1.125E+05)$ must be satisfied. Any change in g_2 moves the constraint parallel to itself changing the optimum solution (design variables and the cost function). Since $u_2 = (5.625E+04)$, Eq. (3.72) gives $\partial f / \partial e_2 = (-5.625E+04)$. It can be verified that the sensitivity coefficient predicts correct changes in the cost function.

It can be verified that the other two solution cases (3 and 5) also give correct values for the sensitivity coefficients.

EXERCISES FOR CHAPTER 3

Section 3.2   Fundamental Concepts

3.1 *Answer True or False*

1. A function can have several local minimum points in a small neighborhood of $\mathbf{x}^*$.

2. A function cannot have more than one global minimum point.
3. The value of the function having global minimum at several points must be the same.
4. A function defined on an open set cannot have global minimum.
5. Gradient of a function $f(\mathbf{x})$ at a point is normal to the surface defined by the level surface $f(\mathbf{x}) = $ constant.
6. Gradient of a function at a point gives a local direction of maximum decrease in the function.
7. Hessian matrix of a continuously differentiable function can be asymmetric.
8. Hessian matrix for a function is calculated using only the first derivatives of the function.
9. Taylor series expansion for a function at a point uses function value and its derivatives.
10. Taylor series expansion can be written at a point where the function is discontinuous.
11. Taylor series expansion of a complicated function replaces it with a polynomial function at the point.
12. Linear Taylor series expansion of a complicated function at a point is only a good local approximation for the function.
13. A quadratic form can have first-order terms in the variables.
14. For a given $\mathbf{x}$, the quadratic form defines a vector.
15. Every quadratic form has a symmetric matrix associated with it.
16. A symmetric matrix is positive definite if its eigenvalues are nonnegative.
17. A matrix is positive semidefinite if some of its eigenvalues are negative and others are nonnegative.
18. All eigenvalues of a negative definite matrix are strictly negative.
19. Quadratic form appears as one of the terms in Taylor's expansion of a function.
20. A positive definite quadratic form must have positive value for any $\mathbf{x} \neq \mathbf{0}$.

Write Taylor series expansion for the following functions up to quadratic terms:

3.2 $\cos x$ about the point $x^* = \pi/4$
3.3 $\cos x$ about the point $x^* = \pi/3$
3.4 $\sin x$ about the point $x^* = \pi/6$
3.5 $\sin x$ about the point $x^* = \pi/4$
3.6 e^x about the point $x^* = 0$
3.7 e^x about the point $x^* = 2$
3.8 $f(x_1, x_2) = 10x_1^4 - 20x_1^2 x_2 + 10x_2^2 + x_1^2 - 2x_1 + 5$ about the point $(1, 1)$. Compare approximate and exact values of the function at the point $(1.2, 0.8)$.

Determine the form of the following quadratic functions:

3.9 $F(\mathbf{x}) = x_1^2 + 4x_1 x_2 + 2x_1 x_3 - 7x_2^2 - 6x_2 x_3 + 5x_3^2$
3.10 $F(\mathbf{x}) = 2x_1^2 + 2x_2^2 - 5x_1 x_2$

3.11 $F(\mathbf{x}) = x_1^2 + x_2^2 + 3x_1x_2$

3.12 $F(\mathbf{x}) = 3x_1^2 + x_2^2 - x_1x_2$

3.13 $F(\mathbf{x}) = x_1^2 - x_2^2 + 4x_1x_2$

3.14 $F(\mathbf{x}) = x_1^2 - x_2^2 + x_3^2 - 2x_2x_3$

3.15 $F(\mathbf{x}) = x_1^2 - 2x_1x_2 + 2x_2^2$

3.16 $F(\mathbf{x}) = x_1^2 - x_1x_2 - x_2^2$

Section 3.3 Unconstrained Optimum Design Problems

3.17 *Answer True or False*

1. If the first order necessary condition at a point is satisfied for an unconstrained problem, it can be a local maximum point for the function.

2. A point satisfying first-order necessary conditions for an unconstrained function may not be a local minimum point.

3. A function can have negative value at its maximum point.

4. If a constant is added to a function, the location of its minimum point is changed.

5. If a function is multiplied by a positive constant, the location of its minimum point is unchanged.

6. If the curvature of an unconstrained function of a single variable at the point x^* is zero, then it is a local maximum point for the function.

7. The curvature of an unconstrained function of a single variable at its local minimum point is negative.

8. The Hessian of an unconstrained function at its local minimum point must be positive semidefinite.

9. The Hessian of an unconstrained function at its minimum point is negative definite.

10. If Hessian of an unconstrained function is indefinite at a candidate point, the point may be a local maximum or minimum.

Find stationary points for the following functions (use a numerical method such as the Newton-Raphson method in Appendix C, if needed). Also determine local minimum, local maximum and inflection points for the functions (inflection points are those stationary points that are neither minimum nor maximum):

3.18 $f(x_1, x_2) = 3x_1^2 + 2x_1x_2 + 2x_2^2 + 7$

3.19 $f(x_1, x_2) = x_1^2 + 4x_1x_2 + x_2^2 + 3$

3.20 $f(x_1, x_2) = x_1^3 + 12x_1x_2^2 + 2x_2^2 + 5x_1^2 + 3x_2$

3.21 $f(x_1, x_2) = 5x_1 - \frac{1}{16}x_1^2x_2 + \dfrac{1}{4x_1}x_2^2$

3.22 $f(x) = \cos x$

3.23 $f(x_1, x_2) = x_1^2 + x_1x_2 + x_2^2$

3.24 $f(x) = x^2 e^{-x}$

3.25 $f(x_1, x_2) = x_1 + \dfrac{10}{x_1 x_2} + 5x_2$

3.26 $f(x_1, x_2) = x_1^2 - 2x_1 + 4x_2^2 - 8x_2 + 6$

3.27 $f(x_1, x_2) = 3x_1^2 - 2x_1 x_2 + 5x_2^2 + 8x_2$

3.28 The annual operating cost U for an electrical line system is given by the following expression

$$U = \frac{(21.9E+07)}{V^2 C} + (3.9E+06)C + (1.0E+03)V$$

where $V =$ line voltage in kilovolts and $C =$ line conductance in mhos. Find stationary points for the function, and determine V and C to minimize the operating cost.

Section 3.4   Constrained Optimum Design Problems

3.29 *Answer True or False*

1. A regular point of the feasible region is defined as a point where cost function gradient is independent of the gradients of active constraints.

2. A point satisfying K–T conditions for a general optimum design problem can be a local max-point for the cost function.

3. At the optimum point the number of active independent constraints is always more than the number of design variables.

4. In the general optimum design problem formulation, the number of independent equality constraints must be "$\leq$" to the number of design variables.

5. In the general optimum design problem formulation, the number of inequality constraints cannot exceed the number of design variables.

6. At the optimum point, Lagrange multipliers for the "$\leq$ type" inequality constraints must be nonnegative.

7. At the optimum point, the Lagrange multiplier for a "$\leq$ type" constraint can be zero.

8. While solving an optimum design problem by K–T conditions, each case defined by the switching conditions can have multiple solutions.

9. In optimum design problem formulation, "$\geq$ type" constraints cannot be treated.

10. Optimum design points for constrained optimization problems give stationary value to the Lagrange function with respect to design variables.

11. Optimum design points having at least one active constraint give stationary value to the cost function.

12. At a constrained optimum design point that is regular, the cost function gradient is linearly dependent on the gradients of the active constraint functions.

13. If a slack variable has zero value at the optimum, the inequality constraint is inactive.

14. Gradients of inequality constraints that are active at the optimum point must be zero.

15. Design problems with equality constraints have gradient of the cost function as zero at the optimum point.

Find points satisfying Kuhn–Tucker necessary conditions for the following problems. Check for regularity of the points:

3.30 Miminize $\quad f(x_1, x_2) = 4x_1^2 + 3x_2^2 - 5x_1x_2 - 8x_1$
subject to $\quad x_1 + x_2 = 4$

3.31 Minimize $\quad f(x_1, x_2) = 9x_1^2 + 18x_1x_2 + 13x_2^2 - 4$
subject to $\quad x_1^2 + x_2^2 + 2x_1 = 16$

3.32 Minimize $\quad f(x_1, x_2) = (x_1 - 1)^2 + (x_2 - 1)^2$
subject to $\quad x_1 + x_2 - 4 = 0$

3.33 Consider the following problem with equality constraints:
Minimize $\quad (x_1 - 1)^2 + (x_2 - 1)^2$
subject to $\quad x_1 + x_2 - 4 = 0$
$\qquad\qquad x_1 - x_2 - 2 = 0$
1. Is it a valid optimization problem? Explain.
2. Explain how would you solve the problem? Are necessary conditions needed to find the optimum solution?

3.34 Minimize $\quad f(x_1, x_2) = 2x_1 + 3x_2 - x_1^3 - 2x_2^2$
subject to $\quad x_1 + 3x_2 \leqq 6$
$\qquad\qquad 5x_1 + 2x_2 \leqq 10$
$\qquad\qquad x_1, x_2 \geqq 0$

3.35 Minimize $\quad f(x_1, x_2) = 4x_1^2 + 3x_2^2 - 5x_1x_2 - 8x_1$
subject to $\quad x_1 + x_2 \leqq 4$

3.36 Minimize $\quad f(x_1, x_2) = x_1^2 + x_2^2 - 4x_1 - 2x_2 + 6$
subject to $\quad x_1 + x_2 \geqq 4$

3.37 Minimize $\quad f(x_1, x_2) = 2x_1^2 - 6x_1x_2 + 9x_2^2 - 18x_1 + 9x_2$
subject to $\quad x_1 + 2x_2 \leqq 10$
$\qquad\qquad 4x_1 - 3x_2 \leqq 20; \qquad x_i \geqq 0; \ i = 1, 2$

3.38 Minimize $\quad f(x_1, x_2) = (x_1 - 1)^2 + (x_2 - 1)^2$
subject to $\quad x_1 + x_2 - 4 \leqq 0$

3.39 Minimize $\quad f(x_1, x_2) = (x_1 - 1)^2 + (x_2 - 1)^2$
subject to $\quad x_1 + x_2 - 4 \leqq 0$
$\qquad\qquad x_1 - x_2 - 2 \leqq 0$

3.40 Minimize $\quad f(x_1, x_2) = (x_1 - 1)^2 + (x_2 - 1)^2$
subject to $\quad x_1 + x_2 - 4 \leqq 0$
$\qquad\qquad 2 - x_1 \leqq 0$

3.41 Minimize $\quad f(x_1, x_2) = 9x_1^2 - 18x_1x_2 + 13x_2^2 - 4$
subject to $\quad x_1^2 + x_2^2 + 2x_1 \geqq 16$

3.42 Minimize $\quad f(x_1, x_2) = (x_1 - 3)^2 + (x_2 - 3)^2$
subject to $\quad x_1 + x_2 \leqq 4$
$\qquad\qquad x_1 - 3x_2 = 1$

3.43 Minimize $\quad f(x_1, x_2) = x_1^3 - 16x_1 + 2x_2 - 3x_2^2$
subject to $\quad x_1 + x_2 \leqq 3$

3.44 Minimize $f(x_1, x_2) = 3x_1^2 - 2x_1x_2 + 5x_2^2 + 8x_2$
subject to $x_1^2 - x_2^2 + 8x_2 \leq 16$

3.45 Consider the design of a can formulated in Section 2.6.2:

Minimize $f(D, H) = \pi DH + \dfrac{\pi}{2}D^2,$ cm^2

subject to $\dfrac{\pi}{4}D^2H \geq 400,$ cm^3

$3.5 \leq D \leq 8,$ cm
$8 \leq H \leq 18,$ cm

Write K–T conditions and solve them. Interpret the necessary conditions at the solution point graphically.

3.46 A minimum weight tubular column design problem is formulated in Section 2.6.7 as find mean radius R and thickness t to minimize

$$\text{mass} = 2\rho l \pi R t$$

subject to the stress constraint

$$\frac{P}{2\pi R t} \leq \sigma_a$$

the buckling load constraint

$$P \leq \frac{\pi^3 E R^3 t}{4l^2}$$

and non-negativity of design variables

$$R \geq 0, \; t \geq 0$$

Solve the K–T conditions for the problem imposing an additional constraint $R/t \leq 50$ for the following data: $P = 50\,\text{kN}$, $l = 5.0\,\text{m}$, $E = 210$ GPa, $\sigma_a = 250$ MPa and $\rho = 7850\,\text{kg/m}^3$. Interpret the necessary conditions at the solution point graphically.

3.47 A minimum weight tubular column design problem is formulated in Section 2.6.7 as find outer radius R_o and inner radius R_i to minimize

$$\text{mass} = \pi \rho l (R_o^2 - R_i^2)$$

subject to stress constraint

$$\frac{P}{\pi(R_o^2 - R_i^2)} \leq \sigma_a$$

the buckling load constraint

$$P \leq \frac{\pi^3 E}{16l^2}(R_o^4 - R_i^4)$$

and nonnegativity of design variables

$$R_o \geq 0, \qquad R_i \geq 0$$

Solve the K–T conditions for the problem imposing an additional constraint

$0.5(R_o + R_i)/(R_o - R_i) \le 50$. Use the same data as in Exercise 3.46. Interpret the necessary conditions at the solution point graphically.

3.48 An engineering design problem is formulated as

minimize $f(x_1, x_2) = x_1^2 + 320x_1x_2$

subject to $\dfrac{1}{60x_2}x_1 - 1 \le 0$

$1 - \frac{1}{3600}x_1(x_1 - x_2) \le 0$

$x_1, x_2 \ge 0$

Write K–T necessary conditions and solve for the candidate minimum designs. Verify the solutions graphically. Interpret the K–T conditions on the graph for the problem.

3.49 A 100×100 m lot is available to construct a multistory office building. At least $20\,000$ m^2 total floor space is needed. According to a zoning ordinance, the maximum height of the building can only be 21 m, and the area for parking outside the building must be at least 25% of the total floor area. It has been decided to fix the height of each story at 3.5 m. The cost of the building in millions of dollars is estimated at $(0.6h + 0.001A)$ where A is the cross-sectional area of the building per floor and h is the height of the building. Formulate and solve the minimum cost design problem graphically (same as Exercise 2.1). Verify the K–T conditions at the solution points and show gradients of cost and constraint functions on the graph.

3.50 A refinery has two crude oils:

Crude A costs \$30/barrel (bbl) and $20\,000$ bbl are available
Crude B costs \$36/bbl and $30\,000$ bbl are available

The company manufactures gasoline and lube oil from the crudes. Yield and sale price per barrel of the product and markets are shown in Table 3.3. How much crude oils should the company use to maximize its profit? Formulate and solve the optimum design problem graphically (same as Exercise 2.2). Verify the K–T conditions at the solution points and show gradients of cost and constraint functions on the graph.

TABLE 3.3
Data for refinery operations

Product	Yield/bbl		Sale price per bbl	Market (bbl)
	Crude A	Crude B		
Gasoline	0.6	0.8	\$50	20 000
Lube oil	0.4	0.2	\$120	10 000

3.51 Design a beer mug shown in Fig. E3.51 to hold as much beer as possible. The height and radius of the mug should be no more than 20 cm. The mug must be at least 5 cm in radius. The surface area of the sides must not be greater than 900 cm^2 (ignore the area of the bottom of the mug and ignore the mug handle – see Fig. E3.51). Formulate and solve the optimum design problem graphically (same as Exercise 2.3). Verify the K–T conditions and show gradients of cost and constraint functions on the graph.

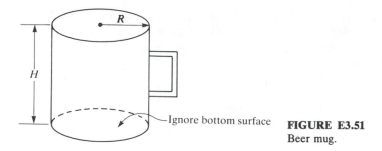

—Ignore bottom surface

FIGURE E3.51
Beer mug.

3.52 A Company is redesigning its parallel flow heat exchanger of length l to increase its heat transfer. An end view of the unit is shown in Fig. E3.52 There are certain limitations on the design problem. The smallest available conducting tube has a radius of 0.5 cm and all tubes must be of the same size. Further, the total cross-sectional area of all the tubes cannot exceed 2000 cm² to ensure adequate space inside the outer shell. Formulate and solve the problem graphically to determine the number of tubes, and the radius of each tube to maximize the surface area of the tubes in the exchanger (same as Exercise 2.4). Verify K–T conditions at the solution points and show gradients of cost and constraint functions on the graph.

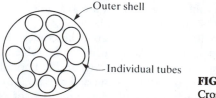

—Outer shell

—Individual tubes

FIGURE E3.52
Cross-section of a heat exchanger.

3.53 Proposals for a parking ramp having been defeated, and we plan to build a parking lot in the downtown urban renewal section. The cost of land is $200W + 100D,$ where W is the width along the street, and D the depth of the lot in meters. The available width along the street is 100 m, while the maximum depth available is 200 m. We want to have at least 10 000 m² in the lot. To avoid unsightly lots, the city requires the longer dimension of any lot be no more than twice the shorter dimension. Formulate the minimum cost design problem and solve it graphically (same as Exercise 2.5). Verify K–T conditions at the solution points and show gradients of cost and constraint functions on the graph.

3.54 A manufacturer sells products A and B. Profit from A is $10/kg and from B $8/kg. Available raw materials for the products are: 100 kg of C and 80 kg of D. To produce 1 kg of A, 0.4 kg of C and 0.6 kg of D are needed. To produce 1 kg of B, 0.5 kg of C and 0.5 kg of D are needed. The market for the products is 70 kg for A and 110 kg for B. How much of A and B should be produced to maximize profit? Formulate the design optimization problem and solve it graphically (same as Exercise 2.6). Verify K–T conditions at the solution points and show gradients of cost and constraint functions on the graph.

3.55 Design a diet of bread and milk to get at least 5 units of vitamin A and 4 units of vitamin B each day. The amount of vitamins A and B in 1 kg of each food and the cost of food are given in Table 3.4. Formulate the design optimization problem and solve it graphically so that we get at least the basic requirements of vitamins at the minimum cost (same as Exercise 2.7). Verify K–T conditions at the solution points and show gradients of cost and constraint functions on the graph.

TABLE 3.4
Data for the diet problem

Vitamin	Bread	Milk
A	1	2
B	3	2
Cost/kg	2	1

3.56 Enterprising chemical engineering students have set up a still in a bathtub. They can produce 225 bottles of pure alcohol each week. They bottle two products from alcohol: (i) wine, 20 proof, and (ii) whiskey, 80 proof.

Recall that pure alcohol is 200 proof. They have an unlimited supply of water, but can only obtain 800 empty bottles per week because of stiff competition. The weekly supply of sugar is enough for 600 bottles of wine or 1200 bottles of whiskey. They make $1.0 profit on each bottle of wine and $2.00 profit on each bottle of whiskey. They can sell whatever they produce. How many bottles of wine and whiskey should they produce each week to maximize profit. Formulate the design optimization problem and solve it graphically (same as Exercise 2.8). Verify K–T conditions at the solution points and show gradients of cost and constraint functions on the graph.

3.57 Design a can closed at one end using the smallest area of sheet metal for a specified interior volume of 600 cm^3. The can is a right circular cylinder with an interior height h and radius r. The ratio of height to diameter must not be less than 1.0 and not greater than 1.5. The height cannot be more than 20 cm. Formulate the design optimization problem and solve it graphically (same as Exercise 2.9). Verify K–T conditions at the solution points and show gradients of cost and constraint functions on the graph.

3.58 Design a shipping container with dimensions $b \times b \times h$ to minimize the ratio: (round-trip cost of shipping the container only)/(one-way cost of shipping the contents only). Use the following data:

mass of the container/area: 80 kg/m^2
Maximum b: 10 m
Maximum h: 18 m
One-way shipping cost,
    full or empty: $18/kg gross mass
Mass of the contents: 150 kg/m^3

Formulate the design optimization problem and solve it graphically (same as

Exercise 2.10). Verify K–T conditions at the solution points and show gradients of cost and constraint functions on the graph.

3.59 Design a circular tank closed at both ends to have a volume of $150 \, \text{m}^3$. The fabrication cost is proportional to the surface area of the sheet metal and is $\$400/\text{m}^2$. The tank is to be housed in a shed with sloping roof. Therefore, height H of the tank is limited by the relation $H \leq 10 - D/2$, where D is the diameter of the tank. Formulate the minimum cost design problem and solve it graphically (same as Exercise 2.12). Verify K–T conditions at the solution points and show gradients of cost and constraint functions on the graph.

3.60 Two electrical generators are interconnected to provide total power to meet the load. Each generator's cost is a function of the power output, as shown in Fig. E3.60. All costs and power are expressed on per unit basis. The total power need is at least 60 units. Formulate a minimum cost design problem and solve it graphically (same as Exercise 2.14). Verify K–T conditions at the solution points and show gradients of cost and constraint functions on the graph.

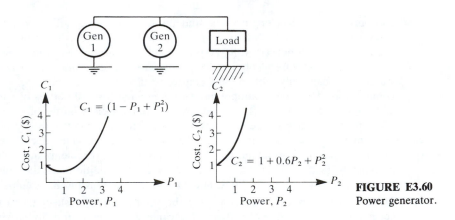

FIGURE E3.60 Power generator.

Section 3.5 Global Optimality

3.61 *Answer True or False*

1. A linear inequality constraint always defines a convex feasible region.
2. A linear equality constraint always defines a convex feasible region.
3. A nonlinear equality constraint cannot give a convex feasible region.
4. A function is convex if and only if its Hessian is positive definite everywhere.
5. An optimum design problem is convex if all constraints are linear and cost function is convex.
6. A convex programming problem always has an optimum solution.
7. An optimum solution for a convex programming problem is always unique.
8. A nonconvex programming problem cannot have global optimum solution.

9. For a convex design problem, Hessian of the cost function must be positive semidefinite everywhere.

10. Check for convexity of a function can actually identify a domain over which the function may be convex.

3.62 Using the definition of a line segment given in Eq. (3.58), show that the following set is convex

$$S = \{\mathbf{x} \mid x_1^2 + x_2^2 - 1.0 \leq 0\}$$

3.63 Find the domain for which the following functions are convex: (i) $\sin x$, (ii) $\cos x$.

Check for convexity of the following functions. If the function is not convex everywhere, then determine the domain (set S) over which the function is convex:

3.64 $f(x_1, x_2) = 3x_1^2 + 2x_1x_2 + 2x_2^2 + 7$

3.65 $f(x_1, x_2) = x_1^2 + 4x_1x_2 + x_2^2 + 3$

3.66 $f(x_1, x_2) = x_1^3 + 12x_1x_2^2 + 2x_2^2 + 5x_1^2 + 3x_2$

3.67 $f(x_1, x_2) = 5x_1 - \frac{1}{16}x_1^2x_2^2 + \dfrac{1}{4x_1} x_2^2$

3.68 $f(x_1, x_2) = x_1^2 + x_1x_2 + x_2^2$

3.69 Check convexity of the problem (Exercise 3.45):

minimize $f(D, H) = \pi D H + \dfrac{\pi}{2} D^2$

subject to $\dfrac{\pi}{4} D^2 H \geq 400$

$$3.5 \leq D \leq 8.0$$
$$8.0 \leq H \leq 18.0$$

3.70 Check convexity of the function (Exercise 3.28):

$$U(V, C) = \frac{(21.9\text{E}+07)}{(V^2 C)} + (3.9\text{E}+06)C + (1.0\text{E}+03)V$$

Formulate and check convexity for the following problems:

3.71 Exercise 3.49

3.72 Exercise 3.51

3.73 Exercise 3.52

3.74 Exercise 3.53

3.75 Exercise 3.57

3.76 Exercise 3.58

3.77 Exercise 3.59

3.78 Exercise 3.60

3.79 Minimum weight design of the symmetric three-bar truss of Fig. 2.5 is defined as follows:

minimize $f(x_1, x_2) = 2\sqrt{2}x_1 + x_2$

subject to the constraints

$$g_1 \equiv \frac{1}{\sqrt{2}} \left[\frac{P_u}{x_1} + \frac{P_v}{(x_1 + \sqrt{2} x_2)} \right] - 20\,000 \leq 0$$

$$g_2 \equiv \frac{\sqrt{2} P_v}{(x_1 + \sqrt{2} x_2)} - 20\,000 \leq 0$$

$$g_3 \equiv -x_1 \leq 0$$

$$g_4 \equiv -x_2 \leq 0$$

where x_1 is the cross-sectional area of members 1 and 3 (symmetric structure) and x_2 is the cross-sectional area of member 2, $P_u = P \cos \theta$, $P_v = P \sin \theta$ with $P > 0$ and $0 \leq \theta \leq 90$. Check for convexity of the problem for $\theta = 60°$.

3.80 For the three-bar truss problem of Exercise 3.79, consider the case of K–T conditions with g_1 as the only active constraint. Solve the conditions for optimum solution and determine the range for the load angle θ for which the solution is valid.

3.81 For the three-bar truss problem of Exercise 3.79, consider the case of K–T conditions with only g_1 and g_2 as active constraints. Solve the conditions for optimum solution and determine range for the load angle θ for which the solution is valid.

3.82 For the three-bar truss problem of Exercise 3.79, consider the case of K–T conditions with g_2 as the only active constraint. Solve the conditions for optimum solution and determine the range for the load angle θ for which the solution is valid.

3.83 For the three-bar truss problem of Exercise 3.79, consider the case of K–T conditions with g_1 and g_4 as active constraints. Solve the conditions for optimum solution and determine the range for the load angle θ for which the solution is valid.

Section 3.6 Second-order Conditions for Constrained Optimization

3.84 *Answer True or False*

1. A convex programming problem always has a unique global minimum point.
2. For a convex programming problem, K–T necessary conditions are also sufficient.
3. Hessian of the Lagrange function must be positive definite at constrained minimum points.
4. For a constrained problem, if the sufficiency condition of Theorem 3.12 is violated, the candidate point x^* may still be a minimum point.
5. If Hessian of the Lagrange function at x^*, $\nabla^2 L(x^*)$, is positive definite, the optimum design problem is convex.
6. For a constrained problem, the sufficient condition at x^* is satisfied if there are no feasible directions in a neighborhood of x^* along which the cost function reduces.

Solve the following problems graphically. Check necessary and sufficient conditions for candidate local minimum points and verify them on the graph for the problem:

3.85 Minimize $f(x_1, x_2) = 4x_1^2 + 3x_2^2 - 5x_1x_2 - 8x_1$
subject to $x_1 + x_2 = 4$

3.86 Minimize $f(x_1, x_2) = 9x_1^2 + 18x_1x_2 + 13x_2^2 - 4$
subject to $x_1^2 + x_2^2 + 2x_1 = 16$

3.87 Minimize $f(x_1, x_2) = 2x_1 + 3x_2 - x_1^3 - 2x_2^2$
subject to $x_1 + 3x_2 \leqq 6$
$5x_1 + 2x_2 \leqq 10$
$x_1, x_2 \geqq 0$

3.88 Minimize $f(x_1, x_2) = 4x_1^2 + 3x_2^2 - 5x_1x_2 - 8x_1$
subject to $x_1 + x_2 \leqq 4$

3.89 Minimize $f(x_1, x_2) = 9x_1^2 - 18x_1x_2 + 13x_2^2 - 4$
subject to $x_1^2 + x_2^2 + 2x_1 \geqq 16$

3.90 Minimize $f(x_1, x_2) = (x_1 - 3)^2 + (x_2 - 3)^2$
subject to $x_1 + x_2 \leqq 4$
$x_1 - 3x_2 = 1$

3.91 Minimize $f(x_1, x_2) = x_1^3 - 16x_1 + 2x_2 - 3x_2^2$
subject to $x_1 + x_2 \leqq 3$

3.92 Minimize $f(x_1, x_2) = 3x_1^2 - 2x_1x_2 + 5x_2^2 + 8x_2$
subject to $x_1^2 - x_2^2 + 8x_2 \leqq 16$

3.93 A cantilever beam is subjected to the point load P (kN), as shown in Fig. E3.93. Maximum bending moment in the beam is Pl (kN · m) and maximum shear is P (kN). Formulate the minimum mass design problem and solve it graphically using hollow circular cross-section. Check the necessary and sufficient conditions at the optimum point.

    The material should not fail under bending stress or shear stress. The maximum bending stress is calculated as

$$\sigma = \frac{PlR_o}{I}$$

where I = moment of inertia of the cross-section. The maximum shearing stress is calculated as

$$\tau = \frac{P}{3I}(R_0^2 + R_0R_i + R_i^2)$$

The data for the problem are $P = 10$ kN; $l = 5$ m; modulus of elasticity, $E = 210$ GPa; allowable bending stress, $\sigma_a = 250$ MPa; allowable shear stress, $\tau_a = 90$ MPa; and mass density, $\rho = 7850$ kg/m³.

$$0 \leqq R_0 \leqq 20 \text{ cm}; \qquad 0 \leqq R_i \leqq 20 \text{ cm}$$

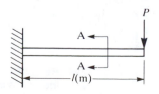

Section A–A

FIGURE E3.93
Cantilever beam.

3.94 Design a hollow circular beam-column shown in Fig. E3.94 for two conditions: when $P = 50$ (kN), the axial stress σ should be less than σ_a, and when $P = 0$, deflection δ due to self-weight should satisfy $\delta \leq 0.001l$. Limits for dimensions are, $t = 0.10–1.0$ cm, $R = 2.0–20.0$ cm and $R/t \geq 20$. Formulate the minimum weight design problem and solve it graphically. Check the necessary and sufficient conditions for the solution points and verify them on the graph. The data for the problem are $\delta = \dfrac{5wl^4}{384EI}$; $w = $ self weight force/length (N/m); $\sigma_a = 250$ MPa; modulus of elasticity, $E = 210$ GPa; mass density, $\rho = 7800$ kg/m^3; and $\sigma = P/A$; gravitational constant, $g = 9.80$ m/s^2.

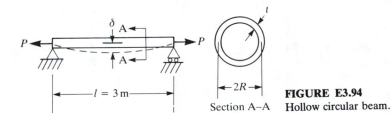

$l = 3$ m	$\leftarrow 2R \rightarrow$
	Section A–A

FIGURE E3.94
Hollow circular beam.

3.95 In the design of a closed-end, thin-walled cylindrical pressure vessel shown in Fig. E3.95, the design objective is to select the mean radius R and wall thickness t to minimize the total mass. The vessel should contain at least 25 m^3 of gas at a pressure of 3.5 MPa. It is required that the circumferential stress in the pressure vessel not exceed 210 MPa and the circumferential strain not exceed (1.0E$-$03). The circumferential stress and strain are calculated from the equations

$$\sigma_c = \frac{PR}{t}, \qquad \varepsilon_c = \frac{PR(2 - v)}{2Et}$$

where $\rho = $ mass density (7850 kg/m^3), $\sigma_c = $ circumferential stress (Pa), $\varepsilon_c = $ circumferential strain, $P = $ internal pressure (Pa), $E = $ Young's modulus (210 GPa) and $v = $ Poisson's ratio (0.3).

1. Formulate the optimum design problem and solve the problem graphically.
2. Check necessary and sufficient conditions for the solutions points and verify them graphically.

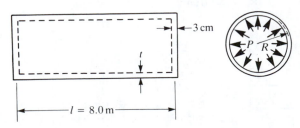

FIGURE E3.95
Cylindrical pressure vessel.

Find optimum solutions for the following problems graphically. Check necessary and sufficient conditions for the solution points and verify them on the graph for the problem:

3.96 Exercise 3.46
3.97 Exercise 3.47
3.98 Exercise 3.45
3.99 Exercise 3.49
3.100* Exercise 2.48
3.101* Exercise 2.49
3.102* Exercise 2.50
3.103* Exercise 2.65

Section 3.7   Physical Meaning of Lagrange Multipliers

Solve the following problems graphically, verify necessary and sufficient conditions for the solution points and study the effect on the cost function of variations in the right-hand side of constraints:

3.104 Exercise 3.85
3.105 Exercise 3.86
3.106 Exercise 3.87
3.107 Exercise 3.88
3.108 Exercise 3.89
3.109 Exercise 3.90
3.110 Exercise 3.91
3.111 Exercise 3.92

Section 3.8   Engineering Design Examples

3.112 *Answer True or False*

1. Candidate minimum points for a constrained problem that do not satisfy second order sufficiency conditions can be global minimum designs.
2. Lagrange multipliers may be used to calculate the sensitivity coefficient for the cost function with respect to the right-hand side parameters even if Theorem 3.14 cannot be used.
3. Relative magnitudes of the Lagrange multipliers provide useful information for practical design problems.

3.113 A circular tank that is closed at both ends is to be fabricated to have a volume of 250π m^3. The fabrication cost is found to be proportional to the surface area of the sheet metal needed for fabrication of the tank and is \$400/m^2. The tank is to be housed in a shed with a sloping roof which limits the height of the tank by the relation $H \leqq 8D$, where H is the height and D is the diameter of the tank. The problem is formulated as minimize

$$f = 400(0.5\pi D^2 + \pi DH)$$

subject to the constraints

$$\frac{\pi}{4} D^2 H = 250\pi, \quad \text{and } H \leq 8D$$

Ignore any other constraints.

1. Check for convexity of the problem.
2. Write Kuhn–Tucker necessary conditions.
3. Solve Kuhn–Tucker necessary conditions for local minimum points. Check sufficient conditions and verify the conditions graphically.
4. What will be the change in cost if volume requirement is changed to 255π m^3 in place of 250π m^3?

3.114 A symmetric (area of member 1 is the same as area of member 3) three-bar truss is to be designed to support a load P as shown in Fig. 2.5. The following notation is used:

$S_1 = P \cos \theta$ (N)
$S_2 = P \sin \theta$ (N)
$A_1 = $ cross-sectional area of members 1 and 3 (m^2)
$A_2 = $ cross-sectional area of member 2 (m^2)
$\sigma_1 = $ stress in Member 1 (Pa)
$\sigma_2 = $ stress in Member 2 (Pa)
$\sigma_a = $ allowable stress (Pa)
$\rho = $ material mass density

Stresses in members are calculated as

$$\sigma_1 = \frac{S_1}{\sqrt{2}A_1} + \frac{S_2}{\sqrt{2}(A_1 + \sqrt{2}A_2)}$$

$$\sigma_2 = \frac{\sqrt{2}S_2}{A_1 + \sqrt{2}A_2}$$

1. Formulate the minimum mass design problem treating A_1 and A_2 as design variables.
2. Check for convexity of the problem.
3. Write Kuhn–Tucker necessary conditions for the problem.
4. Solve the optimum design problem using the data: $P = 50$ kN, $\theta = 30°$, $\rho = 7800$ kg/m^3, $\sigma_a = 150$ MPa. Verify the solution graphically and interpret the necessary conditions on the graph for the problem.
5. What will be the effect on the cost function if σ_a is increased to 152 MPa?

Formulate and solve the following problems graphically; check necessary and sufficient conditions at the solution points, verify the conditions on the graph for the problem and study the effect of variations in constraint limits on the cost function:

3.115 Exercise 3.49	**3.116** Exercise 3.51
3.117 Exercise 3.52	**3.118** Exercise 3.53
3.119 Exercise 3.57	**3.120** Exercise 3.58
3.121 Exercise 3.59	**3.122** Exercise 3.60
3.123 Exercise 3.93	**3.124** Exercise 3.94
3.125 Exercise 3.46	**3.126** Exercise 3.97
3.127 Exercise 3.45	**3.128** Exercise 3.95
3.129* Exercise 2.48	**3.130*** Exercise 2.49
3.131* Exercise 2.50	**3.132*** Exercise 2.53
3.133* Exercise 2.54	**3.134*** Exercise 2.55
3.135* Exercise 2.57	**3.136*** Exercise 2.58
3.137* Exercise 2.59	**3.138*** Exercise 2.60
3.139* Exercise 2.61	**3.140*** Exercise 2.62
3.141* Exercise 2.63	**3.142*** Exercise 2.64
3.143* Exercise 2.65	

CHAPTER
4

LINEAR PROGRAMMING METHODS FOR OPTIMUM DESIGN

4.1 INTRODUCTION

An optimum design problem having linear cost and constraint functions in the design variables is called a *linear programming problem*. We shall use the abbreviation LP for linear programming problems, or simply for linear programs. LP problems arise in many fields of engineering such as water resources, systems engineering, traffic flow control, the management of resources, transportation engineering, and electrical engineering. In areas of aerospace, automotive, structural, or mechanical system design, most problems are not linear. However, structural design problems utilizing plastic or limit analysis concepts can be formulated as linear programming problems. In addition, one way of solving nonlinear programming problems is to transform them to a sequence of linear programs (Chapter 6). Many other nonlinear programming methods also solve a linear programming problem at each iteration. Thus, linear programming methods are useful in many fields and must be clearly understood. This chapter describes the basic theory and concepts for solving such problems.

In Section 2.7, a general mathematical model for optimum design was defined to find a design variable vector **x** to minimize a cost function

$$f(\mathbf{x}) = f(x_1, x_2, \ldots, x_n) \tag{4.1}$$

subject to the equality constraints

$$h_j(\mathbf{x}) \equiv h_j(x_1, \ldots, x_n) = 0; \qquad j = 1 \text{ to } p \tag{4.2}$$

and inequality constraints

$$g_i(\mathbf{x}) \equiv g_i(x_1, \ldots, x_n) \leqq 0; \qquad i = 1 \text{ to } m \tag{4.3}$$

In Chapter 3, a general theory of optimum design to treat the foregoing model was described. The theory can also be used to solve LP problems. However, more efficient and elegant numerical methods are available to solve the LP problem directly. Since there are numerous LP problems in the real world, it is worthwhile to discuss special methods for them in detail.

Any linear function $f(\mathbf{x})$ of n variables has the following form:

$$f(\mathbf{x}) = c_1 x_1 + c_2 x_2 + \ldots + c_n x_n \equiv \sum_{i=1}^{n} c_i x_i = \mathbf{c}^T \mathbf{x} \tag{4.4}$$

where c_i, $i = 1$ to n are constants. All functions of an LP problem can be represented in the preceding form. Therefore, the general optimization model of Eqs (4.1) to (4.3) is replaced by a linear model in this chapter and a standard form for the model is defined. The two-phase Simplex algorithm to solve LP problems is developed and illustrated with simple numerical examples. The theory of dual linear programming and post-optimal analysis are also discussed.

It is noted here that the subject of linear programming is well developed and several excellent full-length textbooks and journal articles are available on the subject [Hadley, 1961; Ackoff and Sasieni, 1968; Randolph and Meeks, 1978]. These references may be consulted for in-depth treatment and proofs of the results used in this chapter.

Details of the Simplex method are described to show numerical steps needed to solve LP problems. Before attempting to implement the algorithm into computer codes, the existence of standard packages for solving LP problems must be checked. Most computer center libraries have at least one software package to treat such problems, e.g. LINDO [Schrage, 1981]. It is more economical to use the available software rather than developing a new one.

The following is an outline of this chapter.

Section 4.2 Definition of a Standard Linear Programming (LP) Problem. A standard form for the LP problem requiring minimization of a cost subject to the equality constraints and nonnegativity of design variables is defined. Any other LP problem can be transformed into the standard form. Such transformations are explained with examples.

Section 4.3 Properties of Linear Programming Problems. Some fundamental properties of LP problems are discussed. Several terms used later in the chapter are defined. It is shown that optimum solution for an LP problem always lies on the boundary of the feasible region (constraint set). In addition, it is at least at one of the vertices of the convex feasible region (convex polyhedral constraint set). Some LP theorems are stated and their significance is discussed. Geometrical meaning of the optimum solution is given.

Section 4.4 The Simplex Method. Basics of the Simplex method for solving LP problems are described. Ideas of a canonical form, pivot step, pivot row, pivot column and pivot element are introduced. The Simplex tableau is introduced and its notation is explained. The method is described as an extension of the standard Gauss–Jordan elimination process for solving a system of linear equations $\mathbf{Ax} = \mathbf{b}$ where $\mathbf{A}$ is an $m \times n$ ($m < n$) matrix, $\mathbf{x}$ is an n-vector and $\mathbf{b}$ is an m-vector. The method is developed and illustrated for "$\leqq$ type" constraints.

Section 4.5 Initial Basic Feasible Solution – Artificial Variables. The basic Simplex method of Section 4.4 is extended to handle "$\geqq$ type" and equality constraints. A basic feasible solution is needed to initiate the solution process. Such a solution is immediately available if only "$\leqq$ type" constraints are present. However, for the "$\geqq$ type" and equality constraints we must introduce artificial variables, define an auxiliary minimization LP problem and solve it. The standard Simplex method can be used to solve the auxiliary problem. This is called Phase I of the Simplex procedure. At the end of Phase I, a basic feasible solution for the original problem is known. Phase II then continues to find a solution to the LP problem.

An alternate form of the Simplex method that does not require two-phase procedure is also described. It modifies the original cost function by adding a penalty term and is sometimes called the *Big-M Method.*

It is noted that the Simplex method gives a complete answer to an LP problem. It tells us (1) whether a solution to the problem exists (infeasible or feasible problem), (2) whether the problem is unbounded, (3) if a solution exists, the method finds it, and (4) whether there are multiple solutions to the problem. The method is illustrated with examples.

Section 4.6 Post-optimality Analysis. In engineering design problems, there is always some uncertainty about parameters of the problem. Therefore, it is useful to study sensitivity of the solution to changes in parameters. This type of analysis is the topic of Section 4.6. *Lagrange multipliers* for the constraints can be recovered from the final LP tableau. The constraint variation sensitivity Theorem 3.14 of Chapter 3 shows that the Lagrange multipliers are implicit derivatives of the cost function with respect to the right-hand side parameters of the constraints (resource limits). The theorem can be used to study the effect on cost function as the resource limits change. Methods are also described to find ranges for the right-hand side parameters, cost coefficients and the coefficient matrix. The final tableau can be used to find all the ranges. If variations in the parameters remain within the determined ranges, then the information at the known solution can be used to find the new solution. Otherwise, the problem has to be re-solved. The methods are illustrated with examples.

Section 4.7 Duality in Linear Programming. Associated with every LP problem is another problem called the *dual.* The original LP is called the primal problem. Theory of duality plays an important role in the sensitivity analysis. Some theorems related to dual and primal problems are stated and explained. Dual variables are related to Lagrange multipliers of the primal constraints. Solution of the dual problem can be recovered from the final primal solution. Or, solution of the primal problem can be recovered from the final dual solution. Therefore, only one of the two problems needs to be solved. This is illustrated with examples.

4.2 DEFINITION OF A STANDARD LINEAR PROGRAMMING PROBLEM

Linear programming problems may have equality as well as inequality constraints. Also, many problems require maximization of a function whereas others require minimization. The standard LP problem can be defined in several equivalent ways and the same numerical procedure can be adapted to treat different formulations. In this text, *we formulate LP problems as minimization of a function with equality constraints and nonnegativity of design variables*. This form is not as restrictive as it may appear because all other LP problems can be readily transcribed into it. We shall explain the process of transcribing a given LP problem into the standard form.

4.2.1 Linear Constraints

The ith linear inequality or equality constraint involving k design variables, y_j, $j = 1$ to k is written in one of the following three forms:

$$a_{i1} y_1 + a_{i2} y_2 + \ldots + a_{ik} y_k \left\{ \lesseqgtr \right\} b_i \tag{4.5}$$

where a_{ij} and b_i are known constants. Also b_i, called the *resource limits*, are assumed to be always nonnegative, i.e. $b_i \geqq 0$. b_i's can always be made nonnegative by multiplying both sides of Eq. (4.5) by -1 if necessary. Note, however, that multiplication by -1 changes the sense of the original inequality, i.e. "$\leqq$ type" becomes "$\geqq$ type" and vice versa. For example, a constraint $y_1 + 2y_2 \leqq -2$ must be transformed as $-y_1 - 2y_2 \geqq 2$ to have a positive right-hand side.

Since only equality constraints are treated in the standard LP, the inequalities in Eq. (4.5) must be converted to equalities. This is no real restriction since any inequality can be converted into equality by introducing a nonnegative *slack or surplus variable* as explained in the following paragraphs. Note also that since b_i's are required to be nonnegative in Eq. (4.5), it is not always possible to convert "$\geqq$" inequalities to the "$\leqq$ form" and keep $b_i \geqq 0$. In Chapters 2 and 3 this was done where a standard optimization problem was defined with only "$\leqq$ type" constraints. However, in this chapter, we will have to explicitly treat "$\geqq$ type" linear inequalities. It will be seen later that "$\geqq$ type" constraints do require a special treatment in LP methods.

For the ith "$\leqq$ type" constraint, we introduce a nonnegative *slack variable* $s_i \geqq 0$ and convert it to an equality as

$$a_{i1} y_1 + a_{i2} y_2 + \ldots + a_{ik} y_k + s_i = b_i \qquad (4.6)$$

We also introduced the idea of slack variables in Chapter 3. There s_i^2 was used as a slack variable instead of s_i. That was done to avoid the additional constraint $s_i \geqq 0$. However, in LP problems we cannot use s_i^2 as a slack variable because it makes the problem nonlinear. Therefore, we will use s_i as a slack variable along with the additional constraint $s_i \geqq 0$. For example, a constraint $2y_1 - y_2 \leqq 4$ will be transformed as $2y_1 - y_2 + s_1 = 4$ with $s_1 \geqq 0$ as its slack variable.

Similarly, the ith "$\geqq$ type" constraint is converted to an equality by subtracting a nonnegative *surplus variable* $s_i \geqq 0$, as

$$a_{i1} y_1 + a_{i2} y_2 + \ldots + a_{ik} y_k - s_i = b_i \qquad (4.7)$$

The idea of a surplus variable is very similar to the slack variable. For the "$\geqq$ type" constraint the left-hand side has to be always greater than or equal to the right-hand side, so we must subtract a nonnegative variable to transform it to an equality. For example, a constraint $-y_1 + 2y_2 \geqq 2$ will be transformed as $-y_1 + 2y_2 - s_1 = 2$ with $s_1 \geqq 0$ as its surplus variable.

Note that slack and surplus variables are *additional unknowns* that must be determined as a part of the solution.

4.2.2 Unrestricted Variables

In addition to the equality constraints, we require all design variables to be nonnegative in the standard LP problem, i.e. $y_i \geqq 0$, $i = 1$ to k. Engineering design variables usually represent some physical quantities such as member cross-sectional area, wire diameter, and material thickness. Therefore, it is reasonable to assume nonnegative values for them. *If a design variable y_j is unrestricted in sign, it can always be written as the difference of two nonnegative variables*, as $y_j = y_j^+ - y_j^-$, with $y_j^+ \geqq 0$ and $y_j^- \geqq 0$. This decomposition is substituted in all equations and y_j^+ and y_j^- are treated as unknowns in the problem. At the optimum, if $y_j^+ \geqq y_j^-$, then y_j is nonnegative, and if $y_j^+ \leqq y_j^-$, then y_j is nonpositive. This treatment for each free variable increases the dimension of the design variable vector by 1.

4.2.3 Standard LP Definition

For notational clarity, let **x** represent an n-vector consisting of k original design variables and $(n - k)$ slack, surplus, or other variables. Now let us define the standard LP problem as find an n-vector **x** to minimize a linear cost function

$$f = c_1 x_1 + c_2 x_2 + \ldots + c_n x_n \qquad (4.8)$$

subject to the equality constraints

$$a_{11}x_1 + a_{12}x_2 + \ldots + a_{1n}x_n = b_1$$
$$a_{21}x_1 + a_{22}x_2 + \ldots + a_{2n}x_n = b_2 \tag{4.9}$$
$$\cdot \quad \cdot \quad \cdots \quad \cdot \quad \cdot$$
$$\cdot \quad \cdot \quad \cdots \quad \cdot \quad \cdot$$
$$a_{m1}x_1 + a_{m2}x_2 + \ldots + a_{mn}x_n = b_m$$

and nonnegativity of design variables

$$x_j \geqq 0; \quad j = 1 \text{ to } n \tag{4.10}$$

The quantities $b_i \geqq 0$, c_j and a_{ij} ($i = 1$ to m and $j = 1$ to n) are assumed to be known constants and m and n are positive integers. Note that only b_i are required to be nonnegative.

The standard LP problem can also be written in the summation notation as

$$\text{minimize } f = \sum_{i=1}^{n} c_i x_i \tag{4.11}$$

subject to the constraints

$$\sum_{j=1}^{n} a_{ij}x_j = b_i; \quad i = 1 \text{ to } m \tag{4.12}$$
$$x_j \geqq 0; \quad j = 1 \text{ to } n$$

Matrix notation may also be used to define the LP problem as

$$\text{minimize } f = \mathbf{c}^T\mathbf{x} \tag{4.13}$$

subject to the constraints

$$\mathbf{A}\mathbf{x} = \mathbf{b} \tag{4.14}$$
$$\mathbf{x} \geqq \mathbf{0} \tag{4.15}$$

where $\mathbf{A} = [a_{ij}]$ is an $m \times n$ matrix, $\mathbf{c}$ and $\mathbf{x}$ are n-vectors, and $\mathbf{b}$ is an m-vector.

The formulations given in Eqs (4.8) to (4.15) are more general than what may appear at first sight, because all LP problems can be transcribed into them. Conversion of "$\leqq$ type" and "$\geqq$ type" inequalities to equalities using slack and surplus variables has previously been explained. Unrestricted variables can be decomposed into the difference of two nonnegative variables. *Maximization of functions* can also be routinely treated. For example, if the objective is to maximize a function (rather than minimize), we simply minimize its negative. Maximization of a function $z = (d_1x_1 + d_2x_2 + \ldots + d_nx_n)$ is equivalent to minimization of its negative, $f = -(d_1x_1 + d_2x_2 + \ldots + d_nx_n)$. *Note that a function that is to be maximized is denoted as z in this chapter.* It is henceforth assumed that the LP problem has been converted into the standard form defined in Eqs (4.8) to (4.15). Recall that the right-hand sides (resource limits) b_i of all equations in the standard LP form must be nonnegative.

Example 4.1 Conversion to standard LP form. Convert the following problem into the standard LP form:

maximize
$$z = 2y_1 + 5y_2$$

subject to
$$3y_1 + 2y_2 \leq 12$$

$$2y_1 + 3y_2 \geq 6$$

$$y_1 \geq 0, \ y_2 \text{ is unrestricted in sign}$$

Solution. To transform the problem into the standard LP form, we take the following steps:

1. Since y_2 is unrestricted in sign, we split it into its positive and negative parts as $y_2 = y_2^+ - y_2^-$ with $y_2^+ \geq 0$, $y_2^- \geq 0$.
2. Substituting this new definition of y_2 into the problem, we get

maximize
$$z = 2y_1 + 5(y_2^+ - y_2^-)$$

subject to
$$3y_1 + 2(y_2^+ - y_2^-) \leq 12$$

$$2y_1 + 3(y_2^+ - y_2^-) \geq 6$$

$$y_1, y_2^+, y_2^- \geq 0$$

3. The right-hand side of both the constraints are nonnegative, so they conform to the standard form; and there is no need to further modify them.
4. Converting to a minimization problem subject to equality constraints, we get the problem in the standard form as

minimize
$$f = -2y_1 - 5(y_2^+ - y_2^-)$$

subject to
$$3y_1 + 2(y_2^+ - y_2^-) + s_1 = 12$$
$$2y_1 + 3(y_2^+ - y_2^-) - s_2 = 6$$
$$y_1, y_2^+, y_2^-, s_1, s_2 \geq 0$$

where $s_1 =$ slack variable for the first constraint and $s_2 =$ surplus variable for the second constraint.

5. We can redefine the solution variables as

$$x_1 = y_1, \ x_2 = y_2^+, \ x_3 = y_2^-, \ x_4 = s_1, \ x_5 = s_2$$

and rewrite the problem in the standard form as

minimize
$$f = -2x_1 - 5x_2 + 5x_3$$

subject to

$$3x_1 + 2x_2 - 2x_3 + x_4 = 12$$

$$2x_1 + 3x_2 - 3x_3 - x_5 = 6$$

$$x_i \geqq 0, \quad i = 1 \text{ to } 5$$

Comparing the preceding equations with Eqs (4.13) to (4.15), we can define the following quantities:

$m = 2$ (the number of equations)

$n = 5$ (the number of variables)

$\mathbf{x} = [x_1 \ x_2 \ x_3 \ x_4 \ x_5]^T$

$\mathbf{c} = [-2 \ -5 \ 5 \ 0 \ 0]^T$

$\mathbf{b} = [12 \ 6]^T$

$$\mathbf{A} = [a_{ij}]_{2 \times 5} = \begin{bmatrix} 3 & 2 & -2 & 1 & 0 \\ 2 & 3 & -3 & 0 & -1 \end{bmatrix}$$

‖

4.3 BASIC CONCEPTS RELATED TO LINEAR PROGRAMMING PROBLEMS

There are several concepts and terms related to linear programming problems. Since they are used throughout this chapter, we define and explain them in this section. It will be prudent to clearly understand and remember them for later use.

4.3.1 Basic Concepts

Since all functions are linear in an LP problem, the feasible region (constraint set) defined by linear equalities or inequalities is *convex* (Section 3.5). Also, the cost function is linear, so it is convex. Therefore, the LP problem is convex, and if an *optimum solution* exists, it is *global* according to Theorem 3.9.

Note also that even when there are inequality constraints in an LP design problem, the *solution always lies on the boundary of the feasible region if it exists*; i.e. some constraints are always active at the optimum. This can be seen by writing necessary conditions of Theorem 3.4 for an unconstrained optimum. These conditions, $\partial f / \partial x_i = 0$ when used for the cost function of Eq. (4.8), give $c_i = 0$ for $i = 1$ to n. This is not possible, as all c_i's are not zero (if all c_i's were zero, there would be no cost function). Therefore, by contradiction, *optimum solution for any LP problem must lie on the boundary of the feasible region*. This is in contrast to the general nonlinear problems where the optimum can be inside or on boundary of the feasible region.

An optimum solution of the LP problem must also satisfy equality constraints in Eqs (4.9), only then the solution can be feasible. Therefore, to have a meaningful optimum design problem, Eqs (4.9) should have more than one solution. We seek a feasible solution among them that has the least cost. To have many solutions the number of linearly independent equations in Eqs (4.9) must be less than n – the number of variables in the LP problem. (Refer to Section B.5 in Appendix B for further discussion on a general solution of m equations in n unknowns.) It is assumed in the following discussion that all the m rows of the matrix $\mathbf{A}$ in Eqs (4.14) (or Eqs (4.9)) are linearly independent and that $m < n$. This means that there are no redundant equations. Therefore, Eqs (4.9) have infinite solutions and we seek a feasible solution that also minimizes the cost function. A method for the solution of simultaneous equations (4.9) based on the *Gaussian elimination* is described in Appendix B. The Simplex method of LP described later in the chapter uses steps of the Gaussian elimination procedure. Therefore, that procedure must be thoroughly reviewed before studying the Simplex method.

We shall use the following example to illustrate the preceding ideas. It shall also be used later to introduce LP terminology and basic steps of the Simplex method.

Example 4.2 Profit maximization problem – characterization of solution for LP problems. As an example of solving constraint equations, we consider the profit maximization problem solved graphically in Section 2.8.1. The problem is to find x_1 and x_2 to minimize

$$f = -400x_1 - 600x_2 \tag{a}$$

subject to

$$x_1 + x_2 \leqq 16 \tag{b}$$

$$\tfrac{1}{28}x_1 + \tfrac{1}{14}x_2 \leqq 1 \tag{c}$$

$$\tfrac{1}{14}x_1 + \tfrac{1}{24}x_2 \leqq 1 \tag{d}$$

$$x_1, x_2 \geqq 0 \tag{e}$$

Solution. Graphical solution for the problem is given in Fig. 4.1. All constraints of Eqs (b) to (e) are plotted and some iso-cost lines are shown. Each point of the region bounded by the polygon ABCDE satisfies all the constraints of Eqs (b) to (d) and the nonnegativity conditions of Eq. (e). It is seen from Fig. 4.1 that the vertex D gives the optimum solution.

Introducing slack variables for constraints of Eqs (b) to (d) and writing the problem in the standard LP form, we have

minimize

$$f = -400x_1 - 600x_2$$

subject to

$$x_1 + x_2 + x_3 = 16$$

$$\tfrac{1}{28}x_1 + \tfrac{1}{14}x_2 + x_4 = 1$$

$$\tfrac{1}{14}x_1 + \tfrac{1}{24}x_2 + x_5 = 1$$

$$x_i \geqq 0, \qquad i = 1 \text{ to } 5$$

$$\tag{f}$$

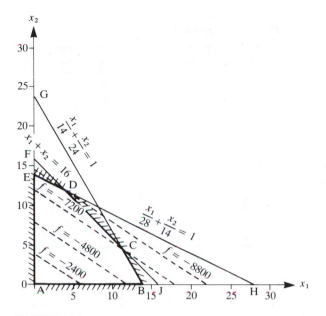

FIGURE 4.1
Graphical solution for profit maximization LP problem. Optimum point = (4, 12). Optimum
Cost = −8800.

where x_3, x_4 and x_5 are slack variables for the first, second and third constraints,
respectively.

Note that all three equations in Eqs (f) are linearly independent. Since the
number of variables (5) exceeds the number of constraint equations (3), a unique
solution cannot exist for Eqs (f) (see Appendix B). Actually there are infinite
solutions. To see this, we write a general solution for the equations by
transferring the terms associated with the variables x_1 and x_2 to the right-hand
side of Eqs (f) as

$$x_3 = 16 - x_1 - x_2$$

$$x_4 = 1 - \tfrac{1}{28}x_1 - \tfrac{1}{14}x_2 \tag{g}$$

$$x_5 = 1 - \tfrac{1}{14}x_1 - \tfrac{1}{24}x_2$$

In these equations, x_1 and x_2 act as independent variables which can be given any
value, and x_3, x_4 and x_5 are dependent on them. Different values for x_1 and x_2
generate different values for x_3, x_4 and x_5.

A solution of particular interest in LP problems is obtained by setting p of
the variables to zero and solving for the rest of them, where p is the difference
between the number of variables (n) and the number of constraint equations (m),
i.e. $p = n - m$ (e.g. $p = 2$ in the case of Eqs (f)). With two variables set to zero, a
unique solution of Eqs (f) exists for the remaining three variables. *A solution
obtained by setting p variables to zero is called the basic solution.* For example, a
basic solution is obtained from Eqs (f) or (g) by setting $x_1 = 0$ and $x_2 = 0$, as

TABLE 4.1
Ten basic solutions for the profit maximization problem

No.	x_1	x_2	x_3	x_4	x_5	f	Location in Fig. 4.1
1	0	0	16	1	1	0	A
2	0	14	2	0	$\frac{5}{12}$	-8400	E
3	0	16	0	$-\frac{2}{7}$	$\frac{1}{3}$	—	infeasible
4	0	24	-8	$-\frac{5}{7}$	0	—	infeasible
5	16	0	0	$\frac{3}{7}$	$-\frac{2}{7}$	—	infeasible
6	14	0	2	$\frac{1}{2}$	0	-5600	B
7	28	0	-12	0	-1	—	infeasible
8	4	12	0	0	$\frac{3}{14}$	-8800	D
9	11.2	4.8	0	$\frac{1}{5}$	0	-7360	C
10	$\frac{130}{17}$	$\frac{168}{17}$	$-\frac{26}{17}$	0	0	—	infeasible

$x_3 = 16$, $x_4 = 1$, $x_5 = 1$. Another basic solution is obtained by setting $x_1 = 0$ and $x_3 = 0$, as $x_2 = 16$, $x_4 = -\frac{2}{7}$ and $x_5 = \frac{1}{3}$.

Table 4.1 shows 10 basic solutions for the present example obtained by using the procedure given in the preceding paragraph. Note that of the 10 solutions, exactly 5 (Nos 1, 2, 6, 8 and 9) correspond to the vertices of the polygon of Fig. 4.1, and the remaining 5 violate the nonnegativity condition. Therefore, only 5 of the 10 basic solutions are feasible. By moving the iso-cost line parallel to itself, it is easily seen that the optimum solution is at point D. Note that the optimum point is at one of the vertices of the *constraint polygon*. This will be observed later as a general property of any LP problem. That is, *if an LP has a solution, it is at least at one of the vertices of the feasible region.* ‖

4.3.2 LP Terminology

We shall now introduce various definitions and terms related to the LP problem. The definitions of convex sets, convex function and the line segment introduced earlier in Section 3.5 will also be used here.

Vertex (Extreme) Point. This is a point of the set that does not lie on a line segment joining two other points of the set. For example, every point on the circumference of a circle and each vertex of the polygon satisfy the requirements for an extreme point.

Feasible Solution. Any solution of the constraint equations satisfying the nonnegativity conditions is a feasible solution. In the profit maximization example of Fig. 4.1, every point bounded by the polygon ABCDE (convex set) is a feasible solution.

Basic Solution. It is a solution of the constraint equations obtained by setting the "redundant number" $(n - m)$ of the variables to zero and solving the equations simultaneously for the remaining variables. The variables set to zero are called *nonbasic,* and the remaining ones are called *basic.* In the profit maximization example, each of the 10 solutions in Table 4.1 is basic (but only A, B, C, D and E are basic and feasible).

Basic Feasible Solution. A basic solution satisfying the nonnegativity conditions on the variables is called a basic feasible solution. Note that solutions 1, 2, 6, 8 and 9 in Table 4.1 are the basic feasible solutions.

Degenerate Basic Solution. If a basic variable has zero value, the solution is said to be a degenerate basic solution.

Degenerate Basic Feasible Solution. If a basic variable has zero value the corresponding basic feasible solution is said to be degenerate.

Optimum solution. A feasible solution minimizing the cost function is called an optimum solution. The point D in Fig. 4.1 corresponds to the optimum solution.

Optimum Basic Solution. It is a basic feasible solution with optimum cost function value. From Table 4.1 and Fig. 4.1 it is clear that only the solution number 8 is the optimum basic solution.

Convex Polyhedron. If the feasible region (constraint set) for an LP problem is bounded, it is called a convex polyhedron.

Basis. Columns of the coefficient matrix **A** of the constraint equations corresponding to the basic variables are said to form a basis for the m-dimensional vector space. Any other m-dimensional vector can be expressed as a linear combination of the basis vectors.

Example 4.3   Determination of basic solutions. Find all basic solutions for the following problem and identify basic feasible solutions on a sketch of the constraint set:

maximize
$$z = 4x_1 + 5x_2$$
subject to
$$-x_1 + x_2 \leqq 4$$
$$x_1 + x_2 \leqq 6$$
$$x_1, x_2 \geqq 0$$

Solution. The feasible region for the problem is shown in Fig. 4.2. Introducing slack variables x_3 and x_4 into the constraint equations and converting maximization of z to minimization, the problem is written in the standard LP form as

minimize
$$f = -4x_1 - 5x_2 \tag{a}$$
subject to
$$-x_1 + x_2 + x_3 = 4 \tag{b}$$
$$x_1 + x_2 + x_4 = 6$$
$$x_i \geqq 0; \quad i = 1 \text{ to } 4 \tag{c}$$

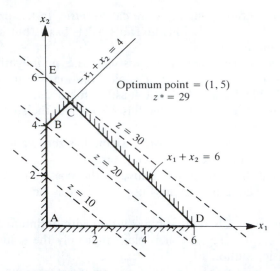

FIGURE 4.2
Graphical solution for the LP problem of Example 4.3. Optimum point $= (1, 5)$. $z^* = 29$.

Since $n = 4$ and $m = 2$, the problem has 6 basic solutions. These solutions are obtained from Eqs (b) by choosing two variables as nonbasic and the remaining two as basic. For example, x_1 and x_2 may be chosen as nonbasic, i.e. $x_1 = 0$, $x_2 = 0$. Then Eqs (b) give $x_3 = 4$, $x_4 = 6$. Also with $x_1 = 0$ and $x_3 = 0$, Eqs (b) give $x_2 = 4$ and $x_4 = 2$ as another basic solution. Similarly, the remaining basic solutions can be obtained.

The six basic solutions for the problem are summarized in Table 4.2 along with the corresponding cost function values. Basic feasible solutions are 1, 2, 5 and 6. These correspond to points $(0, 0)$, $(0, 4)$, $(6, 0)$ and $(1, 5)$ in Fig. 4.2, respectively. The minimum value of the cost function is obtained at the point $(1, 5)$ as $f = -29$ (maximum value of $z = 29$). ‖

TABLE 4.2
Basic solutions for Example 4.3

No.	x_1	x_2	x_3	x_4	f	Location in Fig. 4.2
1	0	0	4	6	0	A
2	0	4	0	2	−20	B
3	0	6	−2	0	——	infeasible
4	−4	0	0	10	——	infeasible
5	6	0	10	0	−24	D
6	1	5	0	0	−29	C

4.3.3   Optimum Solution for LP Problems

Now some important theorems that define optimum solution for LP problems are stated and explained.

> **Theorem 4.1   Extreme points and basic feasible solutions.** The collection of feasible solutions for an LP problem constitutes a convex set whose extreme points correspond to basic feasible solutions.   ‖

This theorem relates *extreme points of the convex polyhedron to the basic feasible solutions*. This is an important result giving geometrical meaning to the basic feasible solutions; they are the vertices of the polyhedron representing the constraint set. As an example, basic feasible solutions in Table 4.1 correspond to vertices in the constraint set of Fig. 4.1.

The following theorem establishes the importance of the basic feasible solutions.

> **Theorem 4.2   Basic theorem of linear programming.** Let the $m \times n$ coefficient matrix **A** of constraint equations have full row rank, i.e. rank(**A**) = m. Then
>
> **1.** if there is a feasible solution, there is a basic feasible solution, and
> **2.** if there is an optimum feasible solution, there is an optimum basic feasible solution.   ‖

Part (1) of the theorem says that if there is any feasible solution to the LP problem, then there must be at least one extreme point or vertex of the *convex feasible region.* Part (2) of the theorem says that if the LP problem has a solution, then it is at least at one of the vertices of the *convex polyhedron* representing feasible solutions. There may be more than one solution if the cost function is parallel to one of the constraints. As noted earlier, the LP problem has an infinite number of feasible designs. We seek a feasible design that minimizes the cost function. Theorem 4.2 says that the optimum solution must be one of the basic feasible solutions, i.e. at one of the extreme points of the convex set. Thus, our task of solving an LP problem has been reduced to the search for optimum only among the basic feasible solutions. For a problem having n variables and m constraints, there are at the most

$$\binom{n}{m} = \frac{n!}{m!\,(n-m)!} \qquad \text{(total number of combinations)}$$

basic solutions. These are only a finite number of possibilities. Thus according to Theorem 4.2, the optimum is at one of these points. We need to systematically search these solutions for the optimum. The Simplex method of the next section is based on searching among the basic feasible solutions to reduce the cost function continuously.

4.4 THE SIMPLEX METHOD

4.4.1 Basic Ideas and Steps of the Simplex Method

Theorem 4.2 guarantees that one of the basic feasible solutions is an optimum for the LP problem. The basic idea of the Simplex method is simply to proceed from one basic feasible solution to another in a way to continually decrease the cost function until the minimum is reached. Thus, to solve an LP problem we need a method to systematically find basic feasible solutions of Eqs (4.9). The *Gauss-Jordan elimination* process (Appendix B, Sections B.4 and B.5) provides such a procedure. Before the Simplex method is developed, the idea of Simplex, canonical forms and pivot step are introduced. These are fundamental in the development of the method.

4.4.1.1 SIMPLEX. A Simplex in two-dimensional space is formed by any three points that do not lie on a straight line. In three-dimensional space, it is formed by four points that do not lie in the same plane. Three points can lie on a plane and the fourth one has to lie outside the plane. In general, a Simplex in the n-dimensional space is a convex hull of any $(n + 1)$ points which do not lie on one hyperplane. A *convex hull* of $(n + 1)$ points is the smallest convex set containing all the points. Thus the Simplex represents a convex set.

4.4.1.2 CANONICAL FORM. The idea of a canonical form is important in the development of the Simplex method. Therefore, we introduce this idea and discuss its use.

An $m \times n$ system of simultaneous equations given in Eqs (4.9) with rank $(\mathbf{A}) = m$ is said to be in *canonical form* if each equation has a variable (with unit coefficient) that does not appear in any other equation. A canonical form in general is written as shown in Eq. (4.16). Note that variables x_1 to x_m

$$
\begin{aligned}
x_1 + a_{1,m+1}x_{m+1} + a_{1,m+2}x_{m+2} + \ldots + a_{1,n}x_n &= b_1 \\
x_2 + a_{2,m+1}x_{m+1} + a_{2,m+2}x_{m+2} + \ldots + a_{2,n}x_n &= b_2 \\
&\vdots \\
x_m + a_{m,m+1}x_{m+1} + a_{m,m+2}x_{m+2} + \ldots + a_{m,n}x_n &= b_m
\end{aligned}
\tag{4.16}
$$

appear in only one of the equations; x_1 appears in the first equation, x_2 in the second equation, and so on. Note also that this sequence of variables x_1 to x_m in Eq. (4.16) is chosen only for convenience. In general, any of the variables x_1 to x_n can be associated with the first equation as long as it does not appear in any other equation. Similarly, the second equation need not be associated with the second variable x_2. This will become clearer when we discuss the Simplex method.

It is possible to write canonical form of Eq. (4.16) as a matrix equation, as also explained in Section B.5 of Appendix B:

$$\mathbf{I}_{(m)}\mathbf{x}_{(m)} + \mathbf{Q}\mathbf{x}_{(n-m)} = \mathbf{b} \tag{4.17}$$

where

$\mathbf{I}_{(m)} = m$-dimensional identity matrix

$\mathbf{x}_{(m)} = [x_1 \ x_2 \ldots x_m]^T$; vector of dimension m

$\boldsymbol{x}_{(n-m)} = [x_{m+1} \ldots x_n]^T$; vector of dimension $(n - m)$

$\mathbf{Q} = m \times (n - m)$ matrix consisting of coefficients of the variables x_{m+1} to x_n in Eq. (4.16)

$\mathbf{b} = [b_1 \ b_2 \ldots b_m]^T$

The Gauss-Jordan elimination process can be used to convert a given system of equations into the canonical form of Eq. (4.16) or Eq. (4.17). If we set $\mathbf{x}_{(n-m)} = \mathbf{0}$ in Eq. (4.17), then $\mathbf{x}_{(m)} = \mathbf{b}$. The variables set to zero in $\mathbf{x}_{(n-m)}$ are called *nonbasic*, and the variables $\mathbf{x}_{(m)}$ solved from Eq. (4.17) are called *basic*. The solution thus obtained is called a basic solution. If the right-hand side parameters b_i are $\geqq 0$, then the canonical form gives a basic feasible solution.

Equations (f) in Example 4.2 of Section 4.3 represent a canonical form. In these equations, the variables x_1 and x_2 are nonbasic, so they have zero value. The variables x_3, x_4 and x_5 are basic and their values are readily obtained from the canonical form as $x_3 = 16$, $x_4 = 1$ and $x_5 = 1$.

Similarly, Eqs (b) of Example 4.3 represent a canonical form giving a basic solution of $x_1 = 0$, $x_2 = 0$, $x_3 = 4$ and $x_4 = 6$.

4.4.1.3 TABLEAU. It is customary to represent the canonical form in a tableau as shown in Table 4.3. A *tableau* is defined as the representation of a scene or a picture. It is a convenient way of representing all the necessary information related to an LP problem. In the Simplex method, tableau consists of detached coefficients of the variables in the cost and constraint functions. The tableau in Table 4.3 does not contain coefficients of the cost function; they can, however, be included, as we shall see later.

TABLE 4.3
Representation of a canonical form in a tableau

Basic ↓	x_1	x_2	.	.	.	x_m	x_{m+1}	x_{m+2}	.	.	.	x_n	RHS
x_1	1	0	.	.	.	0	$a_{1,m+1}$	$a_{1,m+2}$	.	.	.	$a_{1,n}$	b_1
x_2	0	1	.	.	.	0	$a_{2,m+1}$	$a_{2,m+2}$	.	.	.	$a_{2,n}$	b_2
x_3	0	0	.	.	.	0	$a_{3,m+1}$	$a_{3,m+2}$	.	.	.	$a_{3,n}$	b_3
.	.	.	.	.	.	.			.	.	.	.	.
.	.	.	.	.	.	.			.	.	.	.	.
x_m	0	0	.	.	.	1	$a_{m,m+1}$	$a_{m,m+2}$	.	.	.	$a_{m,n}$	b_m

It is important to understand the structure and notation of the tableau as explained in the following because it is used later to develop the Simplex method:

1. The entries of the tableau are obtained by reducing the linear system of equations $\mathbf{Ax} = \mathbf{b}$ associated with the LP problem into the canonical form of Eq. (4.17). In Table 4.3, the first m columns correspond to the identity matrix, the next $(n - m)$ columns correspond to the $\mathbf{Q}$ matrix and the last column corresponds to the right-hand side (RHS) vector $\mathbf{b}$ in Eq. (4.17).

2. Each column of the tableau is associated with a variable; x_1 with the first column, x_2 with the second, and so on. This is due to the fact that the ith column contains a coefficient of the variable x_i in each of the rows in Eq. (4.17).

3. Each row of the tableau contains coefficients of the corresponding row in Eq. (4.17).

4. Each row of the tableau is also associated with a variable as indicated in the left most column in Table 4.3. These variables correspond to the columns of the identity matrix in the tableau. In Table 4.3, x_1 corresponds to the first column, x_2 to the second column, and so on. Note, however, that columns of the identity matrix can appear anywhere in the tableau. They need not be in any sequence either. Since variables associated with the identity matrix are basic, the left most column then identifies a basic variable associated with each row. This will become clearer later when we solve example problems.

5. Since each basic variable appears in only one row, its value is immediately available in the right most column (note that by definition, the nonbasic variables have zero value). For the example of Table 4.3, the basic variables have the values $x_i = b_i$, $i = 1$ to m. If all $b_i \geq 0$, we have a basic feasible solution.

6. The tableau identifies nonbasic variables, basic variables (associated with the columns of the identity matrix) and gives their values, i.e. it gives a basic solution. We shall see later that it can be augmented with the cost function expression, and in that case, it will also immediately give the value of the cost function associated with the basic solution.

7. Columns associated with the basic variables are called basic and others are called nonbasic.

8. *Basic columns* form a basis for the m-dimensional vector space, i.e. any m-dimensional vector can be expressed as a linear combination of the basis vectors (refer to Appendix B, Section B.6).

Example 4.4 Canonical form and tableau. Write the canonical form of Example 4.2 in a tableau.

TABLE 4.4
Tableau for LP problem of Example 4.4

Basic ↓	x_1	x_2	x_3	x_4	x_5	b
x_3	1	1	1	0	0	16
x_4	$\frac{1}{28}$	$\frac{1}{14}$	0	1	0	1
x_5	$\frac{1}{14}$	$\frac{1}{24}$	0	0	1	1

Solution. Table 4.4 shows Eqs (f) of Example 4.2 written in the notation of the tableau of Table 4.3. Note that for this example, the number of equations is three and the number of variables is five, i.e. $m = 3$ and $n = 5$.

The variables x_3, x_4 and x_5 appear in one and only one equation, so columns x_3, x_4 and x_5 define the identity matrix $\mathbf{I}_{(m)}$ of the canonical form of Eq. (4.17). The vectors $\mathbf{x}_{(m)}$ and $\mathbf{x}_{(n-m)}$ are defined as

$$\mathbf{x}_{(m)} = [x_3 \ x_4 \ x_5]^T$$

$$\mathbf{x}_{(n-m)} = [x_1 \ x_2]^T$$

The matrix $\mathbf{Q}$ of Eq. (4.17) is identified as

$$\mathbf{Q} = \begin{bmatrix} 1 & 1 \\ \frac{1}{28} & \frac{1}{14} \\ \frac{1}{14} & \frac{1}{24} \end{bmatrix}$$

If x_1 and x_2 are taken as nonbasic, then the values for the basic variables are obtained from the tableau as

$$x_3 = 16, \qquad x_4 = 1, \qquad x_5 = 1 \qquad \qquad \|$$

4.4.1.4 THE PIVOT STEP. In the Simplex method, we want to systematically search among the basic feasible solutions for the optimum design. Starting from a basic feasible solution, we want to find another one to reduce the cost. This can be done by interchanging a current basic variable with a nonbasic variable. The *pivot step* accomplishes this task and defines a new canonical form, as explained in the following.

Let us select a basic variable x_p ($1 \leq p \leq m$) to be replaced by a nonbasic variable x_q for $(n - m) \leq q \leq n$. We will describe later how to determine x_p and x_q. The pth basic column is to be interchanged with the qth nonbasic column. This is possible only when the pivot element $a_{pq} \neq 0$.

Note that x_q will be basic if it is eliminated from all the equations except the pth one. This can be accomplished by performing a *Gauss–Jordan elimination step* on the qth column of the tableau shown in Table 4.3 using the pth row for elimination. This will give $a_{pq} = 1$ and zeros elsewhere in the qth column.

The row used for the elimination process (pth row) is called the *pivot row*. The column on which the elimination is performed (qth column) is called the *pivot column*. The process of interchanging one basic variable with a nonbasic variable is called the *pivot step*.

Let a'_{ij} denote the new coefficients in the canonical form after the pivot step. Then, the *pivot step* for performing elimination in the qth column using the pth row as the pivot row is described by the following general equations:

$$a'_{pj} = a_{pj}/a_{pq}; \quad j = 1 \text{ to } n \tag{4.18}$$

$$b'_p = b_p/a_{pq} \tag{4.19}$$

$$a'_{ij} = a_{ij} - (a_{pj}/a_{pq})a_{iq}; \quad \begin{cases} i \neq p, \, i = 1 \text{ to } m, \\ j = 1 \text{ to } n \end{cases}; \tag{4.20}$$

$$b'_i = b_i - (b_p/a_{pq})a_{iq}; \quad i \neq p, \, i = 1 \text{ to } m \tag{4.21}$$

In Eqs (4.18) and (4.19), the pth row of the tableau is simply divided by the pivot element a_{pq}. Equations (4.20) and (4.21) perform the elimination step in the qth column of the tableau. Elements above and below the pth row are reduced to zero by proper elimination. These equations may be coded into a computer program to perform the pivot step.

The process of interchanging roles of two variables is illustrated in the following example.

Example 4.5 Pivot step – interchange of basic and nonbasic variables. Assuming x_3 and x_4 as basic variables, Example 4.3 is written in the canonical form as follows;

minimize
$$f = -4x_1 - 5x_2$$
subject to

$$-x_1 + x_2 + x_3 \qquad = 4$$

$$x_1 + x_2 \qquad + x_4 = 6$$

$$x_i \geq 0; \quad i = 1 \text{ to } 4$$

Obtain a new canonical form by interchanging the roles of x_1 and x_4, i.e. make x_1 a basic and x_4 a nonbasic variable.

Solution. The given canonical form can be written in a tableau as shown in Table 4.5. For the canonical form, x_1 and x_2 are nonbasic and x_3 and x_4 are basic, i.e. $x_1 = x_2 = 0$, and $x_3 = 4$ and $x_4 = 6$. This corresponds to point A in Fig. 4.2. In the tableau, the basic variables are identified in the left most column and the right most column gives their values. Also basic variables are identified by examining columns of the tableau. The variables associated with the columns of the identity matrix are basic. Location of the nonzero unit element in a basic column identifies the row whose right-hand side parameter b_i is the current value of the basic variable associated with that column.

To make x_1 basic and x_4 a nonbasic variable, one would like to make $a'_{21} = 1$ and $a'_{11} = 0$. This will replace x_1 with x_4 as the basic variable and a new canonical form will be obtained. The second row is treated as the pivot row, i.e. $a_{21} = 1$

TABLE 4.5

Pivot step to interchange basic variable x_4 with nonbasic variable x_1 for Example 4.5

	Basic ↓	x_1	x_2	x_3	x_4	b	Basic solution
Initial canonical form	x_3	-1	1	1	0	4	$x_1 = 0$, $x_2 = 0$,
	x_4	**1**	1	0	1	6	$x_3 = 4$, $x_4 = 6$

To interchange x_1 with x_4, choose row 2 as pivot row and column 1 as the pivot column. Perform elimination using a_{21} as the pivot element and obtain the second canonical form.

Result of the pivot operation

	Basic ↓	x_1	x_2	x_3	x_4	b	Basic solution
Second canonical form	x_3	0	2	1	1	10	$x_2 = 0$, $x_4 = 0$
	x_1	1	1	0	1	6	$x_1 = 6$, $x_3 = 10$

$(p = 2,\ q = 1)$ is the pivot element. Performing Gauss–Jordan elimination in the first column with $a_{21} = 1$ as the pivot element, we obtain the second canonical form as shown in Table 4.5. For this canonical form $x_2 = x_4 = 0$ are the nonbasic variables and $x_1 = 6$ and $x_3 = 10$ are the basic variables. Thus, referring to Fig. 4.2, this pivot step results in a move from the extreme point A$(0, 0)$ to the extreme point D$(6, 0)$. ||

4.4.1.5 BASIC STEPS OF THE SIMPLEX METHOD. In the Simplex method, we start with a basic feasible solution, i.e. at a vertex of the convex polyhedron. A move is then made to an adjacent vertex while maintaining feasibility as well as reducing the cost function. This is accomplished by replacing a basic variable with a nonbasic variable while *maintaining feasibility of the new solution*. Also, in moving from one extreme point to another the cost function f is to be improved (reduced). In the Simplex method, movements are to adjacent extreme points only. Since there may be several points adjacent to the current extreme point, we naturally wish to choose the one which makes the greatest improvement in the cost function f. If adjacent points make identical improvements in f, the choice becomes arbitrary. An improvement at each step ensures no backtracking. Two basic questions now arise:

1. How to choose a current nonbasic variable that should become basic?
2. Which variable from the current basic set should become nonbasic?

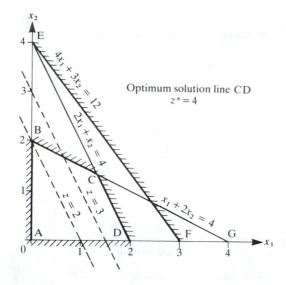

FIGURE 4.3
Graphical solution for the LP problem of Example 4.6. Optimum solution: along line C–D. $z^* = 4$.

The Simplex method answers these questions which require theoretical considerations. We shall discuss these aspects in the next subsection. Here, we consider an example to illustrate the basic steps of the Simplex method.

Example 4.6 Steps of the Simplex method. Solve the following LP problem:

maximize

$$z = 2x_1 + x_2$$

subject to

$$4x_1 + 3x_2 \leq 12$$

$$2x_1 + x_2 \leq 4$$

$$x_1 + 2x_2 \leq 4$$

$$x_1, x_2 \geq 0$$

Solution. The graphical solution for the problem is given in Fig. 4.3. It can be seen that the problem has an infinite number of solutions along the line C–D ($z^* = 4$) because the objective function is parallel to the second constraint.

To illustrate the Simplex method, we use the following steps:

1. We write the problem in the standard LP form, by transforming the maximization of z to minimization of $f = -2x_1 - x_2$, and add slack variables x_3, x_4 and x_5 to the constraints. Thus, the problem becomes

minimize

$$f = -2x_1 - x_2$$

TABLE 4.6
Initial tableau for the LP problem of Example 4.6

Basic ↓	x_1	x_2	x_3	x_4	x_5	b	Ratio $b_i/a_{i1}; a_{i1} > 0$
x_3	4	3	1	0	0	12	$\frac{12}{4} = 3$ x_4 should
$\to x_4$	**$\underline{2}$**	1	0	1	0	4	$\frac{4}{2} = 2 \leftarrow$ become
x_5	1	2	0	0	1	4	$\frac{4}{1} = 4$ nonbasic (pivot row)
Cost function	**$\underline{-2}$** ↑	-1	0	0	0	$f - 0$	

x_4 out, x_1 in

x_1 should become basic (pivot column).
Note: The selected negative cost coefficient and the pivot element are bold faced and underlined throughout.

subject to

$$4x_1 + 3x_2 + x_3 \qquad\qquad = 12$$
$$2x_1 + x_2 \qquad + x_4 \qquad = 4$$
$$x_1 + 2x_2 \qquad\qquad + x_5 = 4$$
$$x_i \geq 0; \qquad i = 1 \text{ to } 5$$

We use the tableau and notation of Table 4.3 which will be augmented with the cost function expression as the last row. The initial tableau for the problem is as shown in Table 4.6.

2. To initiate the Simplex method, a basic feasible solution is needed. This is already available in Table 4.6 with x_1 and x_2 as nonbasic, and x_3, x_4 and x_5 as basic:

basic variables $x_3 = 12$, $x_4 = 4$, $x_5 = 4$

nonbasic variables $x_1 = 0$, $x_2 = 0$

cost function $f = 0$

Note that cost row gives $0 = f - 0$, which gives the current value of the cost function as 0. This solution represents point A in Fig. 4.3 at which none of the constraints is active except nonnegativity constraints on the variables.

3. We *scan the cost row* which should have nonzero entries only in the nonbasic columns, i.e. x_1 and x_2. *If all the nonzero entries are nonnegative, then we have an optimum solution* and the Simplex method can be terminated.

4. In the present example, there are negative entries in nonbasic columns, so the current basic feasible solution is *not optimum*. We select a nonbasic column having negative cost coefficient. In Table 4.6, we select column x_1. This identifies a nonbasic variable (x_1) that should become basic. Thus, eliminations will be performed in the x_1 column.

TABLE 4.7

Second tableau for Example 4.6 making x_1 a basic variable

Basic ↓	x_1	x_2	x_3	x_4	x_5	b
x_3	0	1	1	−2	0	4
x_1	1	0.5	0	0.5	0	2
x_5	0	1.5	0	−0.5	1	2
Cost function	0	0 ↑	0	1 ↑	0	$f+4$

The cost coefficient in nonbasic columns are nonnegative; the tableau gives the optimum solution.

Note that when there is more than one negative entry in the cost row, the variable tapped to become basic is arbitrary among the indicated possibilities. The usual convention is to select a variable associated with the smallest value in the cost row (or, negative element with the largest absolute value).

5. To identify which current basic variable should become nonbasic (i.e. to select the pivot row), we take ratios of the right-hand side parameters with the positive elements in the x_1 column as shown in Table 4.6. We identify the row having the smallest ratio, i.e. the second row. This will make x_4 nonbasic. The pivot element is $a_{21} = 2$ (the intersection of pivot row and pivot column).

6. We perform elimination in column x_1 using row 2 as the pivot row: divide row 2 by 2; multiply new row 2 by 4 and subtract from row 1; subtract new row 2 from row 3; and multiply new row 2 by 2 and add to the cost row. As a result of this elimination step, a new tableau is obtained as shown in Table 4.7. The new basic feasible solution is given as

$$\text{basic variables} \qquad x_3 = 4, \ x_1 = 2, \ x_5 = 2$$

$$\text{nonbasic variables} \quad x_2 = 0, \ x_4 = 0$$

$$\text{cost function} \qquad f + 4 = 0, f = -4$$

7. This solution is identified as point D in Fig. 4.3. We see that the cost function has reduced from 0 to −4. All coefficients in the last row are nonnegative, so no further reduction of the cost function is possible. Thus, the foregoing solution is the optimum. Note that the cost coefficients corresponding to the nonbasic variable x_2 in the last row is zero. This is an indication of *multiple solutions* for the problem. In general, when a cost coefficient in the last row is zero corresponding to a nonbasic variable, the problem may have multiple solutions. We shall discuss this point later in more detail.

Let us see what happens if we do not select a row with the least ratio as the pivot row. Let $a_{31} = 1$ in the third row be the pivot element in Table 4.6. This will interchange nonbasic variable x_1 with the basic variable x_5. Performing the elimination steps in the first column, we obtain the new tableau given in Table 4.8.

TABLE 4.8
Result of improper pivoting in Simplex method for LP problem of Example 4.6

Basic ↓	x_1	x_2	x_3	x_4	x_5	b
x_3	0	−5	1	0	−4	−4
x_4	0	−3	0	1	−2	−4
x_1	1	2	0	0	1	4
Cost function	0	3	0	0	4	$f + 8$

The pivot step making x_1 basic and x_5 nonbasic in Table 4.6 gives a basic solution that is not feasible.

From the tableau, we have

$$\text{basic variables} \quad x_3 = -4, \, x_4 = -4, \, x_1 = 4$$

$$\text{nonbasic variables} \quad x_2 = 0, \, x_5 = 0$$

$$\text{cost function} \quad f + 8 = 0, \, f = -8$$

The foregoing solution corresponds to point G in Fig. 4.3. We see that the basic solution is not feasible as x_3 and x_4 have negative values. Thus, we conclude that if a row with smallest ratio (of right-hand side parameters with positive elements in the pivot column) is not selected, the new basic solution may not be feasible. ‖

Note that for the preceding example, only one iteration of the Simplex method gave the optimum solution. In general, more iterations are needed until all coefficients in the cost row become nonnegative.

Based on the preceding example, *we summarize the basic ideas and procedures of the Simplex method as follows*:

1. *The method needs an initial basic feasible solution (canonical form)* to initiate the search for the optimum solution. Such a form is readily available if all the constraints are "≦ type", as can be seen in the foregoing example. In that case, selecting slack variables as basic and original variables as nonbasic gives the initial basic feasible solution. In this section, we develop basic steps of the Simplex method using only "≦ type" constraints. In the next section, treatment of "≧ type" and equality constraints shall be discussed.

2. For the Simplex method to work, *the cost function must be expressed in terms of only the nonbasic variables*. This is the reason for writing the cost function as the last row in the Simplex tableau so that the basic variable can be eliminated from it during the pivot step.

3. Since the cost function is in terms of the nonbasic variables, cost coefficients

in the last row corresponding to basic variables are zero, and those corresponding to nonbasic variables are nonzero. *If any of the nonzero entries in the last row is negative, we are not at the optimum point*; otherwise an optimum solution has been reached.

4. If the optimum solution is not reached, we need to interchange a current basic variable with a nonbasic variable (i.e. we need to find a new basic feasible solution). To do this we first *select a nonbasic variable that should become basic. A nonbasic variable associated with a column having a negative coefficient in the cost (last) row is selected to become basic*. This identifies the pivot column. To select the current basic variable that should become nonbasic, we take ratios of the right-hand side values (last column) with the positive elements in the pivot column. *We identify the row with the smallest ratio as the pivot row* and the basic variable associated with it that should become nonbasic.

5. If it so happens that all entries in the pivot column are negative, then *the problem is unbounded* as the cost can be reduced to negative infinity.

6. Once the pivot row and column are identified, the *pivot step* can be completed by making the pivot element as one and other elements in the pivot column as zero using the elimination process. The pivot row is used for the elimination process. The pivot step results in eliminating the nonbasic variable associated with the pivot column from all the equations except the pivot equation. Note that the elimination step is also carried out for the last (cost) row. This step gives a *new basic feasible solution* having smaller cost function value.

7. The preceding steps are repeated until an answer to the problem is obtained.

4.4.2* Derivation of the Simplex Method

In the previous subsection, we presented the basic ideas and concepts of the Simplex method. The steps of the method were illustrated in an example. In this subsection, we describe the theory that leads to the steps used in the example problem.

4.4.2.1—SELECTION OF A BASIC VARIABLE THAT SHOULD BECOME NONBASIC.
Derivation of the Simplex method is based on answering the two questions posed in the previous subsection. We will answer the second question first, i.e. which variable from the basic set should become nonbasic? Assume for the moment that x_r is a nonbasic variable tapped to become basic. This indicates that the rth nonbasic column should replace some current basic column. After this interchange, there should be all zero elements in the rth column except the unit element at one location.

To determine a basic variable that should become nonbasic, we need to determine the pivot row for the elimination process. This way the current basic variable associated with that row will become nonbasic after the elimination step. To determine the pivot row, we transfer all the terms associated with the variable x_r to the right-hand side of the canonical form of Eq. (4.16). The system of equations becomes as shown in Eq. (4.22). Since x_r is to become a

$$x_1 + a_{1,m+1}x_{m+1} + \ldots + a_{1,n}x_n = b_1 - a_{1,r}x_r$$

$$x_2 + a_{2,m+1}x_{m+1} + \ldots + a_{2,n}x_n = b_2 - a_{2,r}x_r$$

$$\begin{array}{ccc} \cdot & & \cdot \\ \cdot & & \cdot \\ \cdot & & \cdot \end{array} \tag{4.22}$$

$$x_n + a_{m,m+1}x_{m+1} + \ldots + a_{m,n}x_n = b_m - a_{m,r}x_r$$

basic variable its value should become nonnegative. The new solution must also remain feasible. An examination of the right-hand side of Eq. (4.22) shows that x_r cannot increase arbitrarily. The reason is that if x_r becomes arbitrarily large then some of the new right-hand side parameters $(b_i - a_{i,r}x_r)$, $i = 1$ to m may become negative. Since right-hand side parameters are the new values of the basic variables, this will violate the nonnegativity constraint for the variables and the new basic solution will not be feasible. Thus for the new solution to be basic and feasible, the following constraints must be satisfied by the right-hand sides of Eq. (4.22) in selecting a basic variable that should become nonbasic:

$$b_i - a_{i,r}x_r \geq 0; \qquad i = 1 \text{ to } m \tag{4.23}$$

Any $a_{i,r}$ that are nonpositive pose no limit on how much x_r can be increased since Inequality (4.23) remains satisfied ($b_i \geq 0$). For positive $a_{i,r}$, x_r can be increased from zero until one of the inequalities in Eq. (4.23) becomes active, i.e. one of the right-hand sides of Eq. (4.22) becomes zero. A further increase would violate the nonnegativity conditions of Eq. (4.23). Thus, the maximum value that the incoming variable x_r can take is

$$\frac{b_s}{a_{s,r}} = \min_i \left\{ \frac{b_i}{a_{i,r}}, a_{ir} > 0; i = 1 \text{ to } m \right\} \tag{4.24}$$

where s is the index of the smallest ratio.

Equation (4.24) says that we take ratios of the right-hand side parameters b_i with the positive elements in the rth column ($a_{i,r}$'s) and we select the row index s giving the least ratio. *In the case of a tie, the choice for the index s is arbitrary among the tying indices and in such a case the resulting basic feasible solution will be degenerate.* Thus, Eq. (4.24) identifies a row with the smallest ratio $b_i/a_{i,r}$. The basic variable x_s associated with this row should become nonbasic. *If all $a_{i,r}$ are nonpositive in the rth column, then x_r can be increased indefinitely.* This indicates that the LP problem is *unbounded.* Any practical problem with this situation is not properly constrained, so the problem formulation should be reexamined.

4.4.2.2 SELECTION OF A NONBASIC VARIABLE THAT SHOULD BECOME BASIC. We now know how to select a basic variable that should replace a nonbasic variable. Let us see how we can identify the nonbasic variable that should become basic. This will answer the first question posed earlier. *The main idea of bringing a nonbasic variable into the basic set is to improve the design, i.e. to reduce the current value of the cost function.* A clue to the desired improvement is obtained if we examine the cost function expression. To do this we *need to write the cost function in terms of nonbasic variables only.* We substitute for the current values of basic variables from Eq. (4.16) into the cost function to eliminate them. Current values of the basic variables in terms of nonbasic variables are

$$x_i = b_i - \sum_{j=m+1}^{n} a_{ij} x_j; \qquad i = 1 \text{ to } m \tag{4.25}$$

Substituting Eq. (4.25) into the Eq. (4.11) and simplifying, we obtain an expression for the cost function in terms of the nonbasic variables (x_j, $j = m + 1$ to n) as

$$f = f_0 + \sum_{j=m+1}^{n} c'_j x_j \tag{4.26}$$

where f_0 is the current value of the cost function given as

$$f_0 = \sum_{i=1}^{m} b_i c_i \tag{4.27}$$

and the parameters c'_j are

$$c'_j = c_j - \sum_{i=1}^{m} a_{ij} c_i; \qquad j = (m + 1) \text{ to } n \tag{4.28}$$

The cost coefficients c'_j of the nonbasic variables play a key role in the Simplex method and are called the *reduced or relative cost coefficients.* They are used to identify a nonbasic variable that should become basic to reduce the current value of the cost function. Expressing cost function in terms of the current nonbasic variables is a key step in the Simplex method. We will see later that this is not difficult to accomplish as Gaussian elimination steps can be used routinely on the cost function expression to eliminate basic variables from it. Once this has been done, the reduced cost coefficients c'_j can be readily identified.

In general the reduced cost coefficients c'_j of the nonbasic variables may be positive, negative or zero. Let one of c'_j be negative. Then, note from Eq. (4.26) that if a positive value is assigned to the associated nonbasic variable (i.e. it is made basic), the value of f will reduce. If more than one negative c'_j is present, a widely used rule of thumb is to choose the nonbasic variable associated with the smallest c'_j (i.e. negative c'_j with the largest absolute value) to become basic. Thus, if any c'_j for $(m + 1) \leq j \leq n$ (for nonbasic variables) is negative then it is possible to find a new basic feasible solution (if one exists) that will further reduce the cost function. If a c_j' is zero, then the associated nonbasic variable can be made basic without affecting the cost function value. If all c_j' are nonnegative, then it is not possible to reduce the cost function any further, and the current basic feasible solution is optimum. These results are summarized in the following theorems.

> **Theorem 4.3 Improvement of basic feasible solution.** Given a nondegenerate basic feasible solution with the corresponding cost function f_0, suppose that $c_j' < 0$ for some j. Then there is a feasible solution with $f < f_0$. If the jth nonbasic column associated with c_j' can be substituted for some column in the original basis, the new basic feasible solution will have $f < f_0$. If the jth column cannot be substituted to yield a basic feasible solution, then the feasible region (constraint set) is unbounded and the cost function can be made arbitrarily small (towards negative infinity). ‖

> **Theorem 4.4 Optimum solution for LP problems.** If a basic feasible solution has reduced cost coefficients $c_j' \geqq 0$ for all j, then it is optimum. ‖

Finally, the cost function expression $\mathbf{c}^T \mathbf{x} = f$ may be viewed as another linear equation in the Simplex tableau; for example, the $(m+1)$th row. One performs the pivot step on the entire set of $(m+1)$ equations so that x_1, $x_2, \ldots, x_m$, and f are basic variables. The last row of the tableau represents the cost function expression of Eq. (4.26); that is, with this procedure the cost function is automatically given in terms of the nonbasic variables after each pivot step. The coefficients in the nonbasic columns of the last row are simply the current reduced cost coefficients c_j'.

Note that when all c_j' in the nonbasic columns are positive, the optimum solution is unique. If at least one c_j' (reduced cost coefficient associated with a nonbasic variable) is zero, then there is a possibility of alternate optima. If the nonbasic variable associated with a zero reduced cost coefficient can be made basic according to the foregoing procedure, the extreme point corresponding to alternate optima can be obtained. Since the reduced cost coefficient is zero, the optimum cost function value will not change. Any point on the line segment joining the optimum extreme points also corresponds to an optimum. Note that these optima are global as opposed to local, although there is *no distinct global optimum*. Geometrically, multiple optima for a LP problem imply that the cost function hyperplane is parallel to one of the constraint hyperplanes.

Note that if the nonbasic variable associated with the negative reduced cost coefficient c_j' cannot be made basic (e.g. when all a_{ij} in the c_j' column are negative), then the feasible region is unbounded.

> **Example 4.7 Solution by the Simplex method.** Using the Simplex method, find the optimum (if one exists) for the LP problem of Example 4.3:
>
> minimize
> $$f = -4x_1 - 5x_2$$

TABLE 4.9
Solution of Example 4.7 by the Simplex method

Basic ↓	x_1	x_2	x_3	x_4	b	Ratio b_i/a_{iq}	
x_3 out →x_3	-1	**1**	1	0	4	$\frac{4}{1}=4$	Initial tableau,
x_2 in x_4	1	1	0	1	6	$\frac{6}{1}=6$	pivot: a_{12}
Cost	-4	**-5**	0	0	$f-0$		
x_2	-1	1	1	0	4		Second tableau,
x_4 out →x_4 x_1 in	**2**	0	-1	1	2		pivot: a_{21}
Cost	**-9**	0	5	0	$f+20$		
x_2	0	1	$\frac{1}{2}$	$\frac{1}{2}$	5		Third tableau,
x_1	1	0	$-\frac{1}{2}$	$\frac{1}{2}$	1		optimum point
Cost	0	0	$\frac{1}{2}$	$\frac{9}{2}$	$f+29$		

Reduced cost coefficients in nonbasic columns are nonnegative; the tableau gives optimum solution.

subject to

$$-x_1 + x_2 + x_3 \quad = 4$$

$$x_1 + x_2 \quad\quad + x_4 = 6$$

$$x_i \geq 0; \; i = 1 \text{ to } 4$$

Solution. Writing the problem in the Simplex tableau, we obtain the initial canonical form as shown in Table 4.9. From initial tableau the basic feasible solution is

$$\text{basic variables} \quad x_3 = 4, \; x_4 = 6$$

$$\text{nonbasic variables} \quad x_1 = x_2 = 0$$

From the last row of the tableau, the cost function is $f = 0$.

Note that the cost function in the last row is in terms of nonbasic variables x_1 and x_2. Thus, coefficients in the x_1 and x_2 columns are the reduced cost coefficients c_j'. Scanning the last row, we observe that there are negative c_j' coefficients. Therefore, the current basic solution is not optimum. In the last row, the most negative coefficient of -5 corresponds to the second column. Therefore, we select x_2 to become a basic variable, or elimination should be performed in the second column. This fixes the column index q to 2 in Eq. (4.18). Now taking the ratios of the right-hand side parameters with positive coefficients in the qth column b_i/a_{iq}, we obtain a minimum ratio for the first row as 4. This identifies the first row as the pivot row according to Eq. (4.24). Therefore, the current basic variable associated with the first row, x_3, should become nonbasic.

Now performing the pivot step on column 2 with a_{12} as the pivot element, we obtain the second canonical form as shown in Table 4.9. For this canonical form the basic feasible solution is

$$\text{basic variables} \qquad x_2 = 4, \; x_4 = 2$$

$$\text{nonbasic variables} \quad x_1 = x_3 = 0.$$

The cost function is $f = -20$ ($f + 20 = 0$ from the last row) which is an improvement from 0. Thus, this pivot step results in a move from $(0,0)$ to $(0,4)$ on the convex polyhedron of Fig. 4.2.

The reduced cost coefficient corresponding to the nonbasic column x_1 is negative. Therefore, the cost function can be further improved. Repeating the process we obtain $a_{21} = 2$ as the pivot element, implying that x_1 should become basic and x_4 should become nonbasic. The third canonical form is shown in Table 4.9. For this tableau, all the reduced cost coefficients c_j (corresponding to the nonbasic variables) in the last row are nonnegative. Therefore, the tableau yields the optimum solution as

$$x_1 = 1, \qquad x_2 = 5, \qquad x_3 = 0, \qquad x_4 = 0, \qquad f = -29$$

In Fig. 4.2, this corresponds to the point $(1,5)$, as before. $\qquad\qquad$ ‖

Example 4.8 Solution of profit maximization problem by the Simplex method. Using the Simplex method find the optimum solution for the profit maximization problem of Example 4.2.

Solution. Introducing slack variables in constraints of Eqs (c) through (e) in Example 4.2, we get the LP problem in the standard form as minimize

$$f = -400x_1 - 600x_2$$

subject to

$$x_1 + x_2 + x_3 \qquad\quad = 16$$

$$\tfrac{1}{28}x_1 + \tfrac{1}{14}x_2 + x_4 \;\; = 1$$

$$\tfrac{1}{14}x_1 + \tfrac{1}{24}x_2 + x_5 \;\; = 1$$

$$x_i \geq 0; i = 1 \text{ to } 5$$

Now writing the problem in the standard Simplex tableau, we obtain the initial canonical form as shown in Table 4.10. Thus the initial basic feasible solution is

$$x_1 = 0, \qquad x_2 = 0, \qquad x_3 = 16, \qquad x_4 = x_5 = 1, \qquad f = 0$$

which corresponds to point A in Fig. 4.1. The initial cost function is zero and x_3, x_4 and x_5 are the basic variables.

Using the Simplex procedure we note that $a_{22} = \tfrac{1}{14}$ is the pivot element. This implies that x_4 should be replaced by x_2 in the basic set. Carrying out the pivot operation using the second row as the pivot row, we obtain the second

TABLE 4.10
Solution of Example 4.8 by the Simplex method

	Basic ↓	x_1	x_2	x_3	x_4	x_5	b	Ratio b_i/a_{iq}	
x_4 out → x_2 in	x_3	1	1	1	0	0	16	16	Initial tableau, pivot: a_{22}
	$\to x_4$	$\frac{1}{28}$	$\frac{1}{14}$	0	1	0	1	14	
	x_5	$\frac{1}{14}$	$\frac{1}{24}$	0	0	1	1	24	
	Cost	−400	**−600**	0	0	0	$f-0$		
x_3 out → x_1 in	$\to x_3$	$\frac{1}{2}$	0	1	−14	0	2	4	Second tableau, pivot: a_{11}
	x_2	$\frac{1}{2}$	1	0	14	0	14	28	
	x_5	$\frac{17}{336}$	0	0	$\frac{7}{12}$	1	$\frac{5}{12}$	$\frac{140}{17}$	
	Cost	**−100**	0	0	8400	0	$f+8400$		
	x_1	1	0	2	−28	0	4		Third tableau: optimum solution
	x_2	0	1	−1	28	0	12		
	x_5	0	0	$-\frac{17}{168}$	2	1	$\frac{3}{14}$		
	Cost	0	0	200	5600	0	$f+8800$		
				↑	↑				

Reduced cost coefficients in the nonbasic columns are nonnegative; the tableau gives optimum solution.

canonical form shown in Table 4.10. At this point the basic feasible solution is

$$x_1 = 0, \qquad x_2 = 14, \qquad x_3 = 2, \qquad x_4 = 0, \qquad x_5 = \frac{5}{12}$$

which corresponds to point E on Fig. 4.1. The cost function is reduced to −8400.

The pivot element for the next step is a_{11} implying that x_3 should be replaced by x_1 in the basic set. Carrying out the pivot operation, we obtain the third canonical form shown in Table 4.10. At this point all reduced cost coefficients (corresponding to nonbasic variables) are nonnegative, so according to Theorem 4.4, we have the optimum solution:

$$x_1 = 4, \qquad x_2 = 12, \qquad x_3 = 0, \qquad x_4 = 0, \quad \text{and} \quad x_5 = \frac{3}{14}$$

This corresponds to the D in Fig. 4.1. The optimum value of the cost function is −8800. Note that c_j' corresponding to the nonbasic variables x_3 and x_4 are positive. Therefore, the global optimum is unique, as may also be observed from Fig. 4.1. ‖

4.4.2.3 SIMPLEX ALGORITHM. The basic procedure of the Simplex method is to start with an initial basic feasible solution, i.e. at the vertex of the convex polyhedron. Then a move is made to an adjacent vertex to reduce the cost function until optimum is reached. The steps of the procedure are summarized as follows:

Step 1. Start with an initial basic feasible solution. This is readily obtained if all constraints are "$\leqq$ type" because the slack variables can be selected as basic and the real variables as nonbasic. If there are equality or "$\geqq$ type" constraints, then the two-phase simplex procedure of the next section must be used.

Step 2. The cost function must be in terms of only the nonbasic variables. This is readily available when there are only "$\leqq$ type" constraints. The slack variables are basic and they do not appear in the cost function.

Step 3. If all the reduced cost coefficients for nonbasic variables are nonnegative, we have the optimum solution. If any coefficient is negative, there is a possibility of improving the cost function. We identify a column having negative reduced cost coefficient because the nonbasic variable associated with it can become basic. This is called the pivot column.

Step 4. If all elements in the pivot column are negative, then the problem is unbounded. Design problem formulation should be examined to correct the situation. If there are positive elements in the pivot column, then we take ratios of the right-hand side parameters with the positive elements in the pivot column and identify a row with the smallest ratio. In the case of a tie, any row among the tying ratios can be selected. The basic variable associated with this row should become nonbasic (i.e. becomes zero). The row is called the pivot row and its intersection with the pivot column identifies the pivot element.

Step 5. Complete the pivot step using the Gauss–Jordan elimination procedure and the pivot row identified in Step 4. Elimination must also be performed in the cost function row.

Step 6. Identify basic and nonbasic variables, and their values. Identify the cost function value and go to Step 3.

Example 4.9   LP problem with multiple solutions. Solve the following problem by the Simplex method:

maximize
$$z = x_1 + 0.5x_2$$
subject to
$$2x_1 + 3x_2 \leqq 12$$
$$2x_1 + x_2 \leqq 8$$
$$x_1, x_2 \geqq 0$$

Solution. The problem was solved graphically in Subsection 2.8.2 of Chapter 2. It has *multiple solutions* as may be seen in Fig. 2.8. We will solve the problem using the Simplex method and discuss how multiple solutions can be recognized for general LP problems. The problem is converted to standard LP form as

minimize
$$f = -x_1 - 0.5x_2$$

TABLE 4.11

Solution by the simplex method for Example 4.9

	Basic	x_1	x_2	x_3	x_4	b	
x_4 out $\rightarrow x_4$	x_3	2	3	1	0	12	Initial tableau,
x_1 in	x_4	**2**	1	0	1	8	pivot: a_{21}
		−1	−0.5	0	0	$f - 0$	
x_3 out $\rightarrow x_3$	x_3	0	**2**	1	−1	4	Second tableau,
x_2 in	x_1	1	$\frac{1}{2}$	0	$\frac{1}{2}$	4	pivot: a_{12}
	Cost	0	**0**	0	$\frac{1}{2}$	$f + 4^a$	
	x_2	0	1	$\frac{1}{2}$	$-\frac{1}{2}$	2	Third tableau
	x_1	1	0	$-\frac{1}{4}$	$\frac{3}{4}$	3	
	Cost	0	0	0	$\frac{1}{2}$	$f + 4^b$	

Reduced cost coefficients in nonbasic columns are nonnegative; the tableau gives optimum solution. $c_3' = 0$ indicates possibility of multiple solutions.
[a] First optimum point; [b] Second optimum point.

subject to

$$2x_1 + 3x_2 + x_3 \quad\quad = 12$$
$$2x_1 + x_2 \quad\quad + x_4 = 8$$
$$x_i \geq 0; \quad i = 1 \text{ to } 4$$

Table 4.11 contains iterations of the Simplex method. The optimum point is reached in just one iteration, as all the reduced cost coefficients are nonnegative in the second canonical form. The solution is

$$\text{basic variables} \quad x_1 = 4, \ x_3 = 4$$
$$\text{nonbasic variables} \quad x_2 = x_4 = 0$$
$$\text{optimum cost function} \quad f = -4$$

The solution corresponds to point B in Fig. 2.8.

In the second tableau, the *reduced cost coefficient for the nonbasic variable x_2 is zero*. This means that it is possible to make x_2 basic without any change in the optimum cost function value. This suggests the possibility of existence of *multiple optimum* solutions. Performing the pivot operation on column 2, we find another solution as (Table 4.11):

$$\text{basic variables} \quad x_1 = 3, \ x_2 = 2$$
$$\text{nonbasic variables} \quad x_3 = x_4 = 0$$
$$\text{optimum cost function} \quad f = -4$$

This solution corresponds to point C on Fig. 2.8. Note that any point on the line B–C also gives optimum solution. Multiple solutions can occur when the cost function is parallel to one of the constraints. For the present example, the cost function is parallel to the second constraint. ‖

In general, *if a reduced cost coefficient corresponding to a nonbasic variable is zero in the final tableau, there is a possibility of multiple optimum solutions.* From the practical standpoint, this is not a bad situation. Actually, it may be desirable because it gives choice to the designer; any suitable point on the straight line joining the two optimum designs can be selected to better suit the needs. Note that all optimum design points are global as opposed to local solutions.

Example 4.10 Identification of an unbounded problem with the Simplex method. Solve the LP problem:

maximize

$$z = x_1 - 2x_2$$

subject to

$$2x_1 - x_2 \geqq 0$$

$$-2x_1 + 3x_2 \leqq 6$$

$$x_1, x_2 \geqq 0$$

Solution. The problem has been solved graphically in Subsection 2.8.3. It can be seen from the graphical solution (Fig. 2.9), that the problem is unbounded. We will solve the problem using the Simplex method and see how we can recognize unbounded problems.

Writing the problem in the standard Simplex form, we obtain the initial canonical form shown in Table 4.12 where x_3 and x_4 are the slack variables. (Note that the first constraint has been transformed as $-2x_1 + x_2 \leqq 0$.) The basic feasible

TABLE 4.12
Initial canonical form for Example 4.10 (unbounded problem)

Basic ↓	x_1	x_2	x_3	x_4	b
x_3	-2	1	1	0	0
x_4	-2	3	0	1	6
Cost	**-1** ↑	2	0	0	$f - 0$

c_1' is negative, but no pivot element can be determined (all elements in x_1 column are negative); this implies unbounded solution.

solution is

$$\text{basic variables} \quad x_3 = 0, \, x_4 = 5$$

$$\text{nonbasic variables} \quad x_1 = x_2 = 0$$

$$\text{cost function} \quad f = 0$$

Scanning the last row, we find that the reduced cost coefficient for the nonbasic variable x_1 is negative. Therefore x_1 can become a basic variable. However, a pivot element cannot be selected in the first column because there is no positive element. There is no other possibility of selecting another nonbasic variable to become basic. The reduced cost coefficient for x_2 (the other nonbasic variable) is positive. Therefore, no pivot steps can be performed and yet we are not at the optimum point. Thus, the problem is unbounded.

The foregoing observation will be true in general. For *unbounded problems,* there will be negative reduced cost coefficients for nonbasic variables, but no possibility of pivot steps. ‖

4.5 INITIAL BASIC FEASIBLE SOLUTION – ARTIFICIAL VARIABLES

4.5.1 Two-phase Simplex Method

4.5.1.1 ARTIFICIAL VARIABLES. The foregoing section describes essential steps of the Simplex method for LP problems. An initial basic feasible point is required to initiate the procedure, i.e. an *initial canonical form* is needed. It is possible to obtain such a point by algebraic manipulation to convert the standard LP problem into the canonical form. If the problem consists solely of "$\leq$ type" constraints with nonnegative right-hand side parameters, then the addition of slack variables yields a canonical form. The initial basic feasible solution is immediately available by setting all original design variables to zero (i.e. they are nonbasic). In many design problems, however, there are "$\geq$ type" and "equality" constraints. For such constraints, initial basic feasible solution is not readily available. To obtain such a solution, new nonnegative variables are added to each "$\geq$ type" or "equality" constraint. They are called the *artificial variables* which are different from the surplus variables. They have no physical meaning; however, with their addition we obtain an initial basic feasible solution by treating them as basic.

The artificial variables augment the convex polyhedron of the original problem. The initial basic feasible solution corresponds to an extreme point (vertex) located in the new expanded space. The problem now is to traverse extreme points in the expanded space until one is reached in the original space. When the original space is reached all artificial variables will be nonbasic (i.e. they will have zero value). At this point the augmented space is literally removed so that future movements are only among the extreme points of the original space until the optimum is reached. In short, after creating artificial variables we eliminate them as quickly as possible. The preceding procedure is called the *two-phase Simplex method* of LP.

4.5.1.2 ARTIFICIAL COST FUNCTION. To eliminate the artificial variables, we define an auxiliary function known as the *artificial cost function*. It is simply a sum of all the artificial variables. If each constraint of the standard LP problem needs an artificial variable, then the optimization problem for Phase I is defined as (recall that n is the number of variables and m is the number of constraints in the standard LP problem of Eqs (4.8) to (4.10)):

minimize

$$w = x_{n+1} + x_{n+2} + \ldots + x_{n+m} = \sum_{i=1}^{m} x_{n+i} \tag{4.29}$$

subject to the constraints

$$a_{11}x_1 + a_{12}x_2 + \ldots + a_{1n}x_n + x_{n+1} = b_1$$
$$a_{21}x_1 + a_{22}x_2 + \ldots + a_{2n}x_n + x_{n+2} = b_2$$
$$\cdot \qquad \cdot \qquad \cdots \qquad \cdot \qquad \cdot \qquad \cdot \tag{4.30}$$
$$\cdot \qquad \cdot \qquad \cdots \qquad \cdot \qquad \cdot \qquad \cdot$$
$$a_{m1}x_1 + a_{m2}x_2 + \ldots + a_{mn}x_n + x_{n+m} = b_m$$
$$x_i \geq 0; \qquad i = 1 \text{ to } (n+m)$$

where x_{n+j}; $j = 1$ to m are the artificial variables. The constraints of Eq. (4.30) can be written in the summation notation as

$$\sum_{j=1}^{n} a_{ij}x_j + x_{n+i} = b_i; \qquad i = 1 \text{ to } m \tag{4.31}$$

The preceding minimization problem for Phase I is not in a form suitable for the Simplex method. The reason is that the reduced cost coefficients c_j' are not yet known. They can be identified only if w is in terms of the nonbasic variables. Currently w is in terms of basic variables $x_{n+1}, \ldots, x_{n+m}$. It must be in terms of $x_1, \ldots, x_n$. This can be done using the constraint expressions to eliminate the basic variables from the artificial cost function. Calculating $x_{n+1}, \ldots, x_{n+m}$ from Eqs (4.30) and substituting into Eq. (4.29), we obtain the cost function w in terms of the nonbasic variables x_j; $j = 1$ to n as

$$w = \sum_{i=1}^{m} b_i - \sum_{j=1}^{n} \sum_{i=1}^{m} a_{ij}x_j \tag{4.32}$$

The reduced cost coefficients c_j' are identified as the coefficients of x_j in Eq. (4.32) as

$$c_j' = -\sum_{i=1}^{m} a_{ij}; \qquad j = 1 \text{ to } n \tag{4.33}$$

The standard Simplex procedure can now be employed to solve the auxiliary optimization problem of Phase I. During this phase, the original cost function is treated as a constraint and the elimination step is also executed for it. This way, the real cost function is in terms of the nonbasic variables only at the end of Phase I and the Simplex method can be continued during Phase II. All artificial variables become nonbasic at the end of Phase I. Since w is a sum of all the artificial variables, its minimum value is clearly zero. When $w = 0$, an extreme point of the original convex set is reached. w is then discarded in favor of f and iterations continue until the minimum of f is obtained. Suppose, however, that w cannot be driven to zero. This will be apparent when none of the reduced cost coefficients for the artificial cost function is negative and yet w is greater than zero. Clearly, this means that we cannot reach the original convex set and, therefore, *no feasible solution exists for the original design problem, i.e. it is an infeasible problem.* At this point the designer should re-examine the formulation of the problem which may be over-constrained or improperly formulated.

If there are also "$\leq$ type" constraints in the original problem, these are cast into the standard LP form by adding slack variables that serve as basic variables in Phase I. Therefore, the number of artificial variables is less than m – the total number of constraints. Accordingly, the number of artificial variables required to obtain an initial basic feasible solution is also less than m. This implies that the sums in Eqs (4.32) and (4.33) are not for all the m constraints. They are only over the constraints requiring an artificial variable.

4.5.1.3 PHASE I ALGORITHM. The Simplex method for finding an initial basic feasible solution is described by the following Phase I algorithm.

Step 1. Introduce slack and surplus variables in the constraints and express them in the form of Eq. (4.9). The right-hand side constants of all the constraints must be nonnegative. If the jth design variable y_j is unrestricted in sign, substitute $y_j = y_j^+ - y_j^-$ in all the equations.

Step 2. Define artificial variables for the equality and the "$\geq$ type" constraints, as in Eqs (4.30). This system of equations then gives a canonical form for Phase I. Define an artificial cost function w as the sum of the artificial variables. Eliminate artificial variables from the cost function to express it in terms of only the nonbasic variables as in Eq. (4.32).

Step 3. Write a Simplex tableau for the problem. The artificial cost function is written in the last row and the original cost function in the second last row.

Step 4. Scan the last row of the tableau and note the most negative coefficient. Let its index be r. This implies that x_r should become a basic variable.

Step 5. Calculate the ratios of the right-hand side parameters with the positive coefficients in the rth column and select a row index s according to Eq. (4.24). This implies that x_s should become nonbasic. If Eq. (4.24) does not yield a pivot element (i.e. all elements are negative in the rth column), the *problem is unbounded.*

Step 6. Perform the pivot step on the rows of the tableau with a_{sr} as the pivot element and the sth row as the pivot row.

Step 7. If there are negative entries in the last row, go to Step 4. Otherwise go to Step 8.

Step 8. If all entries in the last row are nonnegative and w is zero, Phase I of the Simplex algorithm is complete and an initial basic feasible solution is obtained for the original problem. If all entries in the last row are nonnegative and $w \neq 0$, then the *problem is infeasible*.

4.5.1.4 PHASE II ALGORITHM. In the final tableau from Phase I, the last row is replaced by the actual cost function equation. The basic variables, however, should not appear in the cost function. Thus, pivot steps need to be performed on the cost function equation to eliminate the basic variables from it. A convenient way of accomplishing this is to treat the cost function as one of the equations in the Phase I tableau, say the second equation from the bottom. Elimination is performed on this equation along with others. In this way, the cost function will be in the correct form to continue with Phase II.

The *Phase II algorithm* for the Simplex method is described by the following steps:

Step 1. Same as Step 4 of Phase I.

Step 2. Same as Step 5 of Phase I.

Step 3. Same as Step 6 of Phase I.

Step 4. If there are negative entries in the last row, go to Step 1. Otherwise, the last tableau yields optimum solution for the problem.

Note that the *Simplex method is quite robust and convergent*. It is guaranteed to yield one of the following results:

1. If the problem is infeasible, the method will indicate that.
2. If the problem is unbounded, the method will indicate that.
3. If there is a solution to the problem, the method will find it. Note that the solution is a global one.
4. If there are multiple solutions, the method will indicate that.

Example 4.11 Use of artificial variable for "$\geq$ type" constraints. Find the optimum solution for the following LP problem using the Simplex method:

maximize
$$z = y_1 + 2y_2$$

subject to
$$3y_1 + 2y_2 \leq 12$$
$$2y_1 + 3y_2 \geq 6$$
$$y_1 \geq 0, \text{ and } y_2 \text{ is free in sign}$$

Solution. The graphical solution for the problem is shown in Fig. 4.4. It can be seen that the optimum solution is at point B. We shall use the two-phase simplex method to verify the solution.

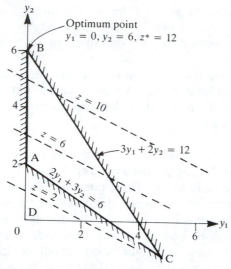

FIGURE 4.4
Graphical solution for Example 4.11.

Since y_2 is free in sign, we decompose it as $y_2 = y_2^+ - y_2^-$. To write the problem in the standard form, we define $x_1 = y_1$, $x_2 = y_2^+$ and $x_3 = y_2^-$, and transform the problem as

minimize
$$f = -x_1 - 2x_2 + 2x_3$$
subject to
$$3x_1 + 2x_2 - 2x_3 + x_4 = 12$$
$$2x_1 + 3x_2 - 3x_3 - x_5 = 6$$
$$x_i \geq 0; \quad i = 1 \text{ to } 5$$

where x_4 is a slack variable for the first constraint and x_5 is a surplus variable for the second constraint. We need to use the two-phase algorithm because the second constraint is "$\geq$ type". Accordingly, we introduce an artificial variable x_6 in the second constraint as

$$2x_1 + 3x_2 - 3x_3 - x_5 + x_6 = 6$$

The artificial cost function is defined as $w = x_6$. Since w should be in terms of nonbasic variables (x_6 is basic), we substitute for x_6 from the foregoing equation and obtain w as

$$w = 6 - 2x_1 - 3x_2 + 3x_3 + x_5$$

The initial tableau for Phase I is shown in Table 4.13. The initial basic variables are $x_4 = 12$ and $x_6 = 6$. The nonbasic variables are $x_1 = x_2 = x_3 = x_5 = 0$. Also $w = 6$ and $f = 0$. This corresponds to the infeasible point D in Fig. 4.4. According to Steps 4 and 5 of the Phase I algorithm, the pivot element is a_{22}, which implies that x_2 should become basic and x_6 should become nonbasic. Performing the pivot step, we obtain the second tableau given in Table 4.13. For

TABLE 4.13
Solution by the two-phase Simplex method for Example 4.11

Basic ↓	x_1	x_2	x_3	x_4	x_5	x_6	b	
x_6 out — x_4 x_2 in — $\rightarrow x_6$	3 2	2 **3**	-2 -3	1 0	0 -1	0 1	12 6	Phase I: initial tableau, pivot: a_{22}
Cost	-1	-2	2	0	0	0	$f-0$	
Artificial cost	-2	**-3**	3	0	1	0	$w-6$	
x_4 out — x_4 x_5 in — x_2	$\frac{5}{3}$ $\frac{2}{3}$	0 1	0 -1	1 0	$\frac{2}{3}$ $-\frac{1}{3}$	$-\frac{2}{3}$ $\frac{1}{3}$	8 2	Second tableau pivot: a_{15}
Cost	$\frac{1}{3}$	0	0	0	$-\frac{2}{3}$	$\frac{2}{3}$	$f+4$	
Artificial cost	0	0	0	0	0	1	$w-0$	End of Phase I
x_5 x_2	$\frac{5}{2}$ $\frac{3}{2}$	0 1	0 -1	$\frac{3}{2}$ $\frac{1}{2}$	1 0	-1 0	12 6	Third tableau: optimum solution
Cost	2 ↑	0	0	1 ↑	0	0	$f+12$	End of Phase II

Reduced cost coefficients in nonbasic columns are nonnegative; the third tableau gives optimum solution.

the second tableau $x_4 = 8$ and $x_2 = 2$ are the basic variables and all others are nonbasic. This corresponds to the feasible point A in Fig. 4.4. Since all the reduced cost coefficients of the artificial cost function are nonnegative and the artificial cost function is zero, an initial basic feasible solution for the original problem is obtained. Therefore, this is the end of Phase I.

For Phase II, column x_6 should be ignored in determining pivots. For the next step, the pivot element is a_{15} according to Steps 1 and 2. This implies that x_4 should be replaced by x_5 as a basic variable. The third tableau is obtained as shown in Table 4.13. The last tableau yields an optimum solution for the problem, which is $x_5 = 12$, $x_2 = 6$, $x_1 = x_3 = x_4 = 0$ and $f = -12$. The solution for the original design problem is then $y_1 = 0$, $y_2 = 6$, and $z = 12$, which agrees with the graphical solution of Fig. 4.4.

Note that the artificial variable column (x_6) in the final tableau is negative of the surplus variable column (x_5). This is true for all "$\geq$ type" constraints. ‖

Example 4.12 Use of artificial variables for equality constraints (infeasible problem). Solve the LP problem

maximize
$$z = x_1 + 4x_2$$

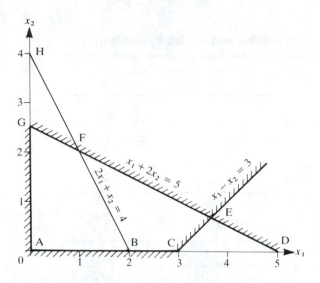

FIGURE 4.5
Constraints for Example 4.12. Infeasible problem.

subject to

$$x_1 + 2x_2 \leqq 5$$

$$2x_1 + x_2 = 4$$

$$x_1 - x_2 \geqq 3$$

$$x_1, x_2 \geqq 0$$

Solution. The constraints for the problem are plotted in Fig. 4.5. It can be seen that the problem has no feasible solution. We will solve the problem using the Simplex method to see how we can recognize an infeasible problem. Writing the problem in the standard LP form, we obtain

minimize
$$f = -x_1 - 4x_2$$
subject to

$$x_1 + 2x_2 + x_3 \qquad\qquad = 5$$

$$2x_1 + x_2 \qquad\quad + x_5 \quad = 4$$

$$x_1 - x_2 \qquad - x_4 \quad + x_6 = 3$$

$$x_i \geqq 0; \qquad i = 1 \text{ to } 6$$

Here x_3 is a slack variable, x_4 is a surplus variable, and x_5 and x_6 are artificial variables. Table 4.14 shows Phase I iterations of the Simplex method. It can be seen that after the first pivot step all the reduced cost coefficients of the artificial cost function for nonbasic variables are positive. However, the artificial cost

TABLE 4.14
Solution for Example 4.12 (infeasible problem)

Basic ↓	x_1	x_2	x_3	x_4	x_5	x_6	b	
x_5 out x_3 $\rightarrow x_5$ x_1 in x_6	1 **2** 1	2 1 −1	1 0 0	0 0 −1	0 1 0	0 0 1	5 4 3	*Initial tableau* *pivot:* a_{21}
Cost	−1	−4	0	0	0	0	$f-0$	
Artificial cost	**−3**	0	0	1	0	0	$w-7$	
x_3 x_1 x_6	0 1 0	$\frac{3}{2}$ $\frac{1}{2}$ $-\frac{3}{2}$	1 0 0	0 0 −1	$-\frac{1}{2}$ $\frac{1}{2}$ $-\frac{1}{2}$	0 0 1	3 2 1	*Second tableau*
Cost	0	$-\frac{7}{2}$	0	0	$\frac{1}{2}$	0	$f+2$	
Artificial cost	0	$\frac{3}{2}$	0	1	$\frac{3}{2}$	0	$w-1$	*End of Phase I*

Artificial variable x_6 is basic at the end of Phase I, i.e. the artificial cost function is not zero. Therefore, no feasible solution for the problem exists.

function is not zero. Therefore there is no feasible solution to the original problem. ‖

Example 4.13 Use of artificial variables (unbounded problem). Solve the LP problem:

maximize
$$z = 3x_1 - 2x_2$$
subject to
$$x_1 - x_2 \geqq 0$$
$$x_1 + x_2 \geqq 2$$
$$x_1, x_2 \geqq 0$$

Solution. The constraints for the problem are plotted in Fig. 4.6. It can be seen that the problem is unbounded. We will solve the problem by the Simplex method and see how to recognize *unboundedness*. Transforming the problem to the standard form, we get

minimize
$$f = -3x_1 + 2x_2$$
subject to
$$-x_1 + x_2 + x_3 \qquad = 0$$
$$x_1 + x_2 - x_4 + x_5 = 2$$
$$x_i \geqq 0; \qquad i = 1 \text{ to } 5$$

TABLE 4.15
Solution for Example 4.13 (unbounded problem)

Basic ↓	x_1	x_2	x_3	x_4	x_5	b	
x_5 out x_3 →x_5 x_1 in	-1	1	1	0	0	0	*Initial tableau,*
	1	1	0	-1	1	2	*pivot:* a_{21}
Cost	-3	2	0	0	0	$f-0$	
Artificial cost	**-1**	-1	0	1	0	$w-2$	
x_3	0	2	1	-1	1	2	*Second tableau*
x_1	1	1	0	-1	1	2	
Cost	0	5	0	**-3**	3	$f+6$	
Artificial cost	0	0	0	0	1	$w-0$	
End of Phase I					*End of Phase II*		

Reduced cost coefficient c_4' is negative but the pivot element cannot be determined, i.e. x_4 cannot be made basic (all elements in the x_4 column are negative in the second tableau). The problem is unbounded.

where x_3 is a slack, x_4 is a surplus and x_5 is an artificial variable. Note that the right-hand side of the first constraint is zero, so it can be treated as either "$\leq$ type" or "$\geq$ type". We will treat it as "$\leq$ type". Note also that the second constraint is "$\geq$ type", so we must use an artificial variable and an artificial cost function to find the initial basic feasible solution.

The solution for the problem is given in Table 4.15. For the initial tableau $x_3 = 0$ and $x_5 = 2$ are basic variables and all others are nonbasic. Note that this is a *degenerate basic feasible solution*. The solution corresponds to point A (the origin) in Fig. 4.6. Scanning the artificial cost row, we observe that there are two possibilities of pivot columns, x_1 and x_2. If x_2 is selected as the pivot column, then the first row must be the pivot row with $a_{12} = 1$ as the pivot element. This will make x_2 basic and x_3 nonbasic. However, x_2 will remain zero and the resulting solution will be degenerate corresponding to point A. One more iteration will be necessary to move from A to D. If we choose x_1 as the pivot column, then $a_{21} = 1$ will be the pivot element making x_1 as basic and x_5 as nonbasic. Carrying out the pivot step, we obtain the second tableau as shown in Table 4.15. The basic feasible solution is $x_1 = 2$, $x_3 = 2$ and other variables as zero. This solution corresponds to point D in Fig. 4.6. This is the basic feasible solution for the original problem as artificial cost function is zero, i.e. $w = 0$. The original cost function has also reduced to -6 from 0. This is the end of Phase I.

Scanning the cost function row, we find that the reduced cost coefficient in the x_4 column is negative. However, since all elements in the column are negative, there is no possibility of any further pivot steps. Thus the problem is unbounded.   ‖

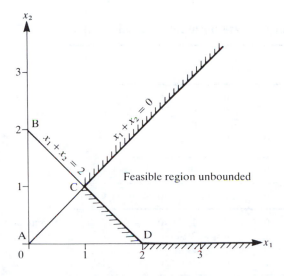

FIGURE 4.6
Constraints for Example 4.13. Unbounded problem.

4.5.1.5 DEGENERATE BASIC FEASIBLE SOLUTION. It is possible that during iterations of the simplex method, a basic variable attains zero value, i.e. the basic feasible solution becomes degenerate. What are the implications of this situation? We shall discuss them with the following example.

Example 4.14 Implications of degenerate basic feasible solution. Solve the following LP problem by the simplex method:

maximize
$$z = x_1 + 4x_2$$
subject to
$$x_1 + 2x_2 \leqq 5$$
$$2x_1 + x_2 \leqq 4$$
$$2x_1 + x_2 \geqq 4$$
$$x_1 - x_2 \geqq 1$$
$$x_1, x_2 \geqq 0$$

Solution. The problem is transcribed into the standard LP form as follows:

minimize
$$f = -x_1 - 4x_2$$
subject to
$$x_1 + 2x_2 + x_3 \qquad\qquad\qquad = 5$$
$$2x_1 + x_2 \qquad + x_4 \qquad\qquad\qquad = 4$$
$$2x_1 + x_2 \qquad\qquad - x_5 \qquad + x_7 \qquad = 4$$
$$x_1 - x_2 \qquad\qquad\qquad - x_6 \qquad + x_8 = 1$$
$$x_i \geqq 0; \qquad i = 1 \text{ to } 8$$

TABLE 4.16
Solution for Example 4.14 (degenerate basic feasible solution)

Basic ↓	x_1	x_2	x_3	x_4	x_5	x_6	x_7	x_8	b	
x_3	1	2	1	0	0	0	0	0	5	*Initial tableau,*
x_4	2	1	0	1	0	0	0	0	4	*pivot:* a_{14}
x_7	2	1	0	0	-1	0	1	0	4	
x_8 out → x_8 x_1 in	**1**	-1	0	0	0	-1	0	1	1	
Cost	-1	-4	0	0	0	0	0	0	$f-0$	
Artificial	**-3**	0	0	0	1	1	0	0	$w-5$	
x_3	0	3	1	0	0	1	0	-1	4	*Second tableau,*
x_4	0	3	0	1	0	2	0	-2	2	*pivot:* a_{23}
x_7 out → x_7 x_2 in	0	**3**	0	0	-1	2	1	-2	2	
x_1	1	-1	0	0	0	-1	0	1	1	
Cost	0	-5	0	0	0	-1	0	1	$f+1$	
Artificial	0	**-3**	0	0	1	-2	0	3	$w-2$	
x_4 out → x_3 x_5 in	0	0	1	0	1	-1	-1	1	2	*Third tableau,*
→ x_4	0	0	0	1	**1**	0	-1	0	0	*pivot:* a_{25}
x_2	0	1	0	0	$-\frac{1}{3}$	$\frac{2}{3}$	$\frac{1}{3}$	$-\frac{2}{3}$	$\frac{2}{3}$	
x_1	1	0	0	0	$-\frac{1}{3}$	$-\frac{1}{3}$	$\frac{1}{3}$	$\frac{5}{3}$	$\frac{5}{3}$	
Cost	0	0	0	0	$-\frac{5}{3}$	$\frac{7}{3}$	$\frac{5}{3}$	$-\frac{7}{3}$	$f+\frac{13}{3}$	
Artificial	0	0	0	0	0	0	1	0	$w-0$	

End of Phase I

Basic	x_1	x_2	x_3	x_4	x_5	x_6	x_7	x_8	b	
x_3	0	0	1	-1	0	-1	0	1	2	*Final tableau*
x_5	0	0	0	1	1	0	-1	0	0	
x_2	0	1	0	$\frac{1}{3}$	0	$\frac{2}{3}$	0	$-\frac{2}{3}$	$\frac{2}{3}$	
x_1	1	0	0	$\frac{1}{3}$	0	$-\frac{1}{3}$	0	$\frac{1}{3}$	$\frac{5}{3}$	
Cost	0	0	0	$\frac{5}{3}$	0	$\frac{7}{3}$	0	$-\frac{7}{3}$	$f+\frac{13}{3}$	

End of Phase II

where x_3 and x_4 are slack variables, x_5 and x_6 are surplus variables, and x_7 and x_8 are artificial variables. The two-phase Simplex procedure takes three iterations to reach the optimum point. These are given in Table 4.16. It can be seen that in the third tableau, the basic variable x_4 has zero value, so the *basic feasible solution is degenerate*. At this iteration, it is determined that x_5 should become basic, so x_5 is the pivot column. We need to determine the pivot row. We take ratios of the right-hand sides with the positive elements in the x_5 column. This determines the second row as the pivot row because it has the least ratio (zero). In general, if the element in the pivot column and the row that gives degenerate basic variable is positive, then that row must always be the pivot row; otherwise, the new solution cannot be feasible. Also, in this case, the new basic feasible solution will be degenerate, as for the final tableau in Table 4.16. The only way the new feasible solution can be nondegenerate is when the element in the pivot column and the degenerate variable row is negative. In that case the new basic feasible solution will be nondegenerate. It is theoretically possible for the Simplex method to fail by cycling between two degenerate basic feasible solutions. However, in practice this usually does not happen.

The final solution for this problem is

basic variables	$x_1 = \frac{5}{3}, \ x_2 = \frac{2}{3}, \ x_3 = 2, \ x_5 = 0$
nonbasic variables	$x_4 = x_6 = x_7 = x_8 = 0$
optimum cost function	$f = -\frac{13}{3}$ or $z = \frac{13}{3}$

‖

4.5.2 Alternate Simplex Method

A slightly different procedure can be used to solve linear programming problems having "$\geqq$ type" and equality constraints. The artificial variables are introduced into the problem as before. However, the artificial cost function is not used. Instead, *the original cost function is augmented by adding artificial variables having large positive constants. The additional terms act as penalties for having artificial variables in the problem.* Since artificial variables are basic, they need to be eliminated from the cost function before the Simplex method can be used. This can easily be done by using the appropriate constraints that contain artificial variables. Once this has been done, the regular simplex method can be used to solve the problem. We illustrate the procedure with an example problem.

Example 4.15 The Big-M method for equality and "$\geqq$ type" constraints. Find the numerical solution for the following problem (Example 4.11) using the alternate Simplex procedure:

maximize
$$z = y_1 + 2y_2$$

subject to

$$3y_1 + 2y_2 \leqq 12$$

$$2y_1 + 3y_2 \geqq 6$$

$y_1 \geqq 0$, y_2 is unrestricted in sign

Solution. Converting the problem to the standard form, we obtain:

minimize
$$f = -x_1 - 2x_2 + 2x_3$$

subject to

$$3x_1 + 2x_2 - 2x_3 + x_4 \qquad\qquad = 12$$
$$2x_1 + 3x_2 - 3x_3 \qquad\quad - x_5 + x_6 = 6$$
$$x_i \geq 0; \qquad i = 1 \text{ to } 6$$

Since y_2 is unrestricted, it has been defined as $y_2 = x_2 - x_3$. x_4 is a slack variable, x_5 is a surplus variable and x_6 is an artificial variable. Following the alternate Simplex procedure, we add Mx_6 (with, say, $M = 10$) to the cost function and define it as

$$f = -x_1 - 2x_2 + 2x_3 + 10x_6$$

Note that if there is a feasible solution to the problem, then all artificial variables will become nonbasic – i.e. zero – and we will recover the original cost function. Also note, that if there are other artificial variables, they will be multiplied by M and added to the cost function. This is sometimes called the *Big-M Method*. Now substituting for x_6 from the second constraint into the foregoing cost function, we get

$$f = -x_1 - 2x_2 + 2x_3$$
$$+ 10(6 - 2x_1 - 3x_2 + 3x_3 + x_5)$$
$$= 60 - 21x_1 - 32x_2 + 32x_3 + 10x_5$$

With this as the cost function, iterations of the Simplex method are shown in Table 4.17. It can be seen that the final solution is the same as given in Table 4.13 and Fig. 4.4. ‖

TABLE 4.17
Solution of Example 4.15 by alternate simplex method

	Basic ↓	x_1	x_2	x_3	x_4	x_5	x_6	b	
x_6 out → x_2 in	x_4 x_6	3 2	2 **3**	-2 -3	1 0	0 -1	0 1	12 6	Initial tableau, pivot: a_{22}
	Cost	-21	**-32**	32	0	10	0	$f - 60$	
x_4 out → x_5 in	x_4 x_2	$\frac{5}{3}$ $\frac{2}{3}$	0 1	0 -1	1 0	$\frac{2}{3}$ $-\frac{1}{3}$	$-\frac{2}{3}$ $\frac{1}{3}$	8 2	Second tableau, pivot: a_{15}
	Cost	$\frac{1}{3}$	0	0	0	$-\frac{2}{3}$	$\frac{32}{3}$	$f + 4$	
	x_5 x_2	$\frac{5}{2}$ $\frac{3}{2}$	0 1	0 -1	$\frac{3}{2}$ $\frac{1}{2}$	1 0	-1 0	12 6	Third tableau
	Cost	2	0	0	1	0	10	$f + 12$	

4.6 POST-OPTIMALITY ANALYSIS

The optimum solution of the LP problem depends on the parameters in vectors **c** and **b**, and matrix **A** defined in Eqs (4.13) to (4.15). These parameters are prone to errors in practical design problems. Thus we are interested not only in the optimum solution but also in how it changes when the parameters change. The changes may be either discrete (e.g. when we are uncertain about which of several choices is the value of a particular parameter) or continuous. *The study of discrete parameter changes is often called sensitivity analysis, and that of continuous changes is called parametric programming.*

There are five basic parametric changes affecting the solution:

1. Changes in the cost function coefficients, c_j.
2. Changes in the resource limits, b_i.
3. Changes in the constraint coefficients, a_{ij}.
4. The effect of including additional constraints.
5. The effect of including additional variables.

A thorough discussion of these changes, while not necessarily difficult, is beyond our scope. In principle, we could imagine solving a new problem for every change. Fortunately, for a small number of changes there are useful shortcuts. Almost all computer programs for LP problems provide some information about parameter variations. We shall study the parametric changes defined in items 1 through 3. *The final tableau contains all the information needed to study these changes.* We shall describe the information contained in the final tableau and its use to study the three parametric changes. For other variations, Randolph and Meeks [1978] may be consulted.

It turns out that the optimum solution of the altered problem can be computed using the optimum solution of the original problem if changes in the parameters are within certain limits. In the following discussion we use a'_{ij}, c'_j and b'_i to represent the corresponding values of the parameters a_{ij}, c_j and b_i in the final tableau.

4.6.1 Changes in Resource Limits

First, we study how the optimum value of the cost function for the problem changes if we change the right-hand side parameters, b_i's (also known as resource limits), of the constraints. The constraint variation sensitivity theorem of Chapter 3 can be used to study the effect of these changes. Use of that theorem requires knowledge of the Lagrange multipliers for the constraints. Therefore, we must determine them. The following theorem gives a way of recovering the multipliers for the constraints of an LP problem from the final tableau.

Theorem 4.5 Lagrange Multiplier Values. In the final tableau, the Lagrange multiplier for the ith constraint equals the reduced cost coefficient in the slack or artificial variable column associated with the ith constraint. A "$\geqq$ type" constraint treated by adding an artificial variable in the Simplex method always has a nonpositive Lagrange multiplier.   ‖

It is important to understand the theorem to properly recover Lagrange multipliers from the final tableau. Let us first consider a "$\leqq$ type" constraint with a nonnegative right-hand side. For such a constraint, the Lagrange multiplier is the reduced cost coefficient (in the final tableau) in the slack variable column associated with the constraint. The *Lagrange multiplier for "$\leqq$ type" constraint is always nonnegative* according to Theorem 3.6.

Next consider an equality constraint with a nonnegative right-hand side. This constraint is treated in the Simplex method by adding an artificial variable. According to Theorem 4.5, its Lagrange multiplier is the reduced cost coefficient of the artificial variable (in the final tableau). *The Lagrange multiplier for an equality constraint may be positive, negative or zero.*

A "$\geqq$ type" constraint is also treated in the Simplex method by adding an artificial variable. Its Lagrange multiplier is the reduced cost coefficient of the artificial variable in the final tableau. *Note that the Lagrange multiplier for a "$\geqq$ type" constraint must be always nonpositive.*

Another way of recovering the Lagrange multiplier for "$\geqq$ type" constraints is to use the surplus variable column. As observed earlier, the surplus variable column in the final tableau is negative of the artificial variable column. Therefore, the reduced cost coefficient of the surplus variable can also be used to recover the Lagrange multiplier for a "$\geqq$ type" constraint. The coefficient is always positive in the final tableau. Therefore, negative of that coefficient gives the Lagrange multiplier for the "$\geqq$ type" constraint.

In Chapter 3, Section 3.7, a *physical meaning of the Lagrange multipliers* was described. There the Lagrange multipliers were related to derivatives of the cost function with respect to the right-hand side parameters (Theorem 3.14). There equality and inequality constraints were treated separately with v_i and u_i as their Lagrange multipliers respectively. In this section, we use a slightly different notation. We use e_i as the right-hand side parameter of any constraint and y_i as its Lagrange multiplier. Using this notation and Theorem 3.14, we obtain the following derivative of the cost function with respect to the right-hand side parameters:

$$\frac{\partial f}{\partial e_i} = -y_i \qquad (4.34)$$

It is remarked here that Theorem 4.5 and Eq. (4.34) are *applicable only if changes in the right-hand side parameters are within certain ranges,* i.e. there are upper and lower limits on changes in the resource limits for which Eq. (4.34) is valid. Calculations for these limits are discussed in Subsection 4.6.2.

Also, *note that Theorem 3.14 and Eq. (4.34) are applicable only for a minimization problem with "$\leqq$ type" or equality constraints.* Therefore, care must be exercised in using the proper cost and constraint function expressions to determine y_i and the effect of changing the resource limits.

As noted in the preceding, *the final tableau for the problem contains all the information to recover Lagrange multipliers for the constraints.* Care must be used to recover proper values and signs for the Lagrange multipliers. Theorem 4.5 describes the process to recover this information from the final tableau which is *summarized* as follows:

1. For the "$\leq$ *type*" ith constraint the Lagrange multiplier $y_i \geq 0$ is the *reduced cost coefficient in the slack variable column* associated with the constraint.

2. For a "$\leq$ type" constraint with a negative right-hand side, we must multiply by -1 to treat it in the Simplex method. This makes it a "$\geq$ *type*" constraint and we must use *surplus and artificial variables* to treat it in the Simplex method. Its Lagrange multiplier $y_i \geq 0$ is the reduced cost coefficient in the surplus variable column. Note, however, that the multiplier is for the constraint written in the "$\leq$ form". As an example, let the reduced cost coefficient of the surplus variable associated with a constraint $x_1 - x_2 \geq 1$ be $\frac{7}{3}$. Therefore, the Lagrange multiplier for the constraint written in the $\leq$ form, $-x_1 + x_2 \leq -1$, is $\frac{7}{3}$. This point is important for the proper use of Eq. (4.34). Without this conversion, use of Eq. (4.34) will be incorrect. Note that for the "$\leq$ type" constraints, the Lagrange multipliers have to be nonnegative (according to the Kuhn–Tucker conditions of Chapter 3). The Lagrange multipliers for the "$\geq$ *type*" *constraints have to be nonpositive.* For the foregoing example, the Lagrange multiplier associated with the constraint $x_1 - x_2 \geq 1$ is $-\frac{7}{3}$.

3. The *Lagrange multipliers must also satisfy the switching conditions* ($u_i s_i = 0$) of Eq. (3.50). Therefore, if the Lagrange multiplier for a constraint is zero, then the constraint must be inactive. Its slack variable must be positive (except for the unusual case where both the Lagrange multiplier and the constraint function have zero value).

4. For the ith *equality constraint, the Lagrange multiplier is the reduced cost coefficient in the artificial variable column* associated with the constraint. Note that the Lagrange multiplier for an equality constraint is *unrestricted in sign.*

Example 4.16 Recovery of Lagrange multipliers from final tableau. Consider the problem

maximize
$$z = 5x_1 - 2x_2$$

subject to
$$2x_1 + x_2 \leq 9$$
$$x_1 - 2x_2 \leq 2$$
$$-3x_1 + 2x_2 \leq 3$$
$$x_1, x_2 \geq 0$$

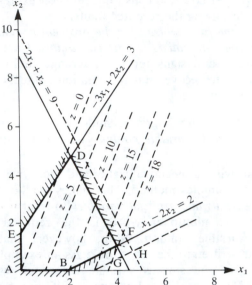

FIGURE 4.7
Graphical solution for Example 4.16.

Solve the problem by the Simplex method. Recover Lagrange multipliers for the constraints.

Solution. Constraints for the problem and cost function contours are plotted in Fig. 4.7. The optimum solution is at point C and is given as $x_1 = 4$, $x_2 = 1$, $z = 18$.

Solving the problem using the Simplex method, we obtain the sequence of calculations given in Table 4.18. From the final tableau,

basic variables	$x_1 = 4$, $x_2 = 1$, $x_5 = 13$
nonbasic variables	$x_3 = 0$, $x_4 = 0$
maximum objective function	$z = 18$ (minimum value is -18)

In the problem formulation, x_3, x_4 and x_5 are the slack variables for the three constraints. Since all constraints are "$\leq$ type", the reduced cost coefficients for the slack variables are the Lagrange multipliers as follows:

first constraint $y_1 = 1.6$ (c_3' in column x_3)

second constraint $y_2 = 1.8$ (c_4' in column x_4)

third constraint $y_3 = 0$ (c_5' in column x_5)

Therefore, Eq. (4.34) gives

$$\frac{\partial f}{\partial e_1} = -1.6; \qquad \frac{\partial f}{\partial e_2} = -1.8; \qquad \frac{\partial f}{\partial e_3} = 0$$

where $f = -(5x_1 - 2x_2)$; Eq. (4.34) is valid for a minimization problem. If the right-hand side of the first constraint changes from 9 to 10, the cost function f

TABLE 4.18
Solution of Example 4.16 by the Simplex method

	Basic ↓	x_1	x_2	x_3	x_4	x_5	b	
x_4 out $\to x_4$ x_1 in	x_3	2	1	1	0	0	9	*Initial tableau,*
	x_4	**1**	−2	0	1	0	2	*pivot: a_{21}*
	x_5	−3	2	0	0	1	3	
	Cost	**−5**	2	0	0	0	$f - 0$	
x_3 out $\to x_3$ x_2 in	x_3	0	**5**	1	−2	0	5	*Second tableau,*
	x_1	1	−2	0	1	0	2	*pivot: a_{12}*
	x_5	0	−4	0	3	1	9	
	Cost	0	**−8**	0	5	0	$f + 10$	
	x_2	0	1	0.2	−0.4	0	1	*Third tableau*
	x_1	1	0	0.4	0.2	0	4	
	x_5	0	0	0.8	1.4	1	13	
	Cost	0	0	1.6	1.8	0	$f + 18$	
		↑	↑	↑	↑	↑		
		c_1'	c_2'	c_3'	c_4'	c_5'		

x_3, x_4, and x_5 are slack variables.

reduces by 1.6, i.e. the new value of f will be −19.6 ($z = 19.6$). Point F in Fig. 4.7 gives the new optimum solution for this case. If the right-hand side of the second constraint changes from 2 to 3, the cost function f reduces by 1.8 to −19.8. Point G in Fig. 4.7 gives the new optimum solution. Note that any small change in the right-hand side of the third constraint will have no effect on the cost function.

When the right-hand side of first and second constraints are changed to 10 and 3 simultaneously, the net change in the cost function is −(1.6 + 1.8), i.e. new f will be −21.4. The new solution is at point H in Fig. 4.7. ‖

It is noted (as in Section 3.7) that the *Lagrange multipliers are very useful* for practical design problems. Their values give the relative effect of changes in the right-hand side parameters of constraints (resource limits). Using their relative values, the designer can determine the most profitable way to adjust the resource limits, if necessary and possible. *The Lagrange multipliers are sometimes called the dual variables* (or, *dual prices*). The concept of duality in linear programming is described in the next section.

Example 4.17 Recovery of Lagrange multipliers from final tableau. Solve the following LP problem and recover proper Lagrange multipliers for the constraints:

maximize
$$z = x_1 + 4x_2$$

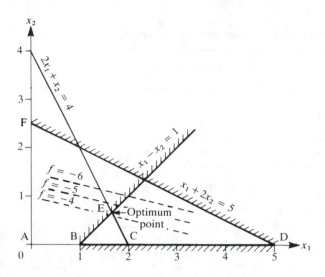

FIGURE 4.8
Constraints for Example 4.17. Feasible region: line E–C.

subject to

$$x_1 + 2x_2 \leqq 5$$

$$2x_1 + x_2 = 4$$

$$x_1 - x_2 \geqq 1$$

$$x_1, x_2 \geqq 0$$

Solution. Constraints for the problem are plotted in Fig. 4.8. It can be seen that line E–C is the feasible region for the problem and point E gives the optimum solution. Converting the problem to standard simplex form, we obtain:

minimize

$$f = -x_1 - 4x_2$$

subject to

$$x_1 + 2x_2 + x_3 \qquad\qquad = 5$$

$$2x_1 + x_2 \qquad\quad + x_5 \quad = 4$$

$$x_1 - x_2 \qquad - x_4 \quad + x_6 = 1$$

$$x_i \geqq 0; \qquad i = 1 \text{ to } 6$$

where x_3 is a slack variable, x_4 is a surplus variable, and x_5 and x_6 are artificial variables. The problem, solved in Table 4.19, takes just two iterations to reach the optimum. The solution from the final tableau is

basic variables $x_1 = \frac{5}{3},\ x_2 = \frac{2}{3},\ x_3 = 2$

nonbasic variables $x_4 = 0,\ x_5 = 0,\ x_6 = 0$

cost function $f = -\frac{13}{3}$

TABLE 4.19
Solution for Example 4.17 with equality constraint

Basic ↓	x_1	x_2	x_3	x_4	x_5	x_6	b	
x_3	1	2	1	0	0	0	5	Initial tableau
x_5	2	1	0	0	1	0	4	
x_6	**1**	-1	0	-1	0	1	1	
Cost	-1	-4	0	0	0	0	$f - 0$	
Artificial	**-3**	0	0	1	0	0	$w - 5$	
x_3	0	3	1	1	0	-1	4	Second tableau
x_5	0	**3**	0	2	1	-2	2	
x_1	1	-1	0	-1	0	1	1	
Cost	0	-5	0	-1	0	1	$f + 1$	
Artificial	0	**-3**	0	-2	0	3	$w - 2$	
x_3	0	0	1	-1	-1	1	2	Third tableau
x_2	0	1	0	$\frac{2}{3}$	$\frac{1}{3}$	$-\frac{2}{3}$	$\frac{2}{3}$	
x_1	1	0	0	$-\frac{1}{3}$	$\frac{1}{3}$	$\frac{1}{3}$	$\frac{5}{3}$	
Cost	0 (c_1')	0 (c_2')	0 (c_3')	$\frac{7}{3}$ (c_4')	$\frac{5}{3}$ (c_5')	$-\frac{7}{3}$ (c_6')	$f + \frac{13}{3}$	
Artificial	0	0	0	0	1	1	$w - 0$	
	End of Phase I				End of Phase II			

x_3, slack variable; x_4, surplus variable; x_5, x_6, artificial variables.

Note that artificial variable column (x_6) is negative of the surplus variable column (x_4) for the third constraint. Using Theorem 4.5, the Lagrange multipliers for the constraints are

first constraint ($x_1 + 2x_2 \leqq 5$)

$$y_1 = 0 \ (c_3' \text{ in the slack variable column } x_3)$$

second constraint ($2x_1 + x_2 = 4$)

$$y_2 = \tfrac{5}{3} \ (c_5' \text{ in the artificial variable column } x_5)$$

third constraint ($x_1 - x_2 \geqq 1$)

$$y_3 = -\tfrac{7}{3} \ (c_6' \text{ in the artificial variable column } x_6)$$

When the third constraint is written in the "$\leqq$ form" ($-x_1 + x_2 \leqq -1$), its Lagrange multiplier is $\tfrac{7}{3}$ which is negative of the preceding value. Note that it is also c_4' in the surplus variable column x_4.

When the right-hand side of the equality constraint is changed to 5 from 4, the cost function $(-x_1 - 4x_2)$ changes by

$$\frac{\partial f}{\partial e_2} = -y_2 = -\tfrac{5}{3}$$

That is, the cost function will reduce by $\tfrac{5}{3}$, from $-\tfrac{13}{3}$ to -6 ($z = 6$). When the right-hand side of the third constraint is changed to 2 (i.e. $-x_1 + x_2 \leqq -2$), the cost function changes by

$$\frac{\partial f}{\partial e_3} = -\tfrac{7}{3}(-1) = \tfrac{7}{3}$$

That is, the cost function will increase by $\tfrac{7}{3}$, from $-\tfrac{13}{3}$ to -2 ($z = 2$). This can be also observed from Fig. 4.8. ‖

4.6.2 Ranging Right-hand Side Parameters

When the right-hand side of a constraint is changed, the constraint boundary moves parallel to itself changing the feasible region for the problem. However, the iso-cost lines do not change. Since the feasible region is changed, the optimum solution may change, i.e. the design variables as well as the cost function may change. There are, however, certain limits on changes for which the optimum solution of the altered problem can be obtained from the information contained in the final tableau. That is, if the changes are within a certain range, the sets of basic and nonbasic variables do not change.

> **Theorem 4.6   Limits on changes in resources.** Let Δ_k be the possible change in the right-hand side b_k of the kth constraint. If Δ_k satisfies the following inequalities, then no more iterations of the Simplex method are required to obtain solution for the altered problem
>
> $$\max_i \{-b_i'/a_{ij}'; \, a_{ij}' > 0\} \leqq \Delta_k \leqq \min_i \{-b_i'/a_{ij}'; \, a_{ij}' < 0\}; \qquad (4.35)$$
>
> where $i = 1$ to m,
>
> b_i' = right-hand side parameter for the ith constraint in the final tableau,
>
> a_{ij}' = parameters in the jth column of the final tableau; the jth column corresponds to x_j which is the slack variable for a "$\leqq$ type" constraint, or the artificial variable for an equality, or "$\geqq$ type" constraint,
>
> Δ_k = possible change in the right-hand side of the kth constraint; the slack or the artificial variable for the kth constraint determines the index j of the column whose elements are used in the Inequalities (4.35).
>
> Furthermore, the new right-hand side parameters b_i'' due to a change of Δ_k in b_k are given as
>
> $$b_i'' = b_i' + \Delta_k a_{ij}'; \qquad i = 1 \text{ to } m \qquad (4.36) \quad ‖$$

Using Eq. (4.36) and the final tableau, new values for the basic variables in each row can be obtained. Equation (4.36) is applicable only if Δ_k is in the range determined from Eq. (4.35). To determine the range, we first determine the column index j according to the rules given in Theorem 4.6. Then in the jth column, we determine the ratios $-b_i'/a_{ij}'$ for all $a_{ij}'>0$. The maximum value among these ratios gives the lower limit on change Δ_k in b_k. If there is no $a_{ij}'>0$, then the said ratios cannot be found. In that case, there is no lower bound on change Δ_k in b_k, i.e. the lower limit is $-\infty$. To compute an upper limit on Δ_k, we determine ratios $-b_i'/a_{ij}'$ for all $a_{ij}<0$. The minimum value among the ratios gives the upper limit on change Δ_k in b_k. If there is no $a_{ij}'<0$, then the said ratios cannot be found. In that case, there is no upper bound on change Δ_k in b_k, i.e. the upper limit is ∞.

Example 4.18 Ranges for resource limits – "$\leq$ type" constraints. Find ranges for the right-hand side parameters of constraints in Example 4.16.

Solution. The graphical solution for the problem is shown in Fig. 4.7. The problem written in the standard form is given as

minimize
$$f = -5x_1 + 2x_2$$
subject to
$$2x_1 + x_2 + x_3 = 9$$
$$x_1 - 2x_2 + x_4 = 2$$
$$-3x_1 + 2x_2 + x_5 = 3$$
$$x_i \geq 0; \quad i = 1 \text{ to } 5$$

where x_3, x_4 and x_5 are the slack variables. The final tableau for the problem is copied from Table 4.18 in Table 4.20.

For the first constraint, x_3 is the slack variable. Therefore, j for Δ_1 is 3 (the third column). Since there is no negative element in the third column, there is no

TABLE 4.20
Final tableau for Example 4.18 (ranging resource limits)

Basic ↓	x_1	x_2	x_3	x_4	x_5	b
x_2	0	1	0.2	-0.4	0	1
x_1	1	0	0.4	0.2	0	4
x_5	0	0	0.8	1.4	1	13
Cost	0	0	1.6	1.8	0	$f+18$

x_3, x_4, and x_5 are slack variables.
For first constraint, use x_3 column, $j=3$ in Eq. (4.35).
For second constraint, use x_4 column, $j=4$ in Eq. (4.35).
For third constraint, use x_5 column, $j=5$ in Eq. (4.35).

upper limit on Δ_1. To find the lower limit, we take the ratios $-b_i'/a_{ij}'$ and find the maximum element:

$$\max\left\{-\frac{1}{0.2}, -\frac{4}{0.4}, -\frac{13}{0.8}\right\} \leq \Delta_1$$

or, $-5 \leq \Delta_1$. Thus, limits for Δ_1 are

$$-5 \leq \Delta_1 \leq \infty$$

and the range on b_1 is obtained by adding the current value of $b_1 = 9$, as

$$4 \leq b_1 \leq \infty$$

For the second constraint ($k = 2$), x_4 is the slack variable. Therefore, we will use elements in the fourth column of the final tableau (a_{i4}', $j = 4$) in the equalities of Eq. (4.35). From the final tableau, we obtain

$$\max\left\{-\frac{4}{0.2}, -\frac{13}{1.4}\right\} \leq \Delta_2 \leq \min\left\{-\frac{1}{(-0.4)}\right\}$$

or, range for Δ_2 is

$$-9.286 \leq \Delta_2 \leq 2.5$$

Therefore, the allowed decrease in b_2 is 9.286 and the allowed increase is 2.5. Adding 2 to the above inequality (the current value of b_2), the range on b_2 is given as

$$-7.286 \leq b_2 \leq 4.5$$

Similarly, for the third constraint, the range for Δ_3 is given as

$$\max\left\{-\tfrac{13}{1}\right\} \leq \Delta_3 \leq \infty, \quad \text{or} \quad -13 \leq \Delta_3 \leq \infty$$

That is, the allowed decrease is 13 and the allowed increase is infinity for b_3. Therefore, range on b_3 is given as

$$-10 \leq b_3 \leq \infty$$

New values of design variables. Let us calculate new values of design variables if the right-hand side of the first constraint is changed from 9 to 10. Note that this change is within the limits determined in the foregoing. In Eq. (4.36), $k = 1$, so $\Delta_1 = 1$. Also, $j = 3$, so we use the third column from Table 4.20 in Eq. (4.36) and obtain new values of the variables as

$$x_2 \equiv b_1'' = b_1' + \Delta_1 a_{13}'$$
$$= 1 + (1)(0.2) = 1.2$$
$$x_1 \equiv b_2'' = b_2' + \Delta_1 a_{23}'$$
$$= 4 + (1)(0.4) = 4.4$$
$$x_5 \equiv b_3'' = b_3' + \Delta_1 a_{33}'$$
$$= 13 + (1)(0.8) = 13.8$$

The other variables remain nonbasic, so they have zero values. The new solution corresponds to point F in Fig. 4.7.

If the right-hand side of the second constraint is changed from 2 to 3, the new values of the variables, using Eq. (4.36) and the x_4 column from Table 4.20, are given as

$$x_2 \equiv b_1'' = b_1' + \Delta_2 a_{14}'$$
$$= 1 + (1)(-0.4) = 0.6$$
$$x_1 \equiv b_2'' = b_2' + \Delta_2 a_{24}'$$
$$= 4 + (1)(0.2) = 4.2$$
$$x_5 \equiv b_3'' = b_3' + \Delta_2 a_{34}'$$
$$= 13 + (1)(1.4) = 14.4$$

This solution corresponds to point G in Fig. 4.7.

When the right-hand sides of two or more constraints are changed simultaneously, we can use Eq. (4.36) to determine new values of the design variables. However, we have to make sure that the new right-hand sides do not change the basic and nonbasic set of variables, i.e. the vertex that gives the optimum solution is not changed. Or, in other words, no new constraint becomes active. As an example, let us calculate the new values of design variables using Eq. (4.36) when the right-hand side of the first and the second constraints are changed to 10 and 3 from 9 and 2, respectively:

$$x_2 \equiv b_1'' = b_1' + \Delta_1 a_{13}' + \Delta_2 a_{14}'$$
$$= 1 + (1)(0.2) + (1)(-0.4) = 0.8$$
$$x_1 \equiv b_2'' = b_2' + \Delta_1 a_{23}' + \Delta_2 a_{24}'$$
$$= 4 + (1)(0.4) + (1)(0.2) = 4.6$$
$$x_5 \equiv b_3'' = b_3' + \Delta_1 a_{33}' + \Delta_2 a_{34}'$$
$$= 13 + (1)(0.8) + (1)(1.4) = 15.2$$

It can be verified that the new solution corresponds to point H in Fig. 4.7. ‖

Example 4.19 Ranges for resource limits – equality and "≧ type" constraints. Find ranges for the right-hand side parameters of the following problem:

maximize
$$z = x_1 + 4x_2$$
subject to
$$x_1 + 2x_2 \leq 5$$
$$2x_1 + x_2 = 4$$
$$x_1 - x_2 \geq 1$$
$$x_1, x_2 \geq 0$$

Solution. This problem is solved in Example 4.17 and the final tableau is copied from Table 4.19 in Table 4.21. The graphical solution for the problem is given in Fig. 4.8. In the tableau, x_3 is a slack variable for the first constraint, x_4 is a surplus

TABLE 4.21

Final tableau for Example 4.19 (Ranging resource limits)

Basic ↓	x_1	x_2	x_3	x_4	x_5	x_6	b
x_3	0	0	1	-1	-1	1	2
x_2	0	1	0	$\frac{2}{3}$	$\frac{1}{3}$	$-\frac{2}{3}$	$\frac{2}{3}$
x_1	1	0	0	$-\frac{1}{3}$	$\frac{1}{3}$	$\frac{1}{3}$	$\frac{5}{3}$
Cost	0	0	0	$\frac{7}{3}$	$\frac{5}{3}$	$-\frac{7}{3}$	$f + \frac{13}{3}$

x_3, slack variable; x_4, surplus variable; x_5, x_6, artificial variables.
For first constraint, use x_3 column, $j = 3$ in Eq. (4.35).
For second constraint, use x_5 column, $j = 5$ in Eq. (4.35).
For third constraint, use x_6 column, $j = 6$ in Eq. (4.35).

variable for the third constraint, x_5 is an artificial variable for the second constraint, and x_6 is an artificial variable for the third constraint.

For the first constraint, x_3 is the slack variable. Therefore, index j for Δ_1 is 3 for use in Inequalities (4.35). There is no negative element in column 3, so there is no upper limit on Δ_1. The lower limit on Δ_1 according to Eq. (4.35) is -2. Therefore, range for Δ_1 is $-2 \leqq \Delta_1 \leqq \infty$. Or, adding current value of $b_1 = 5$, range on b_1 is

$$3 \leqq b_1 \leqq \infty$$

The second constraint is an equality. The index j for use in Inequalities (4.35) is determined by the artificial variable for the constraint which is x_5, i.e. $j = 5$. Accordingly, the range for Δ_2 is given as

$$\max \left\{ - \left(\tfrac{2}{3}\right)/\left(\tfrac{1}{3}\right), \ - \left(\tfrac{5}{3}\right)/\left(\tfrac{1}{3}\right) \right\} \leqq \Delta_2 \leqq \min \left\{ -2/(-1) \right\}$$

Or, $-2 \leqq \Delta_2 \leqq 2$. Range on b_2 can be found by adding the current value of $b_2 = 4$ to above inequality, $2 \leqq b_2 \leqq 6$.

The third constraint is a "$\geqq$ type", so index j for use in Inequalities (4.35) is determined by the artificial variable. This gives $j = 6$. Taking ratios of the right-hand side parameters with the elements in the sixth column, we get the range for Δ_3 as

$$\max \left\{ -\tfrac{2}{1}, \ - \left(\tfrac{5}{3}\right)/\left(\tfrac{1}{3}\right) \right\} \leqq \Delta_3 \leqq \min \left\{ - \left(\tfrac{2}{3}\right)/\left(- \tfrac{2}{3}\right) \right\}$$

or, $-2 \leqq \Delta_3 \leqq 1$. The limits on changes in b_3 are (add current value of $b_3 = 1$ on both sides of the above inequality) $-1 \leqq b_3 \leqq 2$.

New values of design variables. We can use Eq. (4.36) to calculate the new values of the design variables for the right-hand side changes that remain within the previously determined ranges. It can be seen that since the first constraint is not active, it does not affect the optimum solution as long as its right-hand side remains within the range $3 \leqq b_1 \leqq \infty$ that is determined in the foregoing.

Let us determine the new solution when the right-hand side of the second constraint is changed to 5 from 4 (the change is within the range determined previously). The second constraint has x_5 as an artificial variable, so we used column 5 from Table 4.21 in Eq. (4.35) and obtain the new values of the variables as follows:

$$x_3 \equiv b_1'' = b_1' + \Delta_2 a_{15}'$$
$$= 2 + (1)(-1) = 1$$
$$x_2 \equiv b_2'' = b_2' + \Delta_2 a_{25}'$$
$$= \tfrac{2}{3} + (1)(\tfrac{1}{3}) = 1$$
$$x_1 \equiv b_3'' = b_3' + \Delta_2 a_{35}'$$
$$= \tfrac{5}{3} + (1)(\tfrac{1}{3}) = 2$$

The solution can be verified by solving the second and third constraints which remains active.

To determine the new values of design variables when the right-hand side of the third constraint is changed from 1 to 2, we use x_6 column from Table 4.21 in Eq. (4.36) and obtain the new solution as

$$x_3 \equiv b_1'' = b_1' + \Delta_3 a_{16}'$$
$$= 2 + (1)(1) = 3$$
$$x_2 \equiv b_2'' = b_2' + \Delta_3 a_{26}'$$
$$= \tfrac{2}{3} + (1)(-\tfrac{2}{3}) = 0$$
$$x_1 \equiv b_3'' = b_3' + \Delta_3 a_{36}'$$
$$= \tfrac{5}{3} + (1)(\tfrac{1}{3}) = 2$$

It can easily be seen from Fig. 4.8 that the new solution corresponds to point C.

‖

4.6.3 Ranging Cost Coefficients

If a cost coefficient c_k is changed to $c_k + \Delta c_k$, we like to find an admissible range on Δc_k such that the optimum design variables are not changed. *Note that when the cost coefficients are changed, the feasible region for the problem does not change.* However, orientation of the cost function hyperplane and value of the cost function change. Limits on the change Δc_k for the coefficient c_k depend on whether or not x_k is a basic variable at the optimum. Thus, we must consider the two cases separately.

Theorem 4.7   Range for cost coefficient of nonbasic variables. Let c_k be such that x_k is not a basic variable. If this c_k is replaced by any $c_k + \Delta c_k$ where $-c_k' \leqq \Delta c_k \leqq \infty$, then the optimum solution (design variables and the cost function) does not change. Here, c_k' is the reduced cost coefficient corresponding to x_k in the final tableau.

‖

Theorem 4.8   Range for cost coefficient of basic variables. Let c_k be such that x_k^* is a basic variable, and let $x_k^* = b_r'$ (a superscript * is used to indicate optimum value). Then the range on change Δc_k in c_k for which the optimum design variables do not change is given as

$$\max_j \{c_j'/a_{rj}'; a_{rj}'<0\} \leqq \Delta c_k \leqq \min_j \{c_j'/a_{rj}'; a_{rj}'>0\} \qquad (4.37)$$

where $a_{rj}' =$ element in the rth row and the jth column of the final tableau. The index r is determined by the row that determines x_k^*. Index j corresponds to each of the nonbasic columns excluding artificial columns. (*Note*: if no $a_{rj}'>0$, then there is no upper limit; if no $a_{rj}'<0$, then there is no lower limit.)

$c_j' =$ reduced cost coefficient in the jth nonbasic column excluding artificial variable columns.

When Δc_k satisfies Inequality (4.37), the optimum value of the cost function is $f^* + \Delta c_k x_k^*$. ∥

To determine possible changes in the cost coefficient of a basic variable, the first step is to determine the row index r for use in Inequalities (4.37). This represents the row determining the basic variable x_k^*. After r has been determined, we take ratios of the reduced cost coefficients and elements in the rth row according to the rules given in Theorem 4.8. The lower limit on Δc_k is determined by the maximum ratio c_j'/a_{rj}' with $a_{rj}'<0$. The upper limit is determined by the minimum ratio c_j'/a_{rj}' with $a_{rj}'>0$.

Example 4.20 Ranges for cost coefficients – "$\leqq$ type" constraints. Determine ranges for the cost coefficients for the problem:

maximize
$$z = 5x_1 - 2x_2$$
subject to
$$2x_1 + x_2 \leqq 9$$
$$x_1 - 2x_2 \leqq 2$$
$$-3x_1 + 2x_2 \leqq 3$$
$$x_1, x_2 \geqq 0$$

Solution. The problem is solved in Example 4.16. The final tableau is copied from Table 4.18 in Table 4.22. The problem is solved as minimization of the cost function $f = -5x_1 + 2x_2$. Therefore, we will find ranges for the cost coefficients $c_1 = -5$ and $c_2 = 2$. Note that since both x_1 and x_2 are basic variables, Theorem 4.8 will be used.

Since the second row determines the basic variable x_1, $r = 2$ (the row number) for use in Inequalities (4.37). Columns 3 and 4 are nonbasic. Therefore $j = 3, 4$ are the column indices for use in Eq. (4.37). We will take ratios of the reduced cost coefficients with the elements in the second row of the final tableau. This gives the range for Δc_1 as

$$-\infty \leqq \Delta c_1 \leqq \min\left\{\frac{1.6}{0.4}, \frac{1.8}{0.2}\right\}; \quad \text{or} \quad -\infty \leqq \Delta c_1 \leqq 4$$

TABLE 4.22
Final tableau for Example 4.20 (ranging cost coefficients)

Basic ↓	x_1	x_2	x_3	x_4	x_5	b
x_2	0	1	0.2	−0.4	0	1
x_1	1	0	0.4	0.2	0	4
x_5	0	0	0.8	1.4	1	13
Cost	0	0	1.6 (c_3')	1.8 (c_4')	0	$f + 18$

For c_1: x_1 is basic; it is determined by the second row, so use second row in Eq. (4.37) ($r = 2$).
For c_2: x_2 is basic; it is determined by the first row, so use first row in Eq. (4.37) ($r = 1$).

The range for c_1 is obtained by adding the current value of $c_1 = -5$ to both sides of the above inequality,

$$-\infty \leq c_1 \leq -1 \tag{a}$$

Thus, if c_1 changes from −5 to −4, the new cost function value is given as

$$f_{new}^* = f^* + \Delta c_1 x_1^*$$
$$= -18 + (1)(4) = -14$$

That is, the cost function will increase by 4.

For the second cost coefficient, $r = 1$ (the row number) because the first row determines x_2 as a basic variable. Using Inequalities (4.37), we take the ratio of reduced cost coefficients with elements in the first row to obtain,

$$\max\left\{\frac{1.8}{(-0.4)}\right\} \leq \Delta c_2 \leq \min\left\{\frac{1.6}{0.2}\right\}; \quad \text{or} \quad -4.5 \leq \Delta c_2 \leq 8$$

The range for c_2 is obtained by adding the current value of $c_2 = 2$ to both sides of the above inequality,

$$-2.5 \leq c_2 \leq 10 \tag{b}$$

Thus, if c_2 is changed from 2 to 3, the new cost function value is given as

$$f_{new}^* = f^* + \Delta c_2 x_2^*$$
$$= -18 + (1)(1) = -17$$

That is, the cost function will increase by 1.

Note that the range for the coefficients of the maximization function ($z = 5x_1 - 2x_2$) can be obtained from Eqs (a) and (b). To determine these ranges, we multiply Eqs (a) and (b) by −1. Therefore, the range for $c_1 = 5$ is given as $1 \leq c_1 \leq \infty$, and that for $c_2 = -2$ is $-10 \leq c_2 \leq 2.5$. ‖

TABLE 4.23
Final tableau for Example 4.21 (ranging cost coefficients)

Basic ↓	x_1	x_2	x_3	x_4	x_5	x_6	b
x_3	0	0	1	-1	-1	1	2
x_2	0	1	0	$\frac{2}{3}$	$\frac{1}{3}$	$-\frac{2}{3}$	$\frac{2}{3}$
x_1	1	0	0	$-\frac{1}{3}$	$\frac{1}{3}$	$\frac{1}{3}$	$\frac{5}{3}$
Cost	0	0	0	$\frac{7}{3}$ (c_4')	$\frac{5}{3}$	$-\frac{7}{3}$	$f+\frac{13}{3}$

For c_1: x_1 is basic determined by row 3, so use $r=3$ in Eq. (4.37).
For c_2; x_2 is basic determined by row 2, so use $r=2$ in Eq. (4.37).

Example 4.21 Ranges for cost coefficients – equality and "≧ type" constraints. Find ranges for the cost coefficients of the problem,

maximize
$$z = x_1 + 4x_2$$
subject to
$$x_1 + 2x_2 \leqq 5$$
$$2x_1 + x_2 = 4$$
$$x_1 - x_2 \geqq 1$$
$$x_1, x_2 \geqq 0$$

Solution. The problem is solved in Example 4.17 and the final tableau is copied from Table 4.19 in Table 4.23. In the tableau, x_3 is a slack variable for the first constraint, x_4 is a surplus variable for the third constraint, and x_5 and x_6 are artificial variables for the second and third constraints, respectively.

Since both x_1 and x_2 are basic variables, we will use Theorem 4.8 to find ranges for the cost coefficients $c_1 = -1$ and $c_2 = -4$. Note that the problem is solved as minimization of the cost function $f = -x_1 - 4x_2$. Columns 4, 5 and 6 are nonbasic. However, columns 5 and 6 must be excluded for use in Inequalities (4.37) as they correspond to artificial variables. Therefore, the column index j is only 4 for use in Inequalities (4.37).

To find the range for Δc_1, $r = 3$ is used because the third row determines x_1 as a basic variable. Using Inequalities (4.37) with $r = 3$ and $j = 4$, we have

$$\max \{(\tfrac{7}{3})/(-\tfrac{1}{3})\} \leqq \Delta c_1 \leqq \infty; \quad \text{or} \quad -7 \leqq \Delta c_1 \leqq \infty$$

The range for c_1 is obtained by adding the current value of $c_1 = -1$ to both sides of the above inequality,
$$-8 \leqq c_1 \leqq \infty \tag{a}$$

Thus, if c_1 changes from -1 to -2, the new cost function value is given as
$$f^*_{new} = f^* + \Delta c_1 x_1^*$$
$$= -\tfrac{13}{3} + (-1)(\tfrac{5}{3}) = -6$$

For the second cost coefficient, $r = 2$ because the second row determines x_2 as a basic variable. Using Inequalities (4.37) with $r = 2$ and $j = 4$, the range for Δc_2 is obtained as

$$-\infty \leqq \Delta c_2 \leqq 3.5$$

Thus the range for c_2 with current value $c_2 = -4$ is given as

$$-\infty \leqq c_2 \leqq -0.5 \qquad \text{(b)}$$

Thus, if c_2 changes from -4 to -3, the new value of the cost function is given as

$$f^*_{new} = f^* + \Delta c_2 x_2^*$$
$$= -\tfrac{13}{3} + (1)(\tfrac{2}{3}) = -\tfrac{11}{3}$$

From Eqs (a) and (b), the ranges for coefficients of the maximization function ($z = x_1 + 4x_2$) are obtained as

$$-\infty \leqq c_1 \leqq 8, \; 0.5 \leqq c_2 \leqq \infty \qquad \qquad \|$$

4.6.4* Changes in the Coefficient Matrix

Any change in the coefficient matrix $\mathbf{A}$ changes the feasible region for the problem. This may change the optimum solution for the problem depending on whether or not the change is associated with a basic variable. Let a_{ij} be replaced by $a_{ij} + \Delta a_{ij}$. We shall determine limits for Δa_{ij} so that with minor computations the optimum solution for the changed problem can be obtained. We must consider the two cases; (i) when the change is associated with a nonbasic variable, and (ii) when the change is associated with a basic variable.

Theorem 4.9 Change associated with a nonbasic variable. Let j in a_{ij} be such that x_j is not a basic variable, and k be the column index for the slack or artificial variable associated with the ith row. Define a vector

$$\mathbf{c}_B = [c_{B1} \; c_{B2} \ldots \; c_{Bm}]^T \qquad (4.38)$$

where $c_{Bi} = c_j$ if $x_j^* = b_i^*$, $i = 1$ to m (i.e. the index i corresponds to the ith row that determines the optimum value of variable x_j). Also define a scalar

$$R = \sum_{r=1}^{m} c_{Br} a'_{rk} \qquad (4.39)$$

With this notation, if Δa_{ij} satisfies the following inequalities

$$\Delta a_{ij} \geqq c'_j / R \quad \text{when } R < 0, \text{ and } \Delta a_{ij} \leqq \infty \quad \text{when } R \leqq 0 \qquad (4.40)$$

or,

$$\Delta a_{ij} \leqq c'_j / R \quad \text{if } R > 0, \text{ and } \Delta a_{ij} \geqq -\infty \quad \text{if } R \geqq 0 \qquad (4.41)$$

then the optimum solution (design variables and cost function) do not change when a_{ij} is replaced by any $a_{ij} + \Delta a_{ij}$. Also, if $R = 0$, then the solution does not change for any value of Δa_{ij}. $\qquad \|$

To use the theorem, a first step is to determine indices j and k. Then we determine the vector $\mathbf{c}_B$ of Eq. (4.38), and the scalar R of Eq. (4.39). Conditions of Inequalities (4.40) and (4.41) then determine whether or not the given Δa_{ij} will change the optimum solution. If the inequalities are not satisfied, then we have to re-solve the problem to obtain the new solution.

Theorem 4.10   Change associated with a basic variable. Let j in a_{ij} be such that x_j is a basic variable and let $x_j^* = b_t'$ (i.e. t is the row index that determines optimum value of x_j). Let index k and the scalar R be defined as in Theorem 4.9. Let Δa_{ij} satisfy the following inequalities:

$$\max_{r \neq t} \{b_r'/A_r,\ A_r < 0\} \leq \Delta a_{ij} \leq \min_{r \neq t} \{b_r'/A_r,\ A_r > 0\} \tag{4.42}$$

with

$$A_r = b_t' a_{rk}' - b_r' a_{tk}', \qquad r = 1 \text{ to } m;\ r \neq t \tag{4.43}$$

and

$$\max_q \{-c_q'/B_q,\ B_q > 0\} \leq \Delta a_{ij} \leq \min_q \{-c_q'/B_q,\ B_q < 0\} \tag{4.44}$$

with

$$B_q = c_q' a_{tk}' + a_{iq}' R \qquad \text{for all } q \text{ not in the basis} \tag{4.45}$$

and

$$1 + a_{tk}' \Delta a_{ij} > 0 \tag{4.46}$$

Note that the upper and lower limits on Δa_{ij} do not exist if the corresponding denominators do not exist. If Δa_{ij} satisfies the above inequalities, then the optimum solution of the changed problem can be obtained without any further iterations of the Simplex method. If b_r' for $r = 1$ to m is replaced by

$$
\begin{aligned}
b_r'' &= b_r' - \Delta a_{ij} a_{rk}'/(1 + \Delta a_{ij} a_{tk}'), \qquad r = 1 \text{ to } m;\ r \neq t \\
b_t'' &= b_t'/(1 + \Delta a_{ij} a_{tk}')
\end{aligned}
\tag{4.47}
$$

in the final tableau, then the new optimum values for the basic variables can be obtained when a_{ij} is replaced by $a_{ij} + \Delta a_{ij}$. In other words, if $x_j^* = b_t'$, then $x_j' = b_t''$ where x_j' refers to the optimum solution for the changed problem.   ‖

To use the theorem, we need to determine indices j, t and k. Then we determine the constants A_r and B_q from Eqs (4.43) and (4.45). With these, ranges on Δa_{ij} can be determined from Inequalities (4.42) and (4.44). If Δa_{ij} satisfy these inequalities, Eq. (4.47) determines the new solution. If the inequalities are not satisfied, the problem must be re-solved for the new solution.

4.7* DUALITY IN LINEAR PROGRAMMING

Associated with every LP problem is another linear programming problem called its *dual*. The original LP problem is called the *primal*. If primal involves n variables and m constraints, the dual involves n constraints and m variables. The solution of either is sufficient for readily obtaining solution to the other. In this section, we shall develop the important concept of duality in LP.

4.7.1 Standard Primal LP

There are several ways of defining the primal and the corresponding dual problems. However, we shall define a *standard primal problem* as:
Find $x_1, x_2, \ldots, x_n$ to maximize a primal objective function

$$z_p = d_1 x_1 + \ldots + d_n x_n = \sum_{i=1}^{n} d_i x_i \equiv \mathbf{d}^T \mathbf{x} \tag{4.48}$$

subject to the constraints

$$
\begin{aligned}
a_{11} x_1 + \ldots + a_{1n} x_n &\leqq e_1 \\
&\cdots\cdots\cdots \quad (\mathbf{Ax} \leqq \mathbf{e}) \\
a_{m1} x_1 + \ldots + a_{mn} x_n &\leqq e_m \\
x_j \geqq 0; \quad j = 1 \text{ to } n
\end{aligned}
\tag{4.49}
$$

We shall use a subscript "p" on z to indicate the primal objective function. Also z is used as the maximization function. It must be understood that in the standard LP problem defined in Eqs (4.8) to (4.10), all constraints were equalities and right-hand side parameters b_i were nonnegative. However, in the definition of the standard primal problem, all constraints must be "$\leqq$ type" and there is no restriction on the sign of the right-hand side parameters e_i. So, "$\geqq$ type" constraints must be multiplied by -1 to convert them to "$\leqq$ types". Equalities should be also converted to "$\leqq$ type" constraints. This is explained later in this section. In addition, the primal objective function is always maximized. Note that to solve the above primal LP problem by the Simplex method, we must transform it into the standard Simplex form of Eqs (4.8) to (4.10). Only then the two-phase Simplex method can be used.

4.7.2 Dual LP Problem

The dual for the standard primal is defined as follows:
Find dual variables $y_1, y_2, \ldots, y_m$ to minimize a dual objective function

$$f_d = e_1 y_1 + \ldots + e_m y_m \equiv \sum_{i=1}^{m} e_i y_i = \mathbf{e}^T \mathbf{y} \tag{4.50}$$

subject to the constraints

$$a_{11}y_1 + \ldots + a_{m1}y_m \geq d_1$$
$$\cdots\cdots\cdots\cdots$$
$$a_{1n}y_1 + \ldots + a_{mn}y_m \geq d_n$$

$$\quad (\mathbf{A}^T\mathbf{y} \geq \mathbf{d})$$

$$y_i \geq 0; \quad i = 1 \text{ to } m$$

(4.51)

We use a subscript d on f to indicate that it is the cost function for the dual problem. Note the following *relations between the primal and dual problems*:

1. The number of dual variables is equal to the number of primal constraints. Each dual variable is associated with a primal constraint. For example, y_i is associated with the ith primal constraint.
2. The number of dual constraints is equal to the number of primal variables. Each primal variable is associated with a dual constraint. For example, x_i is associated with the ith dual constraint.
3. Primal constraints are "$\leq$ type" inequalities, whereas the dual constraints are "$\geq$ types".
4. The maximization of the primal objective function is replaced by the minimization of the dual cost function.
5. The coefficients d_i of the primal objective function become the right-hand side of the dual constraints. The right-hand side parameters e_i of the primal constraints become coefficients for the dual cost function.
6. The coefficient matrix $[a_{ij}]$ of primal constraints is transposed to $[a_{ji}]$ for dual constraints.
7. The nonnegativity condition applies to both primal and dual variables.

We again emphasize here that the definition of the standard primal is different from the standard LP definition. However, to solve a primal or its dual, we must use the standard form of the Simplex method. An example illustrates this aspect later in the section.

Example 4.22   Dual of an LP problem. Write dual of the problem,

maximize

$$z_p = 5x_1 - 2x_2$$

subject to

$$2x_1 + x_2 \leq 9 \tag{a}$$

$$x_1 - 2x_2 \leq 2 \tag{b}$$

$$-3x_1 + 2x_2 \leq 3 \tag{c}$$

$$x_1, x_2 \geq 0$$

Solution. The problem is already in the standard primal form and the following associated vectors and matrices can be identified:

$$\mathbf{d} = \begin{bmatrix} 5 \\ -2 \end{bmatrix}, \qquad \mathbf{e} = \begin{bmatrix} 9 \\ 2 \\ 3 \end{bmatrix}, \qquad \mathbf{A} = \begin{bmatrix} 2 & 1 \\ 1 & -2 \\ -3 & 2 \end{bmatrix}$$

Since there are three primal constraints, there are three dual variables for the problem. Let y_1, y_2 and y_3 be the dual variables associated with the primal constraints (a), (b) and (c), respectively. Therefore, Eqs (4.50) and (4.51) give dual for the problem as

minimize

$$f_d = 9y_1 + 2y_2 + 3y_3$$

subject to

$$2y_1 + y_2 - 3y_3 \geq 5$$

$$y_1 - 2y_2 + 2y_3 \geq -2$$

$$y_1, y_2, y_3 \geq 0$$

$\|$

4.7.3 Treatment of Equality Constraints

Many design problems have equality constraints. Each equality constraint can be replaced by a pair of inequalities. For example, $2x_1 + 3x_2 = 5$ can be replaced by the pair $2x_1 + 3x_2 \geq 5$ and $2x_1 + 3x_2 \leq 5$. We can multiply the " $\geq$ type" inequality by -1 to convert it into the standard primal form. The following example illustrates treatment of equality and " $\geq$ type" constraints.

Example 4.23 Dual of an LP with equality and " $\geq$ type" constraints. Write dual for the problem:

maximize

$$z_p = x_1 + 4x_2$$

subject to

$$x_1 + 2x_2 \leq 5$$

$$2x_1 + x_2 = 4$$

$$x_1 - x_2 \geq 1$$

$$x_1, x_2 \geq 0$$

Solution. The equality constraint $2x_1 + x_2 = 4$ is equivalent to the two inequalities $2x_1 + x_2 \leq 4$ and $2x_1 + x_2 \geq 4$. The " $\geq$ type" constraints are multiplied by -1 to convert them into the " $\leq$ " form. Thus, the standard primal of the above problem is

maximize

$$z_p = x_1 + 4x_2$$

subject to

$$x_1 + 2x_2 \leq 5$$
$$2x_1 + x_2 \leq 4$$
$$-2x_1 - x_2 \leq -4$$
$$-x_1 + x_2 \leq -1$$
$$x_1, x_2 \geq 0$$

Using Eqs (4.50) and (4.51), dual for the primal is

minimize

$$f_d = 5y_1 + 4(y_2 - y_3) - y_4$$

subject to

$$y_1 + 2(y_2 - y_3) - y_4 \geq 1$$
$$2y_1 + (y_2 - y_3) + y_4 \geq 4$$
$$y_1, y_2, y_3, y_4 \geq 0$$ ||

4.7.4 Alternate Treatment of Equality Constraints

We will show that it is not necessary to replace an equality constraint by a pair of inequalities to write the dual. Note that there are four dual variables for Example 4.23. The variables y_2 and y_3 correspond to the second and third primal constraints written in the standard form. The second and third constraints are actually equivalent to the original equality constraint. Note also that the term $(y_2 - y_3)$ appears in all the expressions of the dual problem. We define $y_5 = y_2 - y_3$. Since it is the difference of two nonnegative variables ($y_2 \geq 0$, $y_3 \geq 0$), it can be positive, negative or zero. Substituting for y_5, the dual problem in Example 4.23 is re-written as

minimize

$$f_d = 5y_1 + 4y_5 - y_4$$

subject to

$$y_1 + 2y_5 - y_4 \geq 1$$
$$2y_1 + y_5 + y_4 \geq 4$$
$$y_1, y_4 \geq 0$$

$y_5 = y_2 - y_3$ is unrestricted in sign

The number of dual variables is now only three. Since the number of dual variables is equal to the number of primal constraints, the dual variable y_5 must be associated with the equality constraint $2x_1 + x_2 = 4$. Thus, we can draw the following conclusion: *if the ith primal constraint is left as an equality, the ith dual variable is unrestricted in sign.*

In a similar manner, we can show that *if the primal variable is unrestricted in sign, then the ith dual constraint is an equality.* This is left as an exercise.

Example 4.24 Recovery of primal from dual. Note that we can convert a dual problem into the standard primal form and write its dual again. It can be shown that dual of this problem gives the primal problem back again. To see this, let us convert the preceding dual problem into standard primal form:

maximize
$$z_p = -5y_1 - 4y_5 + y_4$$
subject to
$$-y_1 - 2y_5 + y_4 \leqq -1$$
$$-2y_1 - y_5 - y_4 \leqq -4$$
$$y_1, y_4 \geqq 0$$

y_5 is unrestricted in sign

Writing dual of the above primal, we obtain

minimize
$$f_d = -x_1 - 4x_2$$
subject to
$$-x_1 - 2x_2 \geqq -5$$
$$-2x_1 - x_2 = -4$$
$$x_1 - x_2 \geqq 1$$
$$x_1, x_2 \geqq 0$$

which is the same as the original problem (Example 4.23). Note that in the above dual problem, the second constraint is an equality because the second primal variable (y_5) is unrestricted in sign. ‖

The following theorem is evident:

Theorem 4.11 Dual of dual. The dual of the dual problem is the primal problem. ‖

4.7.5 Determination of Primal Solution from Dual Solution

It remains to determine how the optimum solution of the primal is obtained from the optimum solution of the dual or vice versa. First, let us multiply each inequality in Eq. (4.51) by $x_1, x_2, \ldots, x_n$, and add them. Since x_j's are restricted to be nonnegative, we get the inequality

$$x_1(a_{11}y_1 + \ldots + a_{m1}y_m) + x_2(a_{12}y_1 + \ldots + a_{m2}y_m)$$
$$+ \ldots + x_n(a_{1n}y_1 + \ldots + a_{mn}y_m) \geqq d_1x_1 + \ldots + d_nx_n$$

In the matrix form the above inequality is written as $\mathbf{x}^T \mathbf{A}^T \mathbf{y} \geq \mathbf{x}^T \mathbf{d}$. Rearranging the equation by collecting terms with $y_1, y_2, \ldots, y_m$ (or taking transpose of the left-hand side as $\mathbf{y}^T \mathbf{A} \mathbf{x}$), we obtain

$$y_1(a_{11}x_1 + a_{12}x_2 + \ldots + a_{1n}x_n) + y_2(a_{21}x_1 + a_{22}x_2 + \ldots + a_{2n}x_n)$$
$$+ \ldots + y_m(a_{m1}x_1 + a_{m2}x_2 + \ldots + a_{mn}x_n) \geq d_1 x_1 + d_2 x_2 + \ldots + d_n x_n \quad (4.52)$$

In the matrix form the above inequality can be written as $\mathbf{y}^T \mathbf{A} \mathbf{x} \geq \mathbf{x}^T \mathbf{d}$. Each quantity in parentheses in Eq. (4.52) is less than the corresponding value of e on the right-hand side of Inequality (4.49). Therefore, substitution from Inequalities (4.49) preserves the inequality of Eq. (4.52),

$$y_1 e_1 + y_2 e_2 + \ldots + y_m e_m \geq d_1 x_1 + d_2 x_2 + \ldots + d_n x_n \quad (4.53)$$

Or, $\mathbf{y}^T \mathbf{e} \geq \mathbf{x}^T \mathbf{d}$. Note in Inequality (4.53) that the left-hand side is the dual cost function and the right-hand side is the primal objective function. Therefore, from Inequality (4.53), $f_d \geq z_p$ for all $(x_1, x_2, \ldots, x_n)$ and $(y_1, y_2, \ldots, y_m)$ satisfying Eqs (4.48) to (4.51). Thus, the vectors $\mathbf{x}$ and $\mathbf{y}$ with $z_p = f_d$ maximize z_p while minimizing f_d. The optimum (minimum) value of the dual cost function is also the optimum (maximum) value of the primal objective function.

The following theorems regarding primal and dual problems can be stated:

> **Theorem 4.12 Relationship between primal and dual.** Let $\mathbf{x}$ and $\mathbf{y}$ be in the constraint sets (i.e. feasible points) of primal (defined in Eqs (4.48) and (4.49)) and dual (defined in Eqs (4.50) and (4.51)) problems, respectively. Then the following conditions hold:
>
> 1. $f_d(\mathbf{y}) \geq z_p(\mathbf{x})$.
> 2. If $f_d = z_p$, then $\mathbf{x}$ and $\mathbf{y}$ are solutions for primal and dual problems, respectively.
> 3. If primal is unbounded, the corresponding dual is infeasible, and vice versa.
> 4. If primal is feasible and dual is infeasible, then primal is unbounded and vice versa.   ‖

> **Theorem 4.13 Primal and dual solutions.** Let both primal and dual have feasible points. Then both have solution in $\mathbf{x}$ and $\mathbf{y}$ respectively and $f_d(\mathbf{y}) = z_p(\mathbf{x})$.
>   ‖

> **Theorem 4.14 Solution of primal from dual.** If the ith dual constraint is a strict inequality at optimum, then the corresponding ith primal variable is nonbasic, i.e. it vanishes. Also, if the ith dual variable is basic, then the ith primal constraint is satisfied at equality.   ‖

The conditions of Theorem 4.14 can be written as

$$\text{if } \sum_{i=1}^{m} a_{ij} y_i > d_j, \quad \text{then} \quad x_j = 0.$$

(*j*th dual constraint is strict inequality, then the *j*th primal variable is nonbasic)

$$\text{if } y_i > 0, \quad \text{then} \quad \sum_{j=1}^{n} a_{ij}x_j = e_i$$

(*i*th dual variable is basic, the *i*th primal constraint is an equality).

These conditions can be used to obtain primal variables using the dual variables. The primal constraints satisfied at equality are identified from values of the dual variables. The resulting linear equations can be solved simultaneously for the primal variables. However, this is not necessary as the final dual tableau can be used directly to obtain primal variables. We shall illustrate the use of these theorems in the following example.

Example 4.25 Primal and dual solutions. Consider the following problem:

maximize
$$z_p = 5x_1 - 2x_2$$
subject to
$$2x_1 + x_2 \leq 9$$
$$x_1 - 2x_2 \leq 2$$
$$-3x_1 + 2x_2 \leq 3$$
$$x_1, x_2 \geq 0$$

Solve the primal and the dual problems and study their final tableaux.

Solution. The problem has been solved using the Simplex method in Example 4.16 and Table 4.18. The final tableau is reproduced from there in Table 4.24. From the final primal tableau,

basic variables	$x_1 = 4, x_2 = 1, x_5 = 13$
nonbasic variables	$x_3 = 0, x_4 = 0$
maximum objective function	$z_p = 18$ (minimum value is -18)

Now, let us write dual for the problem and solve it using the Simplex

TABLE 4.24
Final tableau for Example 4.25 by Simplex method (primal solution)

Basic ↓	x_1	x_2	x_3	x_4	x_5	b
x_2	0	1	0.2	−0.4	0	1
x_1	1	0	0.4	0.2	0	4
x_5	0	0	0.8	1.4	1	13
Cost	0	0	1.6	1.8	0	$f_p + 18$

method. Note that the original problem is already in the standard primal form. There are three primal inequality constraints, so there are three dual variables. There are two primal variables, so there are two dual constraints. Let y_1, y_2 and y_3 be the dual variables. Therefore, dual of the problem is given as

minimize

$$f_d = 9y_1 + 2y_2 + 3y_3$$

subject to

$$2y_1 + y_2 - 3y_3 \geq 5$$

$$y_1 - 2y_2 + 2y_3 \geq -2$$

$$y_1, y_2, y_3 \geq 0$$

Writing the constraints in the standard Simplex form by introducing slack, surplus and artificial variables, we obtain

$$2y_1 + y_2 - 3y_3 - y_4 + y_6 = 5$$

$$-y_1 + 2y_2 - 2y_3 + y_5 = 2$$

$$y_i \geq 0, \ i = 1 \text{ to } 6$$

where y_4 is a surplus variable, y_5 is a slack variable and y_6 is an artificial variable. The two-phase Simplex procedure can be used to solve the problem. Thus we obtain the sequence of calculations for the dual problem shown in Table 4.25.

TABLE 4.25
Solution of dual of Example 4.25

Basic ↓	y_1	y_2	y_3	y_4	y_5	y_6	b
y_6	**2**	1	−3	−1	0	1	5
y_5	−1	2	−2	0	1	0	2
Cost	9	2	3	0	0	0	$f_d - 0$
Artificial	**−2**	−1	3	1	0	0	$w - 5$
y_1	1	0.5	−1.5	−0.5	0	0.5	2.5
y_5	0	**2.5**	−3.5	−0.5	1	0.5	4.5
Cost	0	**−2.5**	16.5	4.5	0	−4.5	$f_d - 22.5$
Artificial	0	0	0	0	0	1	$w - 0$
y_1	1	0	−0.8	−0.4	−0.2	0.4	1.6
y_2	0	1	−1.4	−0.2	0.4	0.2	1.8
Cost	0	0	13.0	4.0	1.0	−4.0	$f_d - 18$

From the final dual tableau, we obtain the following solution:

basic variables	$y_1 = 1.6,\ y_2 = 1.8$
nonbasic variables	$y_3 = 0,\ y_4 = 0,\ y_5 = 0$
minimum value of dual function	$f_d = 18$

Note that at the optimum $f_d = z_p$, which satisfies the conditions of Theorems 4.12 and 4.13. Using Theorem 4.14, we see that the first and second primal constraints must be satisfied at equality since the dual variables y_1 and y_2 associated with the constraints are positive (basic) in Table 4.25. Therefore, primal variables x_1 and x_2 are obtained as a solution of the first two primal constraints satisfied at equality:

$$2x_1 + x_2 = 9$$
$$x_1 - 2x_2 = 2$$

The solution of above equations is given as

$$x_1 = 4, \qquad x_2 = 1$$

which is the same as obtained from the final primal tableau.   ‖

4.7.6 Use of Dual Tableau to Recover Primal Solution

It turns out that we do not need to follow the preceding procedure (use of Theorem 4.14) to recover the primal variables. The final dual tableau contains all the information to recover the primal solution. Similarly, the final primal tableau contains all the information to recover the dual solution. Looking at the final tableau for Example 4.25, we observe that the elements in the last row of the dual tableau match the elements in the last column of the primal tableau. Similarly, the reduced cost coefficients in the final primal tableau match the dual variables. To recover the primal variables from the final dual tableau, we use reduced cost coefficients in columns corresponding to slack or surplus variables. We note that the reduced cost coefficient in column y_4 is precisely x_1 and that in column y_5 is precisely x_2. Therefore, reduced cost coefficients corresponding to slack and surplus variables in the final dual tableau give values of primal variables. Similarly, if we solve the primal problem, we can recover the dual solution from the final primal tableau. The following theorem summarizes this result.

Theorem 4.15 Recovery of primal solution from dual tableau. Let dual of the standard primal defined in Eqs (4.48) and (4.49) (i.e. maximize $\mathbf{d}^T\mathbf{x}$ subject to $\mathbf{A}\mathbf{x} \leq \mathbf{e},\ \mathbf{x} \geq \mathbf{0}$) be solved by the standard Simplex method. Then the value of the ith primal variable equals the reduced cost coefficient of the slack or surplus variable associated with the ith dual constraint in the final dual tableau. In addition, if a dual variable is nonbasic, then its reduced cost coefficient equals the value of slack or surplus variable for the corresponding primal constraint.   ‖

To use the theorem, we must write dual for the standard primal problem. To solve the dual problem by the Simplex method, we must put the problem in the standard form where the right-hand sides are nonnegative. Also, note that the reduced cost coefficients for the slack or surplus variables must be nonnegative in the final dual tableau. Therefore, the corresponding primal variables are also nonnegative.

Note also that if a *dual variable is nonbasic* (i.e. has zero value), then *its reduced cost coefficient equals the value of the slack or surplus variable for the corresponding primal constraint*. In Example 4.25, y_3 the dual variable corresponding to the third primal constraint, is nonbasic. The reduced cost coefficient in the y_3 column is 13. Therefore, the slack variable for the third primal constraint has value 13. This is the same as obtained from the final primal tableau.

We also note that the dual solution can be obtained from the final primal tableau using Theorem 4.15 as

$$y_1 = 1.6, \ y_2 = 1.8, \ y_3 = 0$$

which is the same solution as before.

While using Theorem 4.15, the following additional points should be noted:

1. When the final primal tableau is used to recover the dual solution, the dual variables correspond to the primal constraints expressed in the " $\leq$ " form only. However, the primal constraints must be converted to standard Simplex form while solving the problem. Recall that all the right-hand sides of constraints must be nonnegative for the Simplex method. The dual variables are nonnegative only for the constraints written in the " $\leq$ " form.

2. When a primal constraint is an equality, it is treated in the Simplex method by adding an artificial variable in Phase I. There is no slack or surplus variable associated with an equality. We also know from the previous discussion that the dual variable associated with the equality constraint is unrestricted in sign. Then the question is how to recover its value from the final primal tableau? There are a couple of ways of doing this. The first procedure is to convert the equality constraint into a pair of inequalities, as noted previously. For example, the constraint

$$2x_1 + x_2 = 4 \tag{a}$$

is written as the pair of inequalities

$$2x_1 + x_2 \leq 4 \tag{b}$$

$$-2x_1 - x_2 \leq -4 \tag{c}$$

The two inequalities are treated in a standard way in the Simplex method. The corresponding dual variables are recovered from the final primal tableau using Theorem 4.15. Let $y_2 \geq 0$ and $y_3 \geq 0$ be the dual variables associated with constraints of Eqs (b) and (c), respectively, and y_1 be the dual variable associated with equality constraint of Eq. (a). Then, $y_1 = y_2 - y_3$. Accordingly, y_1 is unrestricted in sign and its value is known using y_2 and y_3.

The second way of recovering dual variable for the equality constraint is to carry along its artificial variable column in Phase II of the Simplex method. Then dual variable for the constraint is the reduced cost coefficient in the artificial variable column in the final primal tableau.

We illustrate these procedures in an example.

Example 4.26   Use of final primal tableau to recover dual solutions. Solve the following LP problem and recover its dual solution from the final primal tableau:

maximize
$$z_p = x_1 + 4x_2$$
subject to
$$x_1 + 2x_2 \leqq 5$$
$$2x_1 + x_2 = 4$$
$$x_1 - x_2 \geqq 1$$
$$x_1, x_2 \geqq 0$$

Solution. We will convert the second equality constraint into a pair of inequalities as in Eqs (b) and (c). Writing the problem into a standard Simplex form, we obtain

minimize
$$f_p = -x_1 - 4x_2$$
subject to
$$x_1 + 2x_2 + x_3 \qquad\qquad\qquad = 5$$
$$2x_1 + x_2 \qquad + x_4 \qquad\qquad = 4$$
$$2x_1 + x_2 \qquad\qquad - x_5 \quad + x_7 \qquad = 4$$
$$x_1 - x_2 \qquad\qquad\qquad - x_6 \quad + x_8 = 1$$
$$x_i \geqq 0; \qquad i = 1 \text{ to } 8$$

where, x_3 and x_4 are the slack variables, x_5 and x_6 are the surplus variables, and x_7 and x_8 are the artificial variables. This problem formulation is the same as for Example 4.14. Various pivot steps for the problem are given in Table 4.16 and the final tableau is reproduced in Table 4.26. The optimum solution is obtained in three iterations as

basic variables	$x_1 = \frac{5}{3},\ x_2 = \frac{2}{3},\ x_3 = 2,\ x_5 = 0$
nonbasic variables	$x_4 = x_6 = x_7 = x_8 = 0$
primal cost function	$f_p = -\frac{13}{3}$

Using Theorem 4.15, the dual variable for the above four constraints are

1. $y_1 = 0$, for the first constraint (reduced cost coefficient of x_3, the slack variable).

TABLE 4.26
Solution for Example 4.26 with the equality constraint converted into two inequalities

Basic ↓	x_1	x_2	x_3	x_4	x_5	x_6	x_7	x_8	b
x_3	0	0	1	-1	0	-1	-0	-1	2
x_5	0	0	0	1	1	0	-1	0	0
x_2	0	1	0	$\frac{1}{3}$	0	$\frac{2}{3}$	0	$-\frac{2}{3}$	$\frac{2}{3}$
x_1	1	0	0	$\frac{1}{3}$	0	$-\frac{1}{3}$	0	$\frac{1}{3}$	$\frac{5}{3}$
Cost	0	0	0	$\frac{5}{3}$	0	$\frac{7}{3}$	0	$-\frac{7}{3}$	$f_p + \frac{13}{3}$

2. $y_2 = \frac{5}{3}$, for the second constraint (reduced cost coefficient of x_4, the slack variable).

3. $y_3 = 0$, for the third constraint (reduced cost coefficient of x_5, the surplus variable).

4. $y_4 = \frac{7}{3}$, for the fourth constraint (reduced cost coefficient of x_6, the surplus variable).

Thus from the above discussion, dual variable for the equality constraint $2x_1 + x_2 = 4$ is $y_2 - y_3 = \frac{5}{3}$. Note also that $y_4 = \frac{7}{3}$ is the dual variable for the fourth constraint written as $-x_1 + x_2 \leq -1$ and not for the constraint $x_1 - x_2 \geq 1$. These observations are important for the topic of sensitivity analysis discussed in Section 4.6.

Now, let us re-solve the same problem with the equality constraint as it is. The standard Simplex form for the problem is given as

minimize
$$f_p = -x_1 - 4x_2$$

subject to

$$x_1 + 2x_2 + x_3 \qquad\qquad = 5$$
$$2x_1 + x_2 \qquad + x_5 \qquad = 4$$
$$x_1 - x_2 \qquad - x_4 \qquad + x_6 = 1$$
$$x_i \geq 0; \qquad i = 1 \text{ to } 6$$

where, x_3 is a slack variable, x_4 is a surplus variable, and x_5 and x_6 are artificial variables. The problem is solved in Table 4.27. It takes two iterations to reach the optimum:

basic variables $x_1 = \frac{5}{3}, \ x_2 = \frac{2}{3}, \ x_3 = 2$

nonbasic variables $x_4 = 0, \ x_5 = 0, \ x_6 = 0$

primal cost function $f_p = -\frac{13}{3}$

TABLE 4.27
Solution for Example 4.26 with equality constraint

Basic ↓	x_1	x_2	x_3	x_4	x_5	x_6	b
x_3	1	2	1	0	0	0	5
x_5	2	1	0	0	1	0	4
x_6	**1**	−1	0	−1	0	1	1
Cost	−1	−4	0	0	0	0	$f_p - 0$
Artificial	**−3**	0	0	1	0	0	$w - 5$
x_3	0	3	1	1	0	−1	4
x_5	0	**3**	0	2	1	−2	2
x_1	1	−1	0	−1	0	1	1
Cost	0	−5	0	−1	0	1	$f_p + 1$
Artificial	0	**−3**	0	−2	0	3	$w - 2$
x_3	0	0	1	−1	−1	1	2
x_2	0	1	0	$\frac{2}{3}$	$\frac{1}{3}$	$-\frac{2}{3}$	$\frac{2}{3}$
x_1	1	0	0	$-\frac{1}{3}$	$\frac{1}{3}$	$\frac{1}{3}$	$\frac{5}{3}$
Cost	0	0	0	$\frac{7}{3}$	$\frac{5}{3}$	$-\frac{7}{3}$	$f_p + \frac{13}{3}$
Artificial	0	0	0	0	1	1	$w - 0$

End of Phase I   *End of Phase II*

Using Theorem 4.15 and the preceding discussion, the dual variables for the above three constraints are

1. $y_1 = 0$, for the first constraint (reduced cost coefficient of x_3, the slack variable).
2. $y_2 = \frac{5}{3}$, for the second constraint (reduced cost coefficient of x_5, the artificial variable).
3. $y_3 = \frac{7}{3}$, for the third constraint (reduced cost coefficient of x_4, the surplus variable).

We see that the two solutions are the same. Therefore, we do not have to replace an equality constraint by two inequalities in the standard Simplex method. The reduced cost coefficient corresponding to the artificial variable associated with the equality constraint gives the value of the dual variable for the constraint. ‖

4.7.7 Dual Variables as Lagrange Multipliers

Section 4.6 describes how the optimum value of the cost function for the problem changes if we change right-hand side parameters, b_i's (resource limits), of constraints. The constraint variation sensitivity theorem of Chapter 3 is used to study this effect. Use of that theorem requires knowledge of the Lagrange multipliers for the constraints which must be determined. It turns out that the dual variables of the problem are related to the Lagrange multipliers. The following theorem gives this relationship.

> **Theorem 4.16 Dual variables as Lagrange Multipliers.** Let **x** and **y** be optimal solutions for the primal and dual problems stated in Eqs (4.48) to (4.51), respectively. Then dual variables **y** are also Lagrange multipliers for primal constraints of Eqs (4.49). ‖

Proof. The theorem can be proved by writing Kuhn–Tucker necessary conditions of Theorem 3.6 for the primal problem defined in Eqs (4.48) and (4.49). To write these conditions, convert the primal problem to a minimization problem and define a Lagrange function as

$$
\begin{aligned}
L &= -\sum_{j=1}^{n} d_j x_j + \sum_{i=1}^{m} y_i \left(\sum_{j=1}^{n} a_{ij} x_j - e_i \right) - \sum_{j=1}^{n} v_j x_j \\
&= -\mathbf{d}^T \mathbf{x} + \mathbf{y}^T (\mathbf{A}\mathbf{x} - \mathbf{e}) - \mathbf{v}^T \mathbf{x}
\end{aligned} \tag{a}
$$

where y_i is the Lagrange multiplier for the ith primal constraint of Eq. (4.49) and v_j is the Lagrange multiplier for the jth nonnegativity constraint for the variable x_j. Write Kuhn–Tucker necessary conditions of Theorem 3.6 as

$$
-d_j + \sum_{i=1}^{m} y_i a_{ij} - v_j = 0; \qquad j = 1 \text{ to } n \quad (\partial L / \partial x_j = 0) \tag{b}
$$

$$
y_i \left(\sum_{j=1}^{n} a_{ij} x_j - e_i \right) = 0, \qquad i = 1 \text{ to } m \tag{c}
$$

$$
v_i x_i = 0, \qquad x_i \geqq 0, \qquad i = 1 \text{ to } n \tag{d}
$$

$$
y_i \geqq 0, \qquad i = 1 \text{ to } m \tag{e}
$$

$$
v_i \geqq 0; \qquad i = 1 \text{ to } n \tag{f}
$$

Rewrite Eq. (b) as

$$
-d_j + \sum_{i=1}^{m} a_{ij} y_i = v_j; \qquad j = 1 \text{ to } n \qquad (-\mathbf{d} + \mathbf{A}^T \mathbf{y} = \mathbf{v})
$$

Using conditions (f) in the preceding equation, we conclude

$$
\sum_{i=1}^{m} a_{ij} y_i \geqq d_j; \qquad j = 1 \text{ to } n \qquad (\mathbf{A}^T \mathbf{y} \geqq \mathbf{d}) \tag{g}
$$

Thus y_i's are feasible solutions for the dual constraints of Eq. (4.51).

Now let x_i represent the optimum solution for the primal problem. Then m of the x_i's are positive (barring degeneracy), and the corresponding v_i are equal to zero from Eq. (d). The remaining x_j's are zero and the corresponding v_j's are greater than zero. Therefore, from Eq. (g), we obtain

$$\text{(i)} \quad v_j > 0, \ x_j = 0, \ \sum_{i=1}^{m} a_{ij} y_i > d_j \tag{h}$$

$$\text{(ii)} \quad v_j = 0, \ x_j > 0, \ \sum_{i=1}^{m} a_{ij} y_i = d_j \tag{i}$$

Now adding the m rows given in Eq. (c), interchanging the sums on the left side and re-arranging, we obtain

$$\sum_{j=1}^{n} x_j \sum_{i=1}^{m} a_{ij} y_i = \sum_{i=1}^{m} y_i e_i \quad (\mathbf{x}^T \mathbf{A}^T \mathbf{y} = \mathbf{y}^T \mathbf{e}) \tag{j}$$

Using Eqs (h) and (i), Eq. (j) can be written as

$$\sum_{j=1}^{n} d_j x_j = \sum_{i=1}^{m} y_i e_i \quad (\mathbf{d}^T \mathbf{x} = \mathbf{y}^T \mathbf{e}) \tag{k}$$

Equation (k) also states that

$$z_p = \sum_{i=1}^{m} y_i e_i = \mathbf{y}^T \mathbf{e} \tag{l}$$

The right-hand side of Eq. (l) represents the dual cost function. According to Theorem 4.12, if primal and dual functions have the same values and if $\mathbf{x}$ and $\mathbf{y}$ are feasible points for the primal and the dual problems, then they are optimum solutions for the respective problems. Thus, the Lagrange multipliers y_i, $i = 1$ to m solve the dual problem defined in Eqs. (4.50) and (4.51).   ‖

EXERCISES FOR CHAPTER 4

Section 4.2   Definition of Standard Linear Programming Problem

4.1 *Answer True or False*

1. A linear programming problem having maximization of a function cannot be transcribed into the standard LP form.
2. A surplus variable must be added to a "$\leq$ type" constraint in the standard LP formulation.
3. A slack variable for an LP constraint can have negative value.
4. A surplus variable for an LP constraint must be nonnegative.
5. If a "$\leq$ type" constraint is active, its slack variable must be positive.
6. If a "$\geq$ type" constraint is active, its surplus variable must be zero.
7. In the standard LP formulation, the resource limits are free in sign.
8. Only "$\leq$ type" constraints can be transcribed into the standard LP form.

9. Variables that are free in sign can be treated in any LP problem.
10. In the standard LP form, all the cost coefficients must be positive.
11. All variables must be nonnegative in the standard LP definition.

Convert the following problems to the standard LP form:

4.2 Minimize $f = 5x_1 + 4x_2 - x_3$

subject to $x_1 + 2x_2 - x_3 \geq 1$

$2x_1 + x_2 + x_3 \geq 4$

$x_1, x_2 \geq 0;$ x_3 is unrestricted in sign

4.3 Maximize $z = x_1 + 2x_2$

subject to $-x_1 + 3x_2 \leq 10$

$x_1 + x_2 \leq 6$

$x_1 - x_2 \leq 2$

$x_1 + 3x_2 \geq 6$

$x_1, x_2 \geq 0$

4.4 Minimize $f = 2x_1 - 3x_2$

subject to $x_1 + x_2 \leq 1$

$-2x_1 + x_2 \geq 2$

$x_1, x_2 \geq 0$

4.5 Maximize $z = 4x_1 + 2x_2$

subject to $-2x_1 + x_2 \leq 4$

$x_1 + 2x_2 \geq 2$

$x_1, x_2 \geq 0$

4.6 Maximize $z = x_1 + 4x_2$

subject to $x_1 + 2x_2 \leq 5$

$x_1 + x_2 = 4$

$x_1 - x_2 \geq 3$

$x_1, x_2 \geq 0$

4.7 Maximize $z = x_1 + 4x_2$

subject to $x_1 + 2x_2 \leq 5$

$2x_1 + x_2 = 4$

$x_1 - x_2 \geq 1$

$x_1, x_2 \geq 0$

4.8 Minimize $f = 9x_1 + 2x_2 + 3x_3$

subject to $-2x_1 - x_2 + 3x_3 \leq -5$

$x_1 - 2x_2 + 2x_3 \geq -2$

$x_1, x_2, x_3 \geq 0$

4.9 Minimize $f = 5x_1 + 4x_2 - x_3$

subject to $x_1 + 2x_2 - x_3 \geq 1$

$2x_1 + x_2 + x_3 \geq 4$

$x_1, x_2 \geq 0;$ x_3 is unrestricted in sign

4.10 Maximize $z = -10x_1 - 18x_2$

subject to $x_1 - 3x_2 \leqq -3$
$2x_1 + 2x_2 \geqq 5$
$x_1, x_2 \geqq 0$

4.11 Minimize $f = 20x_1 - 6x_2$
subject to $3x_1 - x_2 \geqq 3$
$-4x_1 + 3x_2 = -8$
$x_1, x_2 \geqq 0$

4.12 Maximize $z = 2x_1 + 5x_2 - 4.5x_3 + 1.5x_4$
subject to $5x_1 + 3x_2 + 1.5x_3 \leqq 8$
$1.8x_1 - 6x_2 + 4x_3 + x_4 \geqq 3$
$-3.6x_1 + 8.2x_2 + 7.5x_3 + 5x_4 = 15$
$x_i \geqq 0; \quad i = 1 \text{ to } 4$

4.13 Minimize $f = 8x_1 - 3x_2 + 15x_3$
subject to $5x_1 - 1.8x_2 - 3.6x_3 \geqq 2$
$3x_1 + 6x_2 + 8.2x_3 \geqq 5$
$1.5x_1 - 4x_2 + 7.5x_3 \geqq -4.5$
$-x_2 + 5x_3 \geqq 1.5$
$x_1, x_2 \geqq 0; \quad x_3 \text{ is unrestricted in sign}$

4.14 Maximize $z = 10x_1 + 6x_2$
subject to $2x_1 + 3x_2 \leqq 90$
$4x_1 + 2x_2 \leqq 80$
$x_2 \geqq 15$
$5x_1 + x_2 = 25$
$x_1, x_2 \geqq 0$

4.15 Maximize $z = -2x_1 + 4x_2$
subject to $2x_1 + x_2 \geqq 3$
$2x_1 + 10x_2 \leqq 18$
$x_1, x_2 \geqq 0$

4.16 Maximize $z = x_1 + 4x_2$
subject to $x_1 + 2x_2 \leqq 5$
$2x_1 + x_2 = 4$
$x_1 - x_2 \geqq 3$
$x_1 \geqq 0, x_2 \text{ is free in sign}$

4.17 Minimize $f = 3x_1 + 2x_2$
subject to $x_1 - x_2 \geqq 0$
$x_1 + x_2 \geqq 2$
$x_1, x_2 \geqq 0$

4.18 Maximize $z = 3x_1 + 2x_2$
subject to $x_1 - x_2 \geqq 0$
$x_1 + x_2 \geqq 2$
$2x_1 + x_2 \leqq 6$
$x_1, x_2 \geqq 0$

4.19 Maximize $z = x_1 + 2x_2$

subject to $\quad 3x_1 + 4x_2 \leqq 12$

$\qquad x_1 + 3x_2 \geqq 3$

$\qquad\qquad x_1 \geqq 0; \qquad x_2$ is free in sign

Section 4.3 Basic Concepts Related to Linear Programming Problems

4.20 *Answer True or False*

1. In the standard LP definition, the number of constraint equations (i.e. rows in the matrix **A**) must be less than the number of variables.
2. In an LP problem, the number of "$\leqq$ type" constraints cannot be more than the number of design variables.
3. In an LP problem, the number of "$\geqq$ type" constraints cannot be more than the number of design variables.
4. An LP problem has an infinite number of basic solutions.
5. A basic solution must have zero value for some of the variables.
6. A basic solution can have negative values for some of the variables.
7. A degenerate basic solution has exactly m variables with nonzero values where m is the number of equations.
8. A basic feasible solution has all variables with nonnegative values.
9. A basic feasible solution must have m variables with positive values, where m is the number of equations.
10. The optimum point for an LP problem can be inside the feasible region.
11. The optimum point for an LP problem lies at a vertex of the feasible region.
12. The solution to any LP problem is only a local optimum.
13. The solution to any LP problem is a unique global optimum.

Find all the basic solutions for the following LP problems using the Gauss–Jordan elimination method. Identify basic feasible solutions and show them on graph paper.

4.21 Maximize $z = x_1 + 4x_2$

subject to $\quad x_1 + 2x_2 \leqq 5$

$\qquad 2x_1 + x_2 = 4$

$\qquad x_1 - x_2 \geqq 1$

$\qquad x_1, x_2 \geqq 0$

4.22 Maximize $z = -10x_1 - 18x_2$

subject to $\quad x_1 - 3x_2 \leqq -3$

$\qquad 2x_1 + 2x_2 \geqq 5$

$\qquad x_1, x_2 \geqq 0$

4.23 Maximize $z = x_1 + 2x_2$

subject to $\quad 3x_1 + 4x_2 \leqq 12$

$\qquad x_1 + 3x_2 \geqq 3$

$\qquad\qquad x_1 \geqq 0, x_2$ is free in sign

4.24 Minimize $f = 20x_1 - 6x_2$

subject to $\quad 3x_1 - x_2 \geqq 3$

$\qquad -4x_1 + 3x_2 = -8$

$\qquad\qquad x_1,\ x_2 \geqq 0$

4.25 Maximize $z = 5x_1 - 2x_2$

subject to $\quad 2x_1 + x_2 \leqq 9$

$\qquad\qquad x_1 - 2x_2 \leqq 2$

$\qquad -3x_1 + 2x_2 \leqq 3$

$\qquad\qquad x_1,\ x_2 \geqq 0$

4.26 Maximize $z = x_1 + 4x_2$

subject to $\quad x_1 + 2x_2 \leqq 5$

$\qquad\qquad x_1 + x_2 = 4$

$\qquad\qquad x_1 - x_2 \geqq 3$

$\qquad\qquad x_1,\ x_2 \geqq 0$

4.27 Minimize $f = 5x_1 + 4x_2 - x_3$

subject to $\quad x_1 + 2x_2 - x_3 \geqq 1$

$\qquad\qquad 2x_1 + x_2 + x_3 \geqq 4$

$\qquad\qquad x_1,\ x_3 \geqq 0; \qquad x_2$ is free in sign

4.28 Minimize $f = 9x_1 + 2x_2 + 3x_3$

subject to $\quad -2x_1 - x_2 + 3x_3 \leqq -5$

$\qquad\qquad x_1 - 2x_2 + 2x_3 \geqq -2$

$\qquad\qquad x_1,\ x_2,\ x_3 \geqq 0$

4.29 Maximize $z = 4x_1 + 2x_2$

subject to $\quad -2x_1 + x_2 \leqq 4$

$\qquad\qquad x_1 + 2x_2 \geqq 2$

$\qquad\qquad x_1,\ x_2 \geqq 0$

4.30 Maximize $z = 3x_1 + 2x_2$

subject to $\quad x_1 - x_2 \geqq 0$

$\qquad\qquad x_1 + x_2 \geqq 2$

$\qquad\qquad x_1,\ x_2 \geqq 0$

4.31 Maximize $z = 4x_1 + 5x_2$

subject to $\quad -x_1 + 2x_2 \leqq 10$

$\qquad\qquad 3x_1 + 2x_2 \leqq 18$

$\qquad\qquad x_1,\ x_2 \geqq 0$

Section 4.4   The Simplex Method

Solve the following problems by the Simplex method and verify the solution graphically:

4.32 Maximize $z = x_1 + 0.5x_2$

subject to $\quad 6x_1 + 5x_2 \leqq 30$

$$3x_1 + x_2 \leqq 12$$
$$x_1 + 3x_2 \leqq 12$$
$$x_1, x_2 \geqq 0$$

4.33 Maximize $z = 3x_1 + 2x_2$

subject to $\quad 3x_1 + 2x_2 \leqq 6$
$$-4x_1 + 9x_2 \leqq 36$$
$$x_1, x_2 \geqq 0$$

4.34 Maximize $z = x_1 + 2x_2$

subject to $\quad -x_1 + 3x_2 \leqq 10$
$$x_1 + x_2 \leqq 6$$
$$x_1 - x_2 \leqq 2$$
$$x_1, x_2 \geqq 0$$

4.35 Maximize $z = 2x_1 + x_2$

subject to $\quad -x_1 + 2x_2 \leqq 10$
$$3x_1 + 2x_2 \leqq 18$$
$$x_1, x_2 \geqq 0$$

4.36 Maximize $z = 5x_1 - 2x_2$

subject to $\quad 2x_1 + x_2 \leqq 9$
$$x_1 - x_2 \leqq 2$$
$$-3x_1 + 2x_2 \leqq 3$$
$$x_1, x_2 \geqq 0$$

4.37 Minimize $f = 2x_1 - x_2$

subject to $\quad -x_1 + 2x_2 \leqq 10$
$$3x_1 + 2x_2 \leqq 18$$
$$x_1, x_2 \geqq 0$$

Section 4.5   Initial Basic Feasible Solution – Artificial Variables

4.38 *Answer True or False*

1. A pivot step of the Simplex method replaces a current basic variable with a nonbasic variable.
2. The pivot step brings the design point to the interior of the constraint set.
3. The pivot column in the Simplex method is determined by the largest reduced cost coefficient corresponding to a basic variable.
4. The pivot row in the Simplex method is determined by the largest ratio of right-hand side parameters with the positive coefficients in the pivot column.
5. The criterion for a current basic variable to leave the basic set is to keep the new solution basic and feasible.
6. A move from one basic feasible solution to another corresponds to extreme points of the convex polyhedral set.

7. A move from one basic feasible solution to another can increase the cost function value in the Simplex method.

8. The right-hand sides in the Simplex tableau can assume negative values.

9. The right-hand sides in the Simplex tableau can become zero.

10. The reduced cost coefficients corresponding to the basic variables must be positive at the optimum.

11. If a reduced cost coefficient corresponding to a nonbasic variable is zero at the optimum point, there may be multiple solutions to the problem.

12. If all elements in the pivot column are negative, the problem is infeasible.

13. The artificial variables must be positive in the final solution.

14. If artificial variables are positive at the final solution, the artificial cost function is also positive.

15. If artificial cost function is positive at the optimum solution, the problem is unbounded.

Solve the following LP problems by the Simplex method and verify the solution graphically, if possible:

4.39 Maximize $z = x_1 + 2x_2$

subject to $\quad -x_1 + 3x_2 \leq 10$

$$x_1 + x_2 \leq 6$$
$$x_1 - x_2 \leq 2$$
$$x_1 + 3x_2 \geq 6$$
$$x_1, x_2 \geq 0$$

4.40 Maximize $z = 4x_1 + 2x_2$

subject to $\quad -2x_1 + x_2 \leq 4$

$$x_1 + 2x_2 \geq 2$$
$$x_1, x_2 \geq 0$$

4.41 Maximize $z = x_1 + 4x_2$

subject to $\quad x_1 + 2x_2 \leq 5$

$$x_1 + x_2 = 4$$
$$x_1 - x_2 \geq 3$$
$$x_1, x_2 \geq 0$$

4.42 Maximize $z = x_1 + 4x_2$

subject to $\quad x_1 + 2x_2 \leq 5$

$$2x_1 + x_2 = 4$$
$$x_1 - x_2 \geq 1$$
$$x_1, x_2 \geq 0$$

4.43 Minimize $f = 9x_1 + 2x_2 + 3x_3$

subject to $\quad -2x_1 - x_2 + 3x_3 \leq -5$

$$x_1 - 2x_2 + 2x_3 \geq -2$$
$$x_1, x_2, x_3 \geq 0$$

4.44 Minimize $f = 5x_1 + 4x_2 - x_3$

subject to $x_1 + 2x_2 - x_3 \geqq 1$

$2x_1 + x_2 + x_3 \geqq 4$

$x_1, x_2 \geqq 0$; x_3 is unrestricted in sign

4.45 Maximize $z = -10x_1 - 18x_2$

subject to $x_1 - 3x_2 \leqq -3$

$2x_1 + 2x_2 \geqq 5$

$x_1, x_2 \geqq 0$

4.46 Minimize $f = 20x_1 - 6x_2$

subject to $3x_1 - x_2 \geqq 3$

$-4x_1 + 3x_2 = -8$

$x_1, x_2 \geqq 0$

4.47 Maximize $z = 2x_1 + 5x_2 - 4.5x_3 + 1.5x_4$

subject to $5x_1 + 3x_2 + 1.5x_3 \leqq 8$

$1.8x_1 - 6x_2 + 4x_3 + x_4 \geqq 3$

$-3.6x_1 + 8.2x_2 + 7.5x_3 + 5x_4 = 15$

$x_i \geqq 0$; $i = 1$ to 4

4.48 Minimize $f = 8x_1 - 3x_2 + 15x_3$

subject to $5x_1 - 1.8x_2 - 3.6x_3 \geqq 2$

$3x_1 + 6x_2 + 8.2x_3 \geqq 5$

$1.5x_1 - 4x_2 + 7.5x_3 \geqq -4.5$

$-x_2 + 5x_3 \geqq 1.5$

$x_1, x_2 \geqq 0$; x_3 is unrestricted in sign

4.49 Maximize $z = 10x_1 + 6x_2$

subject to $2x_1 + 3x_2 \leqq 90$

$4x_1 + 2x_2 \leqq 80$

$x_2 \geqq 15$

$5x_1 + x_2 = 25$

$x_1, x_2 \geqq 0$

4.50 Maximize $z = -2x_1 + 4x_2$

subject to $2x_1 + x_2 \geqq 3$

$2x_1 + 10x_2 \leqq 18$

$x_1, x_2 \geqq 0$

4.51 Maximize $z = x_1 + 4x_2$

subject to $x_1 + 2x_2 \leqq 5$

$2x_1 + x_2 = 4$

$x_1 - x_2 \geqq 3$

$x_1 \geqq 0$; x_2 is free in sign

4.52 Minimize $f = 3x_1 + 2x_2$

subject to $x_1 - x_2 \geqq 0$

$x_1 + x_2 \geqq 2$

$x_1, x_2 \geqq 0$

4.53 Maximize $z = 3x_1 + 2x_2$

subject to $\quad x_1 - x_2 \geqq 0$

$\qquad x_1 + x_2 \geqq 2$

$\qquad 2x_1 + x_2 \leqq 6$

$\qquad x_1, x_2 \geqq 0$

4.54 Maximize $z = x_1 + 2x_2$

subject to $\quad 3x_1 + 4x_2 \leqq 12$

$\qquad x_1 + 3x_2 \leqq 3$

$\qquad x_1 \geqq 0; \qquad x_2$ is free in sign

4.55 Maximize $z = x_1 + 2x_2$

subject to $\quad -x_1 + 3x_2 \leqq 10$

$\qquad x_1 + x_2 \leqq 6$

$\qquad x_1 - x_2 \leqq 2$

$\qquad x_1 + 3x_2 \geqq 6$

$\qquad x_1, x_2 \geqq 0$

4.56 Maximize $z = 3x_1 + 8x_2$

subject to $\quad 3x_1 + 4x_2 \leqq 20$

$\qquad x_1 + 3x_2 \geqq 6$

$\qquad x_1 \geqq 0; \qquad x_2$ is free in sign

4.57 Minimize $f = 2x_1 - 3x_2$

subject to $\quad x_1 + x_2 \leqq 1$

$\qquad -2x_1 + x_2 \geqq 2$

$\qquad x_1, x_2 \geqq 0$

4.58 Minimize $f = 3x_1 - 3x_2$

subject to $\quad -x_1 + x_2 \leqq 0$

$\qquad x_1 + x_2 \geqq 2$

$\qquad x_1, x_2 \geqq 0$

4.59 A refinery has two crude oils:

1. Crude A costs \$30/barrel (bbl) and 20 000 bbl are available.
2. Crude B costs \$36/bbl and 30 000 bbl are available.

The company manufactures gasoline and lube oil from the crudes. Yield and sale price per barrel of the product and markets are shown in Table 4.28. How much crude oil should the company use to maximize its profit? Formulate and solve the optimum design problem. Verify the solution graphically (same as Exercise 2.2).

4.60 A manufacturer sells products A and B. Profit from A is \$10/kg and from B \$8/kg. Available raw materials for the products are: 100 kg of C and 80 kg of D. To produce 1 kg of A, 0.4 kg of C and 0.6 kg of D are needed. To produce 1 kg of B, 0.5 kg of C and 0.5 kg of D are needed. The market for the products is 70 kg for A and 110 kg for B. How much A and B should be produced to maximize profit? Formulate and solve the design optimization problem. Verify the solution graphically (same as Exercise 2.6).

TABLE 4.28
Data for refinery operations

	Yield/bbl		Sale price	Market
Product	Crude A	Crude B	per bbl	(bbl)
Gasoline	0.6	0.8	$50	20 000
Lube oil	0.4	0.2	$120	10 000

4.61 Design a diet of bread and milk to get at least 5 units of vitamin A and 4 units of vitamin B each day. The amount of vitamins A and B in 1 kg of each food and the cost of the foods are given in Table 4.29. Formulate and solve the design optimization problem so that we get at least the basic requirements of vitamins at the minimum cost. Verify the solution graphically (same as Exercise 2.7).

TABLE 4.29
Data for the diet problem

Vitamin	Bread	Milk
A	1	2
B	3	2
Cost/kg	2	1

4.62 Enterprising chemical engineering students have set up a still in a bathtub. They can produce 225 bottles of pure alcohol each week. They bottle two products from alcohol: (i) wine, 20 proof and (ii) whiskey, 80 proof. Recall that pure alcohol is 200 proof. They have an unlimited supply of water, but can only obtain 800 empty bottles per week because of stiff competition. The weekly supply of sugar is enough for 600 bottles of wine or 1200 bottles of whiskey. They make $1.0 profit on each bottle of wine and $2.00 profit on each bottle of whiskey. They can sell whatever they produce. How many bottles of wine and whiskey should they produce each week to maximize the profit. Formulate and solve the design optimization problem. Verify the solution graphically (same as Exercise 2.8).

4.63* A vegetable oil processor wishes to determine how much shortening, salad oil, and margarine to produce to optimize the use of his current oil stocks. At the current time, he has 250 000 kg of soybean oil, 110 000 kg of cottonseed oil and 2000 kg of milk base substances. The milk base substances are required only in the production of margarine. There are certain processing losses associated with each product; 10% for shortening, 5% for salad oil, and no loss for margarine. The producer's back orders require him to produce at least 100 000 kg of shortening, 50 000 kg of salad oil and 10 000 kg of margarine. In addition, sales forecasts indicate a strong demand for all products in the near future. The profit per kg and the base stock required per kg of each product are given in Table 4.30. Formulate and solve the problem to maximize profit over the next production scheduling period (same as Exercise 2.18).

TABLE 4.30
Data for the vergetable oil processing problem

Product	Profit per kg	Parts per kg of base stock requirements		
		Soybean	Cottonseed	Milk base
Shortening	0.10	2	1	0
Salad oil	0.08	0	1	0
Margarine	0.05	3	1	1

4.64* A trucking firm is considering to purchase several new trucks. It has $2 million to spend. The investment should yield a maximum of capacity in tonnes × kilometers per day. The three available choices are given in Table 4.31. Some additional restrictions should be considered: the company employs 150 drivers, and the labor market does not allow the hiring of additional drivers. Garage and maintenance facilities can handle 30 trucks at the most.

How many trucks of each type should the company purchase? Formulate and solve the design optimization problem (same as Exercise 2.20).

TABLE 4.31
Data for the available trucks

Truck model	Load capacity (tonnes)	Average speed (km/h)	Crew req'd	No. of hours of operation per day (3 shifts)	Initial investment per truck ($)
A	10	55	1	18	40 000
B	20	50	2	18	60 000
C	18	50	2	21	70 000

4.65 Solve the "saw mill" problem formulated in Section 2.6.3.

4.66* Formulate and solve the "steel mill" problem stated in Exercise 2.21.

4.67* Obtain solutions for the three formulations of the "cabinet design" problem given in Section 2.6.4. Compare the three formulations.

Section 4.6 Post-optimality Analysis

4.68 Formulate and solve the "crude oil" problem stated in Exercise 4.59. What is the effect on the cost function if market for lubricating oil suddenly increases to 12 000 barrels? What is the effect on the solution if price of Crude A drops to $24/bbl? Verify the solutions graphically.

4.69 Formulate and solve the problem stated in Exercise 4.60. What are the effects of the following changes?

1. Supply of material C increases to 120 kg.
2. Supply of material D increases to 100 kg.
3. Market for product A decreases to 60.
4. Profit for A reduces to $8/kg.

Verify your solutions graphically.

Find Lagrange multipliers at the optimum for the following problems:

4.70 Maximize $x_1 + 2x_2$
subject to $3x_1 + 2x_2 \leq 12$
$2x_1 + 3x_2 \geq 6$
$x_1 \geq 0;$ x_2 is free in sign

4.71 Exercise 4.39	**4.72** Exercise 4.40
4.73 Exercise 4.41	**4.74** Exercise 4.42
4.75 Exercise 4.43	**4.76** Exercise 4.44
4.77 Exercise 4.45	**4.78** Exercise 4.46
4.79 Exercise 4.47	**4.80** Exercise 4.48
4.81 Exercise 4.49	**4.82** Exercise 4.50
4.83 Exercise 4.51	**4.84** Exercise 4.52
4.85 Exercise 4.53	**4.86** Exercise 4.54
4.87 Exercise 4.55	**4.88** Exercise 4.56
4.89 Exercise 4.57	**4.90** Exercise 4.58

Find ranges for the right-hand side parameters of the following problems:

4.91 Exercise 4.59	**4.92** Exercise 4.60
4.93 Exercise 4.70	**4.94** Exercise 4.39
4.95 Exercise 4.40	**4.96** Exercise 4.41
4.97 Exercise 4.42	**4.98** Exercise 4.43
4.99 Exercise 4.44	**4.100** Exercise 4.45
4.101 Exercise 4.46	**4.102** Exercise 4.47
4.103 Exercise 4.48	**4.104** Exercise 4.49
4.105 Exercise 4.50	**4.106** Exercise 4.51
4.107 Exercise 4.52	**4.108** Exercise 4.53
4.109 Exercise 4.54	**4.110** Exercise 4.55
4.111 Exercise 4.56	**4.112** Exercise 4.57
4.113 Exercise 4.58	

Find the ranges for the cost function coefficients for the following problems:

4.114 Exercise 4.59   **4.115** Exercise 4.60

4.116 Exercise 4.70   **4.117** Exercise 4.39

4.118 Exercise 4.40   **4.119** Exercise 4.41

4.120 Exercise 4.42   **4.121** Exercise 4.43

4.122 Exercise 4.44   **4.123** Exercise 4.45

4.124 Exercise 4.46   **4.125** Exercise 4.47

4.126 Exercise 4.48   **4.127** Exercise 4.49

4.128 Exercise 4.50   **4.129** Exercise 4.51

4.130 Exercise 4.52   **4.131** Exercise 4.53

4.132 Exercise 4.54   **4.133** Exercise 4.55

4.134 Exercise 4.56   **4.135** Exercise 4.57

4.136 Exercise 4.58

4.137 Formulate and solve the "diet" problem stated in Exercise 4.61. Investigate the effect on the optimum solution of the following changes:

1. The cost of milk increases to $1.20/kg.
2. The need for vitamin A increases to 6 units.
3. The need for vitamin B reduces to 3 units.

Verify the solution graphically.

4.138 Formulate and solve the problem stated in Exercise 4.62. Investigate the effect on the optimum solution of the following changes:

1. The supply of empty bottles reduces to 750.
2. The profit on a bottle of wine reduces to $0.80.
3. Only 200 bottles of alcohol can be produced.

4.139* Formulate and solve the problem stated in Exercise 4.63. Investigate the effect on the optimum solution of the following changes:

1. The profit on margarine increases to $0.06/kg.
2. The supply of milk base substances increases to 2500 kg
3. The supply of soybeans reduces to 220 000 kg.

4.140 Solve the "saw mill" problem formulated in Section 2.6.3. Investigate the effect on the optimum solution of the following changes:

1. The transportation cost for the logs increases to $0.16 per kilometer per log.
2. The capacity of Mill A reduces to 200 logs/day.
3. The capacity of mill B reduces to 270 logs/day.

4.141* Formulate and solve the problem stated in Exercise 4.64. Investigate the effect on the optimum solution of the following changes:

1. Due to demand on the capital, the available cash reduces to $1.8 million.
2. The initial investment for truck B increases to $65 000.
3. Maintenance capacity reduces to 28 trucks.

4.142* Formulate and solve the "steel mill" problem stated in Exercise 2.21. Investigate the effect on the optimum solution of the following changes:

1. The capacity of reduction plant 1 increases to 1 300 000.
2. The capacity of reduction plant 2 reduces to 950 000.
3. The capacity of fabricating plant 2 increases to 250 000.
4. The demand for product 2 increases to 130 000.
5. The demand for product 1 reduces to 280 000.

4.143* Obtain solutions for the three formulations of the "cabinet design" problem given in Section 2.6.4. Compare the three formulations. Investigate the effect on the optimum solution of the following changes:

1. Bolting capacity is reduced to 5500/day.
2. The cost of riveting the C_1 component increases to $0.70.
3. The company must manufacture only 95 devices per day.

4.144 Given the following problem:

$$\text{minimize } f = 2x_1 - 4x_2$$
$$\text{subject to} \quad g_1 \equiv 10x_1 + 5x_2 \leq 15$$
$$g_2 \equiv 4x_1 + 10x_2 \leq 36$$
$$x_1 \geq 0, \, x_2 \geq 0$$

Slack variables for g_1 and g_2 are x_3 and x_4, respectively. The final tableau for the problem is

x_1	x_2	x_3	x_4	b
2	1	$\frac{1}{5}$	0	3
-16	0	-2	1	6
10	0	$\frac{4}{5}$	0	$f + 12$

Using the given tableau:

1. Determine the optimum values of f and $\mathbf{x}$.
2. Determine Lagrange multipliers for g_1 and g_2.
3. Determine right-hand-side ranges for g_1 and g_2.
4. What is the smallest value that f can have, with the current basis, if the right-hand side of g_1 is changed? What is the right-hand side of g_1 for that case?

Section 4.7   Duality in Linear Programming

Recover dual variables from the final primal tableau and verify the solution by solving the dual problem for the following exercises:

4.145 Exercise 4.39   **4.146** Exercise 4.40
4.147 Exercise 4.41   **4.148** Exercise 4.42
4.149 Exercise 4.43   **4.150** Exercise 4.44
4.151 Exercise 4.45   **4.152** Exercise 4.46
4.153 Exercise 4.47   **4.154** Exercise 4.48
4.155 Exercise 4.59   **4.156** Exercise 4.60

CHAPTER
5

NUMERICAL METHODS FOR UNCONSTRAINED OPTIMUM DESIGN

5.1 INTRODUCTION

In nonlinear optimization, some or all the functions of the problem (cost function and/or constraint functions) are nonlinear. The term *nonlinear programming* (NLP) is used for numerical methods that solve nonlinear optimization problems. This chapter concentrates on the description of methods for unconstrained problems. Chapter 6 treats constrained problems.

Numerical methods for nonlinear optimization problems are needed because the analytical methods for solving some problems are either too cumbersome or not applicable at all. In analytical methods, described in Chapter 3, we write necessary conditions and solve them for candidate local minimum designs. There are three basic reasons why the methods are inappropriate for many engineering design problems:

1. *The numbers of design variables and constraints can be large.* The necessary conditions give a large number of equations which can be difficult to solve. This is particularly true for constrained problems where the necessary conditions must be solved by considering many cases. In addition, it is difficult to solve equations in a closed form due to nonlinearity.

2. *The functions for the design problem (cost and constraint) can be highly*

nonlinear. Therefore, even if the dimension of the problem is small, the set of necessary conditions is highly nonlinear and intractable.

3. In many engineering applications, *cost and/or constraint functions can be implicit in terms of design variables.* For such functions, explicit functional form in terms of the independent variables is not known. They can be, however, numerically calculated once values of design variables are specified. Such functions are difficult to treat in analytical methods based on writing necessary conditions.

Due to these reasons, we must develop systematic numerical approaches for the optimum design of engineering systems. In such approaches, we estimate an initial design and improve it until optimality conditions are satisfied.

Many *numerical methods* for *nonlinear optimization* have been developed. Some are better than others and research in the area continues to develop still better techniques. Detailed derivations and theory of various methods are beyond the scope of the present text. However, it is important to understand a few basic concepts, ideas and procedures which are used in most algorithms for unconstrained and constrained optimization. Therefore, *the approach followed in this chapter and Chapter 6 is to stress the underlying concepts with example problems. Based on these concepts numerous numerical algorithms can be developed for the optimum design of engineering systems.* We shall describe a few of the procedures.

In the present chapter, unconstrained optimization problems are treated. Numerical details for some algorithms are discussed to give the student a flavor of the type of calculations needed for nonlinear optimization problems. *Most computer center libraries either have some standard computer programs for general use, or they can easily acquire them. Therefore, coding of the algorithms should be attempted only as a last resort.* It is, however, important to understand the methods and the underlying ideas because many of the concepts are applicable (or, can be extended) to constrained optimization problems.

Several numerical problems are solved using the programs given in Appendix D to study the performance of several algorithms and compare them. Many results are given in tabular form. It must be understood that the behavior of any algorithm can be drastically affected by the numerical implementation details. In addition, *numerical performance* of a program can change from one computer to another and from one compiler to another. Therefore, it may not be possible to duplicate exactly the numerical results given in the tables on different computers. The results given in the text are obtained on an Apollo DN560 workstation.

The unconstrained optimization problems are classified as one-dimensional (line search) and multidimensional problems as defined in Fig. 5.1. Numerical methods for solving the problems have been developed over the last several decades. Substantial work was, however, done during the 1950s and 1960s because it was shown that constrained optimization problems could

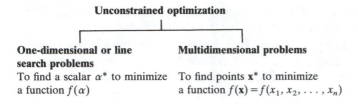

FIGURE 5.1
Classification of unconstrained optimization problems.

be transformed to a sequence of unconstrained problems. Due to this reason the methods gained considerable importance and substantial effort was expended in developing efficient algorithms and computer codes. Therefore, it is important to fully understand the unconstrained optimization methods.

The following is an outline of this chapter.

Section 5.2 General Concepts Related to Numerical Algorithms. The idea of iterative numerical algorithms is introduced. A general search procedure for optimum design problems is given. The algorithm is applicable to both constrained and unconstrained problems. Since we want to minimize the cost function, the idea of a descent step is introduced which simply means that change in design at every iteration must reduce the cost function. Convergence of an algorithm and its rate of convergence are briefly discussed. Most algorithms discussed in the text need gradients of cost and constraint functions. Sometimes analytical gradient evaluation is quite tedious or even impossible. Therefore, gradients must be evaluated numerically using finite differences. This is briefly explained.

Section 5.3 One-dimensional Minimization. Most optimization algorithms are broken down into two phases: search direction and step length determination subproblems. Once the search direction is determined, the step size determination is a one-dimensional minimization problem. Three simple algorithms are described to solve the problem: equal interval search, Golden Sections search and polynomial interpolation. The procedures are illustrated with examples.

Section 5.4 Steepest Descent Method. This is the simplest and the oldest method of computing the search direction. It exploits properties of the gradient of the cost function. However, the algorithm is not very efficient, so it is not recommended for general applications. It is included for historical reasons. In addition, many modern algorithms can be considered as variations of the steepest descent method. The method is illustrated with examples. The scaling of design variables can be very effective in many applications, so it is explained with examples.

Section 5.5 Conjugate Gradient Method. This method is a very simple and effective modification of the steepest descent method. The modification is explained and the method is illustrated with an example. The method is recommended for general applications.

Section 5.6 Newton's Method. This method uses Hessian of the cost function to determine the search direction. It is briefly derived and illustrated with examples. A useful modification of the method is explained.

Section 5.7 Quasi-Newton Methods. Newton's method requires evaluation of second-order derivatives of the cost function. In addition, exact Hessian matrix can be singular or indefinite causing numerical difficulties. Quasi-Newton methods use only first-order information to generate approximate Hessians that remain positive definite at each iteration. Two such methods are described and illustrated.

Section 5.8 Engineering Applications of Unconstrained Methods. There are several useful engineering applications of unconstrained methods. We describe a few of them. The first one involves minimization of total potential energy which gives equilibrium states for structural and mechanical systems. This is illustrated with a simple numerical example. The second application involves the solution of a set of nonlinear equations that are encountered in numerous engineering applications. An unconstrained function involving the original equations is defined. Minimization of the function gives roots for the nonlinear set of equations. This is illustrated with an example.

Section 5.9 Transformation Methods for Optimum Design. It turns out that unconstrained optimization methods can be used to solve constrained problems. This section briefly describes such methods that transform the constrained problem to a sequence of unconstrained problems. The *sequential unconstrained minimization techniques* as well as more modern *multiplier methods* are briefly reviewed.

5.2 GENERAL CONCEPTS RELATED TO NUMERICAL ALGORITHMS

Numerical methods for optimum design are conceptually different from the analytical methods described in Chapter 3. In the analytical approach we write the optimality conditions, and solve them for candidate local minimum designs. Using numerical methods, however, we select a design as an initial estimate for the optimum point and change it iteratively until optimality conditions are satisfied. The process may require several iterations. Thus, with numerical methods we strive to satisfy the optimality conditions using an *iterative process.* In this section, *we describe some basic concepts that are applicable to both constrained and unconstrained numerical optimization methods.*

5.2.1 A General Algorithm

All numerical methods given in the present and subsequent chapters are described by the following iterative prescription:

$$\text{vector form:} \quad \mathbf{x}^{(k+1)} = \mathbf{x}^{(k)} + \Delta\mathbf{x}^{(k)}; \quad k = 0, 1, 2, \ldots \quad (5.1)$$

$$\text{component form:} \quad x_i^{(k+1)} = x_i^{(k)} + \Delta x_i^{(k)}; \quad k = 0, 1, 2, \ldots \quad (5.2)$$

$$i = 1 \text{ to } n$$

In these equations, the superscript k represents the iteration number, subscript i denotes the design variable number, $\mathbf{x}^{(0)}$ is any starting design and $\Delta\mathbf{x}^{(k)}$ represents a small change in the current design. The iterative scheme described in Eq. (5.1) or (5.2) is continued until optimality conditions are satisfied, or an acceptable design is obtained.

The iterative formula is applicable to constrained as well as unconstrained problems. For unconstrained problems, calculations for $\Delta\mathbf{x}^{(k)}$ depend on the cost function and its derivatives at the current design point. For the constrained problem, the constraints must also be considered while computing the change in design $\Delta\mathbf{x}^{(k)}$. Therefore, in addition to the cost function and its derivatives, the constraint functions and their derivatives play a role in determining $\Delta\mathbf{x}^{(k)}$. There are several methods for calculating $\Delta\mathbf{x}^{(k)}$ for unconstrained and constrained problems. We shall describe some of the basic methods later in this chapter.

For optimization methods, the change in design $\Delta\mathbf{x}^{(k)}$ is further decomposed into two parts as

$$\Delta\mathbf{x}^{(k)} = \alpha_k \mathbf{d}^{(k)} \tag{5.3}$$

where $\mathbf{d}^{(k)}$ is a *desirable search direction* of move in the design space and α_k is a *positive scalar called the step size* in that direction. Thus, the process of computing $\Delta\mathbf{x}^{(k)}$ solves two separate subproblems: the direction finding subproblem and the step length determination subproblem (scaling along the direction). The process of moving from one design point to the next is illustrated in Fig. 5.2. In the figure, B is the current design point, $\mathbf{d}^{(k)}$ is the desirable search direction and α is a step length. Therefore, when $\alpha\mathbf{d}^{(k)}$ is added to the current design we reach a new design point C in the design space. The entire process is repeated from point C.

It is observed that there are many procedures for calculating the step size α and the search direction vector $\mathbf{d}^{(k)}$. Various combinations of the procedures can be used to develop different optimization algorithms.

In summary the *basic idea of numerical methods* for nonlinear optimization problems is to start with a reasonable estimate for the optimum design.

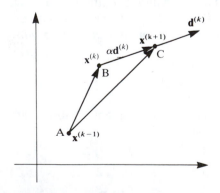

FIGURE 5.2
Conceptual diagram for iterative steps of an optimization method.

Cost and constraint functions and their derivatives are evaluated at that point. Based on them, the design is moved to a new point. The process is continued until either optimality conditions or some other stopping criteria are satisfied. It can be seen that the iterative process represents an organized search through the design space for points that represent local minima. Thus, the procedures are often called the *search techniques* or *direct methods* of optimization.

The preceding iterative process can be summarized as a *general algorithm* applicable to both constrained and unconstrained problems:

Step 1. Estimate a reasonable starting design $\mathbf{x}^{(0)}$. Set the iteration counter $k = 0$.

Step 2. Compute a search direction $\mathbf{d}^{(k)}$ in the design space. This calculation generally requires cost function value and its gradient for the unconstrained problems and, in addition, constraint functions and their gradients for constrained problems.

Step 3. Check for convergence of the algorithm. If it has converged, terminate the iterative process. Otherwise, continue.

Step 4. Calculate a positive step size α_k.

Step 5. Calculate the new design as

$$\mathbf{x}^{(k+1)} = \mathbf{x}^{(k)} + \alpha_k \mathbf{d}^{(k)} \tag{5.4}$$

Set $k = k + 1$ and go to Step 2.

In the remaining sections of this chapter, we shall discuss methods for calculating the step size α and the search direction $\mathbf{d}^{(k)}$ for unconstrained optimization problems.

5.2.2 Descent Step Idea

We have referred to $\mathbf{d}^{(k)}$ as a desirable direction of design change in the iterative process. Now we discuss what we mean by a *desirable direction of design change*.

The objective of the iterative optimization process is to reach a minimum point for the cost function $f(\mathbf{x})$. Let us assume that we are in the kth iteration and we have determined that $\mathbf{x}^{(k)}$ is not a minimum point, i.e. optimality conditions of Theorem 3.4 are not satisfied. If $\mathbf{x}^{(k)}$ is not a minimum point then we should be able to find another point $\mathbf{x}^{(k+1)}$ with a smaller cost function value than the one at $\mathbf{x}^{(k)}$. This statement can be expressed mathematically as

$$f(\mathbf{x}^{(k+1)}) < f(\mathbf{x}^{(k)}) \tag{5.5}$$

Substitute $\mathbf{x}^{(k+1)}$ from Eq. (5.4) into the preceding inequality to obtain

$$f(\mathbf{x}^{(k)} + \alpha_k \mathbf{d}^{(k)}) < f(\mathbf{x}^{(k)}) \tag{5.6}$$

Approximating the left-hand side of Eq. (5.6) by the linear Taylor series expansion about the point $\mathbf{x}^{(k)}$, we get

$$f(\mathbf{x}^{(k)}) + \alpha_k (\mathbf{c}^{(k)} \cdot \mathbf{d}^{(k)}) < f(\mathbf{x}^{(k)}) \tag{5.7}$$

where $\mathbf{c}^{(k)} = \nabla f(\mathbf{x}^{(k)})$ is the gradient of $f(\mathbf{x})$ at the point $\mathbf{x}^{(k)}$ and $(\cdot)$ represents dot product of the two vectors. Since the left-hand side must be smaller than the right-hand side in Eq. (5.7), we conclude that the second term on the left-hand side must be negative, i.e.

$$\alpha_k(\mathbf{c}^{(k)} \cdot \mathbf{d}^{(k)}) < 0 \tag{5.8}$$

Since $\alpha_k > 0$, it may be dropped from the Inequality (5.8). Also, since $\mathbf{c}^{(k)}$ is a known quantity (gradient of the cost function), the search direction $\mathbf{d}^{(k)}$ must be calculated to satisfy the Inequality (5.8). Geometrically, the inequality shows that the angle between the vectors $\mathbf{c}^{(k)}$ and $\mathbf{d}^{(k)}$ must be between 90 and 270°. In other words, any small move in such a direction must decrease the cost function.

We can now define a *desirable direction of design change* as any vector $\mathbf{d}^{(k)}$ satisfying the Inequality (5.8). *Such vectors are also called directions of descent for the cost function and the Inequality (5.8) is called the descent condition.* A step of the iterative optimization method based on these directions is called a *descent step*. There can be several directions of descent at a design point and each optimization algorithm computes it differently.

The descent direction is also sometimes called the "down hill" direction. The problem of minimizing $f(\mathbf{x})$ can be considered as a problem of trying to reach the bottom of a hill from a high point. From there, we find a "down hill" direction and travel along it as far as possible. From the lowest point along the direction, we repeat the process until the bottom is reached. A method based on the idea of a descent step is called a *descent method*. Clearly, such a method will not converge to a maximum point for the function.

The concept of *descent directions* is used in most numerical optimization methods. Therefore, it should be clearly understood.

Example 5.1   Check for the descent condition. Consider the function

$$f(\mathbf{x}) = x_1^2 - x_1 x_2 + 2x_2^2 - 2x_1 + e^{(x_1+x_2)}$$

Verify whether the vector $\mathbf{d} = (1, 2)$ at the point $(0, 0)$ is a descent direction for the function f.

Solution. If $\mathbf{d} = (1, 2)$ is a descent direction, then it must satisfy the Inequality (5.8). To verify this, we calculate the gradient of the function $f(\mathbf{x})$ at $(0, 0)$ which is given as

$$\mathbf{c} = (2x_1 - x_2 - 2 + e^{(x_1+x_2)}, \ -x_1 + 4x_2 + e^{(x_1+x_2)}) = (-1, 1)$$

Therefore,

$$(\mathbf{c} \cdot \mathbf{d}) = (-1, 1)\begin{bmatrix} 1 \\ 2 \end{bmatrix} = -1 + 2 = 1 > 0.$$

Thus, the Inequality (5.8) is not satisfied and given $\mathbf{d}$ is not a descent direction for the function $f(\mathbf{x})$. ‖

5.2.3 Convergence of Algorithms

The central idea behind numerical methods of optimization is to search for the optimum point in an iterative manner generating a sequence of designs. It is important to note that the success of a method depends on the guarantee of convergence of the sequence to the optimum point. *The property of convergence to a local optimum point irrespective of the starting point is called global convergence of the numerical method.* It is desirable to employ convergent numerical methods in practice since they are more reliable. For unconstrained problems, a convergent algorithm must reduce cost function at every iteration until minimum is reached.

5.2.4 Rate of Convergence

In practice, a numerical method may take too many iterations to reach the optimum point. Therefore, it is important to employ methods having a faster rate of convergence. Rate of convergence of an algorithm is usually measured by the numbers of iterations and function evaluations to obtain an acceptable solution. Faster algorithms usually use second-order information about the problem functions. They are known as *Newton methods*. Many algorithms also approximate second-order information using first-order information only. They are known as *quasi-Newton methods*. We shall describe a few of these methods.

5.2.5 Numerical Evaluation of Gradients

It is often cumbersome to calculate the gradient of a complex function analytically. In such cases, *it is possible to approximate the gradient or Hessian matrix of the function by finite differences.* We shall briefly describe such procedures. It is assumed that the function is continuous and differentiable.

The partial derivative of a multivariable function $f(\mathbf{x})$ with respect to a variable x_i is defined as

$$\frac{\partial f}{\partial x_i} = \lim_{\delta x_i \to 0} \frac{f(x_1, \ldots, x_i + \delta x_i, \ldots, x_n) - f(x_1, \ldots, x_i, \ldots, x_n)}{\delta x_i} \tag{5.9}$$

where δx_i is a small perturbation in the variable x_i. Using this definition of the partial derivative, we can evaluate numerically the gradient of the function at a given point. Several expressions can be developed based on Eq. (5.9). We shall describe simple *forward difference, backward difference, and central difference approaches*. In all the approaches, δx_i is replaced by a finite change Δx_i and an approximate derivative value is obtained using the definition of Eq. (5.9). The function is evaluated at two neighboring points and the difference in function values is divided by the change Δx_i. This is also called the *numerical differentiation* of a function.

In the *forward difference* method, the change Δx_i is added to x_i and the partial derivative at the point $(x_1, \ldots, x_n)$ is calculated as

$$\frac{\partial f}{\partial x_i} \approx \frac{f(x_1, \ldots, x_i + \Delta x_i, \ldots, x_n) - f(x_1, \ldots, x_i, \ldots, x_n)}{\Delta x_i} \tag{5.10}$$

Equation (5.10) is used for all x_i, $i = 1$ to n to calculate all the partial derivatives.

In the *backward difference approach*, the change Δx_i is subtracted from x_i and the partial derivative at the point $(x_1, \ldots, x_n)$ is calculated as

$$\frac{\partial f}{\partial x_i} \approx \frac{f(x_1, \ldots, x_i, \ldots, x_n) - f(x_1, \ldots, x_i - \Delta x_i, \ldots, x_n)}{\Delta x_i} \tag{5.11}$$

for $i = 1$ to n.

In the *central difference approach*, the change $\frac{1}{2}\Delta x_i$ is subtracted as well as added to x_i and the partial derivative at the point $(x_1, \ldots, x_n)$ is calculated as

$$\frac{\partial f}{\partial x_i} \approx \frac{f(x_1, \ldots, x_i + \frac{1}{2}\Delta x_i, \ldots, x_n) - f(x_1, \ldots, x_i - \frac{1}{2}\Delta x_i, \ldots, x_n)}{\Delta x_i} \tag{5.12}$$

for $i = 1$ to n.

Note that if the perturbation Δx_i in the variable x_i is too large, then the gradient evaluation is inaccurate. Also, if Δx_i is too small, then due to loss of significant digits the gradient evaluation may be totally inaccurate. There are few practical guidelines to select a reasonable perturbation in the design variables. Usually, if the function is not too nonlinear, a perturbation of 1% works fairly well (i.e., $\Delta x_i = 0.01 |x_i|$). For more details on the specification of perturbation in the variables, consult Gill, Murray and Wright [1981].

The main disadvantage of finite difference calculation of the gradients is that the number of function evaluations increases significantly making an optimization algorithm inefficient. This is due to the fact that for each increment Δx_i, the functions must be re-evaluated. Note also that the central difference approach requires twice the number of function evaluations compared to either the forward difference or the backward difference approach.

Example 5.2 Numerical gradient evaluation. Evaluate the gradient of $f(\mathbf{x}) = x_1^2 + x_2$ at a point $(2, 1)$ by the numerical finite difference approach and compare it with the analytic gradient. Use a perturbation of 1% of the design variables.

Solution. The exact gradient of the function at the point $(2, 1)$ is given as

$$\nabla f = (2x_1, 1) = (4, 1)$$

We shall calculate the gradient of the function by all three methods. A 1% change in variables gives $\Delta x_1 = 0.02$ *and* $\Delta x_2 = 0.01$.

Forward Difference Approach:

$$\frac{\partial f}{\partial x_1} \approx \frac{f(2.02, 1) - f(2, 1)}{0.02} = \frac{0.0804}{0.02} = 4.02$$

$$\frac{\partial f}{\partial x_2} \approx \frac{f(2, 1.01) - f(2, 1)}{0.01} = \frac{0.01}{0.01} = 1.00$$

Backward Difference Approach:

$$\frac{\partial f}{\partial x_1} \approx \frac{f(2, 1) - f(1.98, 1)}{0.02} = \frac{0.0796}{0.02} = 3.98$$

$$\frac{\partial f}{\partial x_2} \approx \frac{f(2, 1) - f(2, 0.99)}{0.01} = \frac{0.01}{0.01} = 1.00$$

Central Difference Approch:

$$\frac{\partial f}{\partial x_1} = \frac{f(2.01, 1) - f(1.99, 1)}{0.02} = \frac{0.08}{0.02} = 4.00$$

$$\frac{\partial f}{\partial x_2} = \frac{f(2, 1.005) - f(2, 0.995)}{0.01} = \frac{0.01}{0.01} = 1.00$$

Note that for the given function all three methods give very good approximation to the exact gradient. The central difference approach, however, gives the best approximation. Also note that all three methods give the exact value for $\partial f / \partial x_2$. This is because the function is linear in x_2. For linear functions, the numerical differentiation always gives exact gradients.   ‖

5.3 ONE-DIMENSIONAL MINIMIZATION

5.3.1 The Problem Definition

Unconstrained numerical optimization methods are based on the iterative formula given in Eq. (5.1). As discussed earlier, the problem of obtaining the change in design $\Delta\mathbf{x}$ is usually decomposed into two parts: the direction finding and the step size determination problems, as expressed in Eq. (5.3). We need to discuss numerical methods for solving both the problems. In the following paragraphs, we first discuss the problem of *step size determination*. This is often called the *one-dimensional search (or, line search)* problem. Such problems are simpler to solve. This is one reason for discussing them first. Following one-dimensional minimization methods, several methods are described in Sections 5.4–5.7 for finding a *search direction* **d** in the design space.

For an optimization problem with several variables, the direction finding problem must be solved first. A step size must then be determined by searching for minimum of the function along the given direction in the design space. This is always a one-dimensional minimization problem.

5.3.1.1 REDUCTION TO A FUNCTION OF SINGLE VARIABLE. To see how the line search will be used in multidimensional problems, let us assume for the moment that a desirable direction of design change $\mathbf{d}^{(k)}$ has been found. In Eqs (5.1) and (5.3), scalar α is then the only unknown. Substituting Eq. (5.1) into the cost function $f(\mathbf{x})$, we obtain

$$f(\mathbf{x}^{(k+1)}) = f(\mathbf{x}^{(k)} + \alpha\mathbf{d}^{(k)})$$

Now, since $\mathbf{d}^{(k)}$ is known, the right-hand side becomes a function of the scalar

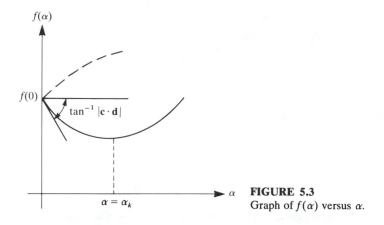

FIGURE 5.3
Graph of $f(\alpha)$ versus α.

parameter α only and we can write the preceding equation as

$$f(\mathbf{x}^{(k+1)}) = f(\mathbf{x}^{(k)} + \alpha \mathbf{d}^{(k)}) \equiv \bar{f}(\alpha) \tag{5.13}$$

where $\bar{f}(\alpha)$ is the new function with α as the only independent variable (later on, we shall drop the "over bar" for functions of single variable). Note that at $\alpha = 0$, $\bar{f}(0) = f(\mathbf{x}^{(k)})$, which is the current value of the cost function. It is important to understand the *reduction of a function of n variables to a function of only one variable* in Eq. (5.13), since this fundamental step is used in almost all optimization methods. It is also important to understand the geometrical significance of Eq. (5.13). We shall elaborate these ideas further in the following paragraphs.

If $\mathbf{x}^{(k)}$ is not a minimum point, then it is possible to find a descent direction $\mathbf{d}^{(k)}$ at the point. Recall that a small move along $\mathbf{d}^{(k)}$ reduces the cost function. Therefore, using Eqs (5.5) and (5.13), the descent condition for the cost function can be expressed as the inequality:

$$f(\alpha) < f(0) \tag{5.14}$$

Since $f(\alpha)$ is a function of single variable, we can plot $f(\alpha)$ versus α on graph paper. To satisfy Inequality (5.14) the curve $f(\alpha)$ versus α must have a negative slope at the point $\alpha = 0$. Such a curve is shown by the solid line in Fig. 5.3. It must be understood that if the direction of search is correct, the graph of $f(\alpha)$ versus α cannot be the one shown by the dashed curve because any positive α would cause the function $f(\alpha)$ to increase, violating the Inequality (5.14). This would also be a contradiction as $\mathbf{d}^{(k)}$ is a direction of descent for the cost function. Therefore, the graph of $f(\alpha)$ versus α must be the solid curve in Fig. 5.3 for all problems. In fact, the slope of the curve $f(\alpha)$ at $\alpha = 0$ is calculated as $f'(0) = \mathbf{c}^{(k)} \cdot \mathbf{d}^{(k)}$ which is negative as seen in Eq. (5.8).

5.3.1.2 ANALYTICAL METHOD TO COMPUTE STEP SIZE. The above discussion shows that if $\mathbf{d}^{(k)}$ is a descent direction, then α *must always be a positive scalar* in Eq. (5.8). Thus, the *one-dimensional minimization problem* is to find $\alpha = \alpha_k$ such that $f(\alpha)$ is minimized. If $f(\alpha)$ is a simple function then we can use the analytical procedure (necessary and sufficient conditions of Section 3.3) to determine α_k. The necessary condition is $df(\alpha_k)/d\alpha = 0$, and the sufficient condition is $d^2f(\alpha_k)/d\alpha^2 > 0$. We shall illustrate the analytical line search procedure with an example problem. Note that differentiation of $f(\mathbf{x}^{(k+1)})$ in Eq. (5.13) with respect to α gives

$$\frac{df(\mathbf{x}^{(k+1)})}{d\alpha} = \frac{\partial f^T(\mathbf{x}^{(k+1)})}{\partial \mathbf{x}} \frac{d\mathbf{x}^{(k+1)}}{d\alpha} = \nabla f(\mathbf{x}^{(k+1)}) \cdot \mathbf{d}^{(k)}$$

Thus, the necessary condition for optimum step size $(df/d\alpha = 0)$ gives

$$\nabla f(\mathbf{x}^{(k+1)}) \cdot \mathbf{d}^{(k)} = 0, \qquad \text{or} \qquad \mathbf{c}^{(k+1)} \cdot \mathbf{d}^{(k)} = 0 \qquad (5.15)$$

which shows that the gradient of the cost function at the new point is orthogonal to the search direction at the kth iteration.

Example 5.3   Analytical step size determination. Let a direction of change for the function

$$f(\mathbf{x}) = 3x_1^2 + 2x_1x_2 + 2x_2^2 + 7 \qquad (a)$$

at the point $(1, 2)$ be given as $(-1, -1)$. Compute the step size α to minimize $f(\mathbf{x})$ in the given direction.

Solution. For the given point $\mathbf{x}^{(k)} = (1, 2)$, $f(\mathbf{x}^{(k)}) = 22$, and $\mathbf{d}^{(k)} = (-1, -1)$. We first check to see if $\mathbf{d}^{(k)}$ is a direction of descent. The gradient of the function at $(1, 2)$ is given as $\mathbf{c}^{(k)} = (10, 10)$ and $\mathbf{c}^{(k)} \cdot \mathbf{d}^{(k)} = 10(-1) + 10(-1) = -20 < 0$. Therefore $(-1, -1)$ is a direction of descent.

The new point $\mathbf{x}^{(k+1)}$ from Eq. (5.1) is given as

$$\begin{bmatrix} x_1 \\ x_2 \end{bmatrix}^{(k+1)} = \begin{bmatrix} 1 \\ 2 \end{bmatrix} + \alpha \begin{bmatrix} -1 \\ -1 \end{bmatrix} \qquad (b)$$

Or, in the component form,

$$x_1^{(k+1)} = 1 - \alpha; \qquad x_2^{(k+1)} = 2 - \alpha \qquad (c)$$

Substituting Eqs (c) into Eq. (a),

$$f(\mathbf{x}^{(k+1)}) = 3(1 - \alpha)^2 + 2(1 - \alpha)(2 - \alpha) + 2(2 - \alpha)^2 + 7$$

Simplifying the above equation,

$$f(\mathbf{x}^{(k+1)}) = 7\alpha^2 - 20\alpha + 22 \equiv f(\alpha) \qquad (d)$$

Therefore, along the given direction $(-1, -1)$, $f(\mathbf{x})$ becomes a function of single variable α. Also observe from Eq. (d) that $f(0) = 22$, which is the cost function value at the current point and that $f'(0) = -20 < 0$ which is the slope of $f(\alpha)$ at $\alpha = 0$ (also recall that $f'(0) = \mathbf{c}^{(k)} \cdot \mathbf{d}^{(k)}$).

Now using the necessary and sufficient conditions of optimality for $f(\alpha)$, we obtain

$$\frac{df}{d\alpha} \equiv 14\alpha_k - 20 = 0; \qquad \alpha_k = \tfrac{10}{7}$$

$$\frac{d^2f}{d\alpha^2} = 14 > 0$$

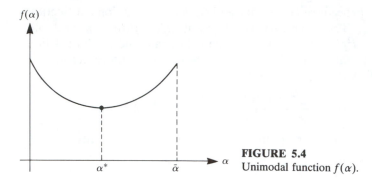

FIGURE 5.4
Unimodal function $f(\alpha)$.

so, the sufficiency condition is satisfied at α_k. Therefore, $\alpha_k = \frac{10}{7}$ minimizes $f(\mathbf{x})$ in the direction $(-1, -1)$. The new design point is

$$\begin{bmatrix} x_1 \\ x_2 \end{bmatrix}^{(k+1)} = \begin{bmatrix} 1 \\ 2 \end{bmatrix} + (\tfrac{10}{7})\begin{bmatrix} -1 \\ -1 \end{bmatrix} = \begin{bmatrix} -\frac{3}{7} \\ \frac{4}{7} \end{bmatrix} \qquad (e)$$

Substituting the new design $(-\frac{3}{7}, \frac{4}{7})$ into Eq. (a), we find the new value of the cost function as $\frac{54}{7}$. This is a substantial improvement (reduction) from the cost function of 22 at the previous design point.   ‖

5.3.1.3 NUMERICAL METHODS TO COMPUTE STEP SIZE.
In the preceding example, *it was possible to simplify expressions* and obtain an explicit form of the function $f(\alpha)$. Also, the functional form of $f(\alpha)$ was quite simple. It was possible to use the necessary and sufficient conditions to find the minimum of $f(\alpha)$, and analytically calculate the step size α_k. For many problems, it is not possible to obtain an explicit expression for $f(\alpha)$. Moreover, even if the functional form of $f(\alpha)$ is known, it may be too complicated to lend itself for analytical solution. Therefore, a *numerical method must be used to find* α_k to minimize $f(\mathbf{x})$ in the direction $\mathbf{d}^{(k)}$.

The numerical line search process is in itself iterative requiring several iterations before a minimum point is reached. Most line search techniques are based on comparing function values at several points along the direction of search. Usually, we must make some assumptions on the form of the line search function to compute step size by numerical methods. For example, it must be assumed that a minimum exists and it is unique in the interval of interest. A function with these properties is called the *unimodal function*. Figure 5.4 shows the graph of such a function which decreases continuously until the minimum point is reached. Comparing Figs 5.3 and 5.4, we observe that $f(\alpha)$ is a unimodal function in some interval. Therefore, it has a unique minimum in it.

Most one-dimensional search methods work for only unimodal functions. This may appear to be a severe restriction on the methods; however, it is not. For functions that are not unimodal, we can think of locating only a local minimum point that is closest to the starting point. This is illustrated in Fig.

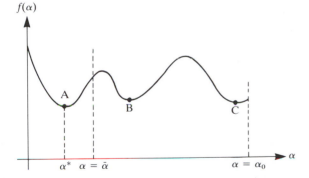

FIGURE 5.5
Nonunimodal function $f(\alpha)$ for $0 \leqq \alpha \leqq \alpha_0$ (unimodal for $0 \leqq \alpha \leqq \bar{\alpha}$).

5.5, where the function $f(\alpha)$ is not unimodal for $0 \leqq \alpha \leqq \alpha_0$. Points A, B and C are all local minima. If we restrict α to lie between 0 and $\bar{\alpha}$, however, there is only one local minimum point A, as the function $f(\alpha)$ is unimodal for $0 \leqq \alpha \leqq \bar{\alpha}$. Thus, the assumption of unimodality is not as restrictive as it appears.

The search problem then is to find α in an interval $0 \leqq \alpha \leqq \bar{\alpha}$ at which the function $f(\alpha)$ has global minimum value. This statement of the problem, however, requires some modification. Since we are dealing with numerical methods, it is not possible to locate exact minimum point α^*. In fact, *what we determine is the interval in which the minimum* lies, i.e. some lower and upper limits α_l and α_u for α^* are determined. The interval (α_l, α_u) is called the *interval of uncertainty* and is designated as $I = \alpha_u - \alpha_l$. Most numerical methods iteratively reduce the interval of uncertainty until it is less than a small specified positive number ε, i.e. $I < \varepsilon$. This is the desired accuracy for locating the minimum. Once the stopping criterion is satisfied, α^* is taken as $0.5(\alpha_l + \alpha_u)$. Methods based on the preceding philosophy are sometimes called the *interval reducing methods*

Several method are available for one-dimensional minimization [Cooper and Steinberg, 1970]. However, we shall discuss only three *zero order* methods. The term zero order emphasizes the fact that only the function values are used to locate the minimum point. The basic procedure of these methods can be divided into two phases. In phase one, the location of the minimum point is bracketed and the initial interval of uncertainty is established. In the second phase, the interval of uncertainty is refined by eliminating regions that cannot contain the minimum. This is done by computing and comparing function values in the interval of uncertainty. We shall describe the two phases in more detail in the following subsections.

5.3.2 Equal Interval Search

The basic idea of any one-dimensional search technique is to reduce successively the interval of uncertainty to a small acceptable value. To clearly discuss the ideas, we start with a very simple-minded approach called the Equal Interval

(a) $f(\alpha)$

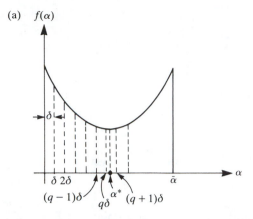

(b) $f(\alpha)$

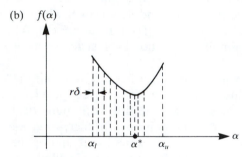

FIGURE 5.6
Equal interval search process. (a) Initial bracketing of minimum; (b) reducing interval of uncertainty.

Search. The idea is quite elementary as illustrated in Fig. 5.6. In the interval $0 \leqq \alpha \leqq \bar{\alpha}$, the function $f(\alpha)$ is evaluated at several points using a uniform grid. To do this, we select a small number δ and evaluate the function at the points δ, 2δ, 3δ, ... , $q\delta$, $(q + 1)\delta$, and so on. We compare values of the function at the two successive points, say q and $(q + 1)$. Then if function at the point q is larger than that at the point $(q + 1)$, i.e.

$$f(q\delta) > f((q + 1)\delta)$$

the minimum point has not been surpassed yet. However, if the function has started to increase, i.e.

$$f(q\delta) < f((q + 1)\delta) \tag{5.16}$$

then the minimum has been surpassed.

Note that once Eq. (5.16) is satisfied for two sucessive points q and $(q + 1)$, the minimum can be between either the points $(q - 1)$ and q, or the points q and $(q + 1)$. To account for both possibilities, we take the minimum to lie between the points $(q - 1)$ and $(q + 1)$. Thus, lower and upper limits for the interval of uncertainty are

$$\alpha_l = (q - 1)\delta, \qquad \alpha_u = (q + 1)\delta \tag{5.17}$$

and the interval of uncertainty I is 2δ.

Once the lower and upper limits have been established, we re-start the search process from $\alpha = \alpha_l$ with some reduced value for the increment in δ, say $r\delta$ where $r \ll 1$. The preceding process is repeated and the minimum is again bracketed. The interval of uncertainty I reduces to $2r\delta$. This is illustrated in Fig. 5.6(b). The value of increment is further reduced and the process is repeated, until the interval of uncertainty is reduced to a sufficiently small value ε to satisfy accuracy requirements. Note that the method is *convergent* for unimodal functions and can be very easily coded into a computer program. Appendix D gives the listing of a computer program for Equal Interval Search.

The efficiency of a method such as the Equal Interval Search depends on the number of function evaluations needed to achieve the desired accuracy. Clearly, the number of function evaluations depends on the initial choice for the value of δ. If δ is very small, the process may take many function evaluations to initially bracket the minimum. An advantage with smaller δ, however, is that the interval of uncertainty at the end of the initial search for minimum is fairly small. Subsequent improvements for the interval of uncertainty require fewer function evaluations. It is usually advantageous to start with a larger value of δ and quickly bracket the minimum point. Then the process is continued until the accuracy requirement is satisfied.

Example 5.4 Equal interval search. For the function $f(\alpha) = 2 - 4\alpha + e^{\alpha}$, use Equal Interval Search to find the minimum within an accuracy of $\varepsilon = 0.001$. Use $\delta = 0.5$.

Solution. Writing necessary conditions of optimality, we get $df/d\alpha = -4 + e^{\alpha} = 0$. Therefore, $\alpha = 1.3863$ is a candidate minimum point. This is indeed a local minimum since $d^2f/d\alpha^2 > 0$ at the point. We shall verify the solution using the equal interval search.

Using $\delta = 0.5$, the function is evaluated at various trial step sizes as shown in Table 5.1. It can be seen that the function starts to increase after the trial step size of 1.5. Therefore, the lower bound on step size is 1.0 and the upper bound is 2.0. The interval of uncertainty is $(2.0 - 1.0) = 1.0$ (i.e. 2δ).

To refine the interval of uncertainty further, let us choose $\delta = 0.05$. Iterations 6–14 in Table 5.1 are with the reduced value of δ. The interval of uncertainty is reduced to 0.10 (i.e. 2δ). The process of refining the interval of uncertainty within a specified accuracy is repetitive and is continued with $\delta = 0.005$ and then with $\delta = 0.0005$.

Appendix D gives the details of a simple computer program for minimizing any function of single variable using the Equal Interval Search. The program defines some data for the algorithm and calls subroutine EQUAL to perform the equal interval line search. After the minimum is found, it prints out the results. The subroutine EQUAL implements the Equal Interval Search algorithm. It calls a subroutine FUNCT to evaluate the function for any given α. The function whose minimum is desired can be coded appropriately in this subroutine. With $\delta = 0.5$ and $\varepsilon = 0.001$, the minimum for the function $f(\alpha) = 2 - 4\alpha + e^{\alpha}$ is obtained at $\alpha^* = 1.3865$ with $f(\alpha^*) = 0.454\,823$ in 37 function evaluations.   ‖

5.3.2.1 ALTERNATE EQUAL INTERVAL SEARCH. A slightly different computational procedure can be followed to reduce the interval of uncertainty

TABLE 5.1
Equal Interval Search for $f(\alpha) = 2 - 4\alpha + e^\alpha$ of Example 5.4

No.	Trial step		Function value	
1		0.000 000	3.000 000	$\delta = 0.5$
2		0.500 000	1.648 721	
3	$\alpha_l \to$	1.000 000	0.718 282	
4		1.500 000	0.481 689	
5	$\alpha_u \to$	2.000 000	1.389 056	
6		1.050 000	0.657 651	Start from
7		1.100 000	0.604 166	$\alpha_l = 1.0$ with
8		1.150 000	0.558 193	$\delta = 0.05$
9		1.200 000	0.520 117	
10		1.250 000	0.490 343	
11		1.300 000	0.469 297	
12	$\alpha_l \to$	1.350 000	0.457 426	
13		1.400 000	0.455 200	
14	$\alpha_u \to$	1.450 000	0.463 115	
15		1.355 000	0.456 761	Start from
16		1.360 000	0.456 193	$\alpha_l = 1.35$ with
17		1.365 000	0.455 723	$\delta = 0.005$
18		1.370 000	0.455 351	
19		1.375 000	0.455 077	
20	$\alpha_l \to$	1.380 000	0.454 902	
21		1.385 000	0.454 826	
22	$\alpha_u \to$	1.390 000	0.454 850	
23		1.380 500	0.454 890	Start from
24		1.381 000	0.454 879	$\alpha_l = 1.380$ with
25		1.381 500	0.454 868	$\delta = 0.0005$
26		1.382 000	0.454 859	
27		1.382 500	0.454 851	
28		1.383 000	0.454 844	
29		1.383 500	0.454 838	
30		1.384 000	0.454 833	
31		1.384 500	0.454 829	
32		1.385 000	0.454 826	
33		1.385 500	0.454 824	
34	$\alpha_l \to$	1.386 000	0.454 823	
35		1.386 500	0.454 823	
36	$\alpha_u \to$	1.387 000	0.454 824	
37		1.386 500	0.454 823	

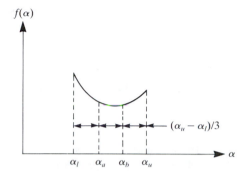

$f(\alpha)$

$(\alpha_u - \alpha_l)/3$

$\alpha_l \quad \alpha_a \quad \alpha_b \quad \alpha_u$

α

FIGURE 5.7
An alternate Equal Interval Search process.

once the minimum has been bracketed. The procedure is to evaluate the function at two points, say α_a and α_b in the interval of uncertainty. The points α_a and α_b are located at a distance of $I/3$ and $2I/3$ (where $I = \alpha_u - \alpha_l$) from the lower limit α_l. That is,

$$\alpha_a = \alpha_l + \tfrac{1}{3}(\alpha_u - \alpha_l) = \tfrac{1}{3}(\alpha_u + 2\alpha_l) \tag{5.18}$$

$$\alpha_b = \alpha_l + \tfrac{2}{3}(\alpha_u - \alpha_l) = \tfrac{1}{3}(2\alpha_u + \alpha_l) \tag{5.19}$$

This is shown in Fig. 5.7. Next the function is evaluated at the two points α_a and α_b. The following two conditions must now be considered:

1. If function at α_a is smaller than that at α_b, i.e. $f(\alpha_a) < f(\alpha_b)$, then the minimum lies between α_l and α_b. The right $\tfrac{1}{3}$ interval between α_b and α_u is discarded. New limits for the interval of uncertainty are $\alpha_l' = \alpha_l$ and $\alpha_u' = \alpha_b$ (the prime on α is used to indicate revised limits for the interval of uncertainty). Therefore, $I' = \alpha_u' - \alpha_l' = \alpha_b - \alpha_l$. The procedure is repeated with the new limits.

2. If function at α_a is larger than that at α_b, i.e. $f(\alpha_a) > f(\alpha_b)$, then the minimum lies between α_a and α_u. The interval between α_l and α_a is discarded. The procedure is repeated with $\alpha_l' = \alpha_a$ and $\alpha_u' = \alpha_u$ ($I' = \alpha_u' - \alpha_l'$).

With the preceding calculations the interval of uncertainty is reduced to $I' = 2I/3$ after every set of two function evaluations. The entire process is continued until the interval of uncertainty is reduced to an acceptable value.

5.3.3 Golden Section Search

5.3.3.1 INITIAL BRACKETING OF MINIMUM. In the preceding methods, the initially selected increment δ is kept fixed to bracket the minimum. This can be an inefficient process if δ happens to be a small number. An alternate procedure is to vary the increment at each step, i.e., multiply it by a constant r greater than 1. This way initial bracketing of the minimum is rapid; however, the length of the initial interval of uncertainty is increased. The *Golden Section* search procedure is such a *variable interval search method*. In this method the value of r is not selected arbitrarily. It is selected as the *Golden Ratio* which is obtained from Fibonacci sequence as 1.618. We shall derive this value of the Golden ratio.

The *Fibonacci sequence* is defined as

$$F_0 = 1; \qquad F_1 = 1; \qquad F_n = F_{n-1} + F_{n-2}, \qquad n = 2, 3, \ldots \qquad (5.20)$$

Any number of the Fibonacci sequence is obtained by adding the previous two numbers, so the sequence is given as

$$1, 1, 2, 3, 5, 8, 13, 21, 34, 55, 89, \ldots$$

Using the sequence, a procedure called Fibonacci search can be developed [Cooper and Steinberg, 1970]. We shall not discuss that procedure. The sequence has the property,

$$\frac{F_n}{F_{n-1}} \to 1.618 \left(\text{or,} \ \frac{F_{n-1}}{F_n} \to 0.618 \right) \quad \text{as} \quad n \to \infty$$

That is, as n becomes large, the ratio between two successive numbers F_n and F_{n-1} in the Fibonacci sequence reaches a constant value of 1.618 or $(\sqrt{5} + 1)/2$. This golden ratio has many other interesting properties that will be exploited in the one-dimensional search procedure.

Figure 5.8 illustrates the foregoing process of initially bracketing the minimum. Starting at $q = 0$, we evaluate $f(\alpha)$ at $\alpha = \delta$, where $\delta > 0$ is a small number. We check to see if the value $f(\delta)$ is smaller than the value $f(0)$. If it is, we then take an increment of 1.618δ in the step size (i.e. the increment is 1.618 times the previous increment). This way we evaluate the function at the following points and compare them:

$$q = 0; \qquad \alpha_0 = \delta$$

$$q = 1; \qquad \alpha_1 = \delta + 1.618\delta = 2.618\delta = \sum_{j=0}^{1} \delta(1.618)^j$$

$$q = 2; \qquad \alpha_2 = 2.618\delta + 1.618(1.618\delta) = 5.236\delta = \sum_{j=0}^{2} \delta(1.618)^j$$

$$q = 3; \qquad \alpha_3 = 5.236\delta + 1.618^3\delta = 9.472\delta = \sum_{j=0}^{3} \delta(1.618)^j$$

$$\vdots$$

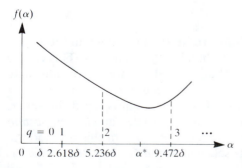

$f(\alpha)$

$q = 0 \ 1 \qquad 2 \qquad\qquad 3 \qquad \cdots$

$0 \quad \delta \ 2.618\delta \ 5.236\delta \qquad \alpha^* \ 9.472\delta$

FIGURE 5.8
Initial bracketing of the minimum point in the Golden Section method.

In general, we continue to evaluate the function at the points

$$\alpha_q = \sum_{j=0}^{q} \delta(1.618)^j; \qquad q = 0, 1, 2, \ldots \tag{5.21}$$

Let us assume that the function at α_{q-1} is smaller than that at α_{q-2} and α_q, i.e.

$$f(\alpha_{q-1}) < f(\alpha_{q-2}), \quad \text{and} \quad f(\alpha_{q-1}) < f(\alpha_q) \tag{5.22}$$

Therefore, the minimum point has been surpassed. Actually the minimum point lies between the previous two intervals, i.e. between α_q and α_{q-2}. Therefore, *upper and lower limits on the interval of uncertainty* are

$$\alpha_u \equiv \alpha_q = \sum_{j=0}^{q} \delta(1.618)^j \tag{5.23}$$

$$\alpha_l \equiv \alpha_{q-2} = \sum_{j=0}^{q-2} \delta(1.618)^j \tag{5.24}$$

and the *initial interval of uncertainty* is given as

$$
\begin{aligned}
I &= \alpha_u - \alpha_l \\
&= \sum_{j=0}^{q} \delta(1.618)^j - \sum_{j=0}^{q-2} \delta(1.618)^j \\
&= \delta(1.618)^{q-1} + \delta(1.618)^q \\
&= \delta(1.618)^{q-1}(1 + 1.618) \\
&= 2.618(1.618)^{q-1}\delta
\end{aligned} \tag{5.25}
$$

5.3.3.2 REDUCTION OF INTERVAL OF UNCERTAINTY. The next task is to start reducing the interval of uncertainty by evaluating and comparing functions at some points in I. *The method uses two function values* within the interval I, just as in the alternate equal interval search of Fig. 5.7. The difference is that the points α_a and α_b are not located at $I/3$ from either end of the interval of uncertainty as shown in Fig. 5.7. Instead, they are located at a distance of $0.382I$ (or $0.618I$) from either end. The factor 0.382 is related to the Golden ratio as we shall see in the following.

To see how the factor 0.618 is determined, consider two points symmetrically located from either end as shown in Fig. 5.9(a) – points α_a and α_b are located at a distance of τI from either end of the interval. Comparing functions values at α_a and α_b, either the left (α_l, α_a) or the right (α_b, α_u) portion of the interval gets discarded. Let us assume that the right portion gets discarded as shown in Fig. 5.9(b), so α_l' and α_u' are the new lower and upper bounds on the minimum. The new interval of uncertainty is $I' = \tau I$. There is one point in the new interval at which the function value is known. It is required that this point be located at a distance of $\tau I'$ from the left end; therefore,

$$\tau I' = (1 - \tau)I \tag{5.26}$$

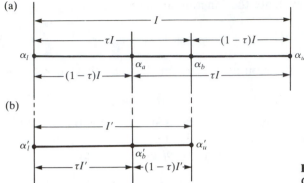

FIGURE 5.9
Golden Section partition.

Since $I' = \tau I$, this gives the equation

$$\tau^2 + \tau - 1 = 0 \tag{5.27}$$

The positive root of this equation is

$$\tau = (-1 + \sqrt{5})/2 = 0.618 \tag{5.28}$$

Thus the two points are located at a distance of $0.618I$ or $0.382I$ from either end of the interval.

The Golden Section search can be initiated once the initial interval of uncertainty is known. If the initial bracketing is done using the variable step increment (with a factor of 1.618 which is 1/0.618), then the function value at one of the points α_{q-1} is already known. It turns out that α_{q-1} is automatically the point α_a. This can be seen by multiplying the initial interval I in Eq. (5.25) by 0.382. If the preceding procedure is not used to initially bracket the minimum, then the points α_a and α_b will have to be calculated by the Golden Section procedure.

5.3.3.3 ALGORITHM FOR ONE-DIMENSIONAL SEARCH BY GOLDEN SECTIONS

Step 1. For a chosen small step size δ in α, let q be the smallest integer to satisfy Eq. (5.22) where α_q, α_{q-1} and α_{q-2} are calculated from Eq. (5.21). The upper and lower bounds on α^* (the optimum value for step size) are given by Eqs (5.23) and (5.24).

Step 2. Compute $f(\alpha_a)$ and $f(\alpha_b)$, where $\alpha_a = \alpha_l + 0.382I$ and $\alpha_b = \alpha_l + 0.618I$ (the interval of uncertainty $I = \alpha_u - \alpha_l$). Note that, at the first iteration, $\alpha_a = \alpha_{q-1}$, so $f(\alpha_a)$ needs no calculation.

Step 3. Compare $f(\alpha_a)$ and $f(\alpha_b)$ and go to Step 4, 5 or 6.

Step 4. If $f(\alpha_a) < f(\alpha_b)$ then optimum point α^* lies between α_l and α_b, i.e. $\alpha_l \le \alpha^* \le \alpha_b$. The new limits for the reduced interval of uncertainty are $\alpha_l' = \alpha_l$ and $\alpha_u' = \alpha_b$. Also, $\alpha_b' = \alpha_a$. Compute $f(\alpha_a')$ where $\alpha_a' = \alpha_l' + 0.382(\alpha_u' - \alpha_l')$ and go to Step 7.

Step 5. If $f(\alpha_a) > f(\alpha_b)$ then optimum point α^* lies between α_a and α_u, i.e. $\alpha_a \leqq \alpha^* \leqq \alpha_u$. Similar to the procedure in Step 4, let $\alpha_l' = \alpha_a$ and $\alpha_u' = \alpha_u$, so that $\alpha_a' = \alpha_b$. Compute $f(\alpha_b')$, where $\alpha_b' = \alpha_l' + 0.618(\alpha_u' - \alpha_l')$ and go to Step 7.

Step 6. If $f(\alpha_a) = f(\alpha_b)$, let $\alpha_l = \alpha_a$ and $\alpha_u = \alpha_b$, and return to Step 2.

Step 7. If the new interval of uncertainty $I' = \alpha_u' - \alpha_l'$ is small enough to satisfy a convergence criterion (i.e. $I' < \varepsilon$), let $\alpha^* = (\alpha_u' + \alpha_l')/2$ and stop. Otherwise, delete the primes on α_l', α_a', α_b' and α_u' and return to Step 3.

Example 5.5 Minimization of a function by Golden Section search. Consider the function $f(\alpha) = 2 - 4\alpha + e^{\alpha}$ as in Example 5.4, and use golden section search to find the minimum within an accuracy of 0.001. Use $\delta = 0.5$. Comment on the efficiency of Golden Section search with respect to the Equal Interval search method.

Solution. Analytically, the optimum value of α is 1.3863. In the Golden Section search, we need to first bracket the minimum point and then reduce successively the interval in which it lies to locate it precisely. Table 5.2 shows various iterations of the method. The minimum point is bracketed in only four iterations as shown in the first part of the table. The initial interval of uncertainty is obtained as $I = (\alpha_u - \alpha_l) = 2.618\,034 - 0.5 = 2.118\,034$ since $f(2.618\,034) > f(1.309\,017)$ in Table 5.2. Note that this interval is considerably larger than the one with Equal Interval search.

Now to reduce the interval of uncertainty, let us calculate α_b as $(\alpha_l + 0.618I)$; also $\alpha_b = \alpha_u - 0.382I$ (calculations are shown in the second part of Table 5.2). Note that α_a and $f(\alpha_a)$ are already known and need no further calculation. This is the main advantage of Golden Section search, that only one additional function evaluation is required once the interval of uncertainty is known. We calculate $\alpha_b = 1.809\,017$ and $f(\alpha_b) = 0.868\,376$. Since $f(\alpha_a) < f(\alpha_b)$, new limits for the reduced interval of uncertainty are $\alpha_l' = 0.5$ and $\alpha_u' = 1.809\,017$. Also, $\alpha_b' = 1.309\,017$ at which the function value is already known. We need to compute only $f(\alpha_a')$ where $\alpha_a' = \alpha_l' + 0.382(\alpha_u' - \alpha_l') = 1.000$. Further refinement of the interval of uncertainty is repetitive and can be accomplished by writing a computer program.

A subroutine GOLD implementing the Golden Section search procedure is given in Appendix D. The minimum for the function f is obtained at $\alpha^* = 1.386\,511$ with $f(\alpha^*) = 0.454\,823$ in 22 function evaluations as shown in Table 5.2. The number of function evaluations is a measure of efficiency of an optimization algorithm. Note that the number of function evaluations for Golden Section search is less than those for the Equal Interval search. This verifies our earlier observation that Golden Section search is the most efficient method for a specified accuracy and initial step length. ‖

It may appear that if the initial step length δ is too large in Equal Interval or Golden Section method, the line search fails, i.e. $f(\delta) > f(0)$. Actually, it indicates that initial δ is not proper and needs to be reduced until $f(\delta) < f(0)$. With this procedure, convergence of zero order methods can be numerically

TABLE 5.2
Golden Section search for $f(\alpha) = 2 - 4\alpha + e^\alpha$ of Example 5.5

Initial Bracketing of Minimum

No.	Trial step	Function value
1	0.000 000	3.000 000
2	$\alpha_l \rightarrow$ 0.500 000	1.648 721
3	$\alpha_a \rightarrow$ 1.309 017	0.466 464
4	$\alpha_u \rightarrow$ 2.618 034	5.236 610

Reducing Interval of Uncertainty

No.	$\alpha_l; [f(\alpha_l)]$	$\alpha_a; [f(\alpha_a)]$	$\alpha_b; [f(\alpha_b)]$	$\alpha_u; [f(\alpha_u)]$	I
1	0.500 000 [1.648 721]	1.309 017 [0.466 464]	1.809 017 [0.868 376]	2.618 034 [5.236 610]	2.118 034
2	0.500 000 [1.648 721]	1.000 000 [0.718 282]	1.309 017 [0.466 464]	1.809 017 [0.868 376]	1.309 017
3	1.000 000 [0.718 282]	1.309 017 [0.466 464]	1.500 000 [0.481 689]	1.809 017 [0.868 376]	0.809 017
4	1.000 000 [0.718 282]	1.190 983 [0.526 382]	1.309 017 [0.466 464]	1.500 000 [0.481 689]	0.500 000
5	1.190 983 [0.526 382]	1.309 017 [0.466 464]	1.381 966 [0.454 860]	1.500 000 [0.481 689]	0.309 017
6	1.309 017 [0.466 464]	1.381 966 [0.454 860]	1.427 051 [0.458 190]	1.500 000 [0.481 689]	0.190 983
7	1.309 017 [0.466 464]	1.354 102 [0.456 873]	1.381 966 [0.454 860]	1.427 051 [0.458 190]	0.118 034
8	1.354 102 [0.456 873]	1.381 966 [0.454 860]	1.399 187 [0.455 156]	1.427 051 [0.458 190]	0.072 949
9	1.354 102 [0.456 873]	1.371 323 [0.455 269]	1.381 966 [0.454 860]	1.399 187 [0.455 156]	0.045 085
10	1.371 323 [0.455 269]	1.381 966 [0.454 860]	1.388 544 [0.454 833]	1.399 187 [0.455 156]	0.027 864
11	1.381 966 [0.454 860]	1.388 544 [0.454 833]	1.392 609 [0.454 902]	1.399 187 [0.455 156]	0.017 221
12	1.381 966 [0.454 860]	1.386 031 [0.454 823]	1.388 544 [0.454 833]	1.392 609 [0.454 902]	0.010 643
13	1.381 966 [0.454 860]	1.384 479 [0.454 829]	1.386 031 [0.454 823]	1.388 544 [0.454 833]	0.006 57
14	1.384 479 [0.454 829]	1.386 031 [0.454 823]	1.386 991 [0.454 824]	1.388 544 [0.454 833]	0.004 065
15	1.384 479 [0.454 829]	1.385 438 [0.454 824]	1.386 031 [0.454 823]	1.386 991 [0.454 824]	0.002 512
16	1.385 438 [0.454 824]	1.386 031 [0.454 823]	1.386 398 [0.454 823]	1.386 991 [0.454 824]	0.001 553
17	1.386 031 [0.454 823]	1.386 398 [0.454 823]	1.386 624 [0.454 823]	1.386 991 [0.454 823]	0.000 960

$\alpha^* = 0.5(1.386\,398 + 1.386\,624) = 1.386\,511; f(\alpha^*) = 0.454\,823.$

enforced. The numerical procedure has been implemented in the EQUAL and GOLD subroutines given in Appendix D.

5.3.4* Polynomial Interpolation

The zero order methods described earlier can require too many function evaluations during line search to determine an appropriate step size. In realistic engineering design problems, the function evaluation requires a significant amount of computational effort. Therefore, zero order methods such as Equal Interval search are inefficient for many practical applications. Instead of evaluating the function at numerous trial steps, we can pass a curve through a limited number of points. Any continuous function on a given interval can be approximated as closely as desired by passing a polynomial of sufficiently high order and its minimum point can be found explicitly. The minimum point of the approximating polynomial is often a good estimate of the exact minimum of the line search function $f(\alpha)$. Thus, polynomial interpolation can be an efficient technique for one-dimensional search.

5.3.4.1 QUADRATIC CURVE FITTING. Many times it is sufficient to approximate the function $f(\alpha)$ on an interval of uncertainty by a quadratic curve (or, second-degree polynomial). Though it is possible to construct polynomials of higher degree for better accuracy, we shall restrict our attention to the simplest case of a quadratic polynomial. To interpolate a function in an interval with quadratic curve, we need to know the function value at three distinct points to determine the three coefficients of the quadratic polynomial. It must also be assumed that the function $f(\alpha)$ is sufficiently smooth and unimodal, and the initial interval of uncertainty (α_l, α_u) is known.

Let α_i be any intermediate point in the interval (α_l, α_u), and $f(\alpha_l)$, $f(\alpha_i)$ and $f(\alpha_u)$ be the function values at the respective points. Figure 5.10 shows the function $f(\alpha)$ and its quadratic approximation $q(\alpha)$ in the interval (α_l, α_u). $\bar{\alpha}$ is the minimum point of the function $q(\alpha)$ whereas α^* is the exact minimum point. An iteration can be used to improve the estimate $\bar{\alpha}$ for α^*.

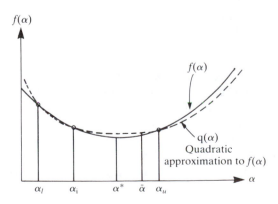

FIGURE 5.10
Quadratic approximation for a function $f(\alpha)$.

Any quadratic function $q(\alpha)$ can be expressed in the general form as

$$q(\alpha) = a_0 + a_1\alpha + a_2\alpha^2 \tag{5.29}$$

where a_0, a_1 and a_2 are the unknown coefficients. Since the function $q(\alpha)$ must have the same value as the function $f(\alpha)$ at the points α_l, α_i and α_u, we get three equations in three unknowns a_0, a_1 and a_2 as follows:

$$a_0 + a_1\alpha_l + a_2\alpha_l^2 = f(\alpha_l)$$
$$a_0 + a_1\alpha_i + a_2\alpha_i^2 = f(\alpha_i)$$
$$a_0 + a_1\alpha_u + a_2\alpha_u^2 = f(\alpha_u)$$

Solving the system of linear simultaneous equations for a_0, a_1 and a_2 we get,

$$a_2 = \frac{1}{(\alpha_u - \alpha_i)}\left[\frac{f(\alpha_u) - f(\alpha_l)}{(\alpha_u - \alpha_l)} - \frac{f(\alpha_i) - f(\alpha_l)}{(\alpha_i - \alpha_l)}\right]$$

$$a_1 = \frac{f(\alpha_i) - f(\alpha_l)}{(\alpha_i - \alpha_l)} - a_2(\alpha_l + \alpha_i) \tag{5.30}$$

$$a_0 = f(\alpha_l) - a_1\alpha_l - a_2\alpha_l^2$$

The minimum point $\bar{\alpha}$ of the quadratic curve $q(\alpha)$ in Eq. (5.29) is calculated by solving the necessary condition $dq(\bar{\alpha})/d\alpha = 0$ and verifying the sufficiency condition $d^2q(\bar{\alpha})/d\alpha^2 > 0$. Thus we get,

$$\bar{\alpha} = -\frac{1}{2a_2}a_1; \quad \text{if } \frac{d^2q}{d\alpha^2} = 2a_2 > 0 \tag{5.31}$$

An iteration can be used to further refine the interval of uncertainty.

The quadratic curve fitting technique may now be given in the form of a computational algorithm:

Step 1. Locate the initial interval of uncertainty (α_l, α_u) by selecting a small number δ. Any zero order method discussed previously may be used.

Step 2. Let α_i be an intermediate point in the interval (α_l, α_u) and $f(\alpha_i)$ be the value of $f(\alpha)$ at α_i.

Step 3. Compute the coefficients a_0, a_1 and a_2 from Eq. (5.30), $\bar{\alpha}$ from Eq. (5.31), and $f(\bar{\alpha})$.

Step 4. Compare α_i and $\bar{\alpha}$. If $\alpha_i < \bar{\alpha}$, go to Step 5a or 5b. Otherwise, go to Step 6a or 6b.

Step 5a. If $f(\alpha_i) < f(\bar{\alpha})$, then $\alpha_l \leq \alpha^* \leq \bar{\alpha}$. The new limits of the reduced interval of uncertainty are $\alpha_l' = \alpha_l$, $\alpha_u' = \bar{\alpha}$ and $\alpha_i' = \alpha_i$ and go to Step 7.

Step 5b. If $f(\alpha_i) > f(\bar{\alpha})$, then $\alpha_i \leq \alpha^* \leq \alpha_u$. The new limits of the reduced interval of uncertainty are $\alpha_l' = \alpha_i$, $\alpha_u' = \alpha_u$ and $\alpha_i' = \bar{\alpha}$ and go to Step 7.

Step 6a. If $f(\alpha_i) < f(\bar{\alpha})$, then $\bar{\alpha} \leq \alpha^* \leq \alpha_u$. The new limits of the reduced interval of uncertainty are $\alpha_l' = \bar{\alpha}$, $\alpha_i' = \alpha_i$ and $\alpha_u' = \alpha_u$ and go to Step 7.

Step 6b. If $f(\alpha_i) > f(\bar{\alpha})$, then $\alpha_l \leq \alpha^* \leq \alpha_i$. The new limits for the reduced interval of uncertainty are $\alpha_l' = \alpha_l$, $\alpha_i' = \bar{\alpha}$ and $\alpha_u' = \alpha_i$ and go to Step 7.

Step 7. If the two successive estimates of the minimum point of $f(\alpha)$ are sufficiently close then stop. Otherwise delete the primes on α_l', α_i' and α_u' and return to Step 2.

Example 5.6   One-dimensional minimization with quadratic interpolation. Find the minimum point of $f(\alpha) = 2 - 4\alpha + e^{\alpha}$ by polynomial interpolation. Use Golden Section search with $\delta = 0.5$ initially to bracket the minimum point.

Solution.
    Iteration 1. From Example 5.5 the following information is known.

$$\alpha_l = 0.5 \qquad\qquad \alpha_i = 1.309\,017 \qquad \alpha_u = 2.618\,034$$

$$f(\alpha_l) = 1.648\,721 \quad f(\alpha_i) = 0.466\,464 \quad f(\alpha_u) = 5.236\,610$$

The coefficients a_0, a_1 and a_2 are calculated from Eqs (5.30) as

$$a_2 = \frac{1}{1.309\,02}\left(\frac{3.5879}{2.1180} - \frac{-1.1823}{0.809\,02}\right) = 2.410$$

$$a_1 = \frac{-1.1823}{0.809\,02} - (2.41)(1.809\,02) = -5.821$$

$$a_0 = 1.648\,271 - (-5.821)(0.50) - 2.41(0.25) = 3.957$$

Therefore, $\bar{\alpha} = 1.2077$ from Eqs (5.31), and $f(\bar{\alpha}) = 0.5149$. Note that $\bar{\alpha} < \alpha_i$ and $f(\alpha_i) < f(\bar{\alpha})$. Thus, new limits of the reduced interval of uncertainty are $\alpha_l' = \bar{\alpha} = 1.2077$, $\alpha_u' = \alpha_u = 2.618\,034$ and $\alpha_i' = \alpha_i = 1.309\,017$.

    Iteration 2. We have the new limits for the interval of uncertainty, the intermediate point and the respective values as

$$\alpha_l = 1.2077, \qquad \alpha_i = 1.309\,017, \qquad \alpha_u = 2.618\,034$$

$$f(\alpha_l) = 0.5149, \qquad f(\alpha_i) = 0.466\,464, \qquad f(\alpha_u) = 5.236\,61$$

The coefficients a_0, a_1 and a_2 are calculated as before, $a_0 = 5.7129$, $a_1 = -7.8339$ and $a_2 = 2.9228$. Thus $\bar{\alpha} = 1.34014$ and $f(\bar{\alpha}) = 0.4590$.

    Comparing with the optimum solution given in Table 5.2, we observe that $\bar{\alpha}$ and $f(\bar{\alpha})$ are quite close to the final solution. One more iteration can give very good approximation to the optimum. Note that only 5 function evaluations are required to obtain a fairly accurate optimum step size for the function $f(\alpha)$. Therefore, the polynomial interpolation approach can be quite efficient for one-dimensional minimization. ||

Example 5.7   One-dimensional minimization with alternate quadratic interpolation. Find the minimum point of $f(\alpha) = 2 - 4\alpha + e^{\alpha}$ using $f(0)$, $f'(0)$ and $f(\alpha_u)$ to fit a quadratic curve, where α_u is an upper bound on the minimum point of $f(\alpha)$.

Solution. Let the general equation for a quadratic curve be $a_0 + a_1\alpha + a_2\alpha^2$, where a_0, a_1 and a_2 are the unknown coefficients. Let us select the upper bound on α^* to be 2.618 034 from Golden Section search. Using the given function $f(\alpha)$, we have $f(0) = 3$, $f(2.618\,034) = 5.236\,61$ and $f'(0) = -3$. Now, as before, we get the following three equations to solve for the unknown coefficients a_0, a_1

and a_2

$$a_0 = f(0) = 3$$
$$a_0 + 2.618\,034a_1 + 6.854a_2 = f(2.618\,034) = 5.236\,61$$
$$a_1 = f'(0) = -3$$

Solving the three equations simultaneously, we get $a_0 = 3$, $a_1 = -3$ and $a_2 = 1.4722$. The minimum point of the parabolic curve using Eq. (5.31) is given as $\bar{\alpha} = 1.0189$ and $f(\bar{\alpha}) = 0.694\,43$. This estimate can be improved using an iteration as before.

Note that in the preceding an estimate of the minimum point of the function $f(\alpha)$ can be found in only two function evaluations. Since the slope $f'(0) = \mathbf{c}^{(k)} \cdot \mathbf{d}^{(k)}$ is known for multidimensional problems, no additional calculations are required to evaluate it at $\alpha = 0$. ‖

5.4 STEEPEST DESCENT METHOD

In the previous section we assumed that a search direction in the design space is known and we tackled the problem of step size determination. In this and subsequent sections we shall address the question of determination of the search direction **d**. *The basic requirement for **d** is that the cost function be reduced if we move a small distance along the direction. This will be called the descent direction.*

Several methods are available for determining a descent direction for unconstrained optimization problems. The steepest descent method or the *gradient method* is the simplest, the oldest and probably the best known numerical method for unconstrained optimization. The philosophy of the method, introduced by Cauchy in 1847, is to find the direction **d** at the current iteration in which the cost function $f(\mathbf{x})$ decreases most rapidly, at least locally. It is due to this philosophy that the method is called the steepest descent search technique. Also, properties of the gradient vector are used in the iterative process which is the reason for its alternate name: the *gradient method*. The steepest descent method is a *first-order method* since only the gradient of the cost function is calculated and used to evaluate the search direction. Later, we shall discuss *second-order methods* where Hessian of the function will be used in determining the search direction. We shall first study properties of the gradient vector of a scalar function before stating an algorithm for the method.

5.4.1 Properties of Gradient Vector

The gradient vector of a scalar function $f(x_1, x_2, \ldots, x_n)$ was defined in Chapter 3. Just as a reminder, we define it again as the column vector:

$$\nabla f = \left[\frac{\partial f}{\partial x_1} \quad \frac{\partial f}{\partial x_2} \cdots \frac{\partial f}{\partial x_n} \right]^T \equiv \mathbf{c} \tag{5.32}$$

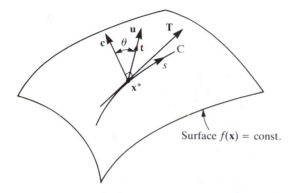

FIGURE 5.11
Gradient vector for the surface $f(\mathbf{x}) = $ constant at the point $\mathbf{x}^*$.

To simplify the notation, we shall use vector $\mathbf{c}$ to represent gradient of the scalar function $f(\mathbf{x})$; that is, $c_i = \partial f / \partial x_i$. We shall use a superscript to denote the point at which this vector is calculated, as

$$\mathbf{c}^{(k)} = \mathbf{c}(\mathbf{x}^{(k)}) = \left[\frac{\partial f(\mathbf{x}^{(k)})}{\partial x_i}\right]^T \tag{5.33}$$

The gradient vector has several *properties* that are used in the steepest descent method. Since proofs of the properties are also quite instructive, they are also given.

Property 1. *The gradient vector* $\mathbf{c}$ *of a function* $f(x_1, x_2, \ldots, x_n)$ *at the point* $\mathbf{x}^* = (x_1^*, x_2^*, \ldots, x_n^*)$ *is orthogonal (normal) to the tangent plane for the surface* $f(x_1, x_2, \ldots, x_n) = constant.$

This is an important property of the gradient vector shown graphically in Fig. 5.11. It shows the surface $f(\mathbf{x}) = $ constant; $\mathbf{x}^*$ is a point on the surface; C is any curve on the surface through the point $\mathbf{x}^*$; $\mathbf{T}$ is a vector tangent to the curve C at the point $\mathbf{x}^*$, $\mathbf{u}$ is any unit vector; and $\mathbf{c}$ is the gradient vector at $\mathbf{x}^*$. According to the above property, vectors $\mathbf{c}$ and $\mathbf{T}$ are normal to each other, i.e. their dot product is zero, $\mathbf{c} \cdot \mathbf{T} = 0$.

Proof. To show this, we take any curve C on the surface $f(x_1, x_2, \ldots, x_n) = $ constant, as shown in Fig. 5.11. Let the curve pass through the point $\mathbf{x}^* = (x_1^*, x_2^*, \ldots, x_n^*)$. Also, let s be any parameter along C. Then a unit tangent vector $\mathbf{T}$ along C at the point $\mathbf{x}^*$ is given as

$$\mathbf{T} = \left[\frac{\partial x_1}{\partial s} \frac{\partial x_2}{\partial s} \cdots \frac{\partial x_n}{\partial s}\right]^T \tag{a}$$

Since $f(\mathbf{x}) = $ constant, the derivative of f along the curve C is zero, i.e.

$$\frac{df}{ds} = 0 \qquad \text{(directional derivative of } f)$$

Or, using the chain rule of differentiation

$$\frac{df}{ds} \equiv \frac{\partial f}{\partial x_1} \frac{\partial x_1}{\partial s} + \ldots + \frac{\partial f}{\partial x_n} \frac{\partial x_n}{\partial s} = 0 \tag{b}$$

Writing Eq. (b) in the vector form after identifying $\partial f/\partial x_i$ and $\partial x_i/\partial s$ (from Eq. a) as components of the gradient and the unit tangent vectors, we obtain

$$\mathbf{c} \cdot \mathbf{T} = 0, \quad \text{or} \quad \mathbf{c}^T \mathbf{T} = 0$$

Since the dot product of the gradient vector $\mathbf{c}$ with the tangential vector $\mathbf{T}$ is zero, the vectors are normal to each other. But $\mathbf{T}$ is any tangent vector at $\mathbf{x}^*$, so $\mathbf{c}$ is orthogonal to the tangent plane for the surface $f(\mathbf{x}) = $ constant at the point $\mathbf{x}^*$.   ‖

Property 2. *The second property is that gradient represents a direction of maximum rate of increase for the function $f(\mathbf{x})$ at the point $\mathbf{x}^*$.*

Proof. To show this, let $\mathbf{u}$ be a unit vector in any direction that is not tangent to the surface. This is shown in Fig. 5.11. Let t be a parameter along $\mathbf{u}$. The derivative of $f(\mathbf{x})$ in the direction $\mathbf{u}$ at the point $\mathbf{x}^*$ (i.e. directional derivative of f) is given as

$$\frac{df}{dt} = \lim_{\varepsilon \to 0} \frac{f(\mathbf{x} + \varepsilon \mathbf{u}) - f(\mathbf{x})}{\varepsilon} \tag{c}$$

where ε is a small number. Using Taylor series expansion

$$f(\mathbf{x} + \varepsilon \mathbf{u}) = f(\mathbf{x}) + \varepsilon \left[u_1 \frac{\partial f}{\partial x_1} + u_2 \frac{\partial f}{\partial x_2} + \ldots + u_n \frac{\partial f}{\partial x_n} \right] + 0(\varepsilon^2)$$

where u_i are components of the unit vector $\mathbf{u}$ and $0(\varepsilon^2)$ are terms of order ε^2. Rewriting the foregoing equation,

$$f(\mathbf{x} + \varepsilon \mathbf{u}) - f(\mathbf{x}) = \varepsilon \sum_{i=1}^{n} u_i \frac{\partial f}{\partial x_i} + 0(\varepsilon^2) \tag{d}$$

Substituting Eq. (d) into Eq. (c) and taking the indicated limit, we get

$$\frac{df}{dt} = \sum_{i=1}^{n} u_i \frac{\partial f}{\partial x_i} = \mathbf{c} \cdot \mathbf{u} = \mathbf{c}^T \mathbf{u} \tag{e}$$

Using the definition of the dot product for Eq. (e),

$$\frac{df}{dt} = \|\mathbf{c}\| \, \|\mathbf{u}\| \cos \theta \tag{f}$$

where θ is the angle between the $\mathbf{c}$ and $\mathbf{u}$ vectors. The right-hand side of Eq. (f) will have extreme value when $\theta = 0$ or $180°$. When $\theta = 0$, vector $\mathbf{u}$ is along $\mathbf{c}$ and $\cos \theta = 1$. Therefore, from Eq. (f), df/dt represents the maximum rate of increase for $f(\mathbf{x})$ when $\theta = 0$. Similarly, when $\theta = 180°$, vector $\mathbf{u}$ points in the negative $\mathbf{c}$ direction. Therefore, from Eq. (f), df/dt represents the maximum rate of decrease for $f(\mathbf{x})$ when $\theta = 180°$.   ‖

According to the foregoing property of the gradient vector, if we need to move away from the surface $f(\mathbf{x}) = $ constant, the function increases most rapidly along the gradient vector compared to a move in any other direction. In Fig. 5.11, a small move along the direction $\mathbf{c}$ will result in a larger increase in the function compared to a similar move along the direction $\mathbf{u}$. Of course, any small move along the direction **T** results in no change in the function, since **T** is tangent to the surface.

Property 3. *The maximum rate of change of $f(\mathbf{x})$ at any point $\mathbf{x}^*$ is the magnitude of the gradient vector.*

Proof. Since **u** is a unit vector, the maximum value of df/dt from Eq. (f) is given as

$$\max \left| \frac{df}{dt} \right| = \|\mathbf{c}\|$$

However, for $\theta = 0$, **u** is in the direction of the gradient vector. Therefore, the magnitude of the gradient represents the maximum rate of change for the function $f(\mathbf{x})$. ‖

These properties show that gradient vector at any point $\mathbf{x}^*$ represents a direction of maximum increase in the function $f(\mathbf{x})$ and the rate of increase is the magnitude of the vector. Gradient is therefore called a direction of steepest ascent for the function $f(\mathbf{x})$.

Example 5.8 Verification of properties of the gradient vector. Verify the properties of the gradient vector for the function $f(\mathbf{x}) = 25x_1^2 + x_2^2$ at the point $\mathbf{x}^{(0)} = (0.6, 4)$.

Solution. Figure 5.12 shows in the $x_1 - x_2$ plane the iso-cost contours of value 25 and 100 for the function f. The value of the function at $(0.6, 4)$ is $f(0.6, 4) = 25$. The gradient of the function at $(0.6, 4)$ is given as

$$\mathbf{c} = \nabla f(0.6, 4) = (\partial f/\partial x_1, \partial f/\partial x_2)$$
$$= (50x_1, 2x_2) = (30, 8)$$
$$\|\mathbf{c}\| = \sqrt{30*30 + 8*8} = 31.048\,35$$

Therefore a unit vector along the gradient is given as

$$\mathbf{C} = \mathbf{c}/\|\mathbf{c}\| = (0.966\,235, 0.257\,663)$$

Using the given function, a vector tangent to the curve at the point $(0.6, 4)$ is given as

$$\mathbf{t} = (-4, 15)$$

This vector is obtained by using the equation for the curve $25x_1^2 + x_2^2 = 25$ and writing the tangent vector as $(\partial x_1/\partial s, \partial x_2/\partial s)$ where s is a parameter along the curve. The unit tangent vector is

$$\mathbf{T} = \mathbf{t}/\|\mathbf{t}\| = (-0.257\,663, 0.966\,235)$$

Property 1. If gradient is normal to the tangent, then $\mathbf{C} \cdot \mathbf{T} = 0$. This is indeed true for the preceding data. We can also use the condition that if two lines are orthogonal, then $m_1 m_2 = -1$, where m_1 and m_2 are the slopes of the two lines. To calculate slope of the tangent we use the equation for the curve $25x_1^2 + x_2^2 = 25$, or

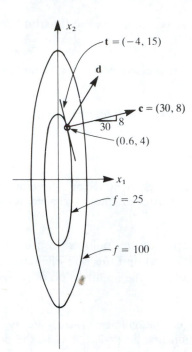

FIGURE 5.12
Iso-cost contours of function $f = 25x_1^2 + x_2^2$ for $f = 25$ and 100.

$x_2 = 5\sqrt{1 - x_1^2}$. Therefore, the slope of the tangent at the point $(0.6, 4)$ is given as

$$m_1 = dx_2/dx_1 = -5x_1/\sqrt{1 - x_1^2} = -\tfrac{1}{3.75}$$

The slope of the gradient vector is $m_2 = \tfrac{30}{8} = 3.75$. Thus $m_1 m_2$ is, indeed, -1, and the two lines are normal to each other.

Property 2. Consider any arbitrary direction

$$\mathbf{d} = (0.501\,034, 0.865\,430)$$

at the point $(0.6, 4)$ as shown in Fig. 5.12. If $\mathbf{C}$ is the direction of steepest ascent, then the function should increase more rapidly along $\mathbf{C}$ than along $\mathbf{d}$. Let us choose a step size $\alpha = 0.1$ and calculate two points, one along $\mathbf{C}$ and the other along $\mathbf{d}$ as

$$\mathbf{x}^{(1)} = \mathbf{x}^{(0)} + \alpha\mathbf{C}$$

$$= \begin{bmatrix} 0.6 \\ 4.0 \end{bmatrix} + 0.1 \begin{bmatrix} 0.966\,235 \\ 0.257\,663 \end{bmatrix} = \begin{bmatrix} 0.696\,623\,5 \\ 4.025\,766\,3 \end{bmatrix}$$

$$\mathbf{x}^{(2)} = \mathbf{x}^{(0)} + \alpha\mathbf{d}$$

$$= \begin{bmatrix} 0.6 \\ 4.0 \end{bmatrix} + 0.1 \begin{bmatrix} 0.501\,034 \\ 0.865\,430 \end{bmatrix} = \begin{bmatrix} 0.650\,103\,4 \\ 4.085\,436\,0 \end{bmatrix}$$

Now, we calculate the function at these points and compare their values:

$$f(\mathbf{x}^{(1)}) = 28.3389$$
$$f(\mathbf{x}^{(2)}) = 27.2657$$

Since $f(\mathbf{x}^{(1)}) > f(\mathbf{x}^{(2)})$, the function increases more rapidly along $\mathbf{C}$ than along $\mathbf{d}$.

Property 3. If the magnitude of the gradient vector represents the maximum rate of change of $f(\mathbf{x})$, then $(\mathbf{c} \cdot \mathbf{c}) > (\mathbf{c} \cdot \mathbf{d})$. $(\mathbf{c} \cdot \mathbf{c}) = 964.0$ and $(\mathbf{c} \cdot \mathbf{d}) = 21.9545$. Therefore, the gradient vector satisfies this property also.

Note that the last two properties are valid only in a local sense, i.e. only in a small neighborhood of the point at which the gradient is evaluated. ‖

5.4.2 Steepest Descent Algorithm

The properties of the gradient vector can be used to define an iterative algorithm for the unconstrained optimization problems. The direction of maximum decrease in the cost function is the negative of its gradient at the given point $\mathbf{x}$. Any small move in the negative gradient direction will result in the maximum local rate of decrease in the cost function. The negative gradient vector then represents a *direction of steepest descent* for the cost function. This result is summarized in the following theorem.

> **Theorem 5.1 Steepest descent direction.** Let $f(\mathbf{x})$ be a differentiable function with respect to $\mathbf{x}$. The direction of steepest descent for $f(\mathbf{x})$ at any point is
>
> $$\mathbf{d} = -\mathbf{c}, \quad \text{or} \quad d_i \equiv -c_i = \frac{-\partial f}{\partial x_i}; \quad i = 1 \text{ to } n \quad (5.34) \; ‖$$

Equation (5.34) gives a direction of change in the design space for use in Eq. (5.4). Based on the preceding discussion, the *steepest descent algorithm* is stated as follows.

Step 1. Estimate a starting design $\mathbf{x}^{(0)}$ and set the iteration counter $k = 0$. Select a convergence parameter $\varepsilon > 0$.

Step 2. Calculate the gradient of $f(\mathbf{x})$ at the point $\mathbf{x}^{(k)}$ as $\mathbf{c}^{(k)} = \nabla f(\mathbf{x}^{(k)})$. Calculate $\|\mathbf{c}^{(k)}\|$. If $\|\mathbf{c}^{(k)}\| < \varepsilon$, then stop the iterative process as $\mathbf{x}^* = \mathbf{x}^{(k)}$ is a minimum point. Otherwise, go to Step 3.

Step 3. Let the search direction at the current point $\mathbf{x}^{(k)}$ be $\mathbf{d}^{(k)} = -\mathbf{c}^{(k)}$.

Step 4. Calculate a step size α_k to minimize $f(\mathbf{x}^{(k)} + \alpha \mathbf{d}^{(k)})$. A one-dimensional search is used to determine α_k.

Step 5. Update the design as $\mathbf{x}^{(k+1)} = \mathbf{x}^{(k)} + \alpha_k \mathbf{d}^{(k)}$. Set $k = k + 1$ and go to Step 2.

The basic idea of the steepest descent method is quite simple. We start with an initial estimate for the minimum design. The direction of steepest descent is computed at that point. If the direction is nonzero, we move as far as possible along it to reduce the cost function. At the new design point we calculate the steepest descent direction again and repeat the entire process.

Note that since $\mathbf{d} = -\mathbf{c}$, *the descent condition of Inequality (5.8) is automatically satisfied as* $\mathbf{c} \cdot \mathbf{d} = -\|\mathbf{c}\|^2 < 0$.

It is interesting to note that the successive directions of steepest descent are normal to one another, i.e.

$$\mathbf{c}^{(k)} \cdot \mathbf{c}^{(k+1)} = 0 \tag{5.35}$$

This can be shown quite easily by using the necessary conditions to determine the optimum step size. In Step 4 of the algorithm, it is required to compute α_k to minimize $f(\mathbf{x}^{(k)} + \alpha \mathbf{d}^{(k)})$. The necessary condition for this is $df/d\alpha = 0$. Using the chain rule of differentiation, we get

$$\frac{df(\mathbf{x}^{(k+1)})}{d\alpha} = \left[\frac{\partial f(\mathbf{x}^{(k+1)})}{\partial \mathbf{x}} \right]^T \frac{\partial \mathbf{x}^{(k+1)}}{\partial \alpha}$$

which gives

$$\mathbf{c}^{(k+1)} \cdot \mathbf{d}^{(k)} = 0, \quad \text{or} \quad \mathbf{c}^{(k+1)} \cdot \mathbf{c}^{(k)} = 0 \tag{5.36}$$

since

$$\mathbf{c}^{(k+1)} = \frac{\partial f(\mathbf{x}^{(k+1)})}{\partial \mathbf{x}} \quad \text{and} \quad \frac{\partial \mathbf{x}^{(k+1)}}{\partial \alpha} = \frac{\partial}{\partial \alpha} (\mathbf{x}^{(k)} + \alpha \mathbf{d}^{(k)}) = \mathbf{d}^{(k)}$$

In the two-dimensional case, $\mathbf{x} = (x_1, x_2)$. Figure 5.13 is a view of the design variable space. The closed curves in the figure are contours of the cost function $f(\mathbf{x})$. The figure shows several steepest descent directions that are orthogonal to each other.

5.4.2.1 LINE SEARCH TERMINATION CRITERION.
The numerical methods of one-dimensional minimization are often used to perform line search in multidimensional problems. Many times the numerical methods will give an approximate or inexact value of the step size. Thus, line search termination criterion is useful to decide the accuracy of a numerical method in the step size calculation. For the exact value of the step size, Eq. (5.36) must hold, i.e.

$$\mathbf{c}^{(k+1)} \cdot \mathbf{d}^{(k)} = 0 \tag{5.37}$$

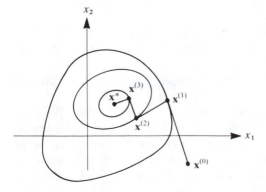

FIGURE 5.13
Orthogonal steepest descent paths.

where $\mathbf{c}^{(k+1)}$ is the gradient at $\mathbf{x}^{(k+1)}$ and $\mathbf{d}^{(k)}$ is the direction of travel in the previous iteration. Due to round off and truncation errors in computer calculations, the line search termination criterion of Eq. (5.37) cannot be satisfied precisely; nevertheless, it gives an indication of the accuracy in numerical evaluation of the step size. Note that the line search termination criterion does not depend on how the direction of descent $\mathbf{d}$ is calculated.

Example 5.9 Use of steepest descent algorithm. Minimize $f(x_1, x_2) = x_1^2 + x_2^2 - 2x_1x_2$ using the steepest descent method starting from the point $(1, 0)$.

Solution. To solve the problem, we follow steps of the steepest descent algorithm.

1. The starting design is estimated as $\mathbf{x}^{(0)} = (1, 0)$.
2. $\mathbf{c}^{(0)} = (2x_1 - 2x_2, 2x_2 - 2x_1) = (2, -2)$; $\|\mathbf{c}^{(0)}\| = 2\sqrt{2} \neq 0$.
3. Set $\mathbf{d}^{(0)} = -\mathbf{c}^{(0)} = (-2, 2)$.
4. Calculate α to minimize $f(\mathbf{x}^{(0)} + \alpha\mathbf{d}^{(0)})$ where $\mathbf{x}^{(0)} + \alpha\mathbf{d}^{(0)} = (1 - 2\alpha, 2\alpha)$

$$f(\mathbf{x}^{(0)} + \alpha\mathbf{d}^{(0)}) = (1 - 2\alpha)^2 + (2\alpha)^2 - 2(1 - 2\alpha)(2\alpha)$$
$$= 16\alpha^2 - 8\alpha + 1 \equiv f(\alpha)$$

Since this is a simple function of α, we can use necessary and sufficient conditions to solve for the optimum step length. In general, numerical one-dimensional search will have to be used to calculate α. Using the analytic approach to solve for α, we get

$$\frac{df(\alpha)}{d\alpha} = 0; \qquad 32\alpha - 8 = 0 \text{ or } \alpha_0 = 0.25$$

$$\frac{d^2f(\alpha)}{d\alpha^2} = 32 > 0.$$

Therefore, the sufficiency condition for minimum is satisfied.

5. Updating the design $(\mathbf{x}^{(0)} + \alpha_0\mathbf{d}^{(0)})$:

$$x_1^{(1)} = 1 - 0.25(2) = 0.5$$
$$x_2^{(1)} = 0 + 0.25(2) = 0.5$$

Solving for $\mathbf{c}^{(1)}$ from the expression in Step 2, we see that $\mathbf{c}^{(1)} = (0, 0)$ which satisfies the stopping criterion. Therefore, $(0.5, 0.5)$ is a minimum point for the given problem. ||

The preceding problem is quite simple and an optimum point is obtained in only one iteration. This is because the condition number of the Hessian of the cost function is one (condition number is a scalar associated with the given matrix; refer to Section B.8 in Appendix B). In such a case, the steepest descent method will converge in just one iteration with any starting point. In general, the algorithm will require several iterations before an acceptable optimum is reached.

Example 5.10   Use of steepest descent algorithm. Minimize $f(x_1, x_2, x_3) = x_1^2 + 2x_2^2 + 2x_3^2 + 2x_1x_2 + 2x_2x_3$ using the steepest descent method with a starting design as $(2, 4, 10)$. Select the convergence parameter ε as 0.005. Perform a line search by Golden Section search with initial step length $\delta = 0.05$ and an accuracy of 0.0001.

Solution.

1. Let $\mathbf{c} \equiv \nabla f = (2x_1 + 2x_2, 4x_2 + 2x_1 + 2x_3, 4x_3 + 2x_2)$.   Now, $\mathbf{c}^{(0)} = (12, 40, 48)$ and $\|\mathbf{c}^{(0)}\| = \sqrt{4048} = 63.6 > \varepsilon$.
2. $\mathbf{d}^{(0)} = -\mathbf{c}^{(0)} = (-12, -40, -48)$.
3. Calculate α_0 by Golden Section search to minimize $f(\mathbf{x}^{(0)} + \alpha \mathbf{d}^{(0)})$; $\alpha_0 = 0.1587$.
4. Update the design as $\mathbf{x}^{(1)} = \mathbf{x}^{(0)} + \alpha_0 \mathbf{d}^{(0)}$:

$$\mathbf{x}^{(1)} = (0.0956, -2.348, 2.381)$$

5. $\mathbf{c}^{(1)} = (-4.5, -4.438, 4.828)$, $\|\mathbf{c}^{(1)}\| = 7.952 > \varepsilon$.

   Note that $\mathbf{c}^{(1)} \cdot \mathbf{d}^{(0)} = 0$ which verifies the line search termination criterion. The steps in steepest descent algorithm should be repeated until the convergence criterion is satisfied. Appendix D contains the computer program and user supplied subroutines FUNCT and GRAD to implement steps of the steepest descent algorithm. The iterative history for the problem with the program is given in Table 5.3. The optimum cost function value is 0.0 and the optimum point is $(0, 0, 0)$. Note that a large number of iterations and function evaluations are needed to reach the optimum.   ‖

The method of steepest descent is quite simple and robust (it is convergent). However, it has several drawbacks. These are:

1. Even if convergence of the method is guaranteed, a large number of iterations may be required for the minimization of even positive definite quadratic forms, i.e. the method can be quite slow to converge to the minimum point.
2. Information calculated at the previous iterations is not used. Each iteration is started independent of others, which is inefficient.
3. Only first-order information about the function is used at each iteration to determine the search direction. This is one reason that convergence of the method is slow. It can further deteriorate if only inaccurate line search is used. Moreover, the rate of convergence depends on the condition number of the Hessian of the cost function at the optimum point. If the condition number is large, the rate of convergence of the method is slow.
4. Practical experience with the method has shown that a substantial decrease in the cost function is achieved in the initial few iterations and the cost function decreases quite slowly in later iterations.
5. The direction of steepest descent (direction of most rapid decrease in the cost function) may be good in a local sense (in a small neighborhood) but not in a global sense.

TABLE 5.3

Optimum solution for Example 5.10 with steepest descent program

No.	x_1	x_2	x_3	f(x)	α	‖c‖
1	2.000 00E + 00	4.000 00E + 00	1.000 00E + 01	3.320 00E + 02	1.587 18E − 01	6.362 39E + 01
2	9.538 70E − 02	−2.348 71E + 00	2.381 55E + 00	1.075 03E + 01	3.058 72E − 01	7.959 22E + 00
3	1.473 84E + 00	−9.903 42E − 01	9.045 62E − 01	1.059 36E + 00	1.815 71E − 01	2.061 42E + 00
4	1.298 26E + 00	−1.134 77E + 00	6.072 28E − 01	6.737 57E − 01	6.499 89E − 01	8.139 10E − 01
5	1.085 73E + 00	−6.615 14E − 01	5.036 40E − 01	4.585 34E − 01	1.905 88E − 01	1.217 29E + 00
6	9.240 28E − 01	−7.630 36E − 01	3.718 42E − 01	3.172 18E − 01	5.880 53E − 01	5.631 54E − 01
7	7.346 84E − 01	−4.922 94E − 01	3.946 01E − 01	2.240 07E − 01	1.938 77E − 01	8.193 77E − 01
8	6.406 97E − 01	−5.484 01E − 01	2.794 74E − 01	1.589 46E − 01	5.725 54E − 01	3.991 41E − 01
9	5.350 08E − 01	−3.461 39E − 01	2.673 96E − 01	1.133 73E − 01	1.946 60E − 01	5.775 45E − 01
10	4.614 78E − 01	−3.890 14E − 01	1.939 50E − 01	8.091 74E − 02	5.697 67E − 01	2.848 37E − 01
11	3.789 02E − 01	−2.493 07E − 01	1.952 19E − 01	5.783 10E − 02	1.946 01E − 01	4.118 95E − 01
12	3.284 64E − 01	−2.786 95E − 01	1.402 91E − 01	4.131 41E − 02	5.720 88E − 01	2.033 39E − 01
13	2.715 19E − 01	−1.772 81E − 01	1.381 32E − 01	2.949 40E − 02	1.944 39E − 01	2.947 10E − 01
14	2.348 72E − 01	−1.987 04E − 01	9.963 96E − 02	2.104 99E − 02	5.726 50E − 01	1.451 12E − 01
15	1.934 49E − 01	−1.266 69E − 01	9.898 06E − 02	1.502 33E − 02	1.944 57E − 01	2.104 30E − 01
16	1.674 77E − 01	−1.418 72E − 01	7.125 41E − 02	1.071 97E − 02	5.717 77E − 01	1.035 80E − 01
17	1.381 96E − 01	−9.039 73E − 02	7.052 67E − 02	7.653 62E − 03	1.945 34E − 01	1.500 77E − 01
18	1.195 99E − 01	−1.012 63E − 01	5.081 80E − 02	5.463 41E − 03	5.708 32E − 01	7.395 59E − 02
19	9.866 59E − 02	−6.460 51E − 02	5.039 27E − 02	3.901 56E − 03	1.945 71E − 01	1.070 16E − 01
20	8.541 14E − 02	−7.232 89E − 02	3.631 34E − 02	2.786 93E − 03	5.721 47E − 01	5.280 78E − 02
21	7.044 12E − 02	−4.608 67E − 02	3.597 26E − 02	1.989 46E − 03	1.943 20E − 01	7.653 97E − 02
22	6.097 61E − 02	−5.162 11E − 02	2.592 30E − 02	1.419 91E − 03	5.743 72E − 01	3.766 50E − 02
23	5.022 96E − 02	−3.284 70E − 02	2.566 47E − 02	1.012 41E − 03	1.942 24E − 01	5.469 15E − 02
24	4.347 74E − 02	−3.680 93E − 02	1.848 53E − 02	7.219 38E − 04	5.744 09E − 01	2.685 89E − 02
25	3.581 70E − 02	−2.341 88E − 02	1.830 00E − 02	5.148 07E − 04	1.943 79E − 01	3.901 03E − 02
26	3.099 71E − 02	−2.624 87E − 02	1.317 57E − 02	3.670 49E − 04	5.714 30E − 01	1.916 85E − 02
27	2.557 04E − 02	−1.673 48E − 02	1.305 84E − 02	2.621 04E − 04	1.944 75E − 01	2.776 37E − 02
28	2.213 38E − 02	−1.874 14E − 02	9.409 25E − 03	1.871 31E − 04	5.725 54E − 01	1.368 15E − 02
29	1.824 92E − 02	−1.193 96E − 02	9.321 00E − 03	1.335 49E − 04	1.944 75E − 01	1.983 48E − 02
30	1.579 51E − 02	−1.337 52E − 02	6.714 12E − 03	9.530 54E − 05	5.718 73E − 01	9.766 61E − 03
31	1.302 74E − 02	−8.524 32E − 03	6.653 48E − 03	6.804 63E − 05	1.944 39E − 01	1.415 34E − 02
32	1.127 62E − 02	−9.547 93E − 03	4.793 62E − 03	4.856 93E − 05	5.726 20E − 01	6.970 01E − 03
33	9.296 91E − 03	−6.082 44E − 03	4.748 60E − 03	3.466 11E − 05	1.942 83E − 01	1.010 55E − 02

Optimum design variables: 8.047 87E − 03, −6.813 19E − 03, 3.421 74E − 03.

Optimum cost function value: 2.473 47E − 05.

Norm of gradient at optimum: 4.970 71E − 03.

Total no. of function evaluations: 753.

The methods discussed in later sections try to overcome some of these difficulties.

5.4.3* Scaling of Design Variables

The rate of convergence of the steepest descent method is at the most linear even for a quadratic cost function. It is possible to accelerate the rate of convergence of the steepest descent method by scaling the design variables.

TABLE 5.4

Optimum solution for Example 5.11 with steepest descent method: $f(\mathbf{x}) = 25x_1^2 + x_2^2$; $\mathbf{x}^{(0)} = (1, 1)$

No.	x_1	x_2	$f(\mathbf{x})$	α	$\|\mathbf{c}\|$
1	1.000 00E + 00	1.000 00E + 00	2.600 00E + 01	2.003 29E − 02	5.004 00E + 01
2	−1.644 37E − 03	9.599 34E − 01	9.215 41E − 01	4.789 76E − 02	1.921 63E + 00
3	3.773 64E − 02	4.036 27E − 02	3.723 01E − 02	2.003 29E − 02	1.888 55E + 00
4	−6.205 22E − 05	3.874 55E − 02	1.501 31E − 03	4.815 11E − 01	7.755 31E − 02
5	1.431 89E − 03	1.432 69E − 03	5.331 03E − 05	2.003 29E − 02	7.165 18E − 02

Optimum design variables: −2.354 50E − 06, 1.375 29E − 03.

Optimum cost function value: 1.891 57E − 06.

Norm of gradient at optimum: 2.753 10E − 03.

Total no. of function evaluations: 111.

For a quadratic cost function it is possible to scale the design variables such that the condition number of the Hessian matrix in the new design variables is unity. The steepest descent method converges in only one iteration for a positive definite quadratic function with a unit condition number. To obtain the optimum point with the original design variables we could then unscale the transformed design variables. We shall demonstrate the advantage of scaling the design variables by the following two examples.

Example 5.11   Effect of scaling of design variables. Minimize $f(x_1, x_2) = 25x_1^2 + x_2^2$ with a starting design $(1, 1)$ by the steepest descent method. How would you scale the design variables to accelerate the rate of convergence?

Solution. Let us solve the problem by the computer program for the steepest descent method given in Appendix D. The history of the iterative process is given in Table 5.4. Note the inefficiency of the method on such a simple quadratic cost function. Figure 5.14 shows the iso-cost contours of the cost function and progress of the method from the initial design. The Hessian of $f(x_1, x_2)$ is

$$\mathbf{H} = \begin{bmatrix} 50 & 0 \\ 0 & 2 \end{bmatrix}$$

Now let us introduce new design variables y_1 and y_2 such that

$$\mathbf{x} = \mathbf{D}\mathbf{y} \quad \text{where} \quad \mathbf{D} = \begin{bmatrix} \dfrac{1}{\sqrt{50}} & 0 \\ 0 & \dfrac{1}{\sqrt{2}} \end{bmatrix}$$

Note that, in general, we may use $D_{ii} = 1/\sqrt{H_{ii}}$ for $i = 1$ to n if Hessian is a diagonal matrix. Thus, $x_1 = y_1/\sqrt{50}$ and $x_2 = y_2/\sqrt{2}$ and

$$f(y_1, y_2) = \tfrac{1}{2}(y_1^2 + y_2^2)$$

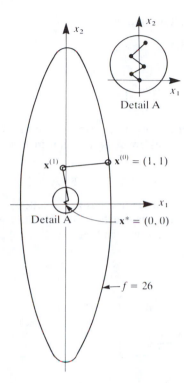

FIGURE 5.14
Iteration history for Example 5.11 with the steepest descent method.

The minimum point of $f(y_1, y_2)$ can be found in one iteration by the steepest descent method. The optimum point is $(0, 0)$ in the new design variable space. To obtain the minimum point in the original design space, we have to unscale the transformed design variables as

$$x_1^* = y_1/\sqrt{50} = 0 \quad \text{and} \quad x_2^* = y_2/\sqrt{2} = 0 \qquad \|$$

Example 5.12 Effect of scaling of design variables. Minimize $f(x_1, x_2) = 6x_1^2 - 6x_1x_2 + 2x_2^2 - 5x_1 + 4x_2 + 2$ with a starting design $(-1, -2)$ by the steepest descent method. Scale the design variables to have a condition number of unity for the Hessian matrix of function f in the new design variables.

Solution. Note that the function f in this problem contains the cross term x_1x_2 unlike the previous example. Thus, we need to compute the eigenvalues and eigenvectors of the Hessian matrix to find a suitable scaling or transformation of the design variables. The Hessian **H** of the function f is given as

$$\mathbf{H} = \begin{bmatrix} 12 & -6 \\ -6 & 4 \end{bmatrix}$$

The eigenvalues of the Hessian can be calculated as 0.7889 and 15.211. The corresponding eigenvectors are $(0.4718, 0.8817)$ and $(-0.8817, 0.4718)$. Now let us define new variables y_1 and y_2 by the following transformation

$$x = Qy \tag{a}$$

where

$$Q = \begin{bmatrix} 0.4718 & -0.8817 \\ 0.8817 & 0.4718 \end{bmatrix}$$

Note that the columns of Q are the eigenvectors of the Hessian matrix H. The transformation of variables defined by Eq. (a) gives the function in terms of y_1 and y_2 as

$$f(y_1, y_2) = 0.5(0.7889y_1^2 + 15.211y_2^2) + 1.1678y_1 + 6.2957y_2 + 2$$

The condition number of the Hessian matrix in the new design variables y_1 and y_2 is still not unity. To achieve the condition number equal to unity for the Hessian, we must define another transformation of y_1 and y_2 such that

$$y = Dz$$

where

$$D = \begin{bmatrix} \dfrac{1}{\sqrt{0.7889}} & 0 \\ 0 & \dfrac{1}{\sqrt{15.211}} \end{bmatrix}$$

Finally, the new design variables are given by

$$y_1 = \frac{z_1}{\sqrt{0.7889}} \quad \text{and} \quad y_2 = \frac{z_2}{\sqrt{15.211}}$$

and $f(z_1, z_2) = 0.5(z_1^2 + z_2^2) + 1.3148z_1 + 1.6142z_2$. Note that the condition number of the Hessian of f in the variables z_1 and z_2 is unity. The steepest descent method converges to the solution of $f(z_1, z_2)$ in one iteration as $(-1.3158, -1.6142)$. The minimum point in the original design space is found by defining the inverse transformation as

$$x = QDz$$

This gives the minimum point in the original design space as $(-\frac{1}{3}, -\frac{3}{2})$. ||

5.5 CONJUGATE GRADIENT METHOD

The method of conjugate gradients, due to Fletcher and Reeves [1964], is a very simple and effective modification of the steepest descent method. It was noted in the previous section that the steepest descent directions at two consecutive steps are orthogonal to each other. This tends to slow down the steepest descent method although it is convergent. The conjugate gradient directions are not orthogonal to each other. Rather, these directions tend to cut diagonally through the orthogonal steepest descent directions. Therefore, they improve the rate of convergence of the steepest descent method considerably. Actually, the *conjugate gradient directions* $\mathbf{d}^{(i)}$ are orthogonal with respect to a symmetric and positive definite matrix $\mathbf{A}$, i.e. $\mathbf{d}^{(i)^T}\mathbf{A}\mathbf{d}^{(j)} = 0$ for all i and j and $i \neq j$.

The *conjugate gradient algorithm* is stated as follows.

Step 1. Estimate a starting design as $\mathbf{x}^{(0)}$. Set the iteration counter $k = 0$. Select the convergence parameter ε. Calculate

$$\mathbf{d}^{(0)} = -\mathbf{c}^{(0)} \equiv -\nabla f(\mathbf{x}^{(0)})$$

Check to see if $\|\mathbf{c}^{(0)}\| < \varepsilon$. If it is, then stop. Otherwise, go to Step 4 (note that Step 1 of the conjugate gradient and the steepest descent methods is the same).

Step 2. Compute the gradient of the cost function as

$$\mathbf{c}^{(k)} = \nabla f(\mathbf{x}^{(k)})$$

Calculate $\|\mathbf{c}^{(k)}\|$. If $\|\mathbf{c}^{(k)}\| < \varepsilon$, then stop; otherwise continue.

Step 3. Calculate the new conjugate direction as

$$\mathbf{d}^{(k)} = -\mathbf{c}^{(k)} + \beta_k \mathbf{d}^{(k-1)} \tag{5.38}$$

where

$$\beta_k = (\|\mathbf{c}^{(k)}\| / \|\mathbf{c}^{(k-1)}\|)^2 \tag{5.39}$$

Step 4. Compute $\alpha_k = \alpha$ to minimize $f(\mathbf{x}^{(k)} + \alpha \mathbf{d}^{(k)})$.

Step 5. Change the design as

$$\mathbf{x}^{(k+1)} = \mathbf{x}^{(k)} + \alpha_k \mathbf{d}^{(k)} \tag{5.40}$$

Set $k = k + 1$ and go to Step 2.

Note that the conjugate direction in Eq. (5.38) satisfies the descent condition of Inequality (5.8). This can be shown by substituting $\mathbf{d}^{(k)}$ from Eq. (5.38) into Inequality (5.8), and using the step size determination condition given in Eq. (5.15).

The first step of the conjugate gradient method is just the steepest descent step. The only difference between the conjugate gradient and steepest descent methods is in Eq. (5.38). In this step the current steepest descent direction is modified by adding a scaled direction used in the previous iteration. The scale factor is determined using lengths of the gradient vector for the cost function at the two iterations as shown in Eq. (5.39). Thus the conjugate direction is nothing but a deflected steepest descent direction. This is an extremely simple modification that requires little additional calculation. It is, however, very effective in substantially improving the rate of convergence of the steepest descent method. Therefore, *the conjugate gradient method should always be preferred over the steepest descent method.* In the next section an example is discussed to compare the rate of convergence of the steepest descent, conjugate gradient and Newton's methods. We shall see there that the method performs quite well compared to the other two methods.

The conjugate gradient algorithm finds the minimum in n iterations for positive definite quadratic forms having n design variables. For general functions, it is recommended that the iterative process be re-started every $(n + 1)$ iterations for computational stability, if the minimum has not been found by then. That is, set $\mathbf{x}^{(0)} = \mathbf{x}^{(n+1)}$ and re-start the process from Step 1 of the algorithm. The algorithm is very simple to program and works very well for general unconstrained minimization problems. It is a convergent algorithm and is available in the IDESIGN software system [Arora and Tseng, 1987] where an inexact line search is used.

> **Example 5.13 Use of conjugate gradient algorithm.** Consider the example problem solved in Example 5.10:
>
> $$\text{minimize } f(x_1, x_2, x_3) = x_1^2 + 2x_2^2 + 2x_3^2 + 2x_1x_2 + 2x_2x_3$$
>
> Carry out two iterations of the conjugate gradient method starting from the design $(2, 4, 10)$.
>
> **Solution.** The first iteration of the conjugate gradient method is the same as given in Example 5.10:
>
> $$\mathbf{c}^{(0)} = (12, 40, 48); \quad \|\mathbf{c}^{(0)}\| = 63.6, \quad f(\mathbf{x}^{(0)}) = 332.0$$
> $$\mathbf{x}^{(1)} = (0.0956, -2.348, 2.381)$$
>
> The second iteration starts from Step 2 of the conjugate gradient algorithm:
>
> $$\mathbf{c}^{(1)} = (-4.5, -4.438, 4.828), \quad f(\mathbf{x}^{(1)}) = 10.75$$
> $$\|\mathbf{c}^{(1)}\| = 7.952 > \varepsilon, \text{ so continue.}$$
> $$\beta_1 = [\|\mathbf{c}^{(1)}\|/\|\mathbf{c}^{(0)}\|]^2$$
> $$= (7.952/63.3)^2 = 0.015\,633$$
> $$\mathbf{d}^{(1)} = -\mathbf{c}^{(1)} + \beta_1\mathbf{d}^{(0)}$$
> $$= \begin{bmatrix} 4.500 \\ 4.438 \\ -4.828 \end{bmatrix} + (0.015\,633) \begin{bmatrix} -12 \\ -40 \\ -48 \end{bmatrix} = \begin{bmatrix} 4.312\,41 \\ 3.812\,68 \\ -5.578\,38 \end{bmatrix}$$
>
> The design is updated as
>
> $$\mathbf{x}^{(2)} = \begin{bmatrix} 0.0956 \\ -2.348 \\ 2.381 \end{bmatrix} + \alpha \begin{bmatrix} 4.312\,41 \\ 3.812\,68 \\ -5.578\,38 \end{bmatrix}$$
>
> Computing the step size to minimize $f(\mathbf{x}^{(1)} + \alpha\mathbf{d}^{(1)})$, we get $\alpha = 0.3156$. Substituting this into the above expression, we get $\mathbf{x}^{(2)} = (1.4566, -1.1447, 0.6205)$. Calculating the gradient at this point we get $\mathbf{c}^{(2)} = (0.6238, -0.4246, 0.1926)$. $|\mathbf{c}^{(2)}\| = 0.7788 > \varepsilon$, so we need to continue the iterations. Note that $\mathbf{c}^{(2)} \cdot \mathbf{d}^{(1)} = 0$.
>
> The problem is solved using the conjugate gradient method available in the IDESIGN software system with $\varepsilon = 0.005$. Table 5.5 shows the iteration history from the program. It can be seen that a very precise optimum is obtained in only 4 iterations and 10 function evaluations. Comparing these with the steepest

TABLE 5.5

Iteration history for Example 5.13 with conjugate gradient method: $f(\mathbf{x}) = x_1^2 + 2x_2^2 + 2x_3^2 + 2x_1x_2 + 2x_2x_3$

No.	x_1	x_2	x_3	$f(\mathbf{x})$	$\|\mathbf{c}\|$
1	$2.0000E+00$	$4.0000E+00$	$1.0000E+01$	$3.3200E+02$	$6.3624E+01$
2	$9.5358E-02$	$-2.3488E+00$	$2.3814E+00$	$1.0750E+01$	$7.9593E+00$
3	$1.4578E+00$	$-1.1452E+00$	$6.2143E-01$	$7.5823E-01$	$7.7937E-01$
4	$-6.4550E-10$	$-5.8410E-10$	$-1.3150E-10$	$6.8520E-20$	$3.0512E-05$

descent method results given in Table 5.3, we conclude that the conjugate gradient method is much superior. ‖

5.6 NEWTON'S METHOD

In the steepest descent method, only first-order derivative information is used in the representation of the cost function at a point to determine the direction of travel. If second-order derivatives were available, we could use them to represent the cost surface more accurately, and a better direction of travel could be found. With the inclusion of second-order information we could expect a better rate of convergence. For example, Newton's method which uses Hessian matrix for the function has a *quadratic* rate of convergence. In other words, for any positive definite quadratic function, the method will converge in only one iteration, and step size will be one.

The *basic idea* of the Newton's method is to use second-order Taylor's series expansion of the function about the current design point. This gives a quadratic expression for the change in design $\Delta\mathbf{x}$. The necessary conditions for minimization of this function then give an explicit calculation for the direction of travel in the design space. Proper step length in this direction then completes one iteration of the method.

In the following, we shall omit the argument $\mathbf{x}^{(k)}$ from all functions, as the derivation applies to any design iteration. Using second-order Taylor series expansion for the function $f(\mathbf{x})$, we obtain

$$f(\mathbf{x} + \Delta\mathbf{x}) = f(\mathbf{x}) + \mathbf{c}^T\Delta\mathbf{x} + 0.5\Delta\mathbf{x}^T\mathbf{H}\Delta\mathbf{x} \qquad (5.41)$$

where $\Delta\mathbf{x}$ is a small change in design and $\mathbf{H}$ is Hessian of f at the point $\mathbf{x}$ (sometimes denoted as $\nabla^2 f$). Equation (5.41) is a quadratic function in terms of $\Delta\mathbf{x}$. The theory of convex programming problems in Chapter 3 guarantees that if $\mathbf{H}$ is positive semidefinite, then there is a $\Delta\mathbf{x}$ that gives a global minimum for the function of Eq. (5.41). In addition, if $\mathbf{H}$ is positive definite, then the minimum for Eq. (5.41) is unique. Writing optimality conditions $(\partial f/\partial(\Delta\mathbf{x}) = \mathbf{0})$ for the function of Eq. (5.41),

$$\mathbf{c} + \mathbf{H}\Delta\mathbf{x} = \mathbf{0} \qquad (5.42)$$

Assuming $\mathbf{H}$ to be nonsingular, we get an expression for $\Delta\mathbf{x}$ as

$$\Delta\mathbf{x} = -\mathbf{H}^{-1}\mathbf{c} \tag{5.43}$$

Using this value for $\Delta\mathbf{x}$, the new estimate for the design is given as

$$\mathbf{x}^{(1)} = \mathbf{x}^{(0)} + \Delta\mathbf{x} \tag{5.44}$$

Since Eq. (5.41) is just an approximation for f at the point $\mathbf{x}^{(0)}$, $\mathbf{x}^{(1)}$ will probably not be the precise minimum point of $f(\mathbf{x})$. Therefore, the process will have to be repeated to obtain improved estimates until the minimum is reached.

Each iteration of Newton's method requires computation of the Hessian of the cost function which is a symmetric matrix. Therefore, it needs computation of $n(n+1)/2$ second-order derivatives of $f(\mathbf{x})$ (recall that n is the number of design variables). This can require considerable computational effort and we should try to improve efficiency of the method before recalculating the second-order information. An easy way of improving the method is to use the step length parameter with the direction $\Delta\mathbf{x}$. The step length can be calculated to minimize the cost function in the direction $\Delta\mathbf{x}$. Any of the one-dimensional search procedures may be used for this purpose. In addition to improving efficiency, the use of a step size makes the method very stable and guarantees its convergence to a local minimum point starting from any design point provided the Hessian $\mathbf{H}$ remains positive definite at all iterations. The classical Newton's method uses the step size as one in the direction $\Delta\mathbf{x}$. However, a full step along $\Delta\mathbf{x}$ may not result in a descent step for the cost function, i.e. $f(\mathbf{x} + \Delta\mathbf{x})$ may not be smaller than $f(\mathbf{x})$. Therefore, *classical Newton's method is not guaranteed to converge.*

The *modified Newton's algorithm* is stated as follows.

Step 1. Make an engineering estimate for a starting design $\mathbf{x}^{(0)}$. Set iteration counter $k = 0$. Select a tolerance ε for the stopping criterion.

Step 2. Calculate $c_i^{(k)} = \partial f(\mathbf{x}^{(k)})/\partial x_i$; $i = 1$ to n. If $\|\mathbf{c}^{(k)}\| < \varepsilon$, stop the iterative process. Otherwise, continue.

Step 3. Calculate the Hessian matrix as

$$\mathbf{H}(\mathbf{x}^{(k)}) = \left[\frac{\partial^2 f}{\partial x_i\, \partial x_j}\right]; \qquad i = 1 \text{ to } n; \qquad j = 1 \text{ to } n$$

Step 4. Calculate the direction of travel in the design space using Eq. (5.43) as

$$\mathbf{d}^{(k)} = -\mathbf{H}^{-1}\mathbf{c}^{(k)} \tag{5.45}$$

Note that the calculation of $\mathbf{d}^{(k)}$ in the above equation is symbolic. For computational efficiency, a system of linear simultaneous equation is solved instead of evaluating the inverse of Hessian matrix.

Step 5. Update the design as $\mathbf{x}^{(k+1)} = \mathbf{x}^{(k)} + \alpha_k \mathbf{d}^{(k)}$, where α_k is calculated to minimize $f(\mathbf{x}^{(k)} + \alpha \mathbf{d}^{(k)})$. A one-dimensional search procedure may be used to calculate α. To start with, $\alpha = 1$ is a good initial estimate for the step size.

Step 6. Set $k = k + 1$ and go to Step 2.

It is emphasized here that unless $\mathbf{H}$ is positive definite, the direction $\mathbf{d}^{(k)}$ determined from Eq. (5.45) will not be that of descent for the cost function. To see this we substitute $\mathbf{d}^{(k)}$ from Eq. (5.45) into the descent condition of Eq. (5.8) to obtain

$$-\mathbf{c}^{(k)^T}\mathbf{H}^{-1}\mathbf{c}^{(k)} < 0$$

The foregoing condition will always be satisfied if $\mathbf{H}$ is positive definite. If $\mathbf{H}$ is negative definite or negative semidefinite the condition is always violated. With $\mathbf{H}$ as indefinite or positive semidefinite, the condition may or may not be satisfied, so we must check for it. If the direction obtained in Step 4 is not that of descent for the cost function, then we should stop there because a positive step size cannot be determined.

Example 5.14 Use of modified Newton's method. Minimize $f(\mathbf{x}) = 3x_1^2 + 2x_1x_2 + 2x_2^2 + 7$ using the modified Newton's algorithm starting from the point $(5, 10)$. Use $\varepsilon = 0.0001$ as the stopping criterion.

Solution. We will follow the steps of the modified Newton's method.

1. $\mathbf{x}^{(0)}$ is given as $(5, 10)$.
2. The gradient vector $\mathbf{c}^{(0)}$ at the point $(5, 10)$ is given as

$$\mathbf{c}^{(0)} = (6x_1 + 2x_2,\ 2x_1 + 4x_2) = (50,\ 50)$$
$$\|\mathbf{c}^{(0)}\| = \sqrt{50^2 + 50^2} = 50\sqrt{2} > \varepsilon$$

Therefore, the convergence criterion is not satisfied.
3. The Hessian matrix at the point $(5, 10)$ is given as

$$\mathbf{H}^{(0)} = \begin{bmatrix} 6 & 2 \\ 2 & 4 \end{bmatrix}$$

Note that the Hessian is positive definite.
4. The direction of design change is

$$\mathbf{d}^{(0)} = -\mathbf{H}^{-1}\mathbf{c}^{(0)} = \frac{-1}{20}\begin{bmatrix} 4 & -2 \\ -2 & 6 \end{bmatrix}\begin{bmatrix} 50 \\ 50 \end{bmatrix} = \begin{bmatrix} -5 \\ -10 \end{bmatrix}$$

Note that since $\mathbf{H}$ is positive definite, the direction $\mathbf{d}^{(0)}$ is that of descent for the cost function.
5. Step size α is calculated to minimize $f(\mathbf{x}^{(0)} + \alpha\mathbf{d}^{(0)})$:

$$\mathbf{x}^{(1)} = \mathbf{x}^{(0)} + \alpha\mathbf{d}^{(0)}$$

$$= \begin{bmatrix} 5 \\ 10 \end{bmatrix} + \alpha\begin{bmatrix} -5 \\ -10 \end{bmatrix} = \begin{bmatrix} 5 - 5\alpha \\ 10 - 10\alpha \end{bmatrix}$$

$$\frac{df}{d\alpha} = 0; \quad \text{or} \quad \nabla f(\mathbf{x}^{(1)}) \cdot \mathbf{d}^{(0)} = 0$$

Using the Step 2 calculations, $\nabla f(\mathbf{x}^{(1)})$ is given as

$$\nabla f(\mathbf{x}^{(1)}) = \begin{bmatrix} 6(5-5\alpha) + 2(10-10\alpha) \\ 2(5-5\alpha) + 4(10-10\alpha) \end{bmatrix} = \begin{bmatrix} 50 - 50\alpha \\ 50 - 50\alpha \end{bmatrix}$$

Therefore,

$$\nabla f(\mathbf{x}^{(1)}) \cdot \mathbf{d}^{(0)} = (50 - 50\alpha, \ 50 - 50\alpha) \begin{bmatrix} -5 \\ -10 \end{bmatrix} = 0$$

Or, $-5(50 - 50\alpha) - 10(50 - 50\alpha) = 0$

Solving the preceding equation, we get $\alpha = 1$. Note that the Golden Section search also gives $\alpha = 1$. Therefore

$$\mathbf{x}^{(1)} = \begin{bmatrix} 5 - 5\alpha \\ 10 - 10\alpha \end{bmatrix} = \begin{bmatrix} 0 \\ 0 \end{bmatrix}$$

The gradient of the cost function at $\mathbf{x}^{(1)}$ is

$$\mathbf{c}^{(1)} = \begin{bmatrix} 50 - 50\alpha \\ 50 - 50\alpha \end{bmatrix} = \begin{bmatrix} 0 \\ 0 \end{bmatrix}$$

Since $\|\mathbf{c}^{(1)}\| < \varepsilon$, the Newton's method has given the solution in just one iteration. This is an unusual situation because the function is a positive definite quadratic form (Hessian of f is positive definite everywhere).   ‖

A computer program based on the modified Newton's method is given in Appendix D which needs three user supplied subroutines FUNCT, GRAD and HASN. These subroutines evaluate cost function, the gradient and Hessian matrix of the cost function, respectively.

Example 5.15   Use of modified Newton's method. Minimize $f(\mathbf{x}) = 10x_1^4 - 20x_1^2x_2 + 10x_2^2 + x_1^2 - 2x_1 + 5$ using the computer program for the modified Newton's method given in Appendix D from the point $(-1, 3)$. Golden Section search may be used for step size determination with $\delta = 0.05$ and line search accuracy equal to 1.0E $-$ 04. For the stopping criterion, select $\varepsilon = 0.005$.

Solution. Note that $f(\mathbf{x})$ is not a quadratic function in terms of the design variables. Thus we cannot expect the Newton's method to converge in one iteration. The gradient of $f(\mathbf{x})$ is given as

$$\mathbf{c} = \nabla f(\mathbf{x}) = (40x_1^3 - 40x_1x_2 + 2x_1 - 2, \ -20x_1^2 + 20x_2)$$

and the Hessian matrix of $f(\mathbf{x})$ is

$$\mathbf{H} = \nabla^2 f(\mathbf{x}) = \begin{bmatrix} 120x_1^2 - 40x_2 + 2 & -40x_1 \\ -40x_1 & 20 \end{bmatrix}$$

The iteration history of the Newton's method for this example problem is given in Table 5.6. The optimum point is $(1, 1)$ and the optimum value of $f(\mathbf{x})$ is 4.0. Newton's method has converged in seven iterations to the optimum solution. Figure 5.15 shows the iso-cost contours for the function and the progress of the method from the starting design $(-1, 3)$. Note that the step size is approximately

TABLE 5.6

Optimum solution for Example 5.15 with modified Newton's method:
$f(\mathbf{x}) = 10x_1^4 - 20x_1^2x_2 + 10x_2^2 + x_1^2 - 2x_1 + 5;\ \mathbf{x}^{(0)} = (-1, 3)$

No.	x_1	x_2	$f(\mathbf{x})$	α	$\|\mathbf{c}\|$
1	$-1.000\,00E + 00$	$3.000\,00E + 00$	$4.800\,00E + 01$	$9.961\,01E - 01$	$8.588\,36E + 01$
2	$-1.051\,08E + 00$	$1.109\,96E + 00$	$8.207\,21E + 00$	$1.901\,81E - 01$	$3.885\,43E + 00$
3	$-6.158\,44E - 01$	$1.940\,34E - 01$	$6.954\,06E + 00$	$1.599\,29E + 00$	$8.630\,19E + 00$
4	$-6.655\,13E - 02$	$-1.862\,86E - 01$	$5.501\,25E + 00$	$3.113\,32E + 00$	$4.639\,26E + 00$
5	$6.231\,68E - 01$	$3.156\,68E - 01$	$4.194\,81E + 00$	$2.146\,04E + 00$	$1.797\,58E + 00$
6	$9.527\,89E - 01$	$8.824\,41E - 01$	$4.008\,66E + 00$	$1.113\,63E + 00$	$1.009\,13E + 00$
7	$9.876\,69E - 01$	$9.771\,56E - 01$	$4.000\,18E + 00$	$9.572\,47E - 01$	$9.641\,31E - 02$

Optimum design variables: $9.998\,80E - 01, 9.996\,81E - 01$.
Optimum cost function value: $4.000\,00E + 00$.
Norm of gradient at optimum: $3.268\,83E - 03$.
Total no. of function evaluations: 198.

equal to one in the last phase of the iterative process. This is because any function resembles a quadratic function sufficiently close to the optimum point and step size is equal to unity for a quadratic function. ‖

The *drawbacks* of the Newton method for general applications are:

1. It requires calculations of second-order derivatives at each iteration which is usually quite time consuming. In some applications, it may not even be possible to calculate them. Also, a system of simultaneous linear equations needs to be solved. Therefore each iteration of the method requires substantially more calculations compared to the steepest descent method.

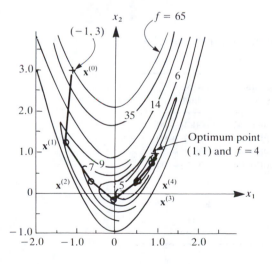

FIGURE 5.15
Iteration history for Example 5.15 with Newton's method.

2. Hessian of the function may be singular at some iterations. Thus, Eq. (5.43) cannot be used to compute the search direction. Also, unless the Hessian is positive definite, the Newton direction cannot be guaranteed to be that of descent for the cost function.

3. Each iteration is started afresh. Previously calculated gradients and Hessians are not used. So, it is a memoryless method.

4. The method is not convergent unless Hessian remains positive definite and a step size determination scheme is used. However, the method has a high (quadratic) rate of convergence. For a strictly convex quadratic function, the method will converge in only one iteration from any starting design.

Example 5.16 Comparison of steepest descent, Newton and conjugate gradient methods. Minimize $f(\mathbf{x}) = 50(x_2 - x_1^2)^2 + (2 - x_1)^2$ starting from the point $(5, -5)$. Use the steepest descent, Newton and conjugate gradient methods and compare their performance.

Solution. The optimum solution for the problem is known as $(2, 4)$ with $f(2, 4) = 0$. We use exact gradient expressions and $\varepsilon = 0.005$ to solve the problem using the steepest descent and Newton's method programs given in Appendix D and the conjugate gradient method available in IDESIGN. Table 5.7 summarizes final results with the three methods. For the steepest descent method, $\delta_0 = 0.05$ and line search termination criterion of $0.000\,01$ are used and, for the Newton's method, they are 0.05 and 0.0001 respectively. Golden Section search is used in both the methods. It can be observed again that the steepest descent method is the most inefficient and the conjugate gradient is the most efficient method for the present example. Therefore, the conjugate gradient method is recommended for general applications.

   Tables 5.8 and 5.9 show iteration histories for the conjugate gradient and Newton's methods. ||

TABLE 5.7

Comparative evaluation of three methods for Example 5.16: $f(\mathbf{x}) = 50(x_2 - x_1^2)^2 + (2 - x_1)^2$

	Steepest descent	Conjugate gradient	Newton's method
x_1	1.9941E + 00	2.0000E + 00	2.0000E + 00
x_2	3.9765E + 00	3.9998E + 00	3.9999E + 00
f	3.4564E − 05	1.0239E − 08	2.5054E − 10
$\|\mathbf{c}\|$	3.3236E − 03	1.2860E − 04	9.0357E − 04
No. of function evaluations	138 236	65	349
No. of iterations	9670	22	13

TABLE 5.8

Iteration history for Example 5.16 with conjugate gradient method in IDESIGN: $f(\mathbf{x}) = 50(x_2 - x_1^2)^2 + (2 - x_1)^2$

No.	x_1	x_2	$f(\mathbf{x})$	$\|\mathbf{c}\|$
1	5.0000E + 00	−5.0000E + 00	4.500 90E + 04	3.015 56E + 04
2	8.3032E − 01	−4.5831E + 00	1.391 35E + 03	1.020 07E + 03
3	−2.6902E − 01	−3.9193E + 00	8.018 38E + 02	4.554 50E + 02
4	1.9543E − 01	4.2482E − 02	3.257 37E + 00	3.800 95E + 00
5	2.3653E − 01	2.0314E − 02	3.173 30E + 00	4.011 01E + 00
6	4.4392E − 01	1.2086E − 01	2.711 74E + 00	8.451 04E + 00
7	8.8952E − 01	7.2438E − 01	1.456 72E + 00	1.176 08E + 01
8	8.8169E − 01	7.7490E − 01	1.250 93E + 00	1.818 53E + 00
9	1.0894E + 00	1.1451E + 00	9.162 47E − 01	8.382 04E + 00
10	1.2857E + 00	1.5986E + 00	6.580 01E − 01	1.367 65E + 01
11	1.3261E + 00	1.7676E + 00	4.581 43E − 01	3.838 76E + 00
12	1.4782E + 00	2.1584E + 00	3.077 39E − 01	7.330 75E + 00
13	1.5795E + 00	2.4589E + 00	2.416 86E − 01	1.113 96E + 01
14	1.6437E + 00	2.7122E + 00	1.325 14E − 01	4.303 67E + 00
15	1.7595E + 00	3.0783E + 00	7.329 38E − 02	5.972 70E + 00
16	1.8439E + 00	3.3749E + 00	5.579 89E − 02	9.279 73E + 00
17	1.8946E + 00	3.5934E + 00	1.186 16E − 02	1.725 34E + 00
18	1.9453E + 00	3.7794E + 00	4.200 64E − 03	1.873 03E + 00
19	1.9950E + 00	3.9760E + 00	7.814 25E − 04	1.589 99E + 00
20	1.9923E + 00	3.9692E + 00	5.951 53E − 05	2.794 13E − 02
21	2.0000E + 00	3.9998E + 00	1.023 87E − 08	5.385 26E − 02
22	2.0000E + 00	3.9998E + 00	2.111 92E − 09	1.285 96E − 04

TABLE 5.9

Iteration history for Example 5.16 with Newton's method:
$f(\mathbf{x}) = 50(x_2 - x_1^2)^2 + (2 - x_1)^2$

No.	x_1	x_2	$f(\mathbf{x})$	α	$\|\mathbf{c}\|$
1	5.000 00E + 00	−5.000 00E + 00	4.500 90E + 04	9.999 89E − 01	3.015 56E + 04
2	4.999 00E + 00	2.498 97E + 01	8.994 01E + 00	1.033 37E − 01	6.338 85E + 00
3	4.699 31E + 00	2.199 34E + 01	7.692 35E + 00	1.414 99E + 00	9.054 99E + 01
4	4.317 82E + 00	1.853 55E + 01	5.956 92E + 00	2.538 87E + 00	9.860 86E + 01
5	3.819 68E + 00	1.450 82E + 01	3.645 37E + 00	1.994 32E + 00	6.659 34E + 01
6	3.424 14E + 00	1.164 95E + 01	2.310 69E + 00	2.417 33E + 00	5.484 39E + 01
7	3.019 93E + 00	9.063 13E + 00	1.201 82E + 00	2.212 92E + 00	3.681 41E + 01
8	2.682 27E + 00	7.149 53E + 00	5.670 42E − 01	2.511 19E + 00	2.593 46E + 01
9	2.371 13E + 00	5.593 55E + 00	1.789 43E − 01	2.504 77E + 00	1.464 02E + 01
10	2.130 97E + 00	4.526 55E + 00	2.763 54E − 02	2.398 58E + 00	6.593 78E + 00
11	2.002 64E + 00	4.014 35E + 00	7.218 96E − 04	9.862 71E − 01	1.555 90E + 00
12	1.998 45E + 00	3.993 85E + 00	2.451 27E − 06	9.947 86E − 01	1.718 63E − 02
13	2.000 00E + 00	3.999 99E + 00	2.505 38E − 10	—	9.035 71E − 04

5.6.1* Marquardt Modification

As noted before the modified Newton's method has several drawbacks that can cause numerical difficulties. For example, if Hessian of the cost function $\mathbf{H}$ is not positive definite the direction found from Eq. (5.45) may not be that of descent for the cost function. In that case, a step cannot be executed along the direction.

Marquardt [1963] suggested a modification to the direction finding process that has desirable features of the steepest descent and Newton methods. It turns out that far away from the solution point the method behaves like the steepest descent method which is quite good there. Near the solution point, it behaves like the Newton method which is very effective there. In the modified procedure, the Hessian is modified as $(\mathbf{H} + \lambda \mathbf{I})$ where λ is a positive constant. The search direction is then computed from Eq. (5.45) as

$$\mathbf{d}^{(k)} = -(\mathbf{H} + \lambda \mathbf{I})^{-1}\mathbf{c}^{(k)} \tag{5.46}$$

Note that when λ is large, the effect of $\mathbf{H}$ essentially gets neglected and $\mathbf{d}^{(k)}$ is essentially $-(1/\lambda)\mathbf{c}^{(k)}$, which is the steepest descent direction with $(1/\lambda)$ as the step size. As the algorithm proceeds, λ is reduced (i.e. step size is increased). When λ becomes sufficiently small, then the effect of $\lambda \mathbf{I}$ is essentially neglected and the Newton direction is obtained from Eq. (5.46). If the direction $\mathbf{d}^{(k)}$ of Eq. (5.46) does not reduce the cost function, then λ is increased (step size is reduced) and direction is recomputed.

Marquardt's algorithm is given in the following steps.

Step 1. Make an engineering estimate for starting design $\mathbf{x}^{(0)}$. Set iteration counter $k = 0$. Select a tolerance ε as the stopping criterion, and λ_0 as a large constant (say 1.0E+04).

Step 2. Calculate $c_i^{(k)} = \partial f(\mathbf{x}^{(k)})/\partial x_i$; $i = 1$ to n. If $\|\mathbf{c}^{(k)}\| < \varepsilon$, stop the iterative process. Otherwise continue.

Step 3. Calculate the Hessian matrix as

$$\mathbf{H}(\mathbf{x}^{(k)}) = \left[\frac{\partial^2 f}{\partial x_i \, \partial x_j}\right]; \quad i = 1 \text{ to } n; \quad j = 1 \text{ to } n$$

Step 4. Calculate the direction of change as

$$\mathbf{d}^{(k)} = -(\mathbf{H} + \lambda_k \mathbf{I})^{-1}\mathbf{c}^{(k)}$$

Step 5. If $f(\mathbf{x}^{(k)} + \mathbf{d}^{(k)}) < f(\mathbf{x}^{(k)})$, then go to Step 6. Otherwise, let $\lambda_k = 2\lambda_k$ and go to Step 4.

Step 6. Set $\lambda_{k+1} = 0.5\lambda_k$, $k = k + 1$ and go to Step 2.

5.7* QUASI-NEWTON METHODS

In Section 5.4 the steepest descent method was described. Some of the drawbacks of that method were pointed out. It was noted that the method has a poor rate of convergence because only first-order information is used. This flaw was corrected with Newton's method where second-order derivatives were used. Newton's method has very good convergence properties. However, the method can be inefficient because it requires calculation of $n(n+1)/2$ second-order derivatives (recall that n is the number of design variables). For most engineering design problems, calculation of second-order derivatives can be not only tedious but also impossible. In addition, both the steepest descent and Newton's methods are not learning processes. That is, in both the methods each iteration is started with the new design variables without using any other information from the previous iterations. Also, Newton's method runs into difficulties if Hessian of the function is singular at any iteration.

The methods presented in this subsection require the computation of only first derivatives. By making use of information obtained from previous iterations, however, convergence towards the minimum is speeded up. One of the methods generates an approximation to the matrix of second derivatives of the cost function. Therefore, these methods are learning processes as they accumulate the information from previous iterations. In this regard the methods presented here have desirable features of both the steepest descent and the Newton's methods. They are also called quasi-Newton or *update methods*. They use first-order derivatives to generate approximations for the Hessian matrix.

The methods were initially developed for positive definite quadratic function. For such functions they converge to the exact optimum in at the most n iterations, where n is the number of design variables. This ideal behavior does not carry over to general cost functions. A general cost function, however, looks very much like a positive definite quadratic function near the minimum point. Therefore, similar behavior could be expected for such functions once we are near the optimum point. For a nonconvex function, there is no guarantee of convergence in n iterations. If the method does not converge, it is generally re-started at every $(n+1)$th iteration.

There are several ways to approximate Hessian or its inverse. The basic idea is to update the current approximation using changes in design and the gradient vector. While updating the properties of symmetry and positive definiteness are preserved. Positive definiteness is essential, because without that the search direction may not be that of descent for the cost function. We shall describe two of the most popular methods in the class of update methods.

5.7.1* Davidon-Fletcher-Powell Method

This method initially proposed by Davidon [1959] was modified by Fletcher and Powell [1963] and that method is presented here. This is one of the most powerful methods for the minimization of a general function $f(\mathbf{x})$. The method builds up *approximate inverse of the Hessian* of $f(\mathbf{x})$ using only the first derivatives. The method is often called the DFP (Davidon, Fletcher and Powell) method:

Step 1. Estimate an initial design $\mathbf{x}^{(0)}$. Choose a symmetric positive definite matrix $\mathbf{A}^{(0)}$ as an estimate for inverse of the Hessian of the cost function. In the absence of more information, $\mathbf{A}^{(0)} = \mathbf{I}$ may be chosen. Also, specify a convergence parameter ε. Set $k = 0$. Compute the gradient vector as

$$\mathbf{c}^{(0)} = \nabla f(\mathbf{x}^{(0)})$$

Step 2. Calculate the norm of the gradient vector as $\|\mathbf{c}^{(k)}\|$. If $\|\mathbf{c}^{(k)}\| < \varepsilon$, then stop the iterative process. Otherwise continue.

Step 3. Calculate the search direction as

$$\mathbf{d}^{(k)} = -\mathbf{A}^{(k)}\mathbf{c}^{(k)}$$

Step 4. Compute optimum step size $\alpha_k = \alpha$ to minimize $f(\mathbf{x}^{(k)} + \alpha \mathbf{d}^{(k)})$.

Step 5. Update the design as

$$\mathbf{x}^{(k+1)} = \mathbf{x}^{(k)} + \alpha_k \mathbf{d}^{(k)}$$

Step 6. Update the matrix $\mathbf{A}^{(k)}$ – approximation for the inverse of the Hessian of the cost function as

$$\mathbf{A}^{(k+1)} = \mathbf{A}^{(k)} + \mathbf{B}^{(k)} + \mathbf{C}^{(k)}; \qquad n \times n \text{ matrix}$$

where the correction matrices $\mathbf{B}^{(k)}$ and $\mathbf{C}^{(k)}$ are calculated as

$$\mathbf{B}^{(k)} = \frac{\mathbf{s}^{(k)}\mathbf{s}^{(k)^T}}{(\mathbf{s}^{(k)} \cdot \mathbf{y}^{(k)})}; \qquad \mathbf{C}^{(k)} = \frac{-\mathbf{z}^{(k)}\mathbf{z}^{(k)^T}}{(\mathbf{y}^{(k)} \cdot \mathbf{z}^{(k)})}$$

with

$$\mathbf{s}^{(k)} = \alpha_k \mathbf{d}^{(k)} \qquad \text{(change in design)}$$

$$\mathbf{y}^{(k)} = \mathbf{c}^{(k+1)} - \mathbf{c}^{(k)} \qquad \text{(change in gradient)}$$

$$\mathbf{c}^{(k+1)} = \nabla f(\mathbf{x}^{(k+1)})$$

$$\mathbf{z}^{(k)} = \mathbf{A}^{(k)}\mathbf{y}^{(k)}$$

Step 7. Set $k = k + 1$ and go to Step 2.

Note that the first iteration of the method is the same as that for the steepest descent method.

Fletcher and Powell (1963) prove that this algorithm has the following properties:

1. The matrix $\mathbf{A}^{(k)}$ is positive definite for all k. This implies that the method will always converge to a local minimum point, since

$$\frac{d}{d\alpha} f(\mathbf{x}^{(k)} + \alpha \mathbf{d}^{(k)})|_{\alpha=0} = -\mathbf{c}^{(k)^T}\mathbf{A}^{(k)}\mathbf{c}^{(k)} < 0$$

   as long as $\mathbf{c}^{(k)} \neq 0$. This means that $f(\mathbf{x}^{(k)})$ may be decreased by choosing $\alpha > 0$, if $\mathbf{c}^{(k)} \neq \mathbf{0}$ (i.e. $\mathbf{d}^{(k)}$ is a direction of descent).

2. When this method is applied to a positive definite quadratic form, $\mathbf{A}^{(k)}$ converges to inverse of the Hessian of the quadratic form.

Example 5.17 Application of DFP method. Execute two iterations of the DFP method for the problem

$$\text{minimize } f(\mathbf{x}) = 5x_1^2 + 2x_1x_2 + x_2^2 + 7$$

starting from the point $(1, 2)$.

Solution. We shall follow steps of the algorithm.

Iteration 1

1. $\mathbf{x}^{(0)} = (1, 2)$; $\mathbf{A}^{(0)} = \mathbf{I}$, $k = 0$, $\varepsilon = 0.001$
   $\mathbf{c}^{(0)} = (10x_1 + 2x_2, 2x_1 + 2x_2) = (14, 6)$
2. $\|\mathbf{c}^{(0)}\| = \sqrt{14^2 + 6^2} = 15.232 > \varepsilon$, so continue
3. $\mathbf{d}^{(0)} = -\mathbf{c}^{(0)} = (-14, -6)$
4. $\mathbf{x}^{(1)} = \mathbf{x}^{(0)} + \alpha\mathbf{d}^{(0)} = (1 - 14\alpha, 2 - 6\alpha)$

$$f(\mathbf{x}^{(1)}) \equiv f(\alpha) = 5(1 - 14\alpha)^2 + 2(1 - 14\alpha)(2 - 6\alpha) + (2 - 6\alpha)^2 + 7$$

$$\frac{df}{d\alpha} = 5(2)(-14)(1 - 14\alpha) + 2(-14)(2 - 6\alpha) + 2(-6)(1 - 14\alpha) + 2(-6)(2 - 6\alpha) = 0$$

$$\alpha = 0.0988$$

$$\frac{d^2f}{d\alpha^2} = 2348 > 0$$

Therefore, step size $\alpha = 0.0988$.

5. $\mathbf{x}^{(1)} = \mathbf{x}^{(0)} + \alpha\mathbf{d}^{(0)} = (-0.386, 1.407)$
6. $\mathbf{s}^{(0)} = \alpha_0\mathbf{d}^{(0)} = (-1.386, -0.593)$
   $\mathbf{c}^{(1)} = (-1.046, 2.042)$
   $\mathbf{y}^{(0)} = \mathbf{c}^{(1)} - \mathbf{c}^{(0)} = (-15.046, -3.958)$
   $\mathbf{z}^{(0)} = \mathbf{y}^{(0)} = (-15.046, -3.958)$
   $\mathbf{s}^{(0)} \cdot \mathbf{y}^{(0)} = 23.20$
   $\mathbf{y}^{(0)} \cdot \mathbf{z}^{(0)} = 242.05$

$$\mathbf{s}^{(0)}\mathbf{s}^{(0)T} = \begin{bmatrix} 1.921 & 0.822 \\ 0.822 & 0.352 \end{bmatrix}$$

$$\mathbf{B}^{(0)} = \frac{\mathbf{s}^{(0)} \cdot \mathbf{s}^{(0)T}}{\mathbf{s}^{(0)} \cdot \mathbf{y}^{(0)}} = \begin{bmatrix} 0.0828 & 0.0354 \\ 0.0354 & 0.0152 \end{bmatrix}$$

$$\mathbf{z}^{(0)}\mathbf{z}^{0T} = \begin{bmatrix} 226.40 & 59.55 \\ 59.55 & 15.67 \end{bmatrix}$$

$$\mathbf{C}^{(0)} = -\frac{\mathbf{z}^{(0)}\mathbf{z}^{0T}}{\mathbf{y}^{(0)} \cdot \mathbf{z}^{0)}} = \begin{bmatrix} -0.935 & -0.246 \\ -0.246 & -0.065 \end{bmatrix}$$

$$\mathbf{A}^{(1)} = \mathbf{A}^{(0)} + \mathbf{B}^{(0)} + \mathbf{C}^{(0)} = \begin{bmatrix} 0.148 & -0.211 \\ -0.211 & 0.950 \end{bmatrix}$$

Iteration 2

2. $\|\mathbf{c}^{(1)}\| = 2.29 > \varepsilon$, so continue
3. $\mathbf{d}^{(1)} = -\mathbf{A}^{(1)}\mathbf{c}^{(1)} = (0.586, -1.719)$

4. Step size determination: minimize $f(\mathbf{x}^{(1)} + \alpha \mathbf{d}^{(1)})$; $\alpha = 0.776$

5. $\mathbf{x}^{(2)} = \mathbf{x}^{(1)} + \alpha \mathbf{d}^{(1)}$
$\qquad = (-0.386, 1.407) + (0.455, -1.334) = (0.069, 0.073)$

6. $\mathbf{s}^{(1)} = \alpha_1 \mathbf{d}^{(1)} = (0.455, -1.334)$
$\quad \mathbf{c}^{(2)} = (0.836, 0.284)$
$\quad \mathbf{y}^{(1)} = \mathbf{c}^{(2)} - \mathbf{c}^{(1)} = (1.882, -1.758)$
$\quad \mathbf{z}^{(1)} = \mathbf{A}^{(1)} \mathbf{y}^{(1)} = (0.649, -2.067)$
$\quad \mathbf{s}^{(1)} \cdot \mathbf{y}^{(1)} = 3.201$
$\quad \mathbf{y}^{(1)} \cdot \mathbf{z}^{(1)} = 4.855$

$$\mathbf{s}^{(1)}\mathbf{s}^{(1)T} = \begin{bmatrix} 0.207 & -0.607 \\ -0.607 & 1.780 \end{bmatrix}$$

$$\mathbf{B}^{(1)} = \frac{\mathbf{s}^{(1)}\mathbf{s}^{(1)T}}{\mathbf{s}^{(1)} \cdot \mathbf{y}^{(1)}} = \begin{bmatrix} 0.0647 & -0.19 \\ -0.19 & 0.556 \end{bmatrix}$$

$$\mathbf{z}^{(1)}\mathbf{z}^{(1)T} = \begin{bmatrix} 0.421 & -1.341 \\ -1.341 & 4.272 \end{bmatrix}$$

$$\mathbf{C}^{(1)} = -\frac{\mathbf{z}^{(1)}\mathbf{z}^{(1)T}}{\mathbf{y}^{(1)} \cdot \mathbf{z}^{(1)}} = \begin{bmatrix} -0.0867 & 0.276 \\ 0.276 & -0.880 \end{bmatrix}$$

$$\mathbf{A}^{(2)} = \mathbf{A}^{(1)} + \mathbf{B}^{(1)} + \mathbf{C}^{(1)} = \begin{bmatrix} 0.126 & -0.125 \\ -0.125 & 0.626 \end{bmatrix}$$

It can be verified that the matrix $\mathbf{A}$ is quite close to the inverse of the Hessian of the cost function. One more iteration of the DFP method will yield the optimum solution of $(0, 0)$.

$\qquad\qquad\qquad\qquad\qquad\qquad\qquad\qquad\qquad\qquad\qquad\qquad\qquad$ ‖

5.7.2*   Direct Update Methods

In these methods, the Hessian rather than its inverse is updated at every iteration. Several updating methods can be described. We shall present a method that is most popular and has proved to be most effective in applications. Detailed derivation of the method is given in Gill *et al.* [1981]. It is known as the *BFGS (Broyden-Fletcher-Goldfarb-Shanno) method* described as follows.

Step 1. Estimate an initial design $\mathbf{x}^{(0)}$. Choose a symmetric positive definite matrix $\mathbf{H}^{(0)}$ as an estimate for the Hessian of the cost function. In the absence of more information, let $\mathbf{H}^{(0)} = \mathbf{I}$. Choose a convergence parameter ε. Set $k = 0$, and compute the gradient vector as

$$\mathbf{c}^{(0)} = \nabla f(\mathbf{x}^{(0)})$$

Step 2. Calculate the norm of the gradient vector as $\|\mathbf{c}^{(k)}\|$. If $\|\mathbf{c}^{(k)}\| < \varepsilon$ then stop the iterative process; otherwise continue.

Step 3. Solve the following linear system of equations to obtain the search direction:

$$\mathbf{H}^{(k)}\mathbf{d}^{(k)} = -\mathbf{c}^{(k)}$$

Step 4. Compute optimum step size $\alpha_k = \alpha$ to minimize $f(\mathbf{x}^{(k)} + \alpha \mathbf{d}^{(k)})$.

Step 5. Update the design as

$$\mathbf{x}^{(k+1)} = \mathbf{x}^{(k)} + \alpha_k \mathbf{d}^{(k)}$$

Step 6. Update the Hessian approximation for the cost function as

$$\mathbf{H}^{(k+1)} = \mathbf{H}^{(k)} + \mathbf{D}^{(k)} + \mathbf{E}^{(k)}$$

where the correction matrices $\mathbf{D}^{(k)}$ and $\mathbf{E}^{(k)}$ are given as

$$\mathbf{D}^{(k)} = \frac{\mathbf{y}^{(k)}\mathbf{y}^{(k)^T}}{(\mathbf{y}^{(k)} \cdot \mathbf{s}^{(k)})}; \qquad \mathbf{E}^{(k)} = \frac{\mathbf{c}^{(k)}\mathbf{c}^{(k)^T}}{(\mathbf{c}^{(k)} \cdot \mathbf{d}^{(k)})}$$

with

$$\mathbf{s}^{(k)} = \alpha_k \mathbf{d}^{(k)} \qquad \text{(change in design)}$$

$$\mathbf{y}^{(k)} = \mathbf{c}^{(k+1)} - \mathbf{c}^{(k)} \qquad \text{(change in gradient)}$$

$$\mathbf{c}^{(k+1)} = \nabla f(\mathbf{x}^{(k+1)})$$

Step 7. Set $k = k + 1$ and go to Step 2.

Note again that the first iteration of the method is the same as that for the steepest descent method.

It can be shown that the BFGS update formula keeps the Hessian approximation positive definite if exact line search is used. This is important to know as the search direction is guaranteed to be that of descent for the cost function only if $\mathbf{H}^{(k)}$ is positive definite. In numerical calculation, difficulties can arise because Hessian can become singular or indefinite due to inexact line search and round-off and truncation errors. Therefore, some safeguards against the numerical difficulties must be implemented into computer programs for stable and convergent calculations. Another numerical procedure that is extremely useful is to update decomposed factors (Cholesky factors) of the Hessian rather than the Hessian itself [Gill *et al.*, 1981]. This way the matrix can be numerically guaranteed to be positive definite.

Example 5.18   Application of the BFGS method. Execute two iterations of the BFGS method for the problem

$$\text{minimize } f(\mathbf{x}) = 5x_1^2 + 2x_1x_2 + x_2^2 + 7$$

starting from the point $(1, 2)$.

Solution. We shall follow steps of the algorithm. Note that the first iteration gives steepest descent step for the cost function.

Iteration 1

1. $\mathbf{x}^{(0)} = (1, 2)$, $\mathbf{H}^{(0)} = \mathbf{I}$, $\varepsilon = 0.001$, $k = 0$
   $\mathbf{c}^{(0)} = (10x_1 + 2x_2, 2x_1 + 2x_2) = (14, 6)$
   $\|\mathbf{c}^{(0)}\| = \sqrt{14^2 + 6^2} = 15.232 > \varepsilon$, so continue
3. $\mathbf{d}^{(0)} = -\mathbf{c}^{(0)} = (-14, -6)$; since $\mathbf{H}^{(0)} = \mathbf{I}$
4. Step size determination – same as Example 5.17; $\alpha = 0.099$

5. $\mathbf{x}^{(1)} = \mathbf{x}^{(0)} + \alpha \mathbf{d}^{(0)}$
   $= (-0.386, 1.407)$

6. $\mathbf{s}^{(0)} = \alpha_0 \mathbf{d}^{(0)} = (-1.386, -0.593)$
   $\mathbf{c}^{(1)} = (-1.046, 2.042)$
   $\mathbf{y}^{(0)} = \mathbf{c}^{(1)} - \mathbf{c}^{(0)} = (-15.046, -3.958)$
   $\mathbf{y}^{(0)} \cdot \mathbf{s}^{(0)} = 23.20$
   $\mathbf{c}^{(0)} \cdot \mathbf{d}^{(0)} = -232.0$

$$\mathbf{y}^{(0)}\mathbf{y}^{(0)T} = \begin{bmatrix} 226.40 & 59.55 \\ 59.55 & 15.67 \end{bmatrix}$$

$$\mathbf{D}^{(0)} = \frac{\mathbf{y}^{(0)}\mathbf{y}^{(0)T}}{\mathbf{y}^{(0)} \cdot \mathbf{s}^{(0)}} = \begin{bmatrix} 9.760 & 2.567 \\ 2.567 & 0.675 \end{bmatrix}$$

$$\mathbf{c}^{(0)}\mathbf{c}^{(0)T} = \begin{bmatrix} 196 & 84 \\ 84 & 36 \end{bmatrix}$$

$$\mathbf{E}^{(0)} = \frac{\mathbf{c}^{(0)}\mathbf{c}^{(0)T}}{\mathbf{c}^{(0)} \cdot \mathbf{d}^{(0)}} = \begin{bmatrix} -0.845 & -0.362 \\ -0.362 & -0.155 \end{bmatrix}$$

$$\mathbf{H}^{(1)} = \mathbf{H}^{(0)} + \mathbf{D}^{(0)} + \mathbf{E}^{(0)} = \begin{bmatrix} 9.915 & 2.205 \\ 2.205 & 0.520 \end{bmatrix}$$

Iteration 2 ($k = 1$)

2. $\|\mathbf{c}^{(1)}\| = 2.29 > \varepsilon$, so continue

3. $\mathbf{H}^{(1)}\mathbf{d}^{(1)} = -\mathbf{c}^{(1)}$; or, $\mathbf{d}^{(1)} = (17.20, -76.77)$

4. Step size determination: $\alpha = 0.018\ 455$

5. $\mathbf{x}^{(2)} = \mathbf{x}^{(1)} + \alpha \mathbf{d}^{(1)}$
   $= (-0.0686, -0.0098)$

6. $\mathbf{s}^{(1)} = \alpha_1 \mathbf{d}^{(1)} = (0.317, -1.417)$
   $\mathbf{c}^{(2)} = (-0.706, -0.157)$
   $\mathbf{y}^{(1)} = \mathbf{c}^{(2)} - \mathbf{c}^{(1)}$
   $= (0.340, -2.199)$
   $\mathbf{y}^{(1)} \cdot \mathbf{s}^{(1)} = 3.224$
   $\mathbf{c}^{(1)} \cdot \mathbf{d}^{(1)} = -174.76$

$$\mathbf{y}^{(1)}\mathbf{y}^{(1)T} = \begin{bmatrix} 0.1156 & -0.748 \\ -0.748 & 4.836 \end{bmatrix}$$

$$\mathbf{D}^{(1)} = \frac{\mathbf{y}^{(1)}\mathbf{y}^{(1)T}}{\mathbf{y}^{(1)} \cdot \mathbf{s}^{(1)}} = \begin{bmatrix} 0.036 & -0.232 \\ -0.232 & 1.500 \end{bmatrix}$$

$$\mathbf{c}^{(1)}\mathbf{c}^{(1)T} = \begin{bmatrix} 1.094 & -2.136 \\ -2.136 & 4.170 \end{bmatrix}$$

$$\mathbf{E}^{(1)} = \frac{\mathbf{c}^{(1)}\mathbf{c}^{(1)T}}{\mathbf{c}^{(1)} \cdot \mathbf{d}^{(1)}} = \begin{bmatrix} -0.0063 & 0.0122 \\ 0.0122 & -0.0239 \end{bmatrix}$$

$$\mathbf{H}^{(2)} = \mathbf{H}^{(1)} + \mathbf{D}^{(1)} + \mathbf{E}^{(1)} = \begin{bmatrix} 9.945 & 1.985 \\ 1.985 & 1.996 \end{bmatrix}$$

It can be verified that $\mathbf{H}^{(2)}$ is quite close to the Hessian of the given cost function. One more iteration of the BFGS method will yield the optimum solution of $(0, 0)$.

$\|$

5.8 ENGINEERING APPLICATIONS OF UNCONSTRAINED METHODS

There are several engineering problems where unconstrained optimization methods have been used. For example, linear as well as nonlinear simultaneous equations can be solved with unconstrained optimization methods; response of structural and mechanical systems can be computed by minimization of the total potential energy. These problems can be formulated and solved by unconstrained methods. Some of the methods have also been incorporated into the commercially available codes. In addition, unconstrained optimization methods can be used to solve constrained problems. This is described in Section 5.9. In this section, we describe two applications of the unconstrained methods.

5.8.1* Minimization of Total Potential Energy

The equilibrium states of structural and mechanical systems are characterized by the stationary points of the total potential energy of the system. This is known as the *principle of stationary potential energy*. If at a stationary point the potential energy actually has a minimum value, the equilibrium state is called stable. In structural mechanics, these principles are of fundamental importance and form the basis for many numerical methods for structural analysis.

To demonstrate the principle, we consider the symmetric two-bar truss shown in Fig. 5.16. The structure is subjected to a load of W at the node C. Under the action of this load, node C moves to a point C'. The problem is to compute the displacements x_1 and x_2 of node C. This can be done by writing the total potential energy of the truss in terms of x_1 and x_2 and then minimizing it. Once displacements x_1 and x_2 are known, member forces and stresses can be calculated using them. Let

$E =$ modulus of elasticity (N/m^2) (this is the property of a material which relates stresses in the material to strains).

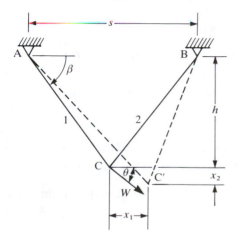

FIGURE 5.16
Two-bar truss.

s = span of the truss (m).
h = height of the truss (m).
A_1 = cross-sectional area of member 1 (m²).
A_2 = cross-sectional area of member 2 (m²).
θ = angle at which load W is applied (degrees).
L = length of the members $(L^2 = h^2 + 0.25s^2)$ (m).
W = load (N).
x_1 = horizontal displacement (m).
x_2 = vertical displacement (m).

Thus the total potential energy of the system, assuming small displacements is given as

$$P(x_1, x_2) = \frac{EA_1}{2L}(x_1 \cos \beta + x_2 \sin \beta)^2 + \frac{EA_2}{2L}(-x_1 \cos \beta + x_2 \sin \beta)^2$$
$$- Wx_1 \cos \theta - Wx_2 \sin \theta, \qquad \text{N} \cdot \text{m} \qquad \text{(a)}$$

where the angle β is shown in Fig. 5.16. Minimization of P with respect to x_1 and x_2 gives the displacements x_1 and x_2 for the equilibrium state of the two-bar truss.

Example 5.19   Minimization of total potential energy of a two-bar truss. For the two-bar truss problem use the following numerical data: $A_1 = A_2 = 1.0\text{E} - 05 \text{ m}^2$, $h = 1.0 \text{ m}$, $s = 1.5 \text{ m}$, $W = 10 \text{ kN}$, $\theta = 30°$, $E = 207 \text{ GPa}$. Minimize the total potential energy given in Eq. (a) by (i) the graphical method, (ii) the analytical method, and (iii) the conjugate gradient method.

Solution. Substituting the above data into Eq. (a) and simplifying, we get (note that $\cos \beta = s/2L$ and $\sin \beta = h/L$):

$$P(x_1, x_2) = \frac{EA}{L}(s/2L)^2 x_1^2 + \frac{EA}{L}(h/L)^2 x_2^2 - Wx_1 \cos \theta - Wx_2 \sin \theta$$
$$= (5.962\text{E}+06)x_1^2 + (1.0598\text{E}+06)x_2^2 - (8.66\text{E}+03)x_1 - (5.00\text{E}+03)x_2, \qquad \text{N} \cdot \text{m}$$

Iso-cost contours for the function are shown in Fig. 5.17. The optimum solution from the graph is read as

$$x_1 = (7.2634\text{E}{-}03) \text{ m}; \qquad x_2 = (2.3359\text{E}{-}03) \text{ m}; \qquad P = -37.348 \text{ N} \cdot \text{m}$$

Using the necessary conditions, we get

$$2(5.962\text{E}+06)x_1 - (8.66\text{E}+03) = 0, \qquad x_1 = (7.2629\text{E}{-}03) \text{ m}$$
$$2(1.0598\text{E}+06)x_2 - (5.00\text{E}+03) = 0, \qquad x_2 = (2.3589\text{E}{-}03) \text{ m}$$

The conjugate gradient method given in IDESIGN also converges to the same solution. ‖

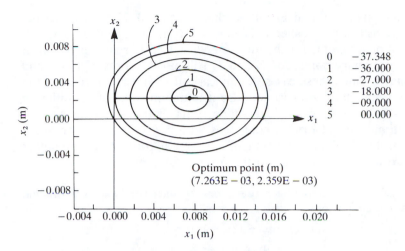

0	-37.348
1	-36.000
2	-27.000
3	-18.000
4	-09.000
5	00.000

Optimum point (m)
(7.263E - 03, 2.359E - 03)

FIGURE 5.17
Iso-cost contours of the potential energy function $P(x_1, x_2)$ for a two-bar truss ($P = 0$, -9.0, -18.0, -27.0, -36.0 and -37.348, N.m).

5.8.2 Solution of Nonlinear Equations

Unconstrained optimization methods can be used to find roots of a nonlinear system of equations. To demonstrate this, we consider the following 2×2 system:

$$F_1(x_1, x_2) = 0 \tag{a}$$

$$F_2(x_1, x_2) = 0 \tag{b}$$

We define a function which is the sum of the squares of the functions F_1 and F_2 as

$$f(x_1, x_2) = F_1^2(x_1, x_2) + F_2^2(x_1, x_2) \tag{c}$$

Note that if x_1 and x_2 are roots of Eqs (a) and (b), then $f = 0$ in Eq. (c). If x_1 and x_2 are not roots then the function $f > 0$ represents the sum of the squares of the errors in the equations $F_1 = 0$ and $F_2 = 0$. Thus, the optimization problem is to find x_1 and x_2 to minimize the function $f(x_1, x_2)$ of Eq. (c). We need to show that the necessary conditions for minimization of $f(\mathbf{x})$ give roots for the nonlinear system of equations. The necessary conditions give

$$\frac{\partial f}{\partial x_1} \equiv 2F_1 \frac{\partial F_1}{\partial x_1} + 2F_2 \frac{\partial F_2}{\partial x_1} = 0 \tag{d}$$

$$\frac{\partial f}{\partial x_2} = 2F_1 \frac{\partial F_1}{\partial x_2} + 2F_2 \frac{\partial F_2}{\partial x_2} = 0 \tag{e}$$

Note that the necessary conditions are satisfied if $F_1 = F_2 = 0$, i.e. x_1 and x_2 are roots of the equations $F_1 = 0$ and $F_2 = 0$. At this point $f = 0$. Note also that the necessary conditions can be satisfied if $\partial F_i / \partial x_j = 0$ for $i, j = 1, 2$. If $\partial F_i / \partial x_j = 0$, x_1 and x_2 are stationary points for the functions F_1 and F_2. For most problems it is unlikely that stationary points for F_1 and F_2 will also be roots of $F_1 = 0$ and $F_2 = 0$, so we may exclude these cases. In any case, if x_1 and x_2 are roots of the equations, then f must have zero value. Also if optimum value of f is different from zero ($f \neq 0$), then x_1 and x_2 cannot be roots of the nonlinear system. Thus if the optimization algorithm converges with $f \neq 0$, then the optimum point for the problem of minimization of f is not a root of the nonlinear system. The algorithm should be re-started from a different point.

Example 5.20   Roots of nonlinear equations by unconstrained minimization. Find roots of the equations

$$F_1(\mathbf{x}) \equiv 3x_1^2 + 12x_2^2 + 10x_1 = 0$$
$$F_2(\mathbf{x}) \equiv 24x_1 x_2 + 4x_2 + 3 = 0$$

Solution. We define the error function $f(\mathbf{x})$ as

$$f(\mathbf{x}) = F_1^2 + F_2^2 = (3x_1^2 + 12x_2^2 + 10x_1)^2 + (24x_1 x_2 + 4x_2 + 3)^2$$

To minimize this function, we can use any of the methods discussed previously. Table 5.10 shows the iteration history with the conjugate gradient method available in the IDESIGN software system [Arora and Tseng, 1987]. A root of the equations is $x_1 = -0.3980$, $x_2 = 0.5404$ starting from the point $(-1, 1)$. Starting from the point $(-50, 50)$ another root is found as $(-3.331, 0.039\,48)$. However, starting from another point $(2, 3)$, the program converges to $(0.020\,63, -0.2812)$ with $f = 4.351$. Since $f \neq 0$, this point is not a root of the given system of equations. When this happens, we start from a different point and re-solve the problem.   ‖

Note that the preceding procedure can be generalized to a system of n equations in n unknowns. In this case, the error function $f(\mathbf{x})$ will be defined as

$$f(\mathbf{x}) = \sum_{i=1}^{n} F_i^2(\mathbf{x})$$

TABLE 5.10
Root of nonlinear equations of Example 5.20: the conjugate gradient method

No.	x_1	x_2	F_1	F_2	f
0	−1.0000	1.0000	5.0000	−17.0000	314.0000
1	−0.5487	0.4649	−1.9900	−1.2626	5.5530
2	−0.4147	0.5658	0.1932	−0.3993	0.1968
3	−0.3993	0.5393	−0.0245	−0.0110	7.242E − 4
4	−0.3979	0.5403	−9.377E − 4	−1.550E − 3	2.759E − 6
5	−0.3980	0.5404	−4.021E − 4	−3.008E − 4	1.173E − 8

5.9* TRANSFORMATION METHODS FOR OPTIMUM DESIGN

It turns out that unconstrained optimization methods can be used to solve constrained problems. The basic idea is to construct a composite function using the cost and constraint functions. It also contains certain parameters – called the penalty parameters – that penalize the composite function for violation of constraints. The larger the violation, the larger is the penalty. Once the composite function is defined for a set of penalty parameters, it is minimized using any of the unconstrained optimization techniques. The penalty parameters are then adjusted based on certain conditions and the composite function is redefined and minimized. The process is continued until there is no significant improvement in the estimate for the optimum point.

Methods based on the foregoing philosophy have been generally called Sequential Unconstrained Minimization Techniques, or in short SUMT. It can be seen that the basic idea of SUMT is quite straightforward. Because of their simplicity, there was considerable enthusiasm for using the methods in engineering design during the 1960s and 1970s. However, more efficient constrained methods have been developed in the late 1970s and 1980s, and they have superseded SUMTs. More recently, however, there have been attempts to revive the methods and research is continuing to further refine the methods. A very brief discussion of the basic concepts and philosophy of the methods is included in the text to give the students a flavor for the techniques. For more detailed presentations, texts by Gill *et al.* [1981], Reklaitis, Ravindran and Ragsdell [1983], and Haftka and Kamat [1985] should be consulted.

The term "transformation method" is used to describe any method that solves the constrained optimization problem by transforming it into one or more unconstrained problems. They include the so-called penalty (exterior) and barrier (interior) function methods as well as multiplier (augmented Lagrangian) methods. To remind the reader of the original constrained problem that we are trying to solve, we re-state it as follows: Find an n-vector $\mathbf{x} = (x_1, x_2, \ldots, x_n)$ to minimize a cost function

$$f = f(\mathbf{x}) \tag{5.47}$$

subject to the equality constraints

$$h_i(\mathbf{x}) = 0; \quad i = 1 \text{ to } p \tag{5.48}$$

and the inequality constraints

$$g_i(\mathbf{x}) \leqq 0; \quad i = 1 \text{ to } m \tag{5.49}$$

All transformation methods convert the constrained optimization problem defined in Eqs (5.47) to (5.49) into an unconstrained problem for the *transformation function*:

$$\phi(\mathbf{x}, \mathbf{r}) = f(\mathbf{x}) + P(\mathbf{h}(\mathbf{x}), \mathbf{g}(\mathbf{x}), \mathbf{r}) \tag{5.50}$$

where $\mathbf{r}$ is a vector of controlling (penalty) parameters and P is a real valued function whose action of imposing the penalty is controlled by $\mathbf{r}$. The form of penalty function P depends on the method used. The basic procedure is to choose an initial design estimate $\mathbf{x}^{(0)}$, and define the function ϕ of Eq. (5.50). The controlling parameters $\mathbf{r}$ are also initially selected. The function ϕ is minimized for $\mathbf{x}$ keeping $\mathbf{r}$ fixed. The parameters $\mathbf{r}$ are then adjusted and the procedure is repeated until no further improvement is possible.

5.9.1 Sequential Unconstrained Minimization Techniques

Sequential unconstrained minimization techniques consist of two basically different types of penalty functions. The first one is called the *penalty function* method and the second is called the *barrier function* method. The basic idea of the penalty function approach is to define the function P in Eq. (5.50) in such a way that if there are constraint violations, a larger penalty gets added to the cost function. Several penalty functions can be defined.

The most popular one is called the quadratic loss function defined as

$$P(\mathbf{h}(\mathbf{x}), \mathbf{g}(\mathbf{x}), r) = r \left\{ \sum_{i=1}^{p} [h_i(\mathbf{x})]^2 + \sum_{i=1}^{m} [g_i^+(\mathbf{x})]^2 \right\} \qquad (5.51)$$

where $g_i^+(\mathbf{x}) = \max(0, g_i(\mathbf{x}))$, and r is a scalar. Note that $g_i^+(\mathbf{x}) \geq 0$; it is zero if inequality is inactive ($g_i(\mathbf{x}) < 0$) and it is positive if inequality is violated. It can be seen that if the equality constraint is not satisfied, i.e. $h_i(\mathbf{x}) \neq 0$, or inequality is violated, i.e. $g_i(\mathbf{x}) > 0$, Eq. (5.51) gives a positive value to the function P, and the cost function is penalized, as seen in Eq. (5.50). The starting design point for the method can be arbitrary. The methods based on the philosophy of penalty functions are sometimes called the *exterior methods* because they iterate through the infeasible region.

The *barrier function methods* are applicable only to the inequality constrained problems. Popular barrier functions are

1. Inverse Barrier Function

$$P(\mathbf{g}(\mathbf{x}), r) = (1/r) \sum_{i=1}^{m} [-1/g_i(\mathbf{x})] \qquad (5.52)$$

2. Log Barrier Function

$$P(\mathbf{g}(\mathbf{x}), r) = (-1/r) \sum_{i=1}^{m} \log(-g_i(\mathbf{x})) \qquad (5.53)$$

These are called the barrier function methods because a large barrier is constructed around the feasible region. In fact the function P becomes infinite if any of the inequalities is active. Thus when the iterative process is started from a feasible point, it cannot go into the infeasible region because the iterative process cannot cross the huge barrier.

For both methods, it can be shown that as $r \to \infty$, $\mathbf{x}(r) \to \mathbf{x}^*$ where $\mathbf{x}(r)$ is the minimum of the transformed function $\phi(\mathbf{x}, r)$ of Eq. (5.50) and $\mathbf{x}^*$ a solution of the original constrained optimization problem.

The advantages and disadvantages of the penalty function method are:

1. It is applicable to general constrained problems, i.e. equalities as well as inequalities can be treated.
2. The starting design point can be arbitrary.
3. The method iterates through the infeasible region where the problem functions may be undefined.
4. If the iterative process terminates prematurely, the final design may not be feasible and hence not usable.

The advantages and disadvantages of the barrier function method are:

1. The method is applicable to inequality constrained problems only.
2. The starting design point must be feasible. It turns out, however, that the method itself can be used to determine the starting point [Haug and Arora, 1979].
3. The method always iterates through the feasible region, so if it terminates prematurely, the final design is feasible and hence usable.

The above two methods are referred to as the Sequential Unconstrained Minimization Technique or SUMT in the literature [Fiacco and McCormick, 1968]. These methods have certain weaknesses that are most serious when r is large. The penalty and barrier functions tend to be ill-behaved near the boundary of the feasible region (constraint set) where the optimum points usually lie. There is also a problem of selecting the sequence $r^{(k)}$. The choice of $r^{(0)}$ and the rate at which $r^{(k)}$ tends to infinity can seriously affect the computational effort to find a solution. Furthermore, the Hessian matrix of the unconstrained function becomes ill-conditioned as $r \to \infty$. Some of these difficulties have been overcome with the use of *extended-barrier functions* [Haftka and Kamat, 1985].

5.9.2 Multiplier (Augmented Lagrangian) Methods

To alleviate some of the difficulties of the methods presented in the previous section, a different class of transformation methods have been developed in the literature. These are called the *multiplier or augmented Lagrangian methods*. In these methods, there is no need for the controlling parameters r to go to infinity. As a result the transformation function ϕ has good conditioning with no singularities. The multiplier methods are convergent as are the SUMTs. That is, they converge to a local minimum starting from any point. They have been proved to possess a faster rate of convergence than the previous two methods in Section 5.9.1. In multiplier methods, the penalty function is given as

$$P(\mathbf{h}(\mathbf{x}), \mathbf{g}(\mathbf{x}), \mathbf{r}, \boldsymbol{\theta}) = \tfrac{1}{2} \sum_{i=1}^{p} r_i'(h_i + \theta_i')^2 + \tfrac{1}{2} \sum_{i=1}^{m} r_i[(g_i + \theta_i)^+]^2 \qquad (5.54)$$

where $\theta_i > 0$, $r_i > 0$, θ_i' and r_i' are parameters associated with the ith inequality and equality constraints. If $\theta_i = \theta_i' = 0$, $r_i = r_i' = r$, then Eq. (5.54) reduces to the well-known quadratic loss function given in Eq. (5.51), where convergence is enforced by letting $r \to \infty$. However, the objective of the multiplier methods is to keep each r_i and r_i' finite. The idea of multiplier methods is to start with some r_i, r_i', θ_i' and θ_i and minimize the transformation function of Eq. (5.50). The parameters r_i, r_i', θ_i' and θ_i are then adjusted using some rule and the entire process is repeated until optimality conditions are satisfied. For a more detailed discussion and applications of the methods, the reader should consult Belegundu and Arora [1984b] and the references cited earlier.

EXERCISES FOR CHAPTER 5

Section 5.2   General Concepts Related to Numerical Algorithms

5.1 *Answer True or False*

1. All optimum design algorithms require a starting point to initiate the iterative process.
2. A vector of design changes must be computed at each iteration of the iterative process.
3. The design change calculation can be divided into step size determination and direction finding subproblems.
4. The search direction requires evaluation of gradient of the cost function.
5. Step size along the search direction is always negative.
6. Step size along the search direction can be zero.
7. In unconstrained optimization, cost function can increase for an arbitrary small step along the descent direction.
8. A descent direction always exists, if the current point is not a local minimum.
9. In unconstrained optimization, a direction of descent can be found at a point where the gradient of the cost function is zero.
10. The descent direction makes an angle of 0–90° with the gradient of the cost function.

Determine if the given direction at the point is that of descent for the following functions (show all the calculations):

5.2 $f(\mathbf{x}) = 3x_1^2 + 2x_1 + 2x_2^2 + 7;$ $\quad \mathbf{d} = (-1, 1)$ at $\mathbf{x} = (2, 1)$

5.3 $f(\mathbf{x}) = x_1^2 + x_2^2 - 2x_1 - 2x_2 + 4;$ $\quad \mathbf{d} = (2, 1)$ at $\mathbf{x} = (1, 1)$

5.4 $f(\mathbf{x}) = x_1^2 + 2x_2^2 + 2x_3^2 + 2x_1x_2 + 2x_2x_3;$ $\quad \mathbf{d} = (-3, 10, -12)$ at $\mathbf{x} = (1, 2, 3)$

5.5 $f(\mathbf{x}) = 0.1x_1^2 + x_2^2 - 10;$ $\quad \mathbf{d} = (1, 2)$ at $\mathbf{x} = (4, 1)$

5.6 $f(\mathbf{x}) = (x_1 - 2)^2 + (x_2 - 1)^2;$ $\quad \mathbf{d} = (2, 3)$ at $\mathbf{x} = (4, 3)$

5.7 $f(\mathbf{x}) = 10(x_2 - x_1^2)^2 + (1 - x_1)^2;$ $\quad \mathbf{d} = (162, -40)$ at $\mathbf{x} = (2, 2)$

5.8 $f(\mathbf{x}) = (x_1 - 2)^2 + x_2^2;$ $\quad \mathbf{d} = (-2, 2)$ at $\mathbf{x} = (1, 1)$

5.9 $f(\mathbf{x}) = 0.5x_1^2 + x_2^2 - x_1x_2 - 7x_1 - 7x_2;$ $\quad \mathbf{d} = (7, 6)$ at $\mathbf{x} = (1, 1)$

5.10 $f(\mathbf{x}) = (x_1 + x_2)^2 + (x_2 + x_3)^2;$ $\quad \mathbf{d} = (4, 8, 4)$ at $\mathbf{x} = (1, 1, 1)$

5.11 $f(\mathbf{x}) = x_1^2 + x_2^2 + x_3^2;$ $\quad \mathbf{d} = (2, 4, -2)$ at $\mathbf{x} = (1, 2, -1)$

5.12 $f(\mathbf{x}) = (x_1 + 3x_2 + x_3)^2 + 4(x_1 - x_2)^2;$ $\quad \mathbf{d} = (-2, -6, -2)$ at $\mathbf{x} = (-1, -1, -1)$

5.13 $f(\mathbf{x}) = 9 - 8x_1 - 6x_2 - 4x_3 + 2x_1^2 + 2x_2^2 + x_3^2 + 2x_1x_2 + 2x_2x_3;$
$\quad \mathbf{d} = (-2, 2, 0)$ at $\mathbf{x} = (1, 1, 1)$

5.14 $f(\mathbf{x}) = (x_1 - 1)^2 + (x_2 - 2)^2 + (x_3 - 3)^2 + (x_4 - 4)^2;$
$\quad \mathbf{d} = (2, -2, 2, -2)$ at $\mathbf{x} = (2, 1, 4, 3)$

Section 5.3 One-dimensional Minimization

5.15 *Answer True or False*

1. Step size determination is always a one-dimensional problem.
2. In unconstrained optimization, the slope of the cost function along the descent direction at zero step size is always positive.
3. The optimum step lies outside the interval of uncertainty.
4. After initial bracketing, the Golden Section search requires two function evaluations to reduce the interval of uncertainty.

5.16 Find the minimum of the function $f(\alpha) = 7\alpha^2 - 20\alpha + 22$ using the Equal Interval search method within an accuracy of 0.001. Use $\delta = 0.05$.

5.17 For the function $f(\alpha) = 7\alpha^2 - 20\alpha + 22$, use the Golden Section method to find the minimum within an accuracy of 0.005 (final interval of uncertainty should be less than 0.005). Use $\delta = 0.05$.

5.18 Write a FORTRAN program to implement the alternate Equal Interval search process shown in Fig. 5.7 for any given function $f(\alpha)$. For the function $f(\alpha) = 2 - 4\alpha + e^\alpha$, use your FORTRAN program to find the minimum within an accuracy of 0.001. Use $\delta = 0.50$.

5.19 Consider the function $f(x_1, x_2, x_3) = x_1^2 + 2x_2^2 + 2x_3^2 + 2x_1x_2 + 2x_2x_3$. Verify whether the vector $\mathbf{d} = (-12, -40, -48)$ at the point $(2, 4, 10)$ is a descent direction for f. What is the slope of the function at the given point? Find an optimum step size along $\mathbf{d}$ by any numerical method.

5.20 Consider the function $f = x_1^2 + x_2^2 - 2x_1 - 2x_2 + 4$. At the point $(1, 1)$, let a search direction be defined as $\mathbf{d} = (1, 2)$. Express f as a function of one variable at the given point along $\mathbf{d}$. Find an optimum step size along $\mathbf{d}$ analytically.

For the following functions, direction of change at a point is given. Derive the function of one variable (line search function) that can be used to determine optimum step size (show all the calculations):

5.21 $f(\mathbf{x}) = 0.1x_1^2 + x_2^2 - 10$; $\mathbf{d} = (-1, -2)$ at $\mathbf{x} = (5, 1)$

5.22 $f(\mathbf{x}) = (x_1 - 2)^2 + (x_2 - 1)^2$; $\mathbf{d} = (-4, -6)$ at $\mathbf{x} = (4, 4)$

5.23 $f(\mathbf{x}) = 10(x_2 - x_1^2)^2 + (1 - x_1)^2$; $\mathbf{d} = (-162, 40)$ at $\mathbf{x} = (2, 2)$

5.24 $f(\mathbf{x}) = (x_1 - 2)^2 + x_2^2$; $\mathbf{d} = (2, -2)$ at $\mathbf{x} = (1, 1)$

5.25 $f(\mathbf{x}) = 0.5x_1^2 + x_2^2 - x_1x_2 - 7x_1 - 7x_2$; $\mathbf{d} = (7, 6)$ at $\mathbf{x} = (1, 1)$

5.26 $f(\mathbf{x}) = (x_1 + x_2)^2 + (x_2 + x_3)^2$; $\mathbf{d} = (-4, -8, -4)$ at $\mathbf{x} = (1, 1, 1)$

5.27 $f(\mathbf{x}) = x_1^2 + x_2^2 + x_3^2$; $\mathbf{d} = (-2, -4, 2)$ at $\mathbf{x} = (1, 2, -1)$

5.28 $f(\mathbf{x}) = (x_1 + 3x_2 + x_3)^2 + 4(x_1 - x_2)^2$; $\mathbf{d} = (1, 3, 1)$ at $\mathbf{x} = (-1, -1, -1)$

5.29 $d(\mathbf{x}) = 9 - 8x_1 - 6x_2 - 4x_3 + 2x_1^2 + 2x_2^2 + x_3^2 + 2x_1x_2 + 2x_2x_3$;
     $\mathbf{d} = (2, -2, 0)$ at $\mathbf{x} = (1, 1, 1)$

5.30 $f(\mathbf{x}) = (x_1 - 1)^2 + (x_2 - 2)^2 + (x_3 - 3)^2 + (x_4 - 4)^2$;
     $\mathbf{d} = (-2, 2, -2, 2)$ at $\mathbf{x} = (2, 1, 4, 3)$

For the following problems, calculate the optimum step size α^* using the Equal Interval search with $\delta = 0.05$ and $\varepsilon = 0.001$ at the given point and the search direction (show all calculations):

5.31 Exercise 5.21   **5.32** Exercise 5.22

5.33 Exercise 5.23   **5.34** Exercise 5.24

5.35 Exercise 5.25   **5.36** Exercise 5.26

5.37 Exercise 5.27   **5.38** Exercise 5.28

5.39 Exercise 5.29   **5.40** Exercise 5.30

For the following problems, calculate the optimum step size α^* using the Golden Section search with $\delta = 0.05$, $\varepsilon = 0.001$ at the given point and the search direction (show all calculations):

5.41 Exercise 5.21   **5.42** Exercise 5.22

5.43 Exercise 5.23   **5.44** Exercise 5.24

5.45 Exercise 5.25   **5.46** Exercise 5.26

5.47 Exercise 5.27   **5.48** Exercise 5.28

5.49 Exercise 5.29   **5.50** Exercise 5.30

5.51 Write a Fortran program to implement the polynomial interpolation with a quadratic curve fitting. Choose a function $f(\alpha) = 7\alpha^2 - 20\alpha + 22$. Use the Golden Section method to initially bracket the minimum point of $f(\alpha)$ with $\delta = 0.05$. Use your program to find the minimum point of $f(\alpha)$. Comment on the accuracy of the solution.

5.52 For the function $f(\alpha) = 7\alpha^2 - 20\alpha + 22$, use two function values, $f(0)$ and $f(\alpha_u)$, and the slope of f at $\alpha = 0$ to fit a quadratic curve. Here α_u is any upper bound on the minimum point of $f(\alpha)$. What is the estimate of the minimum point from the above quadratic curve? How many iterations will be required to find α^*? Why?

5.53 Under what situation can the polynomial interpolation approach not be used for one-dimensional minimization?

5.54 Given

$$f(\mathbf{x}) = 10 - x_1 + x_1 x_2 + x_2^2$$
$$\mathbf{x}^{(0)} = (2, 4); \qquad \mathbf{d}^{(0)} = (-1, -1)$$

For the one-dimensional search, three values of α, $\alpha_l = 0$, $\alpha_i = 2$ and $\alpha_u = 4$ are tried. Using quadratic polynomial interpolation

1. At what α is the function a minimum? Prove that this is a minimum point and not a maximum.
2. At what values of α is $f(\alpha) = 15$?

Section 5.4   Steepest Descent Method

5.55 *Answer True or False*

1. The steepest descent method is convergent.
2. The steepest descent method can converge to a local maximum point starting from a point where the gradient of the function is nonzero.
3. Steepest descent directions are orthogonal to each other.
4. Steepest descent direction is orthogonal to the cost surface.

For the following problems, complete two iterations of the steepest descent method starting from the given design point:

5.56 $f(x_1, x_2) = x_1^2 + 2x_2^2 - 4x_1 - 2x_1 x_2;$ starting design $(1, 1)$

5.57 $f(x_1, x_2) = 12.096x_1^2 + 21.504x_2^2 - 1.7321x_1 - x_2;$ starting design $(1, 1)$

5.58 $f(x_1, x_2) = 6.983x_1^2 + 12.415x_2^2 - x_1;$ starting design $(2, 1)$

5.59 $f(x_1, x_2) = 12.096x_1^2 + 21.504x_2^2 - x_2;$ starting design $(1, 2)$

5.60 $f(x_1, x_2) = 25x_1^2 + 20x_2^2 - 2x_1 - x_2;$ starting design $(3, 1)$

5.61 $f(x_1, x_2, x_3) = x_1^2 + 2x_2^2 + 2x_3^2 + 2x_1 x_2 + 2x_2 x_3;$ starting design $(1, 1, 1)$

5.62 $f(x_1, x_2) = 8x_1^2 + 8x_2^2 - 80\sqrt{x_1^2 + x_2^2 - 20x_2 + 100}$
$$ - 80\sqrt{x_1^2 + x_2^2 + 20x_2 + 100} - 5x_1 - 5x_2 $$

Starting design $(4, 6)$; the step size may be approximated or calculated using a computer program.

5.63 $f(x_1, x_2) = 9x_1^2 + 9x_2^2 - 100\sqrt{x_1^2 + x_2^2 - 20x_2 + 100}$
$$ - 64\sqrt{x_1^2 + x_2^2 + 16x_2 + 64} - 5x_1 - 41x_2 $$

Starting design $(5, 2)$; the step size may be approximated or calculated using a computer program.

5.64 $f(x_1, x_2) = 100(x_2 - x_1^2)^2 + (1 - x_1)^2;$ starting design $(5, 2)$

5.65 $f(x_1, x_2, x_3, x_4) = (x_1 - 10x_2)^2 + 5(x_3 - x_4)^2 + (x_2 - 2x_3)^4 + 10(x_1 - x_4)^4$
Let the starting design be $(1, 2, 3, 4)$.

5.66 Solve Exercises 5.56 to 5.65 using the computer program given in Appendix D for the steepest descent method.

5.67 Consider the following three functions:

$$f_1 = x_1^2 + x_2^2 + x_3^2; \qquad f_2 = x_1^2 + 10x_2^2 + 100x_3^2; \qquad f_3 = 100x_1^2 + x_2^2 + 0.1x_3^2$$

Minimize f_1, f_2 and f_3 using the program for the steepest descent method given in Appendix D. Choose the starting design to be $(1, 1, 2)$ for all functions. What do you conclude from observing the performance of the method on the foregoing functions? How would you scale the design variables for the functions f_2 and f_3 to improve the rate of convergence of the method?

5.68 Calculate the gradient of the following functions at the given points by the forward, backward and central difference approaches with a 1% change in the point and compare them with the exact gradient:

1. $f(\mathbf{x}) = 12.096x_1^2 + 21.504x_2^2 - 1.7321x_1 - x_2$ at $(5, 6)$
2. $f(\mathbf{x}) = 50(x_2 - x_1^2)^2 + (2 - x_1)^2$ at $(1, 2)$
3. $f(\mathbf{x}) = x_1^2 + 2x_2^2 + 2x_3^2 + 2x_1x_2 + 2x_2x_3$ at $(1, 2, 3)$

5.69 Consider the following optimization problem

$$\text{maximize} \sum_{i=1}^{n} u_i \frac{\partial f}{\partial x_i} \qquad (\mathbf{c} \cdot \mathbf{u})$$

subject to the constraint

$$\sum_{i=1}^{n} u_i^2 = 1$$

Here $\mathbf{u} = (u_1, u_2, \ldots, u_n)$ are components of a unit vector. Solve this optimization problem and show that $\mathbf{u}$ that maximizes the above cost function is indeed in the direction of the gradient $\mathbf{c}$.

Section 5.5   Conjugate Gradient Method

5.70 *Answer True or False*

1. The conjugate gradient method usually converges faster than the steepest descent method.
2. Conjugate directions are computed from gradients of the cost function.
3. Conjugate directions are normal to each other.
4. The conjugate direction at the kth point is orthogonal to the gradient of the cost function at the $(k + 1)$th point when exact step size is calculated.
5. The conjugate direction at the kth point is orthogonal to the gradient of the cost function at the $(k - 1)$th point.

For the following problems, complete two iterations of the conjugate gradient method:

5.71 Exercise 5.56   **5.72** Exercise 5.57
5.73 Exercise 5.58   **5.74** Exercise 5.59
5.75 Exercise 5.60   **5.76** Exercise 5.61
5.77 Exercise 5.62   **5.78** Exercise 5.63

5.79 Exercise 5.64   **5.80** Exercise 5.65

5.81 Write a FORTRAN program to implement the conjugate gradient method (actually modify the steepest descent program given in Appendix D). Solve Exercises 5.71 to 5.80 using the program.

Section 5.6   Newton's Method

5.82 *Answer True or False*

1. In Newton's method, it is always possible to calculate a search direction at any point.
2. The Newton direction is always that of descent for the cost function.
3. Newton's method is convergent starting from any point with step size as one.
4. Newton's method needs only gradient information at any point.

For the following problems, complete one iteration of the modified Newton's method; also check for the descent condition:

5.83 Exercise 5.56   **5.84** Exercise 5.57

5.85 Exercise 5.58   **5.86** Exercise 5.59

5.87 Exercise 5.60   **5.88** Exercise 5.61

5.89 Exercise 5.62   **5.90** Exercise 5.63

5.91 Exercise 5.64   **5.92** Exercise 5.65

5.93 Write a FORTRAN computer program to implement the modified Newton's algorithm. Use Equal Interval search for line search. Solve Exercises 5.83 to 5.92 using the program.

Section 5.7   Quasi-Newton Methods

5.94 *Answer True or False: for the unconstrained problems*

1. The DFP method generates approximation to the inverse of the Hessian.
2. The DFP method generates positive definite approximation to the inverse of the Hessian.
3. The DFP method always gives a direction of descent for the cost function.
4. The BFGS method generates positive definite approximation to the Hessian of the cost function.
5. The BFGS method always gives direction of descent for the cost function.
6. The BFGS method always converges to the Hessian of the cost function.

For the following problems, complete two iterations of the Davidon-Fletcher-Powell and BFGS methods:

5.95 Exercise 5.56   **5.96** Exercise 5.57

5.97 Exercise 5.58   **5.98** Exercise 5.59

5.99 Exercise 5.60 **5.100** Exercise 5.61

5.101 Exercise 5.62 **5.102** Exercise 5.63

5.103 Exercise 5.64 **5.104** Exercise 5.65

5.105 Write a computer program to implement the Davidon-Fletcher-Powell method. Solve Exercises 5.95 to 5.104 using the program.

5.106 Write a computer program to implement the BFGS method. Solve Exercises 5.95 to 5.104 using the program.

Section 5.8 Engineering Applications of Unconstrained Methods

Find the equilibrium configuration for the two-bar structure of Fig. 5.16 using the following numerical data:

5.107 $A_1 = 1.5 \text{ cm}^2$, $A_2 = 2.0 \text{ cm}^2$, $h = 100 \text{ cm}$, $s = 150 \text{ cm}$,
$W = 100\,000 \text{ N}$, $\theta = 45°$, $E = 21 \text{ MN/cm}^2$

5.108 $A_1 = 100 \text{ mm}^2$, $A_2 = 200 \text{ mm}^2$, $h = 1000 \text{ mm}$, $s = 1500 \text{ mm}$,
$W = 50\,000 \text{ N}$, $\theta = 60°$, $E = 210\,000 \text{ N/mm}^2$

Find roots of the following nonlinear equations using the conjugate gradient method:

5.109 $F(x) \equiv 3x - e^x = 0$

5.110 $F(x) \equiv \sin x = 0$

5.111 $F(x) \equiv \cos x = 0$

5.112 $F(x) \equiv \dfrac{2x}{3} - \sin x = 0$

5.113 $F_1(\mathbf{x}) \equiv 1 - \dfrac{10}{(x_1^2 x_2)} = 0$, $\qquad F_2(\mathbf{x}) \equiv 1 - \dfrac{2}{(x_1 x_2^2)} = 0$

5.114 $F_1(\mathbf{x}) \equiv 5 - \frac{1}{8}x_1 x_2 - \dfrac{1}{4x_1^2}x_2^2 = 0$, $\qquad F_2(\mathbf{x}) \equiv -\frac{1}{16}x_1^2 + \dfrac{1}{2x_1}x_2 = 0$

NUMERICAL
METHODS FOR
CONSTRAINED
OPTIMUM
DESIGN

6.1 INTRODUCTION

In the previous chapter, the constrained problem was transformed into a sequence of unconstrained problems. The solutions of the unconstrained problems converged to the solution of the original constrained problem. Any one of the unconstrained numerical optimization methods could be used to solve the transformed constrained problem.

In this chapter, we describe numerical methods – sometimes called the *primal methods* – to directly solve the original constrained optimization problem. The constrained problem we are trying to solve was formulated in Section 2.7. For convenience of reference, it is restated as: find $\mathbf{x} = (x_1, \ldots, x_n)$, a design variable vector of dimension n, to minimize a cost function

$$f = f(\mathbf{x}) \tag{6.1}$$

subject to the equality constraints

$$h_i(\mathbf{x}) = 0; \quad i = 1 \text{ to } p \tag{6.2}$$

the inequality constraints

$$g_i(\mathbf{x}) \leqq 0; \quad i = 1 \text{ to } m \tag{6.3}$$

and explicit bounds on design variables

$$x_{il} \leqq x_i \leqq x_{iu}; \qquad i = 1 \text{ to } n \tag{6.4}$$

where x_{il} and x_{iu} are respectively the smallest and largest allowed values for the ith design variable x_i. Note that the explicit design variable bound constraints of Eq. (6.4) are quite simple and easy to treat in actual numerical implementations. It is usually efficient to treat them in that manner. However, in the discussion and illustration of the numerical methods, we shall assume that they are included in the inequality constraints in Eq. (6.3). Note also that we shall discuss only the methods that can treat the general constrained problem defined in Eqs (6.1) to (6.4).

Just as for unconstrained problems, several methods have been investigated for the preceding model of general constrained optimization problems. Most methods follow the two-phase approach as before: *search direction* and *step length* determination phases. The approach followed here will be to describe the underlying *ideas* and *concepts* of the methods. A comprehensive coverage of all the methods giving their advantages and disadvantages will be avoided. Only a few simple and generally applicable methods will be described and illustrated with examples.

In Section 5.4 we described the steepest descent method for solving unconstrained optimization problems. That method is quite straightforward. It is, however, not directly applicable to constrained problems. One reason is that we must consider constraints while computing the search direction. In this chapter, we shall describe a *constrained steepest descent method* that computes the direction of design change considering local behavior of cost and constraint functions. The methods (and most others) are based on linearization of the problem about the current estimate of the optimum design. Therefore, linearization of the problem is quite important and will be discussed in detail. Once the problem has been linearized, it is natural to ask if it can be solved using linear programming methods. Therefore, we shall first describe a method that is a simple extension of the Simplex method for linear programming. Then we shall discuss extension of the steepest descent method to constrained problems.

The following is an outline of this chapter.

Section 6.2 Basic Concepts and Ideas. This section contains basic concepts and ideas, and a definition of the terms used in numerical methods for constrained optimization. The status of a constraint at a design point is defined. Active, inactive, violated and ε-active constraints (equality as well as inequalities) are defined. Normalization of constraints and its advantages are explained with examples. The ideas of a "potential constraint strategy", descent function, and convergence of algorithms are explained.

Section 6.3 Linearization of the Problem. Use of Taylor series expansion to linearize a nonlinear constrained optimization problem is a fundamental step in most numerical optimization methods. This is explained and demonstrated with examples.

Section 6.4 Sequential Linear Programming Algorithm. Once the linearized problem has been defined, the Simplex method of linear programming can be used to solve for the search direction. This is demonstrated with examples. The idea of move limits and their needs are explained. A Sequential Linear Programming (SLP) algorithm is defined. Advantages and disadvantages of the method are discussed.

Section 6.5 Quadratic Programming subproblem. The linearized subproblem is transformed to a quadratic programming (QP) subproblem having quadratic cost function and linear constraints. The advantages of the QP subproblem over an LP subproblem are explained. A method for solving QP problems, based on extensions of the Simplex method for LP is described and illustrated.

Section 6.6 Constrained Steepest Descent Method. Using the QP subproblem a constrained steepest descent algorithm is given and illustrated with examples. A descent function and descent condition are defined. A simple bisection procedure is given to determine the proper step size. The method is very general as it can treat equality as well as inequality constraints, and the initial design point can be arbitrary, i.e. it can be feasible or infeasible.

Section 6.7 Constrained Quasi-Newton Methods. In this section, the Constrained Steepest Descent algorithm is extended to include Hessian of the Lagrange function in definition of the QP subproblem. Derivation of the subproblem is given and the procedure for updating the approximate Hessian is explained. The idea of constrained quasi-Newton methods is quite simple and straightforward, but very effective in their numerical performance. The method is illustrated with an example and numerical aspects are discussed. The method is used in Chapters 7 and 8 to solve several design problems.

Section 6.8 Other Methods. Several other methods for constrained optimization problems have been developed and evaluated. In addition, many variations of the methods have been discussed. We briefly describe the basic ideas of three methods – the feasible directions, gradient projection and generalized reduced gradient – that have been used to solve some engineering design problems.

6.2 BASIC CONCEPTS AND IDEAS

6.2.1 Basic Concepts Related to Algorithms for Constrained Problems

In the *direct numerical (search) methods,* we select a design to initiate the iterative process, as for the unconstrained methods described in Chapter 5. The iterative process is continued until no further moves are possible and the optimality conditions are satisfied. Most of the general concepts of iterative numerical algorithms discussed in Section 5.2, also apply to methods for constrained optimization problems. Therefore, those concepts should be thoroughly reviewed again.

All numerical methods discussed in this chapter are based on the following iterative prescription as also given in Eqs (5.1) and (5.2) for unconstrained problems:

vector form: $\mathbf{x}^{(k+1)} = \mathbf{x}^{(k)} + \Delta\mathbf{x}^{(k)};$ $k = 0, 1, 2, \ldots$ (6.5)

component form: $x_i^{(k+1)} = x_i^{(k)} + \Delta x_i^{(k)};$ $k = 0, 1, 2, \ldots$ (6.6)

$$i = 1 \text{ to } n$$

The superscript k represents the iteration or design cycle number, subscript i refers to the ith design variable, $\mathbf{x}^{(0)}$ is the starting design estimate, and $\Delta\mathbf{x}^{(k)}$ represents a small change in the current design. As in the unconstrained numerical methods, the change in design $\Delta\mathbf{x}^{(k)}$ is decomposed as

$$\Delta\mathbf{x}^{(k)} = \alpha_k \mathbf{d}^{(k)} \tag{6.7}$$

where α_k is a step size in the search direction $\mathbf{d}^{(k)}$. Thus, the design improvement involves the solution of the search direction and step size determination subproblems. Solution of both the subproblems can involve values of cost and constraint functions as well as their gradients at the current design point.

Conceptually, algorithms for unconstrained and constrained optimization problems are based on the same iterative philosophy. There is one important difference, however; constraints must be considered while determining the search direction as well as the step size for the constrained problems. A different procedure for determining either one can give a different optimization algorithm. We shall describe, in general terms, a couple of ways in which the algorithms may proceed in the design space.

All algorithms need a design estimate to initiate the iterative process. The starting design can be feasible or infeasible. If it is inside the feasible region as Point A in Fig. 6.1, then there are two possibilities:

1. The gradient of the cost function vanishes at the point, so it is an unconstrained stationary point. We need to check the sufficient condition for optimality of the point.
2. If the current point is not stationary, then we can reduce the cost function by moving along a descent direction, say, the steepest descent direction $(-\mathbf{c})$ as shown in Fig. 6.1. We continue such iterations until either a constraint is encountered or unconstrained minimum point is reached.

For the remaining discussion, we assume that the optimum point is on the boundary of the constraint set, i.e. some constraints are active at the optimum.

Once the constraint boundary is encountered at Point B, one strategy is to travel along a tangent to the boundary such as direction B–C in Fig. 6.1. This results in an infeasible point from where the constraints are corrected to again reach a feasible point such as D in Fig. 6.1. From there the preceding steps are repeated until the optimum point is reached. Another strategy is to

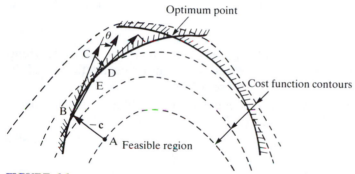

FIGURE 6.1
Conceptual steps of constrained optimization algorithms initiated from a feasible point.

deflect the tangential direction B–C towards the feasible region by certain angle θ. Then a line search is performed through the feasible region to reach the boundary point E, as shown in Fig. 6.1. The procedure is then repeated from there.

When the starting point is infeasible, as Point A in Fig. 6.2, then one strategy is to correct constraints to reach the constraint boundary at Point B. From there, the strategies described in the preceding paragraph can be followed to reach the optimum point. This is shown in Path 1 in Fig. 6.2. The second strategy is to iterate through the infeasible region by computing directions that take successive design points closer to the optimum point, shown as Path 2 in Fig. 6.2.

Several algorithms based on the strategies described in the foregoing have been developed and evaluated. Some algorithms are better for a certain class of problems than others. A few algorithms work well if the problem has only inequality constraints whereas others can treat both equality and inequality constraints simultaneously. In this text, we shall concentrate mostly on general algorithms that have no restriction on the form of the functions or

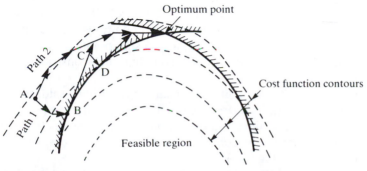

FIGURE 6.2
Conceptual steps of constrained optimization algorithms initiated from an infeasible point.

the constraints. Most of the algorithms that we shall describe will be based on the following *four basic steps*.

1. *Linearization of cost and constraint functions* about the given (current) estimate for optimum design. Some methods require linearization of all the constraints while others only for a subset of them.
2. *Definition of a search direction determination subproblem* using the linearized cost and constraint functions (either all or a subset of them).
3. *Solution of the subproblem* which gives a search direction in the design space.
4. *Calculation of a step size to minimize an appropriate descent function in the search direction* (explained in more detail in Sections 6.2.5 and 6.6.1). For the unconstrained problems, the cost function is used as the descent function. For constrained problems several descent functions have been used including the cost function. Some descent functions augment the cost function with penalty terms that include the effect of constraint violations.

6.2.2   Constraint Status at a Design Point

An inequality constraint can be either active, ε-active, violated or inactive at a design point. On the other hand, an equality constraint is either active or violated at a design point. The precise definitions of the status of a constraint at a design point are needed in the development and discussion of numerical methods.

   Active Constraint. An inequality constraint $g_i(\mathbf{x}) \leqq 0$ is said to be *active* (or *tight*) at a design point $\mathbf{x}^{(k)}$ if it is satisfied as an equality at that point, i.e. $g_i(\mathbf{x}^{(k)}) = 0$.

   Inactive constraint. An inequality constraint $g_i(\mathbf{x}) \leqq 0$ is said to be *inactive* at a design point $\mathbf{x}^{(k)}$ if it has negative value at that point, i.e. $g_i(\mathbf{x}^{(k)}) < 0$.

   Violated Constraint. An *inequality* constraint $g_i(\mathbf{x}) \leqq 0$ is said to be *violated* at a design point $\mathbf{x}^{(k)}$ if it has positive value there, i.e. $g_i(\mathbf{x}^{(k)}) > 0$. An *equality* constraint $h_i(\mathbf{x}) = 0$ is *violated* at a design point $\mathbf{x}^{(k)}$ if it has nonzero value there, i.e. $h_i(\mathbf{x}^{(k)}) \neq 0$. Note that by these definitions, an equality constraint is always either active or violated for any design point.

   We now introduce the ε-active constraints. To define such constraints, we select a small positive number ε.

   ε-Active Constraint. Any inequality constraint $g_i(\mathbf{x}) \leqq 0$ is said to be ε-active at the point $\mathbf{x}^{(k)}$ if $g_i(\mathbf{x}^{(k)}) < 0$ but $g_i(\mathbf{x}^{(k)}) + \varepsilon \geqq 0$, where $\varepsilon > 0$ is a small number. An ε-active constraint for a design point simply means that the design point is arbitrarily close to the constraint boundary on the feasible side (within the ε-band as shown in Fig. 6.3).

   To understand the idea of the status of a constraint, we refer to Fig. 6.3. Consider the ith inequality constraint $g_i(\mathbf{x}) \leqq 0$. The constraint boundary

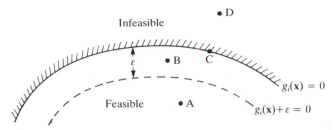

FIGURE 6.3
Status of a constraint at design points A, B, C and D.

(surface in n-dimensional space), $g_i(\mathbf{x}) = 0$, is plotted, and feasible and infeasible sides for the constraint are identified. An artificial boundary at a distance of ε from the boundary $g_i(\mathbf{x}) = 0$ and inside the feasible region is also plotted. We consider four design points A, B, C and D as shown in Fig. 6.3. For design point A, the constraint $g_i(\mathbf{x})$ is negative and even $g_i(\mathbf{x}) + \varepsilon < 0$. Thus, the constraint is *inactive* for design point A. For design point B, $g_i(\mathbf{x})$ is strictly less than zero, so it is inactive. However, $g_i(\mathbf{x}) + \varepsilon > 0$, so the constraint is ε-*active* for design point B. For design point C, $g_i(\mathbf{x}) = 0$ as shown in Fig. 6.3. Therefore, the constraint is *active* there. For design point D, $g_i(\mathbf{x})$ is greater than zero, so the constraint is *violated*.

6.2.3 Constraint Normalization

In numerical calculations, it is desirable to normalize all the constraint functions. As noted earlier, ε-active, active and violated constraints are used in computing a desirable direction of design change. Usually one value for ε (say 0.10) is used for all constraints. Since different constraints involve different orders of magnitude, it is not proper to use the same ε for all the constraints unless they are normalized. For example, consider a stress constraint as

$$\sigma \leqq \sigma_a, \quad \text{or} \quad \sigma - \sigma_a \leqq 0 \tag{6.8}$$

and a displacement constraint as

$$\delta \leqq \delta_a, \quad \text{or} \quad \delta - \delta_a \leqq 0 \tag{6.9}$$

where σ = calculated stress in a member, σ_a = an allowable stress (working stress), δ = calculated deflection at a point and δ_a = an allowable deflection.

Note that the units for the two constraints are different. Constraint of Eq. (6.8) involves stress which have units of Pascals (Pa, N/m²). For example, allowable stress for steel is 250 MPa. The other constraint in Eq. (6.9) involves deflections of the structure which may be only a few centimeters. For example, allowable deflection δ_a may be only 2 cm. Thus, the values of the two constraints are of widely differing orders of magnitude. If the constraints are violated, it is difficult to judge severity of their violation. We can, however, normalize the constraints by dividing them by their respective allowable values to obtain the normalized constraint as

$$R - 1.0 \leqq 0 \tag{6.10}$$

where $R = \sigma/\sigma_a$ for the stress constraint, and $R = \delta/\delta_a$ for the deflection constraint. Here, both σ_a and δ_a are assumed to be positive; otherwise, sense of the inequality will change. For normalized constraints, it is easy to check for ε-active constraint using the same value of ε for both of them.

There are other constraints that must be written in the form

$$1.0 - R \leqq 0 \tag{6.11}$$

when normalized with respect to their nominal value. For example, the fundamental vibration frequency ω of a structure or a structural element must be above a given threshold value of ω_a, i.e. $\omega \geqq \omega_a$. When the constraint is normalized and converted to the standard "less than" form, it is given as in Eq. (6.11) with $R = \omega/\omega_a$. In subsequent discussions, it is assumed that all equality as well as inequality constraints have been converted to the normalized form of either Eq. (6.10) or Eq. (6.11).

There are some constraints that cannot be normalized. For these constraints the allowable values may be zero. For example, the lower bound on some design variables may be zero. Such constraints cannot be normalized with respect to lower bounds as division by a zero is not possible. These constraints may be kept in the original form.

Example 6.1   Constraint normalization and status at a point. Consider the two constraints, $\bar{h} \equiv x_1^2 + x_2/2 = 18$ and $\bar{g} \equiv 500x_1 - 30\,000x_2 \leqq 0$. At the design points $(1, 1)$ and $(-4.5, -4.5)$, investigate whether the constraints are active, violated, ε-active or inactive. Use $\varepsilon = 0.1$ to check ε-active constraints.

Solution. Let us normalize the constraint $\bar{h}$ and express it in the standard form as

$$h \equiv \tfrac{1}{18}x_1^2 + \tfrac{1}{36}x_2 - 1.0 = 0$$

Evaluating the constraint at the two points we get, $h(1, 1) = -0.9166$, and $h(-4.5, -4.5) = 0$. Therefore, the equality constraint is violated at $(1, 1)$ and active at $(-4.5, -4.5)$.

The inequality constraint $\bar{g}$ cannot be normalized by dividing it by $30\,000x_2$ because x_2 has negative value for the second point. We must treat this situation very carefully. For the point $(1, 1)$, we can divide the constraint by $30\,000x_2$ and write it as follows, since x_2 is strictly positive,

$$g \equiv \frac{x_1}{60x_2} - 1.0 \leqq 0$$

At the point $(1, 1)$, $g = -0.9833 < 0$, so the constraint is inactive. Even $g + \varepsilon < 0$, so the constraint is not ε-active.

The second point for the inequality constraint can be treated in two ways. In the first method, since we know that x_2 has negative value, when we divide by $30\,000x_2$, the sense of the inequality is changed and it becomes $x_1/60x_2 - 1.0 \geqq 0$.

We can now convert this to a " $\leq$ type" constraint by multiplying it by -1 as

$$g \equiv \frac{-x_1}{60x_2} + 1.0 \leq 0$$

Now $g(-4.5, -4.5) = 0.983 > 0$, so the constraint is violated.

Another approach is to divide the constraint using the absolute value of x_2, i.e. $30\,000\,|x_2|$. In this case the constraint becomes

$$g \equiv \frac{x_1}{60\,|x_2|} - \frac{x_2}{|x_2|} \leq 0.$$

We can define a sign function as

$$\text{sign}\,(x_2) = \begin{cases} 1, & \text{if } x_2 > 0 \\ -1, & \text{if } x_2 < 0 \end{cases}$$

Then the constraint becomes

$$g \equiv \frac{x_1}{60\,|x_2|} - \text{sign}\,(x_2) \leq 0$$

This form of the constraint can be used for both the given points. It can be verified that with this form $g(1, 1) = -0.9833 < 0$, and $g(-4.5, -4.5) = 0.983 > 0$.

‖

For some types of constraints, it may not be desirable to normalize them in actual numerical calculations. For example, when the constraint $500x_1 - 30\,000x_2 \leq 0$ is replaced by $x_1/60|x_2| - \text{sign}\,(x_2) \leq 0$, the nature of the constraint function is changed. The original constraint function is linear but the normalized constraint function is nonlinear. Linear constraints are numerically easier to treat and implement than the nonlinear ones. Care and judgement needs to be exercised while normalizing the constraints in numerical calculations. If one normalization does not work, another form of normalization should be tried. In some cases, it may be better to use unnormalized constraints, especially the equality constraints. Thus, in numerical calculations, some experimentation with normalization of constraints is suggested for some problems.

6.2.4 Potential Constraint Strategy

To evaluate the search direction in numerical methods for constrained optimization, one needs to know the cost and constraint functions and their gradients at a design iteration. The numerical algorithms for constrained optimization can be classified based on whether gradients of all the constraints or only a subset of them are required at a design iteration. The numerical algorithms that use gradients of only a subset of the constraints are said to use *potential constraint strategy*. The potential constraint set, in general, is comprised of active, ε-active and violated constraints at the current iteration. There can be different potential constraint strategies. For example, we can define a very simple potential constraint strategy by defining an index set I_k at the kth iteration as follows:

$$I_k = \begin{bmatrix} \{j \mid j = 1 \text{ to } p \text{ for equalities}\} \text{ and} \\ \{i \mid g_i(\mathbf{x}^{(k)}) + \varepsilon \geq 0, \ i = 1 \text{ to } m\} \end{bmatrix} \tag{6.12}$$

Note that the potential constraint index set I_k defined by Eq. (6.12) includes all the active, ε-active and violated constraints. *The equality constraints are always included in the set I_k by definition.* Also, note that I_k simply contains a list of constraints in the potential set.

Example 6.2   Determination of potential constraint set. Consider the following six constraints:

$$g_1 \equiv \tfrac{1}{18} x_1^2 + \tfrac{1}{36} x_2 - 1.0 \leq 0$$

$$g_2 \equiv x_2 - 10 \leq 0$$

$$g_3 \equiv -x_2 - 2 \leq 0$$

$$g_4 \equiv -\frac{x_1}{60 x_2} + 1.0 \leq 0$$

$$g_5 \equiv -\frac{x_2}{x_1} + 1.0 \leq 0$$

$$g_6 \equiv -x_1 \leq 0$$

Let $\mathbf{x}^{(k)} = (-4.5, -4.5)$ and $\varepsilon = 0.1$. Form the potential constraint index set I_k defined by Eq. (6.12).

Solution. After normalization, constraints g_2 and g_3 are given as $\bar{g}_2 \equiv x_2/10 - 1.0 \leq 0$ and $\bar{g}_3 \equiv -x_2/2 - 1.0 \leq 0$. Evaluating the constraints at the given point, we obtain

$$g_1 = \tfrac{1}{18}(-4.5)^2 + \tfrac{1}{36}(-4.5) - 1.0 = 0 \text{ (active)}$$

$$\bar{g}_2 = -\frac{4.5}{10} - 1.0 = -1.45 < 0 \text{ (inactive)}$$

$$\bar{g}_3 = -\frac{-4.5}{2} - 1.0 = 1.25 > 0 \text{ (violated)}$$

$$g_4 = -\frac{-4.5}{60(-4.5)} + 1.0 = 0.983 > 0 \text{ (violated)}$$

$$g_5 = -\frac{-4.5}{-4.5} + 1.0 = 0 \text{ (active)}$$

$$g_6 = -(-4.5) = 4.5 > 0 \text{ (violated)}$$

Therefore, we see that g_1 and g_5 are active and g_3, g_4 and g_6 are violated. Since $\bar{g}_2 + \varepsilon < 0$, so it is not ε-active. Thus

$$I_k = \{1, 3, 4, 5, 6\}$$

6.2.5 Descent Function

For unconstrained optimization, each algorithm in Chapter 5 required reduction in the cost function at every design iteration. With that requirement, a descent towards the minimum point was maintained. *A function used to monitor progress towards the minimum is called the descent function* or the *merit function*. The cost function is used as the descent function in unconstrained optimization problems.

The idea of a descent function is very important in constrained optimization as well. Use of the cost function, however, as a descent function for constrained optimization is quite cumbersome. Therefore, many other descent functions have been proposed and used. We shall discuss some of the functions later in this chapter. At this point, the purpose of the descent function should be well understood. The basic idea is to compute a search direction $\mathbf{d}^{(k)}$ and then a step size along it such that the descent function is reduced. With this requirement proper progress towards the minimum point is maintained. The descent function also has the property that its minimum value is the same as that of the original cost function.

6.2.6 Convergence of an algorithm

The idea of convergence of an algorithm is very important in constrained optimization problems. We first define and then discuss its importance, and how to achieve it.

An algorithm is said to be convergent if it reaches a minimum point starting from an arbitrary design. An algorithm that has been proven to converge starting from an arbitrary point is called a *robust* method. In practical applications of optimization, such *reliable algorithms* are highly desirable. Many engineering design problems require considerable numerical effort to evaluate functions and their gradients. Failure of the algorithm can have disastrous effects with respect to wastage of valuable resources as well as morale of designers. Thus, it is extremely important to develop convergent algorithms for practical applications. It is equally important to enforce convergence in numerical implementation of algorithms in general purpose design optimization software.

A convergent algorithm satisfies the following two requirements:

1. There is a descent function for the algorithm. The idea is that the descent function must decrease at each iteration. This way, progress towards the minimum point can be monitored.
2. The direction of design changes $\mathbf{d}^{(k)}$ is a continuous function of the design variables. This is also an important requirement. It implies that a proper direction of descent can be found in the design space at each iteration. The direction can be found such that proper descent towards the minimum point can be maintained. This requirement also avoids "oscillations", or "zig-zagging" in the descent function.

In addition to the preceding two requirements, the assumption of the feasible region (constraint set) for the problem being closed and bounded must be satisfied for the guarantee of the algorithms to be convergent. The algorithm may or may not converge if the two conditions are not satisfied. Closedness of a set means that all the boundary points be included in the set. Boundedness implies that there are upper and lower bounds on the elements in the set. These requirements are satisfied if all functions of the problem are continuous. The preceding assumptions are not unreasonable for many engineering design applications.

6.3   LINEARIZATION OF THE CONSTRAINED PROBLEM

At each iteration, most numerical methods for constrained optimization compute design change by solving an approximate subproblem which is obtained by writing linear Taylor series expansions for the cost and constraint functions. This idea of approximate or linearized subproblems is central to the development of numerical optimization methods and should be thoroughly understood.

All search methods start with a design estimate and iteratively improve it. Let $\mathbf{x}^{(k)}$ be the design estimate at the kth iteration and $\Delta\mathbf{x}^{(k)}$ be the desired change in design. We write Taylor series expansion of the cost and constraint functions about the point $\mathbf{x}^{(k)}$ to obtain the approximate subproblem as

minimize $\qquad f(\mathbf{x}^{(k)} + \Delta\mathbf{x}^{(k)}) \cong f(\mathbf{x}^{(k)}) + \nabla f^T(\mathbf{x}^{(k)})\Delta\mathbf{x}^{(k)}$ $\qquad$ (6.13)

subject to the linearized equality constraints

$$h_j(\mathbf{x}^{(k)} + \Delta\mathbf{x}^{(k)}) \cong h_j(\mathbf{x}^{(k)}) + \nabla h_j^T(\mathbf{x}^{(k)})\Delta\mathbf{x}^{(k)} = 0; \qquad j = 1 \text{ to } p \qquad (6.14)$$

and the linearized inequality constraints

$$g_j(\mathbf{x}^{(k)} + \Delta\mathbf{x}^{(k)}) \cong g_j(\mathbf{x}^{(k)}) + \nabla g_j^T(\mathbf{x}^{(k)})\Delta\mathbf{x}^{(k)} \leqq 0; \qquad j = 1 \text{ to } m \qquad (6.15)$$

where ∇f, ∇h_j and ∇g_j are gradients of the cost function, jth equality constraint, and jth inequality constraint respectively, and "$\cong$" implies approximate equality.

In the following discussion, we introduce some simplified notations as follows:

$$f_k = f(\mathbf{x}^{(k)}) \qquad (6.16)$$

i.e. cost function value at the current design $\mathbf{x}^{(k)}$,

$$e_j = -h_j(\mathbf{x}^{(k)}) \qquad (6.17)$$

i.e. negative of the jth equality constraint function value at the current design $\mathbf{x}^{(k)}$,

$$b_j = -g_j(\mathbf{x}^{(k)}) \qquad (6.18)$$

i.e. negative of the jth inequality constraint function value at the current design $\mathbf{x}^{(k)}$,

$$c_i = \partial f(\mathbf{x}^{(k)})/\partial x_i \qquad (6.19)$$

i.e. ith component of the gradient of the cost function at the current design $\mathbf{x}^{(k)}$, i.e. derivative of the cost function with respect to the ith design variable x_i,

$$n_{ij} = \partial h_j(\mathbf{x}^{(k)})/\partial x_i \qquad (6.20)$$

i.e. ith component of the gradient of the jth equality constraint at the current design $\mathbf{x}^{(k)}$, or derivative of the jth equality constraint function with respect to ith design variable x_i,

$$a_{ij} = \partial g_j(\mathbf{x}^{(k)})/\partial x_i \qquad (6.21)$$

i.e. ith component of the gradient of the jth inequality constraint at the current design $\mathbf{x}^{(k)}$, or derivative of the jth inequality constraint function with respect to ith design variable x_i, and

$$d_i = \Delta x_i^{(k)} \qquad (6.22)$$

i.e. $\mathbf{d}$ is the n dimensional design change vector or the search direction.

Note also that the linearization of the problem is done at any design iteration, so the argument $\mathbf{x}^{(k)}$ as well as the superscript k indicating the iteration number shall be omitted for some quantities.

Using these notations, the approximate subproblem given in Eqs (6.13) to (6.15) is defined as follows:

minimize $\qquad \bar{f} = \sum_{i=1}^{n} c_i d_i \qquad (\bar{f} = \mathbf{c}^T \mathbf{d}) \qquad (6.23)$

subject to the linearized equality constraints

$$\sum_{i=1}^{n} n_{ij} d_i = e_j; \qquad j = 1 \text{ to } p \qquad (\mathbf{N}^T \mathbf{d} = \mathbf{e}) \qquad (6.24)$$

and the linearized inequality constraints

$$\sum_{i=1}^{n} a_{ij} d_i \leq b_j; \qquad j = 1 \text{ to } m \qquad (\mathbf{A}^T \mathbf{d} \leq \mathbf{b}) \qquad (6.25)$$

where columns of the matrix $\mathbf{N}$ ($n \times p$) are the gradients of equality constraints and the columns of the matrix $\mathbf{A}$ ($n \times m$) are the gradients of the inequality constraints. Note that since f_k is a constant which does not affect solution of the linearized subproblem, it is dropped from Eq. (6.23). Therefore, $\bar{f}$ represents the linearized change in the original cost function.

Let $\mathbf{n}^{(j)}$ and $\mathbf{a}^{(j)}$ represent the gradients of the j equality and jth inequality constraints respectively. Therefore, they are given as

$$\mathbf{n}^{(j)} = \left(\frac{\partial h_j}{\partial x_1}, \frac{\partial h_j}{\partial x_2}, \ldots, \frac{\partial h_j}{\partial x_n} \right) \qquad (6.26)$$

$$\mathbf{a}^{(j)} = \left(\frac{\partial g_j}{\partial x_1}, \frac{\partial g_j}{\partial x_2}, \ldots, \frac{\partial g_j}{\partial x_n} \right) \qquad (6.27)$$

The matrices **N** and **A** are also given as

$$\mathbf{N} = [\mathbf{n}^{(j)}]_{(n \times p)}; \qquad \mathbf{A} = [\mathbf{a}^{(j)}]_{(n \times m)} \tag{6.28}$$

Example 6.3   Definition of linearized subproblem. Consider the optimization problem of Example 3.28,

minimize $f(\mathbf{x}) = x_1^2 + x_2^2 - 3x_1 x_2$

subject to

$$g_1(\mathbf{x}) \equiv \tfrac{1}{6}x_1^2 + \tfrac{1}{6}x_2^2 - 1.0 \leq 0$$

$$g_2(\mathbf{x}) \equiv -x_1 \leq 0$$

$$g_3(\mathbf{x}) \equiv -x_2 \leq 0$$

Linearize the cost and constraint functions about the point $\mathbf{x}^{(0)} = (1, 1)$ and write the approximate problem given by Eqs (6.23) to (6.25).

Solution. The graphical solution for the problem is shown in Fig. 6.4. It can be seen that the optimum solution is at the point $(\sqrt{3}, \sqrt{3})$ with cost function as -3. The given point $(1, 1)$ is inside the feasible region. The gradients of cost and constraint functions are,

$$\nabla f = (2x_1 - 3x_2, \, 2x_2 - 3x_1)$$

$$\nabla g_1 = (\tfrac{2}{6}x_1, \, \tfrac{2}{6}x_2)$$

$$\nabla g_2 = (-1, 0)$$

$$\nabla g_3 = (0, -1)$$

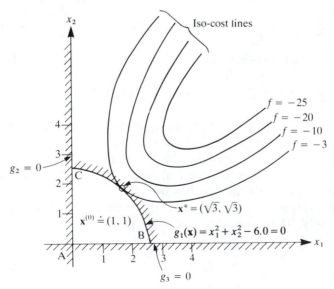

FIGURE 6.4
Graphical representation of the cost and constraints for Example 6.3.

Evaluating the cost and constraint functions and their gradients at the point $(1, 1)$, we get

$$f(\mathbf{x}^{(0)}) = -1.0$$

$$b_1 \equiv -g_1(\mathbf{x}^{(0)}) = \tfrac{2}{3}$$

$$b_2 \equiv -g_2(\mathbf{x}^{(0)}) = 1$$

$$b_3 \equiv -g_3(\mathbf{x}^{(0)}) = 1$$

$$\nabla f(\mathbf{x}^{(0)}) = (-1, -1), \quad \text{i.e. vector } \mathbf{c} = (-1, -1)$$

$$\nabla g_1(\mathbf{x}^{(0)}) = (\tfrac{1}{3}, \tfrac{1}{3})$$

Note that the given design point $(1, 1)$ is in the feasible region since all the constraints are satisfied. The matrix $\mathbf{A}$ and vector $\mathbf{b}$ of Eqs (6.25) are defined as

$$\mathbf{A} = \begin{bmatrix} \tfrac{1}{3} & -1 & 0 \\ \tfrac{1}{3} & 0 & -1 \end{bmatrix}; \quad \mathbf{b} = \begin{bmatrix} \tfrac{2}{3} \\ 1 \\ 1 \end{bmatrix}$$

Now the linearized subproblem of Eqs (6.23) to (6.25) can be written as,

minimize
$$\bar{f} = [-1 \quad -1] \begin{bmatrix} d_1 \\ d_2 \end{bmatrix} \tag{6.29}$$

subject to

$$\begin{bmatrix} \tfrac{1}{3} & \tfrac{1}{3} \\ -1 & 0 \\ 0 & -1 \end{bmatrix} \begin{bmatrix} d_1 \\ d_2 \end{bmatrix} \leq \begin{bmatrix} \tfrac{2}{3} \\ 1 \\ 1 \end{bmatrix} \tag{6.30}$$

Or, in the expanded notation, we get

minimize
$$\bar{f} = -d_1 - d_2$$

subject to

$$\tfrac{1}{3}d_1 + \tfrac{1}{3}d_2 \leq \tfrac{2}{3}$$

$$-d_1 \leq 1; \quad -d_2 \leq 1$$

The last two constraints in the subproblem ensure nonnegativity of the design variables required in the original optimization problem. Note that unless we enforce limits on the design changes d_i, $i = 1, 2$, the foregoing subproblem may be unbounded.

Note also that the preceding linearized subproblem is in terms of the variables d_1 and d_2, i.e. changes in the design variables. *We may also write the subproblem in terms of the original variables x_1 and x_2.* To do this we replace the original functions of the problem with their linear approximations at the given point. For the foregoing problem, linearization of the original problem at the point $\mathbf{x}^{(0)} = (1, 1)$ using linear Taylor series expansion of Eq. (3.11), is given as follows:

$$\bar{f}(x_1, x_2) = f(\mathbf{x}^{(0)}) + \nabla f \cdot (\mathbf{x} - \mathbf{x}^{(0)})$$

$$= -1 + [-1 \quad -1]\begin{bmatrix}(x_1 - 1)\\(x_2 - 1)\end{bmatrix}$$

$$= -x_1 - x_2 + 1$$

$$\bar{g}_1(x_1, x_2) \equiv g_1(\mathbf{x}^{(0)}) + \nabla g_1 \cdot (\mathbf{x} - \mathbf{x}^{(0)})$$

$$\equiv -\tfrac{2}{3} + [\tfrac{1}{3} \quad \tfrac{1}{3}]\begin{bmatrix}(x_1 - 1)\\(x_2 - 1)\end{bmatrix}$$

$$\equiv \tfrac{1}{3}(x_1 + x_2 - 4) \leq 0$$

$$\bar{g}_2 \equiv -x_1 \leq 0$$

$$\bar{g}_3 = -x_2 \leq 0$$

In the foregoing expressions, "overbar" for a function indicates linearized approximation. The feasible regions for the linearized problem at the point $(1, 1)$ and the original problem are shown in Fig. 6.5. Since the linearized cost function is parallel to the linearized first constraint $\bar{g}_1$, the optimum solution for the linearized problem is any point on the line D–E in Fig. 6.5.

It is important to note that the linear approximations for the functions of the problem change from point to point. Therefore, the feasible region for the linearized problem will change with the point at which the linear approximations are written. ‖

Example 6.4   Linearization of rectangular beam design problem. Linearize the rectangular beam design problem formulated in Section 2.8.5 at the point $(50, 200)$ mm.

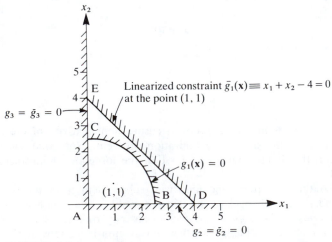

FIGURE 6.5
Graphical representation of the linearized feasible region for Example 6.3.

Solution. The problem, after normalization, is defined as follows. Find width b and depth d to minimize

$$f(b, d) = bd$$

subject to

$$g_1 \equiv \frac{(2.40E+07)}{bd^2} - 1.0 \leq 0$$

$$g_2 \equiv \frac{(1.125E+05)}{bd} - 1.0 \leq 0$$

$$g_3 \equiv \frac{d}{2b} - 1.0 \leq 0$$

$$g_4 \equiv -b \leq 0$$

$$g_5 \equiv -d \leq 0$$

At the given point the problem functions are evaluated as

$$f(50, 200) = (1.00E+04)$$

$$g_1(50, 200) = 11.00 > 0 \text{ (violation)}$$

$$g_2(50, 200) = 10.25 > 0 \text{ (violation)}$$

$$g_3(50, 200) = 1.00 > 0 \text{ (violation)}$$

$$g_4(50, 200) = -50 < 0 \text{ (inactive)}$$

$$g_5(50, 200) = -200 < 0 \text{ (inactive)}$$

In the following calculations, we shall assume that constraints g_4 and g_5 will remain satisfied, i.e. the design will remain in the first quadrant.
   The gradients of the functions are evaluated as

$$\nabla f(50, 200) = (d, b)$$
$$= (200, 50)$$

$$\nabla g_1(50, 200) = (2.40E+07)\left(\frac{-1}{b^2d^2}, \frac{-2}{bd^3}\right)$$
$$= (-0.24, -0.12)$$

$$\nabla g_2(50, 200) = (1.125E+05)\left(\frac{-1}{b^2d}, \frac{-1}{bd^2}\right)$$
$$= (-0.225, -0.05625)$$

$$\nabla g_3(50, 200) = \frac{1}{2}\left(\frac{-1}{b^2}d, \frac{1}{b}\right)$$
$$= (-0.04, 0.01)$$

   Using the function values and their gradients, the linear Taylor series expansions give the linearized subproblem at the point $(50, 200)$ in terms of the original variables as

$$\bar{f}(b, d) = 200b + 50d - 10\,000$$

$$\bar{g}_1(b, d) = -0.24b - 0.12d + 47.00 \leqq 0$$

$$\bar{g}_2(b, d) = -0.225b - 0.056\,25d + 32.75 \leqq 0$$

$$\bar{g}_3(b, d) = -0.04b + 0.01d + 1.00 \leqq 0$$

The linearized constraint functions are plotted in Fig. 6.6 and their feasible region is identified. The feasible region for the original constraints is also identified. It can be observed that the two regions are quite different. Since the linearized cost function is parallel to constraint $\bar{g}_2$, the optimum solution lies on the line I–J. If point I is selected as the solution for the linearized subproblem, then the new point is given as

$$b = 95.28, \qquad d = 201.10, \qquad \bar{f} = 19{,}111$$

For any point on line I–J all the original constraints are still violated. Apparently, for nonlinear constraints, several iterations are needed to correct constraint violations and reach the feasible region. This will be observed in several example problems later in this chapter and in Chapter 8.

One interesting observation concerns the third constraint; the original constraint $d - 2b \leqq 0$ is normalized as $d/2b - 1 \leqq 0$. The normalization does not change the constraint boundary; thus the graphical representation for the problem remains the same as may be verified in Fig. 6.6. However, the normalization changes the form of the constraint function that affects its linearization. If the constraint is not normalized, its linearization will give the same functional form as the original constraint for all design points, i.e.

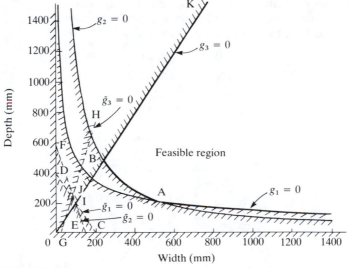

FIGURE 6.6
Feasible region for the original and the linearized constraints of the rectangular beam design problem of Example 6.4.

$d - 2b \leqq 0$. This is shown as line O–K in Fig. 6.6. The linearized form of the normalized constraint changes; it gives the line G–H for the point $(50, 200)$. This is quite different from the original constraint. The iterative process with and without the normalized constraint can lead to different paths to the optimum point. In conclusion, we must be careful while normalizing the constraints so as not to change functional form for the constraints as far as possible.   ‖

6.4 SEQUENTIAL LINEAR PROGRAMMING ALGORITHM

6.4.1 The Basic Idea

Note that all the functions in Eqs (6.23) to (6.25) are linear in the variables d_i. Therefore, linear programming methods can be used to solve for d_i. Such procedures where linear programming is used at each iteration to compute design change are called *Sequential Linear Programming* methods or in short SLP. In this section, we shall briefly describe such procedures.

To solve the LP by the standard Simplex method, the right-hand side parameters e_j and b_j in Eqs (6.17) and (6.18) must be nonnegative. If any b_j is negative, we must multiply the corresponding constraint by -1 to make the right-hand side nonnegative. This will change the sense of the inequality in Eq. (6.25), i.e. it will become a "$\geqq$ type" constraint. It must be noted that the problem defined in Eqs (6.23) to (6.25) may not have a bounded solution or the changes in design may become too large, thus invalidating the linear approximations. Therefore, limits must be imposed on changes in design. Such constraints are usually called *move limits* in the optimization literature. Move limits can be expressed as

$$-\Delta_{il}^{(k)} \leqq d_i \leqq \Delta_{iu}^{(k)}; \qquad i = 1 \text{ to } n \tag{6.31}$$

where $\Delta_{il}^{(k)}$ and $\Delta_{iu}^{(k)}$ are the maximum allowed decrease and increase in the ith design variable, respectively at the kth iteration. The problem is still linear in terms of d_i, so LP methods can be used to solve it. Note that the iteration counter k is used to specify Δ_{il} and Δ_{iu}. That is, the move limits may change every iteration. Figure 6.7 shows the effect of imposing the move limits on changes in the design $\mathbf{x}^{(k)}$; the new design estimate is required to stay in the rectangular area ABCD.

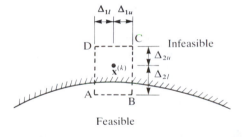

Feasible

FIGURE 6.7
Linear move limits on design changes.

Selection of proper move limits is of critical importance because it can mean success or failure of the SLP algorithm. Their specification, however, requires some experience with the method as well as intuition and judgement. Therefore, the user should not hesitate to try different move limits if one specification leads to failure or improper design. Many times lower and upper bounds are specified on the real design variables x_i. Therefore, move limits must be selected to remain within the specified bounds. Also, since linear approximations for the functions are used, the design changes should not be very large, and the move limits should not be excessively large. Usually Δ_{il} and Δ_{iu} are *selected as some fraction of the current design variable values* (this may vary from 1 to 100%). If the resulting LP problem turns out to be infeasible, the move limits will have to be relaxed (i.e. allow larger changes in design) and the subproblem solved again. Usually, a certain amount of experience with the problem is necessary to select proper move limits and adjust them at every iteration to solve the problem successfully.

Another point must also be noted before an SLP algorithm can be stated. This concerns the sign of the variables d_i (or Δx_i) which can be positive or negative, i.e. current values of design variables can increase or decrease. To allow for such a change, we must treat the LP variables d_i as *free* in sign. This can be done as explained in Section 4.2. Each free variable d_i is replaced by $d_i = d_i^+ - d_i^-$ in all the expressions. The LP subproblem defined in Eqs (6.23) to (6.25) is then transformed to the standard form to use the Simplex method.

We must specify the following stopping criteria for the SLP method:

1. All constraints must be satisfied. This can be expressed as

$$g_i \leqq \varepsilon_1; \ i = 1 \text{ to } m \quad \text{and} \quad |h_i| \leqq \varepsilon_1; \ i = 1 \text{ to } p$$

   where $\varepsilon_1 > 0$ is a specified small number defining tolerance for constraint violation.
2. The changes in design should be almost zero; that is, $\|\mathbf{d}\| \leqq \varepsilon_2$, where $\varepsilon_2 > 0$ is a specified small number.

Ideally, $\varepsilon_1 = \varepsilon_2 = 0$ is the best choice for locating the precise minimum point but in practice in the computer implementation of optimization algorithms a small positive tolerance for these numbers must be specified.

6.4.2 An SLP Algorithm

The *sequential linear programming* algorithm is stated as follows:

Step 1. Estimate a starting design as $\mathbf{x}^{(0)}$. Set $k = 0$. Specify two small positive numbers, ε_1 and ε_2.

Step 2. Evaluate cost and constraint functions at the current design $\mathbf{x}^{(k)}$, i.e. calculate f_k, b_j; $j = 1$ to m, and e_j; $j = 1$ to p as defined in Eqs (6.16) to (6.18).

Step 3. Evaluate the cost and constraint function gradients at the current design $\mathbf{x}^{(k)}$, i.e. evaluate

$$c_i = \frac{\partial f}{\partial x_i}; \qquad i = 1 \text{ to } n$$

$$n_{ij} = \frac{\partial h_j}{\partial x_i}; \qquad j = 1 \text{ to } p; \qquad i = 1 \text{ to } n$$

$$a_{ij} = \frac{\partial g_j}{\partial x_i}; \qquad j = 1 \text{ to } m; \qquad i = 1 \text{ to } n$$

(Note that the potential constraint strategy is not used because gradients of all constraints are calculated. It can be, however, incorporated easily to make the method more efficient.)

Step 4. Select the proper move limits $\Delta_{il}^{(k)}$ and $\Delta_{iu}^{(k)}$ as some fraction of the current design.

Step 5. Define the LP subproblem of Eqs (6.23) to (6.25).

Step 6. If needed, convert the LP subproblem to the standard Simplex form (refer to Section 4.2), and solve it for $\mathbf{d}^{(k)}$.

Step 7. Check for convergence. If $g_i \leq \varepsilon_1$; $i = 1$ to m; $|h_i| \leq \varepsilon_1$; $i = 1$ to p; and $\|\mathbf{d}^{(k)}\| \leq \varepsilon_2$, then stop. Otherwise, continue.

Step 8. Update the design as

$$\mathbf{x}^{(k+1)} = \mathbf{x}^{(k)} + \mathbf{d}^{(k)}$$

Set $k = k + 1$ go to Step 2.

It is interesting to note here that the LP problem defined in Eqs (6.23) to (6.25) can be *transformed to be in the original variables* by substituting $d_i = x_i - x_i^{(k)}$. This was demonstrated in Examples 6.3 and 6.4. The move limits on d_i of Eq. (6.31) can also be transformed to be in the original variables. This way the solution of the LP problem directly gives the estimate for the next design point.

Example 6.5 Study of sequential linear programming algorithm. Consider the problem given in Example 6.3. Define the linearized subproblem at the point $(3, 3)$, and discuss its solution imposing proper move limits.

Solution. To define the linearized subproblem, the following quantities are calculated at the given point $(3, 3)$:

$$f(3, 3) = 3^2 + 3^2 - 3(3)(3) = -9$$

$$g_1(3, 3) \equiv \tfrac{1}{6}(3^2) + \tfrac{1}{6}(3^2) - 1 = 2 > 0 \text{ (violation)}$$

$$g_2(3, 3) \equiv -x_1 = -3 < 0 \text{ (inactive)}$$

$$g_3(3, 3) \equiv -x_2 = -3 < 0 \text{ (inactive)}$$

$$\mathbf{c} \equiv \nabla f = (2x_1 - 3x_2, \ 2x_2 - 3x_1) = (-3, -3)$$

$$\nabla g_1 = (\tfrac{2}{6}x_1, \ \tfrac{2}{6}x_2) = (1, 1)$$

$$\nabla g_2 = (-1, 0)$$

$$\nabla g_3 = (0, -1)$$

The given point is in the infeasible region as the first constraint is violated. The linearized subproblem is defined according to Eqs (6.23) to (6.25) as

minimize
$$\bar{f} = [-3 \quad -3]\begin{bmatrix} d_1 \\ d_2 \end{bmatrix}$$
subject to
$$\begin{bmatrix} 1 & 1 \\ -1 & 0 \\ 0 & -1 \end{bmatrix}\begin{bmatrix} d_1 \\ d_2 \end{bmatrix} \leq \begin{bmatrix} -2 \\ 3 \\ 3 \end{bmatrix}$$

The subproblem has only two variables, so it can be solved using the graphical solution procedure, as shown in Fig. 6.8. This figure when superimposed on Fig. 6.4 represents a linearized approximation for the original problem at the point $(3, 3)$. The feasible solution for the linearized subproblem must lie in the region ABC in Fig. 6.8. The cost function is parallel to the line B–C, thus any point on the line minimizes the function. We may choose $d_1 = -1$ and $d_2 = -1$ as the solution which satisfies all the linearized constraints (note that the linearized change in cost $\bar{f}$ is 6).

If 100% move limits are selected, i.e.

$$-3 \leq d_1 \leq 3; \qquad -3 \leq d_2 \leq 3$$

then the solution to the LP subproblem must lie in the region ADEF. If the move limits are set as 20% of the current value of design variables, then the solution must satisfy

$$-0.6 \leq d_1 \leq 0.6; \qquad -0.6 \leq d_2 \leq 0.6$$

In this case the solution must lie in the region $A_1D_1E_1F_1$. It can be seen that there is no feasible solution to the linearized subproblem because region $A_1D_1E_1F_1$ does not intersect the line B–C. We must enlarge the region $A_1D_1E_1F_1$ by increasing the move limits. Thus, we note that if the move limits are too restrictive, the linearized subproblem may not have any solution.

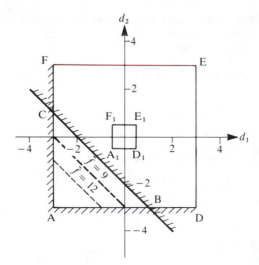

FIGURE 6.8
Graphical solution for the linearized subproblem of Example 6.5.

If we choose $d_1 = -1$ and $d_2 = -1$, then the improved design is given as $(2, 2)$. This is still an infeasible point, as can be seen in Fig. 6.4. Therefore, although the linearized constraint is satisfied with $d_1 = -1$ and $d_2 = -1$, the original nonlinear constraint g_1 is still violated. ‖

Example 6.6 Use of sequential linear programming. Consider the problem given in Example 6.3. Perform one iteration of the SLP algorithm. Use $\varepsilon_1 = \varepsilon_2 = 0.001$ and choose move limits such that a 15% design change is permissible. Let $\mathbf{x}^{(0)} = (1, 1)$ be the starting design.

Solution. The given point represents a feasible solution for the problem as may be seen in Fig. 6.4. The linearized subproblem with the appropriate move limits on design changes d_1 and d_2 at the point $\mathbf{x}^{(0)}$ is obtained in Example 6.3 as

minimize
$$\bar{f} = -d_1 - d_2$$

subject to

$$\tfrac{1}{3}d_1 + \tfrac{1}{3}d_2 \leqq \tfrac{2}{3}$$

$$-(1 + d_1) \leqq 0$$

$$-(1 + d_2) \leqq 0$$

$$-0.15 \leqq d_1 \leqq 0.15$$

$$-0.15 \leqq d_2 \leqq 0.15$$

Graphical solution for the linearized subproblem is given in Fig. 6.9. Move limits of 15% define the solution region to be DEFG. The optimum solution for the problem is at point F where $d_1 = 0.15$ and $d_2 = 0.15$. It can be seen that much larger move limits are possible in the present case.

We shall solve the problem using the Simplex method as well. Note that in the linearized subproblem, the design changes d_1 and d_2 are free in sign. If we wish to solve the problem by the Simplex method we must define new variables,

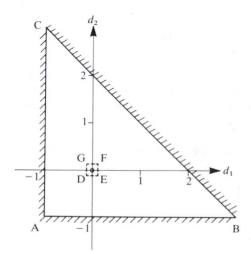

FIGURE 6.9
Graphical solution for the linearized subproblem of Example 6.6.

A, B, C and D such that

$$d_1 = A - B, \; d_2 = C - D \text{ and A, B, C and D} \geqq 0$$

Therefore, substituting these decompositions into the foregoing equations, we get the following LP problem written in standard form as

minimize
$$\bar{f} = -A + B - C + D$$

subject to
$$\tfrac{1}{3}(A - B + C - D) \leqq \tfrac{2}{3}$$

$$-A + B \leqq 1.0$$

$$-C + D \leqq 1.0$$

$$A - B \leqq 0.15$$

$$B - A \leqq 0.15$$

$$C - D \leqq 0.15$$

$$D - C \leqq 0.15$$

$$A, B, C, D \geqq 0$$

The solution to the foregoing LP problem with the Simplex method is $A = 0.15$, $B = 0$, $C = 0.15$ and $D = 0$. Therefore,

$$d_1 = A - B = 0.15$$

$$d_2 = C - D = 0.15$$

After updating the design we get,

$$\mathbf{x}^{(1)} = \mathbf{x}^{(0)} + \mathbf{d}^{(0)}$$

$$= (1.15, 1.15)$$

At the new design $(1.15, 1.15)$ we have,

$$f(\mathbf{x}^{(1)}) = -1.3225$$

$$g_1(\mathbf{x}^{(1)}) = -0.6166$$

Note that the cost function value has decreased for the new design $\mathbf{x}^{(1)}$ without violating the constraint. This indicates that the new design is an improvement over the previous one. Since the norm of the design change, $\|\mathbf{d}\| = 0.212$ is larger than the permissible tolerance (0.001), we need to go through more iterations to satisfy the stopping criteria.

Note also that the linearized subproblem at the point $(1, 1)$ can be written in the original variables. This was done in Example 6.3 and the linearized subproblem was obtained as

minimize
$$\bar{f} = -x_1 - x_2 + 1$$

subject to
$$\bar{g}_1 \equiv \tfrac{1}{3}(x_1 + x_2 - 4) \leqq 0$$

$$\bar{g}_2 \equiv -x_1 \leqq 0$$

$$\bar{g}_3 \equiv -x_2 \leqq 0$$

The 15% move limits can be also transformed to be in the original variables using $-\Delta_{il} \leqq x_i - x_i^{(0)} \leqq \Delta_{iu}$ as

$$-0.15 \leqq (x_1 - 1) \leqq 0.15 \quad \text{or} \quad 0.85 \leqq x_1 \leqq 1.15$$

$$-0.15 \leqq (x_2 - 1) \leqq 0.15 \quad \text{or} \quad 0.85 \leqq x_2 \leqq 1.15$$

Solving the subproblem, we obtain the same solution $(1.15, 1.15)$ as before. ‖

6.4.3 Observations on the SLP Algorithm

The sequential linear programming algorithm is a simple and straightforward approach to solve constrained optimization problems. The algorithm has been applied to solve various engineering design problems. However, many superior methods have emerged and replaced the SLP algorithm. The following observations highlight the important limitations of the SLP method:

1. *The method cannot be used as a black box approach for engineering design problems.* The selection of move limits is a trial and error process and can be best achieved in an interactive mode. The foregoing examples show that the move limits can be too restrictive resulting in no solution for the LP subproblem. They can also slow down the rate of convergence from a feasible point.

2. *The method is not convergent* since no descent function is defined, and line search is not performed along the search direction. Therefore, the method is not reliable for arbitrary starting points.

3. *The rate of convergence and performance of the SLP method* depend to a large extent on the selection of the move limits.

4. *It is possible to implement a potential constraint strategy* in the SLP algorithm to achieve efficiency. The algorithm discussed in this section did not include any such strategy to maintain simplicity in the method.

5. *The method can cycle between two points* if the optimum solution is not a vertex of the constraint set.

6. Although the method is quite simple (conceptually as well as numerically) *it is not recommended as a technique for general design optimization environment due to lack of robustness* and uncertainty in the specification of move limits.

6.5 QUADRATIC PROGRAMMING SUBPROBLEM

As observed in the previous section the SLP algorithm is a simple extension of linear programming to solve general constrained optimization problems. However, the method has several drawbacks, the major one being the lack of robustness. To correct the drawbacks, we shall develop a method where a quadratic programming (QP) subproblem is solved iteratively and the step size is found by minimizing a descent function along the search direction. In this section, we shall discuss the definition of a QP subproblem and a method of solving it.

6.5.1 Quadratic Step Size Constraint

Performance of the sequential linear programming method depends quite heavily on the selection of proper move limits on design changes. The method cannot be proved to converge to a local minimum from an arbitrary starting design. In addition, the move limits must be adjusted at each design iteration. To overcome these difficulties, other methods have been developed to solve for design changes. Most of the methods utilize linear approximations of Eqs (6.23) to (6.25) for the nonlinear programming problem. However, the linear move limits of Eq. (6.31) are abandoned in favor of a step size constraint of the form

$$\|\mathbf{d}\| \leqq \xi \qquad (6.32)$$

where $\|\mathbf{d}\|$ is the length of the search direction and ξ is a specified small positive number. Equation (6.32) imposes a constraint on the length of the design change vector. We shall see later that the parameter ξ will not have to be specified in Eq. (6.32). Using the definition of the length of a vector and squaring both sides of Eq. (6.32), we obtain the following quadratic step size constraint on $\mathbf{d}$:

$$0.5 \sum_{i=1}^{n} (d_i)^2 \leqq \xi^2 \qquad (0.5\mathbf{d}^T\mathbf{d} \leqq \xi^2) \qquad (6.33)$$

The factor of 0.5 on the left-hand side is introduced to eliminate the factor of 2 during differentiations in later calculations. It does not affect the calculations for the search direction, but affects the step size calculation slightly. The final computational algorithm is not affected. Figure 6.10 shows the quadratic step size constraint on $\mathbf{d}$ which is quite different from the linear move limits shown in Fig. 6.7. It can be seen that the new design is required to be in a hypersphere of radius $\sqrt{2}\,\xi$ with origin at the current point. Thus, the approximate subproblem to be solved at each design iteration is defined as follows:

Compute the design changes $d_1, d_2, \ldots, d_n$ to minimize a linearized cost function

$$\bar{f} = \sum_{i=1}^{n} c_i d_i \qquad (\bar{f} = \mathbf{c}^T \mathbf{d}) \qquad (6.34)$$

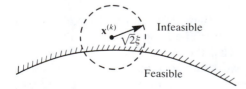

FIGURE 6.10
Quadratic step size constraint on design change.

subject to the linearized equality constraints

$$\sum_{i=1}^{n} n_{ij}d_i = e_j; \qquad j = 1 \text{ to } p \qquad (\mathbf{N}^T\mathbf{d} = \mathbf{e}) \tag{6.35}$$

linearized inequality constraints

$$\sum_{i=1}^{n} a_{ij}d_i \leqq b_j; \qquad j = 1 \text{ to } m \qquad (\mathbf{A}^T\mathbf{d} \leqq \mathbf{b}) \tag{6.36}$$

and the quadratic step size constraint

$$0.5 \sum_{i=1}^{n} (d_i^2) \leqq \xi^2 \qquad (0.5\mathbf{d}^T\mathbf{d} \leqq \xi^2) \tag{6.37}$$

Several different forms and properties of the approximate problem are discussed in the subsequent sections. The subproblem is nonlinear because the step size constraint is nonlinear. Therefore, the Simplex method as presented in Chapter 4 cannot be applied to solve the subproblem. It can be extended, however, as we shall see later.

6.5.2 Quadratic Programming (QP) Subproblem

A quadratic programming problem has quadratic cost function and linear constraints. The nonlinear direction finding subproblem defined in Eqs (6.34) to (6.37) is not in this form, but it can be transformed into it. The subproblem defined in Eqs (6.34) to (6.37) has a drawback in that a solution may not exist if ξ is too small and the current design is infeasible. This can be visualized graphically by referring to Fig. 6.10. If the hypersphere in Fig. 6.10 does not intersect the feasible region (i.e. the radius is too small) then the subproblem is infeasible. On the other hand, the advantage of solving a QP subproblem (defined in the following) is that the solution to it always exists (when constraints are consistent) and is, in fact, unique if the problem is strictly convex. Also, there are many numerical methods to solve a QP problem efficiently and accurately. A modified Simplex method can be used, or the Kuhn–Tucker conditions and procedures of Chapter 3 can be used for simpler problems. Several commercially available software packages are available for solving QP problems, e.g. QPSOL [Gill, Murray, Saunders and Wright, 1984], LINDO [Schrage, 1981], VE06A [Hopper, 1981], and E04NAF [NAG, 1984].

It is possible to formulate the following equivalent QP subproblem whose solution is identical to the subproblem defined by Eqs (6.34) to (6.37):

minimize
$$\bar{f} = \mathbf{c}^T\mathbf{d} + 0.5\mathbf{d}^T\mathbf{d} \tag{6.38}$$

subject to
$$\mathbf{N}^T\mathbf{d} = \mathbf{e} \tag{6.39}$$

$$\mathbf{A}^T\mathbf{d} \leqq \mathbf{b} \tag{6.40}$$

It is left as an exercise for the student to verify by writing the K–T necessary conditions of Chapter 3 that the solution of the preceding QP subproblem is identical to that for the previous subproblem with the Lagrange multiplier for the step size constraint as 1. Note that the QP subproblem is convex and therefore its solution procedure is expected to be stable. Moreover, the solution to the QP subproblem (if one exists) is unique since it is a strict convex programming problem.

Example 6.7 Definition of a QP Subproblem. Consider the following constrained optimization problem:

minimize $$f(\mathbf{x}) = 2x_1^3 + 15x_2^2 - 8x_1x_2 - 4x_1$$

subject to
$$h(\mathbf{x}) = x_1^2 + x_1x_2 + 1.0 = 0$$
$$g(\mathbf{x}) = x_1 - \tfrac{1}{4}x_2^2 - 1.0 \leq 0$$

Linearize the cost and constraint functions about a point $(1, 1)$ and define the QP subproblem.

Solution. Figure 6.11 shows a graphical representation for the problem. The equality constraint is shown as $h = 0$ and inequality as $g = 0$. The feasible region for the inequality constraint is identified and several cost function contours are shown. Since the equality constraint must be satisfied the optimum point must lie on the two curves $h = 0$. Two optimum solutions are identified as

Point A: $\mathbf{x}^* = (1, -2), f(\mathbf{x}^*) = 74$

Point B: $\mathbf{x}^* = (-1, 2), f(\mathbf{x}^*) = 78$

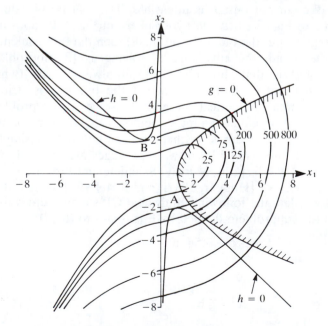

FIGURE 6.11
Graphical representation of Example 6.7.

The gradients of cost and constraint functions are

$$\nabla f = (6x_1^2 - 8x_2 - 4, \; 30x_2 - 8x_1)$$

$$\nabla h = (2x_1 + x_2, \; x_1)$$

$$\nabla g = (1, \; -x_2/2)$$

The cost and constraint function values and their gradients at $(1, 1)$ are

$$f(1, 1) = 5$$

$$h(1, 1) = 3 \neq 0 \; \text{(violation)}$$

$$g(1, 1) = -0.25 < 0 \; \text{(inactive)}$$

$$\mathbf{c} = \nabla f = (-6, 22)$$

$$\nabla h = (3, 1)$$

$$\nabla g = (1, -0.5)$$

Using move limits of 50%, the linear programming subproblem of Eqs (6.23) to (6.25) is defined as

minimize $\qquad\qquad \bar{f} = -6d_1 + 22d_2$

subject to

$$3d_1 + d_2 = -3$$

$$d_1 - 0.5d_2 \leqq 0.25$$

$$-0.5 \leqq d_1 \leqq 0.5$$

$$-0.5 \leqq d_2 \leqq 0.5$$

The QP subproblem of Eqs (6.38) to (6.40) is defined as

minimize $\qquad\qquad \bar{f} = (-6d_1 + 22d_2) + 0.5(d_1^2 + d_2^2)$

subject to

$$3d_1 + d_2 = -3$$

$$d_1 - 0.5d_2 \leqq 0.25$$

Note that if the potential constraint strategy is implemented the inequality constraint need not be included in the definition of the QP subproblem since it is inactive. The gradient of the constraint g need not be calculated at the given design point.

To compare the solutions, the preceding LP and QP subproblems are plotted in Figs 6.12 and 6.13 respectively. In these figures, the solution must satisfy the linearized equality constraint, so it must lie on the line C–D. The feasible region for the linearized inequality constraint is also shown. Therefore, the solution for the subproblem must lie on the line G–C. It can be seen in Fig. 6.12 that with 50% move limits, the linearized subproblem is infeasible. The move limits require the changes to lie in the square HIJK which does not intersect the line G–C. If we relax the move limits to 100%, then point L gives the

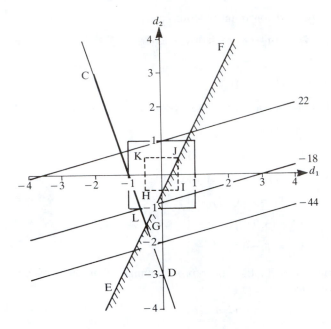

FIGURE 6.12
Solution of the linearized subproblem for Example 6.7 at the point $(1, 1)$.

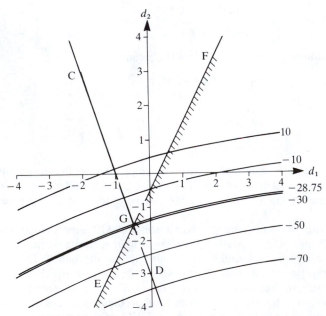

FIGURE 6.13
Solution of the quadratic programming subproblem for Example 6.7 at the point $(1, 1)$.

optimum solution as

$$d_1 = -\tfrac{2}{3}, \qquad d_2 = -1.0, \qquad \bar{f} = -18$$

Thus, we see that the design change with the linearized subproblem is affected by the move limits.

With the QP subproblem, the constraint set remains the same but there is no need for the move limits as shown in Fig. 6.13. The cost function is now a quadratic function which gives the optimum solution at Point G as

$$d_1 = -0.5, \qquad d_2 = -1.5, \qquad \bar{f} = -28.75$$

Note that the design direction determined by the QP subproblem is unique, but it depends on the move limits with the LP subproblem. The two directions determined by LP and QP subproblems are in general different. ‖

6.5.3 Solution of Quadratic Programming Problems

6.5.3.1 DEFINITION OF QP PROBLEM. Quadratic programming (QP) problems are encountered in many real-world applications. In addition, many general nonlinear programming algorithms (discussed in subsequent sections) require solution of a quadratic programming subproblem at each iteration. As seen in Eqs (6.38) to (6.40), the QP subproblem is obtained when a nonlinear problem is linearized and a quadratic step size constraint is imposed. It is extremely important to solve a QP subproblem efficiently so that large-scale problems can be treated. Thus, it is not surprising that substantial research effort has been expended in developing and evaluating many algorithms for solving QP problems [Gill *et al.*, 1981; Luenberger, 1984]. Also, as noted before, many good programs have been developed to solve such problems.

We shall describe a method to solve QP problems that is a simple extension of the *Simplex method*. Many other methods are available that can be considered as variations of that procedure. Some of the available LP codes also have an option to solve QP problems [Schrage, 1981]. The solution procedure in them is usually based on the Simplex method.

Let us define a general QP problem as follows;

minimize
$$q(\mathbf{x}) = \mathbf{c}^T\mathbf{x} + 0.5\mathbf{x}^T\mathbf{H}\mathbf{x} \tag{6.41}$$

subject to linear inequality and equality constraints

$$\mathbf{A}^T\mathbf{x} \leqq \mathbf{b} \tag{6.42}$$

$$\mathbf{N}^T\mathbf{x} = \mathbf{e} \tag{6.43}$$

and

$$\mathbf{x} \geqq \mathbf{0} \tag{6.44}$$

where $\mathbf{c} = n$ dimensional vector of given constants, $\mathbf{x} = n$ dimensional vector of unknowns, $\mathbf{b} = m$ dimensional vector of given constants, $\mathbf{e} = p$ dimensional vector of given constants, $\mathbf{H} = n \times n$ Hessian matrix of given constants, $\mathbf{A} = n \times m$ matrix of given constants, and $\mathbf{N} = n \times p$ matrix of given constants.

Note that all the linear inequality constraints are expressed in the "$\leq$ form". This is needed because we shall write the Kuhn–Tucker necessary conditions of Section 3.4. There we used "$\leq$ form" for the inequalities. Note also that if the matrix $\mathbf{H}$ is positive semidefinite, the QP problem is convex, so any solution (if one exists) represents a global minimum point (which need not be unique). Further, if the matrix $\mathbf{H}$ is positive definite, the problem is strictly convex. Therefore, the problem has a unique global solution (if one exists). We shall assume that the matrix $\mathbf{H}$ is at least positive semidefinite. This is not an unreasonable assumption in practice as many applications satisfy it. For example, in the QP subproblem of Eqs (6.38) to (6.40), $\mathbf{H} = \mathbf{I}$ (an identity matrix), so the Hessian is actually positive definite. Note also that the variables $\mathbf{x}$ are required to be nonnegative in Eq. (6.44). Variables that are free in sign can be easily treated by the method described in Section 4.2.

6.5.3.2 A SOLUTION PROCEDURE – KUHN–TUCKER CONDITIONS FOR THE QP PROBLEM.

A procedure for solving the QP problem of Eqs (6.41) to (6.44) is to first write the Kuhn–Tucker necessary conditions of Section 3.4, and then to transform them into a form that can be treated by Phase I of the Simplex method of Section 4.5. To write the necessary conditions, we introduce slack variables $\mathbf{s}$ for the Inequalities (6.42) and transform them to equalities as

$$\mathbf{A}^T\mathbf{x} + \mathbf{s} = \mathbf{b}; \quad \text{with} \quad \mathbf{s} \geq \mathbf{0} \tag{6.45}$$

Or, slack variable for the jth inequality in Eq. (6.42) can be expressed using Eq. (6.45) as

$$s_j = b_j - \sum_{i=1}^{n} a_{ij}x_i \quad (\mathbf{s} = \mathbf{b} - \mathbf{A}^T\mathbf{x}) \tag{6.46}$$

Note the nonnegativity constraints of Eq. (6.44) (when expressed in the standard form $-\mathbf{x} \leq \mathbf{0}$) do not need slack variables because $x_i \geq 0$ itself is a slack variable.

Let us define the Lagrange function for the QP problem as

$$L = \mathbf{c}^T\mathbf{x} + 0.5\mathbf{x}^T\mathbf{H}\mathbf{x} + \mathbf{u}^T(\mathbf{A}^T\mathbf{x} + \mathbf{s} - \mathbf{b}) - \boldsymbol{\zeta}^T\mathbf{x} + \mathbf{v}^T(\mathbf{N}^T\mathbf{x} - \mathbf{e})$$

where $\mathbf{u}$, $\mathbf{v}$, and $\boldsymbol{\zeta}$ are the Lagrange multiplier vectors for the inequality constraints of Eqs (6.42) or (6.46), equality constraints of Eq. (6.43), and the nonnegativity constraints ($-\mathbf{x} \leq \mathbf{0}$), respectively. The Kuhn–Tucker necessary conditions give

$$\frac{\partial L}{\partial \mathbf{x}} = \mathbf{c} + \mathbf{H}\mathbf{x} + \mathbf{A}\mathbf{u} - \boldsymbol{\zeta} + \mathbf{N}\mathbf{v} = \mathbf{0} \tag{6.47}$$

$$\mathbf{A}^T\mathbf{x} + \mathbf{s} - \mathbf{b} = \mathbf{0} \tag{6.48}$$

$$\mathbf{N}^T\mathbf{x} - \mathbf{e} = \mathbf{0} \tag{6.49}$$

$$u_i s_i = 0; \qquad i = 1 \text{ to } m \qquad (6.50)$$

$$\zeta_i x_i = 0; \qquad i = 1 \text{ to } n \qquad (6.51)$$

$$s_i, u_i \geq 0 \text{ for } i = 1 \text{ to } m; \qquad \zeta_i \geq 0 \text{ for } i = 1 \text{ to } n \qquad (6.52)$$

Since the Lagrange multipliers $\mathbf{v}$ for the equality constraints are free in sign, we may decompose them as

$$\mathbf{v} = \mathbf{y} - \mathbf{z} \text{ with } \mathbf{y}, \mathbf{z} \geq 0 \qquad (6.53)$$

Now, writing the Eqs (6.47), (6.48) and (6.49) into a matrix form, we get

$$\begin{bmatrix} \mathbf{H} & \mathbf{A} & -\mathbf{I}_{(n)} & \mathbf{0}_{(n \times m)} & \mathbf{N} & -\mathbf{N} \\ \mathbf{A}^T & \mathbf{0}_{(m \times m)} & \mathbf{0}_{(m \times n)} & \mathbf{I}_{(m)} & \mathbf{0}_{(m \times p)} & \mathbf{0}_{(m \times p)} \\ \mathbf{N}^T & \mathbf{0}_{(p \times m)} & \mathbf{0}_{(p \times n)} & \mathbf{0}_{(p \times m)} & \mathbf{0}_{(p \times p)} & \mathbf{0}_{(p \times p)} \end{bmatrix} \begin{bmatrix} \mathbf{x} \\ \mathbf{u} \\ \zeta \\ \mathbf{s} \\ \mathbf{y} \\ \mathbf{z} \end{bmatrix} = \begin{bmatrix} -\mathbf{c} \\ \mathbf{b} \\ \mathbf{e} \end{bmatrix} \qquad (6.54)$$

where $\mathbf{I}_{(n)}$ and $\mathbf{I}_{(m)}$ are $n \times n$ and $m \times m$ identity matrices respectively, and $\mathbf{0}$ are zero matrices of the indicated order. In a compact matrix notation, Eq. (6.54) becomes

$$\mathbf{BX} = \mathbf{D} \qquad (6.55)$$

where matrix $\mathbf{B}$ and vectors $\mathbf{X}$ and $\mathbf{D}$ are identified from Eq. (6.54) as

$$\mathbf{B} = \begin{bmatrix} \mathbf{H} & \mathbf{A} & -\mathbf{I}_{(n)} & \mathbf{0}_{(n \times m)} & \mathbf{N} & -\mathbf{N} \\ \mathbf{A}^T & \mathbf{0}_{(m \times m)} & \mathbf{0}_{(m \times n)} & \mathbf{I}_{(m)} & \mathbf{0}_{(m \times p)} & \mathbf{0}_{(m \times p)} \\ \mathbf{N}^T & \mathbf{0}_{(p \times m)} & \mathbf{0}_{(p \times n)} & \mathbf{0}_{(p \times m)} & \mathbf{0}_{(p \times p)} & \mathbf{0}_{(p \times p)} \end{bmatrix}_{[(n+m+p) \times (2n+2m+2p)]}$$

$$(6.56)$$

$$\mathbf{X} = \begin{bmatrix} \mathbf{x} \\ \mathbf{u} \\ \zeta \\ \mathbf{s} \\ \mathbf{y} \\ \mathbf{z} \end{bmatrix}_{(2n+2m+2p)} \qquad \mathbf{D} = \begin{bmatrix} -\mathbf{c} \\ \mathbf{b} \\ \mathbf{e} \end{bmatrix}_{(n+m+p)} \qquad (6.57)$$

The Kuhn–Tucker conditions are now reduced to finding $\mathbf{X}$ as a solution of the linear system in Eqs (6.55) subject to the constraints of Eqs (6.50) to (6.52). In the new variables X_i, the complementary slackness conditions of Eqs (6.50) and (6.51), reduce to

$$X_i X_{n+m+i} = 0; \qquad i = 1 \text{ to } (n+m) \qquad (6.58)$$

and the nonnegativity conditions of Eq. (6.52) reduce to

$$X_i \geq 0; \qquad i = 1 \text{ to } (2n+2m+2p) \qquad (6.59)$$

6.5.3.3 SIMPLEX METHOD FOR SOLVING QP PROBLEM. A solution of the linear system in Eqs (6.55) that satisfies the complementary slackness of Eq. (6.58) and nonnegativity condition of Eq. (6.59) is a solution of the original QP problem. Note that the complementary slackness condition of Eq. (6.58) is nonlinear in the variables X_i. Therefore, it may appear that the Simplex method for LP cannot be used to solve Eq. (6.55). However, a procedure developed by Wolfe [1959] and refined by Hadley [1964] can be used to solve the problem. The procedure converges to a solution in the finite number of steps provided the matrix $\mathbf{H}$ in Eq. (6.41) is positive definite. It can be further shown [Kunzi and Krelle, 1966, p. 123] that the method converges even when $\mathbf{H}$ is positive semidefinite provided the vector $\mathbf{c}$ in Eq. (6.41) is zero.

The method is based on Phase I of the Simplex procedure of Chapter 4 where we introduced an artificial variable for each equality constraint, defined an artificial cost function, and used it to determine an initial basic feasible solution. Following that procedure we introduce an artificial variable Y_i for each of the Eqs (6.55) as

$$\mathbf{BX} + \mathbf{Y} = \mathbf{D} \tag{6.60}$$

where $\mathbf{Y}$ is an $(n + m + p)$ dimensional vector. This way, we initially choose all X_i as nonbasic and Y_j as basic variables. Note that all elements in $\mathbf{D}$ must be nonnegative for the initial basic solution to be feasible. If any of the elements in $\mathbf{D}$ are negative, the corresponding equation in Eqs (6.55) must be multiplied by -1 to have a nonnegative element on the right-hand side.

The artificial cost function for the problem is defined as

$$w = \sum_{i=1}^{n+m+p} Y_i \tag{6.61}$$

To use the Simplex procedure, we need to express the artificial cost function w in terms of nonbasic variables only. We eliminate basic variables Y_i from Eq. (6.61) by substituting Eq. (6.60) into it as

$$w = \sum_{i=1}^{n+m+p} D_i - \sum_{j=1}^{2(n+m+p)} \sum_{i=1}^{n+m+p} B_{ij} X_j$$

$$= w_0 + \sum_{j=1}^{2(n+m+p)} C_j X_j \tag{6.62}$$

where

$$C_j = - \sum_{i=1}^{n+m+p} B_{ij} \quad \text{and} \quad w_0 = \sum_{i=1}^{n+m+p} D_i \tag{6.63}$$

Thus w_0 is the initial value of the artificial cost function and C_j is the initial relative cost coefficient obtained by adding the elements of the jth column of the matrix $\mathbf{B}$ and changing its sign.

Before we can use Phase I of the Simplex method, we need to develop a procedure to impose the complementary slackness condition of Eq. (6.58). The condition is satisfied if both X_i and X_{n+m+i} are not simultaneously basic. Or, if they are, then one of them has zero value (degenerate basic feasible solution). These conditions can easily be checked while determining the pivot element in the Simplex method.

The procedure for solving QP problems is then summarized as:

1. Using the given QP problem of Eqs (6.41) to (6.44) define matrix **B** of Eq. (6.56).
2. Define vector **D** of Eq. (6.57), making sure that all its elements are nonnegative, i.e. $D_i \geqq 0$. If any element is negative, change its sign and multiply by -1 the corresponding row of matrix **B** in Eq. (6.56).
3. Calculate current values of the artificial cost function and relative cost coefficients using Eqs (6.63).
4. Complete Phase I of the Simplex method of Chapter 4. If artificial cost function is nonzero and all relative cost coefficients are nonnegative, or no pivot element can be determined, then the original QP problem is infeasible.
5. If a feasible solution to Eqs (6.55), (6.58) and (6.59) is found, then use the definition of vector **X** in Eq. (6.57) to recover optimum values for the original QP variables **x**, the Lagrange multipliers **u**, **v** and ζ and the slack variable **s**.

The preceding procedure shall be illustrated in an example problem. It is useful to note here that a slightly different procedure to solve the Kuhn–Tucker necessary conditions for the QP problem has been developed by Lemke [1965]. It is known as the *complementary pivot method*. Numerical experiments [Ravindran and Lee, 1981] have shown that method to be computationally more attractive than many other methods for solving QP problems when matrix **H** is positive semidefinite.

Example 6.8   Solution of QP problem. Solve the following QP problem:

minimize
$$f(\mathbf{x}) = (x_1 - 3)^2 + (x_2 - 3)^2 \tag{a}$$

subject to

$$x_1 + x_2 \leqq 4 \tag{b}$$

$$x_1 - 3x_2 = 1 \tag{c}$$

$$x_1, x_2 \geqq 0$$

Solution. Since this is a two variable problem, it can also be solved using the Kuhn–Tucker conditions and procedure of Section 3.4. We shall solve the problem using the Simplex method and verify Kuhn–Tucker conditions for it.

The cost function can be expanded as

$$f(\mathbf{x}) = x_1^2 - 6x_1 + x_2^2 - 6x_2 + 18$$

We shall ignore the constant (18) in the cost function and minimize the following quadratic function expressed in the form of Eq. (6.41):

$$q(\mathbf{x}) = [-6 \quad -6]\begin{bmatrix} x_1 \\ x_2 \end{bmatrix} + 0.5[x_1 \quad x_2]\begin{bmatrix} 2 & 0 \\ 0 & 2 \end{bmatrix}\begin{bmatrix} x_1 \\ x_2 \end{bmatrix} \tag{d}$$

From Eqs (d), (b), and (c), we identify the following quantities:

$$\mathbf{c} = \begin{bmatrix} -6 \\ -6 \end{bmatrix}; \qquad \mathbf{H} = \begin{bmatrix} 2 & 0 \\ 0 & 2 \end{bmatrix}$$

$$\mathbf{A} = \begin{bmatrix} 1 \\ 1 \end{bmatrix}; \qquad \mathbf{b} = [4]$$

$$\mathbf{N} = \begin{bmatrix} 1 \\ -3 \end{bmatrix}; \qquad \mathbf{e} = [1]$$

Using these quantities, matrix $\mathbf{B}$ and vectors $\mathbf{D}$ and $\mathbf{X}$ of Eqs (6.56) and (6.57) are identified as

$$\mathbf{B} = \begin{bmatrix} 2 & 0 & 1 & -1 & 0 & 0 & 1 & -1 \\ 0 & 2 & 1 & 0 & -1 & 0 & -3 & 3 \\ 1 & 1 & 0 & 0 & 0 & 1 & 0 & 0 \\ 1 & -3 & 0 & 0 & 0 & 0 & 0 & 0 \end{bmatrix}$$

$$\mathbf{D} = [6 \quad 6 \mid 4 \mid 1]^T$$

$$\mathbf{X} = [x_1 \quad x_2 \mid u_1 \mid \zeta_1 \quad \zeta_2 \mid s_1 \mid y_1 \quad z_1]^T$$

Note that since all the components of $\mathbf{D}$ are nonnegative, there is no need to multiply any rows of $\mathbf{B}$ by -1 to have nonnegative right-hand sides of Eq. (6.55).

Table 6.1 shows the initial Simplex tableau as well as the four iterations to reach the optimum solution. Note that the relative cost coefficient C_j in the initial tableau is obtained by adding all the elements in the jth column and changing its sign. Also, the complementary slackness condition of Eq. (6.58) requires

$$X_1 X_4 = 0$$

$$X_2 X_5 = 0$$

$$X_3 X_6 = 0$$

implying that X_1 and X_4, X_2 and X_5, and X_3 and X_6 cannot be basic variables simultaneously. We impose these conditions while determining the pivots in Phase I of the Simplex procedure.

After four iterations of the Simplex method, all the artificial variables are nonbasic and artificial cost function is zero. Therefore, the optimum solution is given as

$$X_1 = \tfrac{13}{4}, \quad X_2 = \tfrac{3}{4}, \quad X_3 = \tfrac{3}{4}, \quad X_8 = \tfrac{5}{4}$$

$$X_4 = 0, \quad X_5 = 0, \quad X_6 = 0, \quad X_7 = 0$$

TABLE 6.1
Simplex solution procedure for QP problem of Example 6.8

	X_1	X_2	X_3	X_4	X_5	X_6	X_7	X_8	Y_1	Y_2	Y_3	Y_4	D
Initial													
Y_1	2	0	1	−1	0	0	1	−1	1	0	0	0	6
Y_2	0	2	1	0	−1	0	−3	3	0	1	0	0	6
Y_3	1	1	0	0	0	1	0	0	0	0	1	0	4
Y_4	**1**	−3	0	0	0	0	0	0	0	0	0	1	1
	−4	0	−2	1	1	−1	2	−2	0	0	0	0	$w-17$
1st iteration													
Y_1	0	**6**	1	−1	0	0	1	−1	1	0	0	−2	4
Y_2	0	2	1	0	−1	0	−3	3	0	1	0	0	6
Y_3	0	4	0	0	0	1	0	0	0	0	1	−1	3
X_1	1	−3	0	0	0	0	0	0	0	0	0	1	1
	0	**−12**	−2	1	1	−1	2	−2	0	0	0	4	$w-13$
2nd iteration													
X_2	0	1	$\frac{1}{6}$	$-\frac{1}{6}$	0	0	$\frac{1}{6}$	$-\frac{1}{6}$	$\frac{1}{6}$	0	0	$-\frac{1}{3}$	$\frac{2}{3}$
Y_2	0	0	$\frac{2}{3}$	$\frac{1}{3}$	−1	0	$-\frac{10}{3}$	$\frac{10}{3}$	$-\frac{1}{3}$	1	0	$\frac{2}{3}$	$\frac{14}{3}$
Y_3	0	0	$-\frac{2}{3}$	$\frac{2}{3}$	0	1	$-\frac{2}{3}$	$\frac{2}{3}$	$-\frac{2}{3}$	0	1	$\frac{1}{3}$	$\frac{1}{3}$
X_1	1	0	$\frac{1}{2}$	$-\frac{1}{2}$	0	0	$\frac{1}{2}$	$-\frac{1}{2}$	$\frac{1}{2}$	0	0	0	3
	0	0	0	−1	1	−1	4	**−4**	2	0	0	0	$w-5$
3rd iteration													
X_2	0	1	0	0	0	$\frac{1}{4}$	0	0	0	0	$\frac{1}{4}$	$-\frac{1}{4}$	$\frac{3}{4}$
Y_2	0	0	**4**	−3	−1	−5	0	0	3	1	−5	−1	3
X_8	0	0	−1	1	0	$\frac{3}{2}$	−1	1	−1	0	$\frac{3}{2}$	$\frac{1}{2}$	$\frac{1}{2}$
X_1	1	0	0	0	0	$\frac{3}{4}$	0	0	0	0	$\frac{3}{4}$	$\frac{1}{4}$	$1\frac{3}{4}$
	0	0	**−4**	3	1	5	0	0	−2	0	6	2	$w-3$
4th iteration													
X_2	0	1	0	0	0	$\frac{1}{4}$	0	0	0	0	$\frac{1}{4}$	$-\frac{1}{4}$	$\frac{3}{4}$
X_3	0	0	1	$-\frac{3}{4}$	$-\frac{1}{4}$	$-\frac{5}{4}$	0	0	$\frac{3}{4}$	$\frac{1}{4}$	$-\frac{5}{4}$	$-\frac{1}{4}$	$\frac{3}{4}$
X_8	0	0	0	$\frac{1}{4}$	$-\frac{1}{4}$	$\frac{1}{4}$	−1	1	$-\frac{1}{4}$	$\frac{1}{4}$	$\frac{1}{4}$	$\frac{1}{4}$	$\frac{5}{4}$
X_1	1	0	0	0	0	$\frac{3}{4}$	0	0	0	0	$\frac{3}{4}$	$\frac{1}{4}$	$\frac{13}{4}$
	0	0	0	0	0	0	0	0	1	1	1	1	$w-0$

The optimum solution for the original QP problem is

$$x_1 = \tfrac{13}{4}, \quad x_2 = \tfrac{3}{4}, \quad u_1 = \tfrac{3}{4}, \quad \zeta_1 = 0, \quad \zeta_2 = 0$$

$$s_1 = 0, \quad y_1 = 0, \quad z_1 = \tfrac{5}{4}, \quad v_1 = y_1 - z_1 = -\tfrac{5}{4}$$

$$f(\tfrac{13}{4}, \tfrac{3}{4}) = \tfrac{41}{8}$$

Now let us verify Kuhn–Tucker conditions for the solution. The Lagrangian for the problem using Eqs (a) to (c) is given as

$$L = (x_1 - 3)^2 + (x_2 - 3)^2 + u_1(x_1 + x_2 - 4) + v_1(x_1 - 3x_2 - 1)$$

$$- \zeta_1 x_1 - \zeta_2 x_2$$

Since the slack variable for the inequality constraint is zero, it must be satisfied at equality. Therefore, solving Eqs (b) and (c), we get $x_1 = \tfrac{13}{4}$, $x_2 = \tfrac{3}{4}$. Since $x_1 > 0$, $x_2 > 0$; $\zeta_1 = \zeta_2 = 0$. Differentiating the Lagrangian, we get

$$\frac{\partial L}{\partial x_1} = 2(x_1 - 3) + u_1 + v_1 = 0$$

$$\frac{\partial L}{\partial x_2} = 2(x_2 - 3) + u_1 - 3v_1 = 0$$

Substituting $x_1 = \tfrac{13}{4}$, $x_2 = \tfrac{3}{4}$, $u_1 = \tfrac{3}{4}$, and $v_1 = -\tfrac{5}{4}$ into the preceding equations, we find that they are satisfied. Therefore, the solution obtained using the Simplex procedure satisfies all the Kuhn–Tucker conditions. Also note, that since the cost function is strictly convex, the solution obtained is unique and global. ‖

6.6 CONSTRAINED STEEPEST DESCENT METHOD

The QP subproblem defined in Eqs (6.38) to (6.40) of the previous section can be used to develop a general algorithm for constrained optimization problems. In this section we shall describe such a method. A descent function and a step size determination procedure for the method shall be described. A step-by-step procedure shall be given which can be implemented on a computer.

Note that when there are either no constraints or none is active, minimization of the quadratic function of Eq. (6.38) gives $\mathbf{d} = -\mathbf{c}$ (using the necessary condition, $\partial \bar{f} / \partial \mathbf{d} = \mathbf{0}$). This is just the steepest descent direction of Section 5.4 for the unconstrained problems. When there are constraints, their effect must be included in calculating the search direction. The search direction must satisfy all the linearized constraints. Since the search direction is a modification of the steepest descent direction to satisfy constraints, it is called the *constrained steepest descent direction*. The steps of the resulting constrained steepest descent algorithm will be clear once we define a suitable *descent function and a related line search procedure to calculate the step size* along the search direction. These procedures will be described in detail before stating the algorithm.

The QP subproblem and a Constrained Steepest Descent (CSD) method developed in this section is the most introductory and simple interpretation of the modern and powerful *Sequential Quadratic Programming* (SQP) methods. All features of the algorithms are not discussed here to keep the presentation of the key ideas simple and straightforward. It is important to note that the method works equally well when initiated from feasible or infeasible points. It can also treat equality and inequality constraints in a routine manner.

6.6.1 Descent Function

Recall that in unconstrained optimization methods the cost function is used as the descent function to monitor progress of algorithms towards the optimum point. For constrained problems, the descent function is usually constructed by adding a *penalty* for constraint violations to the current value of the cost function. There exist many descent functions for the constrained optimization problems. In this section, we shall describe one of them and show its use.

One of the properties of a descent function is that its value at the optimum point must be the same as that for the cost function. Also, it should be such that a unit step size is admissible in the neighborhood of the optimum point. A unit step size gives a higher rate of convergence to an optimization algorithm. We shall introduce *Pshenichny's descent function* due to its simplicity and success in solving a large number of engineering design problems [Pshenichny and Danilin, 1978; Belegundu and Arora, 1984]. Other descent functions shall be discussed in Section 6.7.

Pshenichny's descent function Φ at any point $\mathbf{x}$ is defined as

$$\Phi(\mathbf{x}) = f(\mathbf{x}) + RV(\mathbf{x}) \tag{6.64}$$

where R is a positive number called the *penalty parameter* (initially specified by the user), $V(\mathbf{x}) \geq 0$ is either the *maximum constraint violation* among all the constraints or zero (defined later), and $f(\mathbf{x})$ is the cost function value at $\mathbf{x}$. As an example, the descent function at the point $\mathbf{x}^{(k)}$ during the kth iteration is calculated as

$$\Phi_k = f_k + RV_k \tag{6.65}$$

where Φ_k and V_k are the values of $\Phi(\mathbf{x})$ and $V(\mathbf{x})$ at $\mathbf{x}^{(k)}$ as

$$\Phi_k = \Phi(\mathbf{x}^{(k)}); \qquad V_k = V(\mathbf{x}^{(k)}) \tag{6.66}$$

and R is the most current value of the penalty parameter. Note that the penalty parameter may change during the iterative process. Actually, it must be ensured that it is greater than or equal to the sum of all the Lagrange multipliers of the QP subproblem at the point $\mathbf{x}^{(k)}$. This is a necessary condition written as

$$R \geq r_k \tag{6.67}$$

where r_k is a sum of all the Lagrange multipliers at the kth iteration:

$$r_k = \sum_{i=1}^{p} |v_i^{(k)}| + \sum_{i=1}^{m} u_i^{(k)} \tag{6.68}$$

Since Lagrange multiplier $v_i^{(k)}$ for an equality constraint is free in sign, its absolute value is used in Eq. (6.68). $u_i^{(k)}$ is the multiplier for the ith inequality constraint.

The *parameter $V_k \geq 0$ related to the maximum constraint violation* at the kth iteration is determined using the calculated values of the constraint functions at the design point $\mathbf{x}^{(k)}$ as

$$V_k = \max\{0; |h_1|, |h_2|, \ldots, |h_p|; g_1, g_2, \ldots, g_m\} \tag{6.69}$$

Since the equality constraint is violated if it is different from zero, absolute value is used with each h_i in Eq. (6.69). Note that V_k is always nonnegative, i.e. $V_k \geq 0$. If all constraints are satisfied at $\mathbf{x}^{(k)}$, then $V_k = 0$.

Example 6.9   Calculation of descent function. A design problem is formulated as follows:

minimize $\qquad\qquad f(\mathbf{x}) = x_1^2 + 320x_1x_2$

subject to

$$\frac{x_1}{60x_2} - 1 \leq 0$$

$$1 - \frac{x_1(x_1 - x_2)}{3600} \leq 0$$

$$-x_1 \leq 0; \quad -x_2 \leq 0$$

Taking the penalty parameter R as 10 000, calculate the value of the descent function at the point $\mathbf{x}^{(0)} = (40, 0.5)$.

Solution. The cost and constraint functions at the given point $\mathbf{x}^{(0)} = (40, 0.5)$ are evaluated as

$$f_0 \equiv f(40, 0.5) = (40)^2 + 320(40)(0.5) = 8000$$

$$g_1 \equiv \frac{40}{60(0.5)} - 1 = 0.333 \text{ (violation)}$$

$$g_2 \equiv 1 - \frac{40(40 - 0.5)}{3600} = 0.5611 \text{ (violation)}$$

$$g_3 \equiv -40 < 0 \text{ (inactive)}$$

$$g_4 \equiv -0.5 < 0 \text{ (inactive)}$$

Thus, the maximum constraint violation is determined using Eq. (6.69) as

$$V_0 = \max \{0; 0.333, 0.5611, -40, -0.5\} = 0.5611$$

Using Eq. (6.65), the descent function is calculated as

$$\Phi_0 = f_0 + RV_0$$
$$= 8000 + (10\,000)(0.5611)$$
$$= 13\,611 \qquad\qquad\qquad ||$$

6.6.2 Step Size Determination

Before the constrained steepest descent algorithm can be given, a step size determination procedure is needed. We shall next describe such a procedure. In most practical implementations of the algorithm, an *inexact line search* that has worked fairly well is used to determine the step size. We shall describe the procedure and illustrate its use in an example. Note that the exact line search procedures described earlier, such as Golden Section search, can be also used. However, those procedures can be inefficient; therefore, inexact line search is preferred in most constrained optimization methods.

Define a sequence of trial step sizes t_j as follows:

$$t_j = (\tfrac{1}{2})^j; \qquad j = 0, 1, 2, 3, 4, \ldots \qquad (6.70)$$

i.e. the sequence of trial step sizes is given as

$$j = 0; \qquad t_0 = (\tfrac{1}{2})^0 = 1$$
$$j = 1; \qquad t_1 = (\tfrac{1}{2})^1 = \tfrac{1}{2}$$
$$j = 2; \qquad t_2 = (\tfrac{1}{2})^2 = \tfrac{1}{4}$$
$$j = 3; \qquad t_3 = (\tfrac{1}{2})^3 = \tfrac{1}{8}$$
$$j = 4; \qquad t_4 = (\tfrac{1}{2})^4 = \tfrac{1}{16}$$
$$\vdots$$

Thus, we start with the trial step size as $t_0 = 1$. If a certain descent condition (defined in the following paragraph) is not satisfied, the trial step is taken as half of the previous trial, i.e. $t_1 = \tfrac{1}{2}$. If the descent condition is still not satisfied, the trial step size is bisected again. The procedure is continued, until the descent condition is satisfied.

In the following development, we shall *use a second subscript or superscript to indicate values of certain quantities at the trial step sizes*. For example, let t_j be the trial step size at the kth iteration. Then the trial design point for which the descent condition is checked will be expressed as

$$\mathbf{x}^{(k+1,j)} = \mathbf{x}^{(k)} + t_j\mathbf{d}^{(k)} \qquad (6.71)$$

At the kth iteration, we determine an acceptable step size as $\alpha_k = t_j$ with j as the smallest integer to satisfy the *descent condition*

$$\Phi_{k+1,j} \leqq \Phi_k - t_j\beta_k \qquad (6.72)$$

where $\Phi_{k+1,j}$ is the descent function of Eq. (6.65) evaluated at the trial step size t_j and the corresponding design point $\mathbf{x}^{(k+1,j)}$ as

$$\Phi_{k+1,j} = \Phi(\mathbf{x}^{(k+1,j)})$$
$$= f_{k+1,j} + RV_{k+1,j} \qquad (6.73)$$

with $f_{k+1,j} = f(\mathbf{x}^{(k+1,j)})$ and $V_{k+1,j}$ as the maximum constraint violation at the trial design point calculated using Eq. (6.69). Note that in evaluating $\Phi_{k+1,j}$ and Φ_k in Eq. (6.72), the most recent value of the penalty parameter R is used. The constant β_k in Eq. (6.72) is determined using the search direction $\mathbf{d}^{(k)}$ as

$$\beta_k = \gamma \, \|\mathbf{d}^{(k)}\|^2 \qquad (6.74)$$

where γ is a specified constant between 0 and 1. We shall later study the effect of γ on the step size determination process. Note that in the kth iteration β_k defined in Eq. (6.74) is a constant. As a matter of fact t_j is the only variable on the right-hand side of the Inequality (6.72). However, when t_j is changed, the design point is changed, affecting the cost and constraint function values. This way the descent function value on the left-hand side of Inequality (6.72) is changed.

The Inequality (6.72) is called the *descent condition*. It is an important condition that must be satisfied at each iteration to obtain a convergent algorithm. To understand the meaning of condition (6.72), consider Fig. 6.14, where various quantities are plotted as functions of t. For example, the horizontal line A–B represents the constant Φ_k which is the value of the descent function at the current design point $\mathbf{x}^{(k)}$; line A–C represents the function $-t\beta_k$ whose origin has been moved to point A; and the curve AHGD represents the descent function Φ plotted as a function of parameter t and originating from point A. The line A–C and the curve AHGD intersect at point J which corresponds to the point $t = \bar{t}$ on the t-axis. For the descent condition of Inequality (6.72) to be satisfied the curve AHGD must be below the line A–C. This gives only the portion AHJ of the curve AHGD. Thus, we see from the figure that step size larger than $\bar{t}$ does not satisfy the descent condition of Inequality (6.72). To verify this, consider points D and E on the line $t_0 = 1$. Point D represents $\Phi_{k+1,0} = \Phi(\mathbf{x}^{(k+1,0)})$ and point E represents $(\Phi_k - t_0 \beta_k)$. Thus point D represents the left-hand side and point E represents

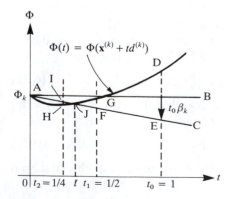

FIGURE 6.14
Geometrical interpretation of the descent condition for determination of step size in constrained steepest descent algorithm.

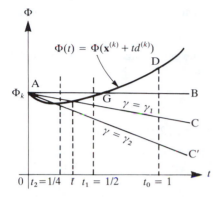

FIGURE 6.15
Effect of parameter γ on step size determination.

the right-hand side of the Inequality (6.72). Since point D is higher than point E, Inequality (6.72) is violated. Similarly, points G and F on the line $t_1 = \frac{1}{2}$ violate the descent condition. Points I and H on the line $t_2 = \frac{1}{4}$ satisfy the descent condition, so the step size α_k at the kth iteration is taken as $\frac{1}{4}$ for the example of Fig. 6.14.

It is important to understand the effect of γ on step size determination. γ is selected as a positive number between 0 and 1. Let us select γ_1 and $\gamma_2 > 0$ with $\gamma_2 > \gamma_1$. Larger γ gives a larger value for the constant β_k in Eq. (6.74). Since β_k is the slope of the line $t\beta_k$, we designate line A–C as $\gamma = \gamma_1$ and A–C′ as $\gamma = \gamma_2$ in Fig. 6.15. Thus, we observe from the figure that a larger γ tends to reduce the step size in order to satisfy the descent condition of Inequality (6.72).

For the purpose of checking the descent condition in actual calculations, it is better to write the inequality of Eq. (6.72) as

$$\Phi_{k+1,j} + t_j \beta_k \leqq \Phi_k; \qquad j = 0, 1, 2 \ldots \tag{6.75}$$

We illustrate the procedure for calculating step size in the following problem example.

Example 6.10 Calculations for step size in constrained steepest descent method. An engineering design problem is formulated as

minimize $\qquad f(\mathbf{x}) = x_1^2 + 320x_1x_2$

subject to

$$g_1(\mathbf{x}) \equiv \frac{x_1}{60x_2} - 1 \leqq 0$$

$$g_2(\mathbf{x}) \equiv 1 - \frac{x_1(x_1 - x_2)}{3600} \leqq 0$$

$$g_3(\mathbf{x}) \equiv -x_1 \leqq 0$$

$$g_4(\mathbf{x}) \equiv -x_2 \leqq 0$$

At a design point $\mathbf{x}^{(0)} = (40, 0.5)$, the search direction is given as $\mathbf{d}^{(0)} = (25.6, 0.45)$. The Lagrange multiplier vector for the constraint is given as

$\mathbf{u} = (4880, 19\,400, 0, 0)$. Choose $\gamma = 0.5$ and calculate the step size for design change using the inexact line search procedure.

Solution. Since the Lagrange multipliers for the constraints are given, the initial value of the penalty parameter is calculated as

$$R = \sum_{i=1}^{4} u_i$$

$$= 4880 + 19\,400 = 24\,280$$

It is important to note that same value of R is to be used on both sides of the descent condition in Eq. (6.72) or Eq. (6.75). The constant β_0 of Eq. (6.74) is calculated as

$$\beta_0 = 0.5(25.6^2 + 0.45^2)$$

$$= 328$$

Calculation of Φ_0. The cost and constraint functions at the starting point $\mathbf{x}^{(0)} = (40, 0.5)$ are calculated as

$$f_0 \equiv f(40, 0.5) = 40^2 + 320(40)(0.5) = 8000$$

$$g_1(40, 0.5) = \frac{40}{60(0.5)} - 1 = 0.333 \text{ (violation)}$$

$$g_2(40, 0.5) = 1 - \frac{40(40 - 0.5)}{3600} = 0.5611 \text{ (violation)}$$

$$g_3(40, 0.5) = -40 < 0 \text{ (inactive)}$$

$$g_4(40, 0.5) = -0.5 < 0 \text{ (inactive)}$$

Maximum constraint violation using Eq. (6.69) is given as

$$V_0 = \max \{0; 0.333, 0.5611, -40, -0.5\}$$

$$= 0.5611$$

Using Eq. (6.65), the current descent function is evaluated as

$$\Phi_0 = f_0 + RV_0$$

$$= 8000 + (24\,280)(0.5611)$$

$$= 21\,624$$

Trial step size $t_0 = 1$. Let $j = 0$ in Eq. (6.70), so the trial step size is $t_0 = 1$. The trial design point is calculated from Eq. (6.71) as

$$x_1^{(1,0)} = x_1^{(0)} + t_0 d_1^{(0)}$$

$$= 40 + (1.0)(25.6) = 65.6$$

$$x_2^{(1,0)} = x_2^{(0)} + t_0 d_2^{(0)}$$

$$= 0.5 + (1.0)(0.45) = 0.95$$

The cost and constraint functions at the trial design point are calculated as

$$f_{1,0} \equiv f(65.6, 0.95)$$

$$= (65.6)^2 + 320(65.6)(0.95) = 24\,246$$

$$g_1(65.6, 0.95) = \frac{65.6}{60(0.95)} - 1$$

$$= 0.151 > 0 \text{ (violation)}$$

$$g_2(65.6, 0.95) = 1 - \frac{65.6(65.6 - 0.95)}{3600}$$

$$= -0.1781 < 0 \text{ (inactive)}$$

$$g_3(65.6, 0.95) = -65.6 < 0 \text{ (inactive)}$$

$$g_4(65.6, 0.95) = -0.95 < 0 \text{ (inactive)}$$

Maximum constraint violation using Eq. (6.69) is given as

$$V_{1,0} = \max \{0; 0.151, -0.1781, -65.6, -0.95\}$$

$$= 0.151$$

The descent function at the trial point is calculated using Eq. (6.73) as

$$\Phi_{1,0} = f_{1,0} + R V_{1,0}$$

$$= 24\,246 + 24\,280(0.151) = 27\,912$$

For the descent condition of Eq. (6.75), we get

$$\text{LHS} = 27\,912 + 328 = 28\,240$$

$$\text{RHS} = 21\,624$$

Since LHS > RHS, the Inequality (6.75) is violated.

Trial step size $t_1 = 0.5$. Let $j = 1$ in Eq. (6.70), so the trial step size is $t_1 = 0.5$. The new trial design point is

$$x_1^{(1,1)} = x_1^{(0)} + t_1 d_1^{(0)}$$

$$= 40 + 0.5(25.6) = 52.8$$

$$x_2^{(1,1)} = x_2^{(0)} + t_1 d_2^{(0)}$$

$$= 0.5 + 0.5(0.45) = 0.725$$

The cost and constraint functions at the new trial design point are calculated as

$$f_{1,1} \equiv f(52.8, 0.725)$$

$$= (52.8)^2 + 320(52.8)(0.725) = 15\,037$$

$$g_1(52.8, 0.725) = \frac{52.8}{60(0.725)} - 1$$

$$= 0.2138 > 0 \text{ (violation)}$$

$$g_2(52.8, 0.725) = 1 - \frac{52.8(52.8 - 0.725)}{3600}$$

$$= 0.2362 > 0 \text{ (violation)}$$

$$g_3(52.8, 0.725) = -52.8 < 0 \text{ (inactive)}$$

$$g_4(52.8, 0.725) = -0.725 < 0 \text{ (inactive)}$$

Maximum constraint violation using Eq. (6.69) is given as

$$V_{1,1} = \max\{0; 0.2138, 0.2362, -52.8, -0.725\}$$
$$= 0.2362$$

The descent function at the trial design point is calculated using Eq. (6.64) as

$$\Phi_{1,1} = f_{1,1} + R V_{1,1}$$
$$= 15\,037 + 24\,280(0.2362) = 20\,772$$

Evaluating the descent condition of Eq. (6.75), we get

$$\text{LHS} = 20\,772 + (0.5)(328) = 20\,936$$

$$\text{RHS} = 21\,624$$

Since LHS < RHS, the Inequality (6.75) is satisfied. Therefore the step size of 0.5 is acceptable. ‖

6.6.3   The CSD Algorithm

We are now ready to state the constrained steepest descent (CSD) algorithm in a step-by-step form. It has been proved [Pshenichny and Danilin, 1978] that the solution point of the sequence $\{\mathbf{x}^{(k)}\}$ generated by the following algorithm is a Kuhn–Tucker point for the general constrained optimization problem defined in Eqs (6.1) to (6.4). The convergence criteria for the algorithm is that $\|\mathbf{d}\| \le \varepsilon$ for a feasible design point. Here ε is a small positive number and $\mathbf{d}$ is the direction vector that is obtained as the solution of the QP subproblem. The constrained steepest descent method is now summarized in the form of a *computational algorithm*.

Step 1. Set $k = 0$. Estimate initial values for design variables as $\mathbf{x}^{(0)}$. Select an appropriate initial value for the penalty parameter R_0, a constant γ between 0 and 1 $(0 < \gamma < 1)$ and two small numbers ε_1 and ε_2 that define the permissible constraint violation and convergence parameter, respectively. $R_0 = 1$ and $\gamma = 0.2$ is a reasonable selection.

Step 2. At $\mathbf{x}^{(k)}$ compute the cost and constraint functions and their gradients. Calculate the maximum constraint violation V_k as defined in Eq. (6.69).

Step 3. Using the cost and constraint function values and their gradients define the QP subproblem given in Eqs (6.38) to (6.40). Solve the QP problem to obtain the search direction $\mathbf{d}^{(k)}$ and Lagrange multipliers $\mathbf{v}^{(k)}$ and $\mathbf{u}^{(k)}$.

Step 4. Check for the following convergence criteria

$$\|\mathbf{d}^{(k)}\| \le \varepsilon_2$$

and the maximum constraint violation $V_k \le \varepsilon_1$. If the convergence criteria are satisfied then stop. Otherwise continue.

Step 5. To check the necessary condition of Eq. (6.67) for the penalty parameter R, calculate the sum r_k of the Lagrange multipliers defined in Eq.

(6.68). Set $R = \max\{R_k, r_k\}$. This will always satisfy the necessary condition of Eq. (6.67).

Step 6. Set $\mathbf{x}^{(k+1)} = \mathbf{x}^{(k)} + \alpha_k \mathbf{d}^{(k)}$, where $\alpha = \alpha_k$ is a proper step size. As for the unconstrained problems, the step size can be obtained by minimizing the descent function of Eq. (6.64) along the search direction $\mathbf{d}^{(k)}$. Any of the procedures discussed in Section 5.3, such as the Golden Section search, can be used to determine optimum step size. However, these procedures are usually inefficient and rarely used in practical applications. Often, only an inexact line search described in Section 6.6.2 is used.

Step 7. Save the current penalty parameter as $R_{k+1} = R$. Update the iteration counter as $k = k + 1$, and go to Step 2.

The CSD algorithm along with the foregoing step size determination procedure is convergent provided second derivatives of all the functions are piece-wise continuous (this is the so-called Lipschitz condition) and the set of design points $\mathbf{x}^{(k)}$ are bounded as follows:

$$\Phi(\mathbf{x}^{(k)}) \leq \Phi(\mathbf{x}^{(0)}); \qquad k = 1, 2, 3, \ldots$$

Example 6.11 Use of constrained steepest descent algorithm. Consider the problem of Example 6.3:

minimize
$$f(\mathbf{x}) = x_1^2 + x_2^2 - 3x_1 x_2$$

subject to
$$g_1(\mathbf{x}) = \tfrac{1}{6}x_1^2 + \tfrac{1}{6}x_2^2 - 1.0 \leq 0$$

$$g_2(\mathbf{x}) = -x_1 \leq 0$$

$$g_3(\mathbf{x}) = -x_2 \leq 0$$

Let $\mathbf{x}^{(0)} = (1, 1)$ be the starting design. Use $R_0 = 10$, $\gamma = 0.5$ and $\varepsilon_1 = \varepsilon_2 = 0.001$ in the constrained steepest descent method. Perform only two iterations.

Solution. The functions of the problem are plotted in Fig. 6.4. The optimum solution for the problem is obtained as $\mathbf{x} = (\sqrt{3}, \sqrt{3})$, $\mathbf{u} = (3, 0, 0)$, $f = -3$.

Iteration 1 $(k = 0)$

Step 1. The initial data are specified as

$$\mathbf{x}^{(0)} = (1, 1); \; R_0 = 10; \; \gamma = 0.5 (0 < \gamma < 1); \; \varepsilon_1 = \varepsilon_2 = 0.001$$

Step 2. To form the QP subproblem, the cost and constraint function values and their gradients must be evaluated at the initial design point $\mathbf{x}^{(0)}$:

$$f(1, 1) = -1$$

$$g_1(1, 1) = -\tfrac{2}{3} < 0 \text{ (inactive)}$$

$$g_2(1, 1) = -1 < 0 \text{ (inactive)}$$

$$g_3(1, 1) = -1 < 0 \text{ (inactive)}$$

$$\nabla f(1, 1) = (-1, -1)$$

$$\nabla g_1(1, 1) = (\tfrac{1}{3}, \tfrac{1}{3})$$

$$\nabla g_2(1, 1) = (-1, 0)$$

$$\nabla g_3(1, 1) = (0, -1)$$

Note that all constraints are inactive at the starting point, so $V_0 = 0$ calculated from Eq. (6.69) as $V_0 = \max \{0; -\tfrac{2}{3}, -1, -1\}$. The linearized constraints are plotted in Fig. 6.5 and the linearized constraint set is plotted in Fig. 6.9.

Step 3. Using the preceding values, the QP subproblem of Eqs (6.38) to (6.40) at (1, 1) is given as

minimize $\bar{f} = (-d_1 - d_2) + 0.5(d_1^2 + d_2^2)$

subject to

$$\tfrac{1}{3}d_1 + \tfrac{1}{3}d_2 \leq \tfrac{2}{3}$$

$$-d_1 \leq 1; \; -d_2 \leq 1$$

Note that the above QP subproblem is strictly convex and thus, has a unique solution. Numerical method must generally be used to solve the subproblem. However, since the present problem is quite simple, it can be solved by writing the Kuhn–Tucker necessary conditions as follows:

$$L = (-d_1 - d_2) + 0.5(d_1^2 + d_2^2) + u_1[\tfrac{1}{3}(d_1 + d_2 - 2) + s_1^2] + u_2(-d_1 - 1 + s_2^2)$$
$$+ u_3(-d_2 - 1 + s_3^2)$$

$$\frac{\partial L}{\partial d_1} \equiv -1 + d_1 + \tfrac{1}{3}u_1 - u_2 = 0$$

$$\frac{\partial L}{\partial d_2} \equiv -1 + d_2 + \tfrac{1}{3}u_1 - u_3 = 0$$

$$\tfrac{1}{3}(d_1 + d_2 - 2) + s_1^2 = 0$$

$$(-d_1 - 1) + s_2^2 = 0; \qquad (-d_2 - 1) + s_3^2 = 0$$

$$u_i s_i = 0; \qquad \text{and} \qquad u_i \geq 0; \qquad i = 1,2,3$$

where u_1, u_2 and u_3 are the Lagrange multipliers for the three constraints and s_1^2, s_2^2 and s_3^2 are the corresponding slack variables. Solving the foregoing K–T conditions, we get the direction vector $\mathbf{d}^{(0)} = (1, 1)$ with $\bar{f} = -1$ and $\mathbf{u}^{(0)} = (0, 0, 0)$. This solution agrees with the graphical solution given in Fig. 6.16. The feasible region for the subproblem is the triangle ABC, and the optimum solution is at Point D.

Step 4. As $\|\mathbf{d}^{(0)}\| = \sqrt{2} > \varepsilon_2$, the convergence criterion is not satisfied.

Step 5. Calculate $r_0 = \sum_{i=1}^{m} u_i^{(0)} = 0$ as defined in Eq. (6.68). To satisfy the necessary condition of Inequality (6.67), let $R = \max \{R_0, r_0\} = \max \{10, 0\} = 10$. It is important to note that $R = 10$ is to be used throughout the first iteration to satisfy the descent condition of Eq. (6.72) or Eq. (6.75).

Step 6. For step size determination, we use the inexact line search described in Section 6.6.2. The current value of the descent function Φ_0 of Eq. (6.65) and

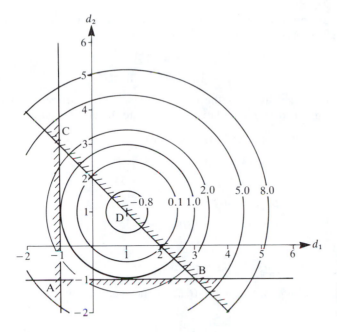

FIGURE 6.16
Solution of quadratic programming subproblem for Example 6.11 at the point (1, 1).

the constant β_0 of Eq. (6.74) are calculated as

$$\Phi_0 = f_0 + RV_0$$
$$= -1 + (10)(0) = -1$$
$$\beta_0 = \gamma \, \|\mathbf{d}^{(0)}\|^2$$
$$= 0.5(1 + 1) = 1$$

Let the trial step size be $t_0 = 1$ and evaluate the new value of the descent function to check the descent condition of Eq. (6.72):

$$\mathbf{x}^{(1,0)} = \mathbf{x}^{(0)} + t_0\mathbf{d}^{(0)}$$
$$= (2, 2)$$

At the trial design point, evaluate the cost and constraint functions, and then evaluate the maximum constraint violation to calculate the descent function:

$$f_{1,0} \equiv f(2, 2) = -4$$
$$V_{1,0} \equiv V(2, 2)$$
$$= \max\{0; \tfrac{1}{3}, -2, -2\} = \tfrac{1}{3}$$
$$\Phi_{1,0} = f_{1,0} + RV_{1,0}$$
$$= -4 + (10)\tfrac{1}{3} = -\tfrac{2}{3}$$
$$\Phi_0 - t_0\beta_0 = -1 - 1 = -2$$

Since $\Phi_{1,0} > \Phi_0 - t_0\beta_0$, the descent condition of Inequality (6.72) is not satisfied. Let us try $j = 1$ (i.e. bisect the step size to $t_1 = 0.5$).

Let the trial step size be $t_1 = 0.5$ and evaluate the new value of the descent function to check the descent condition of Eq. (6.72):

$$\mathbf{x}^{(1,1)} = \mathbf{x}^{(0)} + t_1 \mathbf{d}^{(0)}$$
$$= (1.5, 1.5)$$

At the trial design point, evaluate the cost and constraint functions, and then evaluate the maximum constraint violation to calculate the descent function:

$$f_{1,1} \equiv f(1.5, 1.5) = -2.25$$
$$V_{1,1} \equiv V(1.5, 1.5)$$
$$= \max \{0; -\tfrac{1}{4}, -1.5, -1.5\} = 0$$
$$\Phi_{1,1} = f_{1,1} + RV_{1,1}$$
$$= -2.25 + (10)0 = -2.25$$

and

$$\Phi_0 - t_1\beta_0 = -1 - 0.5 = -1.5$$

Now the descent condition of Inequality (6.72) is satisfied (i.e. $\Phi_{1,1} < \Phi_0 - t_1\beta_0$); thus

$$\alpha_0 = 0.5 \qquad \text{and} \quad \mathbf{x}^{(1)} = (1.5, 1.5)$$

Step 7. Set $R_{0+1} = R_0 = 10$, $k = 1$ and go to Step 2.

Iteration 2 ($k = 1$). For the second iteration, Steps 2 to 7 of the algorithm are repeated as follows:

Step 3. The QP subproblem of Eqs (6.38) to (6.40) at $\mathbf{x}^{(1)} = (1.5, 1.5)$ is defined as follows:

minimize $\qquad\qquad (-1.5d_1 - 1.5d_2) + 0.5(d_1^2 + d_2^2)$

subject to $\qquad\qquad 0.5d_1 + 0.5d_2 \leq 0.25$

$$-d_1 \leq 1.5, \ -d_2 \leq 1.5$$

Since all constraints are inactive, the maximum violation $V_1 = 0$ from Eq. (6.69). The new cost function is given as $f_1 = -2.25$. The solution of the above QP subproblem is $\mathbf{d}^{(1)} = (0.25, 0.25)$ and $\mathbf{u}^{(1)} = (2.5, 0, 0)$.

Step 4. As $\|\mathbf{d}^{(1)}\| = 0.3535 > \varepsilon_2$, the convergence criterion is not satisfied.

Step 5. Let $r_1 = \sum_{i=1}^{m} u_i^{(1)} = 2.5$. Therefore,

$$R = \max \{R_1, r_1\} = \max \{10, 2.5\} = 10.$$

Step 6. For line search, try $j = 0$ in Inequality (6.72) (i.e. $t_0 = 1$):

$$\Phi_1 = f_1 + RV_1$$
$$= -2.25 + 10(0) = -2.25$$
$$\beta_1 = \gamma \|\mathbf{d}^{(1)}\|^2$$
$$= 0.5(0.125) = 0.0625$$

Let the trial step size be $t_0 = 1$ and evaluate the new value of the descent function to check the descent condition of Eq. (6.72):

$$\mathbf{x}^{(2,0)} \equiv \mathbf{x}^{(1)} + t_0 \mathbf{d}^{(1)}$$
$$= (1.75, 1.75)$$
$$f_{2,0} \equiv f(1.75, 1.75)$$
$$= -3.0625$$
$$V_{2,0} \equiv V(1.75, 1.75)$$
$$= \max \{0; 0.0208, -1.75, -1.75\} = 0.0208$$
$$\Phi_{2,0} = f_{2,0} + RV_{2,0}$$
$$= -3.0625 + (10)0.0208 = -2.8541$$

and

$$\Phi_1 - t_0 \beta_1 = -2.25 - (1)(0.0625)$$
$$= -2.3125$$

As the descent condition of Inequality (6.72) is satisfied,

$$\alpha_1 = 1.0 \text{ and } \mathbf{x}^{(2)} = (1.75, 1.75)$$

Step 7. Set $R_2 = R = 10$, $k = 2$ and go to Step 2.

The maximum constraint violation at the current design $\mathbf{x}^{(2)} = (1.75, 1.75)$ is 0.0208, which is greater than the permissible constraint violation. We need to go through more iterations to get back to the feasible region. Note that the optimum point for this problem is (1.732, 1.732). Therefore, the current point is quite close to the optimum with $f_2 = -3.0625$.

Example 6.12   Effect of γ on the performance of CSD algorithm. For the optimum design problem of Example 6.11 study the effect of variations in the parameter γ on the performance of the CSD algorithm.

Solution. In Example 6.11, $\gamma = 0.5$ is used. Let us see what happens if a very small value of γ (say 0.01) is used. All calculations up to Step 6 of Iteration 1 are unchanged. In Step 6, the value of β_0 is changed to $\beta_0 = \gamma \|\mathbf{d}^{(0)}\|^2 = 0.01(2) = 0.02$. Therefore,

$$\Phi_0 - t_0 \beta_0 = -1 - 1(0.02)$$
$$= -1.02$$

which is smaller than $\Phi_{1,0}$, so the descent condition of Inequality (6.72) is violated. Thus, the step size in Iteration 1 will be 0.5 as before.
Calculations in Iteration 2 are unchanged until Step 6 where $\beta_1 = \gamma \|\mathbf{d}^{(1)}\|^2 = 0.02(0.125) = 0.0025$. Therefore,

$$\Phi_1 - t_0 \beta_1 = -2.25 - (1)(0.0025)$$
$$= -2.2525$$

The descent condition of Inequality (6.72) is satisfied. Thus a smaller value of γ has no effect on the first two iterations.

Let us see what happens if a larger value for γ (say 0.9) is chosen. It can be verified that in Iteration 1, there is no difference in calculations. In Step 2, the step size is reduced to 0.5. Therefore, the new design point is $\mathbf{x}^{(2)} = (1.625, 1.625)$. At this point $f_2 = -2.641$, $g_1 = -0.1198$ and $V_1 = 0$. Thus, a larger γ results in a smaller step size and the new design point remains strictly feasible.

$\parallel$

Example 6.13   Effect of penalty parameter R on CSD algorithm. For the optimum design problem of Example 6.11, study the effect of variations in the parameter R on the performance of the CSD algorithm.

Solution. In Example 6.11, initial R is selected as 10. Let us see what happens if R is selected as 1.0. There is no change in the calculations up to Step 5 in Iteration 1. In Step 6,

$$\Phi_{1,0} = -4 + (1)(\tfrac{1}{3})$$

$$= -\tfrac{11}{3}$$

$$\Phi_0 - t_0\beta_0 = -1 + (1)(0)$$

$$= -1$$

Therefore, $\alpha_0 = 1$ satisfies the descent condition of Inequality (6.72) and the new design is given as $\mathbf{x}^{(1)} = (2, 2)$. This is different from what was obtained in Example 6.11.

Iteration 2

Step 3. The QP subproblem of Eqs (6.38) to (6.40) at $\mathbf{x}^{(1)} = (2, 2)$ is defined as follows

minimize $(-2d_1 - 2d_1) + 0.5(d_1^2 + d_2^2)$

subject to

$$\tfrac{2}{3}d_1 + \tfrac{2}{3}d_2 \leq -\tfrac{1}{3}$$

$$-d_1 \leq 2, \qquad -d_2 \leq 2$$

At the point $(2, 2)$, $V_1 = \tfrac{1}{3}$ and $f_1 = -4$. The solution of the QP subproblem is given as $\mathbf{d}^{(1)} = (-0.25, -0.25)$ and $\mathbf{u}^{(1)} = (\tfrac{27}{8}, 0, 0)$.

Step 4. As $\|\mathbf{d}^{(1)}\| = 0.3535 > \varepsilon_2$, the convergence criterion is not satisfied.

Step 5. Let $r_1 = \sum_{i=1}^{3} u_i^{(1)} = \tfrac{27}{8}$. Therefore,

$$R = \max\{R_1, r_1\} = \max\{1, \tfrac{27}{8}\} = \tfrac{27}{8}$$

Step 6. For line search, try $j = 0$ in Inequality (6.72), i.e. $t_0 = 1$:

$$\Phi_1 = f_1 + RV_1 = -4 + (\tfrac{27}{8})(\tfrac{1}{3}) = -2.875$$

$$\Phi_{2,0} = f_{2,0} + RV_{2,0}$$

$$= -3.0625 + (\tfrac{27}{8})(0.0208) = -2.9923$$

$$\beta_1 = \gamma \|\mathbf{d}^{(1)}\|^2$$

$$= 0.5(0.125) = 0.0625$$

$$\Phi_1 - t_0\beta_1 = -2.875 - (1)(0.0625)$$

$$= -2.9375$$

As the descent condition is satisfied, $\alpha_1 = 1.0$ and $\mathbf{x}^{(2)} = (1.75, 1.75)$.

Step 7. Set $R_2 = R_1 = \frac{27}{8}$, $k = 2$ and go to Step 2.

The design at the end of the second iteration is the same as in Example 6.11. This is just a coincidence. We observe that a smaller R gave a larger step size in the first iteration. In general this can change the history of the iterative process. ‖

6.6.4 Observations on the CSD Algorithm

Pshenichny's algorithm given above can be looked upon as the constrained analog of the steepest descent method for unconstrained problems. The following observations are noteworthy for the algorithm:

1. The CSD algorithm is a *first-order* method for constrained optimization. The algorithm converges to a local minimum point starting from an arbitrary point. It can treat equality as well as inequality constraints.

2. The *potential constraint strategy* as suggested by Pshenichny is not introduced in the algorithm for the sake of simplicity. It can be easily incorporated and the algorithm with the potential constraint strategy is considered to be efficient for engineering design problems [Belegundu and Arora, 1984].

3. The Golden Section search may be used to find the step size by minimizing the descent function instead of trying to satisfy the descent condition. Note, however, that the Golden Section search technique is not recommended for numerical algorithms in constrained optimization as it is inefficient.

4. The rate of convergence of the CSD algorithm can be further improved by including higher-order information about the problem functions in the QP subproblem [Lim and Arora, 1986]. This will be discussed in the next section.

5. It is important to note that the step size determined using Eq. (6.70) is not allowed to be greater than one. However in numerical implementations [Lim and Arora, 1986], the algorithm is more efficient for many problems when the step size is allowed to be larger than one. This can be done by allowing the integer j in Eq. (6.70) to have negative values. Therefore in practical applications, it is suggested to investigate the use of a step size that is larger than one.

6. It is also important to note that although the CSD algorithm is theoretically supposed to converge to a local minimum point starting from any point, its numerical behavior can be quite different. This is due to uncertainties in selection of parameters γ and R_0. The step size may become too small for one specification of these parameters. If this happens, some other selection should be tried.

7. The starting point can also affect performance of the algorithm. For example, at some points the QP subproblem may not have any solution.

This need not mean that the original problem is infeasible. Problem may be highly nonlinear, so the linearized constraints may be inconsistent giving infeasible subproblem. This situation can be handled by either temporarily deleting the inconsistant constraints or starting from another point. For more discussion on the implementation of the algorithm, Tseng and Arora [1988] should be consulted.

6.6.5 Use of Potential Constraint Strategy

As noted in the foregoing, the potential set strategy where only a subset of the constraints is included in defining the subproblem, can be incorporated into the CSD algorithm. The procedure is to calculate all the constraint functions for the problem and define a potential set according to Eq. (6.12). There are several alternate ways to define the potential set, and different procedures can lead to different search directions and the path to the optimum point.

The main effect of using a potential set strategy in an algorithm is on the efficiency of the entire iterative process. This is particularly true for large and complex applications where the evaluation of gradients of constraints is a computationally expensive proposition. With the potential set strategy, gradients of only the potential constraints are calculated and used in defining the search direction determination subproblem. The original problem may have hundreds of constraints, but only a few may be in the potential set. Thus with this strategy, not only the number of gradient evaluations is reduced but also the dimension of the subproblem for the search direction is substantially reduced. This can result in additional saving in the computational effort. *Therefore, the potential set strategy is highly beneficial and must be used in practical applications of optimization.*

Later, in Chapter 7, search directions for an example problem are solved by defining the QP subproblems with and without the potential set strategy. It is shown there that the two directions are quite different and both make progress towards the optimum solution.

> **Example 6.14 Minimum area design of a rectangular beam.** For the minimum area beam design problem of Section 2.8.5, find the optimum solution using the CSD algorithm starting from the points $(50, 200)$ mm and $(1000, 1000)$ mm.
>
> **Solution.** The problem is formulated and solved graphically in Section 2.8.5. After normalizing the constraints the problem is defined as follows. Find width b and depth d to minimize the cross-sectional area
>
> $$f(b, d) = bd$$
>
> subject to the bending stress constraint
>
> $$\frac{(2.40\text{E}+07)}{bd^2} - 1.0 \leqq 0$$

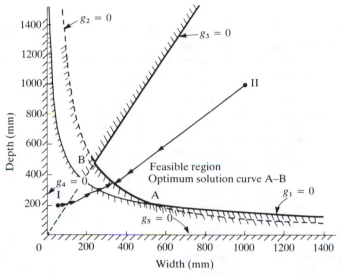

FIGURE 6.17
The history of the iterative process for the rectangular beam design problem.

the shear stress constraint

$$\frac{(1.125E+05)}{bd} - 1.0 \leqq 0$$

the depth constraint

$$\frac{d}{2b} - 1.0 \leqq 0$$

and the explicit bounds on design variables

$$10 \leqq b \leqq 1000$$

$$10 \leqq d \leqq 1000$$

The graphical solution for the problem is given in Fig. 6.17; any point on the curve AB gives an optimum solution.

The problem is solved starting from the given points with the CSD algorithm available in the IDESIGN software package [Arora and Tseng, 1987]. The algorithm has been implemented using the potential set strategy. Initial data to the program is provided and it is allowed to run without interruption until convergence is obtained. Results from the program are summarized in Tables 6.2 and 6.3.

Table 6.2 contains the histories of maximum constraint violation, convergence parameter, cost function and the design variables with the first starting point of (50, 200). The starting point is infeasible with a maximum constraint violation of 1100%. The program finds the optimum solution in eight iterations. Constraint and design variable activities at the optimum point are also given in Table 6.2. The path of the iterative process is traced on the graph of the problem

TABLE 6.2

History of iterative process and optimum solution for rectangular beam design problem, starting point (50, 200) mm

I	Max. vio.	Conv. parm.	Cost	Width	Depth
1	1.100 00E + 01	1.000 00E + 00	1.000 00E + 04	5.000 0E + 01	2.000 0E + 02
2	4.819 71E + 00	1.000 00E + 00	1.957 07E + 04	9.287 6E + 01	2.107 2E + 02
3	1.995 55E + 00	1.000 00E + 00	3.755 57E + 04	1.571 2E + 02	2.390 3E + 02
4	7.023 40E − 01	1.000 00E + 00	6.608 55E + 04	2.302 0E + 02	2.870 7E + 02
5	1.715 08E − 01	2.414 74E + 01	9.603 01E + 04	2.880 1E + 02	3.334 3E + 02
6	1.724 81E − 02	2.979 67E + 00	1.105 92E + 05	3.121 6E + 02	3.542 9E + 02
7	2.180 08E − 04	3.859 68E − 02	1.124 75E + 05	3.151 4E + 02	3.569 9E + 02
8	3.581 47E − 08	6.342 75E − 06	1.125 00E + 05	3.151 7E + 02	3.569 5E + 02

Constraint activity

No.	Active	Value	Lagr. mult
1	No	−4.023 36E − 01	0.000 00E + 00
2	Yes	3.581 47E − 08	1.125 00E + 05
3	No	−4.337 34E − 01	0.000 00E + 00

Design variable activity

No.	Active	Design	Lower	Upper	Lagr. mult.
1	No	3.151 75E + 02	1.000 00E + 01	1.000 00E + 03	0.000 00E + 00
2	No	3.569 45E + 02	1.000 00E + 01	1.000 00E + 03	0.000 00E + 00

Cost function at optimum = 1.125 000E + 05.
No. of calls for cost function evaluation (USERMF) = 8.
No. of calls for evaluation of cost function gradient (USERMG) = 8.
No. of calls for constraint function evaluation (USERCF) = 8.
No. of calls for evaluation of constraint function gradients (USERCG) = 8.
No. of total gradient evaluations = 14.

starting from point I, as shown in Fig. 6.17. The algorithm iterates through the infeasible region to reach the optimum solution which agrees with the one obtained analytically in Section 3.8.2.

Tables 6.2 and 6.3 show that although the first starting point takes more iterations (8) to converge to the optimum point compared to the second point (6), the number of calls for function evaluations is smaller for the first point. The total number of gradient evaluations for the two points are 14 and 3 respectively. Note that if the potential set strategy had not been used, the total number of gradient evaluations would have been 24 and 18 respectively for the two points. These are substantially higher than the actual number of gradient evaluations with IDESIGN which uses potential set strategy. It is clear that for large-scale applications, the potential set strategy can have a substantial impact on the efficiency of calculations for an optimization algorithm.

TABLE 6.3

History of iterative process and optimum solution for rectangular beam design problem, starting point (1000, 1000) mm

I	Max. vio.	Conv. parm	Cost	Width	Depth
1	0.000 00E + 00	1.000 00E + 03	1.000 00E + 06	1.0000E + 03	1.0000E + 03
2	0.000 00E + 00	5.050 00E + 02	2.550 25E + 05	5.0500E + 02	5.0500E + 02
3	0.000 00E + 00	3.812 50E + 02	1.453 52E + 05	3.8125E + 02	3.8125E + 02
4	3.386 21E − 03	5.650 12E − 01	1.121 20E + 05	3.3484E + 02	3.3484E + 02
5	8.551 55E − 06	1.434 12E − 03	1.124 99E + 05	3.3541E + 02	3.3541E + 02
6	5.484 72E − 11	9.198 03E − 09	1.125 00E + 05	3.3441E + 02	3.3541E + 02

Constraint activity

No.	Active	Value	Lagr. mult.
1	No	−3.639 63E − 01	0.000 00E + 00
2	Yes	5.484 72E − 11	1.125 00E + 05
3	No	−5.000 00E − 01	0.000 00E + 00

Design variable activity

No.	Active	Design	Lower	Upper	Lagr. mult.
1	No	3.354 10E + 02	1.000 00E + 01	1.000 00E + 03	0.000 00E + 00
2	No	3.354 10E + 02	1.000 00E + 01	1.000 00E + 03	0.000 00E + 00

Cost function at optimum = 1.125 000E + 05.
No. of calls for cost function evaluation (USERMF) = 12.
No. of calls for evaluation of cost function gradient (USERMG) = 6.
No. of calls for constraint function evaluation (USERCF) = 12.
No. of calls for evaluation of constraint function gradients (USERCG) = 3.
No. of total gradient evaluations = 3.

6.7* CONSTRAINED QUASI-NEWTON METHODS

Thus, far we have used only linear approximation for the cost and constraint functions in defining the search direction determination subproblem. The rate of convergence of algorithms based on such subproblems can be slow. It is possible to use quadratic approximations for the cost and constraint functions in defining the subproblem. This is likely to improve the rate of convergence of the algorithms because curvature information for the functions is used in determining the search direction. However, it turns out that the subproblem defined with quadratic approximations for the functions is as difficult to solve as the original nonlinear optimization problem. Therefore, there is essentially no advantage in pursuing this line of investigation.

It turns out that the QP subproblem, defined in Section 6.5.2, can be modified slightly to introduce curvature information for the Lagrange function into the quadratic cost function of Eq. (6.38) [Wilson, 1963]. The original nonlinear constraints are still approximated by linear constraints as in Eqs (6.39) and (6.40). Thus, to define the new QP subproblem, we need to evaluate Hessian matrix for the Lagrange function. This procedure still leads to two difficulties:

1. Second-order derivatives of all constraints and cost function must be evaluated, which is usually a very tedious calculation.
2. Lagrange multiplier estimates for all constraints must be available to calculate Hessian of the Lagrange function. These are usually known after the QP subproblem has been solved.

A great breakthrough in the methods based on the foregoing philosophy occurred when it was realized that the Hessian of the Lagrange function could be approximated using only first-order information [Han, 1976, 1977; Powell, 1978a, 1978b]. The idea is quite similar to that used in unconstrained quasi-Newton methods described in Section 5.7. We use the gradient of the Lagrange function at the two points and the change in design to update approximation to the Hessian of the Lagrange function. We shall call these constrained Quasi-Newton methods. They have been also called Constrained Variable Metric (CVM), Sequential Quadratic Programming (SQP), or Recursive Quadratic Programming (RQP) methods in the literature. Several variations of the methods can be generated. However, we shall describe a method based on the above philosophy.

6.7.1 Derivation of Quadratic Programming Subproblem

There are several ways to derive the quadratic programming (QP) subproblem that has to be solved at each optimization iteration. Understanding of the detailed derivation of the QP subproblem is not necessary in using the constrained quasi-Newton methods. Therefore, the reader who is not interested in the derivation can skip this subsection.

It is customary to derive the QP subproblem by considering only the equality constrained design optimization problem as

minimize $\qquad f(\mathbf{x})$ subject to $h_i(\mathbf{x}) = 0$; $\qquad i = 1$ to p $\qquad$ (6.76)

Later on, the inequality constraints are easily incorporated into the subproblem. The procedure for the derivation of the QP subproblem is to write Kuhn–Tucker necessary conditions for the problem defined in Eq. (6.76) and then attempt to solve them by Newton's method for nonlinear equations. Each iteration of Newton's method can be then interpreted as being equivalent to the solution of a QP subproblem. In the following derivations, we assume that all functions are twice continuously differentiable, and gradients of all constraints are linearly independent.

The Lagrange function for the design optimization problem defined in Eq. (6.76) is given as

$$L(\mathbf{x}, \mathbf{v}) = f(\mathbf{x}) + \sum_{i=1}^{p} v_i h_i(\mathbf{x}) = f(\mathbf{x}) + \mathbf{v} \cdot \mathbf{h}(\mathbf{x}) \qquad (6.77)$$

where v_i is the Lagrange multiplier for the ith equality constraint $h_i(\mathbf{x}) = 0$. Note that v_i is free in sign. The Kuhn–Tucker necessary conditions give

$$\nabla L(\mathbf{x}, \mathbf{v}) = \mathbf{0}, \qquad \nabla f(\mathbf{x}) + \sum_{i=1}^{p} v_i \nabla h_i(\mathbf{x}) = \mathbf{0} \qquad (6.78)$$

and

$$h_i(\mathbf{x}) = 0; \qquad i = 1 \text{ to } p \qquad (6.79)$$

Note that Eq. (6.78) actually represents n equations, since the dimension of the design variable vector is n. These equations along with the p equality constraints in Eq. (6.79) give $(n + p)$ equations in $(n + p)$ unknowns (n design variables in $\mathbf{x}$ and p Lagrange multipliers in $\mathbf{v}$). These are nonlinear equations, so the Newton–Raphson method of Appendix C can be used to solve them.

Let us write Eqs (6.78) and (6.79) in a compact notation as

$$\mathbf{F}(\mathbf{y}) = \mathbf{0} \qquad (6.80)$$

where $\mathbf{F}$ and $\mathbf{y}$ are identified as

$$\mathbf{F} = \begin{bmatrix} \nabla L \\ \mathbf{h} \end{bmatrix}_{(n+p \times 1)} \quad \text{and} \quad \mathbf{y} = \begin{bmatrix} \mathbf{x} \\ \mathbf{v} \end{bmatrix}_{(n+p \times 1)} \qquad (6.81)$$

Now using the iterative procedure of Section C.2 of Appendix C, we assume that $\mathbf{y}^{(k)}$ at the kth iteration is known and a change $\Delta \mathbf{y}^{(k)}$ is desired. Using linear Taylor series expansion for Eq. (6.80), $\Delta \mathbf{y}^{(k)}$ is given as a solution of the linear system (refer to Eq. C.12 in Appendix C):

$$\nabla \mathbf{F}^T(\mathbf{y}^{(k)}) \Delta \mathbf{y}^{(k)} = -\mathbf{F}(\mathbf{y}^{(k)}) \qquad (6.82)$$

where $\nabla \mathbf{F}$ is an $(n + p) \times (n + p)$ Jacobian matrix for the nonlinear equations whose ith column is the gradient of the function $F_i(\mathbf{y})$ with respect to the vector $\mathbf{y}$. Substituting definitions of $\mathbf{F}$ and $\mathbf{y}$ from Eq. (6.81) into Eq. (6.82), we obtain

$$\begin{bmatrix} \nabla^2 L & \mathbf{N} \\ \mathbf{N}^T & \mathbf{0} \end{bmatrix}^{(k)} \begin{bmatrix} \Delta \mathbf{x} \\ \Delta \mathbf{v} \end{bmatrix}^{(k)} = - \begin{bmatrix} \nabla L \\ \mathbf{h} \end{bmatrix}^{(k)} \qquad (6.83)$$

where the superscript k indicates that the quantities are calculated at the kth iteration, $\nabla^2 L$ is an $n \times n$ Hessian matrix of the Lagrange function, $\mathbf{N}$ is an $n \times p$ matrix defined in Eq. (6.28) whose ith column is the gradient of the equality constraint h_i, $\Delta \mathbf{x}^{(k)} = \mathbf{x}^{(k+1)} - \mathbf{x}^{(k)}$, and $\Delta \mathbf{v}^{(k)} = \mathbf{v}^{(k+1)} - \mathbf{v}^{(k)}$. Equation (6.83) can be converted to a slightly different form by writing the first row as

$$\nabla^2 L \Delta \mathbf{x}^{(k)} + \mathbf{N} \Delta \mathbf{v}^{(k)} = -\nabla L \qquad (6.84)$$

Substituting for $\Delta \mathbf{v}^{(k)} = \mathbf{v}^{(k+1)} - \mathbf{v}^{(k)}$ and ∇L from Eq. (6.78) into Eq. (6.84), we obtain

$$\nabla^2 L \Delta \mathbf{x}^{(k)} + \mathbf{N}(\mathbf{v}^{(k+1)} - \mathbf{v}^{(k)}) = -\nabla f(\mathbf{x}^{(k)}) - \mathbf{N}\mathbf{v}^{(k)} \tag{6.85}$$

Or, the equation is simplified to

$$\nabla^2 L \Delta \mathbf{x}^{(k)} + \mathbf{N}\mathbf{v}^{(k+1)} = -\nabla f(\mathbf{x}^{(k)}) \tag{6.86}$$

Combining Eq. (6.86) with the second row of Eq. (6.83), we obtain

$$\begin{bmatrix} \nabla^2 L & \mathbf{N} \\ \mathbf{N}^T & \mathbf{0} \end{bmatrix}^{(k)} \begin{bmatrix} \Delta \mathbf{x}^{(k)} \\ \mathbf{v}^{(k+1)} \end{bmatrix} = -\begin{bmatrix} \nabla f \\ \mathbf{h} \end{bmatrix}^{(k)} \tag{6.87}$$

Solution of Eq. (6.87) gives a change in the design $\Delta \mathbf{x}^{(k)}$ and a new value for the Lagrange multiplier vector $\mathbf{v}^{(k+1)}$. The iterative procedure is continued until convergence criteria are satisfied.

It will be now shown that Eq. (6.87) is also the solution of a certain QP problem defined at the kth iteration as

minimize $$\nabla f^T \Delta \mathbf{x} + 0.5 \Delta \mathbf{x}^T \nabla^2 L \Delta \mathbf{x} \tag{6.88}$$

subject to linearized equality constraints

$$h_i + \mathbf{n}^{(i)^T} \Delta \mathbf{x} = 0; \qquad i = 1 \text{ to } p \tag{6.89}$$

where $\mathbf{n}^{(i)}$ is the gradient of the function h_i. The Lagrange function for the problem defined in Eqs (6.88) and (6.89) is given as

$$\bar{L} = \nabla f^T \Delta \mathbf{x} + 0.5 \Delta \mathbf{x}^T \nabla^2 L \Delta \mathbf{x} + \sum_{i=1}^{p} v_i(h_i + \mathbf{n}^{(i)^T} \Delta \mathbf{x}) \tag{6.90}$$

The Kuhn–Tucker necessary conditions treating $\Delta \mathbf{x}$ as the unknown variable give

$$\nabla \bar{L} = \mathbf{0}; \qquad \nabla f + \nabla^2 L \Delta \mathbf{x} + \mathbf{N}\mathbf{v} = \mathbf{0} \tag{6.91}$$

and

$$h_i + \mathbf{n}^{(i)^T} \Delta \mathbf{x} = 0 \tag{6.92}$$

It can be seen that if we combine Eq. (6.91) and (6.92) and write them in a matrix form, we get Eq. (6.87). Thus, the problem of minimizing $f(\mathbf{x})$ subject to $h_i(\mathbf{x}) = 0$; $i = 1$ to p can be solved by iteratively solving the QP subproblem defined in Eqs. (6.88) and (6.89).

Just as in Newton's method for unconstrained problems, the solution $\Delta \mathbf{x}$ should be treated as a search direction and step size determined by minimizing an appropriate descent function to obtain a convergent algorithm. Defining the search direction $\mathbf{d}$ as $\Delta \mathbf{x}$ and including inequality constraints, the QP subproblem for the general constrained optimization problem is defined as

minimize $$\mathbf{c}^T \mathbf{d} + 0.5 \mathbf{d}^T \mathbf{H} \mathbf{d} \tag{6.93}$$

subject to constraints of Eqs (6.39) and (6.40) as

$$\mathbf{n}^{(i)^T} \mathbf{d} = e_i; \qquad i = 1 \text{ to } p \tag{6.94}$$

$$\mathbf{a}^{(i)^T} \mathbf{d} \leq b_i; \qquad i = 1 \text{ to } m \tag{6.95}$$

where the notation defined in Section 6.3 is used, and $\mathbf{H}$ is the Hessian matrix $\nabla^2 L$ or its approximation. Usually, a potential constraint strategy can be used in reducing the number of inequalities in Eq. (6.95) as discussed in Section 6.6.5. We shall further elaborate on this point later.

6.7.2 Quasi-Newton Hessian Approximation

Just as for the quasi-Newton methods of Section 5.7 for unconstrained problems, we can approximate the Hessian of the Lagrange function for the constrained problems. We assume that the approximate Hessian $\mathbf{H}^{(k)}$ at the kth iteration is available and we desire to update it to $\mathbf{H}^{(k+1)}$. The BFGS formula of Section 5.7 for direct updating of the Hessian can be used. It is important to note that the updated Hessian should be kept positive definite because, with this property, the QP subproblem defined in Eqs (6.93) to (6.95) remains strictly convex. Thus a unique search direction is obtained. It turns out that the standard BFGS updating formula can lead to singular or indefinite Hessian. To overcome this difficulty, Powell [1978a] suggested a modification to the standard BFGS formula. Although the modification is based on intuition, it has worked well in most applications. We shall give the modified BFGS formula.

Several intermediate scalars and vectors must be calculated before the final formula can be given. We define these as follows:

$$\mathbf{s}^{(k)} = \alpha_k \mathbf{d}^{(k)}; \quad \text{vector of changes in design} \tag{6.96}$$
$$\text{(note: } \alpha_k \text{ is the step size)}$$

$$\mathbf{z}^{(k)} = \mathbf{H}^{(k)} \mathbf{s}^{(k)}; \quad \text{a vector} \tag{6.97}$$

$$\mathbf{y}^{(k)} = \nabla L(\mathbf{x}^{(k+1)}, \mathbf{u}^{(k)}, \mathbf{v}^{(k)}) - \nabla L(\mathbf{x}^{(k)}, \mathbf{u}^{(k)}, \mathbf{v}^{(k)}); \tag{6.98}$$

difference in the gradients of the Lagrange function at two points

$$\xi_1 = \mathbf{s}^{(k)} \cdot \mathbf{y}^{(k)}; \quad \text{a scalar} \tag{6.99}$$

$$\xi_2 = \mathbf{s}^{(k)} \cdot \mathbf{z}^{(k)}; \quad \text{a scalar} \tag{6.100}$$

$$\theta = 1 \text{ if } \xi_1 \geq 0.2\xi_2, \text{ otherwise } \theta = \frac{0.8\xi_2}{(\xi_2 - \xi_1)}; \quad \text{a scalar} \tag{6.101}$$

$$\mathbf{w}^{(k)} = \theta \mathbf{y}^{(k)} + (1 - \theta)\mathbf{z}^{(k)}; \quad \text{a vector} \tag{6.102}$$

$$\xi_3 = \mathbf{s}^{(k)} \cdot \mathbf{w}^{(k)}; \quad \text{a scalar} \tag{6.103}$$

$$\mathbf{D}^{(k)} = \frac{1}{\xi_3} \mathbf{w}^{(k)} \mathbf{w}^{(k)^T}; \quad \text{an } n \times n \text{ matrix} \tag{6.104}$$

$$\mathbf{E}^{(k)} = \frac{1}{\xi_2} \mathbf{z}^{(k)} \mathbf{z}^{(k)^T}; \quad \text{an } n \times n \text{ matrix} \tag{6.105}$$

With the preceding definition of matrices $\mathbf{D}^{(k)}$ and $\mathbf{E}^{(k)}$, the Hessian is updated as

$$\mathbf{H}^{(k+1)} = \mathbf{H}^{(k)} + \mathbf{D}^{(k)} - \mathbf{E}^{(k)} \qquad (6.106)$$

It turns out that if the scalar ξ_1 in Eq. (6.99) is negative, the original BFGS formula can lead to indefinite Hessian. The use of the modified vector $\mathbf{w}^{(k)}$ given in Eq. (6.102) tends to alleviate this difficulty.

Due to the usefulness of incorporating Hessian into an optimization algorithm, several updating procedures have been developed in the recent literature [Gill *et al.*, 1981]. For example, Cholesky factors of the Hessian can be directly updated. In numerical implementations, it is useful to incorporate such procedures because numerical stability can be guaranteed.

6.7.3 Modified Constrained Steepest Descent Algorithm

The CSD algorithm of Section 6.6 has recently been extended to include Hessian updating and potential set strategy [Belegundu and Arora, 1984a; Lim and Arora, 1986; Thanedar, Arora and Tseng, 1986]. The original algorithm did not use the potential set strategy [Han, 1976, 1977; Powell, 1978a,b,c]. The new algorithm has been extensively investigated numerically and several computational enhancements have been incorporated into it to make it robust as well as efficient. In the following, we describe a very basic algorithm as a simple extension of the CSD algorithm. We refer to the new algorithm that uses a potential set strategy as PLBA (Pshenichny-Lim-Belegundu-Arora):

Step 1. The same as the CSD algorithm of Section 6.6, except also set the initial estimate for the approximate Hessian as identity, i.e. $\mathbf{H}^{(0)} = \mathbf{I}$.

Step 2. Calculate the cost and constraint functions at $\mathbf{x}^{(k)}$ and calculate the gradients of cost and constraint functions. Calculate the maximum constraint violation V_k as defined in Eq. (6.69). If $k > 0$, update Hessian of the Lagrange function using Eqs (6.96) to (6.106). If $k = 0$, skip updating and go to Step 3.

Step 3. Define the QP subproblem of Eqs (6.93) to (6.95) and solve it for the search direction $\mathbf{d}^{(k)}$ and Lagrange multipliers $\mathbf{v}^{(k)}$ and $\mathbf{u}^{(k)}$.

Steps 4–7. Same as for the CSD algorithm of Section 6.6.

Thus, we see that the only difference between the two algorithms is in Steps 2 and 3. We demonstrate use of the algorithm with an example problem.

Example 6.15 Use of constrained quasi-Newton method. Complete two iterations of the PLBA algorithm for Example 6.11:

minimize
$$f(\mathbf{x}) = x_1^2 + x_2^2 - 3x_1x_2$$

subject to
$$g_1(\mathbf{x}) \equiv \tfrac{1}{6}x_1^2 + \tfrac{1}{6}x_2^2 - 1.0 \leqq 0$$

$$g_2(\mathbf{x}) \equiv -x_1 \leqq 0$$

$$g_3(\mathbf{x}) \equiv -x_2 \leqq 0$$

The starting point is $(1, 1)$, $R_0 = 10$, $\gamma = 0.5$, $\varepsilon_1 = \varepsilon_2 = 0.001$.

Solution. The first iteration of the PLBA algorithm is the same as the CSD algorithm. From Example 6.11, results of the first iteration are

$$\mathbf{d}^{(0)} = (1, 1); \quad \alpha_0 = 0.5, \quad \mathbf{x}^{(1)} = (1.5, 1.5)$$

$$\mathbf{u}^{(0)} = (0, 0, 0); \quad R_1 = 10, \quad \mathbf{H}^{(0)} = \mathbf{I}.$$

Iteration 2. At the point $\mathbf{x}^{(1)} = (1.5, 1.5)$, the cost and constraint functions and their gradients are evaluated as

$$f = -6.75$$

$$g_1 = -0.25$$

$$g_2 = -1.5$$

$$g_3 = -1.5$$

$$\nabla f = (-1.5, -1.5)$$

$$\nabla g_1 = (0.5, 0.5)$$

$$\nabla g_2 = (-1, 0)$$

$$\nabla g_3 = (0, -1)$$

To update the Hessian matrix, we define the vectors in Eqs (6.96) and (6.97) as

$$\mathbf{s}^{(0)} = \alpha_0 \mathbf{d}^{(0)}$$

$$= (0.5, 0.5)$$

$$\mathbf{z}^{(0)} = \mathbf{H}^{(0)} \mathbf{s}^{(0)}$$

$$= (0.5, 0.5)$$

Since the Lagrange multiplier vector $\mathbf{u}^{(0)} = (0, 0, 0)$, the gradient of the Lagrangian ∇L is simply the gradient of the cost function ∇f. Therefore, vector $\mathbf{y}^{(0)}$ of Eq. (6.98) is given as

$$\mathbf{y}^{(0)} = \nabla f(\mathbf{x}^{(1)}) - \nabla f(\mathbf{x}^{(0)})$$

$$= (-0.5, -0.5)$$

Also, the scalars in Eqs (6.99) and (6.100) are given as

$$\xi_1 = \mathbf{s}^{(0)} \cdot \mathbf{y}^{(0)}$$

$$= -0.5$$

$$\xi_2 = \mathbf{s}^{(0)} \cdot \mathbf{z}^{(0)}$$

$$= 0.5$$

Since $\xi_1 < 0.2\xi_2$, θ in Eq. (6.101) is calculated as

$$\theta = 0.8(0.5)/(0.5 + 0.5)$$

$$= 0.4$$

The vector $\mathbf{w}^{(0)}$ in Eq. (6.102) is given as

$$\mathbf{w}^{(0)} = 0.4(-0.5, -0.5) + (1 - 0.4)(0.5, 0.5)$$

$$= (0.1, 0.1)$$

The scalar ξ_3 in Eq. (6.103) is given as $(0.5, 0.5) \cdot (0.1, 0.1) = 0.1$. The two correction matrices in Eqs (6.104) and (6.105) are calculated as

$$\mathbf{D}^{(0)} = \begin{bmatrix} 0.1 & 0.1 \\ 0.1 & 0.1 \end{bmatrix}; \quad \mathbf{E}^{(0)} = \begin{bmatrix} 0.5 & 0.5 \\ 0.5 & 0.5 \end{bmatrix}$$

Finally, from Eq. (6.106), the updated Hessian is given as

$$\mathbf{H}^{(1)} = \begin{bmatrix} 1 & 0 \\ 0 & 1 \end{bmatrix} + \begin{bmatrix} 0.1 & 0.1 \\ 0.1 & 0.1 \end{bmatrix} - \begin{bmatrix} 0.5 & 0.5 \\ 0.5 & 0.5 \end{bmatrix}$$

$$= \begin{bmatrix} 0.6 & -0.4 \\ -0.4 & 0.6 \end{bmatrix}$$

Step 3. With the updated Hessian and other data previously calculated, the QP subproblem of Eqs (6.93) to (6.95) is defined as

minimize $\quad -1.5d_1 - 1.5d_2 + 0.5(0.6d_1^2 - 0.8d_1d_2 + 0.6d_2^2)$

subject to

$$0.5d_1 + 0.5d_2 \leqq 0.25$$

$$-d_1 \leqq 1.5, \ -d_2 \leqq 1.5$$

The QP subproblem is strictly convex and, thus, has a unique solution. Using $K-T$ conditions the solution for the QP subproblem is obtained as

$$\mathbf{d}^{(1)} = (0.25, 0.25), \quad \mathbf{u}^{(1)} = (2.9, 0, 0)$$

This solution is the same as in Example 6.11. Therefore, the rest of the steps have the same calculations.

It is seen that in this example, inclusion of approximate Hessian does not actually change the search direction at the second iteration. In general, it will give different directions and better convergence. ‖

6.7.4 Observations on the Constrained Quasi-Newton Methods

The quasi-Newton methods have been developed recently. They are considered to be most efficient, reliable and generally applicable. Schittkowski and co-workers [1980, 1981, 1983] have extensively analyzed the methods and evaluated them against several other methods using a set of nonlinear programming test problems. Their conclusion is that quasi-Newton methods are far superior to others. Lim and Arora [1986], Thanedar et al. [1986], Thanedar, Arora, Tseng, Lim and Park [1987] and Arora and Tseng [1987b] have evaluated the methods for a class of engineering design problems. Gabrielle and Beltracchi [1987] have also discussed several enhancements of Pshenichny's Constrained Steepest Descent (CSD) algorithm including incorporation of quasi-Newton updates of the Hessian of the Lagrangian. In general, these investigations have shown the quasi-Newton methods to be superior. Therefore the methods are recommended for general engineering design applications.

Numerical implementation of an algorithm is an art. Considerable care, judgement, safeguards and user-friendly features must be designed and incorporated into the software. Numerical calculations must be robustly implemented. Each step of the algorithm must be analyzed and proper numerical procedures developed to implement the intent of the step. The software must be properly evaluated for performance by solving many different problems. Many aspects of numerical implementation of algorithms are discussed by Gill *et al.* [1981].

The steps of the PLBA algorithm have recently been analyzed [Tseng and Arora, 1988]. Various potential constraint strategies have been incorporated and evaluated. Several descent functions have been investigated. Procedures to resolve inconsistencies in the QP subproblem have been developed and evaluated. As a result of these enhancements and evaluations, a very powerful algorithm for engineering design applications has become available.

6.7.4.1 DESCENT FUNCTIONS. Descent functions play an important role in the constrained quasi-Newton methods, so we shall discuss them briefly. The descent function of Eq. (6.64) can be derived from the Lagrange function for the problem defined in Eqs (6.1) to (6.3):

$$L(\mathbf{x}) = f(\mathbf{x}) + \sum_{i=1}^{p} v_i h_i + \sum_{i=1}^{m} u_i g_i \qquad (6.107)$$

where v_i are the Lagrange multipliers for the equality constraints that are free in sign and $u_i \geq 0$ are the Lagrange multipliers for the inequality constraints. Using absolute values for $v_i h_i$, we can derive the descent function of Eq. (6.64) from Eq. (6.107) as

$$L(\mathbf{x}) \leq f(\mathbf{x}) + \sum_{i=1}^{p} |v_i h_i| + \sum_{i=1}^{m} u_i g_i$$

$$\leq f(\mathbf{x}) + \left(\sum_{i=1}^{p} |v_i| + \sum_{i=1}^{m} u_i \right) V(\mathbf{x})$$

$$\leq f(\mathbf{x}) + RV = \Phi(\mathbf{x}) \qquad (6.108)$$

Thus the Lagrange function is a lower bound on the descent function $\Phi(\mathbf{x})$ of Eq. (6.64). Note that the function $\Phi(\mathbf{x})$ is nondifferentiable.

Another nondifferentiable descent function has been proposed by Han [1977] and Powell [1978c]. We shall denote this as Φ_H and define it as follows at the kth iteration:

$$\Phi_H = f(\mathbf{x}^{(k)}) + \sum_{i=1}^{p} r_i^{(k)} |h_i| + \sum_{i=1}^{m} \mu_i^{(k)} \max\{0, g_i\} \qquad (6.109)$$

where $r_i^{(k)} \geq |v_i^{(k)}|$ are the penalty parameters for equality constraints and $\mu_i^{(k)} \geq u_i^{(k)}$ are the penalty parameters for inequality constraints. The penalty parameters sometimes become very large, so Powell [1978c] suggested a procedure to adjust them as follows:

First iteration $\quad r_i^{(0)} = |v_i^{(0)}|; \quad \mu_i^{(0)} = u_i^{(0)}$ (6.110)

Subsequent iterations: $\quad r_i^{(k)} = \max \{|v_i^{(k)}|, \tfrac{1}{2}(r_i^{(k-1)} + |v_i^{(k)}|)\}$

$$\mu_i^{(k)} = \max \{u_i^{(k)}, \tfrac{1}{2}(\mu_i^{(k-1)} + u_i^{(k)})\}$$ (6.111)

Schittkowski (1981) has suggested using the following augmented Lagrangian function Φ_A as the descent function:

$$\Phi_A = f(\mathbf{x}) + P_1(\mathbf{v}, \mathbf{h}) + P_2(\mathbf{u}, \mathbf{g})$$ (6.112)

where $\quad\quad P_1(\mathbf{v}, \mathbf{h}) = \sum_{i=1}^{p} (v_i h_i + \tfrac{1}{2} r_i h_i^2)$ (6.113)

$$P_2(\mathbf{u}, \mathbf{g}) = \sum_{i=1}^{m} \begin{cases} (u_i g_i + \tfrac{1}{2}\mu_i g_i^2), & \text{if } (g_i + u_i/\mu_i) \geqq 0 \\ \tfrac{1}{2} u_i^2 / \mu_i, & \text{otherwise} \end{cases}$$ (6.114)

where the penalty parameters r_i and μ_i have been defined previously in Eqs (6.110) and (6.111). One good feature of Φ_A is that the function and its gradient are continuous.

6.8* OTHER METHODS

Many other methods and their variations for constrained optimization have been developed and evaluated in the literature. For details, Gill *et al.* [1981], Luenberger [1984] and Reklaitis *et al.* [1983] should be consulted. In this section, we shall briefly discuss the basic ideas of three methods that have been used quite successfully for engineering design problems.

6.8.1 Method of Feasible Directions

The method of feasible directions is one of the earliest primal methods for solving constrained optimization problems. The basic idea of the method is to move from one feasible design to an improved feasible design. Thus given a feasible design $\mathbf{x}^{(k)}$, an "improving feasible direction" $\mathbf{d}^{(k)}$ is determined such that for a sufficiently small step size $\alpha > 0$, the following two properties are satisfied: (i) the new design, $\mathbf{x}^{(k+1)} = \mathbf{x}^{(k)} + \alpha \mathbf{d}^{(k)}$ is feasible, and (ii) the new cost function is smaller than the old one, i.e. $f(\mathbf{x}^{(k+1)}) < f(\mathbf{x}^{(k)})$. Once $\mathbf{d}^{(k)}$ is determined, a line search is performed to determine how far to proceed along $\mathbf{d}^{(k)}$. This leads to a new feasible design $\mathbf{x}^{(k+1)}$, and the process is repeated from there.

The method is based on the general algorithm described in Section 6.2.1, where the design change determination is decomposed into search direction and step size determination subproblems. The direction is determined by defining a linearized subproblem at the current feasible point, and step size is determined to reduce the cost function as well as maintain feasibility. Since linear approximations are used, it is difficult to maintain feasibility with respect to the equality constraints. Therefore, the method has been developed and applied mostly to inequality constrained problems. Some procedures have been developed to treat equality constraints in these methods. However, we shall describe the method for problems with only inequality constraints.

Now we define a subproblem that yields improving feasible direction at the current design point. An *improving feasible direction* is defined as the one that reduces the cost function as well as remains strictly feasible for a small step size. Thus it is a direction of descent for the cost function as well as having its point on the inside of the feasible region. The improving feasible direction $\mathbf{d}$ satisfies the conditions $\mathbf{c}^T\mathbf{d} < 0$ and $\mathbf{a}^{(i)^T}\mathbf{d} < 0$ for $i \in I_k$, where I_k is a potential constraint set at the current point as defined in Eq. (6.12). It can be obtained by minimizing the maximum of $\mathbf{c}^T\mathbf{d}$ and $\mathbf{a}^{(i)^T}\mathbf{d}$ for $i \in I_k$. Denoting this maximum by β, the direction finding subproblem is defined as

$$\text{minimize } \beta \tag{6.115}$$

subject to

$$\mathbf{c}^T\mathbf{d} \leqq \beta \tag{6.116}$$

$$\mathbf{a}^{(i)^T}\mathbf{d} \leqq \beta \quad \text{for } i \in I_k \tag{6.117}$$

$$-1 \leqq d_j \leqq 1; \quad j = 1 \text{ to } n \tag{6.118}$$

The normalization constraint of Eq. (6.118) has been introduced to obtain a bounded solution. Other forms of normalization constraints can also be used. Let $(\beta, \mathbf{d})$ be an optimum solution for the above problem. If $\beta < 0$, then $\mathbf{d}$ is an improving feasible direction. If $\beta = 0$, then the current design point satisfies the Kuhn–Tucker necessary conditions.

There are many different line search algorithms that may be used to determine the appropriate step size along the search direction. Also, to estimate a better feasible direction $\mathbf{d}^{(k)}$, the constraints of Eq. (6.117) can be expressed as $\mathbf{a}^{(i)^T}\mathbf{d} \leqq \theta_i\beta$ where $\theta_i > 0$ are the "push-off" factors. The greater the value of θ_i, the greater is the direction vector $\mathbf{d}$ pushed into the feasible region. The reason for introducing θ_i is to prevent the iterations from repeatedly hitting the constraint boundary and slowing down the convergence. Figure 6.18 shows the physical significance of θ_i in the direction-finding subproblem. It depicts a two-variable design space with one active constraint. If θ_i is taken as zero, then the right-hand side of Eq. (6.117) ($\theta_i\beta$) becomes zero. The direction $\mathbf{d}$ in this case tends to follow the active constraint, i.e. it is tangent to the constraint surface. On the other hand, if θ_i is very large, the direction $\mathbf{d}$ tends to follow the cost function contour. Thus, a small value of θ_i will result in a direction which rapidly reduces the cost function. It may, however, rapidly encounter the same constraint surface due to nonlinearities. Larger values of θ_i will reduce the risk of re-encountering the same constraint, but will not reduce the cost function as fast. A value of $\theta_i = 1$ yields acceptable results for most problems.

The disadvantages of the method are (i) a feasible starting point is

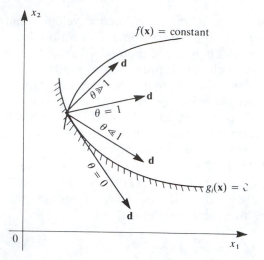

FIGURE 6.18
The effect of push-off factor θ_i on search direction **d** in the feasible directions method.

needed – special algorithms must be used to obtain such a point if it is not known – and (ii) equality constraints are difficult to impose and require special procedures for their implementation.

6.8.2   Gradient Projection Method

The gradient projection method was developed by Rosen in 1961. Just as in the feasible directions method, the method also uses first-order information about the problem at the current point. The feasible directions method requires the solution of an LP at each iteration to find the search direction. In some applications, this can be an expensive calculation. Thus, Rosen was motivated to develop a method that does not require the solution of an LP. His idea was to develop a procedure in which the direction vector could be calculated easily, although it may not be as good as the one obtained from the feasible directions approach. Thus, in the gradient projection method, an explicit expression for the search direction is available.

The basic procedure for the gradient projection method is to start with an initial point. If the point is inside the feasible region, the steepest descent direction for the cost function is used until a constraint boundary is encountered. If the starting point is infeasible then the constraint correction step is used to reach the constraint set. When the point is on the boundary, a direction that is tangent to the constraint surface is used to change the design. This direction is computed by projecting the steepest descent direction for the cost function on to the tangent plane. This was termed the constrained steepest descent (CSD) direction in Section 6.6. A step is executed in the negative projected gradient direction. Since the direction is at a tangent to the constraint surface, the new point will be infeasible. Therefore, a series of correction steps need to be executed to reach the feasible region.

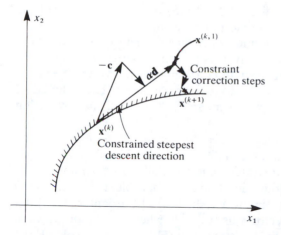

FIGURE 6.19
The steps of the gradient projection method.

The iterative process of the gradient projection method is illustrated in Fig. 6.19. At the point $\mathbf{x}^{(k)}$, $-\mathbf{c}^{(k)}$ is the steepest descent direction and $\mathbf{d}^{(k)}$ is the negative projected gradient (constrained steepest descent) direction. An arbitrary step takes the point $\mathbf{x}^{(k)}$ to $\mathbf{x}^{(k+1)}$ from where constraint correction steps are executed to reach the feasible point $\mathbf{x}^{(k,1)}$. Comparing the gradient projection method and the constrained steepest descent method of Section 6.6, we observe that at a feasible point where some constraints are active, the two methods have identical directions. The only difference is in the step size determination.

Philosophically, the idea of the gradient projection method is quite good, i.e. the search direction is easily computable, although it may not be as good as the feasible direction. However, numerically the method has considerable uncertainty. The step size specification is arbitrary; the constraint correction process is quite tedious. A serious drawback is that convergence of the algorithm is tedious to enforce. For example, during the constraint correction steps, it must be ensured that $f(\mathbf{x}^{(k+1)}) < f(\mathbf{x}^{(k)})$. If this condition cannot be satisfied or constraints cannot be corrected, then the step size must be reduced and the entire process must be repeated. This can be quite tedious and inefficient. Despite these drawbacks, the method has been applied quite successfully to a wide variety of engineering design problems [Haug and Arora, 1979]. In addition, many variations of the method have been investigated in the literature [Gill *et al.* 1981; Luenberger, 1984; Belegundu and Arora, 1985]. The method and its variations are quite suitable for implementation in an interactive environment. We shall elaborate on this aspect in Chapter 7.

6.8.3 Generalized Reduced Gradient Method

In 1967, Wolfe developed the reduced gradient method based on a simple variable elimination technique for equality constrained problems [Abadie,

1970]. The generalized reduced gradient (GRG) method is an extension of the reduced gradient method to accommodate nonlinear inequality constraints. In this method, a search direction is found, such that for any small move, the current active constraints remain precisely active. If some active constraints are not precisely satisfied due to nonlinearity of constraint functions, the Newton–Raphson method is used to return to the constraint boundary. Thus, the GRG method can be considered somewhat similar to the gradient projection method.

Since inequality constraints can always be converted to equalities by adding slack variables, we can form an equality constrained NLP model. Also, we can employ potential constraint strategy and treat all the constraints in the subproblem as equalities. The direction-finding subproblem in the GRG method can be defined in the following way [Abadie and Carpenter, 1969]: let us partition the design variable vector $\mathbf{x}$ as $[\mathbf{y}^T, \mathbf{z}^T]^T$ where $\mathbf{y}_{(n-p)}$ and $\mathbf{z}_{(p)}$ are vectors of independent and dependent design variables, respectively. First-order changes in the cost and constraint functions (treated as equalities) are given as

$$\Delta f = \frac{\partial f^T}{\partial \mathbf{y}} \Delta \mathbf{y} + \frac{\partial f^T}{\partial \mathbf{z}} \Delta \mathbf{z} \tag{6.119}$$

$$\Delta h_i = \frac{\partial h_i^T}{\partial \mathbf{y}} \Delta \mathbf{y} + \frac{\partial h_i^T}{\partial \mathbf{z}} \Delta \mathbf{z} \tag{6.120}$$

Since we started with a feasible design, any change in the variables must keep the current equalities satisfied at least to first order, i.e. $\Delta h_i = 0$. Therefore, Eq. (6.120) can be written in the matrix form as

$$\mathbf{A}^T \Delta \mathbf{y} + \mathbf{B}^T \Delta \mathbf{z} = \mathbf{0}, \qquad \text{or} \qquad \Delta \mathbf{z} = -(\mathbf{B}^{-T} \mathbf{A}^T) \Delta \mathbf{y} \tag{6.121}$$

where columns of matrices $\mathbf{A}_{((n-p) \times p)}$ and $\mathbf{B}_{(p \times p)}$ contain gradients of equality constraints with respect to $\mathbf{y}$ and $\mathbf{z}$ respectively. Equation (6.121) can be viewed as the one that determines $\Delta \mathbf{z}$ (change in the dependent variable) when $\Delta \mathbf{y}$ (change in the independent variable) is specified. Substituting $\Delta \mathbf{z}$ from Eq. (6.121) into Eq. (6.119) we can identify $df/d\mathbf{y}$ as

$$\Delta f = \left(\frac{\partial f^T}{\partial \mathbf{y}} - \frac{\partial f^T}{\partial \mathbf{z}} \mathbf{B}^{-T} \mathbf{A}^T \right) \Delta \mathbf{y}$$

or

$$\frac{df}{d\mathbf{y}} = \frac{\partial f}{\partial \mathbf{y}} - \mathbf{A} \mathbf{B}^{-1} \frac{\partial f}{\partial \mathbf{z}} \tag{6.122}$$

This is commonly known as the generalized reduced gradient and can be viewed as the gradient of an unconstrained function.

In line search, the cost function is treated as the descent function. For a trial value of α, the design variables are updated using $\Delta \mathbf{y} = -\alpha \, df/d\mathbf{y}$ and $\Delta \mathbf{z}$ from Eq. (6.121). If the trial design is not feasible, then independent design variables are considered to be fixed and dependent variables are changed iteratively by applying the Newton–Raphson (NR) method (Eq. (6.121)) until we get a feasible design point. If the new feasible design satisfies the descent condition, then line search is terminated; otherwise, the previous trial step size is discarded and the procedure is repeated with a reduced step size. It can be observed that when $df/d\mathbf{y} = \mathbf{0}$ in Eq. (6.122), the Kuhn–Tucker conditions of optimality are satisfied for the original NLP problem.

The main computational burden associated with the GRG algorithm arises from the Newton–Raphson (NR) iterations during line search. Strictly speaking, the gradients of constraints need to be recalculated and the Jacobian matrix $\mathbf{B}$ needs to be inverted at every NR iteration during the line search. This is prohibitively expensive. Towards this end, many efficient numerical schemes have been suggested, e.g. the use of a quasi-Newton formula to update $\mathbf{B}^{-1}$ without recomputing gradients but requiring only constraint function values. This can cause problems if the set of independent variables changes every iteration. Another difficulty is to select a feasible starting point. Special algorithms must be used to handle arbitrary starting points, as in the feasible directions method.

There is some confusion in the literature on the relative merits and demerits of the reduced gradient method. For example, the method has been declared superior to the gradient projection method, whereas the two methods are considered essentially the same by Sargeant [1974]. The confusion arises when studying the reduced gradient method in the context of solving inequality constrained problems; some algorithms convert the inequalities into equalities by adding nonnegative slack variables while others adopt potential constraint strategy. It turns out that if a potential constraint strategy is used, the reduced gradient method becomes essentially the same as the gradient projection method [Belegundu and Arora, 1985]. On the other hand, if inequalities are converted to equalities, it behaves quite differently from the gradient projection method. Unfortunately, the inequality constrained problem in most engineering applications must be solved using a potential constraint strategy, as the addition of slack variables to inequalities implies that all constraints are active at every iteration and must, therefore, be differentiated. This is ruled out for large-scale applications due to the enormous computation and storage of information involved. Therefore, as observed by Belegundu and Arora [1985], we need not differentiate between gradient projection and reduced gradient methods when solving most engineering optimization problems.

EXERCISES FOR CHAPTER 6

Section 6.2   Basic Concepts and Ideas

6.1 *Answer True or False*

1. The basic numerical iterative philosophy for solving constrained and unconstrained problems is the same.
2. Step size determination is a one-dimensional problem for unconstrained problems.

3. Step size determination is a multidimensional problem for constrained problem.
4. An inequality constraint $g_i(\mathbf{x}) \leq 0$ is violated at $\mathbf{x}^{(k)}$ if $g_i(\mathbf{x}^{(k)}) > 0$.
5. An inequality constraint $g_i(\mathbf{x}) \leq 0$ is active at $\mathbf{x}^{(k)}$ if $g_i(\mathbf{x}^{(k)}) > 0$.
6. An equality constraint $h_i(\mathbf{x}) = 0$ is violated at $\mathbf{x}^{(k)}$ if $h_i(\mathbf{x}^{(k)}) > 0$.
7. An equality constraint is always active at the optimum.
8. Constraint normalization is useful in checking ε-active constraints.
9. Potential constraint strategy means that only a subset of the constraints is used in the direction-finding subproblem.
10. In constrained optimization problems, search direction is found using the cost gradient only.
11. In constrained optimization problems, search direction is found using the constraint gradients only.
12. In constrained problems, the descent function is used to calculate the search direction.
13. In constrained problems, the descent function is used to calculate a feasible point.
14. Cost function can be used as a descent function in constrained problems.
15. One-dimensional search on a descent function is needed for convergence of algorithms.
16. A robust algorithm guarantees convergence.
17. A constraint set must be closed and bounded to guarantee convergence of algorithms.
18. Efficient algorithms use potential constraint strategy.
19. A constraint $x_1 + x_2 \leq -2$ can be normalized as $(x_1 + x_2)/(-2) \leq 1.0$.
20. A constraint $x_1^2 + x_2^2 \leq 9$ is active at $x_1 = 3$ and $x_2 = 3$.

Section 6.3   Linearization of the Constrained Problem

6.2 *Answer True or False*

1. Linearization of cost and constraint functions is a basic step for solving nonlinear optimization problems.
2. General constrained problems cannot be solved by solving a sequence of linear programming subproblems.
3. In general, the linearized subproblem without move limits may be unbounded.
4. The sequential linear programming method for general constrained problems is guaranteed to converge.
5. Move limits are essential in the sequential linear programming procedure.
6. Equality constraints can be treated in the sequential linear programming algorithm.

For the following problems, create a linear approximation at the given point and show it on the graph of the original problem:

6.3 Consider the beam design problem of Section 2.8.5:

minimize $\qquad\qquad\qquad f(b, d) = bd$

subject to

$$g_1 \equiv \frac{(2.40\text{E}+08)}{bd^2} - 10 \leqq 0$$

$$g_2 \equiv \frac{(2.25\text{E}+05)}{bd} - 2 \leqq 0$$

$$g_3 \equiv d - 2b \leqq 0$$

$$10 \leqq b \leqq 1000; \qquad 10 \leqq d \leqq 1000$$

where b is the width and d is the depth of the beam (mm). Create a linear approximation for the problem at the point $(250, 300)$ mm. Plot the linearized subproblem and the original problem on the same graph sheet.

6.4 A minimum mass tubular column design problem is formulated in Section 2.6.7 as find the mean radius R and thickness t to minimize

$$\text{mass} = 2\rho l \pi R t$$

subject to the stress constraint

$$\frac{P}{2\pi R t} \leqq \sigma_a$$

the buckling load constraint

$$P \leqq \frac{\pi^3 E R^3 t}{4l^2}$$

and simple bounds on the design variables as

$$5 \leqq R \leqq 100 \text{ cm}; \qquad 0.5 \leqq t \leqq 5 \text{ cm}$$

Let $P = 50$ kN, $E = 210$ GPa, $l = 500$ cm, $\sigma_a = 250$ MPa and $\rho = 7850$ kg/m³. Create a linear approximation for the problem at the point $R = 12$ cm and $t = 4$ cm. Plot the linearized subproblem and the original problem on the same graph sheet.

6.5 A minimum volume wall bracket problem is formulated in Section 3.8.1 as find the cross-sectional areas A_1 and A_2 (cm²) to minimize

$$\text{volume} = 50A_1 + 40A_2$$

subject to the stress constraints

$$\frac{125}{A_1} - 1 \leqq 0$$

$$\frac{100}{A_2} - 1 \leqq 0$$

and nonnegativity of the design variables

$$-A_1 \leqq 0; \qquad -A_2 \leqq 0$$

Create a linear approximation for the problem at the point $A_1 = 150 \text{ cm}^2$ and $A_2 = 150 \text{ cm}^2$. Plot the linearized subproblem and the original problem on the same graph sheet.

6.6 A 100×100 m lot is available to construct a multistory office building. At least 20 000 m² of total floor space is needed. According to a zoning ordinance, the maximum height of the building can be only 21 m, and the area for parking outside the building must be at least 25% of the total floor area. It has been decided to fix the height of each story at 3.5 m. The cost of the building in millions of dollars is estimated at $(0.6h + 0.001A)$, where A is the cross-sectional area of the building per floor and h is the height of the building. Formulate the minimum cost design problem (same as Exercise 2.1), transcribe it into the standard normalized form, and create a linear approximation at the point $h = 12$, $A = 4000$. Plot the linearized subproblem and the original problem on the same graph sheet.

6.7 Design a beer mug shown in Fig. E6.7 to hold as much beer as possible (same as Exercise 2.3). The height and radius of the mug should be no more than 20 cm. The mug must be at least 5 cm in radius. The surface area of the sides must not be greater than 900 cm² (ignore the area of the bottom of the mug and ignore the mug handle). Formulate the optimum design problem, transcribe it into the standard normalized form, and create a linear approximation at the point radius = 6 cm and height = 15 cm. Plot the linearized subproblem and the original problem on the same graph sheet.

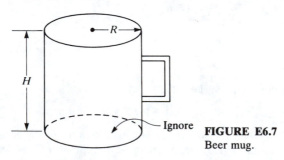

FIGURE E6.7
Beer mug.

6.8 A Company is redesigning its parallel flow heat exchanger of length l to increase its heat transfer. An end view of the unit is shown in Fig. E6.8. There are certain limitations on the design problem. The smallest available conducting tube has a radius of 0.5 cm and all tubes must be of the same size. Further, the total cross-sectional area of all the tubes cannot exceed 2000 cm² to ensure adequate space inside the outer shell. Formulate the problem to determine the number of tubes, and the radius of each tube to maximize the surface area of the

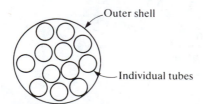

FIGURE E6.8
Cross section of a heat exchanger.

tubes in the exchanger (same as Exercise 2.4). Transcribe the problem into the standard normalized form and create a linear approximation at the point, radius = 2 and number of tubes = 100. Plot the linearized subproblem and the original problem on the same graph sheet.

6.9 After proposals for a parking ramp have been defeated, we plan to build a parking lot in the downtown urban renewal section. The cost of land is $200W + 100D$, where W is the width along the street, and D the depth of the lot in meters. The available width along the street is 100 m, while the maximum depth available is 200 m. We want to have at least 10 000 m² in the lot. To avoid unsightly lots, the city requires the longer dimension of any lot be no more than twice the shorter dimension. Formulate the minimum cost design problem (same as Exercise 2.5), transcribe it into the standard normalized form, and create a linear approximation at the point $W = 100$ and $D = 100$. Plot the linearized subproblem and the original problem on the same graph sheet.

6.10 Design a can closed at one end using the smallest area of sheet metal for a specified interior volume of 600 cm³. The can is a right circular cylinder with an interior height = h, and radius = r. The ratio of height to diameter must not be less than 1.0 and not greater than 1.5. The height cannot be more than 20 cm. Formulate the design optimization problem (same as Exercise 2.9), transcribe it into the standard normalized form and create a linear approximation at the point $r = 6$, $h = 16$. Plot the linearized subproblem and the original problem on the same graph sheet.

6.11 Design a shipping container closed at both ends with dimension $b \times b \times h$ to minimize the ratio: (round-trip cost of shipping the container only)/(one-way cost of shipping the contents only). Use the following data:

Mass of the container/area: 80 kg/m²
Maximum b: 10 m
Maximum h: 18 m
One-way shipping cost,
  full or empty: $18/kg gross mass
Mass of the contents: 150 kg/m³

Formulate the design optimization problem (same as Exercise 2.10), transcribe it into the standard normalized form, and create a linear approximation at the point $b = 5$ and $h = 10$. Plot the linearized subproblem and the original problem on the same graph sheet.

6.12 Certain mining operations require an open top rectangular container to transport materials. The data for the problem are:

Construction costs:
  sides: $50/m²
  ends: $60/m²
  bottom: $90/m²
Salvage value: 25% of the construction cost
Useful life; 20 years
Yearly maintenance: $12/m² of outside surface area
Minimum volume needed: 150 m³
Interest rate: 12% per annum

Formulate the problem of determining the container dimensions for minimum cost (same as Exercise 2.11). Transcribe the problem into the standard normalized form and create a linear approximation for it at the point width = 5, depth = 5 and height = 5.

6.13 Design a circular tank closed at both ends to have a volume of $150\,\text{m}^3$. The fabrication cost is proportional to the surface area of the sheet metal and is $400/\text{m}^2$. The tank is to be housed in a shed with sloping roof. Therefore, height H of the tank is limited by the relation $H \leq 10 - D/2$, where D is the diameter of the tank. Formulate the minimum cost design problem (same as Exercise 2.12), transcribe it into the standard normalized form and create a linear approximation at the point $D = 4$ and $H = 8$. Plot the linearized subproblem and the original problem on the same graph sheet.

6.14 Design the steel framework shown in Fig. E6.14 at a minimum cost. The cost of all horizontal members in one direction is $20w$ and in the other direction it is $30d$. The cost of a vertical column is $50h$. The frame must enclose a total volume of at least $600\,\text{m}^3$. Formulate the design optimization problem (same as Exercise 2.13), transcribe it into the standard normalzzed form and create a linear approximation at the point $w = 10$, $d = 10$ and $h = 4$.

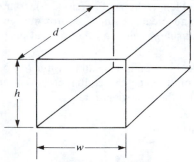

FIGURE E6.14
Steel frame.

6.15 Two electrical generators are interconnected to provide total power to meet the load. Each generator's cost is a function of the power output, as shown in Fig. E6.15. All costs and power are expressed on a per unit basis. The total power need is at least 60 units. Formulate a minimum cost design problem (same as Exercise 2.14), transcribe it into the standard normalized form, and create a linear approximation for it at the point $P_1 = 2$ and $P_2 = 1$. Plot the linearized subproblem and the original problem on the same graph sheet.

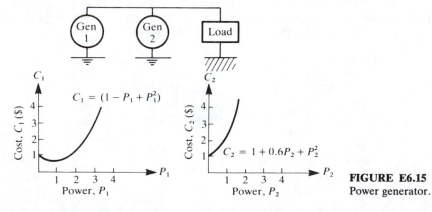

FIGURE E6.15
Power generator.

Section 6.4   Sequential Linear Programming Algorithm

Complete two iterations of the sequential linear programming algorithm for the following problems (try 50% move limits and adjust them if necessary):

6.16 Exercise 6.3   **6.17** Exercise 6.4

6.18 Exercise 6.5   **6.19** Exercise 6.6

6.20 Exercise 6.7   **6.21** Exercise 6.8

6.22 Exercise 6.9   **6.23** Exercise 6.10

6.24 Exercise 6.11   **6.25** Exercise 6.12

6.26 Exercise 6.13   **6.27** Exercise 6.14

6.28 Exercise 6.15

Section 6.5   Quadratic Programming Subproblem

Solve the following QP problems using the extended Simplex method:

6.29 Minimize $f(\mathbf{x}) = (x_1 - 3)^2 + (x_2 - 3)^2$

subject to $x_1 + x_2 \leq 5$

$x_1, x_2 \geq 0$

6.30 Minimize $f(\mathbf{x}) = (x_1 - 1)^2 + (x_2 - 1)^2$

subject to $x_1 + 2x_2 \leq 6$

$x_1, x_2 \geq 0$

6.31 Minimize $f(\mathbf{x}) = (x_1 - 1)^2 + (x_2 - 1)^2$

subject to $x_1 + 2x_2 \leq 2$

$x_1, x_2 \geq 0$

6.32 Minimize $f(\mathbf{x}) = x_1^2 + x_2^2 - x_1 x_2 - 3x_1$

subject to $x_1 + x_2 \leq 3$

$x_1, x_2 \geq 0$

6.33 Minimize $f(\mathbf{x}) = (x_1 - 1)^2 + (x_2 - 1)^2 - 2x_2 + 2$

subject to $x_1 + x_2 \leq 4$

$x_1, x_2 \geq 0$

6.34 Minimize $f(\mathbf{x}) = 4x_1^2 + 3x_2^2 - 5x_1 x_2 - 8x_1$

subject to $x_1 + x_2 = 4$

$x_1, x_2 \geq 0$

6.35 Minimize $f(\mathbf{x}) = x_1^2 + x_2^2 - 2x_1 - 2x_2$

subject to $x_1 + x_2 - 4 = 0$

$x_1 - x_2 - 2 = 0$

$x_1, x_2 \geq 0$

6.36 Minimize $f(\mathbf{x}) = 4x_1^2 + 3x_2^2 - 5x_1x_2 - 8x_1$

   subject to $x_1 + x_2 \leq 4$

   $x_1, x_2 \geq 0$

6.37 Minimize $f(\mathbf{x}) = x_1^2 + x_2^2 - 4x_1 - 2x_2$

   subject to $x_1 + x_2 \geq 4$

   $x_1, x_2 \geq 0$

6.38 Minimize $f(\mathbf{x}) = 2x_1^2 - 6x_1x_2 + 9x_2^2 - 18x_1 + 9x_2$

   subject to $x_1 - 2x_2 \leq 10$

   $4x_1 - 3x_2 \leq 20$

   $x_1, x_2 \geq 0$

6.39 Minimize $f(\mathbf{x}) = x_1^2 + x_2^2 - 2x_1 - 2x_2$

   subject to $x_1 + x_2 - 4 \leq 0$

   $2 - x_1 \leq 0$

   $x_1, x_2 \geq 0$

6.40 Minimize $f(\mathbf{x}) = 2x_1^2 + 2x_2^2 + x_3^2 + 2x_1x_2 - x_1x_3 - 0.8x_2x_3$

   subject to $1.3x_1 + 1.2x_2 + 1.1x_3 \geq 1.15$

   $x_1 + x_2 + x_3 = 1$

   $x_1 \leq 0.7$

   $x_2 \leq 0.7$

   $x_3 \leq 0.7$

   $x_1, x_2, x_3 \geq 0$

For the following exercises, obtain linear and quadratic programming subproblems, plot them on a graph sheet, obtain the search direction for the subproblems, and show them on the graphical representation of the original problem:

6.41 Exercise 6.3   **6.42** Exercise 6.4

6.43 Exercise 6.5   **6.44** Exercise 6.6

6.45 Exercise 6.7   **6.46** Exercise 6.8

6.47 Exercise 6.9   **6.48** Exercise 6.10

6.49 Exercise 6.11   **6.50** Exercise 6.12

6.51 Exercise 6.13   **6.52** Exercise 6.14

6.53 Exercise 6.15

Section 6.6   Constrained Steepest Descent Method

6.54 *Answer True or False*

   1. The constrained steepest descent (CSD) method, when there are active constraints, is based on using the cost function gradient as the search direction.

2. The constrained steepest descent method solves two subproblems: the search direction and step size determination.

3. The cost function is used as the descent function in the CSD method.

4. The QP subproblem in the CSD method is strictly convex.

5. The search direction if one exists is unique for the QP subproblem in the CSD method.

6. Constraint violations play no role in step size determination in the CSD method.

7. Lagrange multipliers of the subproblem play a role in step size determination in the CSD method.

8. Constraints must be evaluated during line search in the CSD method.

9. Inexact line search is used in the CSD method.

10. Potential constraint strategy can be used in the CSD method.

For the following problems, complete two iterations of the constrained steepest descent method for the given starting point (let $R_0 = 1$ and $\gamma = 0.5$):

6.55 Exercise 6.3	**6.56** Exercise 6.4
6.57 Exercise 6.5	**6.58** Exercise 6.6
6.59 Exercise 6.7	**6.60** Exercise 6.8
6.61 Exercise 6.9	**6.62** Exercise 6.10
6.63 Exercise 6.11	**6.64** Exercise 6.12
6.65 Exercise 6.13	**6.66** Exercise 6.14
6.67 Exercise 6.15	

Section 6.7 Constrained Quasi-Newton Methods

Complete two iterations of the constrained quasi-Newton method and compare the search directions with the ones obtained with the CSD algorithm (note that the first iteration is the same for both methods; let $R_0 = 1$, $\gamma = 0.5$):

6.68 Exercise 6.3	**6.69** Exercise 6.4
6.70 Exercise 6.5	**6.71** Exercise 6.6
6.72 Exercise 6.7	**6.73** Exercise 6.8
6.74 Exercise 6.9	**6.75** Exercise 6.10
6.76 Exercise 6.11	**6.77** Exercise 6.12
6.78 Exercise 6.13	**6.79** Exercise 6.14
6.80 Exercise 6.15	

CHAPTER
7

INTERACTIVE
DESIGN
OPTIMIZATION

7.1 INTRODUCTION

Optimization techniques can – to some extent – automate the tedious trial-and-error aspects of the design process, thus allowing the engineer to concentrate on the more creative aspects. They harness a computer's speed with computational algorithms to methodically generate efficient (not just satisfactory) designs which are needed in today's competitive world. Optimization techniques can be applied to virtually any engineering design situation, such as the design of aircraft structures, buildings, automotive structures, engine components, heat exchangers, land developments, water reclamation projects, chemical processes, electronic circuits, and many more.

 Most numerical methods for design optimization involve considerable repetitive calculations. They must be transcribed into proper computer software. Performance and robustness of the methods is affected by the round-off and truncation errors in computer calculations. *Most optimization algorithms are proved to converge only in the limit,* i.e. there is no algorithm for general optimization problems that is proved to converge in a finite number of iterations. In addition, they lack preciseness in their computational steps because a given algorithm can be implemented in several different ways depending on the programmer's knowledge, experience and preferences. To compound matters further, the same software when installed on different computers, or when different compilers are used for the same computer, the performance of the software can alter. All these difficulties and the lack of preciseness point to the need to somehow monitor the progress of an algorithm and interact with it for practical applications.

In order to monitor progress of the optimum design process, proper hardware and software are needed. *The software must have proper interactive facilities for the designer to change the course of the design process if necessary.* The design information must be displayed in a comprehensible form. Proper help facilities should also be available. The graphical display of various data and information can facilitate the interactive decision making process, so it should be available.

In this chapter, we describe *the interactive design optimization process*. The role of designer interaction and algorithms for interaction are described. Desired interactive capabilities and decision making facilities are discussed and simple examples are used to demonstrate their use in the design process. These discussions essentially lay out the specifications for an interactive design optimization software.

The following is an outline of this chapter.

Section 7.2 Role of Interaction in Design Optimization. The interactive design optimization process is defined and the role of proper computer hardware and software is delineated. The advantages of designer interaction and why it is useful are discussed.

Section 7.3 Interactive Design Optimization Algorithms. Several algorithms suitable for designer interaction are described and illustrated with examples. Using these algorithms, the designer can find feasible designs, request a certain reduction in the cost function, and request constraint correction at no change in cost or a specified increase in cost. These are very useful procedures and, when properly implemented, can be used to actually guide the iterative design process towards better designs and ultimately optimum designs.

Section 7.4 Desired Interactive Capabilities. Capabilities useful for designer interaction are described. These include data preparation, decision making, and graphics facilities.

Section 7.5 Interactive Design Optimization Software. In this section, a software package having various interactive capabilities is described. User interfaces for the program are explained.

Section 7.6 Examples of Interactive Design Optimization. To demonstrate the use of various interactive capabilities, an example of the design of coil springs is selected. The problem is formulated in detail. Various facilities of the software package (batch as well as interactive) are exploited to solve the problem. The use of interactive decision making and graphics is explained and demonstrated.

7.2 ROLE OF INTERACTION IN DESIGN OPTIMIZATION

7.2.1 What is Interactive Design Optimization?

In Chapter 1 we described the engineering design process. The differences between the conventional and the optimum design process were explained. The optimum design process requires sophisticated computational algorithms. However, most algorithms have uncertainties in their computational steps. Therefore, it is prudent to interactively monitor their progress and guide the optimum design process. *Interactive design optimization algorithms are based on utilizing the designer's input during the iterative process. They are in some sense open-ended algorithms in which the designer can specify what needs to be done depending on the current design conditions. They must be implemented into an interactive software having capabilities to interrupt the iterative process and report the status of the design to the user.* Relevant data and conditions must be displayed at the designer's command at a graphics workstation. Various options should be available to the designer to facilitate decision making and change design data. It should be possible to restart or terminate the process. With such facilities designers have complete control over the design optimization process. They can guide it to obtain better designs and ultimately the best design.

It is clear that for interactive design optimization, proper algorithms must be implemented in a highly flexible and user-friendly software. It must be possible for the designer to interact with the algorithm and change its course of calculations. We describe later in Section 7.3 algorithms that are suitable for designer interaction. Figure 7.1 is a conceptual flow diagram for the interactive design optimization process. It is a modification of Fig. 1.3, in which an interactive block has been added. The designer interacts with the design process through this block. We shall discuss the desired interactive capabilities and their use later in this chapter.

7.2.2 Role of Computers in Interactive Design Optimization

As we have discussed earlier, *the conventional trial-and-error design philosophy is changing with the emergence of fast computers and computational algorithms*. The new design methodology is characterized by the phrase *model and analyze*. Once the design problem is properly formulated, numerical methods can be used to optimize the system. The methods are iterative and generate a sequence of design points before converging to the optimum solution. They are best suited for computer implementation to exploit the speed of computers for performing repetitive calculations.

It is extremely important to select only robust optimization algorithms for

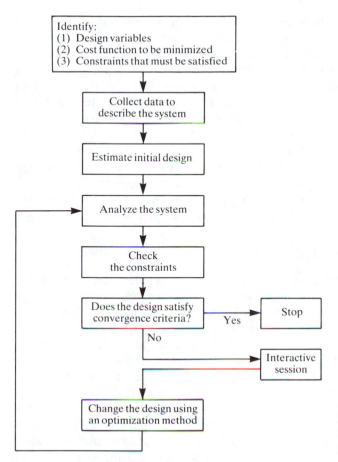

FIGURE 7.1
Interactive optimum design process.

practical applications. Otherwise, failure of the design process will undoubtedly result in the waste of computer resources and, more importantly, the loss of the designer's time and morale.

An optimization algorithm involves a limiting process, because some parameters go to zero or infinity as the optimum design is approached. The representation of such limiting processes is difficult in *computer implementation* as it may lead to underflow or overflow. In other words, the limiting processes can never be satisfied exactly on a computer and quantities such as zero and infinity must be redefined as very small and large numbers respectively on the computer. These quantities are relative and machine-dependent.

Often, the proof of convergence or rate of convergence of an iterative optimization algorithm is based on *exact arithmetic* and under restrictive conditions. That is why the theoretical behavior of an algorithm may no longer be valid in practice due to inexact arithmetic causing round-off and truncation errors in computer representation of numbers. This discussion highlights the fact that proper coding and interactive monitoring of theoretically convergent algorithms is also equally important.

Even if there is an abundance of optimization theory to solve nonlinear programming problems, it does not necessarily tell us how to implement certain steps into the computer program. *Thus, we have to develop guidelines and rules based on heuristics* and past numerical experience. A wealth of information is generated during the iterative optimization process. This information can be used to extrapolate design and generate additional informaton about the problem which may be otherwise difficult to calculate. For example, in the quasi-Newton methods of unconstrained optimization, approximate higher-order information is generated from the computed data. To fully exploit all the information generated during the iterative design optimization process, the present trend is to develop *knowledge-based expert systems* that emulate the decision making power of an expert. Such software systems can considerably facilitate the interactive design process.

7.2.3   Why Interactive Design Optimization?

The design process can be quite complex. Often the problem cannot be stated in a precise form for complete analysis and there are uncertainties in the design data. The solution to the problem need not exist. On many occasions, *the formulation of the problem must be developed as part of the design process.* Therefore, it is neither desirable nor useful to optimize an inexact problem to the end in a batch environment. It would be a complete waste of valuable resources to find out at the end that wrong data were used or a constraint was inadvertently omitted. It is essential to have an interactive algorithm and software capable of designer interaction. Such a capability can be extremely useful in a practical design environment. Using the capability, not only can better designs be obtained but more *insights into the problem behavior can be gained. The problem formulation can be refined,* and inadequate and absurd designs can be avoided. We shall describe some interactive algorithms and other suitable capabilities to demonstrate the usefulness of designer interaction in the design process.

7.3   INTERACTIVE DESIGN OPTIMIZATION ALGORITHMS

It is clear from the preceding discussion that for a useful interactive capability proper algorithms must be implemented into a well designed software. Some optimization algorithms are not suitable for designer interaction. For example, the constrained steepest descent method of Section 6.6 and the quasi-Newton method of Section 6.7 are not suitable for the interactive environment. Their steps are in some sense closed-ended allowing little opportunity for the designer to change course from the iterative design process. However, it turns out that the *QP subproblem and the basic concepts discussed there can be utilized to devise algorithms suitable for the interactive environment.* We shall describe these algorithms and illustrate them with examples.

Depending on the design condition at the current iteration, the *designer may want to ask any of the following four questions*:

1. If the current design is feasible but not optimum, can the cost function be reduced by $\gamma\%$?
2. If the starting design is infeasible, can a feasible design be obtained at any cost?
3. If the current design is infeasible, can a feasible design be obtained without increasing the cost?
4. If the current design is infeasible, can a feasible design be obtained with only $\delta\%$ penalty on the cost?

We shall describe algorithms to answer these questions. It will be seen that the algorithms are conceptually quite simple and easy to implement. As a matter of fact, they are modifications of the constrained steepest descent (CSD) and quasi-Newton methods of Sections 6.6 and 6.7. It should be also clear that if an interactive software with commands to execute the foregoing steps is available, the designer can actually use the commands to guide the process to successively better designs and ultimately an optimum design.

7.3.1 Cost Reduction Algorithm

A subproblem for the cost reduction algorithm can be defined with or without the approximate Hessian **H**. Without Hessian updating, the problem is defined in Eqs (6.38) to (6.40) and, with Hessian updating, it is defined in Eqs (6.93) to (6.95). Although Hessian updating can be used, we shall define the cost reduction subproblem without it to keep the discussion and the presentation simple. Since the cost reduction problem is solved from a feasible or almost feasible point, the right-hand side vector **e** in Eq. (6.40) is zero. Thus, the cost reduction QP subproblem is defined as

minimize $\qquad\qquad\qquad \mathbf{c}^T\mathbf{d} + 0.5\mathbf{d}^T\mathbf{d}$ $\qquad\qquad\qquad$ (7.1)

subject to

$$\mathbf{N}^T\mathbf{d} = \mathbf{0} \qquad\qquad (7.2)$$

$$\mathbf{A}^T\mathbf{d} \leqq \mathbf{b} \qquad\qquad (7.3)$$

Columns of matrices **N** and **A** contain gradients of equality and inequality constraints respectively, and **c** is the gradient of the cost function. Equation (7.2) gives the dot product of **d** with all the columns of **N** as zero. Therefore, **d** is orthogonal to gradients of all the equality constraints. Since gradients in the matrix **N** are normal to the corresponding constraint surfaces, the search direction **d** lies in a plane tangent to the equality constraints. The right-hand side vector **b** for the inequality constraints in Eq. (7.3) contains zero elements corresponding to the active constraints and positive elements corresponding to the inactive constraints. If an active constraint remains satisfied at equality (i.e. $\mathbf{a}^{(i)} \cdot \mathbf{d} = 0$), the direction $\mathbf{d}$ is in a plane tangent to that constraint. Otherwise, it must point into the feasible region for the constraint.

The QP subproblem defined in Eqs (7.1) to (7.3) can incorporate the potential constraint strategy as explained in Section 6.6. The subproblem can be solved for the cost reduction direction by any of the available subroutines cited in Section 6.5.3. In the example problems, however, we shall solve the QP subproblem using Kuhn–Tucker conditions. We shall call this procedure of reducing cost from a feasible point the *Cost Reduction (CR) Algorithm.*

After the direction has been determined, the step size can be calculated by a line search on the proper descent function. Or, we can require a certain reduction in the cost function and determine the step size that way. For example, we can require a fractional reduction γ in the cost function (for a 5% reduction, $\gamma = 0.05$), and calculate a step size based on it. Let α be the step size along $\mathbf{d}$. Then the first-order change in the cost using linear Taylor series expansion is given as $\alpha \, |\mathbf{c} \cdot \mathbf{d}|$. Equating this to the required reduction in cost $|\gamma f|$, the step size is calculated as

$$\alpha = \frac{|\gamma f|}{|\mathbf{c} \cdot \mathbf{d}|} \tag{7.4}$$

Note that $\mathbf{c} \cdot \mathbf{d}$ should not be zero in Eq. (7.4) to give a reasonable step size.

Example 7.1   Cost reduction step. Consider the design optimization problem

minimize $f(\mathbf{x}) = x_1^2 - 3x_1x_2 + 4.5x_2^2 - 10x_1 - 6x_2$

subject to

$$x_1 - x_2 \leq 3$$

$$x_1 + 2x_2 \leq 12$$

$$x_1, x_2 \geq 0$$

From the feasible point $(4, 4)$, calculate the cost reduction direction and the new design point requiring a cost reduction of 10%.

Solution. The constraints can be written in the standard form as

$$g_1 \equiv \tfrac{1}{3}(x_1 - x_2) - 1.0 \leq 0$$

$$g_2 \equiv \tfrac{1}{12}(x_1 + 2x_2) - 1.0 \leq 0$$

$$g_3 \equiv -x_1 \leq 0$$

$$g_4 \equiv -x_2 \leq 0$$

The optimum solution for the problem is calculated using Kuhn–Tucker conditions as

$$\mathbf{x}^* = (6, 3); \qquad \mathbf{u}^* = (17, 16, 0, 0); \qquad f(\mathbf{x}^*) = -55.5$$

At the given point $(4, 4)$,

$$f(4, 4) = -24$$

$$g_1 = -1.0 < 0 \text{ (inactive)}$$

$$g_2 = 0 \text{ (active)}$$

$$g_3 = -4.0 < 0 \text{ (inactive)}$$

$$g_4 = -4.0 < 0 \text{ (inactive)}$$

Therefore, constraint g_2 is active, and all the others are inactive. The cost function is much larger than the optimum value. The constraints for the problem are plotted in Fig. 7.2. The feasible region is identified as OABC. Several cost function contours are shown there. The optimum solution is at the point B $(6, 3)$. The given point $(4, 4)$ is identified as D on the line B–C in Fig. 7.2.

The gradients of cost and constraint functions at the point D $(4, 4)$ are calculated as

$$\mathbf{c} = (2x_1 - 3x_2 - 10, \ -3x_1 + 9x_2 - 6)$$

$$= (-14, 18)$$

$$\mathbf{a}^{(1)} = (\tfrac{1}{3}, -\tfrac{1}{3})$$

$$\mathbf{a}^{(2)} = (\tfrac{1}{12}, \tfrac{1}{6})$$

$$\mathbf{a}^{(3)} = (-1, 0)$$

$$\mathbf{a}^{(4)} = (0, -1)$$

These gradients are shown at point D in Fig. 7.2. Each constraint gradient points

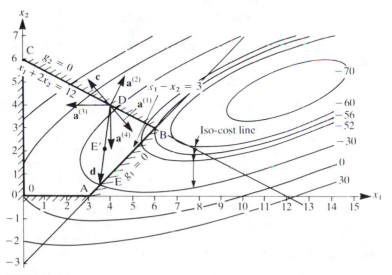

FIGURE 7.2
Feasible region for Example 7.1. Cost reduction step from point D.

in a direction in which the constraint function value increases. Using these quantities the QP subproblem of Eqs (7.1) to (7.3) is defined as

minimize $\qquad\qquad (-14d_1 + 18d_2) + 0.5(d_1^2 + d_2^2)$

subject to

$$\begin{bmatrix} \frac{1}{3} & -\frac{1}{3} \\ \frac{1}{12} & \frac{1}{6} \\ -1 & 0 \\ 0 & -1 \end{bmatrix} \begin{bmatrix} d_1 \\ d_2 \end{bmatrix} \leq \begin{bmatrix} 1 \\ 0 \\ 4 \\ 4 \end{bmatrix}$$

The solution for the QP subproblem using Kuhn–Tucker conditions or the simplex method of Section 6.5 is

$$\mathbf{d} = (-0.5, -3.5); \qquad \mathbf{u} = (43.5, 0, 0, 0)$$

At the solution, only the first constraint is active having a positive Lagrange multiplier. The direction $\mathbf{d}$ is shown in Fig. 7.2. Since the second constraint is inactive, $\mathbf{a}^{(2)} \cdot \mathbf{d}$ must be negative according to Eq. (7.3) and it is (-0.625). Therefore, direction $\mathbf{d}$ points towards the feasible region with respect to the second constraint which can be observed in Fig. 7.2.

The step size is calculated from Eq. (7.4) based on a 10% reduction ($\gamma = 0.1$) of the cost function as

$$\alpha = \frac{|0.1(-24)|}{|(14, -18) \cdot (-0.5, -3.5)|} = 0.3/7$$

Thus, the new design point is given as

$$\mathbf{x}^{(1)} = \begin{bmatrix} 4 \\ 4 \end{bmatrix} + \frac{0.3}{7} \begin{bmatrix} -0.5 \\ -3.5 \end{bmatrix} = \begin{bmatrix} 3.979 \\ 3.850 \end{bmatrix}$$

which is quite close to point D along direction $\mathbf{d}$. The cost function at this point is calculated as

$$f(\mathbf{x}^{(1)}) = -26.304$$

which is approximately 10% less than the one at the current point $(4, 4)$. It may be checked that all constraints are inactive at the new point.

Direction $\mathbf{d}$ points into the feasible region at point $\mathbf{D}$ as can be seen in Fig. 7.2. Any small move along $\mathbf{d}$ results in a feasible design. If the step size is taken as one (which would be obtained if inexact line search of Section 6.6 was performed), then the new point is given as $(3.5, 0.5)$ which is marked as E in Fig. 7.2. At point E constraint g_1 is active and the cost function has a value of -29.875 which is smaller than the previous value of -26.304. If we perform exact line search then α is computed as 0.5586 and the new point is given as $(3.7207, 2.0449)$ – identified as point E′ in Fig. 7.2. The cost function at this point is -39.641 which is still better than the one with step size as unity. ‖

Example 7.2   Cost reduction step with potential constraints. For Example 7.1, calculate the cost reduction step by considering the potential inequality constraints only.

Solution. In some algorithms, only the potential inequality constraints at the current point are considered while defining the direction finding subproblem, as discussed previously in Section 6.6. The direction determined with this sub-problem can be different from that obtained by including all constraints in the subproblem.

For the present problem, only the second constraint is active ($g_2 = 0$) at the point $(4, 4)$. The QP subproblem with this active constraint is defined as

minimize $$(-14d_1 + 18d_2) + 0.5(d_1^2 + d_2^2)$$

subject to

$$\tfrac{1}{12}d_1 + \tfrac{1}{6}d_2 \leqq 0$$

Solving the problem by Kuhn–Tucker conditions, we get

$$\mathbf{d} = (14, -18); \qquad u = 0$$

Since the Lagrange multiplier for the constraint is zero, it is not active, so $\mathbf{d} = -\mathbf{c}$ is the solution to the subproblem. This search direction points into the feasible region along the negative cost function gradient direction, as seen in Fig. 7.2. An appropriate step size can be calculated along the direction.

If we require the constraint to remain active (i.e. $d_1/12 + d_2/6 = 0$), then the solution to the subproblem is given as

$$\mathbf{d} = (18.4, -9.2); \qquad u = -52.8$$

This direction is tangent to the constraint, i.e. along the line D–B in Fig. 7.2. ‖

7.3.2 Constraint Correction Algorithm

If at a design point, constraint violations are very large, it may be useful to find out if a feasible design can be obtained. Several algorithms can be used to correct constraint violations. We shall describe a procedure that is a minor variation of the constrained steepest descent method of Section 6.6. A QP subproblem that gives constraint correction can be obtained from Eqs (6.38) to (6.40) by neglecting the term related to the cost function. In other words, we do not put any restriction on the changes in the cost function, and define the QP subproblem as

minimize $$0.5\mathbf{d}^T\mathbf{d} \tag{7.5}$$

subject to

$$\mathbf{N}^T\mathbf{d} = \mathbf{e} \tag{7.6}$$

$$\mathbf{A}^T\mathbf{d} \leqq \mathbf{b} \tag{7.7}$$

A solution to the subproblem gives a direction with the shortest distance to the constraint boundary (linear approximation) from an infeasible point. Equation (7.5) essentially says: *find a direction* $\mathbf{d}$ *having the shortest path to the linearized feasible region from the current point.* Equations (7.6) and (7.7) impose the requirement of constraint corrections. Note that the potential set strategy as described in Section 6.6 can also be used here. After the direction has been found, a step size can be determined to make sure that the constraint violations are improved. We shall call this procedure the *Constraint Correction (CC) Algorithm*.

Note that constraint correction usually results in an increase in cost. However, there can be some unusual cases where constraint correction can be accompanied by a reduction in the cost function.

Example 7.3 Constraint correction step. For Example 7.1, calculate the constraint correction step from the infeasible point $(9, 3)$.

Solution. The feasible region for the problem and the starting point (F) are shown in Fig. 7.3. The constraint and cost gradients are also shown there. At the point $(9, 3)$, the following data are calculated:

$$f(9, 3) = -67.5$$

$$g_1 = 1 > 0 \text{ (violation)}$$

$$g_2 = 0.25 > 0 \text{ (violation)}$$

$$g_3 = -9 < 0 \text{ (inactive)}$$

$$g_4 = -3 < 0 \text{ (inactive)}$$

$$\mathbf{c} = (-1, -6)$$

and gradients of the constraints are the same as in Example 7.1. Cost and constraint gradients are shown at point F in Fig. 7.3. Thus, the constraint

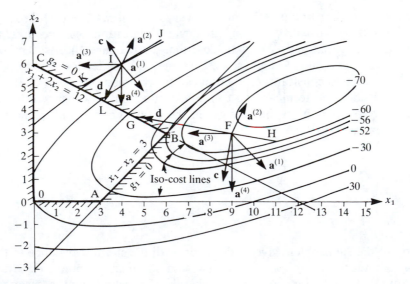

FIGURE 7.3
Feasible region for Example 7.1. Constraint correction and constant cost steps from point F; constant cost step from point I.

correction QP subproblem of Eqs (7.5) to (7.7) is defined as

minimize $$0.5(d_1^2 + d_2^2)$$

subject to

$$\begin{bmatrix} \frac{1}{3} & -\frac{1}{3} \\ \frac{1}{12} & \frac{1}{6} \\ -1 & 0 \\ 0 & -1 \end{bmatrix} \begin{bmatrix} d_1 \\ d_2 \end{bmatrix} \leq \begin{bmatrix} -1 \\ -0.25 \\ 9 \\ 3 \end{bmatrix}$$

Using the Kuhn–Tucker necessary conditions, the solution for the QP subproblem is given as

$$\mathbf{d} = (-3, 0); \qquad \mathbf{u} = (6, 12, 0, 0)$$

Note that the shortest path from Point F to the feasible region is along the line F–B and the QP subproblem actually gives this solution. The new design point is given as

$$\mathbf{x}^{(1)} = \begin{bmatrix} 9 \\ 3 \end{bmatrix} + \begin{bmatrix} -3 \\ 0 \end{bmatrix} = \begin{bmatrix} 6 \\ 3 \end{bmatrix}$$

which is point B in Fig. 7.3. At the new point, constraints g_1 and g_2 are active, and g_3 and g_4 are inactive. Thus, a single step corrects both the violations precisely. *This is due to linearity of all the constraints in the present example.* In general *several iterations* may be needed to correct the constraint violations. Note that the new point actually represents the optimum solution. ‖

7.3.3 Algorithm for Constraint Correction at Constant Cost

In some instances, the constraint violations are not very large. It is useful to know whether a feasible design can be obtained without any increase in the cost. This shall be called a constant cost subproblem which can be defined by adding another constraint to the QP subproblem given in Eqs (7.5) to (7.7). The additional constraint simply requires the current linearized cost to either remain constant or reduce; that is, the linearized change in cost $(\mathbf{c} \cdot \mathbf{d})$ be nonpositive which is expressed as

$$\mathbf{c} \cdot \mathbf{d} \leq 0 \qquad (7.8)$$

The constraint imposes the condition that the direction $\mathbf{d}$ be either orthogonal to the gradient of the cost function (i.e. $\mathbf{c} \cdot \mathbf{d} = 0$), or make an angle between 90 and 270° with it (i.e. $\mathbf{c} \cdot \mathbf{d} < 0$). We shall see this in the example problem discussed later.

If the Inequality (7.8) is active (i.e. dot product is zero, so $\mathbf{d}$ is orthogonal to $\mathbf{c}$), then there is no change in the linearized cost function value. However, there may be some change in the original cost function due to nonlinearities. If the constraint is inactive, then there is actually some reduction in the linearized cost function along with correction of the constraints. This is a desirable situation. Thus, we observe that a constant cost problem is also a QP subproblem defined in Eqs (7.5) to (7.8). *It seeks a shortest path to the feasible region that either reduces the linearized cost function or keeps it unchanged.* We shall call this procedure the *Correction at Constant Cost (CCC) Algorithm.*

Note that the constant cost QP subproblem can be infeasible if the current cost function contour does not intersect the feasible region. This can happen in practice, so a QP subproblem should be solved properly. If it turns out to be infeasible, then the constraint of Eq. (7.8) must be relaxed, and the linearized cost function must be allowed to increase to obtain a feasible point. This will be discussed in the next subsection.

Example 7.4   Constraint correction at constant cost. For Example 7.3, calculate the constant cost step from the infeasible point $(9, 3)$.

Solution. To obtain the constant cost step from point F in Fig. 7.3, we impose an additional constraint of Eq. (7.8) as $(\mathbf{c} \cdot \mathbf{d}) \leqq 0$ on the QP subproblem given in Example 7.3. Substituting for $\mathbf{c}$, the constraint is given as

$$-d_1 - 6d_2 \leqq 0 \tag{7.9}$$

which imposes the condition that the linearized cost function either remain constant at -67.5 or reduce further. From the graphical representation for the problem in Fig. 7.3, we observe that the cost function value of -67.5 at the given point $(9, 3)$ is below the optimum cost of -55.5. Therefore, the current cost function value represents a lower bound on the optimum cost function value. However, the linearized cost function line, shown as G–H in Fig. 7.3, intersects the feasible region. Thus, the QP subproblem of Example 7.3 with the preceding additional constraint has feasible solutions. The Inequality of Eq. (7.8) imposes the condition that direction $\mathbf{d}$ be either on the line G–H (if the constraint is active) or above it (if the constraint is inactive). In case it is inactive, the angle between $\mathbf{c}$ and $\mathbf{d}$ will be between 90 and 270°. If it is below the line G–H, it violates the Inequality (7.8). Note that the shortest path from F to the feasible region is along the line F–B. But this path is below the line G–H and thus not feasible for the preceding QP subproblem.

Solving the problem using Kuhn–Tucker conditions, we obtain the solution for the preceding QP subproblem as

$$\mathbf{d} = (-4.5, 0.75); \quad \mathbf{u} = (0, 83.25, 0, 0, 2.4375)$$

Thus the new point is given as

$$\mathbf{x} = (4.5, 3.75) \quad \text{with } f = -34.6$$

At the new point (G in Fig. 7.3), all the constraints are inactive except the second one (g_2). The constant cost condition of Eq. (7.8) is also active which implies that the direction $\mathbf{d}$ is orthogonal to the cost gradient vector $\mathbf{c}$. As seen in Fig. 7.3, this is indeed true. Note that due to the highly nonlinear nature of the cost function at the point F $(9, 3)$, the new cost function actually increases substantially. Thus the direction $\mathbf{d}$ is not truly a constant cost direction. Although the new point corrects all the violations, the cost function increases beyond the optimum point which is undesirable. Actually, from point F it is better to solve just the constraint correction problem, as in Example 7.3. The increase in cost is smaller for that direction. Thus, in certain cases, it is better to solve just the constraint correction subproblem. ||

Example 7.5 Constraint correction at constant cost. Consider another starting point as $(4, 6)$ for Example 7.3, and calculate the constant cost step from there.

Solution. The starting point is identified as I in Fig. 7.3. The following data are calculated at the point $(4, 6)$:

$$f(4, 6) = 30$$

$$g_1 = -\tfrac{5}{3} \text{ (inactive)}$$

$$g_2 = \tfrac{1}{3} > 0 \text{ (violated)}$$

$$g_3 = -4 < 0 \text{ (inactive)}$$

$$g_4 = -6 < 0 \text{ (inactive)}$$

$$\mathbf{c} = (-20, 36)$$

and the constraint gradients are the same as in Example 7.1. Cost and constraint gradients are shown in Fig. 7.3 at point I. Note that the cost function at point I is above the optimum value. Therefore, the constant cost constraint of Eq. (7.8) may not be active for the solution of the subproblem, i.e. we may be able to correct constraints and reduce the cost function at the same time.

The constant cost QP subproblem given in Eqs (7.5) to (7.8) is defined as

minimize $\qquad\qquad 0.5(d_1^2 + d_2^2)$

$$\begin{bmatrix} \tfrac{1}{3} & -\tfrac{1}{3} \\ \tfrac{1}{12} & \tfrac{1}{6} \\ -1 & 0 \\ 0 & -1 \end{bmatrix} \begin{bmatrix} d_1 \\ d_2 \end{bmatrix} \leqq \begin{bmatrix} \tfrac{5}{3} \\ -\tfrac{1}{3} \\ 4 \\ 6 \end{bmatrix}$$

$$-20d_1 + 36d_2 \leqq 0$$

Writing the Kuhn–Tucker conditions for the QP subproblem we obtain the solution as

$$\mathbf{d} = (-0.8, -1.6); \qquad \mathbf{u} = (0, 9.6, 0, 0, 0)$$

Note that only the second constraint is active. Therefore, the new point should be precisely on the constraint equation $x_1 + 2x_2 = 12$, shown as point L in Fig. 7.3. The constant cost constraint is not active (the direction $\mathbf{d}$ lies below the line J–K in Fig. 7.3 making an angle greater than 90° with $\mathbf{c}$). Thus the new cost function value should reduce at the new point $(3.2, 4.4)$, and it does; -3.28 versus 30, which is a substantial reduction. ||

7.3.4 Algorithm for Constraint Correction at Specified Increase in Cost

As observed in the previous subsection, the constant cost subproblem can be infeasible. In that case, the current value of the cost function must be allowed to increase. This can be done quite easily by slightly modifying the Inequality (7.8) as

$$\mathbf{c}^T \mathbf{d} \leq \Delta \qquad (7.10)$$

where Δ is a specified limit on the increase in cost. The increase in cost can be specified based on the condition that the new cost based on the linearized expression does not go beyond the previous cost at a feasible point, if known. Note again that the QP subproblem in this case can be infeasible if the increase in cost specified in Δ is not enough. Therefore, Δ may have to be adjusted. We shall call this procedure the *Constraint Correction at Specified Cost (CCS) Algorithm*.

> **Example 7.6   Constraint correction at specified increase in cost.** For Example 7.4, calculate the constraint correction step from the infeasible point $(9, 3)$ with a 10% increase in cost.
>
> **Solution.** Since the current value of the cost function is -67.5, the 10% increase in cost gives Δ as 6.75 in Eq. (7.10). Therefore, using $\mathbf{c} = (-1, -6)$ as calculated in Example 7.3, the constraint of Eq. (7.10) becomes
>
> $$-d_1 - 6d_2 \leq 6.75$$
>
> Other constraints and the cost function are the same as defined in Example 7.3. Solving the problem using Kuhn–Tucker conditions, we obtain the solution of the subproblem as
>
> $$\mathbf{d} = (-3, 0); \qquad \mathbf{u} = (6, 12, 0, 0, 0)$$
>
> At the solution, first two constraints are active, and the constraint of Eq. (7.10) is inactive. Note that this is the same solution as obtained in Example 7.3. Thus the new point represents the optimum solution.   ‖

7.3.5   Constraint Correction with Minimum Increase in Cost

It is possible to define a subproblem that minimizes the increase in cost and at the same time corrects constraints. The subproblem is defined as

$$\text{minimize} \quad f_L = \mathbf{c}^T \mathbf{d}$$

subject to the constraints of Eqs (7.6) and (7.7), where f_L is the linearized change in the cost function. This problem may be unbounded, so we impose the following move limits

$$-\Delta_i \leq d_i \leq \Delta_i; \qquad i = 1 \text{ to } n$$

to obtain a bounded problem. Here Δ_i represents the maximum and minimum value for d_i. The preceding subproblem is linear, so any LP code can be used to solve it. A line search can be performed along the direction $\mathbf{d}$ to find the proper step size.

Example 7.7 Constraint correction with minimum increase in cost. For Example 7.4, solve the constraint correction problem with the minimum increase in cost from point F (9, 3) shown in Fig. 7.3.

Solution. At the point (9, 3) the following data are calculated in Example 7.3.3:

$$f(9, 3) = -67.5$$

$$g_1 = 1 > 0 \text{ (violation)};$$

$$g_2 = 0.25 > 0 \text{ (violation)};$$

$$g_3 = -9 < 0 \text{ (inactive)}$$

$$g_4 = -3 < 0 \text{ (inactive)}$$

$$\mathbf{c} = (-1, -6).$$

Therefore, the subproblem is defined as

minimize
$$f_L = -d_1 - 6d_2$$
subject to

$$\tfrac{1}{3}d_1 - \tfrac{1}{3}d_2 \leqq -1$$

$$\tfrac{1}{12}d_1 + \tfrac{1}{6}d_2 \leqq -0.25$$

$$-d_1 \leqq 9$$

$$-d_2 \leqq 3$$

$$-\Delta_1 \leqq d_1 \leqq \Delta_1; \qquad -\Delta_2 \leqq d_2 \leqq \Delta_2$$

The linear programming subproblem can be solved using the Simplex method in Chapter 4. We solve the problem using the program LINDO [Schrage, 1981]. With $\Delta_1 = \Delta_2 = 1$, the problem is infeasible, the move limits are too restrictive and the feasible point cannot be found. Since the problem has two variables, one can easily plot all the problem functions on a graph sheet and verify that there is no solution to the preceding linearized subproblem. When $\Delta_1 = \Delta_2 = 3$, the following solution is obtained:

$$d = -3, \qquad d_2 = 0, \qquad f_L = 3$$

and the second constraint is active with the Lagrange multiplier as 36. The lower limit on d_1 is also active with the Lagrange multiplier as 2.0. When d_1 and d_2 are added to the starting point (9, 3), the new point is given as (6, 3). This is actually the optimum point with the cost function value as -55.5. Note that since $f_L = 3$, the cost function was supposed to increase by only 3 from -67.5. However, due to nonlinearity, it has actually increased by 12.

Note that since the Lagrange multiplier for the lower bound constraint on d_1 is 2, the Constraint Sensitivity Theorem 3.14 predicts that f_L will reduce by 2 to 1 if Δ_1 is changed to 4. This is indeed the case. With $\Delta_1 = \Delta_2 = 4$, the solution of the subproblem is obtained as

$$d_1 = -4, \qquad d_2 = 0.5, \qquad f_L = 1.0$$

and the second constraint is active with the Lagrange multiplier as 36. The lower limit on d_1 is still active with the Lagrange multiplier as 2.0. The new point is given as $(5, 3.5)$ with the cost function as -43.375. For this point the cost function is actually increased by 24.125 rather than just 1 as predicted by the solution of the linearized subproblem. ‖

7.3.6 Observations on Interactive Algorithms

We have discussed several algorithms that are useful for interactive design optimization. They are demonstrated for a problem that has linear constraints and quadratic cost function. There are certain limitations of these algorithms that should be clearly understood:

1. *All the algorithms use linear approximations for the cost and constraint functions.* For highly nonlinear problems, the solution of the subproblems are therefore valid for a small region around the current point.
2. *The step size calculated in Eq. (7.4) using the desired reduction γ in the cost function is based on the linear approximation for the cost function.* With the calculated step size, the *actual reduction* in the cost function may be *smaller* or *larger* than γ depending on the nonlinearity of the cost function.
3. In the constraint correction problem of Example 7.3, only one step is needed to correct all the constraints. This is due to linearity of all the constraints. When the constraints are nonlinear, *several constraint correction steps are usually needed* to reach the feasible region. The spring design problem solved later in the chapter demonstrates this fact.
4. *Constraint correction is most often accompanied by an increase in the cost function.* However, in certain cases it may also result in a decrease in the cost function. This is rare and depends on the nonlinearity of functions and the starting point.
5. The *constant cost condition* of Eq. (7.8) is based on the linearized cost function. Even if this constraint is active at the solution for the subproblem, *there may be changes to the original cost function at the new point.* This is due to *nonlinearity* of the cost and constraint functions. We have observed this phenomena in Examples 7.4–7.7.
6. For some infeasible points, *it is better to solve the constraint correction subproblem rather than the constant cost subproblem.*
7. As seen in Examples 7.1 and 7.2, *there are several cost reduction directions at a given feasible point.* They depend on the definition of the QP subproblem. It is difficult to determine the best possible direction.
8. *The Lagrange multipliers evaluated during the solution of QP subproblems can be quite different* from their values at the optimum solution to the original problem. This can be observed in the solution of Examples 7.1–7.7.

7.4 DESIRED INTERACTIVE CAPABILITIES

Interactive software for design optimization should be flexible and user-friendly. Help facilities should be available in the program which can be menu-driven, command-driven or both. We shall describe several desirable capabilities of such an interactive software.

First of all, the program should be able to treat general nonlinear programming as well as unconstrained problems. It should be able to treat equality, inequality and design variable bound constraints. It should have the choice of a few good algorithms that are robustly implemented. It should trap user's mistakes and not abort abnormally.

7.4.1 Interactive Data Preparation

The software should have a module for interactive data preparation and editing. The commands for data entry should be explicit. Only *the minimum amount of data* should be required. The user should be able to edit any data that have been entered previously. The step-by-step procedure should be to display the *menu* for data selection and entry. Or, it should be possible to enter data in a simple *question/answer* session. The system should be set up in a way so that it is *protected* from any of the designer's mistakes. If data mismatch is found, *messages* should be given in detail. The interactive input procedure should be *simple* so that even a beginner can follow it easily.

7.4.2 Interactive Capabilities

As observed earlier, it is prudent to allow designer interaction in the computer-aided design process. Such a dialogue can be very beneficial, saving computer and human resources. For the use of the interactive software system in engineering, two questions arise: (i) what are the *advantages and disadvantages* of the interaction and (ii) what *type of interactive capability* needs to be provided? We shall address both these questions.

All general-purpose design optimization software need the following information about the problem to be solved: (i) input data such as number of design variables, number of constraints, etc., (ii) the cost and constraint functions, and (iii) the gradients of cost and constraint functions. If the gradients are not available then the system should automatically approximate them by a finite difference method as explained in Chapter 5. If there is a mistake in the input data or problem definition, error will occur in the problem-solving procedure. The optimization system should take care of such mistakes as much as possible.

It is also useful to monitor the optimum process through the interactive session. *Histories of the cost function, constraint functions, design variables, maximum constraint violation, and convergence parameter should be monitored. When these histories are graphically displayed, they can be of great help in certain decision making.* If the design process is not proceeding satisfactorily (there could be inaccuracies or errors in the problem formulation and modeling), it is necessary to terminate it and check the formulation of the problem. This will save human as well as computer resources. Also, if one algorithm is not progressing satisfactorily, a switch should be made to another one. The system should be able to give suggestions for design change based on the analysis of the trends. Therefore, monitoring the iterative process interactively is an important capability that should be available in a design optimization software.

The designer should also be able to guide the problem-solving process. For example, the program can be run for a certain number of iterations and interrupted to see if the process is progressing satisfactorily. If it is not progressing as expected, *a decision to change the course of calculations can be made. If there are constraint violations,* the designer may want to know whether they can be *corrected without any penalty* on the cost function. If this cannot be done, the *penalty* on the cost function to correct the constraints should be made available. When the design is in the feasible region, the system should have the capability to perform calculations and determine if the cost function can be reduced by a certain percentage and still remain feasible. *If the iterative process does not progress well, then the designer should be able to restart the program from any previous iteration or any other design.* At the optimum point, the penalty to tighten a constraint or the gain to relax it should be displayed. This information is available from the Lagrangian multipliers for the constraints. In practical optimization, these interactive capabilities can be quite useful.

It should be possible to change the input data for a design problem during the iterative process. After monitoring the process for a few iterations it may be necessary to *change the problem or program parameters.* This should be possible without terminating the program. Design sensitivity coefficients of the cost function and potential constraints should be displayed in a convenient form, e.g. as normalized bar charts. This information will show *relative sensitivity* of the design variables. The designer should also be able to determine the *status of the design variables* and change it interactively if desired. *The trend information* when displayed graphically can aid the designer in gaining insights into the problem behavior, so this capability should be available.

It should also be possible to utilize the interactive design optimization software in the batch environment with a minimum of input data. The system should have default values for the best parameters determined through expertise and numerical experimentation.

7.4.3 Interactive Decision Making

When the program is run interactively, a wide range of options should be available to the designer. The following is a list of possible capabilities that can aid the designer in decision making:

1. The designer may want to *re-examine the problem formulation* or design data. Therefore, it should be possible to exit the program at any iteration.
2. It should be possible at any iteration to display the *status of the design,*

such as current values of variables, cost function, maximum constraint violation and other such data.

3. It should be possible to *change data* at any iteration, such as design variables and their limits, convergence criteria and other data.

4. The designer should be able to *fix design variables* to any value. It should also be possible to release the fixed design variables.

5. The designer should be able to *run the algorithm one iteration* at a time or several iterations.

6. It should be possible to *restart* the program from any iteration.

7. It should be possible to *change the algorithm* during the iterative process.

8. The designer should be able to *request a reduction in the cost function* by $x\%$ from a feasible point.

9. The designer should be able to *request a constraint correction* at any iteration.

10. The designer should be able to *request a constant cost step*.

11. The designer should be able to *request a constraint correction* with an $x\%$ limit on the increase in cost.

12. The designer should be able to *request various graphical displays*.

7.4.4 Interactive Graphics

Graphical display of data is a convenient way to interpret results and draw conclusions. Interactive graphics can play a major role in design decision making during the iterative optimization process. Possible graphical displays are:

1. Plots of cost function, convergence parameters and maximum constraint violation *histories*. These show the progress of the iterative process towards the optimum point.

2. *Histories of design variables*. These can be used to observe the trend in design variables and possibly extrapolate their values.

3. *Constraint function histories* can be displayed. This can show constraints that are not playing any role in the design process. It can also show dominant constraints.

4. *Sensitivity coefficients* for the cost and constraint functions can be displayed in the form of bar charts. These are nothing but normalized gradients of cost and constraint functions. They show sensitive or insensitive variables and functions.

It can be seen that using interactive graphics capabilities, designers can observe the progress of the optimization process. They can *learn* more about the behavior of the design problem and perhaps *refine* its formulation. We shall discuss these capabilities further with example problems.

7.5 INTERACTIVE DESIGN OPTIMIZATION SOFTWARE

The preceding paragraphs essentially describe specifications for a general-purpose interactive design optimization software. Based on them, a software system can be designed and implemented. It can be observed that to implement all the flexibilities and capabilities, the software will be quite large and complex. The most modern software design and data management techniques will have to be utilized to achieve the stated goals. The entire process of software design, implementation and evaluation can be quite costly and time-consuming, requiring the equivalent of several man-years.

In this section, we shall briefly describe a software that has some of the previously stated capabilities. Other available software may be similarly extended to have similar capabilities. The present program is called IDESIGN which stands for *I*nteractive *Design* Optimization of Engineering Systems. It has interactive and graphics facilities suitable for computer-aided optimization and design [Arora, 1984; Arora and Tseng, 1987]. With the IDESIGN program, the computer and the designer's experience can be utilized to adjust design variables so as to improve the design objective while satisfying the constraints. It contains four state-of-the-art nonlinear optimization algorithms. Efficient and reliable implementations of the algorithms have been developed over several years of testing. The simpler cases of linear and unconstrained optimization can also be handled.

IDESIGN has several facilities that permit the engineer to interact with and control the optimization process. One can backtrack to any previous design or manually input a new trial design. Design information can be displayed in a variety of ways or represented in graphs. The system has been designed to accommodate both experienced users and beginners. The beginner can respond to one menu at a time as guided by on-line instruction. The expert can prepare an input data file and thus bypass immediate menus. The software identifies and helps the user correct improper responses. Input and output can be echoed to a "dialogue" file for the user's reference. Input can also be received from a file for batch mode operation for large-scale problems.

IDESIGN is written in structured, double precision FORTRAN77. Possible hardware and operating system dependencies are isolated in a handful of subroutines. The same version of the code has been successfully compiled and executed on a variety of computers (IBM, PRIME, VAX, APOLLO, CDC) and under a variety of operating systems (PRIMOS, VMS, AEGIS). Using the program, we shall optimize several design examples in Section 7.6 and Chapter 8.

7.5.1 User Interface for IDESIGN

IDESIGN consists of a main program and several standard subroutines that need not be changed by the user. In order to solve a design problem (refer to

Sections 6.1 and 6.2.3 for a proper definition of the problem for IDESIGN), the user must prepare additional subroutines for the program. The input data, such as the initial design, lower and upper limits on design variables, problem parameters and the parameter values to invoke various options available in the program, must also be provided. The input data and options available in the program are described in the User's manual [Arora and Tseng, 1987a].

The user must describe the design problem by coding the following four FORTRAN subroutines:

> **USERMF**: **M**inimization (cost) **F**unction evaluation subroutine
> **USERCF**: **C**onstraint **F**unctions evaluation subroutine
> **USERMG**: **M**inimization (cost) function **G**radient evaluation subroutines
> **USERCG**: **C**onstraint functions **G**radient evaluation subroutine

A fifth subroutine USEROU may also be provided by the user to perform post-optimality analyses for the optimum solution and obtain more output using specialized FORMATS.

Figure 7.4 shows a conceptual layout of the interactive design optimization environment with the program IDESIGN. To create a design system for a particular application, the designer needs to develop FORTRAN subroutines that define the problem – cost and constraint functions as well as gradient evaluation subroutines. The designer has all the flexibilities to develop these subroutines as long as the "argument" requirements to interface with IDESIGN are satisfied. For example, additional arrays may be declared, external subroutines or programs may be called, and additional input data may be entered. Through these subroutines, the designer may also incorporate more interactive commands that are specific to the domain of the application.

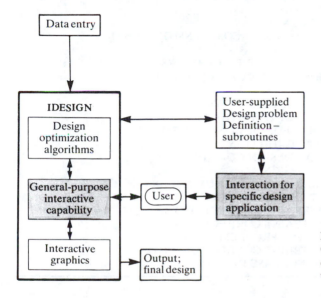

FIGURE 7.4
Conceptual layout of interactive design optimization environment with IDESIGN.

General-purpose interactive capability is available in IDESIGN as shown on the left-hand portion of Fig. 7.4. In this part, interactive commands that are not connected to any specific area of application are available. Table 7.1 contains a list of commands that are currently available. Using these commands, the designer can interactively guide the process towards acceptable designs. The command CH/XXX is particularly useful, as it allows the designer to change design variable values and their upper and lower limits, algorithm, and convergence criteria. It can also be used for obtaining advice from IDESIGN for the best changes to design variables to correct constraints. The PLOT commands can be used to observe trends in the design variables, determine critical constraints, and determine sensitive and insensitive variables with respect to the cost and constraint functions.

It can be seen that the foregoing interactive facilities can be utilized to gain insights into the behavior of a particular design problem. Having gained this knowledge, the designer can perhaps develop alternate design concepts that may be more efficient and economical.

TABLE 7.1
Interactive commands available in IDESIGN

Command	Purpose
CON	CONTINUE IDESIGN
DIS	DISPLAY THE DATA
HELP	HELP THE USER
QUIT	STOP IDESIGN
PLOT/NO.	NO = 1 COST HISTORY
	NO = 2 CONVERGENCE PARAMETER HISTORY
	NO = 3 MAX CONSTRAINT VIOLATION HISTORY
	NO = 4 DESIGN VARIABLES
	NO = 5 CONSTRAINTS
	NO = 6 CONSTRAINT AND COST SENSITIVITY BAR CHARTS
CH/XXX	CHANGE VARIABLES OR PARAMETERS XXX = ABBREVIATION OF PARAMETER
OPT	GO TO OPTIMUM
OK/XX	READY TO CONTINUE IDESIGN XX = NUMBER OF ITERATIONS; IF/XX IS OMITTED, IDESIGN GOES TO THE NEXT DESIGN POINT FOR OK/FEA, IDESIGN GOES TO NEXT FEASIBLE DESIGN
RS/XXX	RESTART FROM ITERATION NUMBER XXX
CR	SOLVES COST REDUCTION PROBLEM
CC	SOLVES CONSTRAINT CORRECTION PROBLEM
CCC	SOLVES CONSTRAINT CORRECTION AT CONSTANT COST PROBLEM
CCS	SOLVES CONSTRAINT CORRECTION WITH BOUND ON INCREASE IN COST

7.5.2 Capabilities of IDESIGN

IDESIGN is a general-purpose optimization program that has been under development since the late 1970s. Several state-of-the-art nonlinear programming algorithms and other capabilities have been added. It has a wide range of options and user-friendly features that are invoked through the input data. The *program has been used* to solve several classes of optimal design problems:

1. *Small-scale engineering design problems* having explicit cost and constraint functions, such as the ones described earlier in this text.
2. *Structural design problems* modeled using finite elements, such as trusses, frames, mixed finite elements, bridges, industrial buildings, high-rise buildings, plate girders, machine elements and many others [Arora and Haug, 1979; Arora and Thanedar, 1986]. More details of applications in this area are also given in Chapter 8.
3. *Dynamic response optimization* applications, such as vibration isolation, steady-state response, designs for earthquake resistance, worst case design and transient response problems [Hsieh and Arora, 1984; Lim and Arora, 1987; Tseng and Arora, 1987].
4. *Biomechanics applications,* such as muscle force distribution and contact force determination problems [Pederson *et al.*, 1987].
5. *Optimal control of systems* – structural, mechanical and aerospace applications. More details of applications in this area are discussed in Chapter 8.
6. *System identification problems,* such as environmental and material modeling problems.

Problem Type and Algorithms. The program can solve any general nonlinear programming problem formulated as given in Eqs (6.1) to (6.4), linear programming problems, and unconstrained problems. Although the program has the option of solving linear programming problems, the algorithm used is not as efficient as the Simplex method. So, for large linear programming problems, it is suggested to use a program based on the Simplex method. The following *algorithms* are available:

1. *Cost function bounding* algorithm [Arora, 1984].
2. *Pshenichny's linearization method* [Section 6.6; Belegundu and Arora, 1984].
3. *Sequential quadratic programming* algorithm that generates and uses approximate second-order information for the Lagrange function [Section 6.7; Lim and Arora, 1986].
4. A *hybrid method* that combines the cost function bounding and the sequential quadratic programming algorithms [Thanedar *et al.*, 1986].
5. *Conjugate gradient* method for unconstrained problems (Section 5.5).

If an algorithm is not specified by the user, the program automatically uses the best algorithm.

Interactive Capability. A wide range of interactive capabilities as discussed earlier are available. For example, input data can be entered and edited, problem parameters and program options can be altered, consulting support to improve the starting design is available, and various graphs and charts can be drawn on the screen.

Gradient Evaluation. The following capabilities to evaluate gradients and check gradient expressions are available:

1. If the user does not program gradient expressions in USERMG and USERCG subroutines, the program has an option to *automatically calculate* them. The finite difference method (forward, backward, or central) described in Section 5.4 is employed using the specified value of δ (input data).
2. An option is available in IDESIGN to determine the *optimum value of δ* for the finite difference gradient evaluation of cost and constraint functions.
3. If the user has programmed *gradient expressions* in USERMG and USERCG subroutines, an option is available to *verify* them, i.e. the gradient evaluation is checked using the finite difference approach. If the gradient expressions are in error, an option is available to either stop the program or continue its execution.

These options have proved to be extremely useful in practical applications.

Output. Several levels of output can be obtained from the program. This is specified in the input data. The minimum output giving the final design, design variables and constraint activities, and histories of cost function, convergence parameter and maximum constraint violation, can be obtained. More detailed information at each iteration, such as the gradient matrix and other intermediate results, can also be obtained. The detailed output is used primarily for debugging the program.

7.6 EXAMPLES OF INTERACTIVE DESIGN OPTIMIZATION

In this section we shall demonstrate the use of some of the interactive capabilities by solving a spring design optimization problem. A detailed formulation of the problem will be described. Given numerical data will be used to solve the problem using batch and interactive capabilities of IDESIGN. Capabilities of the program are also demonstrated using several design problems in Chapter 8 and in several other sources [Arora and Baenziger, 1986, 1987; Arora and Tseng, 1987, 1988; Tseng and Arora, 1987].

The use of coil springs is encountered in numerous practical applications. Detailed methods for analyzing and designing such mechanical components

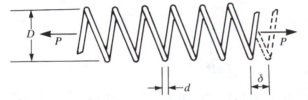

FIGURE 7.5
A coil spring.

have been developed over the years (see, e.g. Spott, 1953; Wahl, 1963; Shigley, 1977; Haug and Arora, 1979). As an example of optimum design, consider a coil spring loaded in tension or compression as shown in Fig. 7.5. The problem is to design a minimum mass spring to carry given loads without material failure while satisfying other performance requirements.

7.6.1 Formulation of the Spring Design Problem

To formulate the problem of designing coil springs, the following notation and data are defined:

Number of inactive coils	$Q = 2$
Applied load	$P = 10\,\text{lbs}$
Shear modulus	$G = (1.15\text{E}+07)\,\text{lb/in}^2$
Deflection along the axis of the spring	δ
Minimum spring deflection	$\Delta = 0.5\,\text{in}$
Weight density of spring material	$\gamma = 0.285\,\text{lb/in}^3$
Gravitational constant	$g = 386\,\text{in/sec}^2$
Mass density of material $(\rho = \gamma/g)$	$\rho = (7.383\,42\text{E}-04)\,\text{lb-sec}^2/\text{in}^4$
Allowable shear stress	$\tau_a = (8.0\text{E}+04)\,\text{lb/in}^2$
Frequency of surge waves	ω
Lower limit on surge wave frequency	$\omega_0 = 100\,\text{Hz}$
Limit on the outer diameter	$\bar{D} = 1.5\,\text{in}$

The three design variables for the problem are defined as d = wire diameter (in), D = mean coil diameter (in) and N = number of active coils.

For a spring under tension or compression, the wire experiences twisting. Therefore, the shear stress constraint should be imposed. We have the following design expressions for the spring:

Load deflection equation: $\qquad P = K\delta$ (7.11)

Spring constant: $\qquad K = \dfrac{d^4 G}{8D^3 N}$ (7.12)

Shear stress: $\qquad \tau = \dfrac{8kPD}{\pi d^3}$ (7.13)

Wahl stress concentration factor: $\quad k = \dfrac{(4D - d)}{(4D - 4d)} + \dfrac{0.615d}{D}$ $\qquad$ (7.14)

Frequency of surge waves: $\qquad\qquad \omega = \dfrac{d}{2\pi D^2 N}\sqrt{\dfrac{G}{2\rho}}$ $\qquad$ (7.15)

The expression for the Wahl stress concentration factor k in Eq. (7.14) has been determined experimentally to account for unusually high stresses at certain points of the spring. These expressions can be used to define constraints for the problem.

Cost function. The problem is to *minimize the mass* of the spring (volume × mass density) which is given as

$$\text{mass} = \tfrac{1}{4}(N + Q)\pi^2 D d^2 \rho \qquad (7.16)$$

Deflection constraint. It is often required that *deflection* under a load P be at least Δ. Therefore, the constraint is that the calculated deflection δ must be greater than or equal to Δ. This form of the constraint is common to spring design. In many applications, the function of the spring is to provide a modest restoring force as parts go through large displacements in carrying out kinematic functions. Mathematically, this performance requirement ($\delta \geqq \Delta$) is stated in an inequality form using Eqs (7.11) and (7.12) as

$$\frac{8PD^3N}{d^4G} \geqq \Delta \qquad (7.17)$$

Shear stress constraint. To prevent material overstressing, *shear stress* in the wire must be no greater than τ_a. Using Eqs (7.13) and (7.14) the constraint is expressed in mathematical form as

$$\frac{8PD}{\pi d^3}\left[\frac{(4D - d)}{(4D - 4d)} + \frac{0.615d}{D}\right] \leqq \tau_a \qquad (7.18)$$

Constraint on frequency of surge waves. We also wish to avoid resonance in dynamic applications. Resonance can be avoided if we make the *frequency of surge waves* (along the spring) as large as possible. In the present problem, we require the frequency of surge waves for the spring to be at least ω_0 (Hertz). Using Eq. (7.15), this constraint ($\omega \geqq \omega_0$) is expressed in a mathematical form as

$$\frac{d}{2\pi D^2 N}\sqrt{\frac{G}{2\rho}} \geqq \omega_0 \qquad (7.19)$$

Diameter constraint. The *outer diameter* of the spring should not be greater than $\bar{D}$, so

$$D + d \leqq \bar{D} \qquad (7.20)$$

Explicit bounds on design variables. To avoid practical difficulties, we put *minimum and maximum size* limits on the wire diameter, coil diameter, and the number of turns:

$$d_{min} \leqq d \leqq d_{max}; \qquad D_{min} \leqq D \leqq D_{max}; \qquad N_{min} \leqq N \leqq N_{max} \qquad (7.21)$$

Thus the minimum mass spring design problem is to select the design variables d, D and N to minimize the mass of Eq. (7.16), while satisfying the ten inequality constraints of Eqs (7.17) through (7.21).

Standard definition of the problem. After *normalizing* the constraints, using the defined data and writing them in the standard form of Section 2.7, we obtain the optimum design *formulation for the spring problem* as

minimize $\qquad\qquad f = (N+2)Dd^2 \qquad\qquad\qquad (7.22)$

subject to the deflection constraint

$$g_1 \equiv 1.0 - \frac{D^3 N}{(71875d^4)} \leqq 0 \qquad (7.23)$$

the shear stress constraint

$$g_2 \equiv \frac{D(4D-d)}{12566d^3(D-d)} + \frac{2.46}{12566d^2} - 1.0 \leqq 0 \qquad (7.24)$$

the surge wave frequency constraint

$$g_3 \equiv 1.0 - \frac{140.54d}{D^2 N} \leqq 0 \qquad (7.25)$$

and the outer diameter constraint

$$g_4 \equiv \frac{D+d}{1.5} - 1.0 \leqq 0 \qquad (7.26)$$

The lower and upper bounds on the design variables are selected as follows:

$$0.05 \leqq d \leqq 0.20 \text{ in}$$

$$0.25 \leqq D \leqq 1.30 \text{ in} \qquad (7.27)$$

$$2 \leqq N \leqq 15$$

Note that the constant $\pi^2 \rho / 4$ in the cost function of Eq. (7.22) has been neglected. This simply scales the cost function value without affecting the final optimum solution. The problem has three design variables and 10 inequality constraints in Eqs (7.23) to (7.27). If we attempt to solve the problem analytically using the Kuhn–Tucker conditions of Section 3.4, we will have to consider 2^{10} cases which will be quite tedious and time-consuming.

7.6.2 Optimum Solution for the Spring Design Problem

Any suitable program can be used to solve the problem defined in Eqs (7.22) to (7.27). We solve the problem using the sequential quadratic programming

TABLE 7.2
History of the iterative optimization process for the spring design problem in batch environment

Iter.	Max. vio.	Conv. parm	Cost	d	D	N
1	$9.61791E-01$	$1.00000E+00$	$2.08000E-01$	$2.0000E-01$	$1.3000E+00$	$2.0000E+00$
2	$2.48814E+00$	$1.00000E+00$	$1.30122E-02$	$5.0000E-02$	$1.3000E+00$	$2.0038E+00$
3	$6.89874E-01$	$1.00000E+00$	$1.22613E-02$	$5.7491E-02$	$9.2743E-01$	$2.0000E+00$
4	$1.60301E-01$	$1.42246E-01$	$1.20798E-02$	$6.2522E-02$	$7.7256E-01$	$2.0000E+00$
5	$1.23963E-02$	$8.92216E-03$	$1.72814E-02$	$6.8435E-02$	$9.1481E-01$	$2.0336E+00$
6	$1.97357E-05$	$6.47793E-03$	$1.76475E-02$	$6.8770E-02$	$9.2373E-01$	$2.0396E+00$
7	$9.25486E-06$	$3.21448E-02$	$1.76248E-02$	$6.8732E-02$	$9.2208E-01$	$2.0460E+00$
8	$2.27139E-04$	$7.68889E-02$	$1.75088E-02$	$6.8542E-02$	$9.1385E-01$	$2.0782E+00$
9	$5.14338E-03$	$8.80280E-02$	$1.69469E-02$	$6.7635E-02$	$8.7486E-01$	$2.2346E+00$
10	$8.79064E-02$	$8.87076E-02$	$1.44839E-02$	$6.3848E-02$	$7.1706E-01$	$2.9549E+00$
11	$9.07017E-02$	$6.66881E-02$	$1.31958E-02$	$6.0328E-02$	$5.9653E-01$	$4.0781E+00$
12	$7.20705E-02$	$7.90647E-02$	$1.26517E-02$	$5.7519E-02$	$5.1028E-01$	$5.4942E+00$
13	$6.74501E-02$	$6.86892E-02$	$1.22889E-02$	$5.4977E-02$	$4.3814E-01$	$7.2798E+00$
14	$2.81792E-02$	$4.50482E-02$	$1.24815E-02$	$5.3497E-02$	$4.0092E-01$	$8.8781E+00$
15	$1.57825E-02$	$1.94256E-02$	$1.25465E-02$	$5.2424E-02$	$3.7413E-01$	$1.0202E+01$
16	$5.85935E-03$	$4.93063E-03$	$1.26254E-02$	$5.1790E-02$	$3.5896E-01$	$1.1113E+01$
17	$1.49687E-04$	$2.69244E-05$	$1.26772E-02$	$5.1698E-02$	$3.5692E-01$	$1.1289E+01$
18	$0.00000E+00$	$9.76924E-08$	$1.26787E-02$	$5.1699E-02$	$3.5695E-01$	$1.1289E+01$

Constraint activity

No.	Active	Value	Lagr. mult.
1	Yes	$-4.66382E-09$	$1.07717E-02$
2	Yes	$-2.46286E-09$	$2.44046E-02$
3	No	$-4.04792E+00$	$0.00000E+00$
4	No	$-7.27568E-01$	$0.00000E+00$

Design variable activity

No.	Active	Design	Lower	Upper	Lagr. mult.
1	Lower	$5.16987E-02$	$5.00000E-02$	$2.00000E-01$	$0.00000E+00$
2	Lower	$3.56950E-01$	$2.50000E-01$	$1.30000E+00$	$0.00000E+00$
3	No	$1.12895E+01$	$2.00000E+00$	$1.50000E+01$	$0.00000E+00$

Cost function at optimum = $1.267868E-02$.

No. of calls for cost function evaluation (USERMF) = 18.

No. of calls for evaluation of cost function gradient (USERMG) = 18.

No. of calls for constraint function evaluation (USERCF) = 18.

No. of calls for evaluation of constraint function gradients (USERCG) = 18.

No. of total gradient evaluations = 34.

algorithm of Section 6.7 available in the IDESIGN software package. The history of the iterative design process is shown in Table 7.2. The table shows iteration number (Iter.), maximum constraint violation (Max. vio.), convergence parameter (Conv. parm.), cost function (Cost), and design variable values at each iteration. It also gives constraint activity at the optimum point indicating whether a constraint is active or not, constraint function values and their Lagrange multipliers. Design variable activity is shown at the optimum point, and the final cost function value and the number of calls to user routines are also given.

The following stopping criteria are used for the present problem:

1. The maximum constraint violation (Max. vio.) should be less than ε_1, i.e. $V \leqq \varepsilon_1$ in Step 4 of the algorithm given in Section 6.7. ε_1 is taken as $1.00\mathrm{E}-04$.
2. The length of the direction vector (Conv. parm.) should be less than ε_2, i.e. $\|\mathbf{d}\| \leqq \varepsilon_2$ in Step 4 of the algorithm given in Section 6.7. ε_2 is taken as $1.00\mathrm{E}-03$.

The starting design estimate is $(0.2, 1.3, 2.0)$ where the maximum constraint violation is 96.2% and the cost function value is 0.208. At the 6th iteration, a feasible design (maximum constraint violation is $1.97\mathrm{E}-05$) is obtained at a cost function value of $(1.764\,75\mathrm{E}-02)$. Note that in this example, the constraint correction is accompanied by a substantial reduction (by a factor of 10) in the cost function. However, most often, the constraint correction will result in an increase in cost. The program takes another 12 iterations to reach the optimum design. At the optimum point, the deflection and shear stress constraints of Eqs (7.23) and (7.24) are active. The Lagrange multiplier values are $(1.077\mathrm{E}-02)$ and $(2.4405\mathrm{E}-02)$. Design variable one (wire diameter) is close to its lower bound.

7.6.3 Interactive Solution for Spring Design Problem

In the previous subsection, the spring design problem was *solved in the batch environment where the designer had no control over the iterative process*. The program took 18 iterations to converge to the optimum point. We shall solve the problem interactively starting from the same design point. The procedure will be to *interrupt* the program at every iteration, analyze the design conditions, and give *interactive commands* to execute a particular step. In the current application, only the cost function value and maximum constraint violation are monitored and used to make decisions. In more advanced applications, histories of design variables and other graphic facilities explained in the next subsection can also be used to make decisions. For example, design variable values can be extrapolated based on the observation of trends. This will be demonstrated in the next subsection.

Table 7.3 contains histories of design variables, maximum constraint violation, convergence parameter and the cost function. It also shows the interactive algorithm used at each iteration. The initial objective is to obtain a feasible design, so the constraint correction (CC) algorithm is executed for the first six iterations. A feasible design is obtained at the 7th iteration. Note that during the first six iterations, constraint correction is accompanied by a reduction in the cost function. At the 7th iteration, the cost reduction (CR) algorithm is executed with a 20% reduction in the cost function. At the 8th iteration the cost function is reduced but constraint violtion again appears. For the next two iterations, constraint correction at constant cost (CCC) is sought and a nearly feasible design is obtained at the 10th iteration. At the 10th iteration, constraint correction at a specified increase in cost (CCS) is sought.

TABLE 7.3
Interactive solution process for the spring design problem

Iter.	Algor.	Max. vio.	Conv. parm.	Cost	d	D	N
1	CC	9.617 91E − 01	1.000 00E + 00	2.080 00E − 01	2.0000E − 01	1.3000E + 00	2.0000E + 00
2	CC	2.488 14E + 00	1.000 00E + 00	1.301 22E − 02	5.0000E − 02	1.3000E + 00	2.0038E + 00
3	CC	6.898 74E − 01	1.000 00E + 00	1.226 13E − 02	5.7491E − 02	9.2743E − 01	2.0000E + 00
4	CC	1.603 01E − 01	1.000 00E + 00	1.207 98E − 02	6.2522E − 02	7.7256E − 01	2.0000E + 00
5	CC	3.705 54E − 01	1.000 00E + 00	1.033 15E − 02	5.8477E − 02	5.1558E − 01	3.8601E + 00
6	CC	5.060 54E − 01	1.000 00E + 00	7.968 02E − 03	5.0000E − 02	2.9195E − 01	8.9170E + 00
7	CR	0.000 00E + 00	1.676 23E − 02	1.473 52E − 02	5.5455E − 02	4.3230E − 01	9.0837E + 00
8	CCC	3.533 58E − 02	1.676 23E − 02	1.196 85E − 02	5.2692E − 02	3.8896E − 01	9.0828E + 00
9	CCC	4.249 50E − 04	1.676 23E − 02	1.272 98E − 02	5.3485E − 02	4.0151E − 01	9.0831E + 00
10	CCS	1.089 57E − 04	1.676 23E − 02	1.272 90E − 02	5.3395E − 02	3.9916E − 01	9.1854E + 00
11	CR	0.000 00E + 00	5.490 55E − 05	1.273 00E − 02	5.3396E − 02	3.9918E − 01	9.1854E + 00

Constraint activity

No.	Active	Value	Lagr. mult.
1	Yes	−2.946 70E − 09	1.095 81E − 02
2	Yes	−1.361 88E − 09	2.457 45E − 02
3	No	−4.123 84E + 00	0.000 00E + 00
4	No	−6.982 84E − 01	0.000 00E + 00

Design variable activity

No.	Active	Design	Lower	Upper	Lagr. mult.
1	Lower	5.339 56E − 02	5.000 00E − 02	2.000 00E − 01	0.000 00E + 00
2	No	3.991 78E − 01	2.500 00E − 01	1.300 00E + 00	0.000 00E + 00
3	No	9.185 39E + 00	2.000 00E + 00	1.500 00E + 01	0.000 00E + 00

Cost function at optimum = 1.273 000E − 02.
No. of calls for cost function evaluation (USERMF) = 11
No. of calls for evaluation of cost function gradient (USERMG). = 11
No. of calls for constraint function evaluation (USERCF) = 11
No. of calls for evaluation of constraint function gradients (USERCG) = 11
No. of total gradient evaluations = 20
CC, constraint correction step; CR, cost reduction step; CCC, constraint correction at constant cost; and CCS, constraint correction at specified cost.

At the 11th iteration, all constraints are satisfied and the convergence parameter is quite small, so the program is terminated. The cost function is fairly close to the true optimum. However, the design point is somewhat different. It turns out that there are several near optimum designs in the neighborhood of the true optimum for this example problem.

7.6.4 Use of Interactive Graphics

The graphical display of a large amount of data is an extremely convenient way to interpret results and draw conclusions. Interactive graphics can play a major role in decision making during the design process. We demonstrate the possible use of interactive graphics during the design optimization process using the spring design problem as an example. We execute the spring design problem for 10 iterations starting from the point (0.2, 1.3, 2.0). At the 10th iteration the program is stopped and the execution control is transferred to the designer. Various plotting commands available in the IDESIGN program are used to display the data on the screen. All calculations are performed on the APOLLO DN460 workstation. In the following, we will explain various graphics facilities and their possible use in the practical design environment.

Design variable trend. The history of design variables when displayed on the screen can show their trend. For example, Fig. 7.6 shows the variation of design variables as the iterations progress. It shows that design variable 1 reduces at the 1st iteration and then remains almost constant. If the information were displayed at an intermediate iteration, the variable could be assigned a fixed value since it was not changing very much. Design variable 2

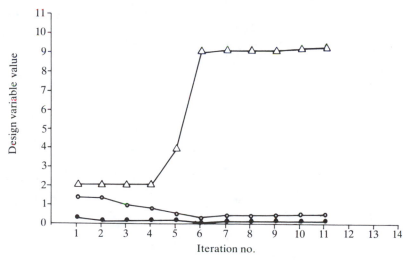

FIGURE 7.6
History of design variables for the spring design problem. ●, d; ○, D; △, N.

reduces for the first few iterations and then remains almost constant. Variable 3 does not change for the first three iterations and then increases rapidly for the next two iterations. Using the trend information, the designer can extrapolate the value of a design variable manually.

We conclude that by using a design variable history, we can make the following decisions:

1. Based on the displayed trend, we can *extrapolate* the value of a *design variable*.

2. If a *design variable* is not changing, we can *fix* it for a few iterations and optimize only the remaining ones.

Maximum constraint violation history. Figure 7.7 shows a plot of maximum constraint violation versus the iteration number for the spring design problem. Using this graph, we can locate feasible designs. For example, designs after iteration 7 are feasible. Designs at all previous iterations had some violation of constraints.

Cost function history. Figure 7.8 shows the history of the cost function for the first 10 iterations for the spring design problem. It shows that the cost function reduces substantially at the 1st iteration. After that it is changing slowly and appears to be quite close to the optimum. The iterative process could have been terminated at a feasible design.

Convergence parameter history. Figure 7.9 shows the convergence parameter history for the spring design problem. This parameter is supposed to go to zero as the optimum is reached. It can be seen that the parameter is close to

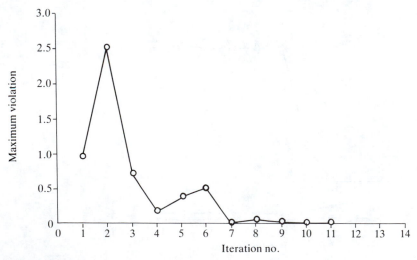

FIGURE 7.7
History of the maximum constraint violation for the spring design problem.

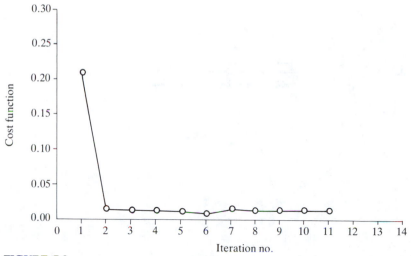

FIGURE 7.8
History of the cost function for the spring design problem.

zero at the 7th iteration, so the solution is quite close to the optimum and the
iterative process could be terminated.

    Constraint function history. Figure 7.10 shows the history of the four
constraints for the spring design problem. A value of less than zero indicates
the constraint to be inactive and greater than zero indicates the constraint to
be violated. It can be seen that the 1st and 4th constraints are violated in the
beginning, but during later iterations the first and the second constraints are
active. The third constraint is never active or violated and may be ignored. The

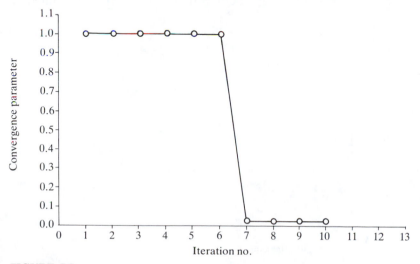

FIGURE 7.9
Convergence parameter history for the spring design problem.

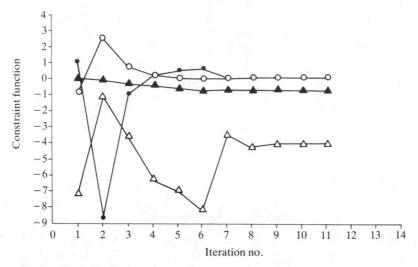

FIGURE 7.10
History of the constraint functions for the spring design problem. ●, g_1; ○, g_2; △, g_3; ▲, g_4.

constraint does not influence the solution or the optimization process. *Thus, using the history of constraint functions, we can identify constraints that are critical in determining the optimum solution.* The designer can further analyze these constraints to determine whether they can be adjusted to improve the optimum solution.

Cost function sensitivity chart. Figure 7.11 shows a normalized bar chart for the cost function sensitivity to design variables. The chart is obtained by plotting the relative values of the gradient components of the cost function

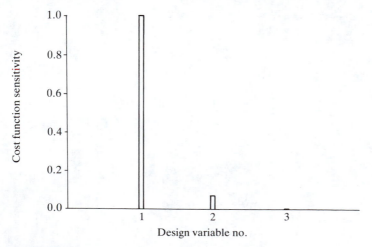

FIGURE 7.11
Cost function sensitivity to design variables for the spring design problem.

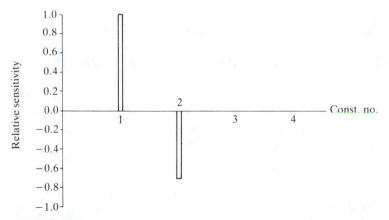

FIGURE 7.12
Sensitivity of constraints to design variable 1 for the spring design problem.

(derivatives with respect to the design variables). For the spring design problem, the cost function is most sensitive to the first design variable and least sensitive to the third one. Knowing this, the designer can decide to fix the third variable and optimize only the first and the second ones.

Design variable sensitivity for various constraints. It may be useful to know what happens to various constraints if a design variable is changed. Figure 7.12 shows such a normalized bar chart for design variable 1 (wire diameter) for the spring design problem. It shows how the four constraints change if design variable 1 is changed slightly. For example, a small increase in the variable increases the value of the first constraint and reduces it for constraint 2. The bar chart is obtained by plotting the normalized derivatives of all the active constraints with respect to the first design variable.

Constraint sensitivity chart. Figure 7.13 is a plot for the normalized

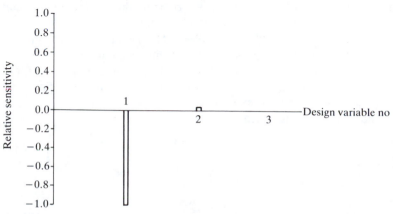

FIGURE 7.13
Sensitivity of constraint no. 2 to design variables for the spring design problem.

gradient components for the second constraint for the spring design problem. It indicates what happens to the constraint function value if any design variable is changed. For example, an increase in variable 1 (wire diameter) reduces the value of the constraint quite rapidly. An increase in the values of design variable 2 increases the value of constraint 2.

Concluding remark. It can be seen that various graphs and bar charts give extremely useful information about the design problem. Such information can be used in accelerating the optimum design process as well as in learning more about the behavior of the system. The insights gained can lead to new concepts and better designs for the system.

Exercises for Chapter 7

Section 7.3   Interactive Design Optimization Algorithms

7.1 Consider the following constrained problem:

minimize $$f(\mathbf{x}) = 2x_1 + 3x_2 - x_1^3 - 2x_2^2$$

subject to

$$x_1 + 3x_2 \leqq 6$$

$$5x_1 + 2x_2 \leqq 10$$

$$x_1, x_2 \geqq 0$$

Plot the problem functions on a graph sheet and identify the feasible region. Locate the optimum point.

7.2 At the feasible point $(0, 2)$ for Exercise 7.1, compute the cost reduction direction and obtain a new point requiring a 10% reduction in the cost function. Show the direction on the graph sheet and explain your solution.

7.3 At the infeasible point $(3, 3)$ for Exercise 7.1, compute the constraint correction direction and show it on the graph sheet and explain your solution.

7.4 At the infeasible point $(2, 0.8)$ for Exercise 7.1, compute the constant cost direction and show it on the graph sheet and explain your solution.

7.5 At the infeasible point $(2, 0.8)$ for Exercise 7.1, compute the constraint correction direction allowing only a 10% increase in the cost function. Show the direction on the graph sheet and explain your solution.

7.6 Consider the following constrained problem:

maximize $$2x_1 + 4x_2 - x_1^2 - x_2^2$$

subject to

$$2x_1 + 3x_2 \leqq 6$$

$$x_1 + 4x_2 \leqq 5$$

$$x_1, x_2 \geqq 0$$

Plot the problem functions on a graph sheet and identify the feasible region. Locate the optimum point.

7.7 At the feasible point $(0, 1.25)$ for Exercise 7.6, compute the cost reduction direction and obtain a new point requiring a 10% reduction in the cost function. Show the direction on the graph sheet and explain your result.

7.8 At the infeasible point $(4, 2)$ for Exercise 7.6, compute the constraint correction direction and show it on the graph sheet. Explain your solution.

7.9 At the infeasible point $(1, 2)$ for Exercise 7.6, compute the constant cost direction and show it on the graph sheet. Explain your solution.

7.10 At the infeasible point $(2, 2)$ for Exercise 7.6, compute the constraint correction direction allowing only a 10% increase in the cost function. Show the direction on the graph sheet and explain your solution.

CHAPTER
8

PRACTICAL
DESIGN
OPTIMIZATION

8.1 INTRODUCTION

Thus far we have considered simple engineering design problems to describe optimization concepts and computational methods. *Explicit expressions for all the functions* of the problem in terms of the design variables were assumed to be known. Whereas some problems can be formulated with explicit functions, there are numerous other *complex applications for which explicit dependence of the problem functions on design variables is not known.* In addition, complex systems require large and more sophisticated analysis models. The number of design variables and constraints can be quite large. A check for convexity of the problem is almost impossible. The existence of even feasible designs is not guaranteed, much less the optimum solution. The calculation of functions of the problem can require large computational effort. In many cases large special purpose codes must be used to compute problem functions. We have also seen in previous chapters that modern computational algorithms require gradients of cost and constraint functions. When an explicit form of the problem functions is not known, *gradient evaluation requires special procedures* which must be developed and implemented into proper software. Finally, various software components must also be integrated properly to create an optimum design capability for a particular class of design problems.

In this chapter, issues of the optimum design of complex practical engineering systems are addressed. Formulation of the problem, gradient evaluation, and other practical issues, such as algorithm and software selection are discussed. The important problem of interfacing a particular application to a design optimization software is discussed and several engineering design applications are described.

The following is an outline of this Chapter.

Section 8.2 Formulation of Practical Design Optimization Problems. This section describes general guidelines for the proper formulation of a design problem. Most practical problems are complex, having implicit cost and constraint functions. The ideas are elaborated with an example problem.

Section 8.3 Gradient Evaluation. This section concentrates on procedures for the evaluation of gradients of implicit constrains. A numerical example is used to illustrate one of the procedures.

Section 8.4 Issues in Practical Applications. The attributes of a good algorithm for practical applications are described. These consist of robustness, efficiency, generality and ease of use.

Section 8.5 Use of General Purpose Software. Practical design problems need large software components for their solution. Issues involving different architectures for general-purpose design optimization software are discussed. The problem of interfacing an application to a general purpose optimizer is delineated.

Sections 8.6–8.9 Practical Applications. These sections describe various design applications. Problems are defined, formulated and solved to illustrate the power of numerical optimization techniques.

8.2 FORMULATION OF PRACTICAL DESIGN OPTIMIZATION PROBLEMS

8.2.1 General Guidelines

The problem formulation for a design task is an important step which must define a realistic model for the engineering system under consideration. The mathematics of optimization methods can easily give rise to situations that are absurd or violate the laws of physics. So, to transcribe a design task correctly into a mathematical model, the designers must use intuition, skill and experience. The following points can serve as guiding principles to generate a mathematical model that is faithful to the real world design task.

1. Once the conceptual design is ready we must concentrate on the *detailed design of the system* to perform the given task. To start with, all the possible parameters or unknowns should be looked upon as potential design variables which should be independent of each other as far as possible. Also, a variety of failure criteria and other technological requirements must be taken as constraints for the safe performance of the

system. In short, considerable flexibility and freedom must be allowed before analyzing different possibilities. As we gain more knowledge about the problem, redundant or unnecessary design variables can be fixed or eliminated from the model. Finally, only significant cost and constraint functions can be retained in the design optimization model.

2. The *existence of an optimum solution* to a design optimization model depends on its formulation. If the constraints are too restrictive, there may not be any feasible solution for the problem. In such a case, the constraints must be relaxed by allowing larger resource limits for inequality constraints. The question of uniqueness of the global solution for the problem depends on the strict convexity of the cost and constraint functions. In reality, most of the *engineering design problems are not convex, thus global optimality of a local solution cannot be guaranteed. Usually there are multiple local optimum solutions.* This, however, is not necessarily an undesirable situation, since it offers additional freedom to the designer to choose a suitable solution among the many available ones.

3. In *numerical computations,* it is sometimes easier to find a feasible design with respect to the inequality constraints than it is with respect to the equality constraints. This, of course, depends on the problem structure and nonlinearity of functions. A constraint expressed as an inequality defines a much larger feasible region than the same constraint expressed as an equality. In case the number of equality constraints is greater than the number of design variables in a problem, there will not exist any solution unless some of the constraints are dependent.

4. The representation of engineering design problems by the standard nonlinear programming design optimization model with a single real valued objective function subject to a set of equality and inequality constraints is not as restrictive as it may appear. The problem of optimizing more than one objective function simultaneously (*multiobjective problems*) can be transformed into the standard problem by assigning weighting factors to different objective functions to combine them into a single objective function. Or, the most important criterion can be treated as the cost function and the remaining ones as constraints. By varying the limits for the ones treated as constraints, trade-off curves can be generated and used for the final design of the system.

5. The idea of *design variable linking* is useful to reduce the number of design variables in an optimization model. If one of the design variables can be expressed in terms of others then that variable can be eliminated from the model. Also, if the designer can identify any symmetry in the system, it can help in reducing the number of design variables.

6. The *potential cost functions* for many structural, mechanical, automotive and aerospace systems are weight, volume, mass, stress at a point, performance, reliability of a system, etc. The constraints can be placed on

stresses, strains, displacements, vibration frequencies, manufacturing limitations and other performance criteria.

7. It is important to have *continuous and differentiable cost and constraint functions*. If a function is not continuous or differentiable then conventional optimization theory is not adequate. In certain instances, it may be possible to replace a nondifferentiable function such as $|x|$ with a smooth function x^2 without changing the problem definition drastically.

8. In general it is *difficult to determine dependent constraints* and eliminate them from the formulation. Modern optimization algorithms and the associated software are capable of handling difficulties arising from the dependent constraints. Also, equality constraints can be used to reduce the number of design variables by expressing one variable in terms of the others. However, such an approach is appropriate for only small-scale problems where explicit expressions for the constraints are available. In more complex applications, equality constraints must be retained and treated in the optimization algorithm.

9. In engineering design problems, *lower and upper limits on the design variables* are often imposed due to practical limitations. If there is no lower limit on a design variable then a large negative number may be taken as the lower limit, and similarly a large positive number may be prescribed as the upper limit if no upper limit is given in the problem definition.

10. For nonlinear programming problems, the design variables are often assumed to be continuous. In practice, however, *discrete and integer variables* often arise. For example, due to manufacturing limitations structural elements and spare parts for many engineering systems are available only in fixed shapes and sizes. Therefore, once we obtain the optimum solution, we can select members that have dimensions nearest to the optimum values. Or, the adaptive optimization procedure described in Section 2.7 can be used to obtain practical solutions.

11. In general, *it is desirable to normalize all the constraints with respect to their limit values,* as discussed in Section 6.2.3. In numerical computations, this procedure leads to more stable behavior. Therefore, as far as possible, all constraints should be normalized in practical applications.

8.2.2 Example of a Practical Design Optimization Problem

Optimum design formulation of complex engineering systems requires more general tools and procedures than the ones discussed previously. We shall demonstrate this by considering a class of problems that has a wide range of applications in automotive, aerospace, mechanical and structural engineering. This important application area is chosen to *demonstrate the procedure of problem formulation and explain the treatment of implicit constraints.* Evaluation of constraint functions and their gradients shall be explained. Readers unfamiliar with this application area should use the material as *guiding principles* for their area of interest because similar analyses and procedures will have to be used in other practical applications.

The application area that we have chosen to investigate is the optimum design of systems modeled by the *finite element techniques.* It is common practice to analyze complex structural systems using the technique that is also available in many commercial software packages. Displacements, stresses and strains at various points, vibration frequencies, and buckling loads for the system can be computed and constraints imposed on them. We shall describe an optimum design formulation for this application area.

Let **x** represent an *n* component vector containing *design parameters for the system.* This may contain thicknesses of members, cross-sectional areas, parameters describing the shape of the system, and stiffness and material properties of elements. Once **x** is specified, a design of the system is known. To analyze the system (calculate stresses, strains and frequencies, buckling load and displacements), the procedure is to first calculate displacements at some key points – called the grid points or nodal points – of the finite element model. From these displacements, strains (relative displacement of the material particles) and stresses at various points of the system are available in many textbooks [Cook, 1981; Huebner and Thornton, 1982; Grandin, 1986].

Let **U** be a vector having *l* components representing generalized displacements at key points of the system. The basic equation that determines the displacement vector **U** for a linear elastic system – called the equilibrium equation in terms of displacements – is given as

$$\mathbf{K}(\mathbf{x})\mathbf{U} = \mathbf{F}(\mathbf{x}) \tag{8.1}$$

where $\mathbf{K}(\mathbf{x})$ is an $l \times l$ matrix called the stiffness matrix and $\mathbf{F}(\mathbf{x})$ is an effective load vector having *l* components. The stiffness matrix $\mathbf{K}(\mathbf{x})$ is a *property of the structural system* which depends explicitly on the design variables, material properties and geometry of the system. Systematic procedures have been developed to automatically calculate the matrix with different finite elements. The load vector $\mathbf{F}(\mathbf{x})$, in general, can also depend on design variables. We shall not discuss procedures to calculate $\mathbf{K}(\mathbf{x})$ because that is beyond the scope of the present text. Our objective is to demonstrate how the design can be optimized once a finite element model for the problem (meaning Eq. (8.1)) has been developed. We shall pursue that objective assuming that the finite element model for the system has been developed.

It can be seen that once the design **x** is specified, the displacements **U** can be calculated by solving the linear system of Eq. (8.1). Note that a different **x** will give, in general, different values for the displacements **U**. Thus **U** is a function of **x**; however, its explicit functional form cannot be written. That is, **U** is an *implicit function* of the design variables **x**. The stress σ_i at the *i*th point is calculated using the displacements and is an explicit function of **U** and **x** as

$\sigma_i(\mathbf{U}, \mathbf{x})$. However, since $\mathbf{U}$ is an implicit function of $\mathbf{x}$, σ_i also becomes an implicit function of the design variables $\mathbf{x}$. The stress and displacement related constraints can be written in a functional form as

$$g_i(\mathbf{x}, \mathbf{U}) \leqq 0 \qquad (8.2)$$

In many automotive, aerospace, mechanical and structural engineering applications, the amount of material used must be minimized for efficient and cost-effective systems. Thus, the usual cost function for this class of applications is the weight, mass or material volume of the system. This is usually an explicit function of the design variables $\mathbf{x}$. Implicit cost functions, such as stress, displacement, vibration frequencies, etc., can also be treated by introducing artificial design variables [Haug and Arora, 1979]. We shall assume that this has been done and treat only explicit cost functions.

In summary, a general formulation for the design problem involving explicit and implicit functions of design variables is defined as: find an n-dimensional vector $\mathbf{x}$ of design variables to minimize a cost function $f(\mathbf{x})$ satisfying the implicit design constraints of Eq. (8.2) with $\mathbf{U}$ satisfying the system of Eq. (8.1), and other practical limitations. Note that equality constraints, if present, can be routinely included as in the previous chapters. We illustrate the procedure of problem formulation with an example problem.

Example 8.1 Design of a two-member frame. Consider the design of a two-member frame subjected to out-of-plane loads as shown in Fig. 8.1. Such frames are encountered in numerous automotive, aerospace, mechanical and structural engineering applications. We wish to formulate the problem of minimizing the volume of the frame subject to stress and size limitations [Bartel, 1969].

Solution. Since the optimum structure will be symmetric, the two members of the frame are identical. Also, it has been determined that hollow rectangular sections shall be used as members with three design variables defined as $d =$ width

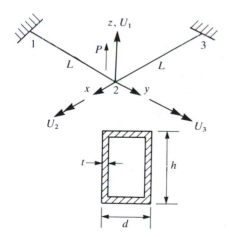

FIGURE 8.1
Two-member frame.

of the member (in), $h =$ height of the member (in) and $t =$ wall thickness (in). Thus, the design variable vector is $\mathbf{x} = (d, h, t)$.

The volume for the structure is taken as the cost function which is an explicit function of the design variables given as

$$f(\mathbf{x}) = 2L(2dt + 2ht - 4t^2) \tag{8.3}$$

To calculate stresses, we need to solve the analysis problem. The members are subjected to both bending and torsional stresses and the combined stress constraint needs to be imposed at points 1 and 2. Let σ and τ be the maximum bending and shear stresses in the member, respectively. The failure criterion for the member is based on a combined stress theory, known as the von Mises (or maximum distortion energy) yield condition [Crandall *et al.*, 1978]. With this criterion, the effective stress σ_e is given as $\sqrt{\sigma^2 + 3\tau^2}$ and the *stress constraint* is written in a normalized form as

$$\frac{1}{\sigma_a^2}(\sigma^2 + 3\tau^2) - 1.0 \leqq 0 \tag{8.4}$$

where σ_a is the allowable design stress.

The stresses are calculated from the member-end moments and torques which are calculated using the finite element procedure. The three generalized nodal displacements (deflections and rotations) for the finite element model shown in Fig. 8.1 are defined as $U_1 =$ vertical displacement at node 2, $U_2 =$ rotation about line 3–2 and $U_3 =$ rotation about line 1–2. Using these, the equilibrium equation (Eq. (8.1)) for the finite element model that determines the displacements U_1, U_2 and U_3, is given as (for details of the procedure to obtain the equation using individual member equilibrium equations, refer to texts by Cook, 1981; Haug and Arora, 1979; Huebner and Thornton, 1982; Grandin, 1986):

$$\frac{EI}{L^3}\begin{bmatrix} 24 & -6L & 6L \\ -6L & \left(4L^2 + \dfrac{GJ}{EI}L^2\right) & 0 \\ 6L & 0 & \left(4L^2 + \dfrac{GJ}{EI}L^2\right) \end{bmatrix}\begin{bmatrix} U_1 \\ U_2 \\ U_3 \end{bmatrix} = \begin{bmatrix} P \\ 0 \\ 0 \end{bmatrix} \tag{8.5}$$

where $E =$ modulus of elasticity,   (3.0E+07) psi
$L =$ member length,   100 in
$G =$ shear modulus,   (1.154E+07) psi
$P =$ load at node 2,   −10 000 lbs

$I =$ moment of inertia $= \frac{1}{12}[dh^3 - (d - 2t)(h - 2t)^3]$,   in⁴ $\tag{8.6}$

$J =$ polar moment of inertia $= \dfrac{2t(d - t)^2(h - t)^2}{(d + h - 2t)}$,   in⁴ $\tag{8.7}$

$A =$ area for calculation of torsional shear stress
$= (d - t)(h - t)$,   in²

From Eq. (8.5), the stiffness matrix $\mathbf{K}(\mathbf{x})$ and the load vector $\mathbf{F}(\mathbf{x})$ of Eq. (8.1) can be identified. Note that in the present example, the load vector $\mathbf{F}$ does not depend on the design variables.

It can be seen from Eq. (8.5) that $\mathbf{U}$ is an implicit function of $\mathbf{x}$. If $\mathbf{K}$ can be inverted explicitly in terms of the design variables $\mathbf{x}$, then $\mathbf{U}$ can be written as an explicit function of $\mathbf{x}$. This is possible in the present example; however, we shall deal with the implicit form to illustrate the procedures for evaluating constraints and their gradients.

For a given design, once the displacements U_1, U_2 and U_3 have been calculated from Eq. (8.5), the torque, and bending moment at points 1 and 2 for member 1–2 are calculated as

$$T = -\frac{GJ}{L}U_3, \qquad \text{lb-in} \tag{8.8}$$

$$M_1 = \frac{2EI}{L^2}(-3U_1 + U_2 L), \qquad \text{lb-in (moment at end 1)} \tag{8.9}$$

$$M_2 = \frac{2EI}{L^2}(-3U_1 + 2U_2 L), \qquad \text{lb-in (moment at end 2)} \tag{8.10}$$

Using these moments, the torsional shear and bending stresses are calculated as

$$\tau = \frac{T}{2At}, \qquad \text{psi} \tag{8.11}$$

$$\sigma_1 = \frac{1}{2I}M_1 h, \qquad \text{psi (bending stress at end 1)} \tag{8.12}$$

$$\sigma_2 = \frac{1}{2I}M_2 h, \qquad \text{psi (bending stress at end 2)} \tag{8.13}$$

Thus the stress constraints of Eq. (8.4) at points 1 and 2 are given as

$$g_1(\mathbf{x}, \mathbf{U}) = \frac{1}{\sigma_a^2}(\sigma_1^2 + 3\tau^2) - 1.0 \leq 0 \tag{8.14}$$

$$g_2(\mathbf{x}, \mathbf{U}) = \frac{1}{\sigma_a^2}(\sigma_2^2 + 3\tau^2) - 1.0 \leq 0 \tag{8.15}$$

It can be observed that since moments T, M_1 and M_2 are implicit functions of design variables, the stresses are also implicit functions. They are also explicit functions of design variables, as seen in Eqs (8.12) and (8.13). Therefore, the stress constraints of Eqs (8.14) and (8.15) are implicit as well as explicit functions of design variables. This observation is important because gradient evaluation for implicit constraint functions requires special procedures, which are explained in the next section.

In addition to the two stress constraints, the following upper and lower bound constraints on design variables are imposed:

$$2.5 \leq d \leq 10.0$$

$$2.5 \leq h \leq 10.0$$

$$0.1 \leq t \leq 1.0$$

It can easily be observed that the explicit forms of the constraint functions g_1 and g_2 in terms of the design variables d, h and t are quite difficult to obtain even for this simple problem. We will need an explicit form for the displacements U_1, U_2 and U_3 in Eqs (8.8) to (8.10) in order to have an explicit form for the stress τ, σ_1

and σ_2. To have an explicit form for U_1, U_2 and U_3, we will have to explicitly invert the coefficient matrix for the equilibrium equation (8.5). Although this is not impossible for the present example, it is quite impossible to do, in general. Thus we observe that the constraints are implicit functions of the design variables.

To illustrate the procedure, we select a design point as $(2.5, 2.5, 0.1)$ and calculate the displacements and stresses. Using the given data, we calculate the following quantities that are needed in further calculations:

$$I = \tfrac{1}{12}[2.5^4 - 2.3^4] = 0.9232 \text{ in}^4$$

$$J = \frac{1}{4.8}[2(0.1)(2.4)^2(2.4)^2] = 1.3824 \text{ in}^4$$

$$A = (2.4)(2.4) = 5.76 \text{ in}^2$$

$$GJ = (1.154E+07)(1.3824) = (1.5953E+07)$$

$$EI = (3.0E+07)(0.9232) = (2.7696E+07)$$

$$4L^2 + \frac{GJ}{EI}L^2 = \left(4 + \frac{1.5953}{2.7696}\right)100^2 = (4.576E+04)$$

Using the foregoing data, the equilibrium equation (Eq. 8.5), is given as

$$27.696 \begin{bmatrix} 24 & -600 & 600 \\ -600 & 45\,760 & 0 \\ 600 & 0 & 45\,760 \end{bmatrix} \begin{bmatrix} U_1 \\ U_2 \\ U_3 \end{bmatrix} = \begin{bmatrix} -10\,000 \\ 0 \\ 0 \end{bmatrix}$$

Solving the preceding equation, the three generalized displacements of node 2 are given as

$$U_1 = -43.681\,90 \text{ in}$$
$$U_2 = -0.572\,75$$
$$U_3 = 0.572\,75$$

Using Eqs (8.8) to (8.10), torque in the member and bending moments at points 1 and 2 are

$$T = -\frac{1.5953E+07}{100}(0.572\,75) = -(9.1371E+04) \text{ lb-in}$$

$$M_1 = \frac{2(2.7696E+07)}{(100)(100)}[-3(-43.681\,90) - 0.572\,75(100)]$$

$$= (4.086\,31E+05) \text{ lb-in}$$

$$M_2 = \frac{2(2.769E+07)}{(100)(100)}[-3(-43.6819) - 2(0.572\,75)(100)]$$

$$= (9.1373E+04) \text{ lb-in}$$

Since $M_1 > M_2$, σ_1 will be larger than σ_2 as observed from Eqs (8.12) and (8.13). Therefore, only the g_1 constraint of Eq. (8.14) needs to be imposed.

The torsional shear stress and bending stress at point 1 are calculated from Eqs (8.11) and (8.12) as

$$\tau = \frac{-(9.137\,31E+04)}{2(5.76)(0.1)} = -(7.931\,7E+04) \text{ psi}$$

$$\sigma_1 = \frac{(4.086\,31\mathrm{E}+05)(2.5)}{2(0.9232)} = (5.532\,81\mathrm{E}+05)\ \mathrm{psi}$$

Taking the allowable stress σ_a as 40 000 psi, the effective stress constraint of Eq. (8.14) is given as

$$g_1 = \frac{1}{(4.0\mathrm{E}+04)^2}\,[(5.532\,81\mathrm{E}+05)^2 + 3(-7.9317\mathrm{E}+04)^2] - 1$$

$$= (2.0212\mathrm{E}+02) > 0$$

Therefore, the constraint is very severely violated at the given design. ‖

8.3 GRADIENT EVALUATION

To use the modern optimization method, we need to evaluate gradients of constraint functions. When the constraint functions are implicit in design variables, we need to develop and utilize special procedures for gradient evaluation. We shall develop a procedure using the finite element application of Section 8.2.

Let us consider the constraint function $g_i(\mathbf{x}, \mathbf{U})$ of Eq. (8.2). Using the chain rule of differentiation, the total derivative of g_i with respect to the jth design variable is given as

$$\frac{dg_i}{dx_j} = \frac{\partial g_i}{\partial x_j} + \frac{\partial g_i^T}{\partial \mathbf{U}}\frac{d\mathbf{U}}{dx_j} \tag{8.16}$$

where

$$\frac{\partial g_i}{\partial \mathbf{U}} = \left[\frac{\partial g_i}{\partial U_1}\ \frac{\partial g_i}{\partial U_2}\ \cdots\ \frac{\partial g_i}{\partial U_l}\right]^T \tag{8.17}$$

and

$$\frac{d\mathbf{U}}{dx_j} = \left[\frac{\partial U_1}{\partial x_j}\ \frac{\partial U_2}{\partial x_j}\ \cdots\ \frac{\partial U_l}{\partial x_j}\right]^T \tag{8.18}$$

Therefore, to calculate the gradient of a constraint, we need to calculate the partial derivatives $\partial g_i/\partial x_j$ and $\partial g_i/\partial \mathbf{U}$, and the total derivatives $d\mathbf{U}/dx_j$. The partial derivatives $\partial g_i/\partial x_j$ and $\partial g_i/\partial \mathbf{U}$ are quite easy to calculate using the form of the function $g_i(\mathbf{x}, \mathbf{U})$. To calculate $d\mathbf{U}/dx_j$, we differentiate the equilibrium Eq. (8.1) to obtain

$$\frac{\partial \mathbf{K}(\mathbf{x})}{\partial x_j}\mathbf{U} + \mathbf{K}(\mathbf{x})\frac{d\mathbf{U}}{dx_j} = \frac{\partial \mathbf{F}}{\partial x_j} \tag{8.19}$$

Or, the equation can be rearranged as

$$\mathbf{K}(\mathbf{x})\frac{d\mathbf{U}}{dx_j} = \frac{\partial \mathbf{F}}{\partial x_j} - \frac{\partial \mathbf{K}(\mathbf{x})}{\partial x_j}\mathbf{U} \tag{8.20}$$

The equation can be used to calculate dU/dx_j. The derivative of the stiffness matrix $\partial K(x)/\partial x_j$ can easily be calculated if the explicit dependence of K on x is known. Note that Eq. (8.20) needs to be solved for each design variable. Once dU/dx_j are known, the gradient of the constraint is calculated from Eq. (8.16). The derivative vector in Eq. (8.16) is often called the *design gradient*. We shall illustrate the procedure with an example problem.

It should be noted that substantial work has been done in developing and implementing efficient procedures for calculating derivatives of implicit functions with respect to the design variables [Arora and Haug, 1979; Adelman and Haftka, 1986]. The subject is generally known as *design sensitivity analysis*. For efficiency considerations and proper numerical implementations, the foregoing literature should be consulted. The procedures have been programmed into general-purpose software for automatic computation of design gradients.

Example 8.2 Gradient evaluation for a two-member frame. Calculate the gradient of the stress constraint $g_1(x, U)$ for the two-member frame of Example 8.1 at the design point $(2.5, 2.5, 0.1)$.

Solution. The problem has been formulated in Example 8.1. The finite element model has been defined there and nodal displacements and member stresses have been calculated. We shall use Eqs (8.16) and (8.20) to evaluate the gradient of the stress constraint of Eq. (8.14).

The partial derivatives of the constraint of Eq. (8.14) with respect to x and U are given as

$$\frac{\partial g_1}{\partial x} = \frac{1}{\sigma_a^2} \left[2\sigma_1 \frac{\partial \sigma_1}{\partial x} + 6\tau \frac{\partial \tau}{\partial x} \right] \tag{8.21}$$

$$\frac{\partial g_1}{\partial U} = \frac{1}{\sigma_a^2} \left[2\sigma_1 \frac{\partial \sigma_1}{\partial U} + 6\tau \frac{\partial \tau}{\partial U} \right] \tag{8.22}$$

Using Eqs (8.8) to (8.13), partial derivatives of τ and σ_1 with respect to x and U are calculated as follows.

Partial derivatives of shear stress. Differentiating the expression for shear stress in Eq. (8.11) with respect to the design variables x, we get

$$\frac{\partial \tau}{\partial x} = \frac{1}{2At} \frac{\partial T}{\partial x} - \frac{T}{2A^2 t} \frac{\partial A}{\partial x} - \frac{T}{2At^2} \frac{\partial t}{\partial x} \tag{8.23}$$

where partial derivatives of the torque T with respect to design variable x are given as

$$\frac{\partial T}{\partial x} = -\frac{GU_3}{L} \frac{\partial J}{\partial x} \tag{8.24}$$

with $\partial J/\partial x$ calculated as

$$\frac{\partial J}{\partial d} = \frac{4t(d-t)(h-t)^2(d+h-2t) - 2t(d-t)^2(h-t)^2}{(d+h-2t)^2} = 0.864$$

$$\frac{\partial J}{\partial h} = \frac{4t(d-t)^2(h-t)(d+h-2t) - 2t(d-t)^2(h-t)^2}{(d+h-2t)^2} = 0.864$$

$$\frac{\partial J}{\partial t} = \frac{2(d-t)^2(h-t)^2 - 4t(d-t)(h-t)^2 - 4t(d-t)^2(h-t)}{(d+h-2t)}$$

$$- \frac{2t(d-t)^2(h-t)^2(-2)}{(d+h-2t)^2} = 12.096$$

Therefore, $\partial J/\partial \mathbf{x}$ is assembled as

$$\frac{\partial J}{\partial \mathbf{x}} = \begin{bmatrix} 0.864 \\ 0.864 \\ 12.096 \end{bmatrix}$$

and Eq. (8.24) gives $\partial T/\partial \mathbf{x}$ as

$$\frac{\partial T}{\partial \mathbf{x}} = -\frac{(1.154\mathrm{E}+07)}{100}(0.572\,75) \begin{bmatrix} 0.864 \\ 0.864 \\ 12.096 \end{bmatrix}$$

$$= -(6.610\mathrm{E}+04) \begin{bmatrix} 0.864 \\ 0.864 \\ 12.096 \end{bmatrix}$$

Other quantities needed to complete calculations in Eq. (8.23) are $\partial A/\partial \mathbf{x}$ and $\partial t/\partial \mathbf{x}$ which are calculated as

$$\frac{\partial A}{\partial \mathbf{x}} = \begin{bmatrix} (h-t) \\ (d-t) \\ -(h-t)-(d-t) \end{bmatrix} = \begin{bmatrix} 2.4 \\ 2.4 \\ -4.6 \end{bmatrix}$$

$$\frac{\partial t}{\partial \mathbf{x}} = \begin{bmatrix} 0 \\ 0 \\ 1 \end{bmatrix}$$

Substituting various quantities into Eq. (8.23), we get the partial derivative of τ with respect to $\mathbf{x}$ as

$$\frac{\partial \tau}{\partial \mathbf{x}} = \frac{1}{2At}\left[\frac{\partial T}{\partial \mathbf{x}} - \frac{T}{A}\frac{\partial A}{\partial \mathbf{x}} - \frac{T}{t}\frac{\partial t}{\partial \mathbf{x}}\right] = \begin{bmatrix} -1.653\mathrm{E}+04 \\ -1.653\mathrm{E}+04 \\ 3.580\mathrm{E}+04 \end{bmatrix}$$

Differentiating the expression for the shear stress τ in Eq. (8.11) with respect to the generalized displacements $\mathbf{U}$, we get

$$\frac{\partial \tau}{\partial \mathbf{U}} = \frac{1}{2At}\frac{\partial T}{\partial \mathbf{U}}$$

where

$$\frac{\partial T}{\partial \mathbf{U}} = \begin{bmatrix} 0 \\ 0 \\ -GJ/L \end{bmatrix} = \begin{bmatrix} 0 \\ 0 \\ -1.5953\mathrm{E}+05 \end{bmatrix}$$

Therefore, $\partial \tau / \partial \mathbf{U}$ is given as

$$\frac{\partial \tau}{\partial \mathbf{U}} = \begin{bmatrix} 0 \\ 0 \\ -1.3848\text{E}+05 \end{bmatrix}$$

Partial derivatives of bending stress. Differentiating the expression for σ_1 given in Eq. (8.12) with respect to design variables $\mathbf{x}$, we get

$$\frac{\partial \sigma_1}{\partial \mathbf{x}} = \frac{h}{2I}\frac{\partial M_1}{\partial \mathbf{x}} + \frac{M_1}{2I}\frac{\partial h}{\partial \mathbf{x}} - \frac{M_1 h}{2I^2}\frac{\partial I}{\partial \mathbf{x}} \tag{8.25}$$

where $\partial M_1 / \partial \mathbf{x}$, $\partial I / \partial \mathbf{x}$ and $\partial h / \partial \mathbf{x}$ are given as

$$\frac{\partial M_1}{\partial \mathbf{x}} = \frac{2E}{L^2}(-3U_1 + U_2 L)\frac{\partial I}{\partial \mathbf{x}}$$

$$\frac{\partial I}{\partial d} = \tfrac{1}{12}[h^3 - (h - 2t)^3]$$

$$= 0.288167$$

$$\frac{\partial I}{\partial h} = \tfrac{1}{4}[dh^2 - (d - 2t)(h - 2t)^2]$$

$$= 0.8645$$

$$\frac{\partial I}{\partial t} = \frac{(h - 2t)^3}{6} + \frac{(h - 2t)^2(d - 2t)}{2}$$

$$= 8.11133$$

$$\frac{\partial h}{\partial \mathbf{x}} = \begin{bmatrix} 0 \\ 1 \\ 0 \end{bmatrix}$$

Substituting various quantities into Eq. (8.25),

$$\frac{\partial \sigma_1}{\partial \mathbf{x}} = \begin{bmatrix} 0 \\ 2.2131\text{E}+05 \\ 0 \end{bmatrix}$$

Differentiating the expression for σ_1 in Eq. (8.12) with respect to generalized displacements $\mathbf{U}$, we get

$$\frac{\partial \sigma_1}{\partial \mathbf{U}} = \frac{h}{2I}\frac{\partial M_1}{\partial \mathbf{U}}$$

where $\partial M_1 / \partial \mathbf{U}$ is given from Eq. (8.9) as

$$\frac{\partial M_1}{\partial \mathbf{U}} = \frac{2EI}{L^2}\begin{bmatrix} -3 \\ L \\ 0 \end{bmatrix}$$

Therefore, $\partial\sigma_1/\partial\mathbf{U}$ is given as

$$\frac{\partial\sigma_1}{\partial\mathbf{U}} = \frac{Eh}{L^2}\begin{bmatrix} -3 \\ L \\ 0 \end{bmatrix} = \begin{bmatrix} -2.25\text{E}+05 \\ 7.50\text{E}+05 \\ 0 \end{bmatrix}$$

Substituting various quantities into Eqs (8.21) and (8.22), we obtain the partial derivatives of constraints as

$$\frac{\partial g_1}{\partial\mathbf{x}} = \begin{bmatrix} 4.9167 \\ 157.973 \\ -10.648 \end{bmatrix}$$

$$\frac{\partial g_1}{\partial\mathbf{U}} = \begin{bmatrix} -15.561 \\ 518.700 \\ 41.19 \end{bmatrix}$$

Derivatives of the displacements. To calculate derivatives of the displacements, we use Eq. (8.20). Since the load vector does not depend on design variables, $\partial\mathbf{F}/\partial x_j = 0$ for $j = 1, 2, 3$ in Eq. (8.20). To calculate $\partial(\mathbf{KU})/\partial x_j$ on the right-hand side of Eq. (8.20), we differentiate Eq. (8.5) with respect to the design variables. For example, differentiation of Eq. (8.5) with respect to d gives the following vector:

$$\frac{\partial\mathbf{K}(\mathbf{x})}{\partial d}\mathbf{U} = \begin{bmatrix} -3.1214\text{E}+03 \\ -2.8585\text{E}+04 \\ 2.8585\text{E}+04 \end{bmatrix}$$

Similarly, by differentiating with respect to h and t, we obtain

$$\frac{\partial\mathbf{K}(\mathbf{x})}{\partial\mathbf{x}}\mathbf{U} = (1.0\text{E}+3)\begin{bmatrix} -3.1214 & -9.3642 & -87.861 \\ -28.585 & 28.454 & 3.300 \\ 28.585 & -28.454 & -3.300 \end{bmatrix}$$

Since $\mathbf{K}(\mathbf{x})$ is already known in Example 8.1, we use Eq. (8.20) to calculate $d\mathbf{U}/d\mathbf{x}$ as

$$\frac{d\mathbf{U}}{d\mathbf{x}} = \begin{bmatrix} 16.909 & 37.645 & 383.42 \\ 0.2443 & 0.4711 & 5.0247 \\ -0.2443 & -0.4711 & -5.0247 \end{bmatrix}$$

Finally, substituting all the quantities in Eq. (8.16), we obtain the design gradient for the effective stress constraint of Eq. (8.14) as

$$\frac{dg_1}{d\mathbf{x}} = \begin{bmatrix} -141.55 \\ -202.87 \\ -3577.30 \end{bmatrix}$$

As noted in Example 8.1, the stress constraint of Eq. (8.14) is severely violated at the given design. The signs of the foregoing design derivatives indicate that all the variables will have to be increased to reduce the constraint violation at the point $(2.5, 2.5, 0.1)$. ‖

8.4 ISSUES IN PRACTICAL DESIGN OPTIMIZATION

Several issues need to be considered for practical design optimization. For example, careful consideration needs to be given to the selection of an algorithm and the software. Improper choice of either one can mean failure of the optimum design process and frustration with optimization techniques. In this section we shall discuss some of the issues that can have a significant impact on practical applications of the optimization methodology.

8.4.1 Selection of An Algorithm

Many algorithms have been developed and evaluated for practical design optimization. We need to consider several aspects while selecting an algorithm for practical applications, such as robustness, efficiency, generality and ease of use.

8.4.1.1 ROBUSTNESS. Characteristics of a robust algorithm are discussed in Section 6.2.6. For practical applications it is extremely important to use a method that is theoretically guaranteed to converge. *A method having such a guarantee starting from any initial design estimate is called robust.* Robust algorithms usually require a few more calculations during each iteration compared to algorithms that have no proof of convergence. Such approximate algorithms usually require considerable tuning for each problem before a reasonable optimum solution is obtained, and many times they converge to nonoptimum solutions. Thus, although the approximate algorithms consume slightly less computer time during each iteration they have considerable uncertainty in their performance. They usually need considerable experimentation and tuning to make them work for a particular problem. This can be an unnecessary distraction requiring the designer's time. Therefore, in the overall sense, such approximate algorithms are actually more costly to use than the robust algorithms. It is suggested that only robust algorithms be used in practical applications, i.e. robustness must be given higher priority over efficiency while selecting an algorithm and the associated software.

8.4.1.2 POTENTIAL CONSTRAINT STRATEGY. To evaluate the search direction in numerical methods for constrained optimization, one needs to know the cost and constraint functions and their gradients. *The numerical algorithms can be classified into two categories based on whether gradients of all the constraints or only a subset of them are required during a design iteration.* The numerical algorithms that need the gradients of only a subset of the constraints are said to use *potential constraint strategy*. The potential constraint set, in general, is comprised of active, nearly active and violated constraints at the current iteration. For a further discussion on the topic of potential set strategy, refer to Section 6.2.4.

8.4.2 Attributes of a Good Optimization Algorithm

Based on the preceding discussion, attributes of a good algorithm for practical design applications are defined as follows.

1. *Reliability*. The algorithm must be reliable for general design applications because such algorithms converge to a minimum point starting from any initial design estimate. Reliability of an algorithm is guaranteed if it is theoretically proven to converge.

2. *Generality*. The algorithm must be general, which implies that it should be able to treat equality as well as inequality constraints. In addition, it should not impose any restrictions on the form of functions of the problem.

3. *Ease of use*. The algorithm must be easy to use by the experienced as well as inexperienced designer. From a practical standpoint, this is an extremely important requirement because an algorithm requiring selection of tuning parameters is quite difficult to use. The proper specification of the parameters usually requires not only a complete knowledge and under-standing of the mathematical structure of the algorithm but also experimen-tation with each problem. Such an algorithm is unsuitable for practical design applications.

4. *Efficiency*. The algorithm must be efficient for general engineering applica-tions. An efficient algorithm has (i) faster rate of convergence to the minimum point, and (ii) least number of calculations within one design iteration. The *rate of convergence* can be accelerated by incorporating the second-order information about the problem into the algorithm. Incorpora-tion of the second-order information, however, requires additional calcula-tions during an iteration. Therefore, there is a trade-off between efficiency of calculation within an iteration and the rate of convergence. Some existing algorithms use second-order information whereas others do not. Efficiency within an iteration implies the minimum of calculations for the search direction and the step size. One way to achieve efficiency is to use a *potential constraint strategy* in calculating the search direction. There are some algorithms that use this strategy in their calculations while others do not. When the potential constraint strategy is used, the direction finding subproblem needs gradients of only potential constraints. Otherwise, gradients of all constraints are needed which is inefficient in most practical applications.

   Another consideration for improving efficiency within an iteration is to keep the number of function evaluations for step size determination to a minimum. This can be achieved by using step size determination procedures requiring fewer calls for function evaluations, e.g. inexact line search.

The designer needs to ask the following questions before selecting an optimization algorithm for practical applications:

1. Does the algorithms have proof of convergence? That is, is it theoretically guaranteed to converge to an optimum point starting from any initial design estimate? Can the starting design be infeasible?
2. Can the algorithm solve a general optimization problem without any restrictions on the constraint functions? Can it treat equality as well as inequality constraints?
3. Is the algorithm easy to use? In other words, does it require tuning for each problem?
4. Does the algorithm incorporate a potential constraint strategy?

8.5   USE OF GENERAL-PURPOSE SOFTWARE

As we have seen in previous sections, practical systems require considerable computer analysis before optimum solutions are obtained. For a particular application, problem functions and gradient evaluation software as well as optimization software must be integrated to create an optimum design capability. Depending on the application, each of the software components can be very large. Therefore, to create a design optimization capability, the most sophisticated and modern computer facilities must be used to integrate the software components.

For the example of structures modeled by the finite elements discussed in Section 8.2.2, large analysis packages must be used to analyze the structure. From the calculated response, constraint functions must be evaluated and programs must be developed to calculate gradients. All the software components must be integrated to create the optimum design capability for structures modeled by finite elements.

In this section we shall discuss the issues involved in selecting a general-purpose optimization software. Also interfacing of the software with a particular application shall be discussed.

8.5.1   Software Selection

Several issues need to be explored before a general-purpose optimization software is selected for integration with other application-dependent software. The most important question pertains to the optimization algorithm and how well it is implemented. The attributes of a good algorithm are given in Section 8.4.2. The software must contain at least one algorithm that satisfies all the requirements stated there. The algorithm should also be robustly implemented because a good algorithm when badly implemented is not very useful. The proof of convergence of most algorithms is based on certain assumptions. These need to be strictly adhered to while implementing the algorithm. In addition, most algorithms have numerical uncertainties in their steps which need to be recognized and proper procedures developed for numerical implementation. It is also important that the software is well tested on a range of applications of varying difficulty.

Several other user-friendly features are desirable. For example, the possibility of interaction during the iterative process, interactive graphics and other aids to facilitate design decision-making are highly desirable. The topic of interactive design optimization and facilities is covered in more detail in Chapter 7. Documentation for the software is also very important. How good is the user's manual? What sample problems are available with the program and how well are they documented? How easy is it to install the program on different computer systems? All these questions should be investigated before selecting the software.

8.5.2 Integration of an Application into a General-purpose Software

Each general-purpose program for optimization requires that the particular design application be somehow integrated into the software. Ease of integration of the software components for various applications can influence its selection. Also, the amount of data preparation needed to use the program is important.

Some general-purpose libraries are available that contain various subroutines implementing different algorithms. The user is required to write a main program to call the subroutine. Subprograms for function and gradient evaluation must also be written. Several exits are made from the subroutine and it is re-entered with different conditions. The main program becomes quite complex and is prone to user errors. In addition, there is little chance for user interaction in this environment because the user must develop his own interactive capability.

The other approach is to develop a computer program with options of various algorithms and facilities. Each application is implanted into the program through a standard interface which consists of "subroutine calls". The user prepares a few subroutines to describe the design problem only. All the data flow between the program and the subroutines is through the subroutine arguments. For example, design variable data are sent to the subroutines and the expected output is the constraint function values and their gradients. Most interactive capabilities, graphics and other user friendly features are built into the program. A key feature of this approach is that the users do not get involved with the algorithmic idiosyncrasies in selecting various parameters and trying to make the algorithm work. They are relieved of these duties by the software developers, and can concentrate on their design problem formulation and its description for the program.

Both the procedures described in the foregoing have been successfully used in the past for many practical applications of optimization. In most cases, the choice of the procedure has been dictated by the availability of the software. We shall use the program IDESIGN, which is based on the second

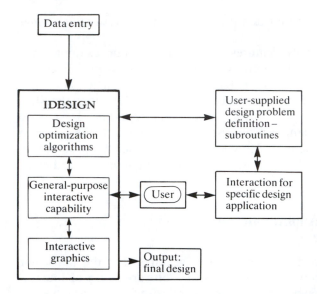

FIGURE 8.2
Implantation of a design application into IDESIGN. IDESIGN is combined with user-supplied subroutines to create an executable module.

approach, to solve several design optimization problems. Figure 8.2 shows how an application can be implanted into the program IDESIGN. As explained in Section 7.5, the program is combined with the user-supplied subroutines to create an executable module. The user-supplied subroutines can be quite simple for problems having explicit functions, and complex for problems having implicit functions. External subroutines or programs may need to be called upon to generate the function values and their gradients needed by IDESIGN. This has been done and several complex problems have been solved and reported in the literature [Tseng and Arora, 1987, 1988].

Besides the general-purpose interactive capability contained in IDESIGN, the user can write additional interactive facilities for the specific application, as shown in Fig. 8.2. Input data for the application are supplied to the program to execute it. Using this arrangement several practical problems from different fields have been solved with the program. In the remaining sections of this chapter, we shall describe several of the design problems and solve them with the foregoing arrangement for the program IDESIGN.

8.6 OPTIMUM DESIGN OF A TWO-MEMBER FRAME WITH OUT-OF-PLANE LOADS

Figure 8.1 shows a two-member frame subjected to out-of-plane loads. The members of the frame are subjected to torsional, bending and shearing loads. The objective of the problem is to design the structure having minimum volume without material failure due to applied loads. The problem has been formulated in Section 8.2.2 using the finite element approach. In defining the stress constraint, von Mises yield criterion is used and the shear stress due to the transverse load is neglected.

The formulation and equations given in Sections 8.2.2 and 8.3 are used to develop appropriate subroutines for the program IDESIGN. The data given there are used to optimize the problem. Two widely separated starting designs, $(2.5, 2.5, 0.1)$ and $(10, 10, 1)$, are tried to observe their effect on the convergence rate. For the first starting point all the variables are at their lower bounds and for the second point they are all at their upper bounds. Both starting points converge to the same optimum solution with almost the same values for design variables and Lagrange multipliers for active constraints, as shown in Tables 8.1 and 8.2 (refer to Section 7.6 for an explanation of the notations used in the tables). However, the numbers of iterations, and the numbers of calls for function and gradient evaluations are quite different. For the first starting point, the stress constraint is severely violated (by 20 212%). Several iterations are expended to bring the design close to the feasible region. For the second starting point, the stress constraint is satisfied and the program takes only six iterations to find the optimum solution. Note that both the solutions reported in Tables 8.1 and 8.2 are obtained by using the Sequential Quadratic Programming Option (SQP) available in IDESIGN. Also, a very severe convergence criterion is used to obtain the precise optimum point.

The preceding discussion shows that the starting design estimate for the iterative process can have a substantial impact on the rate of convergence of an algorithm. In many practical applications, a good starting design is available or can be obtained after some preliminary analyses. It is desirable to use such a starting design for the optimization algorithm because this can save substantial computational effort.

8.7 OPTIMUM DESIGN OF A THREE-BAR STRUCTURE FOR MULTIPLE PERFORMANCE REQUIREMENTS

In the previous section, we discussed design of a structural system for one performance requirement – the material must not fail under the applied loads. In this section, we discuss a similar application where the system must perform safely under several operating environments. The problem that we have chosen is the three-bar structure that is formulated in Section 2.6.8. The structure is shown in Fig. 2.5. The design requirement is to minimize the weight of the structure and satisfy the constraints of member stress, deflection at node 4, buckling of members, vibration frequency and explicit bounds on the design variables. We shall optimize symmetric and asymmetric structures and compare the solutions. A very strict convergence criterion shall be used to obtain the precise optimum designs.

TABLE 8.1

History of the iterative process and optimum solution for a two-member frame, starting point (2.5, 2.5, 0.10)

I	Max. vio.	Conv. parm	Cost	d	h	t
1	2.021 19E + 02	1.000 00E + 00	1.920 00E + 02	2.5000E + 00	2.5000E + 00	1.0000E − 01
2	8.578 97E + 01	1.000 00E + 00	2.318 57E + 02	2.5000E + 00	3.4964E + 00	1.0000E − 01
3	3.587 17E + 01	1.000 00E + 00	2.854 19E + 02	2.5000E + 00	4.8355E + 00	1.0000E − 01
4	2.329 76E + 01	1.000 00E + 00	3.506 78E + 02	5.7019E + 00	3.2651E + 00	1.0000E − 01
5	1.238 49E + 01	1.000 00E + 00	3.689 07E + 02	4.0394E + 00	5.3833E + 00	1.0000E − 01
6	1.186 48E + 01	1.000 00E + 00	3.738 86E + 02	2.5000E + 00	7.0472E + 00	1.0000E − 01
7	2.715 33E + 01	1.000 00E + 00	3.834 11E + 02	7.2853E + 00	2.5000E + 00	1.0000E − 01
8	1.252 10E + 01	1.000 00E + 00	4.037 73E + 02	6.5039E + 00	3.7904E + 00	1.0000E − 01
9	6.626 98E + 00	1.000 00E + 00	4.238 16E + 02	4.5737E + 00	6.2217E + 00	1.0000E − 01
10	5.932 94E + 00	1.000 00E + 00	4.396 60E + 02	2.5000E + 00	8.6915E + 00	1.0000E − 01
11	1.720 46E + 01	1.000 00E + 00	4.573 49E + 02	9.1337E + 00	2.5000E + 00	1.0000E − 01
12	7.639 27E + 00	1.000 00E + 00	4.810 71E + 02	8.5316E + 00	3.6952E + 00	1.0000E − 01
13	3.532 62E + 00	1.000 00E + 00	5.038 13E + 02	7.1404E + 00	5.6550E + 00	1.0000E − 01
14	2.256 09E + 00	1.000 00E + 00	5.244 50E + 02	3.7077E + 00	9.6036E + 00	1.0000E − 01
15	2.982 13E + 00	1.000 00E + 00	5.748 22E + 02	1.0000E + 01	4.5705E + 00	1.0000E − 01
16	1.275 17E + 00	1.000 00E + 00	5.975 47E + 02	8.4379E + 00	6.7008E + 00	1.0000E − 01
17	6.788 24E − 01	1.000 00E + 00	6.144 56E + 02	5.5614E + 00	1.0000E + 01	1.0000E − 01
18	1.589 21E − 01	6.222 70E − 01	6.762 20E + 02	7.1055E + 00	1.0000E + 01	1.0000E − 01
19	1.472 60E − 02	7.012 49E − 02	7.011 11E + 02	7.7278E + 00	1.0000E + 01	1.0000E − 01
20	1.560 97E − 04	7.593 55E − 04	7.039 16E + 02	7.7979E + 00	1.0000E + 01	1.0000E − 01

Constraint activity

No.	Active	Value	Lagr. mult.
1	Yes	1.560 97E − 04	1.946 31E + 02

Design variable activity

No.	Active	Design	Lower	Upper	Lagr. mult.
1	No	7.797 91E + 00	2.500 00E + 00	1.000 00E + 01	0.000 00E + 00
2	Upper	1.000 00E + 01	2.500 00E + 00	1.000 00E + 01	7.897 73E + 01
3	Lower	1.000 00E − 01	1.000 00E − 01	1.000 00E + 00	3.190 90E + 02

Cost function at optimum = 7.039 163E + 02.
No. of calls for cost function evaluation (USERMF) = 20.
No. of calls for evaluation of cost function gradient (USERMG) = 20.
No. of calls for constraint function evaluation (USERCF) = 20.
No. of calls for evaluation of constraint function gradients (USERCG) = 20.
No. of total gradient evaluations = 20.

8.7.1 Symmetric Three-bar Structure

A detailed formulation for the symmetric structure where members 1 and 3 are similar is discussed in Section 2.6.8. In the present application, the structure is designed to withstand three loading conditions and the foregoing constraints. Table 8.3 contains all the data used for designing the structure. All the

TABLE 8.2
History of the iterative process and optimum solution for a two-member frame, starting point (10, 10, 1)

I	Max. vio.	Conv. parm	Cost	d	h	t
1	$0.000\,00E+00$	$6.400\,00E+03$	$7.200\,00E+03$	$1.0000E+01$	$1.0000E+01$	$1.0000E+00$
2	$0.000\,00E+00$	$2.278\,73E+01$	$7.875\,00E+02$	$9.9438E+00$	$9.9438E+00$	$1.0000E-01$
3	$1.250\,20E-02$	$1.319\,93E+00$	$7.130\,63E+02$	$9.0133E+00$	$9.0133E+00$	$1.0000E-01$
4	$2.199\,48E-02$	$1.036\,43E-01$	$6.997\,34E+02$	$7.6933E+00$	$1.0000E+01$	$1.0000E-01$
5	$3.441\,15E-04$	$1.673\,49E-03$	$7.038\,80E+02$	$7.7970E+00$	$1.0000E+01$	$1.0000E-01$
6	$9.404\,69E-08$	$4.305\,13E-07$	$7.039\,47E+02$	$7.7987E+00$	$1.0000E+01$	$1.0000E-01$

Constraint activity

No.	Active	Value	Lagr. mult.
1	Yes	$9.404\,69E-08$	$1.946\,30E+02$

Design variable activity

No.	Active	Design	Lower	Upper	Lagr. mult.
1	No.	$7.798\,67E+00$	$2.500\,00E+00$	$1.000\,00E+01$	$0.000\,00E+00$
2	Upper	$1.000\,00E+01$	$2.500\,00E+00$	$1.000\,00E+01$	$7.897\,67E+01$
3	Lower	$1.000\,00E-01$	$1.000\,00E-01$	$1.000\,00E+00$	$3.190\,90E+02$

Cost function at optimum $= 7.039\,466E+02$.
No. of calls for cost function evaluation (USERMF) $= 9$.
No. of calls for evaluation of cost function gradient (USERMG) $= 6$.
No. of calls for constraint function evaluation (USERCF) $= 9$.
No. of calls for evaluation of constraint function gradients (USERCG) $= 4$.
No. of total gradient evaluations $= 4$.

TABLE 8.3
Design data for a three-bar structure

Allowable stress:	Members 1 and 3, $\sigma_{1a} = \sigma_{3a} = 5000$ psi
	Member 2, $\sigma_{2a} = 20\,000$ psi
Allowable displacements:	$u_a = 0.005$ in
	$v_a = 0.005$ in
Modulus of elasticity:	$E = (1.00E+07)$ psi
Weight density:	$\gamma = (1.00E-01)$ lb/in^3
Constant:	$\beta = 1.0$
Lower limit on design:	$= (0.1, 0.1, 0.1)$ in^2
Upper limit on design:	$= (100, 100, 100)$ in^2
Starting design:	$= (1, 1, 1)$ in^2
Lower limit on frequency:	$= 2500$ Hz
Loading conditions:	$= 3$

	Angle, θ (degrees)		
	45	90	135
Load, P (lb)	40 000	30 000	20 000

expressions programmed for IDESIGN are given in Section 2.6.8. The constraint functions are appropriately normalized and expressed in the standard form.

To study the effect of imposing more performance requirements, the following three design cases are defined (note that explicit design variable bound constraints are also imposed in all cases):

Case 1. Stress constraints only (total constraints = 13).

Case 2. Stress and displacement constraints (total constraints = 19).

TABLE 8.4
History of the iterative process and final solution for a symmetric three-bar structure, Case 1 – stress constraints

I	Max. vio.	Conv. parm	Cost	$A_1 = A_3$	A_2
1	4.656 80E + 00	1.000 00E + 00	3.828 43E + 00	1.0000E + 00	1.0000E + 00
2	2.145 31E + 00	1.000 00E + 00	6.720 82E + 00	1.9528E + 00	1.1973E + 00
3	8.742 38E − 01	1.000 00E + 00	1.126 58E + 01	3.3986E + 00	1.6531E + 00
4	2.784 05E − 01	1.090 96E + 00	1.652 72E + 01	5.0449E + 00	2.2580E + 00
5	4.979 00E − 02	2.611 72E − 01	2.012 46E + 01	6.1359E + 00	2.7697E + 00
6	2.289 84E − 03	5.786 37E − 02	2.107 24E + 01	6.3971E + 00	2.9788E + 00
7	2.899 32E − 05	3.885 94E − 02	2.111 63E + 01	6.3874E + 00	3.0501E + 00
8	2.204 83E − 04	3.972 59E − 03	2.110 68E + 01	6.3140E + 00	3.2482E + 00
9	1.586 18E − 06	5.341 72E − 05	2.111 14E + 01	6.3094E + 00	3.2657E + 00

Constraint activity

No.	Active	Value	Lagr. mult.
1	Yes	1.586 18E − 06	2.111 14E + 01
2	No	−7.320 69E − 01	0.000 00E + 00
3	No	−8.169 82E − 01	0.000 00E + 00
4	No	−6.117 57E − 01	0.000 00E + 00
5	No	−6.117 57E − 01	0.000 00E + 00
6	No	−8.058 79E − 01	0.000 00E + 00
7	No	−8.660 35E − 01	0.000 00E + 00
8	No	−4.999 99E − 01	0.000 00E + 00
9	No	−9.084 91E − 01	0.000 00E + 00

Design variable activity

No.	Active	Design	Lower	Upper	Lagr. mult.
1	No	6.309 42E + 00	1.000 00E − 01	1.000 00E + 02	0.000 00E + 00
2	No	3.265 69E + 00	1.000 00E − 01	1.000 00E + 02	0.000 00E + 00

Cost function at optimum = 2.111 143E + 01.
No. of calls for cost function evaluation (USERMF) = 9.
No. of calls for evaluation of cost function gradient (USERMG) = 9.
No. of calls for constraint function evaluation (USERCF) = 9.
No. of calls for evaluation of constraint function gradients (USERCG) = 9.
No. of total gradient evaluations = 19.

Case 3. All constraints – stress, displacement, member buckling and frequency (total constraints = 29).

Tables 8.4 to 8.6 contain the history of the iterative process, constraint and design variable activities at the final design, and the optimum cost function for

TABLE 8.5

History of the iterative process and final solution for a symetric three-bar structure, Case 2 – stress and displacement constraints

I	Max. vio.	Conv. parm	Cost	$A_1 = A_3$	A_2
1	6.999 92E + 00	1.000 00E + 00	3.828 43E + 00	1.0000E + 00	1.0000E + 00
2	3.266 63E + 00	1.000 00E + 00	6.905 98E + 00	1.8750E + 00	1.6027E + 00
3	1.416 50E + 00	1.000 00E + 00	1.145 04E + 01	3.3105E + 00	2.0868E + 00
4	5.234 73E − 01	1.000 00E + 00	1.693 91E + 01	5.2511E + 00	2.0868E + 00
5	1.338 70E − 01	1.000 00E + 00	2.204 25E + 01	7.0554E + 00	2.0868E + 00
6	1.413 63E − 02	2.802 30E − 01	2.339 85E + 01	7.8884E + 00	1.0868E + 00
7	1.242 55E − 02	2.058 97E − 02	2.289 35E + 01	7.9984E + 00	2.7066E − 01
8	1.506 50E − 04	3.054 85E − 04	2.296 95E + 01	7.9999E + 00	3.4230E − 01
9	2.268 86E − 08	4.538 76E − 08	2.297 04E + 01	7.9999E + 00	3.4320E − 01

Constraint activity

No.	Active	Value	Lagr. mult.
1	Yes	−2.860 00E − 02	0.000 00E + 00
2	No	−9.714 00E − 01	0.000 00E + 00
3	No	−7.643 00E − 01	0.000 00E + 00
4	Yes	1.332 27E − 15	1.697 04E + 01
5	Yes	−5.720 00E − 02	0.000 00E + 00
6	No	−5.000 00E − 01	0.000 00E + 00
7	No	−5.000 00E − 01	0.000 00E + 00
8	No	−7.500 00E − 01	0.000 00E + 00
9	No	−1.000 00E + 00	0.000 00E + 00
10	Yes	2.268 86E − 08	6.000 00E + 00
11	No	−9.857 00E − 01	0.000 00E + 00
12	No	−5.143 00E − 01	0.000 00E + 00
13	No	−8.821 50E − 01	0.000 00E + 00
14	No	−5.000 00E − 01	0.000 00E + 00
15	No	−5.286 00E − 01	0.000 00E + 00

Design variable activity

No.	Active	Design	Lower	Upper	Lagr. mult.
1	No	7.999 92E + 00	1.000 00E − 01	1.000 00E + 02	0.000 00E + 00
2	No	3.432 00E − 01	1.000 00E − 01	1.000 00E + 02	0.000 00E + 00

Cost function at optimum = 2.297 040E + 01.
No. of calls for cost function evaluation (USERMF) = 9.
No. of calls for evaluation of cost function gradient (USERMG) = 9.
No. of calls for constraint function evaluation (USERCF) = 9.
No. of calls for evaluation of constraint function gradients (USERCG) = 9.
No. of total gradient evaluations = 48.

TABLE 8.6

History of the iterative process and final solution for a symmetric three-bar structure, Case 3 – all constraints

I	Max. vio.	Conv. parm	Cost	$A_1 = A_3$	A_2
1	6.999 92E + 00	1.000 00E + 00	3.828 43E + 00	1.000 0E + 00	1.0000E + 00
2	2.888 48E + 00	1.000 00E + 00	7.292 79E + 00	2.057 3E + 00	1.4738E + 00
3	1.231 13E + 00	1.000 00E + 00	1.203 39E + 01	3.585 6E + 00	1.8923E + 00
4	4.377 71E − 01	1.694 16E + 00	1.763 00E + 01	5.564 1E + 00	1.8923E + 00
5	1.021 80E − 01	6.728 96E − 01	2.142 17E + 01	7.258 3E + 00	8.9226E − 01
6	8.669 19E − 03	4.809 65E − 02	2.282 44E + 01	7.931 2E + 00	3.9172E − 01
7	7.387 41E − 05	3.187 76E − 04	2.296 91E + 01	7.999 3E + 00	3.4362E − 01
8	5.456 57E − 09	2.315 29E − 08	2.297 04E + 01	7.999 9E + 00	3.4320E − 01

Constraint activity

No.	Active	Value	Lagr. mult.
1	No	−1.569 67E − 01	0.000 00E + 00
2	Yes	−2.860 00E − 02	0.000 00E + 00
3	No	−9.714 00E − 01	0.000 00E + 00
4	No	−7.643 00E − 01	0.000 00E + 00
5	Yes	5.456 57E − 09	1.697 04E + 01
6	Yes	−5.720 00E − 02	0.000 00E + 00
7	No	−5.000 00E − 01	0.000 00E + 00
8	No	−5.000 00E − 01	0.000 00E + 00
9	No	−7.500 00E − 01	0.000 00E + 00
10	No	−1.000 00E + 00	0.000 00E + 00
11	Yes	0.000 00E + 00	6.000 00E + 00
12	No	−9.857 00E − 01	0.000 00E + 00
13	No	−5.143 00E − 01	0.000 00E + 00
14	No	−8.821 50E − 01	0.000 00E + 00
15	No	−5.000 00E − 01	0.000 00E + 00
16	No	−5.286 00E − 01	0.000 00E + 00

Design variable activity

No.	Active	Design	Lower	Upper	Lagr. mult.
1	No	7.999 92E + 00	1.000 00E − 01	1.000 00E + 02	0.000 00E + 00
2	No	3.432 00E − 01	1.000 00E − 01	1.000 00E + 02	0.000 00E + 00

Cost function at optimum = 2.297 040E + 01.
No. of calls for cost function evaluation (USERMF) = 8.
No. of calls for evaluation of cost function gradient (USERMG) = 8.
No. of calls for constraint function evaluation (USERCF) = 8.
No. of calls for evaluation of constraint function gradients (USERCG) = 8.
No. of total gradient evaluations = 50.

the three cases with the SQP method in IDESIGN. The active constraints at the optimum point and their Lagrange multipliers (for normalized constraints) are:

Case 1. Stress in member 1 under loading condition 1, 21.11.

Case 2. Stress in member 1 under loading condition 1, 0.0; horizontal displacement under loading condition 1, 16.97; vertical displacement under loading condition 1, 6.00; horizontal displacement under loading condition 2, 0.0.

Case 3. Same as for Case 2.

Note that the cost function value at the optimum point increases for Case 2 as compared to Case 1. This is consistent with the hypothesis that more constraints for the system imply a smaller feasible region, thus giving a higher value for the optimum cost function. There is no difference between the solutions for Cases 2 and 3 because none of the additional constraints for Case 3 is active.

8.7.2 Asymmetric Three-bar Structure

When the symmetry condition for the structure (member 1 same as member 3) is relaxed, we get three design variables for the problem compared to only two for the symmetric case, i.e. areas A_1, A_2 and A_3 for members 1, 2 and 3, respectively. With this the design space becomes expanded, so we can expect better optimum designs compared to the previous cases. The data used for the problem are the same as given in Table 8.3. The structure is optimized for the following three cases (note that the explicit design variable bound constraints are imposed in all cases):

Case 4. Stress constraints only (total constraints = 15).

Case 5. Stress and displacement constraints (total constraints = 21).

Case 6. All constraints – stress, displacement, buckling, and frequency (total constraints = 31).

The structure can be analyzed by considering either the equilibrium of node 4 or the general finite element procedures [Cook, 1981]. By following the general procedures, the following expressions for displacements, member stresses, and fundamental vibration frequency are obtained (note that the notations are defined in Section 2.6.8):

Displacements:

$$u = \frac{l}{E}\left[\frac{(A_1 + 2\sqrt{2}A_2 + A_3)P_u + (A_1 - A_3)P_v}{A_1A_2 + \sqrt{2}A_1A_3 + A_2A_3}\right], \quad \text{in}$$

$$v = \frac{l}{E}\left[\frac{-(A_1 - A_3)P_u + P_v(A_1 + A_3)}{A_1A_2 + \sqrt{2}A_1A_3 + A_2A_3}\right], \quad \text{in}$$

Member stresses:

$$\sigma_1 = \frac{(\sqrt{2}A_2 + A_3)P_u + A_3P_v}{A_1A_2 + \sqrt{2}A_1A_3 + A_2A_3}, \qquad \text{psi}$$

$$\sigma_2 = \frac{-(A_1 - A_3)P_u + (A_1 + A_3)P_v}{A_1A_2 + \sqrt{2}A_1A_3 + A_2A_3}, \qquad \text{psi}$$

$$\sigma_3 = \frac{-(A_1 + \sqrt{2}A_2)P_u + A_1P_v}{A_1A_2 + \sqrt{2}A_1A_3 + A_2A_3}, \qquad \text{psi}$$

Lowest eigenvalue:

$$\zeta = \frac{3E}{2\sqrt{2}\rho l^2}\left[\frac{A_1 + \sqrt{2}A_2 + A_3 - [(A_1 - A_3)^2 + 2A_2^2]^{1/2}}{\sqrt{2}(A_1 + A_3) + A_2}\right]$$

Fundamental frequency:

$$\omega = \frac{1}{2\pi}\sqrt{\zeta}, \qquad \text{Hz}$$

Tables 8.7–8.9 contain the history of the iterative process, constraint and design variable activities at the final design, and the optimum cost for the three cases with the SQP method in IDESIGN. The active constraints at the optimum point and their Lagrange multipliers (for normalized constraints) are

Case 4. Stress in member 1 under loading condition 1, 11.00; stress in member 3 under loading condition 3, 4.97.

Case 5. Horizontal displacement under loading condition 1, 11.96; vertical displacement under loading condition 2, 8.58.

Case 6. Frequency constraint, 6.73; horizontal displacement under loading condition 1, 13.28; vertical displacement under loading condition 2, 7.77.

Note that the optimum weight for Case 5 is higher than that for Case 4, and for Case 6 it is higher than that for Case 5. This is consistent with the previous observation; the number of constraints for Case 5 is larger than that for Case 4, and for Case 6 it is larger than that for Case 5.

8.7.3   Comparison of Solutions

Table 8.10 contains a comparison of solutions for all six cases. Since an asymmetric structure has a larger design space, the optimum solutions should be better than those for the symmetric case, and they are; Case 4 is better than Case 1, Case 5 is better than Case 2, and Case 6 is better than Case 3. *These results show that for better practical solutions, more flexibility should be allowed in the design process by defining more design variables.*

TABLE 8.7

History of the iterative process and final solution for an asymmetric three-bar structure, Case 4 – stress constraints

I	Max. vio.	Conv. parm	Cost	A_1	A_2	A_3
1	4.656 80E + 00	1.000 00E + 00	3.828 43E + 00	1.0000E + 00	1.0000E + 00	1.0000E + 00
2	2.106 35E + 00	1.000 00E + 00	6.514 95E + 00	1.9491E + 00	1.4289E + 00	1.6473E + 00
3	8.511 50E − 01	1.000 00E + 00	1.008 09E + 01	3.4599E + 00	1.9566E + 00	2.2849E + 00
4	2.687 61E − 01	1.313 59E + 00	1.352 03E + 01	5.2718E + 00	2.4566E + 00	2.5513E + 00
5	4.698 96E − 02	2.307 91E − 01	1.548 78E + 01	6.5854E + 00	2.5261E + 00	2.5799E + 00
6	2.124 93E − 03	8.464 22E − 02	1.595 71E + 01	6.9464E + 00	2.4233E + 00	2.6233E + 00
7	1.174 23E − 04	5.201 74E − 02	1.597 28E + 01	6.9820E + 00	2.3237E + 00	2.6694E + 00
8	4.031 39E − 04	2.524 83E − 03	1.596 20E + 01	7.0220E + 00	2.1322E + 00	2.7572E + 00
9	4.809 86E − 07	6.270 73E − 05	1.596 84E + 01	7.0236E + 00	2.1383E + 00	2.7558E + 00

Constraint activity

No.	Active	Value	Lagr. mult.
1	Yes	4.809 86E − 07	1.100 20E + 01
2	No	−8.385 71E − 01	0.000 00E + 00
3	No	−6.457 16E − 01	0.000 00E + 00
4	No	−6.575 54E − 01	0.000 00E + 00
5	No	−6.961 93E − 01	0.000 00E + 00
6	No	−1.272 18E − 01	0.000 00E + 00
7	No	−8.228 58E − 01	0.000 00E + 00
8	No	−7.942 85E − 01	0.000 00E + 00
9	Yes	4.809 18E − 07	4.966 50E + 00

Design variable activity

No.	Active	Design	Lower	Upper	Lagr. mult.
1	No	7.023 59E + 00	1.000 00E − 01	1.000 00E + 02	0.000 00E + 00
2	No	2.138 31E + 00	1.000 00E − 01	1.000 00E + 02	0.000 00E + 00
3	No	2.755 79E + 00	1.000 00E − 01	1.000 00E + 02	0.000 00E + 00

Cost function at optimum = 1.596 844E + 01.
No. of calls for cost function evaluation (USERMF) = 9.
No. of calls for evaluation of cost function gradient (USERMG) = 9.
No. of calls for constraint function evaluation (USERCF) = 9.
No. of calls for evaluation of constraint function gradients (USERCG) = 9.
No. of total gradient evaluations = 26.

8.8 DISCRETE VARIABLE OPTIMUM DESIGN

In many practical applications of optimization, design variables for a problem must be selected from a given set of values. For example, structural elements must be chosen from those that are already commercially available. This is called discrete variable optimization which is essential to economize on

TABLE 8.8
History of the iterative process and final solution for an asymmetric three-bar structure, Case 5 – stress and displacement constraints

I	Max. vio.	Conv. parm	Cost	A_1	A_2	A_3
1	$6.999\,92E + 00$	$1.000\,00E + 00$	$3.828\,43E + 00$	$1.000\,0E + 00$	$1.0000E + 00$	$1.0000E + 00$
2	$3.265\,89E + 00$	$1.000\,00E + 00$	$6.773\,40E + 00$	$1.963\,4E + 00$	$1.6469E + 00$	$1.6616E + 00$
3	$1.410\,09E + 00$	$1.000\,00E + 00$	$1.102\,22E + 01$	$3.603\,5E + 00$	$2.2775E + 00$	$2.5799E + 00$
4	$5.106\,70E - 01$	$1.000\,00E + 00$	$1.578\,74E + 01$	$5.805\,2E + 00$	$2.3814E + 00$	$3.6743E + 00$
5	$1.223\,08E - 01$	$8.954\,54E - 01$	$1.944\,06E + 01$	$7.783\,7E + 00$	$2.0911E + 00$	$4.4842E + 00$
6	$9.560\,69E - 03$	$7.643\,35E - 02$	$2.043\,20E + 01$	$8.679\,2E + 00$	$1.7974E + 00$	$4.4975E + 00$
7	$2.035\,55E - 04$	$7.951\,90E - 02$	$2.054\,81E + 01$	$8.808\,4E + 00$	$1.8341E + 00$	$4.4244E + 00$
8	$2.367\,02E - 04$	$1.448\,77E - 02$	$2.053\,92E + 01$	$8.884\,8E + 00$	$1.9071E + 00$	$4.2900E + 00$
9	$2.187\,02E - 05$	$3.830\,28E - 04$	$2.054\,32E + 01$	$8.910\,8E + 00$	$1.9299E + 00$	$4.2508E + 00$
10	$6.721\,42E - 09$	$1.425\,07E - 06$	$2.054\,36E + 01$	$8.910\,6E + 00$	$1.9295E + 00$	$4.2516E + 00$

Constraint activity

No.	Active	Value	Lagr. Mult.
1	No	$-1.954\,61E - 01$	$0.000\,00E + 00$
2	No	$-8.477\,31E - 01$	$0.000\,00E + 00$
3	No	$-8.045\,39E - 01$	$0.000\,00E + 00$
4	Yes	$6.721\,42E - 09$	$1.196\,42E + 01$
5	No	$-3.909\,23E - 01$	$0.000\,00E + 00$
6	No	$-6.769\,85E - 01$	$0.000\,00E + 00$
7	No	$-7.500\,00E - 01$	$0.000\,00E + 00$
8	No	$-3.230\,15E - 01$	$0.000\,00E + 00$
9	No	$-6.460\,30E - 01$	$0.000\,00E + 00$
10	Yes	$6.721\,41E - 09$	$8.579\,42E + 00$
11	No	$-9.022\,69E - 01$	$0.000\,00E + 00$
12	No	$-8.404\,35E - 01$	$0.000\,00E + 00$
13	No	$-2.640\,08E - 01$	$0.000\,00E + 00$
14	No	$-1.662\,77E - 01$	$0.000\,00E + 00$
15	No	$-3.617\,39E - 01$	$0.000\,00E + 00$

Design variable activity

No.	Active	Design	Lower	Upper	Lagr. mult.
1	No	$8.910\,58E + 00$	$1.000\,00E - 01$	$1.000\,00E + 02$	$0.000\,00E + 00$
2	No	$1.929\,54E + 00$	$1.000\,00E - 01$	$1.000\,00E + 02$	$0.000\,00E + 00$
3	No	$4.251\,57E + 00$	$1.000\,00E - 01$	$1.000\,00E + 02$	$0.000\,00E + 00$

Cost function at optimum $= 2.054\,363E + 01$.
No. of calls for cost function evaluation (USERMF) = 10.
No. of calls for evaluation of cost function gradient (USERMG) = 10.
No. of calls for constraint function evaluation (USERCF) = 10.
No. of calls for evaluation of constraint function gradients (USERCG) = 10.
No. of total gradient evaluations = 43.

TABLE 8.9
History of the iterative process and final solution for an asymmetric three-bar structure, Case 6 – all constraints

I	Max. vio.	Conv. parm	Cost	A_1	A_2	A_3
1	6.999 92E + 00	1.000 00E + 00	3.828 43E + 00	1.0000E + 00	1.0000E + 00	1.0000E + 00
2	2.888 48E + 00	1.000 00E + 00	7.292 79E + 00	2.0573E + 00	1.4738E + 00	2.0573E + 00
3	1.225 37E + 00	1.000 00E + 00	1.175 48E + 01	3.7189E + 00	1.9853E + 00	3.1892E + 00
4	4.274 37E − 01	1.673 31E + 00	1.634 91E + 01	6.0225E + 00	2.1056E + 00	4.0492E + 00
5	9.441 47E − 02	3.900 47E − 01	1.977 44E + 01	7.6958E + 00	1.5955E + 00	5.1586E + 00
6	7.387 41E − 03	3.013 29E − 02	2.093 20E + 01	8.2455E + 00	1.2450E + 00	5.6753E + 00
7	6.754 06E − 05	2.255 16E − 04	2.104 82E + 01	8.2901E + 00	1.2017E + 00	6.7435E + 00
8	6.466 97E − 09	1.881 51E − 08	2.104 94E + 01	8.2905E + 00	1.2013E + 00	5.7442E + 00

Constraint activity

No.	Active	Value	Lagr. mult.
1	Yes	2.997 88E − 09	6.731 33E + 00
2	No	−1.141 25E − 01	0.000 00E + 00
3	No	−8.070 62E − 01	0.000 00E + 00
4	No	−8.858 75E − 01	0.000 00E + 00
5	Yes	6.466 97E − 09	1.327 73E + 01
6	No	−2.282 49E − 01	0.000 00E + 00
7	No	−5.907 14E − 01	0.000 00E + 00
8	No	−7.500 00E − 01	0.000 00E + 00
9	No	−4.092 86E − 01	0.000 00E + 00
10	No	−8.185 73E − 01	0.000 00E + 00
11	Yes	1.234 90E − 09	7.772 13E + 00
12	No	−9.429 38E − 01	0.000 00E + 00
13	No	−8.607 69E − 01	0.000 00E + 00
14	No	−3.860 13E − 01	0.000 00E + 00
15	No	−3.289 51E − 01	0.000 00E + 00
16	No	−4.430 75E − 01	0.000 00E + 00

Design variable activity

No.	Active	Design	Lower	Upper	Lagr. mult.
1	No	8.290 52E + 00	1.000 00E − 01	1.000 00E + 02	0.000 00E + 00
2	No	1.201 30E + 00	1.000 00E − 01	1.000 00E + 02	0.000 00E + 00
3	No	5.744 23E + 00	1.000 00E − 01	1.000 00E + 02	0.000 00E + 00

Cost function at optimum = 2.104 943E + 01.
No. of calls for cost function evaluation (USERMF) = 8.
No. of calls for evaluation of cost function gradient (USERMG) = 8.
No. of calls for constraint function evaluation (USERCF) = 8.
No. of calls for evaluation of constraint function gradients (USERCG) = 8.
No. of total gradient evaluations = 48.

TABLE 8.10
Comparison of optimum costs for six cases of a three-bar structure

	Symmetric structure			Asymmetric structure		
	Case 1	Case 2	Case 3	Case 4	Case 5	Case 6
Optimum weight (lb)	21.11	22.97	22.97	15.97	20.54	21.05
NIT[a]	9	9	8	9	10	8
NCF[a]	9	9	8	9	10	8
NGE[a]	19	48	50	26	43	48

[a] NIT, no. of iterations; NCF, No. of calls for function evaluation; NGE, total number of gradient evaluations.

fabrication costs for the design. The subject is briefly discussed in Section 2.7.6. We shall demonstrate the procedure described there for a simple design problem.

The application area that we have chosen is the optimum design of the aerospace, automotive, mechanical and structural systems, by employing finite element models. The problem is to design a minimum weight system with constraints on various performance specifications. As a sample application, we shall consider the ten-bar cantilever structure shown in Fig. 8.3. The loading and other design data for the problem are given in Table 8.11. The set of discrete values taken from the American Institute of Steel Construction Manual are also given there. The final design for the structure must be selected from this set. The cross-sectional area of each member is treated as a design variable giving a total of ten variables. Constraints are imposed on member stress (10), nodal displacement (8), member buckling (10), vibration frequency (1), and explicit bounds on the design variables (20). This gives a total of 49 constraints. In imposing the member buckling constraint, the moment of inertia is taken as $I = \beta A^2$ where β is a constant and A is the member cross-sectional area. The formulation for the problem is quite similar to the one for the three-bar structure discussed in Section 2.6.8. The only difference is that the explicit form of the constraint function is not known. Therefore, we must use the finite element procedures described in Sections 8.2 and 8.3 for structural analysis and the gradient evaluation of constraints.

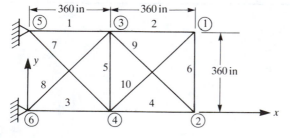

FIGURE 8.3
Ten-bar cantilever truss.

TABLE 8.11
Design data for a ten-bar structure

Modulus of elasticity	$E = (1.00E+07)$ psi
Material weight density	$\gamma = (1.0E-01)$ lb/in^3
Displacement limit	$= \pm 2.0$ in
Stress limit	$= 25\,000$ psi
Frequency limit	$= 22$ Hz
Lower limit on design variables	$= 1.62$ in^2
Upper limit on design variables	$=$ none
Constant β $(I = \beta A^2)$	$= 1.0$
Loading Data:	

Node no.	Load in y-direction (lbs)
1	50 000
2	−150 000
3	50 000
4	−150 000

Available member sizes (in^2): 1.62, 1.80, 1.99, 2.13, 2.38, 2.62, 2.63, 2.88, 2.93, 3.09, 3.13, 3.38, 3.47, 3.55, 3.63, 3.84, 3.87, 3.88, 4.18, 4.22, 4.49, 4.59, 4.80, 4.97, 5.12, 5.74, 7.22, 7.97, 11.50, 13.50, 13.90, 14.20, 15.50, 16.00, 16.90, 18.80, 19.90, 22.00, 22.90, 26.50, 30.00, 33.50.

8.8.1 Continuous Variable Optimization

To compare solutions, the continuous variable optimization problem is solved first. To use the program IDESIGN, USER subroutines are developed using the material of Sections 8.2 and 8.3 to evaluate the functions and their gradients. The optimum solution using a very severe convergence criteria and a uniform starting design of 1.62 in^2, is obtained as

Design variables: 28.28, 1.62, 27.262, 13.737, 1.62, 4.0026, 13.595, 17.544, 19.13, 1.62
Optimum cost function: 5396.5 lbs
Number of iterations: 19
Number of analysis: 21
Maximum constraint violation at optimum: (8.024E−10)
Convergence parameter at optimum: (2.660E−05)
Active constraints at optimum and their Lagrange multipliers
   frequency, 392.4
   stress in member 2, 38.06
   displacement at node 2 in the y direction, 4967
   lower bound for member 2, 7.829
   lower bound for member 5, 205.1
   lower bound for member 10, 140.5

8.8.2 Discrete Variable Optimization

We use the adaptive numerical optimization procedure described in Section 2.7.6 to obtain a discrete variable solution. The procedure is to use the program IDESIGN in an interactive mode. Design conditions are monitored and decisions made to fix design variables that are not changing. The interactive facilities used include design variable histories, maximum constraint violation, and the cost function.

Table 8.12 contains a snapshot of the design conditions at various iterations and the decisions made. It can be seen that for the first five iterations the constraint violations are very large, so the constraint correction (CC) algorithm is used to correct the constraints. At the 6th iteration, it is determined that design variables 5 and 10 are not changing, so they are fixed to their current value. Similarly, at other iterations, variables are assigned values from the available set. At the 14th iteration, all variables have discrete values, the constraint violation is about 1.4% and the structural weight is 5424.69, which is an increase of less than 1% from the true optimum. This is quite a reasonable final solution.

It should be noted that with the discrete variables, several solutions near the true optimum point are possible. A different sequence of fixing variables can give a different solution. For example, starting from the optimum solution with continuous variables, the following acceptable discrete solutions are obtained interactively:

1. 30.0, 1.62, 26.5, 13.9, 1.62, 4.18, 13.5, 18.8, 18.8, 1.62; cost = 5485.6, max. viol. = 4.167% for stress in member 2.
2. Same as (1) except the 8th design variable is 16.9; cost = 5388.9 and max. viol. = 0.58%.

TABLE 8.12
Interactive solution for a ten-member structure with discrete variables

Iteration no.	Maximum violation (%)	Cost function	Algorithm used	Variables fixed to value shown in parentheses
1	1.274E + 04	679.83	CC	All free
2	4.556E + 03	1019.74	CC	All free
3	1.268E + 03	1529.61	CC	All free
4	4.623E + 02	2294.42	CC	All free
5	1.144E + 02	3441.63	CC	All free
6	2.020E + 01	4722.73	CC	5(1.62), 10(1.62)
7	2.418	5389.28	CCC	2(1.80)
11	1.223E − 01	5402.62	SQP	1(30.0), 6(3.84), 7(13.5)
13	5.204E − 04	5411.13	SQP	3(26.5), 9(19.9)
14	1.388	5424.69	—	4(13.5), 8(16.9)

CC, constraint correction algorithm; CCC, constraint correction at constant cost; SQP: sequential quadratic programming.

3. Same as (1) except design variables 2 and 6 are 2.38 and 2.62; cost = 5456.8, max. viol. = 3.74% for stress in member 2.

4. Same as (3) except design variable 2 is 2.62; cost = 5465.4; all constraints are satisfied.

It can be seen that the interactive facilities described in Chapter 7 can be exploited to obtain practical engineering designs.

8.9 OPTIMAL CONTROL OF SYSTEMS BY NONLINEAR PROGRAMMING

8.9.1 A Prototype Optimal Control Problem

Optimal control problems are dynamic in nature as noted in Section 1.6. A brief discussion on differences between optimal control and optimum design problems is given there. It turns out that some optimal control problems can be formulated and solved by the nonlinear programming methods described in Chapters 6 and 7. In this section, we consider a simple optimal control problem that has numerous practical applications. Various formulations of the problem are described and optimal solutions are obtained and discussed.

The application area that we have chosen to demonstrate the use of nonlinear programming methods for this class of problems is the vibration control of systems. This is an important class of problems that is encountered in numerous real-world applications. For example, the control of structures under earthquake and wind loads, vibration control of sensitive instruments to blast loading or shock input, the control of the large space structures, precision control of machines and the alike. To treat these problems we shall consider a simple model of the system to demonstrate the basic formulation and solution procedure. Using the demonstrated procedures, more complex models can be treated to simulate the real-world systems more accurately.

To treat optimal control problems, dynamic response analysis capability must be available. In the present text, we shall assume that students have some background in the vibration analysis of systems. In particular, we shall model systems as single degree of freedom linear spring-mass systems. This leads to a second-order linear differential equation whose closed-from solution is available [Clough and Penzien, 1975]. It will be worthwhile for the students to briefly review the material on the solution of linear differential equations.

To demonstrate the formulation and the solution process, we consider a cantilever structure shown in Fig. 8.4. The data for the problem and various notations used in Fig. 8.4 are defined in Table 8.13. The structure is a highly idealized model of many systems that are used in practice. The length of the structure is L and its cross-section is rectangular with width as b and depth as h. The system is at rest initially at time $t = 0$. It experiences a sudden load due to a shock wave or other such causes. The problem is to control the vibrations of the system such that the displacements are not too large and the system

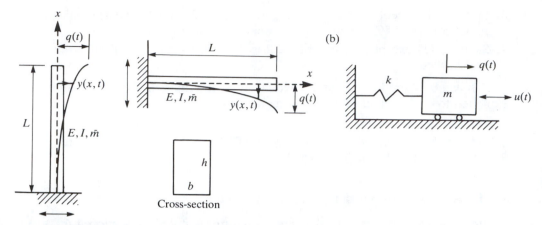

FIGURE 8.4
Model of a system subjected to shock input. (a) Cantilever structures subjected to shock input. (b) Equivalent single degree of freedom model.

comes to rest in a controlled manner. The system has proper sensors and actuators that generate the desired force to suppress the vibrations and bring the system to rest. The control force may also be generated by properly designed dampers or viscoelastic support pads along the length of the structure. We shall not discuss the detailed design of the control force generating mechanisms, but we shall discuss the problem of determining the optimum shape of the control force.

The governing equation that describes the motion of the system is a second-order partial differential equation. To simplify the analysis, we use separation of variables, and express the deflection function $y(x, t)$ as

$$y(x, t) = \psi(x)q(t) \tag{8.26}$$

where $\psi(x)$ is a known function called the shape function, and $q(t)$ is the displacement at the tip of the cantilever, as shown in Fig. 8.4. Several shape

TABLE 8.13
Data for the optimal control problem

Length of the structure	$L = 1.0\,\text{m}$
Width of cross-section	$b = 0.01\,\text{m}$
Depth of cross-section	$h = 0.02\,\text{m}$
Modulus of elasticity	$E = 200\,\text{GPa}$
Mass density	$\rho = 7{,}800\,\text{kg/m}^3$
Moment of inertia	$I = (6.667\text{E}{-}09)\,\text{m}^4$
Mass per unit length	$\bar{m} = 1.56\,\text{kg/m}$
Control function	$u(t) = $ to be determined
Limit on the control function	$u_a = 30\,\text{N}$
Initial velocity	$v_0 = 1.5\,\text{m/s}$

functions can be used; however, we shall use the following one:

$$\psi(x) = \tfrac{1}{2}(3\xi^2 - \xi^3); \qquad \xi = \frac{x}{L} \tag{8.27}$$

Using kinetic and potential energies for the system, $\psi(x)$ of Eq. (8.27), and the data of Table 8.13, the mass and spring constants for an equivalent single degree of freedom system shown in Fig. 8.4 are calculated as follows [Clough and Penzien, 1975]:

Mass

$$\text{kinetic energy} = \frac{1}{2}\int_0^L \bar{m}\dot{y}^2(t)\,dx$$

$$= \frac{1}{2}\left[\int_0^L \bar{m}\psi^2(x)\,dx\right]\dot{q}^2(t)$$

$$= \tfrac{1}{2}m\dot{q}^2(t)$$

where the mass m is identified as

$$m = \int_0^L \bar{m}\psi^2(x)\,dx$$

$$= \tfrac{33}{140}\,\bar{m}L$$

$$= \tfrac{33}{140}\,(1.56)(1.0) = 0.3677 \text{ kg}$$

Spring constant

$$\text{strain energy} = \frac{1}{2}\int_0^L EI[y''(x)]^2\,dx$$

$$= \frac{1}{2}\left[\int_0^L EI(\psi''(x))^2\,dx\right]q^2(t)$$

$$= \tfrac{1}{2}kq^2(t)$$

where the spring constant k is identified as

$$k = \int_0^L EI(\psi''(x))^2\,dx$$

$$= \frac{3EI}{L^3}$$

$$= 3(2.0E{+}11)(6.667E{-}09)/(1.0)^3 = 4000 \text{ N/m}$$

In the foregoing a "dot" over a variable indicates derivatives with respect to time and a "prime" indicates derivatives with respect to the coordinate x.

The equation of motion for the single-degree-of-freedom system along with the initial conditions (initial displacement q_0, initial velocity v_0) are given as

$$m\ddot{q}(t) + kq(t) = u(t) \tag{8.28}$$

$$q(0) = q_0, \qquad \dot{q}(0) = v_0 \tag{8.29}$$

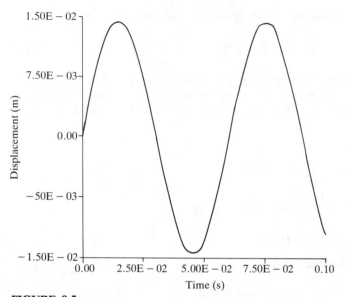

FIGURE 8.5
Displacement response of the equivalent single degree of freedom system to shock loading with no control force.

where $u(t)$ is the control force needed to suppress vibrations due to the initial velocity v_0 (shock loading for the system is transformed to an equivalent initial velocity calculated as impulse of the force divided by the mass). Note that the material damping for the system is neglected. Therefore, if no control force $u(t)$ is used, the system will continue to oscillate. Figures 8.5 and 8.6 show the

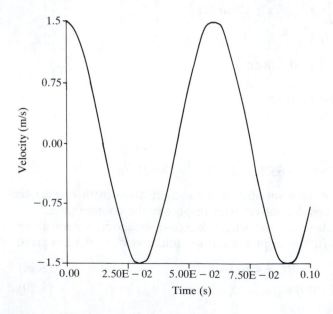

FIGURE 8.6
Velocity response of the equivalent single degree of freedom system to shock loading with no control force.

displacement and velocity response of the system for the initial 0.10 sec when $u(t) = 0$, i.e. no control mechanism is used.

The control problem is to determine the forcing function $u(t)$ such that the system comes to rest in a specified time. We can also pose the problem as follows: determine the control force to minimize the time to bring the system to rest. We shall investigate several of these formulations in the following paragraphs.

We note here that for the preceding simple problem, solution procedures other than the nonlinear programming methods are available [Meirovitch, 1985]. Those procedures may be better for the simple problem. However, we shall use nonlinear programming formulations to solve the problem to demonstrate generality of the method.

8.9.2 Minimization of Error in State Variable

As a first formulation, we define the performance index (cost function) as minimization of error in the state variable (response) in the time interval 0 to T as

$$f_1 = \int_0^T q^2(t) \, dt \tag{8.30}$$

Constraints are imposed on the terminal response, the displacement response and the control force as follows:

Displacement constraint: $|q(t)| \leq q_a$ in the time interval 0 to T (8.31)

Terminal displacement constraint: $q(T) = q_T$ (8.32)

Terminal velocity constraint: $\dot{q}(T) = v_T$ (8.33)

Control force constraint: $|u(t)| \leq u_a$ in the interval $0 \leq t \leq T$ (8.34)

where q_a is the maximum allowed displacement of the system, q_T and v_T are small specified constants, and u_a is the limit on the control force. Thus, the design problem is to compute the control function $u(t)$ in the time interval 0 to T to minimize the performance index of Eq. (8.30) subject to the constraints of Eqs (8.31) to (8.34) and satisfaction of the equations of motion (8.28) and the initial conditions in Eq. (8.29). Note that the constraints of Eqs (8.31) and (8.34) are dynamic in nature and need to be satisfied over the entire time interval 0 to T.

Another performance index can be defined as the sum of the squares of the displacement and the velocity as:

$$f_2 = \int_0^T [q^2(t) + \dot{q}^2(t)] \, dt \tag{8.35}$$

8.9.2.1 FORMULATION FOR NUMERICAL SOLUTION. In order to obtain numerical results, the following data are used:

Allowable time to suppress motion	$T = 0.10\,s$
Initial velocity	$v_0 = 1.5\,m/s$
Initial displacement	$q_0 = 0.0\,m$
Allowable displacement	$q_a = 0.01\,m$
Terminal velocity	$v_T = 0.0\,m/s$
Terminal displacement	$q_T = 0.0\,m$
Limit on the control force	$u_a = 30.0\,N$

For the present example, the equation of motion is quite simple and its analytical solution can be written using Duhamel's integral [Clough and Penzien, 1975] as follows:

$$q(t) = \frac{1}{\omega} v_0 \sin \omega t + q_0 \cos \omega t + \frac{1}{m\omega} \int_0^t u(\eta) \sin \omega(t - \eta)\,d\eta \qquad (8.36)$$

$$\dot{q}(t) = v_0 \cos \omega t - q_0 \omega \sin \omega t + \frac{1}{m} \int_0^t u(\eta) \cos \omega(t - \eta)\,d\eta \qquad (8.37)$$

In more complex applications, the equations of motion will have to be integrated using numerical methods [Shampine and Gordon, 1975; Hsieh and Arora, 1984].

Since explicit forms for the displacement and velocity in terms of the design variable $u(t)$ are known, we can calculate their derivatives explicitly by differentiating Eqs (8.36) and (8.37) with respect to $u(\eta)$, where η is a point between 0 and T, as

$$\frac{dq(t)}{du(\eta)} = \frac{1}{m\omega} \sin \omega(t - \eta) \quad \text{for} \quad t \geq \eta$$

$$= 0 \quad \text{for} \quad t < \eta \qquad (8.38)$$

$$\frac{d\dot{q}(t)}{du(\eta)} = \frac{1}{m} \cos \omega(t - \eta) \quad \text{for} \quad t \geq \eta$$

$$= 0 \quad \text{for} \quad t < \eta \qquad (8.39)$$

In the foregoing expressions, $du(t)/du(\eta) = \delta(t - \eta)$ has been used, where $\delta(t - \eta)$ is the Dirac delta function. The derivative expressions can easily be programmed to impose constraints of the problem. For more general applications, derivatives must be evaluated using numerical computational procedures. Several such procedures developed and evaluated by Hsieh and Arora [1984] and Tseng and Arora [1987] can be used for more complex applications.

Equations (8.36) to (8.39) are used to develop the user-supplied subroutines for the program IDESIGN. Several procedures are needed to solve the problem numerically. First of all, a grid must be used to discretize the time where displacement, velocity and the control force are evaluated.

Interpolation methods, such as cubic splines, B-splines [De Boor, 1978], etc., can be used to evaluate the functions at points other than the grid points.

Another difficulty concerns the dynamic displacement constraint of Eq. (8.31). The constraint must be imposed during the entire time interval 0 to T. Several treatments for such constraints have been investigated [Hsieh and Arora, 1984; Tseng and Arora, 1987]. For example, the constraint can be replaced by several constraints imposed at the local maximum points for the function $q(t)$; it may be replaced by an integral constraint; or it may be imposed at each grid point.

In addition to the foregoing numerical procedures, a numerical integration scheme, such as simple summation, trapezoidal rule, Simpson's rule, Gaussian quadrature, etc., must be selected for evaluating the integrals in Eqs (8.30), (8.36) and (8.37). Based on some preliminary investigations, the following numerical procedures are selected for their simplicity to solve the present problem:

Numerical integration:	Simpson's method
Dynamic constraint:	imposed at each grid point
Design variable (control force):	value at each grid point

8.9.2.2 NUMERICAL RESULTS. Using the foregoing procedures and the numerical data, the problem is solved using the SQP method of Section 6.7 available in the IDESIGN software package. The number of grid points is selected as 41, so there are 41 design variables. The displacement constraint of Eq. (8.31) is imposed at the grid points with its limit set as $q_a = 0.01$ m. As an initial estimate, $u(t)$ is set to zero, so constraints of Eqs (8.31) to (8.33) are violated. The algorithm finds a feasible design in just three iterations. During these iterations, the cost function of Eq. (8.30) is also reduced. The algorithm reaches near to the optimum point at the 11th iteration. Due to severe convergence criteria, it takes another 27 iterations to satisfy the specified convergence criteria. The cost function history is plotted in Fig. 8.7. For all practical purposes the optimum solution is obtained somewhere between the 15th and 30th iterations. Thus, if IDESIGN was being executed in the interactive mode, the designer could terminate the iterative process at the 17th iteration.

The final displacement and velocity responses and the control force history are shown in Figs 8.8–8.10. It can be observed that the displacement and velocity both go to zero at about 0.05 sec, so the system comes to rest at that point. The control force also has zero value after that point and reaches its limit value at several points during that interval. The final cost function value is (8.536E−07).

8.9.2.3 EFFECT OF PROBLEM NORMALIZATION. It turns out that for the present application it is advantageous to normalize the problem and optimize it with normalized variables. We shall briefly discuss these normalizations which

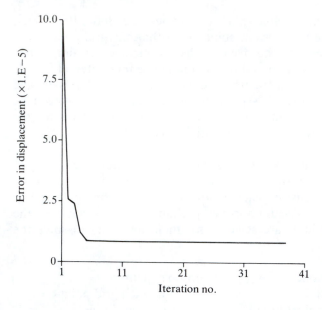

FIGURE 8.7
Cost function history for the optimal control problem of minimizing the error in state variable (cost function f_1).

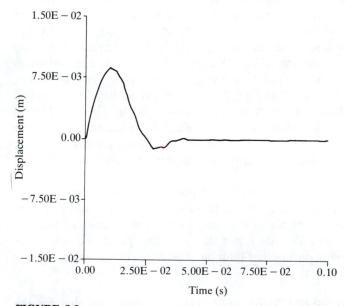

FIGURE 8.8
Displacement response at optimum with minimization of error in state variable as the performance index (cost function f_1).

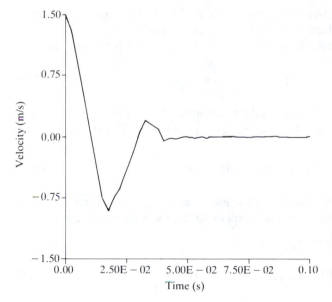

FIGURE 8.9
Velocity response at optimum with minimization of error in state variable as the performance index (cost function f_1).

can also be useful in other applications. Without normalization of the present problem, the cost function and its gradient as well as constraint functions and their gradients have quite small values. The algorithm required a very small value for the convergence parameter (1.0E−09) to converge to the same optimum solution as with the normalized problem. In addition, the rate of convergence without normalization was also quite slow. This apparent numeri-

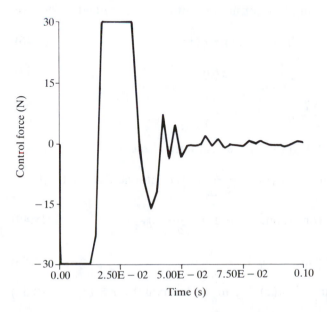

FIGURE 8.10
Optimum control force to minimize error in state variable due to shock input (cost function f_1).

cal difficulty was due to ill-conditioning in the problem which was overcome by the normalization procedure that is described in the following.

The independent variable transformation for the time is defined as

$$t = \tau T \quad \text{or} \quad \tau = \frac{t}{T} \tag{8.40}$$

where τ is the normalized independent variable. With this transformation, when t varies between 0 and T, τ varies between 0 and 1. The displacement is normalized as

$$q(t) = Tq_{max}\bar{q}(\tau) \quad \text{or} \quad \bar{q}(\tau) = \frac{q(t)}{Tq_{max}} \tag{8.41}$$

where $\bar{q}(\tau)$ is the normalized displacement and q_{max} is taken as 0.015. Derivatives of the displacement with respect to time are transformed as

$$\dot{q}(t) = q_{max}\dot{\bar{q}}(\tau)$$

$$\dot{q}(0) = q_{max}\dot{\bar{q}}(0) \quad \text{or} \quad \dot{\bar{q}}(0) = \frac{v_0}{q_{max}} \tag{8.43}$$

$$\ddot{q}(t) = \frac{1}{T}q_{max}\ddot{\bar{q}}(\tau) \tag{8.44}$$

The control force is normalized as

$$u(t) = u_{max}\bar{u}(\tau), \quad \text{or} \quad \bar{u}(\tau) = \frac{u(t)}{u_{max}} \tag{8.45}$$

With this normalization, $\bar{u}(\tau)$ varies between -1 and 1 as $u(t)$ varies between $-u_{max}$ and u_{max}.

Substituting the preceding transformations into Eqs (8.28) and (8.29), we get

$$\bar{m}\ddot{\bar{q}}(\tau) + \bar{k}\bar{q}(\tau) = \bar{u}(\tau) \tag{8.46}$$

$$\bar{q}(0) = \frac{q_0}{Tq_{max}}, \quad \dot{\bar{q}}(0) = \frac{v_0}{q_{max}} \tag{8.47}$$

$$\bar{k} = \frac{kT}{u_{max}}q_{max}, \quad \bar{m} = \frac{m}{Tu_{max}}q_{max}$$

The constraints of Eqs (8.31) to (8.34) are also normalized as

Displacement constraint: $\quad |\bar{q}(\tau)| \leqq \dfrac{q_a}{Tq_{max}}$ in the interval $0 \leqq \tau \leqq 1$ (8.48)

Terminal displacement constraint: $\quad \bar{q}(1) = \dfrac{1}{Tq_{max}}q_T$ (8.49)

Terminal velocity constraint: $\quad \dot{\bar{q}}(1) = \dfrac{1}{q_{max}}v_T$ (8.50)

Control force constraint: $\quad |\bar{u}(\tau)| \leqq 1$ in the interval $0 \leqq \tau \leqq 1$ (8.51)

With the foregoing normalizations, the numerical algorithm behaved considerably better and convergence to the optimum solution as reported earlier was quite rapid. Therefore, for general usage, normalization of the problem is recommended whenever possible. Note that many forms of normalizations of a problem are possible besides the one given in the foregoing. If one form does not work, others should be tried. We shall use the foregoing normalizations in the two formulations discussed in Sections 8.9.3 and 8.9.4.

8.9.2.4 DISCUSSION OF RESULTS. The final solution for the problem can be affected by the number of grid points and convergence criterion. The solution reported previously was obtained using 41 grid points and a convergence criterion of (1.0E−03). A stricter convergence criterion of (1.0E−06) also gave the same solution using a few more iterations.

The number of grid points can also affect the accuracy of the final solution. The use of 21 grid points also gave approximately the same solution. The shape of the final control force was slightly different. The final cost function value was slightly higher than that with 41 grid points.

It is also important to note that the problem can become infeasible if the limit q_a on the displacement in Eq. (8.31) is too severe. For example, when q_a was set to 0.008 m, the problem was infeasible with 41 grid points. However, with 21 grid points a solution was obtained. This also shows that when the number of grid points is small the displacement constraint may actually be violated between the grid points, although it is satisfied at the grid points. Therefore, the number of grid points should be selected judiciously.

The foregoing discussion shows that to impose the constraints more precisely, the exact local max-points should be located and constraint imposed there. To locate the exact max-points, interpolation procedures may be used, or bisection of the interval in which the max-point lies can be used [Hsieh and Arora, 1984]. The gradient of the constraint must be evaluated at the max-points. For the present problem, the preceding procedure is not too difficult to implement because the analytical form for the response is known. For more general applications, the computational as well as programming effort can increase substantially to implement the foregoing procedure.

It is worthwhile to note that several other starting points for the control force such as $u(t) = -30$ N, $u(t) = 30$ N, converged to the same solution as given in Figs 8.8–8.10. The computational effort varied somewhat. The CPU time with 21 grid points was 46 sec and with 41 grid points it was 212 sec on an Apollo DN460 workstation when $u(t) = 0$ was used as the starting point.

It is interesting to note that at the optimum, the dynamic constraint of Eq. (8.31) is not active at any time grid point. It is violated at many intermediate iterations. Also, the terminal response constraints of Eqs (8.32) and (8.33) are satisfied at the optimum with normalized Lagrange multipliers as (−7.97E−04) and (5.51E−05). Since the multipliers are almost zero, the constraints can be somewhat relaxed without affecting the optimal solution.

This can be observed from the final displacement and velocity responses shown in Figs 8.8 and 8.9, respectively. Since the system is essentially at rest after $t = 0.05$ sec, there is no effect of imposing the terminal constraints of Eqs (8.32) and (8.33).

The control force is at its limit value ($u_a = 30$ N) at several grid points; for example, it is at its lower limit at the first six grid points and at the upper limit at the next six. The Lagrange multiplier for the constraint has its largest value initially and gradually decreases to zero after the 13th grid point. According to the Constraint Variation Sensitivity Theorem 3.14, the optimum cost function can be reduced substantially if the limit on the control force is relaxed for a small duration after the system is disturbed.

8.9.3   Minimum Control Effort Problem

Another formulation for the problem is possible where we minimize the total control effort calculated as

$$f_3 = \int_0^T u^2(t) \, dt \qquad (8.52)$$

The constraints are the same as defined in Eqs (8.31) to (8.34) and Eqs (8.28) and (8.29). The numerical procedures for obtaining an optimum solution for the problem are the same as described previously in Section 8.9.2.

This formulation of the problem is quite well behaved. The same optimum solution is obtained quite rapidly (9–27 iterations) with many different starting points. Figures 8.11–8.14 give the cost function history,

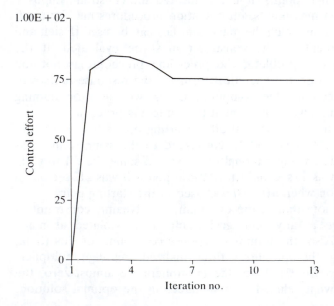

FIGURE 8.11
Cost function history for the optimal control problem of minimization of the control effort (cost function f_3).

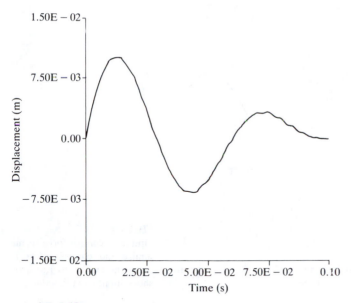

FIGURE 8.12
Displacement response at optimum with minimization of control effort as the performance index (cost function f_3).

displacement and velocity responses, and the control force at the optimum solution which is obtained by starting from $u(t) = 0$ and 41 grid points. The final control effort of 7.481 is much smaller than that for the first case where it was 28.74. The system, however, comes to rest at 0.10 sec compared to 0.05 sec in the previous case. The solution with 21 grid points resulted in a

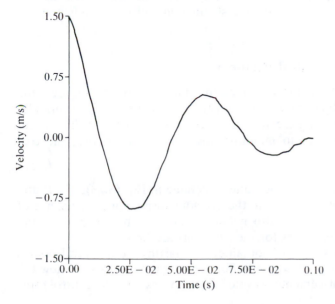

FIGURE 8.13
Velocity response at optimum with minimization of control effort as the performance index (cost function f_3).

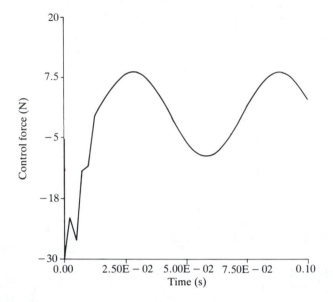

FIGURE 8.14
Optimum control force to minimize the control effort to bring the system to rest after shock input (cost function f_3).

slightly smaller control effort due to the numerical procedures used, as explained earlier.

It is interesting to note that the constraint of Eq. (8.31) is active at the 8th grid point with the normalized Lagrange multiplier as (2.429E−02). The constraints of Eqs (8.32) and (8.33) are also active with the normalized Lagrange multipliers as (−1.040E−02) and (−3.753E−04). In addition, the control force is at its lower limit at the first grid point with the Lagrange multiplier as (7.153E−04). This shows that by increasing or decreasing the limit on the control force, the optimum cost function will not be affected too much.

8.9.4 Minimum Time Control Problem

The idea of this formulation is to minimize the time required to suppress the motion of the system subject to various constraints. In the previous formulations, the desired time to bring the system to rest was specified. In the present formulation, however, we try to minimize the time T. Therefore, the cost function is

$$f_4 = T \tag{8.53}$$

The constraints on the system are the same as defined in Eqs (8.28), (8.29) and (8.31) to (8.34). Note that compared to the previous formulations, gradients of constraints with respect to T are also needed. They can be computed quite easily, since analytical expressions for the functions are known.

The same optimum solution is obtained by starting from several points, such as $T = 0.1$, 0.04, 0.02, and $u(t) = 0$, 30, −30. Figures 8.15–8.18 show the cost function history, displacement and velocity responses, and the control force

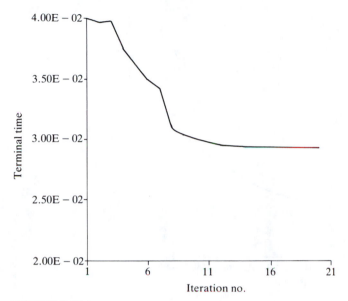

FIGURE 8.15
Cost function history for the optimal control problem of minimization of time to bring the system to rest (cost function f_4).

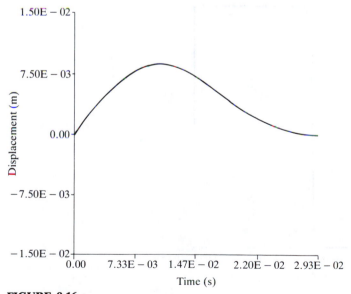

FIGURE 8.16
Displacement response at optimum with minimization of time as the performance index (cost function f_4).

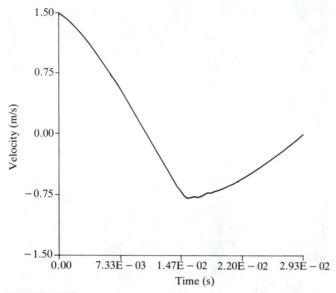

FIGURE 8.17
Velocity response at optimum with minimization of time as the performance index
(cost function f_4).

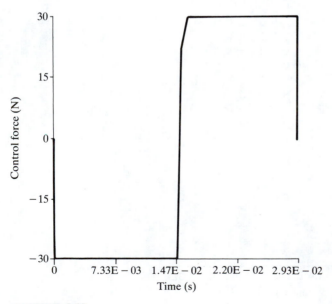

FIGURE 8.18
Optimum control force to minimize time to bring the system to rest after shock input
(cost function f_4).

at the optimum with 41 grid points, $T = 0.04$ and $u(t) = 0$ as the starting point. The CPU time for the entire solution process is 302 sec on the Apollo DN460 workstation. It takes 0.029 33 sec to bring the system to rest. Depending on the starting point, the number of iterations to converge to the final solution vary between 20 and 56.

Constraints of Eqs (8.32) and (8.33) are active with the normalized Lagrange multipliers as $(6.395E{-}02)$ and $(-1.771E{-}01)$ respectively. The control force is at its lower limit for the first 22 grid points and at the upper limit at the remaining points.

It is also interesting to note that even after normalization of the problem, the rate of convergence for the problem could be affected by multiplying the cost function by a positive constant. For example, the number of iterations reduced from 60 to 20 when a factor of 1000 was used. The results reported in the preceding are with the use of this factor of 1000. Apparently, the use of this factor affects the step size determinination process for the algorithm. One conclusion that can be drawn based on the study is that the step size determination process has room for further improvement in the IDESIGN system.

8.9.5 Comparison of Three Formulations for the Optimal Control of System Motion

It is interesting to compare the three formulations for the optimal control of motion of the system shown in Fig. 8.4. Table 8.14 contains a summary of the optimum solutions with the three formulations. All the solutions are obtained with 41 grid points, and $u(t) = 0$ as the starting point using an Apollo DN460 workstation. For the third formulation, $T = 0.04$ sec is used as the starting point.

TABLE 8.14
Summary of optimum solutions for three formulations of optimal control of motion of a system subjected to shock input

	Formulation 1: minimization of error in state variable	Formulation 2: minimization of control effort	Formulation 3: minimization of terminal time
f_1	8.536 07E − 07	2.320 08E − 06	8.644 66E − 07
f_2	1.682 41E − 02	2.735 40E − 02	1.459 66E − 02
f_3	2.873 98E + 01	7.481 04	2.597 61E + 01
f_4	0.10	0.10	2.93 36E − 02
NIT	38	13	20
NCF	38	13	20
NGE	100	68	64
CPU	212	43	302

NIT, number of iterations; NCF, number of calls for function evaluation; NGE, total number of gradients evaluated.

The results of Table 8.14 show that the control effort is largest with the first formulation and the smallest with the second one. The second formulation turns out to be the most efficient as well as convenient to implement. By varying the total time T, this formulation can be used to generate results or Formulation 3. For example, using $T = 0.05$ and 0.029 33 sec, solutions with Formulation 2 were obtained. With $T = 0.029\,33$ sec, the same results as with Formulation 3 were obtained. Also, when $T = 0.025$ sec was used, Formulation 2 resulted in an infeasible problem. For practical applications, Formulation 2 is recommended for the vibration control problems.

EXERCISES FOR CHAPTER 8*

Formulate the following design problems and solve them using a nonlinear programming algorithm starting with a reasonable design estimate. Also solve the problems graphically and trace the history of the iterative process on the graph of the problem:

8.1 Exercise 2.48

8.2 Exercise 2.49

8.3 Exercise 2.50

8.4 Exercise 2.62

8.5 Exercise 2.63

8.6 Exercise 2.64

8.7 Exercise 2.65

8.8 Consider the cantilever beam-mass system shown in Fig. E8.8. Formulate and solve the minimum weight design problem for the rectangular cross-section so that the fundamental vibration frequency is larger than 8 rad/sec and the cross-sectional dimensions satisfy the limitations

$$0.5 \leq b \leq 1.0, \quad \text{in}$$

$$0.2 \leq h \leq 2.0, \quad \text{in}$$

Use a nonlinear programming algorithm to solve the problem. Verify the solution graphically and trace the history of the iterative process on the graph of the problem. Let the starting point be $(0.5, 0.2)$. The data and various equations for the problem are as follows:

Fundamental vibration frequency	$\omega = \sqrt{k_e/m}$	rad/s
Equivalent spring constant, k_e	$\dfrac{1}{k_e} = \dfrac{1}{k} + \dfrac{L^3}{3EI}$	
Mass attached to the spring	$m = \dfrac{W}{g}$	
Weight attached to the spring	$W = 50$ lb	
Length of the beam	$L = 12$ in	
Modulus of elasticity	$E = (3.00\mathrm{E}{+}07)$ psi	
Spring constant	$k = 10$ lb/in	
Moment of inertia	I, in^4	
Gravitational constant	g, in/sec^2	

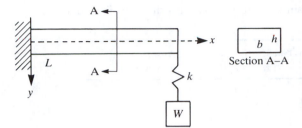

FIGURE E8.8
Cantilever beam with spring-mass at the free end.

8.9 A prismatic steel beam with symmetric I cross-section is shown in Fig. E8.9. Formulate and solve the minimum weight design problem subject to the following constraints:

1. The maximum axial stress due to combined bending and axial load effects should not exceed 100 MPa.
2. The maximum shear stress should not exceed 60 MPa.
3. The maximum deflection should not exceed 15 mm.
4. The beam should be guarded against lateral buckling.
5. Design variables should satisfy the limitations $b \geq 100$ mm, $t_1 \leq 10$ mm, $t_2 \leq 15$ mm, $h \leq 150$ mm

Solve the problem using a numerical optimization method, and verify the solution using K–T necessary conditions for the following data:

Modulus of elasticity	$E = 200$ GPa
Shear modulus	$G = 70$ GPa
Load	$P = 70$ kN
Load angle	$\theta = 45°$
Beam length	$L = 1.5$ m

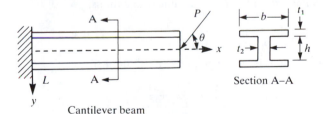

Cantilever beam

Section A–A

FIGURE E8.9
Cantilever I beam. Design variables: b, t_1, t_2 and h.

8.10 *Shape optimization of a structure.* The design objective is to determine the shape of the three-bar structure shown in Fig. E8.10 to minimize its weight [Corcoran, 1970]. The design variables for the problem are the member cross-sectional areas A_1, A_2 and A_3 and the coordinates of nodes A, B, and C (note that x_1, x_2 and x_3 have positive values in the figure; the final values can be positive or negative), so that the truss is as light as possible while satisfying the stress constraints due to the following three loading conditions:

Cond. No. j	Load P_j (lb)	Angle θ_j (degrees)
1	40 000	45
2	30 000	90
3	20 000	135

The stress constraints are written as

$$-5000 \leq \sigma_{1j} \leq 5000, \qquad \text{psi}$$
$$-20\,000 \leq \sigma_{2j} \leq 20\,000, \qquad \text{psi}$$
$$-5000 \leq \sigma_{3j} \leq 5000, \qquad \text{psi}$$

where $j = 1, 2, 3$ represents the index for the three loading conditions and the stresses are calculated from the following expressions:

$$\sigma_{1j} = \frac{E}{L_1}[u_j \cos \alpha_1 + v_j \sin \alpha_1] = \frac{E}{L_1^2}(u_j x_1 + v_j L)$$

$$\sigma_{2j} = \frac{E}{L_2}[u_j \cos \alpha_2 + v_j \sin \alpha_2] = \frac{E}{L_2^2}(u_j x_2 + v_j L)$$

$$\sigma_{3j} = \frac{E}{L_3}[u_j \cos \alpha_3 + v_j \sin \alpha_3] = \frac{E}{L_3^2}(-u_j x_3 + v_j L)$$

where $L = 10$ in and

$$L_1 = \text{length of member 1} = \sqrt{L^2 + x_1^2}$$
$$L_2 = \text{length of member 2} = \sqrt{L^2 + x_2^2}$$
$$L_3 = \text{length of member 3} = \sqrt{L^2 + x_3^2}$$

and u_j and v_j are the horizontal and vertical displacements for the jth loading condition determined from the following linear equations:

$$\begin{bmatrix} k_{11} & k_{12} \\ k_{21} & k_{22} \end{bmatrix} \begin{bmatrix} u_j \\ v_j \end{bmatrix} = \begin{bmatrix} P_j \cos \theta_j \\ P_j \sin \theta_j \end{bmatrix}, \qquad j = 1, 2, 3$$

where the stiffness coefficients are given as ($E = 3.0\text{E} + 07$ psi)

$$k_{11} = E\left(\frac{A_1 x_1^2}{L_1^3} + \frac{A_2 x_2^2}{L_2^3} + \frac{A_3 x_3^2}{L_3^3}\right)$$

$$k_{12} = E\left(\frac{A_1 L x_1}{L_1^3} + \frac{A_2 L x_2}{L_2^3} - \frac{A_3 L x_3}{L_3^3}\right) = k_{21}$$

$$k_{22} = E\left(\frac{A_1 L^2}{L_1^3} + \frac{A_2 L^2}{L_2^3} + \frac{A_3 L^2}{L_3^3}\right)$$

Formulate the design problem and find the optimum solution starting from the point

$$A_1 = 6.0, \ A_2 = 6.0, \ A_3 = 6.0$$
$$x_1 = 5.0, \ x_2 = 0.0, \ x_3 = 5.0$$

Compare the solution with that given in Table 8.7.

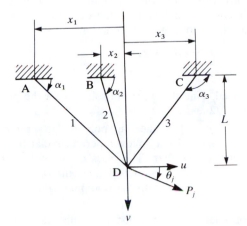

FIGURE E8.10
Three-bar structure–shape optimization.

8.11 *Design synthesis of a nine-speed gear drive.* The arrangement of a nine-speed gear train is shown in Fig. E8.11. The objective of the synthesis is to find the size of all gears from the mesh and speed ratio equations such that the size of the largest gears are kept to a minimum [Osman, Sankar and Dukkipati, 1978]. Due to the mesh and speed ratio equations, it is found that only the following

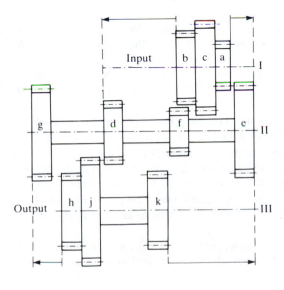

FIGURE E8.11
Schematic arrangement of a nine-speed gear train.

three independent parameters need to be selected:

$$x_1 = \text{gear ratio, } d/a$$
$$x_2 = \text{gear ratio, } e/a$$
$$x_3 = \text{gear ratio, } j/a$$

Due to practical considerations, it is found that the minimization of $|x_2 - x_3|$ results in the reduction of the cost of manufacturing the gear drive.

The gear sizes must satisfy the following mesh equations:

$$\phi^2 x_1(x_1 + x_3 - x_2) - x_2 x_3 = 0$$
$$\phi^3 x_1 - x_2(1 + x_2 - x_1) = 0$$

where ϕ is the step ratio in speed. Find the optimum solution for the problem for two different values of ϕ as $\sqrt{2}$ and $(2)^{1/3}$.

8.12 *Design of a tapered flag pole.* Formulate the flag pole design problem of Exercise 2.63 for the data given there. Use a hollow tapered circular tube with constant thickness as the structural member. The mass of the pole is to be minimized subject to various constraints. Use a numerical optimization method to obtain the final solution and compare it with the optimum solution for the uniform flag pole.

8.13 *Design of a sign support.* Formulate the sign support column design problem described in Exercise 2.64 for the data given there. Use a hollow tapered circular tube with constant thickness as the structural member. The mass of the pole is to be minimized subject to various constraints. Use a numerical optimization method to obtain the final solution and compare it with the optimum solution for the uniform column.

8.14 Repeat the problem of Exercise 8.12 for a hollow square tapered column of uniform thickness.

8.15 Repeat the problem of Exercise 8.13 for a hollow square tapered column of uniform thickness.

8.16 For the optimal control problem of minimization of error in the state variable formulated and solved in Section 8.9.2, study the effect of changing the limit on the control force (u_a) to 25 N or 35 N.

8.17 For the minimum control effort problem formulated and solved in Section 8.9.3, study the effect of changing the limit on the control force (u_a) to 25 N or 35 N.

8.18 For the minimum time control problem formulated and solved in Section 8.9.4, study the effect of changing the limit on the control force (u_a) to 25 N or 35 N.

8.19 For the optimal control problem of minimization of error in the state variable formulated and solved in Section 8.9.2, study the effect of having an additional lumped mass M at the tip of the beam ($M = 0.05$ kg) as shown in Fig. E8.19.

8.20 For the minimum control effort problem formulated and solved in Section 8.9.3, study the effect of having an additional mass M at the tip of the beam ($M = 0.05$ kg).

8.21 For the minimum time control problem formulated and solved in Section 8.9.4, study the effect of having an additional lumped mass M at the tip of the beam ($M = 0.05$ kg).

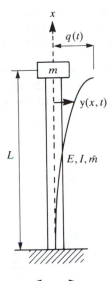

FIGURE E8.19
Cantilever structure with mass at the tip.

8.22 For Exercise 8.19, what will be the optimum solution if th tip mass M is treated as a design variable with limits on it as $0 \leqq M \leqq 0.10$ kg?

8.23 For Exercise 8.20, what will be the optimum solution if the tip mass M is treated as a design varaible with limits on it as $0 \leqq M \leqq 0.10$ kg?

8.24 For Exercise 8.21, what will be the optimum solution if the tip mass M is treated as a design variable with limits on it as $0 \leqq M \leqq 0.10$ kg?

8.25 For the optimal control problem of minimization of error in the state variable formulated and solved in Section 8.9.2 study the effect of including a 1% critical damping in the formulation.

8.26 For the minimum control effort problem formulated and solved in Section 8.9.3, study the effect of including a 1% critical damping in the formulation.

8.27 For the minimum time control problem formulated and solved in Section 8.9.4, study the effect of including a 1% critical damping in the formulation.

8.28 For the spring-mass-damper system shown in Fig. E8.28, formulate and solve the problem of determining the spring constant and damping coefficient to minimize the maximum acceleration of the system over a period of 10 sec when it is subjected to an initial velocity of 5 m/sec. The mass is specified as 5 kg.

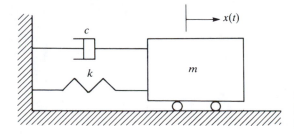

FIGURE E8.28
Damped single degree of freedom system.

The displacement of the mass should not exceed 5 cm for the entire time interval of 10 sec. The spring constant and the damping coefficient must also remain within the limits $1000 \leqq k \leqq 3000$ N/m; $0 \leqq c \leqq 300$ N.S/m. (*Hint*: The objective of minimizing the maximum acceleration is a min–max problem which can be converted to a nonlinear programming problem by introducing an artificial design variable. Let $a(t)$ be the acceleration and A be the artificial variable. Then the objective can be to minimize A subject to an additional constraint $|a(t)| \leqq A$ for $0 \leqq t \leqq 10$).

APPENDIX

A

ECONOMIC ANALYSIS

The main body of this text describes promising analytical and numerical techniques for engineering design optimization. *This appendix departs from the main theme and contains an introduction to engineering decision making based on economic considerations.* More detailed treatment of the subject can be found in texts by Grant, Ireson and Leavenworth [1982], Stark and Nicholls [1972] and Blank and Tarquin [1983].

A.1 TIME VALUE OF MONEY

Engineering systems are designed to perform specific tasks. Usually many alternative designs can perform the same task. The question is, which one of the alternatives is the best? Several factors such as precedents, social environment, and esthetic, economic and psychological values can influence the final selection. This appendix considers only the economic factors influencing the selection of an alternative.

Economic problems are an integral part of engineering because engineers are sensitive to the direct cost of a design. They must anticipate maintenance and operating costs. Future economic conditions must also be taken into account in the decision making process. *We shall discuss ways to measure value of money to enable comparisons of alternative designs.* The following notation

521

is used:

n = number of interest periods, e.g. months, years;

i = *return per dollar per period*; note that i is not the annual interest rate. This shall be further explained in examples;

P = *value or sum of money at the present time, in dollars*;

S_n = *final sum* after n periods or n payments from the present date, in dollars;

R = *a series of consecutive, equal, end-of-period amounts of money –* payment or receipt; e.g. dollars per month, dollars per year, etc.

It is important to understand the notation and the meaning of the symbols for correctly interpreting and solving the examples and the exercises. For example, i must be interpreted as the rate of return per dollar per period and not the annual interest rate, and R is the end-of-period amount and not at the beginning of the period. *It is important to note that we shall quote the annual interest rate in examples and exercises*; using that one can calculate i.

Consider an investment of P dollars which returns i dollars per dollar per period. The return at the end of the first period is iP, and the original investment increases to $(1 + i)P$. This sum is reinvested and returns $i(1 + i)P$ at the end of the next period, so that the original amount is worth $(1 + i)^2P$, etc. If the process is continued for n periods, an original investment P will increase to the final sum S_n, given by

$$S_n = (1 + i)^n P = [\text{spcaf}\,(i, n)]P \tag{A.1}$$

where spcaf (i, n) is called the *single payment compound amount factor*.

Example A.1 **Use of single payment compound amount factor.** Consider an investment of $1000 at an annual interest rate of 9% compounded monthly. Calculate the final sum at the end of 2 and 4 years.

Solution. For the given annual interest rate, the rate of return per dollar per month is $i = 0.09/12 = 0.0075$. The final sum on an investment of $1000 at the end of 2 years ($n = 24$) using the single payment compound amount factor of Eq. (A.1) will be

$$S_{24} = \text{spcaf}\,(0.0075, 24)(1000)$$

$$= (1 + 0.0075)^{24}(1000) = (1.196\,41)(1000) = \$1196.41$$

and at the end of 4 years ($n = 48$) it will become

$$S_{48} = (1 + 0.0075)^{48}(1000)$$

$$= (1.431\,41)(1000) = \$1431.41 \qquad \|$$

A *future payment* S_n made at the end of the nth period has an equivalent present worth P, which can be calculated by inverting Eq. (A.1) as

$$P = (1 + i)^{-n}S_n = [\text{sppwf}\,(i, n)]S_n \tag{A.2}$$

where sppwf (i, n) is called the *single payment present worth factor*. Note from Eq. (A.1) that sppwf (i, n) is the reciprocal of spcaf (i, n).

Example A.2   Use of single payment present worth factor. Consider the case of a person who wants to borrow some money from the bank but can pay back only $10 000 at the end of 2 years. How much can the bank lend if the prevailing annual interest rate is 12% compounded monthly?

Solution. Using the given rate of interest, the rate of return per dollar per period for this example is $i = 0.12/12 = 0.01$. Using the single payment present worth factor of Eq. (A.2), the present worth of $10 000 paid at the end of 2 years $(n = 24)$ is given as $P = [\text{sppwf}\,(0.01, 24)](10\,000)$:

$$P = (1 + 0.01)^{-24}(10\,000)$$
$$= 0.787\,566(10\,000) = \$7876.66$$

Thus, the bank can lend only $7876.66 at the present time.   ‖

Consider a sequence of n *uniform periodic payments R made at the end of each period*. The first payment made at the end of the first period earns interest over $(n - 1)$ periods. Therefore, from Eq. (A.1), it is equivalent to an amount $(1 + i)^{n-1}R$ at the end of the nth period. The second payment made at the end of the second period earns interest over $(n - 2)$ periods and is worth $(1 + i)^{n-2}R$ at the end of the nth period; and so on. This sequence of payments is equivalent to a sum S_n, given by the finite geometric series,

$$S_n = (1 + i)^{n-1}R + \ldots + (1 + i)R + R$$
$$= [(1 + i)^{n-1} + \ldots + (1 + i) + 1]R$$
$$= \frac{1}{i}[(1 + i)^n - 1]R = [\text{uscaf}\,(i, n)]R \tag{A.3}$$

where uscaf (i, n) is the *uniform series compound amount factor*. Likewise, a future sum S_n can be expressed as an equivalent series of uniform payments R by inverting the expression in Eq. (A.3):

$$R = \frac{iS_n}{[(1 + i)^n - 1]} = \text{sfdf}\,(i, n)S_n \tag{A.4}$$

where sfdf (i, n) is called the *sinking fund deposit factor*.

It is important to note in Eqs (A.3) and (A.4), that

1. n is the number of interest periods and the first payment occurs at the end of the first period.
2. The final sum S_n at the end of the nth period includes the final nth payment.

Example A.3   Use of uniform series compound amount factor. Consider the case of a person who decides to deposit $50 at the end of each month for the next 10 years. The prevailing annual rate of interest is 9% compounded monthly. How much will have accumulated at the end of a 10-year period?

Solution. Since the interest is compounded monthly, the value of i for this problem is $0.09/12 = 0.0075$. Using the uniform series compound amount factor of Eq. (A.3), the final sum S_{120} is given as $S_{120} = [\text{uscaf}\,(0.0075, 120)](50)$:

$$S_{120} = \frac{[(1 + 0.0075)^{120} - 1]}{0.0075}(50) \qquad (50)$$

$$= (193.514\,28)(50) = \$9675.71$$

Note that the final sum S_{120} includes the final payment made at the end of the tenth year. ∥

Example A.4   Use of sinking fund deposit factor. A person promises to pay the bank \$10 000 at the end of 2 years. How much money can the bank lend per month if the annual interest rate is 12% compounded monthly?

Solution. Since the annual interest rate is 12%, the rate of return per dollar per period for this problem is $i = 0.12/12 = 0.01$. Using the sinking fund deposit factor of Eq. (A.4), the amount received at the end of each month is given as $R = [\text{sfdf}\,(0.01, 24)](10\,000)$:

$$R = \frac{0.01}{[(1 + 0.01)^{24} - 1]}(10\,000)$$

$$= (0.037\,073)(10\,000) = \$370.73$$

Note that the final payment from the bank occurs at the end of 2 years and at that time a payment of \$10 000 must also be made to the bank. ∥

The sequence of n payments R made at the end of each period can also be expressed as a present worth P. Combining Eqs (A.2) and (A.3) yields

$$P = \frac{1}{i}[1 - (1 + i)^{-n}]R = [\text{uspwf}\,(i, n)]R \qquad (A.5)$$

where uspwf (i, n) is called the *uniform series present worth factor.*

Finally, expressing a present amount P as an equivalent sequence of n uniform payments made at the end of each period gives (from Eq. A.5):

$$R = \frac{iP}{[1 - (1 + i)^{-n}]} = [\text{crf}\,(i, n)]P \qquad (A.6)$$

where crf (i, n) is called the *capital recovery factor.* Table A.1 summarizes all the factors.

Example A.5   Use of uniform series present worth factor. A person promises to pay the bank \$100 per month for the next 2 years. If the annual interest rate is 15% compounded monthly, how much can the bank afford to lend at the present?

Solution. Since the interest is compounded monthly, the value of i for this example is $0.15/12 = 0.0125$. Using the uniform series present worth factor of Eq. (A.5), the present worth of \$100 paid at the end of each month beginning with

TABLE A.1
Interest formulas

To find	Given	Multiply by
S_n	P	Single payment compound amount factor[a] (spcaf), $(1+i)^n$
P	S_n	Single payment present worth factor (sppwf), $(1+i)^{-n}$
S_n	R	Uniform series compound amount factor (uscaf), $\frac{1}{i}[(1+i)^n-1]$
R	S_n	Sinking fund deposit factor (sfdf), $\dfrac{i}{[(1+i)^n-1]}$
P	R	Uniform series present worth factor (uspwf), $\frac{1}{i}[1-(1+i)^{-n}]$
R	P	Capital recovery factor (crf), $\dfrac{i}{[1-(1+i)^{-n}]}$

[a] That is, $S_n = [\text{spcaf}\,(i, n)]P = (1+i)^n P$.

the first one and ending with the 24th, is $P = [\text{uspwf}\,(0.0125, 24)](100)$:

$$P = [1 - (1+0.0125)^{-24}](100)/0.0125$$
$$= (20.6242)(100) = \$2062.42 \qquad \|$$

Example A.6   Use of capital recovery factor. A person puts \$10 000 in the bank and would like to withdraw a fixed sum of money at the end of each month over the next 2 years until the fund is depleted. How much can be withdrawn every month at an annual interest rate of 9% compounded monthly?

Solution. Since the interest is compounded monthly, $i = 0.09/12 = 0.0075$. Using the capital recovery factor of Eq. (A.6), we can find the amount that can be withdrawn at the end of every month for the next 2 years as $R = [\text{crf}\,(0.0075, 24)](10\,000)$:

$$R = \frac{0.0075(10\,000)}{[1 - (1+0.0075)^{-24}]}$$
$$= (0.045\,685)(10\,000) = \$456.85 \qquad \|$$

A.2   CASH FLOW DIAGRAMS

A cash flow diagram is a pictorial representation of cash receipts and disbursements. These diagrams can be helpful in solving problems of economic analysis. Once a correct cash flow diagram for the problem has been drawn, it is a simple matter of using proper interest formulas to perform calculations. In this section, we introduce the idea of a cash flow diagram.

Figure A.1 gives a cash flow diagram from two points of view – the lender's and the borrower's. In the diagram, a person has borrowed a sum of

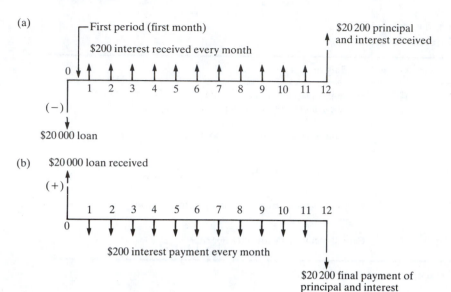

FIGURE A.1
Cash-flow diagrams. (a) Lender's cash flow; (b) borrower's cash flow.

$20 000 and promises to pay it back in 1 year, with simple interest paid every month. The annual interest rate is 12%. Therefore, $200 is paid as interest every month and $20 000 is paid at the end of the 12th month. *Note that vertical lines with arrows pointing downwards imply disbursements, and with arrows pointing upwards imply receipts.* Also, disbursements are shown below (as negative) and receipts (as positive) above the horizontal line.

A.3 ECONOMIC BASES FOR COMPARISON

The formulae given in Table A.1 can be employed in making economic comparisons of alternatives. *Two methods of comparison commonly used are the annual cost (AC) and the present worth (PW) methods.* We shall describe both methods. It is important to realize that the same conditions must be used to compare various alternatives. However, annual base method allows us to compare easily alternatives having different life spans. This will be illustrated with examples. Also, both methods lead to the same conclusion, so either one may be used to compare alternatives.

A note about the *sign convention* used in comparing alternatives is in order. When most of the transactions involve disbursements or costs, positive sign is used for costs and negative sign for receipts in calculating annual costs or present worths. In that case, salvage is considered as a negative cost and any other income is also given a negative sign. With this sign convention, present worth greater than zero actually implies present cost, so an alternative with smaller present worth is to be preferred. If present worth has a negative sign, then it actually represents income. Also, in this case, an alternative with smallest present worth taking into account the algebraic sign is to be preferred, i.e. an alternative with largest numerical value for the present worth. The final selection of an alternative does not depend on the sign convention used in calculations, so any consistent convention can be used. However, we shall use the preceding sign convention in all calculations.

A.3.1 Annual Base Comparisons

An annual base comparison reduces all revenues and expenditures over the selected time to an equivalent annual value. Recall that positive sign will be used for costs and negative sign for income. Therefore, the alternative with lower cost is to be preferred.

> **Example A.7 Alternate Designs.** A design project has two options, A and B. Option A will cost \$280 000 and Option B \$250 000. Annual operating and maintenance paid at the end of each year will be \$8000 for A and \$10 000 for B. Using the *annual cost (AC) method* of comparison with a 12% interest rate, which option should be chosen if both have a 50-year life with no salvage?
>
> *Solution.* The cash flow diagrams for the two options are shown in Fig. A.2. The annual cost of Option A is the sum of the annual maintenance cost and the equivalent uniform payment of the initial cost (\$280 000). The initial cost can be converted to equivalent yearly payment using the capital recovery factor. Thus, the annual cost (AC_A) of Option A is given as
>
> $$AC_A = 280\,000\,\text{crf}\,(0.12, 50) + 8000$$
> $$= \$41\,716.67$$
>
> since crf $(0.12, 50) = 0.120\,42$. Similarly, for Option B, the annual cost is
>
> $$AC_B = 250\,000\,\text{crf}\,(0.12, 50) + 10\,000$$
> $$= \$40\,104.17$$

On this basis Option B is cheaper. ‖

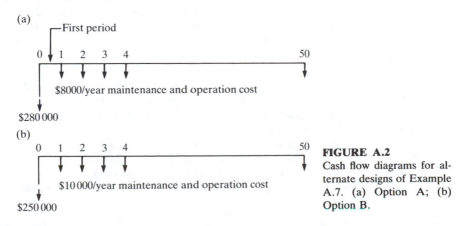

FIGURE A.2
Cash flow diagrams for alternate designs of Example A.7. (a) Option A; (b) Option B.

Example A.8   Alternate power stations. Three companies have submitted bids shown in the following table for the design and operation of a temporary power station which will be used for 4 years. Which design should be used by annual base comparison, if a 15% return is required and if all the equipment can be resold after 4 years for 30% of the initial cost?

	A	B	C
Initial cost ($)	5000	6000	7500
Annual cost ($)	2500	2000	1800

Solution. This example is slightly different from Example A.7 in that the salvage value must also be included while calculating the annual cost. Salvage is an income received at the end of the project. This future sum must be converted to an equivalent yearly income and subtracted from the expenses. A future sum is converted to equivalent yearly value using the sinking fund deposit factor (sfdf). Therefore, the annual cost of Bid A is given as

$$AC_A = 5000 \, \text{crf} \, (0.15, 4) + 2500 - 0.3(5000) \, \text{sfdf} \, (0.15, 4)$$
$$= \$3950.93$$

since crf $(0.15, 4) = 0.350\,27$ and sfdf $(0.15, 4) = 0.200\,27$. Similarly, the annual costs of Bids B and C are

$$AC_B = 6000 \, \text{crf} \, (0.15, 4) + 2000 - 0.3(6000) \, \text{sfdf} \, (0.15, 4)$$
$$= \$3741.11$$
$$AC_C = 7500 \, \text{crf} \, (0.15, 4) + 1800 - 0.3(7500) \, \text{sfdf} \, (0.15, 4)$$
$$= \$3976.39$$

Based on these calculations, Bid B is the cheapest option. ||

Example A.9   Alternate Quarries. A company can purchase either of two mineral quarries. Quarry A costs $600\,000$, is estimated to last 12 years, and would have a land-salvage value at the end of 12 years of $120\,000$. Digging and shipping operations would cost $50\,000$ per year. Quarry B costs $900\,000$, would last 20 years, and would have a salvage value of $60\,000$. Digging and shipping would cost $40\,000$ per year. Which quarry should be purchased? Use the annual base method with an interest rate of 15%. Assume that similar quarries will be available in the future.

Solution. Note for the example that the life spans of Quarries A and B are different. This causes no problem when the annual base method is used. However, with the present worth method of the next section, we shall have to somehow use the same life spans for the two options.

The cash flow diagrams for the quarries are shown in Fig. A.3. To calculate the annual cost of Quarry A, we need to find the annual cost of an initial investment of $600\,000$ using the capital recovery factor (crf), and the equivalent annual income due to salvage of $120\,000$ after 12 years. For this income, the

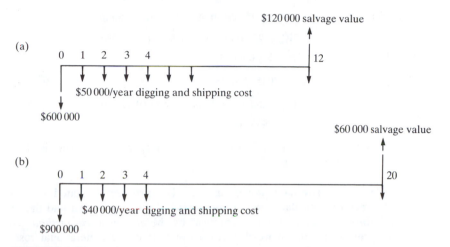

FIGURE A.3
Cash-flow diagrams for the mineral quarries of Example A.9. (a) Quarry A; (b) Quarry B.

sinking fund deposit factor (sfdf) will be used. Therefore,

$$AC_A = 600\,000\ \mathrm{crf}\,(0.15, 12) + 50\,000 - 120\,000\ \mathrm{sfdf}\,(0.15, 12)$$
$$= \$156\,550.77$$

since $\mathrm{crf}\,(0.15, 12) = 0.184\,48$ and $\mathrm{sfdf}\,(0.15, 12) = 0.034\,481$. Similarly,

$$AC_B = 900\,000\ \mathrm{crf}\,(0.15, 20) + 40\,000 - 60\,000\ \mathrm{sfdf}\,(0.15, 20)$$
$$= \$183\,199.64$$

since $\mathrm{crf}\,(0.15, 20) = 0.159\,76$ and $\mathrm{sfdf}\,(0.15, 20) = 0.009\,761\,5$. Based on these calculations, Quarry A is a better investment. ‖

A.3.2 Present Worth Comparisons

In present worth (PW) comparisons, all anticipated revenues and expenditures are expressed by their equivalent present values. The same life spans for all the options must be used for valid comparisons. The same sign convention as before shall be used, i.e. positive sign for costs and negative sign for receipts. Note also, that in most problems, the present worth of a project is actually its total present cost. Therefore, an alternative with lower present worth is to be preferred. We will solve the examples of the previous subsection again using the present worth method.

Example A.10 Alternate designs. The problem is stated in Example A.7. The cash flow diagrams for the problem are shown in Fig. A.2. We will calculate the present worth of the two designs and compare them. To calculate the present worth of Design A, we need to convert the annual maintenance cost of $8000 to its present value. For this we use the uniform series present worth factor (uspwf).

Therefore, the present worth of design A (PW_A) is calculated as

$$PW_A = 280\,000 + 8000 \text{ uspwf } (0.12, 50)$$
$$= \$346\,435.99$$

since uspwf $(0.12, 50) = 8.3045$. Similarly,

$$PW_B = 250\,000 + 10\,000 \text{ uspwf } (0.12, 50)$$
$$= \$333\,044.99$$

Based on these calculations, Design B is a cheaper option. This is the same conclusion as in Example A.7. ∥

Example A.11 Alternate power stations. The problem is stated in Example A.8. We need to calculate the present worths of the annual cost and the salvage value in order to compare the alternatives by the *present worth method*. We use the uniform series present worth factor (uspwf) to convert the annual cost to its present value. The single payment present worth factor (sppwf) is used to convert the salvage value to its present worth. Therefore,

$$PW_A = 5000 + 2500 \text{ uspwf } (0.15, 4) - 0.3\,(5000) \text{ sppwf } (0.15, 4)$$
$$= \$11\,279.82$$

Since uspwf $(0.15, 4) = 2.8550$ and sppwf $(0.15, 4) = 0.571\,75$. Similarly,

$$PW_B = 6000 + 2000 \text{ uspwf } (0.15, 4) - 0.3(6000) \text{ sppwf } (0.15, 4)$$
$$= \$10\,680.80$$
$$PW_C = 7500 + 1800 \text{ uspwf } (0.15, 4) - 0.3\,(7500) \text{ sppwf } (0.15, 4)$$
$$= \$11\,352.52$$

Therefore, Bid B is the cheapest option based on these calculations. This is the same conclusion as in Example A.8. ∥

Example A.12 Alternate quarries. The problem is stated in Example A.9. The cash flow diagrams for the two quarries are shown in Fig. A.3. The life span of the two available options is different. We must somehow use the equivalent present worths to compare the alternatives. Several procedures can be used. We shall demonstrate two of them.

Procedure 1. The basic idea here is to calculate the present worth of Quarry B using a 12-year life span which is the life of Quarry A. We do this by first calculating the annual cost of Quarry B using a 20-year life span. We then calculate the present worth of the annual cost for the first 12 years only and compare it to the present worth of Quarry A:

$$PW_A = 600\,000 + 50\,000 \text{ uspwf } (0.15, 12) - 120\,000 \text{ sppwf } (0.15, 12)$$
$$= \$848\,602.09$$

since uspwf $(0.15, 12) = 5.420\,62$ and sppwf $(0.15, 12) = 0.186\,91$. Calculating the annual cost of B, we get

$$AC_B = 900\,000 \text{ crf } (0.15, 20) + 40\,000 - 60\,000 \text{ sfdf } (0.15, 20)$$
$$= \$183\,199.64$$

since crf $(0.15, 20) = 0.159\,76$ and sfdf $(0.15, 20) = 0.009\,761\,5$. Now, we can calculate the present worth of B using a 12-year life span, as

$$PW_B = 183\,199.64 \text{ uspwf } (0.15, 12)$$
$$= \$993\,055.42$$

Therefore, Quarry A offers a cheaper solution.

Procedure 2. Since it is known that similar quarries would be available in the near future, we can use 5 A and 3 B quarries to get a total life of 60 years for both options. All future investments, expenses and salvage values must be converted to their present worth. Using this procedure, the present worth of Quarries A and B are calculated as

$$PW_A = 600\,000 + 50\,000 \text{ uspwf } (0.15, 60) - 120\,000 \text{ sppwf } (0.15, 12)$$
$$+ 600\,000 \text{ sppwf } (0.15, 12) - 120\,000 \text{ sppwf } (0.15, 24)$$
$$+ 600\,000 \text{ sppwf } (0.15, 24) - 120\,000 \text{ sppwf } (0.15, 36)$$
$$+ 600\,000 \text{ sppwf } (0.15, 36) - 120\,000 \text{ sppwf } (0.15, 48)$$
$$+ 600\,000 \text{ sppwf } (0.15, 48) - 120\,000 \text{ sppwf } (0.15, 60)$$
$$= 600\,000 + 333\,257.30 + 89\,715.43 + 16\,768.46 + 3134.14$$
$$+ 585.79 - 27.37$$
$$= \$1\,043\,433.7$$
$$PW_B = 900\,000 + 40\,000 \text{ uspwf } (0.15, 60) - 60\,000 \text{ sppwf } (0.15, 20)$$
$$+ 900\,000 \text{ sppwf } (0.15, 20) - 60\,000 \text{ sppwf } (0.15, 40)$$
$$+ 900\,000 \text{ sppwf } (0.15, 40) - 60\,000 \text{ sppwf } (0.15, 60)$$
$$= 900\,000 + 266\,605.84 + 51\,324.23 + 3135.93 - 13.69$$
$$= \$1\,221\,052.30$$

Therefore, with this procedure also, Quarry A is a cheaper option. $\parallel$

Use of the annual base or present worth comparison is dependent on the problem and the available information. For some problems the annual base comparison is better suited while others lend themselves to the present worth method. For each problem a suitable method should be selected.

There are many other factors that must also be considered in the comparison of alternatives. For example, the rate of inflation, the effect of nonuniform payments, and variations in the interest rate should also be considered in comparing alternatives. Most real-world problems will require consideration of such factors. However, these are beyond the scope of the present text. Blank and Tarquin [1983] and other texts on the subject of engineering economy may be consulted for a more comprehensive treatment of these factors.

EXERCISES FOR APPENDIX A

A.1 A person decides to deposit $50 per month for the next 18 years at an annual interest rate of 8% compounded monthly. How much money is accumulated at the end of 18 years? How much per month can be withdrawn for the next 4 years after the 18th year?

A.2 A person borrows $4000 at an annual interest rate of 13% compounded monthly to buy a car and promises to pay it back in 30 monthly instalments. What is the monthly instalment? How much money would be needed to pay off the loan after the 14th instalment?

A.3 A person wants to buy a house costing $90 000. The prevailing interest rate is 13% compounded monthly. The down payment must be 20% of the purchase price (i.e., only 80% of the purchase price can be borrowed). What is the monthly instalment if the loan is for 20 years? How much money will be needed at the end of the 5th year (i.e. at the time of the 60th instalment) if the loan has to be paid off at that time?

A.4 A person invests $2000 at an annual interest rate of 9% compounded monthly. How much money will be received at the end of 2 years? How much money will be received if the interest is compounded daily?

A.5 A person borrows $10 000 and agrees to pay $950 per month for the next 12 months. If the interest is compounded monthly, what is the annual interest rate?

A.6 On the day a child is born, the parents decide to save a certain amount of money every year for his college education. They decided to deposit the money on every birthday through the 18th starting with the first one, so that the child can withdraw $10 000 on the 18th, 19th, 20th and 21st birthdays. If the expected rate of return is 9% per year, how much money must be deposited annually?

A.7 A company has two alternative designs for the construction of a bridge. Option A costs $500 000 initially with an annual maintenance cost of $10 000. Option B costs $400 000 initially and its annual maintenance cost is $12 000. Both options have a salvage value of 5% at the end of a 50-year life. Which option should the company adopt? Assume 10% per annum as the rate of return. Use both the present worth and annual cost comparison methods.

A.8. A city in a mountainous region requires an additional 100 mega watt peak power capacity and is considering the following alternatives for the next 20-year period, until a breader reactor meets all needs:

1. Build a new power plant for $10 million and a $1 million per year operating cost. The salvage value at the end of 20 years is $2 million.
2. Build a pumping/generator station that pumps water to a high lake during periods of low power usage and uses the water for power generation during peak load periods. The initial cost is $5 million and the operating costs are $1.5 million per year. There is no salvage value at the end of 20 years.

Which of these alternatives is preferable on a present worth basis? At present interest rates, the following relations hold:

$$S_{20} = 2P, \quad \text{or} \quad P = 0.5S_{20}$$
$$S_{20} = 30R, \quad \text{or} \quad R = \tfrac{1}{30}S_{20}$$

where S_{20} = value at 20 years, P = present worth and R = transaction per year.

A.9 An 80-cm pipeline can be built for $150 000. The annual operating and maintenance cost is estimated at $30 000. The alternate 50-cm line can be built for $120 000. Its operating and maintenance cost is estimated at $35 000 per year. Either line is expected to serve for 25 years with 10% salvage when replaced. Compare the two pipelines on an annual cost and present worth basis assuming a 15% rate of return.

A.10 A company has received two bids for the design and maintenance of a project. Compare the two designs by a present worth analysis using the following data assuming a 10% interest rate. Both designs have an economic life of 40 years with no salvage value.

	A	B
First cost ($)	40 000	50 000
Annual maintenance ($)	1500	500

A.11 A person wants to buy a house costing $100 000. Bank A can lend the money at 12% interest compounded monthly. The bank requires 20% of the purchase price as the down payment and charges 2% of the loan as a loan processing fee. Bank B also requires a 20% down payment. It does not charge a loan processing fee, but its interest rate is 12.5% compounded monthly. Both banks require a monthly mortgage payment on the loan and can lend the money for 20 years. Which bank offers the cheaper option to buy the house? Use both the present worth and annual base comparisons.

A.12 Two rental properties are for sale:

	Property I	Property II
Sale price ($)	600 000	400 000
Annual gross income ($)	80 000	56 000
Annual management cost ($)	4000	3000
Annual maintenance ($)	10 000	6000
Property tax ($)	12 000	8000

Each property requires a minimum down payment of 10%. The prevailing interest rate is 10% compounded annually, and the loan can be obtained for 20 years. Due to the high demand the value of rental properties is expected to double in 20 years. A person has $100 000 which can be kept in the bank giving a return of 6% or invested into the properties.

1. Evaluate the following options using both the annual cost and present worth comparisons assuming a 20-year life for the project:
   Option A: Buy only Property I ($100 000 down payment)
   Option B: Buy only Property II ($100 000 down payment)
   Option C: Buy both properties (total down payment $100 000)
2. If the properties are liquidated at the end of 5 years at 110% of the purchase price, how much money will be received from Options A, B and C?

A.13 A company has two options for an operation. The associated costs are:

	First cost	Annual maintenance cost
Option A ($)	80 000	4000
Option B ($)	110 000	2000

Assume a 40-year life, with zero salvage for either option. Choose the better option, using the present worth comparison with a 15% interest rate.

A.14 To transport its goods, a company has two options: take a slightly longer route to use an existing bridge, or build a new bridge which cuts down the distance and increases the number of trips (which is desirable) for the same cost. The construction of the new bridge will cost $100 000 and will save $100 per day over a 240-day working year. The economic life of the bridge is 40 years and the maintenance cost of the structure is estimated at $1000 per year. Compare the alternatives based on the annual cost and present worth comparisons using a 20% annual rate of return.

A.15 A university is planning to develop a student laboratory. One proposal calls for the construction of the laboratory which would cost $500 000 and would satisfy the need over the next 12 years. The expected annual operating cost would be $40 000. After 12 years, an addition to the laboratory would be constructed for $600 000 with an additional annual operating cost of $30 000.

The alternative plan is to build a single large laboratory now, which would cost $650 000. The annual operating cost would be $42 000 for the first 12 years. At the end of the 12th year, the laboratory would need renovations costing $100 000 and the annual operating costs would be expected to increase to $60 000. Compare the two plans using either the annual base or present worth method with an annual interest rate of 10%.

A.16 A company has three options to correct a problem. The costs of the options are (A) $70 000, (B) $100 000 and (C) $50 000. All options are expected to serve for 60 years without any salvage value. However, Option A requires an expense of $1000 per year, and Option C will require an additional investment of $80 000 at the end of 20 years, which will have a salvage value of $15 000 at 60 years from the present time. Which of the three options should be adopted, assuming a 15% interest rate?

A.17 Your company has decided to buy a new company car. You have been designated to evaluate the following three cars and make your recommendation. Assume the following for all cars:

1. The car will be driven 20 000 miles each year for an expected life of 5 years. A total of 90% of these will be highway miles; the other 10% will be in the city.
2. The price of gasoline is now $1.25 per gallon and will increase at 10% per year for the next 5 years. The gasoline cost will be paid based on the average monthly consumption at the end of each month.
3. At the end of the 5th year, the company will sell the car at a 15% salvage value.
4. Insurance will be 5% of the original car value per year.
5. The annual rate of return is 8%.

The car choices are as follows:

Car A: $6250 list price

Requires:	$150/year maintenance first 2 years
	$300/year maintenance last 3 years
Mileage estimates:	35 m.p.g. highway
	25 m.p.g. city
Financing:	$1000 down payment and the rest in 60 equal monthly payments at 11% annual interest

Car B: $6900 list price

Requires:	$125/year maintenance first 2 years
	$250/year maintenance last 3 years
Mileage estimates:	35 m.p.g. highway
	22 m.p.g. city
Financing:	$1500 down payment and the rest in 60 equal monthly payments at 12% annual interest

Car C: $7200 list price

Requires:	$100/year maintenance first 2 years
	$200/year maintenance last 3 years
Mileage estimates:	38 m.p.g. highway
	28 m.p.g. city
Financing:	$1100 down payment and the rest in 60 equal monthly payments at 10% annual interest

Use the present worth method of comparison (created by G. Jackson).

VECTOR
AND MATRIX
ALGEBRA

B.1 INTRODUCTION

Matrix and vector notation is compact and useful in describing many numerical methods and derivations. Matrix and vector algebra is a basic tool needed in developing methods for the optimum design of systems. The solution of linear optimization problems (linear programming) involves an understanding of the solution process for a system of linear equations. Therefore, it is important to understand operations of vector and matrix algebra and be comfortable with their notation. *The subject is often referred to as linear algebra* and has been well-developed for a long time. It has become a standard tool in almost all engineering and scientific applications. In this appendix, some fundamental properties of vectors and matrices are reviewed. For more comprehensive treatment of the subject, several excellent text books are available and should be consulted [Hohn, 1964; Franklin, 1968; Cooper and Steinberg, 1970; Stewart, 1973; Bell, 1975; Strang, 1976; Jennings, 1977; Deif, 1982; Gere and Weaver, 1983]. In addition, most software libraries have subroutines for linear algebra operations which should be directly utilized.

   After reviewing the basic vector and matrix notations, special matrices, determinant and rank of a matrix, the subject of the solution of a simultaneous system of linear equations is discussed. First an $n \times n$ system and then a rectangular $m \times n$ system are treated. A section on linear independence of vectors is also included. Finally, the eigenvalue problem encountered in many fields of engineering is discussed. Such problems play a prominent role in convex programming problems and sufficiency conditions for optimization.

B.2   DEFINITION OF MATRICES

A matrix is defined as a rectangular array of quantities which can be real numbers, complex numbers, or functions of several variables. The entries in the rectangular array are also called the *elements of the matrix.* Since the solution of simultaneous linear equations is the most common application of matrices, we use them to develop the notion of matrices.

Consider the following system of two simultaneous linear equations in three unknowns:

$$x_1 + 2x_2 + 3x_3 = 6$$
$$-x_1 + 6x_2 - 2x_3 = 3$$
(B.1)

The symbols x_1, x_2, x_3 represent the *solution variables for the system of equations.* Note that the variables x_1, x_2 and x_3 can be replaced by any other variables, say w_1, w_2 and w_3, without affecting the solution. Therefore, they are sometimes called the *dummy variables.* Since they are dummy variables, they can be omitted while writing the equations in a matrix form. For example, Eqs (B.1) can be written in a rectangular array as

$$\left[\begin{array}{ccc|c} 1 & 2 & 3 & 6 \\ -1 & 6 & -2 & 3 \end{array}\right]$$

The entries to the left of the vertical line are coefficients of the variables x_1, x_2 and x_3, and to the right of the vertical line are the numbers on the right-hand side of the equations. It is customary to enclose the array by square brackets as shown. Thus, we see that the system of equations in Eqs (B.1) can be represented by a matrix having two rows and four columns.

An array with m rows and n columns is called a matrix of order "m by n", written as (m, n) or as $m \times n$. To distinguish between matrices and scalars, we shall boldface the variables that represent matrices. In addition, capital letters will be used to represent matrices. For example, a general matrix $\mathbf{A}$ of order $m \times n$ can be represented as

$$\mathbf{A} = \begin{bmatrix} a_{11} & a_{12} & \cdots & a_{1n} \\ a_{21} & a_{22} & \cdots & a_{2n} \\ \vdots & \vdots & & \vdots \\ a_{m1} & a_{m2} & \cdots & a_{mn} \end{bmatrix}$$
(B.2)

The coefficients a_{ij} are called elements of the matrix $\mathbf{A}$; subscripts i and j indicate the row and column numbers for the element a_{ij} (e.g. a_{32} represents the element in the third row and second column). Although the elements can oe real numbers, complex numbers or functions, we shall not deal with complex matrices in the present text. We shall encounter matrices having elements as functions of several variables, e.g. the Hessian matrix of a function discussed in Chapter 3.

It is useful to employ more compact notation for matrices. For example, a matrix $\mathbf{A}$ of order $m \times n$ with a_{ij}'s as its elements is written compactly as

$$\mathbf{A} = [a_{ij}]_{(m \times n)} \tag{B.3}$$

Often, the size of the matrix is also not shown and $\mathbf{A}$ is written as $[a_{ij}]$.

If a matrix has the same number of rows and columns, then it is called the *square matrix*. In Eq. (B.2) or (B.3), if $m = n$, $\mathbf{A}$ is a square matrix. It is called a matrix of order n.

It is important to understand the matrix notation for a set of linear equations because we shall encounter such equations quite often in this text. For example, Eqs (B.1) can be written as

$$\begin{array}{cccc} x_1 & x_2 & x_3 & \mathbf{b} \end{array}$$
$$\begin{bmatrix} 1 & 2 & 3 & | & 6 \\ -1 & 6 & -2 & | & 3 \end{bmatrix}$$

The preceding array containing coefficients of the equations and the right-hand side parameters is called the *augmented matrix*. Note that *each column of the matrix is identified with a variable*; the first column is associated with the variable x_1 because it contains coefficients of x_1 for all equations, the second with x_2, the third with x_3 and the last column with the right-hand side vector which we call $\mathbf{b}$. This interpretation is important while solving linear simultaneous equations (discussed later) or linear programming problems (discussed in Chapter 4).

B.3 TYPE OF MATRICES AND THEIR OPERATIONS

B.3.1 Null Matrix

A matrix having all zero elements is called a null (zero) matrix denoted by the capital letter $\mathbf{0}$. Any *zero matrix* of proper order when pre- or post-multiplied by any other matrix (or scalar) results in a zero matrix.

B.3.2 Vector

A matrix of order $1 \times n$ is called a *row matrix*, or simply *row vector*. Similarly, a matrix of order $n \times 1$ is called a *column matrix*, or simply *column vector*. A vector with n elements is called an n-component vector, or an *n-vector*. In this text, all vectors are considered as column vectors denoted by a lower-case letter in boldface.

B.3.3 Addition of Matrices

If $\mathbf{A}$ and $\mathbf{B}$ are two matrices of the order $m \times n$, then their *sum* is also an $m \times n$ matrix defined as

$$\mathbf{C}_{(m \times n)} = \mathbf{A} + \mathbf{B}; \qquad c_{ij} = a_{ij} + b_{ij} \text{ for all } i \text{ and } j \tag{B.4}$$

Matrix addition satisfies the following properties

$$\mathbf{A} + \mathbf{B} = \mathbf{B} + \mathbf{A} \quad \text{(commutative)} \tag{B.5}$$

If $\mathbf{A}$, $\mathbf{B}$ and $\mathbf{C}$ are three matrices of the same order, then

$$\mathbf{A} + (\mathbf{B} + \mathbf{C}) = (\mathbf{A} + \mathbf{B}) + \mathbf{C} \quad \text{(associative)} \tag{B.6}$$

If $\mathbf{A}$, $\mathbf{B}$ and $\mathbf{C}$ are of same order, then

$$\mathbf{A} + \mathbf{C} = \mathbf{B} + \mathbf{C} \text{ implies } \mathbf{A} = \mathbf{B} \tag{B.7}$$

where $\mathbf{A} = \mathbf{B}$ implies that the matrices are equal. Two matrices $\mathbf{A}$ and $\mathbf{B}$ of order $m \times n$ are equal if $a_{ij} = b_{ij}$ for $i = 1$ to m and $j = 1$ to n.

B.3.4 Multiplication of Matrices

Multiplication of a matrix $\mathbf{A}$ of order $m \times n$ by a scalar k is defined as

$$k\mathbf{A} = [ka_{ij}]_{(m \times n)} \tag{B.8}$$

The multiplication (product) $\mathbf{AB}$ of two matrices $\mathbf{A}$ and $\mathbf{B}$ is defined only if $\mathbf{A}$ and $\mathbf{B}$ are of proper order. The number of columns of $\mathbf{A}$ must be equal to the number of rows of $\mathbf{B}$ for the product $\mathbf{AB}$ to be defined. In that case, the matrices are said to be *conformable* for multiplication. If $\mathbf{A}$ is $m \times n$ and $\mathbf{B}$ is $r \times p$, then the multiplication $\mathbf{AB}$ is defined only when $n = r$, and multiplication $\mathbf{BA}$ is defined only when $m = p$. Multiplication of two matrices of proper order results in a third matrix. If $\mathbf{A}$ and $\mathbf{B}$ are of order $m \times n$ and $n \times p$ respectively, then

$$\mathbf{AB} = \mathbf{C}$$

where $\mathbf{C}$ is a matrix of order $m \times p$. Elements of the matrix $\mathbf{C}$ are determined by multiplying the elements of a row of $\mathbf{A}$ with the elements of a column of $\mathbf{B}$ and adding all the multiplications. Thus

$$\begin{bmatrix} a_{11} & a_{12} & \cdots & a_{1n} \\ a_{21} & a_{22} & \cdots & a_{2n} \\ \vdots & \vdots & & \vdots \\ a_{m1} & a_{m2} & \cdots & a_{mn} \end{bmatrix} \begin{bmatrix} b_{11} & b_{12} & \cdots & b_{1p} \\ b_{21} & b_{22} & \cdots & b_{2p} \\ \vdots & \vdots & & \vdots \\ b_{n1} & b_{n2} & \cdots & b_{np} \end{bmatrix}$$

$$= \begin{bmatrix} c_{11} & c_{12} & \cdots & c_{1p} \\ c_{21} & c_{22} & \cdots & c_{2p} \\ \vdots & \vdots & & \vdots \\ c_{m1} & c_{m2} & \cdots & c_{mp} \end{bmatrix} \tag{B.9}$$

where elements c_{ij} are calculated as

$$c_{ij} = a_{i1}b_{1j} + a_{i2}b_{2j} + \ldots + a_{in}b_{nj} = \sum_{k=1}^{n} a_{ik}b_{kj} \qquad \text{(B.10)}$$

Note that if $\mathbf{B}$ is an $n \times 1$ matrix (i.e. a vector), then $\mathbf{C}$ is an $m \times 1$ matrix. We shall encounter this type of matrix multiplication quite often in this text, e.g. a linear system of equations is represented as $\mathbf{Ax} = \mathbf{b}$, where $\mathbf{x}$ contains the solution variables and $\mathbf{b}$ the right-hand side parameters. Equations (B.1) can be written in this form.

In the product $\mathbf{AB}$ the matrix $\mathbf{A}$ is said to be *post-multiplied* by $\mathbf{B}$ or $\mathbf{B}$ is said to be *pre-multiplied* by $\mathbf{A}$. Whereas the matrix addition satisfies commutative law, matrix multiplication does not, in general, satisfy this law, i.e. $\mathbf{AB} \neq \mathbf{BA}$. Also, even if $\mathbf{AB}$ is well defined, $\mathbf{BA}$ may not be defined.

Example B.1   Multiplication of matrices

$$\mathbf{A} = \begin{bmatrix} 2 & 3 & 1 \\ 6 & 3 & 2 \\ 4 & 2 & 0 \\ 0 & 3 & 5 \end{bmatrix}_{(4 \times 3)} \qquad \mathbf{B} = \begin{bmatrix} 2 & -1 \\ 1 & 0 \\ 3 & -2 \end{bmatrix}_{(3 \times 2)}$$

$$\mathbf{AB} = \begin{bmatrix} 2 & 3 & 1 \\ 6 & 3 & 2 \\ 4 & 2 & 0 \\ 0 & 3 & 5 \end{bmatrix} \begin{bmatrix} 2 & -1 \\ 1 & 0 \\ 3 & -2 \end{bmatrix}$$

$$= \begin{bmatrix} (2 \times 2 + 3 \times 1 + 1 \times 3) & (-2 \times 1 + 3 \times 0 - 1 \times 2) \\ (6 \times 2 + 3 \times 1 + 2 \times 3) & (-6 \times 1 + 3 \times 0 - 2 \times 2) \\ (4 \times 2 + 2 \times 1 + 0 \times 3) & (-4 \times 1 + 2 \times 0 - 0 \times 2) \\ (0 \times 2 + 3 \times 1 + 5 \times 3) & (-0 \times 1 + 3 \times 0 - 5 \times 2) \end{bmatrix}$$

$$= \begin{bmatrix} 10 & -4 \\ 21 & -10 \\ 10 & -4 \\ 18 & -10 \end{bmatrix}_{(4 \times 2)}$$

Note that the product $\mathbf{BA}$ is not defined because the number of columns in $\mathbf{B}$ is not equal to the number of rows in $\mathbf{A}$.   ‖

Example B.2   Multiplication of matrices

$$\mathbf{A} = \begin{bmatrix} 4 & 1 \\ -2 & 0 \\ 8 & 3 \end{bmatrix} \qquad \mathbf{B} = \begin{bmatrix} -3 & 8 & 6 \\ 8 & 3 & -1 \end{bmatrix}$$

$$\mathbf{AB} = \begin{bmatrix} 4 & 1 \\ -2 & 0 \\ 8 & 3 \end{bmatrix} \begin{bmatrix} -3 & 8 & 6 \\ 8 & 3 & -1 \end{bmatrix} = \begin{bmatrix} -4 & 35 & 23 \\ 6 & -16 & -12 \\ 0 & 73 & 45 \end{bmatrix}$$

$$\mathbf{BA} = \begin{bmatrix} -3 & 8 & 6 \\ 8 & 3 & -1 \end{bmatrix} \begin{bmatrix} 4 & 1 \\ -2 & 0 \\ 8 & 3 \end{bmatrix} = \begin{bmatrix} 20 & 15 \\ 18 & 5 \end{bmatrix}$$

Note that for the products **AB** and **BA** to be matrices of the same order, **A** and **B** must be square matrices. ∥

Note that even if matrices **A**, **B** and **C** are properly defined, **AB** = **AC** does not imply **B** = **C**. Also, if **AB** = **0**, it does not imply either **B** = **0** or **A** = **0**. The matrix multiplication, however, satisfies two important laws: the associative and distributive laws. Let matrices **A**, **B**, **C**, **D** and **F** be of proper dimension. Then

1. $(\mathbf{AB})\mathbf{C} = \mathbf{A}(\mathbf{BC})$ (associative law) (B.11)

2a. $\mathbf{B}(\mathbf{C} + \mathbf{D}) = \mathbf{BC} + \mathbf{BD}$

2b. $(\mathbf{C} + \mathbf{D})\mathbf{F} = \mathbf{CF} + \mathbf{DF}$ (distributive laws) (B.12)

2c. $(\mathbf{A} + \mathbf{B})(\mathbf{C} + \mathbf{D}) = \mathbf{AC} + \mathbf{AD} + \mathbf{BC} + \mathbf{BD}$

B.3.5 Transpose of a Matrix

We can write rows of a matrix as columns and obtain another matrix. Such an operation is called transpose of a matrix. If $\mathbf{A} = [a_{ij}]$ is an $m \times n$ matrix, then its transpose denoted as $\mathbf{A}^T$ is an $n \times m$ matrix. It is obtained from **A** by interchanging its rows and columns. The first column of **A** is the first row of $\mathbf{A}^T$; the second column of **A** is the second row of $\mathbf{A}^T$; and so on. Thus, if $\mathbf{A} = [a_{ij}]$, then $\mathbf{A}^T = [a_{ji}]$. The operation of transposing a matrix is illustrated by the following 2×3 matrix:

$$\mathbf{A} = \begin{bmatrix} a_{11} & a_{12} & a_{13} \\ a_{21} & a_{22} & a_{23} \end{bmatrix}; \qquad \mathbf{A}^T = \begin{bmatrix} a_{11} & a_{21} \\ a_{12} & a_{22} \\ a_{13} & a_{23} \end{bmatrix}$$

Some properties of the transpose are

1. $(\mathbf{A}^T)^T = \mathbf{A}$

2. $(\mathbf{A} + \mathbf{B})^T = \mathbf{A}^T + \mathbf{B}^T$

3. $(\alpha\mathbf{A})^T = \alpha\mathbf{A}^T, \quad \alpha = \text{scalar}$

4. $(\mathbf{AB})^T = \mathbf{B}^T\mathbf{A}^T$

B.3.6 Elementary Row–Column Operations

There are three simple but extremely useful operations for rows or columns of a matrix. They are used in later discussions, so we state them here:

1. Interchange any two rows (columns).

2. Multiply any row (column) by a nonzero scalar.

3. Add to any row (column) a scalar multiple of another row (column).

B.3.7 Equivalence of Matrices

A matrix $\mathbf{A}$ is said to be *equivalent* to another matrix $\mathbf{B}$ written as $\mathbf{A} \sim \mathbf{B}$ if $\mathbf{A}$ can be transformed into $\mathbf{B}$ by means of one or more elementary row and/or column operations. If only row (column) operations are used, we say $\mathbf{A}$ is *row (column)* equivalent to $\mathbf{B}$.

B.3.8 Scalar Product–Dot Product of Vectors

A special case of matrix multiplication of particular interest is the multiplication of a row vector by a column vector. If $\mathbf{x}$ and $\mathbf{y}$ are two n-component vectors, then

$$\mathbf{x}^T \mathbf{y} = \sum_{j=1}^{n} x_j y_j \tag{B.13}$$

where

$$\mathbf{x}^T = [x_1 \quad x_2 \ldots x_n] \quad \text{and} \quad \mathbf{y} = [y_1 \quad y_2 \ldots y_n]^T$$

The product in Eq. (B.13) is called the *scalar product or dot product* of $\mathbf{x}$ and $\mathbf{y}$. It is also denoted as $\mathbf{x} \cdot \mathbf{y}$. Note that since the dot product of two vectors is a scalar, $\mathbf{x}^T \mathbf{y} = \mathbf{y}^T \mathbf{x}$.

Associated with any vector $\mathbf{x}$ is a scalar called *norm or length of the vector* defined as

$$(\mathbf{x}^T \mathbf{x})^{1/2} = \left(\sum_{i=1}^{n} x_i^2 \right)^{1/2} \tag{B.14}$$

Often the norm of $\mathbf{x}$ is designated as $\|\mathbf{x}\|$.

B.3.9 Square Matrices

A matrix having the same number of rows and columns is called a *square matrix*; otherwise it is called a rectangular matrix. The elements a_{ii}, $i = 1$ to n are called the *main diagonal elements* and others are called the *off-diagonal elements*. A square matrix having zero entries at all off-diagonal locations is called a *diagonal matrix*. If all main diagonal elements of a diagonal matrix are equal, it is called a *scalar matrix*.

A square matrix $\mathbf{A}$ is called *symmetric* if $\mathbf{A}^T = \mathbf{A}$ and *asymmetric* or *unsymmetric* otherwise. It is called *antisymmetric* if $\mathbf{A}^T = -\mathbf{A}$.

If all the elements below the main diagonal of a square matrix are zero ($a_{ij} = 0$ for $i > j$), it is called an *upper triangular matrix*. Similarly, a *lower triangular matrix* has all zero elements above the main diagonal ($a_{ij} = 0$ for $i < j$). A matrix that has all zero entries except in a band around the main diagonal is called a *banded matrix*.

A square matrix having unit elements on the main diagonal and zeroes elsewhere is called an *identity matrix*. An identity matrix of order n is denoted as $\mathbf{I}_{(n)}$. Identity matrices are useful because their pre- or post-multiplication with another matrix does not change the matrix, e.g. let $\mathbf{A}$ be any $m \times n$ matrix, then

$$\mathbf{I}_{(m)}\mathbf{A} = \mathbf{A} = \mathbf{A}\mathbf{I}_{(n)} \tag{B.15}$$

A scalar matrix $\mathbf{S}_{(n)}$ having diagonal elements as α can be written as

$$\mathbf{S}_{(n)} = \alpha\mathbf{I}_{(n)} \tag{B.16}$$

Note that pre- or post-multiplying any matrix by a scalar matrix of proper order results in multiplying the original matrix by the scalar. This can be proved for any $m \times n$ matrix $\mathbf{A}$ as follows:

$$\mathbf{S}_{(m)}\mathbf{A} = \alpha\mathbf{I}_{(m)}\mathbf{A} = \alpha\mathbf{A} = \alpha(\mathbf{A}\mathbf{I}_{(n)}) = \mathbf{A}(\alpha\mathbf{I}_{(n)}) = \mathbf{A}\mathbf{S}_{(n)} \tag{B.17}$$

B.3.10   Partitioning of Matrices

It is often useful to divide vectors and matrices into a smaller group of elements. This can be done by partitioning the matrix into smaller rectangular arrays called *submatrices* and vectors into subvectors. For example, consider a matrix $\mathbf{A}$ as

$$\mathbf{A} = \begin{bmatrix} 2 & 1 & -6 & 4 & 3 \\ 2 & 3 & 8 & -1 & -3 \\ 1 & -6 & 2 & 3 & 8 \\ -3 & 0 & 5 & -2 & 7 \end{bmatrix}_{(4 \times 5)}$$

A possible partitioning of $\mathbf{A}$ is

$$\mathbf{A} = \left[\begin{array}{ccc|cc} 2 & 1 & -6 & 4 & 3 \\ 2 & 3 & 8 & -1 & -3 \\ \hline 1 & -6 & 2 & 3 & 8 \\ -3 & 0 & 5 & -2 & 7 \end{array} \right]_{(4 \times 5)}$$

Therefore, submatrices of $\mathbf{A}$ are

$$\mathbf{A}_{11} = \begin{bmatrix} 2 & 1 & -6 \\ 2 & 3 & 8 \end{bmatrix}_{(2 \times 3)} \qquad \mathbf{A}_{12} = \begin{bmatrix} 4 & 3 \\ -1 & -3 \end{bmatrix}_{(2 \times 2)}$$

$$\mathbf{A}_{21} = \begin{bmatrix} 1 & -6 & 2 \\ -3 & 0 & 5 \end{bmatrix}_{(2 \times 3)} \qquad \mathbf{A}_{22} = \begin{bmatrix} 3 & 8 \\ -2 & 7 \end{bmatrix}_{(2 \times 2)}$$

where $\mathbf{A}_{ij}$ are matrices of proper order. Thus, $\mathbf{A}$ can be written in terms of submatrices as

$$\mathbf{A} = \left[\begin{array}{c|c} \mathbf{A}_{11} & \mathbf{A}_{12} \\ \hline \mathbf{A}_{21} & \mathbf{A}_{22} \end{array} \right]$$

Note that partitioning of vectors and matrices must be proper so that the operations of addition or multiplication remain defined. To see how two partitioned matrices are multiplied, consider $\mathbf{A}$ an $m \times n$ and $\mathbf{B}$ an $n \times p$ matrix. Let these be partitioned as

$$\mathbf{A} = \left[\begin{array}{c|c} \mathbf{A}_{11} & \mathbf{A}_{12} \\ \hline \mathbf{A}_{21} & \mathbf{A}_{22} \end{array}\right]_{(m \times n)}; \quad \mathbf{B} = \left[\begin{array}{c|c} \mathbf{B}_{11} & \mathbf{B}_{12} \\ \hline \mathbf{B}_{21} & \mathbf{B}_{22} \end{array}\right]_{(n \times p)}$$

Then the product $\mathbf{AB}$ can be written as

$$\mathbf{A} = \left[\begin{array}{c|c} \mathbf{A}_{11} & \mathbf{A}_{12} \\ \hline \mathbf{A}_{21} & \mathbf{A}_{22} \end{array}\right]\left[\begin{array}{c|c} \mathbf{B}_{11} & \mathbf{B}_{12} \\ \hline \mathbf{B}_{21} & \mathbf{B}_{22} \end{array}\right] = \left[\begin{array}{c|c} (\mathbf{A}_{11}\mathbf{B}_{11} + \mathbf{A}_{12}\mathbf{B}_{21}) & (\mathbf{A}_{11}\mathbf{B}_{12} + \mathbf{A}_{12}\mathbf{B}_{22}) \\ \hline (\mathbf{A}_{21}\mathbf{B}_{11} + \mathbf{A}_{22}\mathbf{B}_{21}) & (\mathbf{A}_{21}\mathbf{B}_{12} + \mathbf{A}_{22}\mathbf{B}_{22}) \end{array}\right]_{(m \times p)}$$

Note that the partitioning of matrices $\mathbf{A}$ and $\mathbf{B}$ must be such that the matrix products $\mathbf{A}_{11}\mathbf{B}_{11}$, $\mathbf{A}_{12}\mathbf{B}_{21}$, $\mathbf{A}_{11}\mathbf{B}_{12}$, $\mathbf{A}_{12}\mathbf{B}_{22}$, etc., are proper. In addition, the pairs of matrices $\mathbf{A}_{11}\mathbf{B}_{11}$ and $\mathbf{A}_{12}\mathbf{B}_{21}$, $\mathbf{A}_{11}\mathbf{B}_{12}$ and $\mathbf{A}_{12}\mathbf{B}_{22}$, etc., must be of the same order.

B.4 SOLUTION OF n LINEAR EQUATIONS IN n UNKNOWNS

B.4.1 Introduction

Linear systems of equations are encountered in numerous engineering and scientific applications. Therefore, substantial research and development work has been done to devise several solution procedures. It is critically important to understand the basic ideas and concepts related to linear equations because we use them quite often in this text. In this section we shall describe a basic procedure, known as *Gaussian elimination,* for solution of an $n \times n$ *(square) linear system of equations.* More general methods for solving a *rectangular $m \times n$ system* are discussed in the next section.

It turns out that the idea of determinants is closely related to the solution of a linear system of equations, so first we discuss determinants and their properties. It also turns out that the solution of a square system can be found by inverting the matrix associated with the system, so we describe methods for inverting matrices.

Let us consider the following system of n equations in n unknowns:

$$\mathbf{Ax} = \mathbf{b} \tag{B.18}$$

where $\mathbf{A}$ is an $n \times n$ matrix of specified constants, $\mathbf{x}$ is an n-vector of solution variables, and $\mathbf{b}$ is an n-vector of specified constants known as the right-hand side vector. $\mathbf{A}$ is called the *coefficient matrix* and when the vector $\mathbf{b}$ is added as the $(n+1)$th column of $\mathbf{A}$ as $[\mathbf{A} \,|\, \mathbf{b}]$, the resulting matrix is called the *augmented matrix* for the given system of equations. Note that the left-hand side of Eq. (B.18) consists of multiplication of an $n \times n$ matrix with an n-component vector resulting into another n-component vector. If the right-hand side vector $\mathbf{b}$ is zero, Eq. (B.18) is called a *homogeneous* system; otherwise it is called the *nonhomogeneous system* of equations.

The equation $\mathbf{Ax} = \mathbf{b}$ can also be written in the following summation form:

$$\sum_{j=1}^{n} a_{ij}x_j = b_i; \qquad i = 1 \text{ to } n \tag{B.19}$$

If each row of the matrix $\mathbf{A}$ is interpreted as an n-dimensional row vector $\bar{\mathbf{a}}^{(i)}$, then the left-hand side of Eq. (B.19) can be interpreted as the dot product of two vectors as

$$(\bar{\mathbf{a}}^{(i)} \cdot \mathbf{x}) = b_i; \qquad i = 1 \text{ to } n \tag{B.20}$$

If each column of $\mathbf{A}$ is interpreted as an n-dimensional column vector, then the left-hand side of Eq. (B.18) can be interpreted as the summation of the scaled columns of the matrix $\mathbf{A}$ as

$$\sum_{i=1}^{n} \mathbf{a}^{(i)}x_i = \mathbf{b} \tag{B.21}$$

These interpretations can be useful in devising solution strategies for the system $\mathbf{Ax} = \mathbf{b}$, and in their implementation. For example, Eq. (B.21) shows that the solution variable x_i is simply a scale factor for the ith column of $\mathbf{A}$, i.e. variable x_i is associated with the ith column.

B.4.2 Determinants

To develop the solution strategies for the linear system $\mathbf{Ax} = \mathbf{b}$, we begin by introducing the concept of determinants and study their properties [Hohn, 1964]. The methods for calculating determinants are intimately related to the procedues for solving linear equations, so we shall also discuss them.

Every square matrix has a scalar associated with it, called the determinant calculated from its elements. To introduce the idea of determinants, we set $n = 2$ in Eq. (B.18) and consider the following 2×2 system of simultaneous equations:

$$\begin{bmatrix} a_{11} & a_{12} \\ a_{21} & a_{22} \end{bmatrix} \begin{bmatrix} x_1 \\ x_2 \end{bmatrix} = \begin{bmatrix} b_1 \\ b_2 \end{bmatrix} \tag{a}$$

The number $(a_{11}a_{22} - a_{21}a_{12})$ calculated using elements of the coefficient matrix is called its determinant. To see how this number arises, we shall solve the system in Eqs (a) by the *elimination process*.

Multiplying the first row by a_{22} and the second by a_{12} in Eqs (a), we get

$$\begin{bmatrix} a_{11}a_{22} & a_{12}a_{22} \\ a_{12}a_{21} & a_{12}a_{22} \end{bmatrix} \begin{bmatrix} x_1 \\ x_2 \end{bmatrix} = \begin{bmatrix} a_{22}b_1 \\ a_{12}b_2 \end{bmatrix} \tag{b}$$

Subtracting the second row from the first one in Eqs (b), we eliminate x_2 from the first equation and obtain:

$$(a_{11}a_{22} - a_{12}a_{21})x_1 = a_{22}b_1 - a_{12}b_2 \tag{c}$$

Now repeating the foregoing process to eliminate x_1 from the second row in

Eqs (a) by multiplying the first equation by a_{21} and the second by a_{11} and subtracting, we obtain:

$$(a_{11}a_{22} - a_{12}a_{21})x_2 = a_{11}b_2 - a_{21}b_1 \tag{d}$$

The coefficient of x_1 and x_2 in Eqs (c) and (d) must be nonzero for a unique solution of the system, i.e. $(a_{11}a_{22} - a_{12}a_{21}) \neq 0$, and the values of x_1 and x_2 are calculated as

$$x_1 = \frac{a_{22}b_1 - a_{12}b_2}{a_{11}a_{22} - a_{12}a_{21}}, \qquad x_2 = \frac{a_{11}b_2 - a_{21}b_1}{a_{11}a_{22} - a_{12}a_{21}} \tag{e}$$

The denominator $(a_{11}a_{22} - a_{12}a_{21})$ is identified as the *determinant* of the matrix **A** of Eqs (a). It is denoted by $\det(\mathbf{A})$, or $|\mathbf{A}|$. Thus, for any 2×2 matrix **A**,

$$|\mathbf{A}| = a_{11}a_{22} - a_{12}a_{21} \tag{f}$$

Using the definition of Eq. (f), we can rewrite Eqs (e) as

$$x_1 = \frac{|\mathbf{B}_1|}{|\mathbf{A}|}, \qquad x_2 = \frac{|\mathbf{B}_2|}{|\mathbf{A}|} \tag{g}$$

where $\mathbf{B}_1$ is obtained by replacing the first column of **A** with the right-hand side and $\mathbf{B}_2$ is obtained by replacing the second column of **A** with the right-hand side, as

$$\mathbf{B}_1 = \begin{bmatrix} b_1 & a_{12} \\ b_2 & a_{22} \end{bmatrix}, \qquad \mathbf{B}_2 = \begin{bmatrix} a_{11} & b_1 \\ a_{21} & b_2 \end{bmatrix}$$

Equations (g) are known as *Cramer's rule*. According to this rule, we need to compute only the three determinants – $|\mathbf{A}|$, $|\mathbf{B}_1|$ and $|\mathbf{B}_2|$ – to determine the solution to any 2×2 system of equations. *If $|\mathbf{A}| = 0$, there is no unique solution to Eqs (a). There may be an infinite number of solutions or no solution at all.* These cases are investigated in the next section.

The preceding concept of a determinant can be generalized to $n \times n$ matrices. *For every square matrix **A** of any order, we can associate a unique scalar, called the determinant of **A**.* There are many ways of calculating the determinant of a matrix. These procedures are closely related to the ones used for solving the linear system of equations which we shall discuss subsequently in this section.

B.4.2.1 PROPERTIES OF DETERMINANTS. The determinants have several properties that are useful in devising procedures for their calculation. Therefore, these should be clearly understood.

1. The determinant of any square matrix **A** is also equal to the determinant of the transpose of the matrix, i.e. $|\mathbf{A}| = |\mathbf{A}^T|$.
2. If a square matrix **A** has two identical columns (or rows), then its determinant is zero, i.e. $|\mathbf{A}| = 0$.

3. If a new matrix is formed by interchanging any two columns (or rows) of a given matrix **A** (elementary row–column Operation 1), the determinant of the resulting matrix is negative of the determinant of the original matrix.

4. If a new matrix is formed by adding any multiple of one column (row) to a different column (row) of a given matrix (elementary row–column Operation 3), the determinant of the resulting matrix is equal to the determinant of the original matrix.

5. If a square matrix **B** is identical to a matrix **A**, except some column (or row) is a scalar multiple c of the corresponding column (or row) of **A** (elementary row–column Operation 2), then $|\mathbf{B}| = c\,|\mathbf{A}|$.

6. If elements of a column (or row) of a square matrix **A** are zero, then $|\mathbf{A}| = 0$.

7. If a square matrix **A** is lower or upper triangular, then the determinant of **A** is equal to the product of the diagonal elements:

$$|\mathbf{A}| = a_{11}a_{22} \cdot \ldots \cdot a_{nn} \tag{B.22}$$

8. If **A** and **B** are any two square matrices of the same order, then

$$|\mathbf{AB}| = |\mathbf{A}|\,|\mathbf{B}|$$

9. Let $|\mathbf{A}_{ij}|$ denote the determinant of a matrix obtained by deleting the ith row and the jth column of **A** (yielding a square matrix of order $n-1$); the scalar $|\mathbf{A}_{ij}|$ is called the *minor* of the element a_{ij} of matrix **A**. Then the *cofactor* of a_{ij} is defined as

$$\text{cofac}\,(a_{ij}) = (-1)^{i+j}\,|\mathbf{A}_{ij}| \tag{B.23}$$

The determinant of **A** is calculated in terms of the cofactors as

$$|\mathbf{A}| = \sum_{j=1}^{n} a_{ij}\,\text{cofac}\,(a_{ij}), \qquad \text{for any } i \tag{B.24}$$

Or,

$$|\mathbf{A}| = \sum_{i=1}^{n} a_{ij}\,\text{cofac}\,(a_{ij}), \qquad \text{for any } j \tag{B.25}$$

Note that the cofac (a_{ij}) is also a scalar obtained from the minor $|\mathbf{A}_{ij}|$ but having positive or negative sign determined by the indices i and j as $(-1)^{i+j}$. Equation (B.24) is called the cofactor expansion for $|\mathbf{A}|$ by the ith row; Eq. (B.25) is called the cofactor expansion for $|\mathbf{A}|$ by the jth column. Equations (B.24) and (B.25) can be used to prove Properties 2, 5, 6 and 7 directly.

It is important to note that Eqs (B.24) or (B.25) are difficult to use to calculate the determinant of **A**. These equations require calculation of the cofactors of the elements a_{ij} which are determinants in themselves. However, using the *elementary row and column operations* a square matrix can be converted to either lower or upper triangular form. The determinant is then computed using Eq. (B.22). This will be illustrated in an example later in this section.

A matrix having zero determinant is called a singular matrix; a matrix with a nonzero determinant is called nonsingular. A nonhomogeneous $n \times n$ system of equations has a unique solution if and only if the matrix of coefficients is nonsingular. These properties are discussed and used subsequently to develop methods for solving a system of equations.

B.4.2.2 PRINCIPAL MINOR. Every $n \times n$ square matrix $\mathbf{A}$ has certain scalars associated with it, called the principal minors. They are obtained as determinants of certain submatrices of $\mathbf{A}$. They are useful in determining the "form" of a matrix which is needed in checking sufficiency conditions for optimality as well as the convexity of functions discussed in Chapter 3. Therefore, we discuss the idea of principal minors here.

Let M_k, $k = 1$ to n be called the principal minors of $\mathbf{A}$. Then each M_k is defined as the determinant of the following submatrix:

$$M_k = |\mathbf{A}_{kk}| \tag{B.26}$$

where $\mathbf{A}_{kk}$ is a $k \times k$ submatrix of $\mathbf{A}$ obtained by deleting any $(n - k)$ columns and the corresponding rows. For example, $M_1 = a_{11}$, $M_2 =$ determinant of a 2×2 matrix obtained by deleting all rows and columns of $\mathbf{A}$ except the first two, and so on, are the principal minors of the matrix $\mathbf{A}$.

B.4.3 Gaussian Elimination Procedure

The elimination process described earlier in Section B.4.1 for solving a 2×2 system of equations can be generalized to solve any $n \times n$ system of equations. The entire process can be organized and explained using the matrix notation. The procedure can also be used to calculate the determinant of any matrix. The procedure is known as *Gaussian Elimination,* which we shall describe in detail in the following [Forsythe and Moler, 1967; Franklin, 1968].

Using the three elementary row–column operations defined in Section B.3, the system $\mathbf{Ax} = \mathbf{b}$ of Eq. (B.18) can be transformed to the following form:

$$\begin{bmatrix} 1 & \bar{a}_{12} & \bar{a}_{13} & \cdots & \bar{a}_{1n} \\ 0 & 1 & \bar{a}_{23} & \cdots & \bar{a}_{2n} \\ \vdots & \vdots & \vdots & & \vdots \\ 0 & 0 & 0 & \cdots & 1 \end{bmatrix} \begin{bmatrix} x_1 \\ x_2 \\ \vdots \\ x_n \end{bmatrix} = \begin{bmatrix} \bar{b}_1 \\ \bar{b}_2 \\ \vdots \\ \bar{b}_n \end{bmatrix} \tag{B.27}$$

Or, in expanded form, Eq. (B.27) becomes

$$x_1 + \bar{a}_{12}x_2 + \bar{a}_{13}x_3 + \ldots + \bar{a}_{1n}x_n = \bar{b}_1$$

$$x_2 + \bar{a}_{23}x_3 + \ldots + \bar{a}_{2n}x_n = \bar{b}_2$$

$$x_3 + \ldots + \bar{a}_{3n}x_n = \bar{b}_3 \tag{B.28}$$

$$\vdots \quad \vdots$$

$$x_n = \bar{b}_n$$

Note that we use $\bar{a}_{ij}$ and $\bar{b}_i$ to represent modified elements a_{ij} and b_j of the original system. From the nth equation of the system (B.28), we have $x_n = \bar{b}_n$. If we substitute this value into the $(n-1)$th equation of (B.28), we can solve for x_{n-1}:

$$
\begin{aligned}
x_{n-1} &= \bar{b}_{n-1} - \bar{a}_{n-1,n} x_n \\
&= \bar{b}_{n-1} - \bar{a}_{n-1,n} \bar{b}_n
\end{aligned}
\tag{B.29}
$$

Equation (B.29) can now be substituted into the $(n-2)$th equation of (B.28) and x_{n-2} can be determined. Continuing in this manner, each of the unknowns can be solved in reverse order: $x_n, x_{n-1}, x_{n-2}, \ldots, x_2, x_1$. The procedure of reducing a system of n equations in n unknowns and then solving successively for $x_n, x_{n-1}, x_{n-2}, \ldots, x_2, x_1$ is called the *Gaussian Elimination Procedure or Gauss Reduction*. The latter part of the method (solving successively for $x_n, x_{n-1}, x_{n-2}, \ldots, x_2, x_1$) is called the *backward substitution, or backward pass*.

The Gaussian Elimination procedure uses elementary row–column operations to convert the main diagonal elements of the given coefficient matrix to 1 and the elements below the main diagonal to zero. To carry out these operations we start with the first row and the first column of the given matrix augmented with the right-hand side of the system of equations. To make the diagonal element 1, the first row is divided by the diagonal element. To convert the elements in the first column below the main diagonal to zero, we multiply the first row by the element a_{i1} in the ith row ($i = 2$ to n). The resulting elements of the first row are subtracted from the ith row. This makes the element $\bar{a}_{i1}$ zero in the ith row. These operations are carried out for each row using the first row for elimination each time. Once all the elements below the main diagonal are zero in the first column, the procedure is repeated for the second column using the second row for elimination, and so on. The row used to obtain zero elements in a column is called the *pivot row,* and the column in which elimination is performed is called the *pivot column.* We will illustrate this procedure in an example later.

The foregoing operations of converting elements below the main diagonal to zero can be explained in another way. When we make the elements below the main diagonal in the first column zero, we are eliminating the variable x_1 from all the equations except the first equation (x_1 is associated with the first column). For this elimination step we use equation number 1. In general, when we reduce the elements below the main diagonal in the ith column to zero, we use the ith row as the pivot row. Thus, we eliminate the ith variable from all the equations below the ith row. This explanation is quite straightforward once we realize that each column of the coefficient matrix has a variable associated with it, as noted before.

Example B.3 Solution of equations by Gaussian Elimination. Solve the following 3×3 system of equations

$$
x_1 - x_2 + x_3 = 0
$$

$$
x_1 - x_2 + 2x_3 = 1
\tag{a}
$$

$$
x_1 + x_2 + 2x_3 = 5
$$

Solution. We shall illustrate the Gaussian Elimination procedure in a step-by-step manner using the augmented matrix idea. The augmented matrix for Eqs (a) is defined using the coefficients of the variables in each equation and the right-hand side parameters, as

$$\begin{array}{cccc} x_1 & x_2 & x_3 & \mathbf{b} \end{array}$$
$$\mathbf{B} = \begin{bmatrix} 1 & -1 & 1 & 0 \\ 1 & -1 & 2 & 1 \\ 1 & 1 & 2 & 5 \end{bmatrix}$$

To convert the preceding system to the form of Eq. (B.27), we use the elementary row–column operations as follows:

1. Add -1 times row 1 to row 2 and -1 times row 1 to row 3 (eliminating x_1 from the second and third equations; elementary row operation 3):

$$\begin{array}{cccc} x_1 & x_2 & x_3 & \mathbf{b} \end{array}$$
$$\mathbf{B} \sim \begin{bmatrix} 1 & -1 & 1 & 0 \\ 0 & 0 & 1 & 1 \\ 0 & 2 & 1 & 5 \end{bmatrix} \qquad \begin{array}{l} \text{(recall that symbol "} \sim \text{"} \\ \text{means equivalence} \\ \text{between matrices)} \end{array}$$

2. Since the element at location $(2, 2)$ is zero, interchange rows 2 and 3 to bring a nonzero element at that location (elementary row operation 1). Then dividing the new second row by 2 gives

$$\begin{array}{cccc} x_1 & x_2 & x_3 & \mathbf{b} \end{array}$$
$$\mathbf{B} \sim \begin{bmatrix} 1 & -1 & 1 & 0 \\ 0 & 1 & 0.5 & 2.5 \\ 0 & 0 & 1 & 1 \end{bmatrix}$$

3. Since the element at the location $(3, 3)$ is one and all elements below the main diagonal are zero, the foregoing matrix puts the system of Eqs (a) into the form of Eq. (B.27), as

$$\begin{bmatrix} 1 & -1 & 1 \\ 0 & 1 & 0.5 \\ 0 & 0 & 1 \end{bmatrix} \begin{bmatrix} x_1 \\ x_2 \\ x_3 \end{bmatrix} = \begin{bmatrix} 0 \\ 2.5 \\ 1 \end{bmatrix}$$

Performing the backward substitution, we obtain

$$x_3 = 1 \qquad \text{(from third row)}$$
$$x_2 = 2.5 - 0.5x_3 = 2 \qquad \text{(from second row)}$$
$$x_1 = 0 - x_3 + x_2 = 1 \qquad \text{(from first row)}$$

Therefore, the solution of Eqs (a) is

$$x_1 = 1, \qquad x_2 = 2, \qquad x_3 = 1 \qquad \qquad \|$$

The Gaussian Elimination method can easily be transcribed into a general-purpose computer program that can handle any given system of equations. However, certain modifications must be made to the procedure because numerical calculations on a machine with a finite number of digits introduce round-off errors. These errors can become significantly large if certain precautions are not taken. The modifications primarily involve a reordering of the rows or columns of the augmented matrix in such a way that possible round-off effects tend to be minimized. This reordering must be performed at each step of the elimination process, so that the diagonal element of the pivot row is the absolute largest among the elements of the remaining matrix on the lower right-hand side. This is known as the *total pivoting* procedure. When only the rows are interchanged to bring the absolute largest element from a column to the diagonal location, the procedure is known as *partial pivoting*. Note that many programs are available to solve a system of equations. Thus, before attempting to write a program for Gaussian Elimination, computer center libraries must be searched for the existing ones.

Example B.4 Determinant of a matrix by Gaussian Elimination. The Gaussian Elimination procedure can also be used to calculate the determinant of a matrix. We illustrate the procedure for the following 3×3 matrix:

$$\mathbf{A} = \begin{bmatrix} 2 & 3 & 0 \\ 1 & 2 & 1 \\ 0 & 3 & 4 \end{bmatrix}$$

Solution. Using the Gaussian Elimination procedure, we make the elements below the main diagonal zero, but this time the diagonal elements are not converted to unity. Once the matrix is converted to that form, the determinant is obtained using Eq. (B.22).

$$\begin{bmatrix} 2 & 3 & 0 \\ 1 & 2 & 1 \\ 0 & 3 & 4 \end{bmatrix} \sim \begin{bmatrix} 2 & 3 & 0 \\ 0 & 0.5 & 1 \\ 0 & 3 & 4 \end{bmatrix} \quad \text{(elimination in the first column)}$$

$$\sim \begin{bmatrix} 2 & 3 & 0 \\ 0 & 0.5 & 1 \\ 0 & 0 & -2 \end{bmatrix} \quad \text{(elimination in the second column)}$$

The preceding system is in the canonical form and $|\mathbf{A}|$ is given simply by multiplication of all the diagonal elements, i.e.

$$|\mathbf{A}| = (2)(0.5)(-2) = -2.$$

$\parallel$

B.4.4 Inverse of a Matrix: Gauss-Jordan Elimination

If the multiplication of two square matrices results in an identity matrix, they are called the inverse of each other. Let $\mathbf{A}$ and $\mathbf{B}$ be two square matrices of order n. Then $\mathbf{B}$ is called the inverse of $\mathbf{A}$ if

$$\mathbf{AB} = \mathbf{BA} = \mathbf{I}_{(n)} \tag{B.30}$$

The inverse of $\mathbf{A}$ is usually denoted by $\mathbf{A}^{-1}$. We shall later describe methods to calculate the inverse of a matrix. Every square matrix may not have an inverse. A matrix having no inverse is called a *singular matrix*. If the coefficient matrix of an $n \times n$ system of equations has an inverse, then the system can be solved for the unknown variables. Consider the $n \times n$ system of equations $\mathbf{A}\mathbf{x} = \mathbf{b}$, where $\mathbf{A}$ is the coefficient matrix and $\mathbf{b}$ is the right-hand side vector. Pre-multiplying both sides of the equation by $\mathbf{A}^{-1}$, we get

$$\mathbf{A}^{-1}\mathbf{A}\mathbf{x} = \mathbf{A}^{-1}\mathbf{b}$$

Since $\mathbf{A}^{-1}\mathbf{A} = \mathbf{I}$, the equation reduces to

$$\mathbf{x} = \mathbf{A}^{-1}\mathbf{b}$$

Thus if we know the inverse of matrix $\mathbf{A}$, then the preceding equation can be used to solve for the unknown vector $\mathbf{x}$.

There are a couple of ways to calculate the inverse of a nonsingular matrix. The first procedure is based on the use of the cofactors of $\mathbf{A}$ and its determinant. If $\mathbf{B}$ is the inverse of $\mathbf{A}$, then its elements are given as (called *inverse using cofactors*):

$$b_{ji} = \frac{\text{cofac}\,(a_{ij})}{|\mathbf{A}|}; \qquad i = 1 \text{ to } n; \, j = 1 \text{ to } n \tag{B.31}$$

Note that indices on the left-hand side of the equation are ji and on the right-hand side they are ij. Thus cofactors of the row of matrix $\mathbf{A}$ generate the corresponding column of the inverse matrix $\mathbf{B}$.

The preceding procedure is reasonable for smaller matrices, up to, say 3×3. For larger matrices, it becomes cumbersome and inefficient.

A clue to the second procedure for calculating the inverse is provided by Eq. (B.30). In that equation, elements of $\mathbf{B}$ can be considered as unknowns for the system of equations $\mathbf{A}\mathbf{B} = \mathbf{I}$. Thus the system can be solved using the Gaussian Elimination procedure to obtain the inverse of $\mathbf{A}$. We illustrate the procedure with an example.

Example B.5 Inverse of a matrix by cofactors and Gauss-Jordan reduction. Compute the inverse of the following 3×3 matrix:

$$\mathbf{A} = \begin{bmatrix} 1 & 3 & 0 \\ 1 & 2 & 0 \\ 0 & 3 & 1 \end{bmatrix}$$

Solution.

Inverse by Cofactors. Let $\mathbf{B}$ be a 3×3 matrix that is the inverse of matrix $\mathbf{A}$. In order to use the *cofactors approach* given in Eq. (B.31), we first calculate the determinant of $\mathbf{A}$ as $|\mathbf{A}| = -1$. Using Eq. (B.23), the cofactors of the first row of

A are

$$\text{cofac}\,(a_{11}) = (-1)^{1+1}\begin{vmatrix} 2 & 0 \\ 3 & 1 \end{vmatrix} = 2$$

$$\text{cofac}\,(a_{12}) = (-1)^{1+2}\begin{vmatrix} 1 & 0 \\ 0 & 1 \end{vmatrix} = -1$$

$$\text{cofac}\,(a_{13}) = (-1)^{1+3}\begin{vmatrix} 1 & 2 \\ 0 & 3 \end{vmatrix} = 3$$

Similarly, the cofactors of the 2nd and 3rd rows are

$$-3,\ 1,\ -3;\ 0,\ 0,\ -1$$

Thus, Eq. (B.31) gives the inverse of **A** as

$$\mathbf{B} = \begin{bmatrix} -2 & 3 & 0 \\ 1 & -1 & 0 \\ -3 & 3 & 1 \end{bmatrix}$$

Inverse by Gaussian Elimination. We shall first demonstrate the Gaussian Elimination procedure before presenting the Gauss-Jordan procedure. Since **B** is the inverse of **A**, $\mathbf{AB} = \mathbf{I}$. Or writing this in the expanded form

$$\begin{bmatrix} 1 & 3 & 0 \\ 1 & 2 & 0 \\ 0 & 3 & 1 \end{bmatrix}\begin{bmatrix} b_{11} & b_{12} & b_{13} \\ b_{21} & b_{22} & b_{23} \\ b_{31} & b_{32} & b_{33} \end{bmatrix} = \begin{bmatrix} 1 & 0 & 0 \\ 0 & 1 & 0 \\ 0 & 0 & 1 \end{bmatrix}$$

where b_{ij}'s are the elements of **B**. The foregoing equation can be considered as a system of simultaneous equations having three different right-hand side vectors. We can solve for each unknown column on the left-hand side corresponding to each right-hand side vector by using the Gaussian Elimination procedure. For example, considering the first column of **B** only, we obtain

$$\begin{bmatrix} 1 & 3 & 0 \\ 1 & 2 & 0 \\ 0 & 3 & 1 \end{bmatrix}\begin{bmatrix} b_{11} \\ b_{21} \\ b_{31} \end{bmatrix} = \begin{bmatrix} 1 \\ 0 \\ 0 \end{bmatrix}$$

Using the elimination procedure in the augmented matrix form we obtain

$$\left[\begin{array}{ccc|c} 1 & 3 & 0 & 1 \\ 1 & 2 & 0 & 0 \\ 0 & 3 & 1 & 0 \end{array}\right] \sim \left[\begin{array}{ccc|c} 1 & 3 & 0 & 1 \\ 0 & -1 & 0 & -1 \\ 0 & 3 & 1 & 0 \end{array}\right]$$ (elimination in the first column)

$$\sim \left[\begin{array}{ccc|c} 1 & 3 & 0 & 1 \\ 0 & 1 & 0 & 1 \\ 0 & 0 & 1 & -3 \end{array}\right]$$ (elimination in the second column)

Using back substitution, we obtain the first column of **B** as $b_{31} = -3$, $b_{21} = 1$, $b_{11} = -2$. Similarly, we find $b_{12} = 3$, $b_{22} = -1$, $b_{32} = 3$, $b_{13} = 0$, $b_{23} = 0$ and $b_{33} = 1$. Therefore, the inverse of **A** is given as

$$\mathbf{B} = \begin{bmatrix} -2 & 3 & 0 \\ 1 & -1 & 0 \\ -3 & 3 & 1 \end{bmatrix}$$

Inverse by Gauss-Jordan Elimination. We can organize the procedure for calculating the inverse of a matrix slightly differently. The augmented matrix can be defined with all the three columns of the right-hand side. The Gaussian Elimination process can be carried out below as well as above the main diagonal. With this procedure, the left-hand 3×3 matrix is converted to an identity matrix; and the right-hand 3×3 matrix then contains the inverse of the matrix. When elimination is performed below as well as above the main diagonal, the procedure is called *Gauss-Jordan Elimination*. The process proceeds as follows for calculating the inverse of $\mathbf{A}$:

$$\left[\begin{array}{ccc|ccc} 1 & 3 & 0 & 1 & 0 & 0 \\ 1 & 2 & 0 & 0 & 1 & 0 \\ 0 & 3 & 1 & 0 & 0 & 1 \end{array}\right] \quad \text{(augmented matrix)}$$

$$\sim \left[\begin{array}{ccc|ccc} 1 & 3 & 0 & 1 & 0 & 0 \\ 0 & -1 & 0 & -1 & 1 & 0 \\ 0 & 3 & 1 & 0 & 0 & 1 \end{array}\right] \quad \begin{array}{l}\text{(elimination in the first} \\ \text{column)}\end{array}$$

$$\sim \left[\begin{array}{ccc|ccc} 1 & 0 & 0 & -2 & 3 & 0 \\ 0 & 1 & 0 & 1 & -1 & 0 \\ 0 & 0 & 1 & -3 & 3 & 1 \end{array}\right] \quad \begin{array}{l}\text{(elimination in the second} \\ \text{column)}\end{array}$$

There is no need to perform elimination on the third column since $\bar{a}_{13} = \bar{a}_{23} = 0$ and $\bar{a}_{33} = 1$. We observe from the above matrix that the last three columns give precisely the matrix $\mathbf{B}$ which is the inverse of $\mathbf{A}$. ∥

The Gauss-Jordan procedure of computing the inverse of a 3×3 matrix can be generalized to any nonsingular $n \times n$ matrix. It can also be coded systematically into a general-purpose computer program to compute the inverse of a matrix.

B.5 SOLUTION OF m LINEAR EQUATIONS IN n UNKNOWNS: GENERAL THEORY

In the last section, the concept of determinants was used to determine the existence of a unique solution for any $n \times n$ system of equations. There are many instances in engineering applications where the number of equations is not equal to the number of variables i.e. rectangular systems. In a system of m equations in n unknowns ($m \neq n$), the matrix of coefficients is not square. Therefore, no determinant can be associated with it. Thus, to treat such systems, a more general concept than the determinants is needed. We introduce such a concept in this section.

B.5.1 Rank of a matrix

The general concept needed to develop the solution procedure for a general $m \times n$ system of equations is known as the *rank of the matrix, defined as the order of the largest nonsingular square submatrix of the given matrix.* Using the idea of a rank of a matrix, we can develop a general theory for the solution of a linear system of equations.

Let r be the rank of an $m \times n$ matrix $\mathbf{A}$. Then r satisfies the following conditions:

1. For $m < n$, $r \le m < n$ (if $r = m$, the matrix is said to have full row rank).
2. For $n < m$, $r \le n < m$ (if $r = n$, the matrix is said to have full column rank).
3. For $n = m$, $r \le n$ (if $r = n$, the square matrix is called nonsingular).

In order to determine the rank of a matrix, we need to check the determinants of all the submatrices. This is a cumbersome and time-consuming process. However, it turns out that the Gauss-Jordan elimination process can be used to solve the linear system as well as determine the rank of the matrix. Using the Gauss-Jordan elimination procedure any $m \times n$ matrix $\mathbf{A}$ can be transformed into the following equivalent form (for $m < n$):

$$\mathbf{A} \sim \left[\begin{array}{c|c} \mathbf{I}_{(r)} & \mathbf{0}_{(r \times n-r)} \\ \hline \mathbf{0}_{(m-r \times r)} & \mathbf{0}_{(m-r \times n-r)} \end{array}\right] \tag{B.32}$$

where $\mathbf{I}_{(r)}$ is the $r \times r$ identity matrix. Then r is the rank of the matrix, where r satisfies one of the preceding three conditions. Note that the identity matrix $\mathbf{I}_{(r)}$ is unique for any given matrix.

Example B.6 Rank determination by elementary operations. Determine rank of the following matrix:

$$\mathbf{A} = \begin{bmatrix} 2 & 6 & 2 & 4 \\ -2 & -4 & 2 & 2 \\ 1 & 2 & -1 & -1 \end{bmatrix}$$

Solution. The elementary operations lead to the following matrices:

$$\mathbf{A} \sim \begin{bmatrix} 1 & 3 & 1 & 2 \\ -2 & -4 & 2 & 2 \\ 1 & 2 & -1 & -1 \end{bmatrix} \quad \text{(multiply row 1 by } \tfrac{1}{2}\text{)}$$

$$\sim \begin{bmatrix} 1 & 3 & 1 & 2 \\ 0 & 2 & 4 & 6 \\ 0 & -1 & -2 & -3 \end{bmatrix} \quad \begin{array}{l} \text{(add 2 times row 1 to row 2} \\ \text{and } -1 \text{ times row 1 to row 3)} \end{array}$$

$$\sim \begin{bmatrix} 1 & 0 & 0 & 0 \\ 0 & 2 & 4 & 6 \\ 0 & -1 & -2 & -3 \end{bmatrix} \quad \begin{array}{l} \text{(add } -3 \text{ times column 1 to column 2;} \\ -1 \text{ times column 1 to column 3;} \\ -2 \text{ times column 1 to column 4)} \end{array}$$

$$\sim \begin{bmatrix} 1 & 0 & 0 & 0 \\ 0 & 1 & 2 & 3 \\ 0 & 0 & 0 & 0 \end{bmatrix} \quad \begin{array}{l} \text{(multiply row 2 by } \tfrac{1}{2} \text{ and} \\ \text{add to row 3)} \end{array}$$

$$\sim \left[\begin{array}{cc|cc} 1 & 0 & 0 & 0 \\ 0 & 1 & 0 & 0 \\ \hline 0 & 0 & 0 & 0 \end{array}\right] \quad \begin{array}{l} \text{(add } -2 \text{ times column 2 to column 3;} \\ \text{add } -3 \text{ times column 2 to column 4)} \end{array}$$

This matrix is in the form of Eq. (B.32). The rank of $\mathbf{A}$ is 2, since a 2×2 identity matrix is obtained at the upper left corner. ‖

B.5.2 General Theory for Solution of Linear Equations

Let us now consider solving a system of m simultaneous equations in n unknowns. *The existence of a solution for such a system depends on the rank of the system's coefficient matrix and the augmented matrix.* Let the system be represented as

$$\mathbf{Ax} = \mathbf{b} \tag{B.33}$$

where $\mathbf{A}$ is an $m \times n$ matrix, $\mathbf{b}$ is an m-vector and $\mathbf{x}$ is an n-vector of the unknowns. Note that m may be larger than n, i.e. there may be more equations than the number of unknowns. In that case the system is either *inconsistent* (has no solution) or some of the equations are redundant and may be deleted. The solution process described in the following provides answers to these questions.

Note that if an equation is multiplied by a constant the solution of the system is unchanged. If c times one equation is added to another, the solution of the resulting system is the same as for the original system. Also, if two columns of the coefficient matrix are interchanged (for example, columns i and j), the resulting set of equations is equivalent to the original system; however, the solution variables x_i and x_j are interchanged in the vector $\mathbf{x}$ as follows:

$$\mathbf{x} = [x_1 \quad x_2 \ldots x_{i-1} \quad \overset{\downarrow}{x_j} \quad x_{i+1} \ldots x_{j-1} \quad \overset{\downarrow}{x_i} \quad x_{j+1} \ldots x_n]^T \tag{B.34}$$

This indicates that each column of the coefficient matrix has a variable associated with it as also noted earlier, e.g. x_i and x_j for the ith and the jth columns respectively.

Using the elementary row–column operations, it is always possible to convert a system of m equations in n unknowns in Eq. (B.33) into an equivalent system of the form shown in the following Eq. (B.35). In the equation, a '¯' over each element indicates its new value, obtained by performing row–column operations on the augmented matrix of the original system. The value of subscript r in Eq. (B.35) is the rank of the *coefficient matrix*.

$$\begin{bmatrix}
1 & \bar{a}_{12} & \bar{a}_{13} & \bar{a}_{14} & \ldots & \bar{a}_{1r} & \ldots & \bar{a}_{1n} \\
0 & 1 & \bar{a}_{23} & \bar{a}_{24} & \ldots & \bar{a}_{2r} & \ldots & \bar{a}_{2n} \\
0 & 0 & 1 & \bar{a}_{34} & \ldots & \bar{a}_{3r} & \ldots & \bar{a}_{3n} \\
\cdot & \cdot & \cdot & 1 & & \cdot & & \cdot \\
\cdot & \cdot & \cdot & \cdot & & \cdot & & \cdot \\
\cdot & \cdot & \cdot & \cdot & & \cdot & & \cdot \\
0 & 0 & 0 & 0 & \ldots & 1 & \ldots & \bar{a}_{rn} \\
0 & 0 & 0 & 0 & \ldots & 0 & \ldots & 0 \\
\cdot & \cdot & \cdot & \cdot & & \cdot & & \cdot \\
\cdot & \cdot & \cdot & \cdot & & \cdot & & \cdot \\
\cdot & \cdot & \cdot & \cdot & \ldots & & & \cdot \\
0 & 0 & 0 & 0 & \ldots & 0 & & 0
\end{bmatrix}
\begin{bmatrix}
x_1 \\ x_2 \\ x_3 \\ \cdot \\ \cdot \\ \cdot \\ \cdot \\ \cdot \\ \cdot \\ \cdot \\ \cdot \\ x_n
\end{bmatrix}
=
\begin{bmatrix}
\bar{b}_1 \\ \bar{b}_2 \\ \bar{b}_3 \\ \cdot \\ \cdot \\ \cdot \\ \bar{b}_r \\ \bar{b}_{r+1} \\ \cdot \\ \cdot \\ \cdot \\ \bar{b}_m
\end{bmatrix}
\tag{B.35}$$

Note that if $\bar{b}_{r+1} = \bar{b}_{r+2} = \ldots = \bar{b}_m = 0$ in Eq. (B.35) then the last $(m - r)$ equations become

$$0x_1 + 0x_2 + \ldots + 0x_n = 0$$

These rows can be eliminated from further consideration. However, if any of the last $(m - r)$ components of vector $\bar{\mathbf{b}}$ is not 0, then at least one of the last $(m - r)$ equations is *inconsistent* and the system has no solution. Note also that the rank of the coefficient matrix equals the rank of the augmented matrix if and only if $\bar{b}_i = 0$, $i = (r + 1)$ to m. Thus, a *system of m equations in n unknowns is consistent (i.e. possesses solutions) if and only if the rank of the coefficient matrix equals the rank of the augmented matrix.*

If elementary operations are performed below as well as above the main diagonal to eliminate off-diagonal elements, an equivalent system of the following form is obtained:

$$\left[\begin{array}{c|c} \mathbf{I}_{(r)} & \mathbf{Q}_{(r \times n-r)} \\ \hline \mathbf{0}_{(m-r \times r)} & \mathbf{0}_{(m-r \times n-r)} \end{array}\right] \left[\begin{array}{c} \mathbf{x}_{(r)} \\ \mathbf{x}_{(n-r)} \end{array}\right] = \left[\begin{array}{c} \mathbf{q}_{(r \times 1)} \\ \mathbf{p}_{(m-r \times 1)} \end{array}\right] \tag{B.36}$$

Here $\mathbf{I}_{(r)}$ is an $r \times r$ identity matrix, and $\mathbf{x}_{(r)}$ and $\mathbf{x}_{(n-r)}$ are the r-component and $(n - r)$-component subvectors of vector $\mathbf{x}$. Note that depending on the values of r, n and m, the equation can have several different forms. For example, if $r = n$, the matrices $\mathbf{Q}_{(r \times n-r)}$, $\mathbf{0}_{(m-r \times n-r)}$ and the vector $\mathbf{x}_{(n-r)}$ disappear; similarly, if $r = m$, matrices $\mathbf{0}_{(m-r \times r)}$, $\mathbf{0}_{(m-r \times n-r)}$ and the vector $\mathbf{p}_{(m-r \times 1)}$ disappear. The system of equations (B.33) is consistent only if vector $\mathbf{p} = \mathbf{0}$ in Eq. (B.36). *It must be remembered that for every interchange of columns necessary to produce Eq. (B.36), the corresponding components of* $\mathbf{x}$ *must be interchanged.*

When the system is consistent, the first line of Eq. (B.36) gives

$$\mathbf{I}_{(r)}\mathbf{x}_{(r)} + \mathbf{Q}\mathbf{x}_{(n-r)} = \mathbf{q} \tag{B.37}$$

or

$$\mathbf{x}_{(r)} = \mathbf{q} - \mathbf{Q}\mathbf{x}_{(n-r)} \tag{B.38}$$

Equation (B.38) gives r components of $\mathbf{x}$ in terms of the remaining $(n - r)$ components. If the system is consistent, Eq. (B.38) represents the *general solution* of the system of equations $\mathbf{Ax} = \mathbf{b}$. The last $(n - r)$ components of $\mathbf{x}$ can be assigned arbitrary values; any assignment to $x_{r+1}, \ldots, x_n$ yields a solution. Thus, the system of equations has infinitely many solutions. If $r = n$, the solution is unique. Equation (B.36) is known as the *canonical representation* for the system of equations $\mathbf{Ax} = \mathbf{b}$. This form of equations is very useful in solving linear programming problems in Chapter 4.

The following examples illustrates the Gauss-Jordan elimination procedure.

Example B.7 General solution by Gauss-Jordan reduction. Find a general solution for the set of equations

$$x_1 + x_2 + x_3 + 5x_4 = 6$$

$$x_1 + x_2 - 2x_3 - x_4 = 0$$

$$x_1 + x_2 - x_3 + x_4 = 2$$

Solution. The augmented matrix for the set of equations is given as

$$\mathbf{A} \sim \begin{array}{c} \begin{array}{ccccc} x_1 & x_2 & x_3 & x_4 & \mathbf{b} \end{array} \\ \begin{bmatrix} 1 & 1 & 1 & 5 & 6 \\ 1 & 1 & -2 & -1 & 0 \\ 1 & 1 & -1 & 1 & 2 \end{bmatrix} \end{array} \quad \text{and } \mathbf{x} = \begin{bmatrix} x_1 \\ x_2 \\ x_3 \\ x_4 \end{bmatrix}$$

The following elimination steps are used to transform the system to a canonical form:

1. Subtracting row 1 from rows 2 and 3, we convert elements below the main diagonal in the first column (a_{21} and a_{31}) to zero, i.e. we eliminate x_1 from equations 2 and 3, and obtain

$$\mathbf{A} \sim \begin{array}{c} \begin{array}{ccccc} x_1 & x_2 & x_3 & x_4 & \mathbf{b} \end{array} \\ \begin{bmatrix} 1 & 1 & 1 & 5 & 6 \\ 0 & 0 & -3 & -6 & -6 \\ 0 & 0 & -2 & -4 & -4 \end{bmatrix} \end{array}$$

2. Now, since a_{22} is zero we cannot proceed any further with the elimination process. We must interchange rows and/or columns to bring a nonzero element at location a_{22}. We can interchange either column 3 or column 4 with column 2 to bring a nonzero element in the position a_{22}. (*Note*: The last column can never be interchanged with any other column; it is the right-hand side of the system $\mathbf{Ax} = \mathbf{b}$, so it does not correspond to a variable). Interchanging column 2 with column 3 (elementary column operation 1), we obtain

$$\mathbf{A} \sim \begin{array}{c} \begin{array}{ccccc} x_1 & x_3 & x_2 & x_4 & \mathbf{b} \end{array} \\ \begin{bmatrix} 1 & 1 & 1 & 5 & 6 \\ 0 & -3 & 0 & -6 & -6 \\ 0 & -2 & 0 & -4 & -4 \end{bmatrix} \end{array} \quad \text{and } \mathbf{x} = \begin{bmatrix} x_1 \\ x_3 \\ x_2 \\ x_4 \end{bmatrix}$$

Note that the positions of the variables x_2 and x_3 are also interchanged in the vector $\mathbf{x}$.

3. Now, dividing row 2 by -3, multiplying it by 2 and adding to row 3 gives

$$\mathbf{A} \sim \begin{array}{c} \begin{array}{ccccc} x_1 & x_3 & x_2 & x_4 & \mathbf{b} \end{array} \\ \begin{bmatrix} 1 & 1 & 1 & 5 & 6 \\ 0 & 1 & 0 & 2 & 2 \\ \hline 0 & 0 & 0 & 0 & 0 \end{bmatrix} \end{array} \tag{a}$$

Thus elements below the main diagonal in Eq. (a) are zero and the Gaussian elimination process is complete.

4. To put the equations in the canonical form of Eq. (B.36), we need to perform elimination above the main diagonal also (Gauss-Jordan elimination). Subtracting row 2 from row 1, we obtain

$$
\mathbf{A} \sim \begin{array}{cccc} x_1 & x_3 & x_2 & x_4 \ \ \mathbf{b} \end{array} \left[\begin{array}{cccc|c} 1 & 0 & 1 & 3 & 4 \\ 0 & 1 & 0 & 2 & 2 \\ 0 & 0 & 0 & 0 & 0 \end{array}\right] \tag{b}
$$

5. Using the matrix of Eq. (b), the given system of equations is transformed into a canonical form of Eq. (B.36) as follows:

$$
\begin{bmatrix} 1 & 0 & 1 & 3 \\ 0 & 1 & 0 & 2 \\ 0 & 0 & 0 & 0 \end{bmatrix} \begin{bmatrix} x_1 \\ x_3 \\ x_2 \\ x_4 \end{bmatrix} = \begin{bmatrix} 4 \\ 2 \\ 0 \end{bmatrix}
$$

or

$$
\left[\begin{array}{c|c} \mathbf{I}_{(2)} & \mathbf{Q}_{(2\times2)} \\ \hline \mathbf{0}_{(1\times2)} & \mathbf{0}_{(1\times2)} \end{array}\right] \begin{bmatrix} x_1 \\ x_3 \\ x_2 \\ x_4 \end{bmatrix} = \begin{bmatrix} \mathbf{q}_{(2\times1)} \\ \mathbf{P}_{(1\times1)} \end{bmatrix}
$$

where

$$
\mathbf{Q} = \begin{bmatrix} 1 & 3 \\ 0 & 2 \end{bmatrix}; \qquad \mathbf{q} = \begin{bmatrix} 4 \\ 2 \end{bmatrix}; \qquad \mathbf{p} = 0
$$

$$
\mathbf{x}_{(r)} = (x_1, x_3), \quad \mathbf{x}_{(n-r)} = (x_2, x_4)
$$

6. Since $\mathbf{p} = 0$, the given system of equations is consistent (i.e. it has solutions). Its general solution, in the form of Eq. (B.38) is

$$
\begin{bmatrix} x_1 \\ x_3 \end{bmatrix} = \begin{bmatrix} 4 \\ 2 \end{bmatrix} - \begin{bmatrix} 1 & 3 \\ 0 & 2 \end{bmatrix} \begin{bmatrix} x_2 \\ x_4 \end{bmatrix}
$$

Or, in the expanded notation, the general solution is

$$
\begin{aligned} x_1 &= 4 - x_2 - 3x_4 \\ x_3 &= 2 - 2x_4 \end{aligned} \tag{c}
$$

7. It can be seen that the general solution of Eqs. (c) gives x_1 and x_3 in terms of x_2 and x_4; i.e. x_2 and x_4 are independent variables and x_1 and x_3 are dependent on them. The system has infinite solutions because any specification for x_2 and x_4 gives a solution.

‖

B.5.2.1 BASIC SOLUTIONS. In the preceding general solution, we see that x_2 and x_4 can be given arbitrary values, and the corresponding x_1 and x_3 calculated from the Eq. (c). *Thus the system has an infinite number of solutions.* A particular solution of much interest in linear programming (LP) is obtained by setting $\mathbf{x}_{(n-r)} = 0$ in the general solution of Eq. (B.38). Such a solution is called the *basic solution* of the system of equations $\mathbf{Ax} = \mathbf{b}$. For the present example, a basic solution is $x_1 = 4$, $x_2 = 0$, $x_3 = 2$ and $x_4 = 0$ which is obtained from Eq. (c) by setting $x_2 = x_4 = 0$.

Note that although Eqs (c) give an infinite number of solutions for the system of equations, the number of basic solutions is finite. For example, another basic solution can be obtained by setting $x_2 = x_3 = 0$ and solving for x_1 and x_4. It can be verified that this basic solution is $x_1 = 1$, $x_2 = 0$, $x_3 = 0$ and $x_4 = 1$. The fact that the number of basic solutions is finite is very important for linear programming problems discussed in Chapter 4. The reason is that the *optimum solution for an LP problem is one of the basic solutions.*

Example B.8   Gauss-Jordan reduction process in tabular form. Find a general solution for the following set of equations using a tabular form of the Gauss-Jordan reduction process:

$$-x_1 + 2x_2 - 3x_3 + x_4 = -1$$
$$2x_1 + x_2 + x_3 - 2x_4 = 2$$
$$x_1 - x_2 + 2x_3 + x_4 = 3$$
$$x_1 + 3x_2 - 2x_2 - x_4 = 1$$

Solution. The iterations of the Gauss-Jordan reduction process for the linear system are explained in Table B.1. Three steps are needed to reduce the given system to the canonical form of Eq. (B.36). Note that at the second step, the element a_{33} is zero, so it cannot be used as a pivot element. Therefore, we use element a_{34} as the pivot element and perform elimination in the x_4 column. This effectively means that we interchange column x_3 with column x_4 (as was done in Example B.7).

Re-writing the results from the 3rd step of Table B.1 in the form of Eq. (B.36), we get

$$
\begin{array}{cccc}
x_1 & x_2 & x_4 & x_3
\end{array}
$$
$$
\begin{bmatrix}
1 & 0 & 0 & 1 \\
0 & 1 & 0 & -1 \\
0 & 0 & 1 & 0 \\
0 & 0 & 0 & 0
\end{bmatrix}
\begin{bmatrix}
x_1 \\ x_2 \\ x_4 \\ x_3
\end{bmatrix}
=
\begin{bmatrix}
2 \\ 0 \\ 1 \\ 0
\end{bmatrix}
$$

Since the last equation essentially gives $0 = 0$, the given system of equations is consistent (i.e. it has solutions). Also, since the rank of the coefficient matrix is 3, which is less than the number of equations, there are an infinite number of solutions for the linear system. From the preceding equation the general solution is given as

$$x_1 = 2 - x_3$$
$$x_2 = 0 + x_3$$
$$x_4 = 1$$

A basic solution is obtained by setting x_3 to zero as $x_1 = 2$, $x_2 = 0$, $x_3 = 0$, $x_4 = 1$.   ‖

TABLE B.1

General solution of linear system of equations of Example B.8 by Gauss-Jordan elimination

Step	x_1	x_2	x_3	x_4	b	
Initial	−1	2	−3	1	−1	Divide row 1 by −1 and use it to
	2	1	1	−2	2	perform elimination in column x_1,
	1	−1	2	1	3	e.g. multiply new row 1 by 2
	1	3	−2	−1	1	and subtract it from row 2, etc.
1st step	1	−2	3	−1	1	
	0	5	−5	0	0	Divide row 2 by 5 and perform
	0	1	−1	2	2	elimination in column x_2
	0	5	−5	0	0	
2nd step	1	0	1	−1	1	
	0	1	−1	0	0	Divide row 3 by 2 and perform
	0	0	0	2	2	elimination in column x_4
	0	0	0	0	0	
3rd step	1	0	1	0	2	Canonical form with columns x_1,
	0	1	−1	0	0	x_2 and x_4 containing the identity
	0	0	0	1	1	matrix
	0	0	0	0	0	

To summarize the results of this section, we note that

1. The $m \times n$ system of equations (B.33) is *consistent* if the rank of the coefficient matrix is the same as the rank of the augmented matrix. A consistent system implies that it has a solution.
2. If the number of equations is less than the number of variables ($m < n$) and the system is consistent, having rank less than or equal to m ($r \leq m$), then it has infinitely many solutions.
3. If $m = n = r$, then the system (B.33) has a unique solution.

B.6 CONCEPTS RELATED TO A SET OF VECTORS

In several applications, we come across a set of vectors. It is useful to discuss some concepts related to these sets, such as the linear independence of vectors, and vector spaces. In this section, we briefly discuss these concepts and describe a procedure for checking the linear independence of a set of vectors.

B.6.1 Linear Independence of a Set of Vectors

Consider a set of k vectors each of dimension n:

$$A = \{\mathbf{a}^{(1)}, \mathbf{a}^{(2)}, \ldots, \mathbf{a}^{(k)}\}$$

where a superscript (i) represents the ith vector. A *linear combination* of vectors in the set A is another vector obtained by scaling each vector in A and adding all the resulting vectors. That is, if **b** is a linear combination of a vector in A, then it is defined as

$$\mathbf{b} = x_1\mathbf{a}^{(1)} + x_2\mathbf{a}^{(2)} + \ldots + x_k\mathbf{a}^{(k)} = \sum_{i=1}^{k} x_i\mathbf{a}^{(i)} \qquad \text{(B.39a)}$$

where $x_1, x_2, \ldots, x_k$ are some scalars. The preceding equation can be written compactly in matrix form as

$$\mathbf{b} = \mathbf{A}\mathbf{x} \qquad \text{(B.39b)}$$

where **x** is a k-component vector and **A** is an $n \times k$ matrix with vectors $\mathbf{a}^{(i)}$ as its columns.

To determine if the set of vectors is *linearly independent* or *dependent*, we set the linear combination of Eq. (B.39) to zero as

$$x_1\mathbf{a}^{(1)} + x_2\mathbf{a}^{(2)} + \ldots + x_k\mathbf{a}^{(k)} = \mathbf{0}; \quad \text{or} \quad \mathbf{A}\mathbf{x} = \mathbf{0} \qquad \text{(B.40)}$$

This gives a homogeneous system of equations with x_i's as unknowns. There are n equations in k unknowns. Note that $\mathbf{x} = \mathbf{0}$ satisfies Eq. (B.40). *If $\mathbf{x} = \mathbf{0}$ is the only solution then the set of vectors is linearly independent. In this case rank r of the matrix $\mathbf{A}$ must be equal to k (the number of vectors in the set). If there exists a set of scalars x_i not all zero satisfying Eq. (B.40), then the vectors $\mathbf{a}^{(1)}$, $\mathbf{a}^{(2)}, \ldots, \mathbf{a}^{(k)}$ are said to be linearly dependent. In this case rank r of $\mathbf{A}$ is less than k.*

If a set of vectors is linearly dependent, then one or more vectors are parallel to each other, or there is at least one vector that can be expressed as a linear combination of the rest. That is, at least one of the scalars $x_1, x_2, \ldots, x_k$ must be nonzero. If we assume x_j to be nonzero, then Eq. (B.40) can be written as follows:

$$-x_j\mathbf{a}^{(j)} = x_1\mathbf{a}^{(1)} + x_2\mathbf{a}^{(2)} + \ldots + x_{j-1}\mathbf{a}^{(j-1)} + x_{j+1}\mathbf{a}^{(j+1)}$$

$$+ \ldots + x_k\mathbf{a}^{(k)} = \sum_{i=1}^{k} x_i\mathbf{a}^{(i)}; \qquad i \neq j$$

Or, since $x_j \neq 0$, we can divide both sides by it to obtain

$$\mathbf{a}^{(j)} = -\sum_{i=1}^{k} (x_i/x_j)\mathbf{a}^{(i)}; \qquad i \neq j \qquad \text{(B.41)}$$

in Eq. (B.41) we have expressed $\mathbf{a}^{(j)}$ as a *linear combination* of $\mathbf{a}^{(1)}$, $\mathbf{a}^{(2)}, \ldots, \mathbf{a}^{(j-1)}, \mathbf{a}^{(j+1)}, \ldots, \mathbf{a}^{(k)}$. In general, we see that if a set of vectors is linearly dependent, then at least one of them can be expressed as a linear combination of the rest.

Example B.9 Check for linear independence of vectors. Check linear independence of the following set of vectors:

(i) $\mathbf{a}^{(1)} = \begin{bmatrix} 2 \\ 5 \\ 2 \\ -1 \end{bmatrix}$, $\mathbf{a}^{(2)} = \begin{bmatrix} 3 \\ 2 \\ 1 \\ 0 \end{bmatrix}$, $\mathbf{a}^{(3)} = \begin{bmatrix} 8 \\ 9 \\ 4 \\ -1 \end{bmatrix}$

(ii) $\mathbf{a}^{(1)} = \begin{bmatrix} 2 \\ 6 \\ 2 \\ -2 \end{bmatrix}$, $\mathbf{a}^{(2)} = \begin{bmatrix} 4 \\ 3 \\ 2 \\ 0 \end{bmatrix}$, $\mathbf{a}^{(3)} = \begin{bmatrix} 6 \\ 9 \\ 4 \\ 1 \end{bmatrix}$

Solution. To check for linear independence, we form the linear combination of Eq. (B.39) and set it to zero as in Eq. (B.40). The resulting homogeneous system of equations is solved for the scalars x_i. If all the scalars are zero, then the given set of vectors is linearly independent; otherwise it is dependent.

The vectors in set (i) are linearly dependent, since $x_1 = 1$, $x_2 = 2$ and $x_3 = -1$ give the linear combination of Eq. (B.40) a zero value, i.e.

$$\mathbf{a}^{(1)} + 2\mathbf{a}^{(2)} - \mathbf{a}^{(3)} = \mathbf{0}$$

It may also be checked that the rank of the following matrix whose columns are the given vectors is only 2; so the set of vectors is linearly dependent:

$$\mathbf{A} = \begin{bmatrix} 2 & 3 & 8 \\ 5 & 2 & 9 \\ 2 & 1 & 4 \\ -1 & 0 & -1 \end{bmatrix}$$

For set (ii), let us form a linear combination of the given vectors and set it to zero:

$$x_1 \mathbf{a}^{(1)} + x_2 \mathbf{a}^{(2)} + x_3 \mathbf{a}^{(3)} = \mathbf{0} \tag{a}$$

This is a vector equation which gives the following system when written in the expanded form:

$$2x_1 + 4x_2 + 6x_3 = 0 \tag{b}$$

$$6x_1 + 3x_2 + 9x_3 = 0 \tag{c}$$

$$2x_1 + 2x_2 + 4x_3 = 0 \tag{d}$$

$$-2x_1 \qquad + x_3 = 0 \tag{e}$$

We solve the preceding system of equations by the elimination process. From Eq. (e), we find $x_3 = 2x_1$. Equations (b) to (d) then become

$$14x_1 + 4x_2 = 0 \tag{f}$$

$$24x_1 + 3x_2 = 0 \tag{g}$$

$$10x_1 + 2x_2 = 0 \tag{h}$$

From Eq. (h), we find $x_2 = -5x_1$. Substituting this result into Eqs (f) and (g) gives

$$14x_1 + 4(-5x_1) = -6x_1 = 0 \tag{i}$$

$$24x_1 + 3(-5x_1) = 9x_1 = 0 \tag{j}$$

Equations (i) and (j) imply $x_1 = 0$; therefore, $x_2 = -5x_1 = 0$, $x_3 = 2x_1 = 0$. Thus, the only solution to Eq. (a) is the trivial solution $x_1 = x_2 = x_3 = 0$. The vectors $\mathbf{a}^{(1)}$, $\mathbf{a}^{(2)}$ and $\mathbf{a}^{(3)}$ are therefore linearly independent. ‖

Equation (B.40) may be considered as a set of n simultaneous equations in k unknowns. To see this, define the k vectors as

$$\mathbf{a}^{(1)} = \begin{bmatrix} a_{11} \\ a_{21} \\ a_{31} \\ \cdot \\ \cdot \\ \cdot \\ a_{n1} \end{bmatrix}, \mathbf{a}^{(2)} = \begin{bmatrix} a_{12} \\ a_{22} \\ a_{32} \\ \cdot \\ \cdot \\ \cdot \\ a_{n2} \end{bmatrix} \ldots, \mathbf{a}^{(k)} = \begin{bmatrix} a_{1k} \\ a_{2k} \\ a_{3k} \\ \cdot \\ \cdot \\ \cdot \\ a_{nk} \end{bmatrix}, \mathbf{x} = \begin{bmatrix} x_1 \\ x_2 \\ \cdot \\ \cdot \\ \cdot \\ x_k \end{bmatrix}$$

Also, let $\mathbf{A}_{(n \times k)} = [\mathbf{a}^{(1)}, \mathbf{a}^{(2)}, \ldots, \mathbf{a}^{(k)}]$, i.e. $\mathbf{A}$ is a matrix whose ith column is the ith vector $\mathbf{a}^{(i)}$. Then Eq. (B.40) can be written as

$$\mathbf{Ax} = \mathbf{0} \tag{B.42}$$

The results of Section B.5 show that there is a unique solution to Eq. (B.42) if and only if the rank r of $\mathbf{A}$ is equal to k $(r = k < n)$, the number of columns of $\mathbf{A}$. In that case, the unique solution is $\mathbf{x} = \mathbf{0}$. Therefore, *the vectors* $\mathbf{a}^{(1)}, \mathbf{a}^{(2)}, \ldots, \mathbf{a}^{(k)}$ *are linearly independent if and only if the rank of the matrix* $\mathbf{A}$ *is* k (the number of vectors in the set).

Note that if $k > n$, then the rank of $\mathbf{A}$ cannot exceed n. Therefore, $\mathbf{a}^{(1)}, \mathbf{a}^{(2)}, \ldots, \mathbf{a}^{(k)}$ will always be linearly dependent if $k > n$. Thus, the maximum number of linearly dependent n-component vectors is n. Any set of $(n + 1)$ vectors is always linearly dependent.

Given any set of n linearly independent (n-component) vectors, $\mathbf{a}^{(1)}, \mathbf{a}^{(2)}, \ldots, \mathbf{a}^{(n)}$, any other ($n$-component) vector $\mathbf{b}$ can be expressed as a unique linear combination of these vectors. The problem is to choose a set of scalars $x_1, x_2, \ldots, x_n$ such that

$$x_1\mathbf{a}^{(1)} + x_2\mathbf{a}^{(2)} + \ldots + x_n\mathbf{a}^{(n)} = \mathbf{b}; \quad \text{or} \quad \mathbf{Ax} = \mathbf{b} \tag{B.43}$$

We wish to show that a solution exists for Eq. (B.43) and it is unique. Note that $\mathbf{a}^{(1)}, \mathbf{a}^{(2)}, \ldots, \mathbf{a}^{(n)}$ are linearly independent. Therefore, the rank of the coefficient matrix $\mathbf{A}$ is n, and the rank of the augmented matrix $[\mathbf{A}, \mathbf{b}]$ is also n. It cannot be $(n + 1)$ because the matrix has only n rows. Thus, Eq. (B.43) always possesses a solution for any given $\mathbf{b}$. Moreover, $\mathbf{A}$ is nonsingular, hence the solution is unique.

In summary, we state the following points for a k set of vectors each having n components:

1. If $k > n$, the set of vectors is always linearly dependent, e.g. three vectors each having two components. That is, the number of linearly independent vectors is always less than or equal to n, e.g. for two-component vectors, there are at the most two linearly independent vectors.
2. If there are n linearly independent vectors each of dimension n, then any other n-component vector can be expressed as a unique linear combination of them, e.g. given two linearly independent vectors $\mathbf{a}^{(1)} = (1, 0)$ and $\mathbf{a}^{(2)} = (0, 1)$ of dimension two, any other vector such as $\mathbf{b} = (b_1, b_2)$ can be expressed as a unique linear combination of $\mathbf{a}^{(1)}$ and $\mathbf{a}^{(2)}$.
3. Linear independence of the given set of vectors can be determined in two ways:

   (i) Form the matrix $\mathbf{A}$ of dimension $n \times k$ whose columns are the given vectors. Then, if rank r is equal to k $(r = k)$, the given set is linearly independent; otherwise it is dependent.
   (ii) Set the linear combination of the given vectors to zero as $\mathbf{Ax} = \mathbf{0}$. If $\mathbf{x} = \mathbf{0}$ is the only solution for the resulting system, then the set is independent; otherwise it is dependent.

B.6.2 Vector Spaces

Before defining the concept of a vector space, let us define closure under addition and scalar multiplication:

Definition: Closure under addition. A set of vectors is said to be closed under addition if the sum of any two vectors in the set is also in the set. ‖

Definition: closure under scalar multiplication. A set of vectors is said to be closed under scalar multiplication if the product of any vector in the set by a scalar gives a vector in the set. ‖

Definition: vector space. A nonempty set S of elements (vectors) $\mathbf{x}, \mathbf{y}, \mathbf{z}, \ldots$ is called a vector space if the two algebraic operations (vector addition and multiplication by a real scalar) on them satisfy the following properties:

1. Closure under addition: if $\mathbf{x} \in S$ and $\mathbf{y} \in S$ then $\mathbf{x} + \mathbf{y} \in S$.
2. Commutative in addition: $\mathbf{x} + \mathbf{y} = \mathbf{y} + \mathbf{x}$.
3. Associative in addition: $(\mathbf{x} + \mathbf{y}) + \mathbf{z} = \mathbf{x} + (\mathbf{y} + \mathbf{z})$.
4. Identity in addition: there exists a zero vector $\mathbf{0}$ in the set S such that $\mathbf{x} + \mathbf{0} = \mathbf{x}$ for all $\mathbf{x}$.
5. Inverse in addition: there exists a $-\mathbf{x}$ in the set S such that $\mathbf{x} + (-\mathbf{x}) = \mathbf{0}$ for all $\mathbf{x}$.

6. Closure under scalar multiplication: for real scalars $\alpha, \beta, \ldots$, if $\mathbf{x} \in S$ then $\alpha \mathbf{x} \in S$.

7. Distributive: $(\alpha + \beta)\mathbf{x} = \alpha \mathbf{x} + \beta \mathbf{x}$.

8. Distributive: $\alpha(\mathbf{x} + \mathbf{y}) = \alpha \mathbf{x} + \alpha \mathbf{y}$.

9. Associative in scalar multiplication: $(\alpha \beta)\mathbf{x} = \alpha(\beta \mathbf{x})$.

10. Identity in scalar multiplication: $1\mathbf{x} = \mathbf{x}$. ||

In the preceding section, it was noted that the maximum number of linearly independent vectors in the set of all n-component vectors is n. Thus, for every subset of this set, there exists some maximum number of linearly independent vectors. In particular, every vector space has a maximum number of linearly independent vectors. This number is called the *dimension of the vector space*. If a vector space has dimension k, then any set of k linearly independent vectors in the vector space is called *a basis* for the vector space. Any other vector in the vector space can be expressed as a unique linear combination of the given set of basis vectors.

Example B.10   Check for vector space. Check if the set $S = \{(x_1, x_2, x_3) \mid x_1 = 0\}$ is a vector space.

Solution. To see this consider any two vectors in S as

$$\mathbf{x} = \begin{bmatrix} 0 \\ a \\ b \end{bmatrix} \quad \text{and} \quad \mathbf{y} = \begin{bmatrix} 0 \\ c \\ d \end{bmatrix}$$

where scalars a, b, c and d are completely arbitrary. Then,

$$\mathbf{x} + \mathbf{y} = \begin{bmatrix} 0 \\ a + c \\ b + d \end{bmatrix}$$

Therefore $\mathbf{x} + \mathbf{y}$ is in the set S. Also, for any scalar α,

$$\alpha \mathbf{x} = \begin{bmatrix} 0 \\ \alpha a \\ \alpha b \end{bmatrix}$$

Therefore $\alpha \mathbf{x}$ is in the set S. Thus S is closed under addition and scalar multiplication. All other properties for the definition of a vector space can be proved easily. To show the property (2), we have

$$\mathbf{x} + \mathbf{y} = \begin{bmatrix} 0 \\ a + c \\ b + d \end{bmatrix} = \begin{bmatrix} 0 \\ c + a \\ d + b \end{bmatrix} = \begin{bmatrix} 0 \\ c \\ d \end{bmatrix} + \begin{bmatrix} 0 \\ a \\ b \end{bmatrix} = \mathbf{y} + \mathbf{x}$$

"Associative in addition" is shown as

$$(\mathbf{x}+\mathbf{y})+\mathbf{z} = \begin{bmatrix} 0 \\ a+c \\ b+d \end{bmatrix} + \begin{bmatrix} 0 \\ e \\ f \end{bmatrix} = \begin{bmatrix} 0 \\ a+c+e \\ b+d+f \end{bmatrix}$$

$$= \begin{bmatrix} 0 \\ a \\ b \end{bmatrix} + \begin{bmatrix} 0 \\ c+e \\ d+f \end{bmatrix} = \mathbf{x}+(\mathbf{y}+\mathbf{z}).$$

For identity in addition we have a zero vector in the set S as

$$\mathbf{0} = \begin{bmatrix} 0 \\ 0 \\ 0 \end{bmatrix}$$

such that

$$\mathbf{x}+\mathbf{0} = \begin{bmatrix} 0 \\ a+0 \\ b+0 \end{bmatrix} = \begin{bmatrix} 0 \\ a \\ b \end{bmatrix} = \mathbf{x}$$

Inverse in addition exists if we define $-\mathbf{x}$ as

$$-\mathbf{x} = - \begin{bmatrix} 0 \\ a \\ b \end{bmatrix} = \begin{bmatrix} 0 \\ -a \\ -b \end{bmatrix}$$

such that

$$\mathbf{x}+(-\mathbf{x}) = \begin{bmatrix} 0 \\ a+(-a) \\ b+(-b) \end{bmatrix} = \begin{bmatrix} 0 \\ 0 \\ 0 \end{bmatrix} = \mathbf{0}$$

In a similar way, properties (7) to (10) can easily be shown.

Therefore, the set S is a vector space. Note that the set $V = \{(x_1, x_2, x_3) \mid x_1 = 1\}$ is not a vector space.

Let us now determine the dimension of S. Note that if $\mathbf{A}$ is a matrix whose columns are vectors in S, then $\mathbf{A}$ has three rows, and the first row contains only zeros. Thus, the rank of $\mathbf{A}$ must be less than or equal to 2, and the dimension of S is either 1 or 2. To show that it is in fact 2, we need only find two linearly independent vectors. The following are three such sets of two linearly independent vectors from the set S:

$$(\text{i}) \quad \mathbf{a}^{(1)} = \begin{bmatrix} 0 \\ 1 \\ 0 \end{bmatrix}, \quad \mathbf{a}^{(2)} = \begin{bmatrix} 0 \\ 0 \\ 1 \end{bmatrix}$$

$$(\text{ii}) \quad \mathbf{a}^{(3)} = \begin{bmatrix} 0 \\ 2 \\ 1 \end{bmatrix}, \quad \mathbf{a}^{(4)} = \begin{bmatrix} 0 \\ 0 \\ 1 \end{bmatrix}$$

$$(\text{iii}) \quad \mathbf{a}^{(5)} = \begin{bmatrix} 0 \\ 1 \\ 1 \end{bmatrix}, \quad \mathbf{a}^{(6)} = \begin{bmatrix} 0 \\ 1 \\ -1 \end{bmatrix}$$

Each of these three sets is a basis for S. Any vector in S can be expressed as a linear combination of each of these sets. If $\mathbf{x} = (0, c, d)$ is any element of S, then

(i) $\mathbf{x} = c\mathbf{a}^{(1)} + d\mathbf{a}^{(2)}$

$$\begin{bmatrix} 0 \\ c \\ d \end{bmatrix} = c\begin{bmatrix} 0 \\ 1 \\ 0 \end{bmatrix} + d\begin{bmatrix} 0 \\ 0 \\ 1 \end{bmatrix}$$

(ii) $\mathbf{x} = \dfrac{c}{2}\mathbf{a}^{(3)} + \left(c - \dfrac{d}{2}\right)\mathbf{a}^{(4)}$

$$\begin{bmatrix} 0 \\ c \\ d \end{bmatrix} = (c/2)\begin{bmatrix} 0 \\ 2 \\ 1 \end{bmatrix} + (c - d/2)\begin{bmatrix} 0 \\ 0 \\ 1 \end{bmatrix}$$

(iii) $\mathbf{x} = \left(\dfrac{c+d}{2}\right)\mathbf{a}^{(5)} + \left(\dfrac{c-d}{2}\right)\mathbf{a}^{(6)}$

$$\begin{bmatrix} 0 \\ c \\ d \end{bmatrix} = (c+d)/2\begin{bmatrix} 0 \\ 1 \\ 1 \end{bmatrix} + (c-d)/2\begin{bmatrix} 0 \\ 1 \\ -1 \end{bmatrix}$$

$\parallel$

B.7 EIGENVALUES AND EIGENVECTORS

Given an $n \times n$ matrix $\mathbf{A}$, any nonzero vector $\mathbf{x}$ satisfying

$$\mathbf{Ax} = \lambda\mathbf{x} \tag{B.44}$$

where λ is a scale factor, is called an *eigenvector* (proper or characteristic vector). The scalar λ is called the *eigenvalue* (proper or characteristic value). Since $\mathbf{x} \neq \mathbf{0}$, from Eq. (B.44) we see that λ is given as roots of the characteristic equation

$$|\mathbf{A} - \lambda\mathbf{I}| = 0 \tag{B.45}$$

Equation (B.45) gives an nth degree polynomial in λ. Roots of this polynomial are the required eigenvalues. After eigenvalues have been determined, eigenvectors can be determined from Eq. (B.44).

The coefficient matrix $\mathbf{A}$ may be symmetric or asymmetric. For many applications, $\mathbf{A}$ is a symmetric matrix, so we consider this case in the text. Some properties of eigenvalues and eigenvectors are:

1. Eigenvalues and eigenvectors of a real symmetric matrix are real. They may be complex for real nonsymmetric matrices.
2. Eigenvectors corresponding to distinct eigenvalues of real symmetric matrices are orthogonal to each other (that is, their dot product vanishes).

Example B.11 Calculation of eigenvalues and eigenvectors. Find eigenvalues and eigenvectors of the matrix

$$\mathbf{A} = \begin{bmatrix} 2 & 1 \\ 1 & 2 \end{bmatrix}$$

Solution. The eigenvalue problem is defined as

$$\begin{bmatrix} 2 & 1 \\ 1 & 2 \end{bmatrix}\begin{bmatrix} x_1 \\ x_2 \end{bmatrix} = \lambda \begin{bmatrix} x_1 \\ x_2 \end{bmatrix}$$

The characteristic polynomial is given by $|\mathbf{A} - \lambda\mathbf{I}| = 0$,

$$\begin{vmatrix} 2 - \lambda & 1 \\ 1 & 2 - \lambda \end{vmatrix} = 0$$

or,

$$\lambda^2 - 4\lambda + 3 = 0$$

The roots of this polynomial are

$$\lambda_1 = 3, \lambda_2 = 1$$

Therefore, the eigenvalues are 3 and 1.

The eigenvectors are determined from Eq. (B.44). For $\lambda_1 = 3$, Eq. (B.44) is

$$\begin{bmatrix} (2 - 3) & 1 \\ 1 & (2 - 3) \end{bmatrix}\begin{bmatrix} x_1 \\ x_2 \end{bmatrix} = \begin{bmatrix} 0 \\ 0 \end{bmatrix}$$

or, $x_1 = x_2$. Therefore, a solution of the above equation is $(1, 1)$. After normalization (dividing by its length), the first eigenvector becomes

$$\mathbf{x}^{(1)} = (1/\sqrt{2})\begin{bmatrix} 1 \\ 1 \end{bmatrix}$$

For $\lambda_2 = 1$, Eq. (B.44) is

$$\begin{bmatrix} (2 - 1) & 1 \\ 1 & (2 - 1) \end{bmatrix}\begin{bmatrix} x_1 \\ x_2 \end{bmatrix} = \begin{bmatrix} 0 \\ 0 \end{bmatrix}$$

or $x_1 = -x_2$. Therefore, a solution of the above equation is $(1, -1)$. After normalization, the second eigenvector is

$$\mathbf{x}^{(2)} = (1\sqrt{2})\begin{bmatrix} 1 \\ -1 \end{bmatrix}$$

It may be verified that $\mathbf{x}^{(1)} \cdot \mathbf{x}^{(2)}$ is zero, i.e. $\mathbf{x}^{(1)}$ and $\mathbf{x}^{(2)}$ are orthogonal to each other. ‖

B.8* NORM AND CONDITION NUMBER OF A MATRIX

B.8.1 Norm of Vectors and Matrices

Every n-dimensional vector $\mathbf{x}$ has a scalar-valued function associated with it, denoted as $\|\mathbf{x}\|$. It is called a norm of $\mathbf{x}$ if it satisfies the following three conditions:

1. $\|\mathbf{x}\| > 0$ for $\mathbf{x} = 0$, and $\|\mathbf{x}\| = 0$ only when $\mathbf{x} = \mathbf{0}$.
2. $\|\mathbf{x} + \mathbf{y}\| \leq \|\mathbf{x}\| + \|\mathbf{y}\|$ (triangle inequality).
3. $\|\mathbf{ax}\| = |a| \|\mathbf{x}\|$ where a is a scalar.

The ordinary length of a vector for $n \le 3$ satisfies the foregoing three conditions. The concept of norm is therefore a generalization of ordinary length of a vector in one-, two- or three-dimensional Euclidean space. For example, it can be verified that the Euclidean distance in the n-dimensional space

$$\|\mathbf{x}\| = \sqrt{\mathbf{x}^T \mathbf{x}} = \sqrt{\mathbf{x} \cdot \mathbf{x}} \tag{B.46}$$

satisfies the three norm conditions and hence is a norm.

Every $n \times n$ matrix $\mathbf{A}$ has a *scalar function* associated with it called its norm. It is denoted as $\|\mathbf{A}\|$ and is calculated as

$$\|\mathbf{A}\| = \max_{\mathbf{x} \ne 0} \frac{\|\mathbf{Ax}\|}{\|\mathbf{x}\|} \tag{B.47}$$

Note that since $\mathbf{Ax}$ is a vector, Eq. (B.47) says that the norm of $\mathbf{A}$ is determined by the vector $\mathbf{x}$ that maximizes the ratio $\|\mathbf{Ax}\|/\|\mathbf{x}\|$. The three conditions of the norm can be verified easily for Eq. (B.47), as follows:

1. $\|\mathbf{A}\| > 0$ unless it is a null matrix in which case it is zero.
2. $\|\mathbf{A} + \mathbf{B}\| \le \|\mathbf{A}\| + \|\mathbf{B}\|$.
3. $\|a\mathbf{A}\| = |a| \, \|\mathbf{A}\|$ where a is a scalar.

Other vector norms can also be defined. For example, the summation norm and the max-norm (called the "∞-norm") are defined as

$$\|\mathbf{x}\| = \sum_{i=1}^{n} |x_i|, \quad \text{or} \quad \|\mathbf{x}\| = \max_{1 \le i \le n} |x_i| \tag{B.48}$$

They also satisfy the three conditions of the norm of the vector $\mathbf{x}$.

If λ_1^2 is the largest eigenvalue of $\mathbf{A}^T\mathbf{A}$, then it can be shown using Eq. (B.47) that the norm of $\mathbf{A}$ is also defined as

$$\|\mathbf{A}\| = \lambda_1 > 0$$

Similarly, if λ_n^2 is the smallest eigenvalue of $\mathbf{A}^T\mathbf{A}$, then the norm of $\mathbf{A}^{-1}$ is defined as

$$\|\mathbf{A}^{-1}\| = \lambda_n > 0$$

B.8.2 Condition Number of a Matrix

The condition number is another scalar associated with an $n \times n$ matrix. The idea of a condition number is useful while solving a linear system of equations $\mathbf{Ax} = \mathbf{b}$. Often there is uncertainty in elements of the coefficient matrix $\mathbf{A}$ or the right-hand side vector $\mathbf{b}$. The question then is, how does the solution vector $\mathbf{x}$ change for small perturbations in $\mathbf{A}$ and $\mathbf{b}$? The answer to this question is contained in the condition number of the matrix $\mathbf{A}$.

It can be shown that the condition number of an $n \times n$ matrix $\mathbf{A}$, denoted as cond $(\mathbf{A})$, is given as

$$\text{cond} (\mathbf{A}) = \lambda_1/\lambda_n \geqq 0$$

where λ_1^2 and λ_n^2 are the largest and the smallest eigenvalues of $\mathbf{A}^T\mathbf{A}$. It turns out that a larger condition number indicates that the solution $\mathbf{x}$ is very sensitive to variations in the elements of $\mathbf{A}$ and $\mathbf{b}$. That is, small changes in $\mathbf{A}$ and $\mathbf{b}$ give large changes in $\mathbf{x}$.

A very large condition number for the matrix $\mathbf{A}$ indicates it to be nearly singular. The corresponding system of equations $\mathbf{A}\mathbf{x} = \mathbf{b}$ is called ill-conditioned.

EXERCISES FOR APPENDIX B

Evaluate the following determinants:

B.1 $\begin{vmatrix} 2 & 1 & 3 \\ 1 & 2 & 1 \\ 3 & 1 & 5 \end{vmatrix}$ **B.2** $\begin{vmatrix} 0 & 2 & 3 & 2 \\ 0 & 4 & 5 & 4 \\ 1 & -2 & -2 & 1 \\ 3 & -1 & 2 & 1 \end{vmatrix}$

B.3 $\begin{vmatrix} 0 & 0 & 0 & -2 \\ 0 & 0 & 5 & 3 \\ 0 & 1 & -1 & 1 \\ 2 & 3 & -3 & 2 \end{vmatrix}$

For the following determinants, calculate values of the scalar λ for which determinants vanish:

B.4 $\begin{vmatrix} 2-\lambda & 1 & 0 \\ 1 & 3-\lambda & 0 \\ 0 & 3 & 2-\lambda \end{vmatrix}$ **B.5** $\begin{vmatrix} 2-\lambda & 2 & 0 \\ 1 & 2-\lambda & 0 \\ 0 & 0 & 2-\lambda \end{vmatrix}$

Determine the rank of the following matrices:

B.6 $\begin{bmatrix} 3 & 0 & 1 & 3 \\ 2 & 0 & 3 & 2 \\ 0 & 2 & -8 & 1 \\ -2 & -1 & 2 & -1 \end{bmatrix}$ **B.7** $\begin{bmatrix} 1 & 2 & 2 & 2 & 4 \\ 1 & 6 & 3 & 0 & 3 \\ 2 & 2 & 3 & 3 & 2 \\ 1 & 3 & 2 & 5 & 1 \end{bmatrix}$

B.8 $\begin{bmatrix} 1 & 2 & 3 & 4 \\ 0 & 0 & 0 & 1 \\ 3 & 2 & 3 & 0 \\ 2 & 3 & 1 & 4 \\ 2 & 0 & 6 & 0 \\ 1 & 2 & 1 & 4 \end{bmatrix}$

Obtain solution for the following equations using the Gaussian elimination procedure:

B.9
$$2x_1 + 2x_2 + x_3 = 5$$
$$x_1 - 2x_2 + 2x_3 = 1$$
$$x_2 + 2x_3 = 3$$

B.10
$$x_2 - x_3 = 0$$
$$x_1 + x_2 + x_3 = 3$$
$$x_1 - 3x_2 = -2$$

B.11
$$2x_1 + x_2 + x_3 = 7$$
$$4x_2 - 5x_3 = -7$$
$$x_1 - 2x_2 + 4x_3 = 9$$

B.12
$$2x_1 + x_2 - 3x_3 + x_4 = 1$$
$$x_1 + 2x_2 + 5x_3 - x_4 = 7$$
$$-x_1 + x_2 + x_3 + 4x_4 = 5$$
$$2x_1 - 3x_2 + 2x_3 - 5x_4 = -4$$

B.13
$$3x_1 + x_2 + x_3 = 8$$
$$2x_1 - x_2 - x_3 = -3$$
$$x_1 + 2x_2 - x_3 = 2$$

B.14
$$x_1 + x_2 - x_3 = 2$$
$$2x_1 - x_2 + x_3 = 4$$
$$-x_1 + 2x_2 + 3x_3 = 3$$

B.15
$$-x_1 + x_2 - x_3 = -2$$
$$-2x_1 + x_2 + 2x_3 = 6$$
$$x_1 + x_2 + x_3 = 6$$

B.16
$$-x_1 + 2x_2 + 3x_3 = 4$$
$$2x_1 - x_2 - 2x_3 = -1$$
$$x_1 - 3x_2 + 4x_3 = 2$$

B.17
$$x_1 + x_2 + x_3 + x_4 = 2$$
$$2x_1 + x_2 - x_3 + x_4 = 2$$
$$-x_1 + 2x_2 + 3x_3 + x_4 = 1$$
$$3x_1 + 2x_2 - 2x_3 - x_4 = 8$$

B.18
$$x_1 + x_2 + x_3 + x_4 = -1$$
$$2x_1 - x_2 + x_3 - 2x_4 = 8$$
$$3x_1 + 2x_2 + 2x_3 + 2x_4 = 4$$
$$-x_1 - x_2 + 2x_3 - x_4 = -2$$

Check if the following systems of equations are consistent. If they are, calculate their general solution:

B.19
$$3x_1 + x_2 + 5x_3 + 2x_4 = 2$$
$$2x_1 - 2x_2 + 4x_3 = 2$$
$$2x_1 + 2x_2 + 3x_3 + 2x_4 = 1$$
$$x_1 + 3x_2 + x_3 + 2x_4 = 0$$

B.20
$$x_1 + x_2 + x_3 + x_4 = 10$$
$$-x_1 + x_2 - x_3 + x_4 = 2$$
$$2x_1 - 3x_2 + 2x_3 - 2x_4 = -6$$

B.21
$$x_2 + 2x_3 + x_4 = -2$$
$$x_1 - 2x_2 - x_3 - x_4 = 1$$
$$x_1 - 2x_2 - 3x_3 + x_4 = 1$$

B.22
$$x_1 + x_2 + x_3 + x_4 = 0$$
$$2x_1 + x_2 - 2x_3 - x_4 = 6$$
$$3x_1 + 2x_2 + x_3 + 2x_4 = 2$$

B.23
$$x_2 + 2x_3 + x_4 + 3x_5 + 2x_6 = 9$$
$$-x_1 + 5x_2 + 2x_3 + x_4 + 2x_5 + x_7 = 10$$
$$5x_1 - 3x_2 + 8x_3 + 6x_4 + 3x_5 - 2x_8 = 17$$
$$2x_1 - x_2 + x_4 + 5x_5 - 2x_8 = 5$$

Check the linear independence of the following set of vectors:

B.24 $\mathbf{a}^{(1)} = \begin{bmatrix} 3 \\ 2 \\ 1 \end{bmatrix}$, $\mathbf{a}^{(2)} = \begin{bmatrix} -3 \\ -4 \\ 1 \end{bmatrix}$, $\mathbf{a}^{(3)} = \begin{bmatrix} 2 \\ 3 \\ 0 \end{bmatrix}$, $\mathbf{a}^{(4)} = \begin{bmatrix} 4 \\ 0 \\ 1 \end{bmatrix}$

B.25 $\mathbf{a}^{(1)} = \begin{bmatrix} 1 \\ 2 \\ 3 \\ 4 \\ 5 \end{bmatrix}$, $\mathbf{a}^{(2)} = \begin{bmatrix} -2 \\ 1 \\ 0 \\ 1 \\ -1 \end{bmatrix}$, $\mathbf{a}^{(3)} = \begin{bmatrix} 4 \\ 0 \\ -3 \\ 2 \\ 1 \end{bmatrix}$

Find eigenvalues for the following matrices:

B.26 $\begin{bmatrix} 1 & 2 \\ 2 & 5 \end{bmatrix}$ **B.27** $\begin{bmatrix} 2 & 2 \\ 2 & 4 \end{bmatrix}$

B.28 $\begin{bmatrix} 1 & 1 & 0 \\ 1 & 4 & 0 \\ 0 & 0 & 5 \end{bmatrix}$ **B.29** $\begin{bmatrix} 1 & 0 & 0 \\ 0 & 0 & 1 \\ 0 & 1 & 2 \end{bmatrix}$

B.30 $\begin{bmatrix} 0 & 0 & 0 \\ 0 & 1 & 1 \\ 0 & 1 & 5 \end{bmatrix}$

APPENDIX
C

A NUMERICAL METHOD FOR SOLUTION OF NONLINEAR EQUATIONS

Nonlinear equations are encountered in many fields of engineering. In design optimization, such equations arise when we write the necessary conditions of optimality for unconstrained or constrained problems, discussed in Chapter 3. Roots of that nonlinear set of equations are the candidate minimum designs. Thus the problem of finding *roots of nonlinear equations* must be treated.

The analytical solution of nonlinear equations is almost impossible except in very simple cases where elimination of variables can be carried out. Therefore, numerical methods and digital computers must be used to find solutions of such systems. In this appendix, *we describe a basic numerical method known as the Newton-Raphson method for finding roots of nonlinear equations.* Many variations of the method as well as other methods are available in the literature [Atkinson, 1978]. The method *can find only one root at a time* depending on the initial estimate for the root. Therefore, *several widely separated starting points must be tried* to find different roots. *The method may also fail to converge unless the starting point is in some neighbourhood of the solution. If it fails, a different starting point should be tried.*

We shall describe a basic algorithm that can be coded into a computer program. However, most computer center libraries have several programs for solving a system of nonlinear equations. These programs can be used directly. Therefore, the possibility of utilizing existing programs must be investigated before attempting to code the algorithm.

C.1 SINGLE NONLINEAR EQUATION

To develop the method, let us first consider the problem of finding roots of a general nonlinear equation

$$F(x) = 0 \qquad\qquad (C.1)$$

where $F(x)$ is a nonlinear function of an independent variable x. A simple way of finding roots is to plot $F(x)$ versus x on a graph sheet. Then, the points where the function crosses the x axis are the roots of $F(x) = 0$.

The second method developed by Newton-Raphson is an iterative numerical procedure in which we start with an initial estimate for a root. For the estimate, the function $F(x)$ will generally not have zero value, i.e. the initial estimate is usually not a root of $F(x) = 0$. Therefore, we try to improve the estimate until a root is found. This process, requiring *several cycles* (*or, iterations*) before the root is found, can be described by the following equation:

$$x^{(k+1)} = x^{(k)} + \Delta x^{(k)}, \qquad k = 0, 1, 2, \ldots \qquad (C.2)$$

where superscript k is the iteration number, $x^{(0)}$ is an initial estimate (starting point) and $\Delta x^{(k)}$ is a change in the estimate at the kth iteration. The iterative process is continued until $F(x)$ is reduced to zero or a small acceptable value, say δ (i.e. the stopping criterion is $|F| \le \delta$). Thus the Newton-Raphson method is reduced to somehow computing $\Delta x^{(k)}$ at each iteration of the process. *Note that the iteration concept described by Eq. (C.2) is very general because many other numerical methods are based on the same formula.*

To develop an expression for $\Delta x^{(k)}$, we use a linear (first-order) Taylor series expansion for the function $F(x)$ about the current estimate $x^{(k)}$. Therefore, from Eq. (3.7).

$$F(x^{(k)} + \Delta x^{(k)}) \simeq F(x^{(k)}) + \frac{dF(x^{(k)})}{dx} \Delta x^{(k)} \qquad (C.3)$$

In Eq. (C.3), $x^{(k)}$ is the current estimate for the root, $F(x^{(k)})$ is the current value of the function $F(x)$, and $dF(x^{(k)})/dx$ is the current slope (gradient) of $F(x)$. Our objective is to find $\Delta x^{(k)}$ (an improvement in the current estimate) such that the equation $F(x) = 0$ is satisfied at the new estimate, i.e. $F(x^{(k)} + \Delta x^{(k)}) = 0$. Therefore, setting Eq. (C.3) to zero, $\Delta x^{(k)}$ is given as

$$\Delta x^{(k)} = \frac{-F(x^{(k)})}{(dF/dx)} \qquad\qquad (C.4)$$

For each k, $\Delta x^{(k)}$ is calculated from Eq. (C.4) and a new estimate for the root is obtained from Eq. (C.2). The process is repeated until $F(x)$ goes to zero.

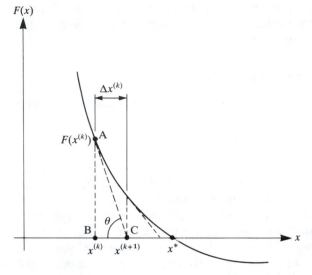

$F(x)$

FIGURE C.1
Graphical representation of the Newton–Raphson method for finding roots of $F(\mathbf{x}) = 0$.

The Newton-Raphson method also has a simple *geometrical representation.* To see this, let us consider the graph of a function $F(x)$ versus x shown in Fig. C.1. Point A on the curve represents the current value of the function $F(x)$ and point B on the x-axis represents the current estimate $x^{(k)}$ for the root of $F(x) = 0$. Point x^* is the root of $F(x) = 0$. The objective of the method is to reach point x^*. The method proceeds by following the tangent to the curve $F(x)$ at point A. Intersection of the tangent with the x-axis gives point C, $x^{(k+1)}$ from where the process is repeated.

The geometry of Fig. C.1 can be used to derive Eq. (C.4). To do this, we consider the triangle ABC:

$$\frac{F(x^{(k)})}{\Delta x^{(k)}} = \tan \theta = -\frac{dF(x^{(k)})}{dx}$$

or

$$\Delta x^{(k)} = -\frac{F(x^{(k)})}{(dF/dx)}$$

which is same as Eq. (C.4).

Steps of the iterative Newton-Raphson method are summarized as follows:

Step 1. Select a starting point $x^{(0)}$ and the parameter δ for stopping the iterative process. Set the iteration counter $k = 0$.

Step 2. Calculate the function F at the current estimate $x^{(k)}$. Check for convergence; if $|F(x^{(k)})| \leq \delta$, then stop the iterative process and accept $x^{(k)}$ as a root of $F(x) = 0$. Otherwise continue.

Step 3. Calculate the derivative of the function dF/dx at the current estimate $x^{(k)}$.

Step 4. Calculate, $\Delta x^{(k)} = -F(x^{(k)})/(dF/dx)$.
Step 5. Update the estimate for the root as $x^{(k+1)} = x^{(k)} + \Delta x^{(k)}$.
Step 6. Set $k = k + 1$ and go to Step 2.
We demonstrate the procedure in the following example.

Example C.1 Roots of a nonlinear equation by the Newton-Raphson method. Find a root of the following equation using the Newton-Raphson method starting from the point $x^{(0)} = 1.0$:

$$F(x) \equiv \frac{2x}{3} - \sin x = 0$$

Solution. The derivative of $F(x)$ for use in Eq. (C.4) is given as

$$\frac{dF}{dx} = \tfrac{2}{3} - \cos x$$

We follow the steps of the foregoing algorithm:

1. $x^{(0)} = 1.0$; let $\delta = 0.001$; $k = 0$.
2. $F(x^{(0)}) = \tfrac{2}{3}(1) - \sin (1) = -0.1750$.
   $|F| = 0.1750 > 0.001$, so $x^{(0)} = 1.0$ is not a root.
3. $\dfrac{dF}{dx} = \tfrac{2}{3} - \cos (1) = 0.1264$.
4. $\Delta x^{(0)} = \dfrac{-F(x^{(0)})}{dF/dx} = -\dfrac{(-0.1750)}{(0.1264)} = 1.3830$.
5. $x^{(0+1)} = x^{(0)} + \Delta x^{(0)}$; or $x^{(1)} = 1.0 + 1.3830 = 2.3830$.
6. $k = 0 + 1 = 1$; go to Step 2.

Results of various iterations of the preceding procedure are summarized in Table C.1. At the 4th iteration $|F(x^{(4)})| = 0.000\,13 < 0.001$ satisfying the specified stopping criteria. Therefore, $x^{(4)} = 1.496$ is taken as an estimate of the root for the given function. To find other roots, we must repeat the preceding process from another starting point. ‖

TABLE C.1
Newton-Raphson iterations for Example C.1: $F(x) = 2x/3 - \sin x$

k	$x^{(k)}$	$F(x^{(k)})$	$dF(x^{(k)})/dx$	$\Delta x^{(k)}$
0	1.000	−0.175	0.1264	1.383
1	2.383	0.900	0.1393	−0.646
2	1.737	0.172	0.832	−0.207
3	1.530	0.021	0.626	−0.034
4	1.496	0.000 13	—	—

C.2   MULTIPLE NONLINEAR EQUATIONS

The foregoing Newton-Raphson procedure can be generalized for the case of n nonlinear equations in n unknowns. We first derive the procedure and then summarize it in a step-by-step algorithm. The reader who is not interested in the derivation can go directly to the algorithm and the example problem.

A set of nonlinear equations can be written in vector form as

$$\mathbf{F}(\mathbf{x}) = \mathbf{0} \tag{C.5}$$

where $\mathbf{F}$ and $\mathbf{x}$ are both n-dimensional vectors. In the iterative procedure, we start with an estimate $\mathbf{x}^{(0)}$ for the root of Eq. (C.5). Just as before, this estimate is improved based on the vector form of Eq. (C.2) as

$$\mathbf{x}^{(k+1)} = \mathbf{x}^{(k)} + \Delta\mathbf{x}^{(k)}, \qquad k = 0, 1, 2, \ldots \tag{C.6}$$

where $\mathbf{x}^{(0)}$ is an initial estimate, k is an iteration number and $\Delta\mathbf{x}^{(k)}$ is a vector of changes in the estimate $\mathbf{x}^{(k)}$. The iterative procedure is continued until $\mathbf{x}^{(k)}$ satisfies Eq. (C.5). Since numerical computations are not exact we need a criterion for judging when a root is found. a common method is to calculate the length of the vector $\mathbf{F}(\mathbf{x})$ as

$$\|\mathbf{F}(\mathbf{x})\| = \left(\sum_{i=1}^{n} \{F_i(\mathbf{x})\}^2 \right)^{1/2} \tag{C.7}$$

and accept $\mathbf{x}^*$ as a root of $\mathbf{F}(\mathbf{x}) = \mathbf{0}$ if

$$\|\mathbf{F}(\mathbf{x}^*)\| \leq \delta \tag{C.8}$$

where $\delta > 0$ is some small specified number. Another stopping criterion might be to require the largest component of $\mathbf{F}(\mathbf{x})$ to satisfy the stopping criterion, i.e. $|F_i|_{max} \leq \delta$.

Now our task is to develop a formula for $\Delta\mathbf{x}^{(k)}$ so that Eq. (C.6) may be used to improve the estimate for the root of Eq. (C.5). To do this, we follow the procedure used for the case of one equation in one variable i.e. we write linear Taylor series expansion of each function. To derive the procedure, we consider the case of two equations in two unknowns in Eq. (C.5) and then generalize the result to the case of n equations in n unknowns. Equation (C.5) for the case of $n = 2$ is given as

$$\begin{aligned} F_1(x_1, x_2) &= 0 \\ F_2(x_1, x_2) &= 0 \end{aligned} \tag{a}$$

and the iterative equation (C.6) becomes

$$\begin{aligned} x_1^{(k+1)} &= x_1^{(k)} + \Delta x_1^{(k)} \\ x_2^{(k+1)} &= x_2^{(k)} + \Delta x_2^{(k)} \end{aligned} \tag{b}$$

To obtain expression for $\Delta x_1^{(k)}$ and $\Delta x_2^{(k)}$, we write linear Taylor series expansions for the functions F_1 and F_2 in Eqs (a) about the current estimate $x^{(k)}$ and, as before, set them to zero:

$$F_1(x_1^{(k)}, x_2^{(k)}) + \frac{\partial F_1}{\partial x_1} \Delta x_1^{(k)} + \frac{\partial F_1}{\partial x_2} \Delta x_2^{(k)} = 0$$

$$F_2(x_1^{(k)}, x_2^{(k)}) + \frac{\partial F_2}{\partial x_1} \Delta x_1^{(k)} + \frac{\partial F_2}{\partial x_2} \Delta x_2^{(k)} = 0$$

(c)

where the partial derivatives $\partial F_i / \partial x_j$ are evaluated at the current estimate $x_1^{(k)}$ and $x_2^{(k)}$. Equations (c) are two linear equations in two unknowns $\Delta x_1^{(k)}$ and $\Delta x_2^{(k)}$; all other quantities are known. Therefore, they are solved for $\Delta x_1^{(k)}$ and $\Delta x_2^{(k)}$, and Eqs (b) are used to update the estimate for the root. The process is repeated until convergence is achieved.

Equations (c) can be written in matrix form as

$$\begin{bmatrix} F_1^{(k)} \\ F_2^{(k)} \end{bmatrix} + \begin{bmatrix} \dfrac{\partial F_1}{\partial x_1} & \dfrac{\partial F_1}{\partial x_2} \\ \dfrac{\partial F_2}{\partial x_1} & \dfrac{\partial F_2}{\partial x_2} \end{bmatrix} \begin{bmatrix} \Delta x_1^{(k)} \\ \Delta x_2^{(k)} \end{bmatrix} = \begin{bmatrix} 0 \\ 0 \end{bmatrix}$$

(d)

where $F_1^{(k)} = F_1(x_1^{(k)}, x_2^{(k)})$ and $F_2^{(k)} = F_2(x_1^{(k)}, x_2^{(k)})$ are the function values at the current estimate $\mathbf{x}^{(k)}$. Or, we can write the equation compactly as

$$\mathbf{F}^{(k)} + \mathbf{J}[\Delta \mathbf{x}^{(k)}] = \mathbf{0}$$

(C.9)

where the vector $\mathbf{F}^{(k)}$, matrix $\mathbf{J}$ and vector $\Delta \mathbf{x}^{(k)}$ are easily identified from Eq. (d). The matrix $\mathbf{J}$ in Eq. (C.9) is usually called the *Jacobian* of the system of equations.

Equation (C.9) allows us to generalize the Newton-Raphson method for n equations in n unknowns. In that case $\mathbf{F}^{(k)}$ and $\Delta \mathbf{x}^{(k)}$ become n dimensional vectors and $\mathbf{J}$ becomes an $n \times n$ matrix of partical derivatives defined as

$$\mathbf{J} = \left[\frac{\partial F_i}{\partial x_j} \right]; \qquad i = 1 \text{ to } n; j = 1 \text{ to } n$$

(C.10)

Using the procedure for calculating the Jacobian for two equations in two unknowns identified in Eqs. (d), we observe that the ith row of $\mathbf{J}$ in Eq. (C.10) is obtained by differentiating the function $F_i(\mathbf{x})$ with respect to all the variables. That is, the first row is obtained by differentiating F_1 with respect to $x_1, x_2, \ldots, x_n$, the second row by differentiating F_2, and so on. Note that the ith row of the Jacobian in Eq. (C.10) can be also considered as a transpose of the gradient vector of $F_i(\mathbf{x})$, i.e. ∇F_i^T.

If the inverse of the matrix $\mathbf{J}$ can be calculated, then an improvement to the estimate for the root is given from Eq. (C.9) as

$$\Delta \mathbf{x}^{(k)} = -\mathbf{J}^{-1} \mathbf{F}^{(k)}$$

(C.11)

In numerical calculations, however, it is inefficient to invert matrices. Therefore, it is recommended to compute $\Delta \mathbf{x}^{(k)}$ by solving the following linear system of equations obtained from Eq. (C.9) as

$$\mathbf{J}[\Delta \mathbf{x}^{(k)}] = -\mathbf{F}^{(k)} \tag{C.12}$$

The *Newton-Raphson algorithm* is then summarized as follows:

Step 1. Select a starting point $\mathbf{x}^{(0)}$ and the parameter δ for stopping the iterative process. Set the iteration counter $k = 0$.

Step 2. Calculate the functions in $\mathbf{F}$ at the current estimate $\mathbf{x}^{(k)}$. Check for convergence; if $\|\mathbf{F}^{(k)}\| \leq \delta$, then stop the iterative process and accept $\mathbf{x}^{(k)}$ as a root of the equation $\mathbf{F}(\mathbf{x}) = \mathbf{0}$. Otherwise, continue.

Step 3. Calculate the Jacobian matrix $\mathbf{J}$ of partial derivatives at the current estimate $\mathbf{x}^{(k)}$ as in Eq. (C.10), i.e. calculate $[\partial F_i / \partial x_j]$; $i = 1$ to n; $j = 1$ to n.

Step 4. Calculate $\Delta \mathbf{x}^{(k)}$ from Eq. (C.12).

Step 5. Update the estimate for the root using Eq. (C.6).

Step 6. Set $k = k + 1$ and go to Step 2.

Note that since the system of equations $\mathbf{F}(\mathbf{x}) = \mathbf{0}$ is nonlinear, it has in general many roots. *The Newton-Raphson algorithm finds only one root at a time.* The method converges to a root depending on the starting estimate $\mathbf{x}^{(0)}$. To find other roots we should re-start the algorithm by selecting a different initial estimate $\mathbf{x}^{(0)}$. *Note also that the method will not work if Jacobian $\mathbf{J}$ is singular at any iteration,* since its inverse cannot be calculated. In addition, the method may fail to converge even if Jacobian $\mathbf{J}$ is nonsingular at all iterations. Several modifications of the basic Newton-Raphson method have been developed to make the method stable and convergent. These extensions, however, are beyond the scope of the present text.

Example C.2 Roots of nonlinear equations by the Newton-Raphson method. Find a root of the following 2×2 system of nonlinear equations using the Newton-Raphson procedure:

$$F_1(x_1, x_2) \equiv 1.0 - \frac{(4.0\text{E}+06)}{x_1^2 x_2} = 0 \tag{e}$$

$$F_2(x_1, x_2) \equiv 250.0 - \frac{(4.0\text{E}+06)}{x_1 x_2^2} = 0 \tag{f}$$

Solution. To use the Newton-Raphson algorithm, we need to compute the Jacobian for the system of Eqs (e) and (f). Using the definition given in Eq. (C.10), the Jacobian is given as

$$\mathbf{J} = \begin{bmatrix} \dfrac{\partial F_1}{\partial x_1} & \dfrac{\partial F_1}{\partial x_2} \\[2mm] \dfrac{\partial F_2}{\partial x_1} & \dfrac{\partial F_2}{\partial x_2} \end{bmatrix} = (4.0\text{E}+06)\begin{bmatrix} 2/x_1^3 x_2 & 1/x_2^2 x_1^2 \\[1mm] 1/x_1^2 x_2^2 & 2/x_1 x_2^3 \end{bmatrix} \tag{g}$$

We use the steps of the algorithm as follows:

1. Let $\mathbf{x}^{(0)} = (500.0, 1.0)$ and $\delta = 0.10$; set $k = 0$.

2. Calculate the function values from Eqs (e) and (f)

$F_1 = 1.0 - (4.0E+06)/(500.0 \times 500.0 \times 1.0) = -15$

$F_2 = 250.0 - (4.0E+06)/(500.0 \times 1.0 \times 1.0) = -7750$

$\|\mathbf{F}\| = \sqrt{15^2 + 7750^2} = 7750 > 0.10$, so $\mathbf{x}^{(0)}$ is not a root; continue the iterative process.

3. Jacobian matrix is calculated using Eq. (g):

$$\mathbf{J} = (4.0E+06) \begin{bmatrix} \dfrac{2}{(500)^3(1.0)} & \dfrac{1}{(500)^2(1.0)^2} \\ \dfrac{1}{(500)^2(1.0)^2} & \dfrac{2}{(500)(1.0)^3} \end{bmatrix}$$

$$= \begin{bmatrix} \frac{8}{125} & 16 \\ 16 & 16\,000 \end{bmatrix}$$

4. Equations (C.9) or (C.12) define the following linear system of equations for $\Delta x_1^{(0)}$ and $\Delta x_2^{(0)}$:

$$\begin{bmatrix} \frac{8}{125} & 16 \\ 16 & 16\,000 \end{bmatrix} \begin{bmatrix} \Delta x_1^{(0)} \\ \Delta x_2^{(0)} \end{bmatrix} = \begin{bmatrix} 15 \\ 7750 \end{bmatrix}$$

Solving the two equations by the elimination process, we get $\Delta x_1^{(0)} = 151.0$, $\Delta x_2^{(0)} = 0.330$.

For this problem it is possible to invert the Jacobian matrix $\mathbf{J}$ and use Eq. (C.11) for calculating $\Delta x_1^{(0)}$ and $\Delta x_2^{(0)}$. $\mathbf{J}^{-1}$ is calculated using the cofactors approach (Appendix B, Section B.5) as

$$\mathbf{J}^{-1} = \frac{1}{|\mathbf{J}|} [\text{cofac}\,(\mathbf{J})]^T$$

$$= \frac{1}{768} \begin{bmatrix} 16\,000 & -16 \\ -16 & \frac{8}{125} \end{bmatrix}$$

Using Eq. (C.11), we obtain the same values for $\Delta x_1^{(0)}$ and $\Delta x_2^{(0)}$ as before.

TABLE C.2
Newton-Raphson iterations for Example C.2

k	$x_1^{(k)}$	$x_2^{(k)}$	$\|\mathbf{F}^{(k)}\|$
0	500.0	1.0000	7750.00
1	651.0	1.3300	3206.00
2	822.0	1.7700	1291.00
3	976.0	2.3500	489.20
4	1047.0	3.0490	161.30
5	1025.0	3.6770	38.50
6	1003.0	3.9630	3.98
7	1000.3	3.9995	0.05

5. Update the estimate for the root using Eq. (C.6) or Eqs (b)

$$x_1^{(0+1)} = x_1^{(0)} + \Delta x_1^{(0)} = 500.0 + 151.0 = 651.0$$
$$x_2^{(0+1)} = x_2^{(0)} + \Delta x_2^{(0)} = 1.00 + 0.33 = 1.33$$

6. Set $k = 0 + 1 = 1$, and go to Step 2.

Results from various iterations of the Newton-Raphson algorithm are summarized in Table C.2. The table is generated by repeating the six steps of the algorithm. At the 7th iteration $\|\mathbf{F}^{(7)}\| = 0.05 < 0.10$. Thus $\mathbf{x}^{(7)} = (1000.3, 3.9995)$ is considered as a root for the system of Eqs (e) and (f) satisfying the desired accuracy. The exact root is $\mathbf{x}^* = (1000.0, 4.0)$. ‖

EXERCISES FOR APPENDIX C

Find roots of the following equations using the Newton-Raphson method and $\delta = 0.001$:

C.1 $F(x) \equiv 3x - e^x = 0$ starting from $x = 0$.
C.2 $F(x) \equiv \sin x = 0$ starting from $x = 10$.
C.3 $F(x) \equiv \cos x = 0$ starting from $x = 2$.
C.4 $F(x) \equiv \dfrac{2x}{3} - \sin x = 0$ starting from $x = -4$.

Complete two iterations of the Newton-Raphson method for the following system of nonlinear equations:

C.5 $F_1(\mathbf{x}) \equiv 1 - \dfrac{10}{x_1^2 x_2} = 0$

$F_2(\mathbf{x}) \equiv 5 - \dfrac{10}{x_1 x_2^2} = 0$; starting point $(4, 1)$

C.6 $F_1(\mathbf{x}) \equiv 5 - \frac{1}{8}x_1 x_2 - \dfrac{1}{4x_1^2} x_2^2 = 0$

$F_2(\mathbf{x}) \equiv -\frac{1}{16}x_1^2 + \dfrac{1}{2x_1} x_2 = 0$; starting point $(10, 10)$

C.7 $F_1(\mathbf{x}) \equiv 3x_1^2 + 12x_2^2 + 10x_1 = 0$

$F_2(\mathbf{x}) \equiv 24x_1 x_2 + 4x_2 + 3 = 0$; starting point $(-5, 0)$

Find all roots of the following system of nonlinear equations using a computer program:

C.8 Exercise C.5
C.9 Exercise C.6
C.10 Exercise C.7

D

SAMPLE
COMPUTER
PROGRAMS

D.1 INTRODUCTION

This appendix contains the *listing of some computer programs based on the algorithms for numerical methods of unconstrained optimization given in Chapter 5.* The objective is to educate the student on how to transform a step-by-step numerical algorithm into a computer program. Note that the given computer programs are not claimed to be the most efficient ones. The key idea is to emphasize the essential numerical aspects of the algorithms in a simple and straightforward way. A beginner in the numerical techniques of optimization is expected to experiment with the computer programs and get a feel for various methods by solving some numerical examples. Thus, a black box usage of the computer programs given in this appendix is highly discouraged. The computer programs given in the following were developed on APOLLO computers.

D.2 EQUAL INTERVAL SEARCH

As discussed in Chapter 5, Equal Interval search is the simplest method of one-dimensional minimization. A computer program based on it is given in Fig. D.1. It is assumed that the one-dimensional function is unimodal,

```
C       MAIN PROGRAM FOR EQUAL INTERVAL SEARCH
        IMPLICIT DOUBLE PRECISION (A-H, O-Z)

        DELTA   = 5.0D-2
        EPSLON  = 1.0D-3
        NCOUNT = 0
        F       = 0.0D0
        ALFA    = 0.0D0
C
C       TO PERFORM LINE SEARCH CALL SUBROUTINE EQUAL
C
        CALL EQUAL(ALFA,DELTA,EPSLON,F,NCOUNT)
        WRITE(*, 10) ' MINIMUM = ', ALFA
        WRITE (*, 10) ' MINIMUM FUNCTION VALUE = ', F
        WRITE (*, *) 'NO. OF FUNCTION EVALUATIONS = ', NCOUNT
10      FORMAT(A,1PE14.5)

        STOP
        END


        SUBROUTINE EQUAL(ALFA,DELTA,EPSLON,F,NCOUNT)
C       ..........................................................................
C       THIS SUBROUTINE IMPLEMENTS EQUAL INTERVAL SEARCH
C         ALFA    = OPTIMUM VALUE ON RETURN
C       DELTA     = INITIAL STEP LENGTH
C       EPSLON    = CONVERGENCE PARAMETER
C       F         = OPTIMUM VALUE OF THE FUNCTION ON RETURN
C       NCOUNT    = NUMBER OF FUNCTION EVALUATIONS ON RETURN
C       ..........................................................................

        IMPLICIT DOUBLE PRECISION (A-H,O-Z)
C
C       ESTABLISH INITIAL DELTA
C
        AL = 0.0D0
        CALL FUNCT(AL,FL,NCOUNT)
10      CONTINUE
        AA = DELTA
        CALL FUNCT(AA,FA,NCOUNT)
        IF (FA .GT. FL) THEN
          DELTA = DELTA * 0.1D0
          GO TO 10
        END IF
C
C       ESTABLISH INITIAL INTERVAL OF UNCERTAINTY
C
20      CONTINUE
        AU = AA + DELTA
        CALL FUNCT(AU,FU,NCOUNT)
        IF (FA .GT. FU) THEN
          AL = AA
```

FIGURE D.1
Program for Equal Interval Search.

```
            AA = AU
            FL = FA
            FA = FU
            GO TO 20
         END IF
C
C        REFINE THE INTERVAL OF UNCERTAINTY FURTHER
C
30       CONTINUE
         IF ((AU − AL) .LE. EPSLON) GO TO 50
         DELTA = DELTA * 0.1D0
         AA = AL
         FA = FL
40       CONTINUE
         AU = AA + DELTA
         CALL FUNCT(AU,FU,NCOUNT)
         IF (FA .GT. FU) THEN
            AL = AA
            AA = AU
            FL = FA
            FA = FU
            GO TO 40
         END IF
         GO TO 30
C
C        MINIMUM IS FOUND
C
50       ALFA = (AU + AL) * 0.5D0
         CALL FUNCT(ALFA,F,NCOUNT)

         RETURN
         END

         SUBROUTINE FUNCT(AL,F,NCOUNT)
C        ...........................................................................
C        CALCULATES THE FUNCTION VALUE
C        AL      = VALUE OF ALPHA, INPUT
C        F       = FUNCTION VALUE ON RETURN
C        NCOUNT  = NUMBER OF CALLS FOR FUNCTION EVALUATION
C        ...........................................................................

         IMPLICIT DOUBLE PRECISION (A–H,O–Z)

         NCOUNT = NCOUNT + 1
         F = 2.0D0 − 4.0D0 * AL + DEXP(AL)

         RETURN
         END
```

FIGURE D.1 (*Continued*)

continuous and has a negative slope in the interval of interest. The initial step length (δ) and line search accuracy (ε) must be specified in the main program. The subroutine EQUAL is called from the main program to perform line search by equal interval search. The three major tasks to be accomplished in the subroutine EQUAL are: (1) to establish initial step length δ such that $f(0) > f(\delta)$, (2) to establish the initial interval of uncertainty, (α_l, α_u), and (3) to reduce the interval of uncertainty such that $(\alpha_u - \alpha_l) \leqq \varepsilon$.

The subroutine EQUAL calls the subroutine FUNCT to evaluate the value of the one-dimensional function at various trial steps. The subroutine FUNCT is supplied by the user. As an example, $f(\alpha) = 2 - 4\alpha + e^{\alpha}$ is chosen as the one-dimensional minimization function. The listing of the program in Fig. D.1 is self explanatory. In the subroutine EQUAL, the following notation is used:

AL = lower limit on α; α_l

AU = upper limit on α; α_u

FL = function value at α_l; $f(\alpha_l)$

FU = function value at α_u; $f(\alpha_u)$

AA = intermediate point α_a

FA = function value at α_a; $f(\alpha_a)$

D.3 GOLDEN SECTION SEARCH

Golden Section search is considered to be one of the efficient methods requiring only function values. The subroutine GOLD, given in Fig. D.2,

```
          SUBROUTINE GOLD(ALFA,DELTA,EPSLON,F,NCOUNT)
C         ...............................................................
C         THIS SUBROUTINE IMPLEMENTS GOLDEN SECTION SEARCH
C         ALFA    = OPTIMUM VALUE OF ALPHA ON RETURN
C         DELTA   = INITIAL STEP LENGTH
C         EPSLON  = CONVERGENCE PARAMETER
C         F       = OPTIMUM VALUE OF THE FUNCTION ON RETURN
C         NCOUNT = NUMBER OF FUNCTION EVALUATIONS ON RETURN
C         ...............................................................

          IMPLICIT DOUBLE PRECISION(A–H,O–Z)

          GR = 0.5D0 * DSQRT(5.0D0) + 0.5D0
C
C         ESTABLISH INITIAL DELTA
C
          AL = 0.0D0
          CALL FUNCT(AL,FL,NCOUNT)
```

FIGURE D.2
Subroutine GOLD for Golden Section Search.

```
10        CONTINUE
          AA = DELTA
          CALL FUNCT(AA,FA,NCOUNT)
          IF (FA .GT. FL) THEN
             DELTA = DELTA * 0.1D0
             GO TO 10
          END IF
C
C         ESTABLISH INITIAL INTERVAL OF UNCERTAINTY
C
          J = 0
20        CONTINUE
          J = J + 1
          AU = AA + DELTA * (GR ** J)
          CALL FUNCT(AU,FU,NCOUNT)
          IF (FA .GT. FU) THEN
             AL = AA
             AA = AU
             FL = FA
             FA = FU
             GO TO 20
          END IF
C
C         REFINE THE INTERVAL OF UNCERTAINTY FURTHER
C
          AB = AL + (AU − AL)/GR
          CALL FUNCT(AB,FB,NCOUNT)
30        CONTINUE
          IF ((AU − AL) .LE. EPSLON) TO GO 80
C
C         IMPLEMENT STEPS 4, 5 OR 6 OF THE ALGORITHM
C
          IF (FA − FB) 40, 60, 50
C
C         FA IS LESS THAN FB (STEP 4)
C
40        AU = AB
          FU = FB
          AB = AA
          FB = FA
          AA = AL + (AU − AL) * (1.0D0 − 1.0D0 / GR)
          CALL FUNCT(AA,FA,NCOUNT)
          GO TO 30
C
C         FA IS GREATER THAN FB (STEP 5)
C
50        AL = AA
          FL = FA
          AA = AB
          FA = FB
          AB = AL + (AU − AL) / GR
          CALL FUNCT(AB,FB,NCOUNT)
          GO TO 30
```

FIGURE D.2 (*Continued*)

```
C
C       FA IS EQUAL TO FB (STEP 6)
C
60      AL = AA
        FL = FA
        AU = AB
        FU = FB
        AA = AL – (1.0D0 – 1.0D0 / GR) * (AU – AL)
        CALL FUNCT (AA,FA, NCOUNT)
        AB = AL + (AU – AL) / GR
        CALL FUNCT (AB,FB,NCOUNT)
        GO TO 30
C
C       MINIMUM IS FOUND
C
80      ALFA = (AU + AL) * 0.5D0
        CALL FUNCT(ALFA,F,NCOUNT)

        RETURN
        END
```

implements the Golden Section search algorithm given in Chapter 5, and is called from the main program given in Fig. D.1; the call to subroutine EQUAL is replaced by a call to the subroutine GOLD. The initial step length and initial interval of uncertainty are established in GOLD, as in the subroutine EQUAL. The interval of uncertainty is reduced further to satisfy the line search accuracy by implementing Steps 4, 5 and 6 of the algorithm given in Chapter 5. The subroutine FUNCT is used to evaluate the function value at a trial step. The following notation is used in the subroutine GOLD: AA = α_a, AB = α_b, AL = α_l, AU = α_u, FA = $f(\alpha_a)$, FB = $f(\alpha_b)$, FL = $f(\alpha_l)$, FU = $f(\alpha_u)$, and GR = Golden Radio, $(\sqrt{5} + 1)/2$.

D.4   STEEPEST DESCENT METHOD

The Steepest Descent Method is the simplest of the gradient-based methods for unconstrained optimization. A computer program for the method is given in Fig. D.3. The basic steps in the algorithm are: (1) to evaluate the gradient of the cost function at the current point, (2) to evaluate an optimum step size along the negative gradient direction, and (3) to update the design, check the convergence criterion and if necessary repeat the preceding steps. The main program essentially follows these steps. The arrays declared in the main program must have dimensions of the design variable vector. Also, the initial data and starting point must be provided by the user. The cost function and its gradient must be provided in subroutines FUNCT and GRAD, respectively. The line search is performed in subroutine GOLDM by Golden Section search for a multivariate problem. As an example, $f(\mathbf{x}) = x_1^2 + 2x_2^2 + 2x_3^2 + 2x_1x_2 + 2x_2x_3$ is chosen as the cost function.

```
C       THE MAIN PROGRAM FOR STEEPEST DESCENT METHOD
C       ............................................................................................
C       DELTA   = INITIAL STEP LENGTH FOR LINE SEARCH
C       EPSLON  = LINE SEARCH ACCURACY
C       EPSL    = STOPPING CRITERION FOR STEEPEST DESCENT METHOD
C       NCOUNT  = NO. OF FUNCTION EVALUATIONS
C       NDV     = NO. OF DESIGN VARIABLES
C       NOC     = NO. OF CYCLES OF THE METHOD
C       X       = DESIGN VARIABLE VECTOR
C       D       = DIRECTION VECTOR
C       G       = GRADIENT VECTOR
C       WK      = WORK ARRAY USED FOR TEMPORARY STORAGE
C       ............................................................................................

        IMPLICIT DOUBLE PRECISION (A–H, O–Z)
        DIMENSION X(3), D(3), G(3), WK(3)
C
C       DEFINE INITIAL DATA
C
        DELTA   = 5.0D-2
        EPSLON  = 1.0D-4
        EPSL    = 5.0D-3
        NCOUNT  = 0
        NDV     = 3
        NOC     = 100
C
C       STARTING VALUES OF THE DESIGN VARIABLES
C
        X(1) = 2.0D0
        X(2) = 4.0D0
        X(3) = 10.0D0

        CALL GRAD(X,G,NDV)
        WRITE(*, 10)
10      FORMAT(' NO.     COST FUNCT STEP SIZE',
     &          ' NORM OF GRAD   ')
        DO 20 K = 1, NOC
           CALL SCALE (G,D,–1.0D0,NDV)
           CALL GOLDM(X,D,WK,ALFA,DELTA,EPSLON,F,NCOUNT,NDV)
           CALL SCALE (D,D,ALFA,NDV)
           CALL PRINT(K,X,ALFA,G,F,NDV)
           CALL ADD(X,D,X,NDV)
           CALL GRAD(X,G,NDV)
           IF(TNORM(G,NDV) .LE. EPSL) GO TO 30
20      CONTINUE
        WRITE (*,*)
        WRITE(*,*)' LIMIT ON NO. OF CYCLES HAS EXCEEDED'
        WRITE(*,*)' THE CURRENT DESIGN VARIABLES ARE:'
        WRITE(*,*) X
        CALL EXIT
```

FIGURE D.3
Computer program for steepest descent method.

```
30      WRITE(*,*)
        WRITE(*,*) 'THE OPTIMAL DESIGN VARIABLES ARE:'
        WRITE (*,40) X
40      FORMAT(3F15.6)
        CALL FUNCT(X,F,NCOUNT,NDV)
        WRITE(*,50)' THE OPTIMUM COST FUNCTION VALUE IS :', F
50      FORMAT(A, F13.6)
        WRITE (*,*) 'TOTAL NO. OF FUNCTION EVALUATIONS ARE', NCOUNT

        STOP
        END


        SUBROUTINE GRAD(X,G,NDV)
C
C       CALCULATES THE GRADIENT OF F(X) IN VECTOR G
C
        IMPLICIT DOUBLE PRECISION (A–H, O–Z)
        DIMENSION X(NDV),G(NDV)

        G(1) = 2.0D0 * X(1) + 2.0D0 * X(2)
        G(2) = 2.0D0 * X(1) + 4.0D0 * X(2) + 2.0D0 * X(3)
        G(3) = 2.0D0 * X(2) + 4.0D0 * X(3)

        RETURN
        END


        SUBROUTINE SCALE (A,X,S,M)
C
C       MULTIPLES VECTOR A(M) BY SCALAR S AND STORES IN X(M)
C
        IMPLICIT DOUBLE PRECISION (A–H, O–Z)
        DIMENSION A(M),X(M)

        D0 10 I = 1, M
          X(I) = S * A(I)
10      CONTINUE

        RETURN
        END

        DOUBLE PRECISION FUNCTION TNORM(X,N)
C
C       CALCULATES NORM OF VECTOR X(N)
C
        IMPLICIT DOUBLE PRECISION (A–H, O–Z)
        DIMENSION X(N)

        SUM = 0.0D0
        DO 10 I = 1, N
          SUM = SUM + X(I) * X(I)
10      CONTINUE
        TNORM = DSQRT(SUM)

        RETURN
        END
```

FIGURE D.3 (*Continued*)

```
      SUBROUTINE ADD(A,X,C,M)
C
C     ADDS VECTORS A(M) AND X(M) AND STORES IN C(M)
C
      IMPLICIT DOUBLE PRECISION (A-H, O-Z)
      DIMENSION A(M), X(M), C(M)

      DO 10 I = 1, M
        C(I) = A(I) + X(I)
10    CONTINUE

      RETURN
      END

      SUBROUTINE PRINT(I,X,ALFA,G,F,M)
C
C     PRINTS THE OUTPUT
C
      IMPLICIT DOUBLE PRECISION (A-H, O-Z)
      DIMENSION X(M),G(M)

      WRITE (*,10) I, F, ALFA, TNORM(G,M)
10    FORMAT(I4, 3F15.6)

      RETURN
      END

      SUBROUTINE FUNCT(X,F,NCOUNT,NDV)
C
C     CALCULATES THE FUNCTION VALUE
C
      IMPLICIT DOUBLE PRECISION (A-H, O-Z)
      DIMENSION X(NDV)

      NCOUNT = NCOUNT + 1
      F = X(1) ** 2 + 2.D0 * (X(2) **2) + 2.D0 * (X(3) ** 2)
   &      + 2.0D0 * X(1) * X(2) + 2.D0 * X(2) * X(3)

      RETURN
      END

      SUBROUTINE UPDATE (XN,X,D,AL,NDV)
C
C     UPDATES THE DESIGN VARIABLE VECTOR
C
      IMPLICIT DOUBLE PRECISION (A-H, O-Z)
      DIMENSION XN(NDV), X(NDV), D(NDV)

      DO 10 I = 1, NDV
        XN(I) = X(I) + AL * D(I)
10    CONTINUE

      RETURN
      END
```

FIGURE D.3 (*Continued*)

```
      SUBROUTINE GOLDM(X,D,XN,ALFA,DELTA,EPSLON,F,NCOUNT,NDV)
C     ..............................................................................
C     IMPLEMENTS GOLDEN SECTION SEARCH FOR MULTIVARIATE PROBLEMS
C     X        = CURRENT DESIGN POINT
C     D        = DIRECTION VECTOR
C     XN       = CURRENT DESIGN + TRIAL STEP * SEARCH DIRECTION
C     ALFA     = OPTIMUM VALUE OF ALPHA ON RETURN
C     DELTA    = INITIAL STEP LENGTH
C     EPSLON   = CONVERGENCE PARAMETER
C     F        = OPTIMUM VALUE OF THE FUNCTION
C     NCOUNT   = NUMBER OF FUNCTION EVALUATIONS ON RETURN
C     ..............................................................................

      IMPLICIT DOUBLE PRECISION (A-H, O-Z)
      DIMENSION X(NDV), D(NDV), XN(NDV)

      GR = 0.5D0 * DSQRT(5.0D0) + 0.5D0
      DELTA1 = DELTA
C
C     ESTABLISH INITIAL DELTA
C
      AL = 0.D0
      CALL UPDATE(XN,X,D,AL,NDV)
      CALL FUNCT(XN,FL,NCOUNT,NDV)
      F = FL
10    CONTINUE
      AA = DELTA1
      CALL UPDATE(XN,X,D,AA,NDV)
      CALL FUNCT (XN,FA,NCOUNT,NDV)
      IF (FA .GT. FL) THEN
         DELTA1 = DELTA1 * 0.1D0
         GO TO 10
      END IF
C
C     ESTABLISH INITIAL INTERVAL OF UNCERTAINTY
C
      J = 0
20    CONTINUE
      J = J + 1
      AU = AA + DELTA1 * (GR ** J)
      CALL UPDATE(XN,X,D,AU,NDV)
      CALL FUNCT(XN,FU,NCOUNT,NDV)
      IF (FA .GT. FU) THEN
         AL = AA
         AA = AU
         FL = FA
         FA = FU
         GO TO 20
      END IF
C
C     REFINE THE INTERVAL OF UNCERTAINTY FURTHER
C
```

FIGURE D.3 (*Continued*)

```
          AB = AL + (AU − AL) / GR
          CALL UPDATE(XN,X,D,AB,NDV)
          CALL FUNCT(XN,FB,NCOUNT,NDV)
30        CONTINUE
          IF((AU−AL) .LE. EPSLON) GO TO 80
C
C         IMPLEMENT STEPS 4, 5 OR 6 OF THE ALGORITHM
C
          IF (FA−FB) 40, 60, 50
C
C         FA IS LESS THAN FB (STEP 4)
C
40        AU = AB
          FU = FB
          AB = AA
          FB = FA
          AA = AL + (1.0D0 − 1.0D0 / GR) * (AU − AL)
          CALL UPDATE(XN,X,D,AA,NDV)
          CALL FUNCT(XN,FA,NCOUNT,NDV)
          GO TO 30
C
C         FA IS GREATER THAN FB (STEP 5)
C
50        AL = AA
          FL = FA
          AA = AB
          FA = FB
          AB = AL + (AU − AL) / GR
          CALL UPDATE(XN,X,D,AB,NDV)
          CALL FUNCT(XN,FB,NCOUNT,NDV)
          GO TO 30
C
C         FA IS EQUAL TO FB (STEP 6)
C
60        AL = AA
          FL = FA
          AU = AB
          FU = FB
          AA = AL + (1.0D0 − 1.0D0 / GR) * (AU − AL)
          CALL UPDATE(XN,X,D,AA,NDV)
          CALL FUNCT(XN,FA,NCOUNT,NDV)
          AB = AL + (AU − AL) / GR
          CALL UPDATE(XN,X,D,AB,NDV)
          CALL FUNCT(XN,FB,NCOUNT,NDV)
          GO TO 30
C
C         MINIMUM IS FOUND
C
80        ALFA = (AU + AL) * 0.5D0

          RETURN
          END
```

FIGURE D.3 (*Continued*)

D.5 MODIFIED NEWTON'S METHOD

The modified Newton's Method evaluates gradient as well as Hessian for the function and thus has a quadratic rate of convergence. Note that even though the method has a superior rate of convergence it may fail to converge due to the singularity or indefiniteness of the Hessian matrix of the cost function. A program for the method is given in Fig. D.4. The cost function, gradient vector and Hessian matrix are calculated in the subroutines FUNCT, GRAD and HASN, respectively. As an example, $f(\mathbf{x}) = x_1^2 + 2x_2^2 + 2x_3^2 + 2x_1x_2 + 2x_2x_3$ is chosen as the cost function. The Newton direction is obtained by solving a system of linear equations in subroutine SYSEQ. It is likely that the Newton direction may not be a descent direction in which the line search will fail to evaluate an appropriate step size. In such a case, the iterative loop is stopped and an appropriate message is printed. The main program for the modified Newton's Method and the related subroutines are given in Fig. D.4.

```
C       THE MAIN PROGRAM FOR MODIFIED NEWTON'S METHOD
C       ........................................................................
C       DELTA   = INITIAL STEP LENGTH FOR LINE SEARCH
C       EPSLON  = LINE SEARCH ACCURACY
C       EPSL    = STOPPING CRITERION FOR MODIFIED NEWTON'S METHOD
C       NCOUNT  = NO. OF FUNCTION EVALUATIONS
C       NDV     = NO. OF DESIGN VARIABLES
C       NOC     = NO. OF CYCLES OF THE METHOD
C       X       = DESIGN VARIABLE VECTOR
C       D       = DIRECTION VECTOR
C       G       = GRADIENT VECTOR
C       H       = HESSIAN MATRIX
C       WK      = WORK ARRAY USED FOR TEMPORARY STORAGE
C       ........................................................................

        IMPLICIT DOUBLE PRECISION (A-H, O-Z)
        DIMENSION X(3), D(3), G(3), H(3,3), WK(3)
C
C       DEFINE INITIAL DATA
C
        DELTA   = 5.0D-2
        EPSLON  = 1.0D-4
        EPSL    = 5.0D-3
        NCOUNT  = 0
        NDV     = 3
        NOC     = 100
C
C       STARTING VALUES OF THE DESIGN VARIABLES
C
        X(1) = 2.0D0
        X(2) = 4.0D0
        X(3) = 10.0D0
```

FIGURE D.4
A program for Newton's method.

```
          CALL GRAD(X,G,NDV)
          WRITE(*, 10)
10        FORMAT(' NO.      COST FUNCT     STEP SIZE',
      &             '    NORM OF GRAD    ')
          DO 20 K = 1, NOC
            CALL HASN(X,H,NDV)
            CALL SCALE (G,D,-1.0D0,NDV)
            CALL SYSEQ(H,NDV,D)
C
C         CHECK FOR THE DESCENT CONDITION
C
          IF (DOT(G,D,NDV) .GE. 1.0E-8) GO TO 60
          CALL GOLDM(X,D,WK,ALFA,DELTA,EPSLON,F,NCOUNT,NDV)
          CALL SCALE(D,D,ALFA,NDV)
          CALL PRINT(K,X,ALFA,G,F,NDV)
          CALL ADD(X,D,X,NDV)
          CALL GRAD(X,G,NDV)
          IF(TNORM(G,NDV) .LE. EPSL) TO TO 30
20        CONTINUE
          WRITE(*,*)
          WRITE(*,*)' LIMIT ON NO. OF CYCLES HAS EXCEEDED'
          WRITE(*,*)' THE CURRENT DESIGN VARIABLES ARE:'
          WRITE(*,*) X
          CALL EXIT

30        WRITE(*,*)
          WRITE(*,*) 'THE OPTIMAL DESIGN VARIABLES ARE :'
          WRITE(*,40) X
40        FORMAT(4X,3F15.6)
          CALL FUNCT(X,F,NCOUNT,NDV)
          WRITE(*,50) ' OPTIMUM COST FUNCTION VALUE IS :', F
50        FORMAT(A, F13.6)
          WRITE(*,*) 'NO. OF FUNCTION EVALUATIONS ARE :   ', NCOUNT
          CALL EXIT

60        WRITE(*,*)
          WRITE(*,*)' DESCENT DIRECTION CANNOT BE FOUND'
          WRITE(*,*)' THE CURRENT DESIGN VARIABLES ARE:'
          WRITE(*,40) X

          STOP
          END

          DOUBLE PRECISION FUNCTION DOT(X,Y,N)
C
C         CALCULATES DOT PRODUCT OF VECTORS X AND Y
C
          IMPLICIT DOUBLE PRECISION (A-H, 0-Z)
          DIMENSION X(N),Y(N)

          SUM = 0.0D0
          DO 10 I = 1, N
          SUM = SUM + X(I) * Y(I)
```

FIGURE D.4 (*Continued*)

```
10      CONTINUE
        DOT = SUM

        RETURN
        END


        SUBROUTINE HASN(X,H,N)
C
C       CALCULATES THE HESSIAN MATRIX H AT X
C
        IMPLICIT DOUBLE PRECISION (A–H,O–Z)
        DIMENSION X(N),H(N,N)


        H(1,1) = 2.0D0
        H(2,2) = 4.0D0
        H(3,3) = 4.0D0
        H(1,2) = 2.0D0
        H(1,3) = 0.0D0
        H(2,3) = 2.0D0
        H(2,1) = H(1,2)
        H(3,1) = H(1,3)
        H(3,2) = H(2,3)


        RETURN
        END


        SUBROUTINE SYSEQ(A,N,B)
C
C       SOLVES AN N × N SYMMETRIC SYSTEM OF LINEAR EQUATIONS AX = B
C       A IS THE COEFFICIENT MATRIX: B IS THE RIGHT HAND SIDE: THESE ARE
C       INPUT
C       B CONTAINS SOLUTION ON RETURN
C
        IMPLICIT DOUBLE PRECISION (A–H,O–Z)
        DIMENSION A(N,N), B(N)
C
C       REDUCTION OF EQUATIONS
C
        M = 0
50      M = M + 1
        MM = M + 1
        B(M) = B(M) / A(M,M)
        IF (M – N) 70, 130, 70
70      DO 80 J = MM, N
           A(M,J) = A(M,J) / A(M,M)
80      CONTINUE
C
C       SUBSTITUTION INTO REMAINING EQUATIONS
C
        DO 120 I = MM, N
           IF(A(I,M)) 90, 120, 90
```

FIGURE D.4 (*Continued*)

```
90        DO 100 J = I, N
             A(I,J) = A(I,J) − A(I,M) * A(M,J)
             A(J,I) = A(I,J)
100       CONTINUE
             B(I) = B(I) − A(I,M) * B(M)
120       CONTINUE
          GO TO 50
C
C         BACK SUBSTITUTION
C
130       M = M − 1
          MM = M + 1
          DO 140 J = MM, N
             B(M) = B(M) − A(M,J) * B(J)
140       CONTINUE
          GO TO 130
150       RETURN
          END
```

FIGURE D.4 (*Continued*)

BIBLIOGRAPHY

Abadie, J. (Ed.) (1970). *Nonlinear Programming,* North-Holland, Amsterdam, Chapter 6.

Abadie, J. and Carpenter, J. (1969). Generalization of the Wolfe Reduced Gradient Method to the Case of Nonlinear Constraints. In *Optimization,* R. Fletcher (Ed.), Academic Press, New York, pp. 37–47.

Ackoff, R. L. and Sasieni, M. W. (1968). *Fundamentals of Operations Research,* John Wiley and Sons, New York.

Adelman, H. and Haftka, R. T. (1986). Sensitivity Analysis of Discrete Structural Systems. *AIAA J.,* Vol. 24, No. 5, pp. 823–832.

Aoki, M. (1971). *Introduction to Optimization Techniques,* Macmillan, New York.

Arora, J. S. (1984). An Algorithm for Optimum Structural Design Without Line Search. In *New Directions in Optimum Structural Design,* E. Atrek, R. H. Gallagher, K. M. Ragsdell and O. C. Zienkiewicz (Eds), John Wiley and Sons, New York, pp. 429–441.

Arora, J. S. and Baenziger, G. (1986). Uses of Artificial Intelligence in Design Optimization. *Computer Methods in Applied Mechanics and Engineering,* Vol. 54, pp. 303–323.

Arora, J. S. and Baenziger, G. (1987). A Nonlinear Optimization Expert System. *Proceedings of the ASCE Structures Congress '87, Computer Applications in Structural Engineering,* D. R. Jenkins (Ed.), pp. 113–125.

Arora, J. S. and Haug, E. J. (1979). Methods of Design Sensitivity Analysis in Structural Optimization. *AIAA J.,* Vol. 17, No. 9, pp. 970–974.

Arora, J. S. and Thanedar, P. B. (1986). Computational Methods for Optimum Design of Large Complex Systems. *Computational Mechanics,* Vol. 1, No. 2, pp. 221–242.

Arora, J. S. and Tseng, C. H. (1987a). *User's Manual for IDESIGN: Version 3.5,* Optimal Design Laboratory, College of Engineering, The University of Iowa, Iowa City, IA 52242, U.S.A. (the program can be acquired by contacting the author).

Arora, J. S. and Tseng, C. H. (1987b). An Investigation of Pshenichnyi's Recursive Quadratic Programming Method for Engineering Optimization – A Discussion. *Journal of Mechanisms, Transmissions and Automation in Design,* Transactions of the ASME, Vol. 109, No. 6, pp. 254–256.

Arora, J. S. and Tseng, C. H. (1988). Interactive Design Optimization. *Engineering Optimization,* Vol. 13, pp. 173–188.

ASTM (1980). *Standard Metric Practice,* No. E380-79, American Society for Testing and Material, 1916 Race St., Philadelphia, Pa 19103.

Atkinson, K. E. (1978). *An Introduction to Numerical Analysis,* John Wiley and Sons, New York.

Bartel, D. L. (1969). *Optimum Design of Spatial Structures.* Ph.D. Dissertation, College of Engineering, The University of Iowa, Iowa City.

Belegundu, A. D. and Arora, J. S. (1984a). A Recursive Quadratic Programming Algorithm with Active Set Strategy for Optimal Design. *International Journal for Numerical Methods in Engineering,* Vol. 20, No. 5, pp. 803–816.

Belegundu, A. D. and Arora, J. S. (1984b). A Computational Study of Transformation Methods for Optimal Design. *AIAA J.,* Vol. 22, No. 4, pp. 535–542.

Belegundu, A. D. and Arora, J. S. (1985). A Study of Mathematical Programming Methods for Structural Optimization. *International Journal for Numerical Methods in Engineering,* Vol. 21, No. 9, pp. 1583–1624.

Bell, W. W. (1975). *Matrices for Scientists and Engineers,* Van Nostrand Reinhold, New York.

Blank, L. and Tarquin, A. (1983). *Engineering Economy,* 2nd ed., McGraw-Hill, New York.

Cauchy, A. (1847). Method Generale Pour La Resolution des Systemes d'Equations Simultanees. *Compt. Rend. Aca. Sci.,* Vol. 25, pp. 536–538.

Clough, R. W. and Penzien, J. (1975). *Dynamics of Structures,* McGraw-Hill, New York.

Cook, R. D. (1981). *Concepts and Applications of Finite Element Analysis,* John Wiley and Sons, New York.

Cooper, L. and Steinberg, D. (1970). *Introduction to Methods of Optimization,* W. B. Saunders, Philadelphia.

Corcoran, P. J. (1970). Configuration Optimization of Structures. *International Journal of Mechanical Sciences,* Vol. 12, pp. 459–462.

Crandall, S. H., Dahl, H. C. and Lardner, T. J. (1978). *Introduction to Mechanics of Solids,* McGraw-Hill, New York.

Dano, S. (1974). *Linear Programming in Industry,* 4th ed., Springer-Verlag, New York.

Day, H. J. and Dolbear, F. (1965). Regional Water Quality Management. *Proceedings of the 1st Annual Meeting of the American Water Resources Association,* University of Chicago, pp. 283–309.

De Boor, C. (1978). *A Practical Guide to Splines,* Vol. 27, Applied Mathematical Sciences, Springer-Verlag, New York.

Deif, A. S. (1982). *Advanced Matrix Theory for Scientists and Engineers,* Halsted Press, New York.

Deininger, R. A. (1975). Water Quality Management – The Planning of Economically Optimal Pollution Control Systems. *Proceedings of the 1st Annual Meeting of the American Water Resources Association,* University of Chicago, pp. 254–282.

Drew, D. (1968). *Traffic Flow Theory and Control,* McGraw-Hill, New York.

Fletcher, R. and Powell, M. J. D. (1963). A Rapidly Convergent Descent Method for Minimization. *The Computer Journal,* Vol. 6, pp. 163–180.

Fletcher, R. and Reeves, R. M. (1964). Function Minimization by Conjugate Gradients. *The Computer Journal,* Vol. 7, pp. 149–160.

Forsythe, G. E. Moler, C. B. (1967). *Computer Solution of Linear Algebraic Systems,* Prentice-Hall, Englewood Cliffs.

Franklin, J. N. (1968). *Matrix Theory,* Prentice-Hall, Englewood Cliffs.

Gabrielle, G. A. and Beltracchi, T. J. (1987). An Investigation of Pschenichnyi's Recursive Quadratic Programming Method for Engineering Optimization. *Journal of Mechanisms,*

Transmissions and Automation in Design, Transactions of the ASME, Vol. 109, No. 6, pp. 248–253.

Gere, J. M. and Weaver, W. (1983). *Matrix Algebra for Engineers,* Brooks/Cole Engineering Division, Monterey.

Gill, P. E., Murray, W. and Wright, M. H. (1981). *Practical Optimization,* Academic Press, New York.

Gill, P. E., Murray, W., Saunders, M. A. and Wright, M. H. (1984). *User's Guide for QPSOL: Version 3.2,* Systems Optimization Laboratory, Department of Operations Research, Stanford University, Stanford, California 94305, USA.

Grandin, H. (1986). *Fundamentals of The Finite Element Method,* Macmillan, New York.

Grant, E. L., Ireson, W. G. and Leavenworth, R. S. (1982). *Principles of Engineering Economy,* 7th ed., John Wiley and Sons, New York.

Hadley, G. (1961). *Linear Programming,* Addison-Wesley, Reading, MA.

Hadley, G. (1964). *Nonlinear and Dynamic Programming,* Addison-Wesley, Reading, MA.

Haftka, R. T. and Kamat, M. P. (1985). *Elements of Structural Optimization,* Martinus Nijhoff, Dordrecht.

Han, S. P. (1976). Superlinearly Convergent Variable Metric Algorithms for General Nonlinear Programming. *Math. Programming,* Vol. 11, pp. 263–282.

Han, S. P. (1977). A Globally Convergent Method for Nonlinear Programming. *Journal of Optimization Theory and Applications,* Vol. 22, pp. 297–309.

Haug, E. J. and Arora, J. S. (1979). *Applied Optimal Design,* Wiley Interscience, New York.

Hock, W. and Schittkowski, K. (1980). *Test Examples for Nonlinear Programming Codes,* Lecture Notes in Economics and Mathematical Systems, 187, Springer-Verlag, New York.

Hock, W. and Schittkowski, K. (1983). A Comparative Performance Evaluation of 27 Nonlinear Programming Codes. *Computing 30,* Springer-Verlag, New York, pp. 335–358.

Hohn, F. E. (1964). *Elementary Matrix Algebra,* Macmillan, New York.

Hopper, M. J. (1981). *Harwell Subroutine Library,* Computer Science and Systems Division, AERE Harwell, Oxfordshire.

Hsieh, C. C. and Arora, J. S. (1984). Design Sensitivity Analysis and Optimization of Dynamic Response. *Computer Methods in Applied Mechanics and Engineering,* Vol. 43, pp. 195–219.

Huebner, K. H. and Thornton, E. A. (1982). *The Finite Element Method for Engineers,* John Wiley and Sons, New York.

Iyengar, N. G. R. and Gupta, S. K. (1980). *Programming Methods in Structural Design,* John Wiley and Sons, New York.

Jennings, A. (1977). *Matrix Computations for Engineers,* John Wiley and Sons, New York.

Kunzi, H. P. and Krelle, W. (1966). *Nonlinear Programming,* Blaisdell Publishing Co., Waltham, MA.

Lemke, C. E. (1965). Bimatrix Equilibrium Points and Mathematical Programming. *Management Science,* Vol. 11, pp. 681–689.

Lim, O. K. and Arora, J. S. (1986). An Active Set RQP Algorithm for Optimal Design. *Computer Methods in Applied Mechanics and Engineering,* Vol. 57, pp. 51–65.

Lim, O. K. and Arora, J. S. (1987). Dynamic Response Optimization Using an Active Set RQP Algorithm. *International Journal for Numerical Methods in Engineering,* Vol. 24, No. 10, pp. 1827–1840.

Luenberger, D. G. (1984). *Linear and Nonlinear Programming,* Addison-Wesley, Reading, MA.

Lynn, W. R. (1964). State Development of Wastewater Treatment Works. *J. Water Pollution Control Federation,* pp. 722–751.

Marquardt, D. W. (1963). An Algorithm for Least Squares Estimation of Nonlinear Parameters. *SIAM J.,* Vol. 11, pp. 431–441.

McCormick, G. P. (1967). Second Order Conditions for Constrained Optima. *SIAM J. Applied Mathematics,* Vol. 15, pp. 641–652.

Meirovitch, L. (1985). *Introduction to Dynamics and Controls,* John Wiley and Sons, New York.

NAG (1984). "FORTRAN Library Manual". Numerical Algorithms Group, 1101 31st Street, Suite 100, Downers Grove, Illinois 60515, USA.

Osman, M. O. M., Sankar, S. and Dukkipati, R. V. (1978). Design Synthesis of a Multi-speed Machine Tool Gear Transmission Using Multiparameter Optimization. *Journal of Mechanical Design,* Transactions of ASME, Vol. 100, April, pp. 303–310.

Pederson, D. R., Brand, R. A., Cheng, C. and Arora, J. S. (1987). Direct Comparison of Muscle Force Predictions Using Linear and Nonlinear Programming. *Journal of Biomechanical Engineering,* Transactions of the ASME, Vol. 109, No. 3, pp. 192–199.

Powell, M. J. D. (1978a). A Fast Algorithm for Nonlinearity Constrained Optimization Calculations. *Lecture Notes in Mathematics,* G. A. Watson *et al.* (Eds), Springer Verlag, Berlin. (Also in *Numerical Analysis,* Proceedings of the Biennial Conference held at Dundee, June 1977.)

Powell, M. J. D. (1978b). The Convergence of Variable Metric Methods for Nonlinearity Constrained Optimization Calculations. In *Nonlinear Programming 3,* O. L. Mangasarian, R. R. Meyer and S. M. Robinson (Eds), Academic Press, New York.

Powell, M. J. D. (1978c). Algorithms for Nonlinear Functions that Use Lagrange Functions. *Mathematical Programming,* Vol. 14, pp. 224–248.

Pshenichny, B. N. (1978). Algorithms for the General Problem of Mathematical Programming. *Kibernetica,* No. 5.

Randolph, P. H. and Meeks, H. D. (1978). *Applied Linear Optimization,* Grid Inc., 4666 Indianola Avenue, Columbus, OH.

Ravindran, A. and Lee, H. (1981). Computer Experiments on Quadratic Programming Algorithms. *European Journal of Operations Research,* Vol. 8, No. 2, pp. 166–174.

Reklaitis, G. V., Ravindran, A. and Ragsdell, K. M. (1983). *Engineering Optimization: Methods and Applications,* John Wiley and Sons, New York.

Roark, R. J. and Young, W. C. (1975). *Formulas for Stress and Strain,* 5th ed., McGraw-Hill, New York.

Rosen, J. B. (1961). The Gradient Projection Method for Nonlinear Programming. *Journal of the Society for Industrial and Applied Mathematics,* Vol. 9, pp. 514–532.

Rubinstein, M. F. and Karagozian, J. (1966). Building Design Under Linear Programming. *Proc. ASCE,* Vol. 92, No. ST6, pp. 223–245.

Sargeant, R. W. H. (1974). Reduced-gradient and Projection Methods for Nonlinear Programming. In *Numerical Methods for Constrained Optimization,* P. E. Gill and W. Murray (Eds), Academic Press, New York, pp. 149–174.

Sasieni, M., Yaspan, A. and Friedman, L. (1960). *Operations-Methods and Problems,* John Wiley and Sons, New York.

Schittkowski, K. (1981). The Nonlinear Programming Method of Wilson, Han and Powell with an Augmented Lagrangian Type Line Search Function, Part 1: Convergence Analysis, Part 2: An Efficient Implementation with Linear Least Squares Subproblems. *Numerische Mathematik,* Vol. 38, pp. 83–127.

Schmit, L. A. (1960). Structural Design by Systematic Synthesis. *Proceedings of the Second ASCE Conference on Electronic Computations,* Pittsburgh, pp. 105–122.

Schrage, L. (1981). *User's Manual for LINDO,* The Scientific Press, Palo Alto, CA.

Shampine, L. F. and Gordon, M. K. (1975). *Computer Simulation of Ordinary Differential Equations: The Initial Value Problem.* W. H. Freeman, San Francisco.

Shigley, J. E. (1977). *Mechanical Engineering Design,* McGraw-Hill, New York.

Siddall, J. N. (1972). *Analytical Decision-Making In Engineering Design,* Prentice Hall, Englewood Cliffs.

Spotts, M. F. (1953). *Design of Machine Elements,* 2nd ed., Prentice Hall, Englewood Cliffs.

Stark, R. M. and Nicholls, R. L. (1972). *Mathematical Foundations for Design: Civil Engineering Systems,* McGraw-Hill, New York.

Stewart, G. (1973). *Introduction to Matrix Computations,* Academic Press, New York.

Stoecker, W. F. (1971). *Design of Thermal Systems,* McGraw-Hill, New York.

Strang, G. (1976). *Linear Algebra and Its Applications,* Academic Press, New York.

Sun, P. F., Arora, J. S. and Haug, E. J. (1975). *Fail-Safe Optimal Design of Structures,* Technical Report No. 19 (also Ph.D. Dissertation of P. F. Sun), Department of Civil and Environmental Engineering, The University of Iowa, Iowa City.

Thanedar, P. B., Arora, J. S. and Tseng, C. H. (1986). A Hybrid Optimization Method and Its Role in Computer Aided Design. *Computers and Structures,* Vol. 23, No. 3, pp. 305–314.

Thanedar, P. B., Arora, J. S., Tseng, C. H., Lim, O. K. and Park, G. J. (1987), Performance of Some SQP Algorithms on Structural Design Problems. *International Journal for Numerical Methods in Engineering,* Vol. 23, No. 12, pp. 2187–2203.

Tseng, C. H. and Arora, J. S. (1987). Optimal Design for Dynamics and Control Using a Sequential Quadratic Programming Algorithm. Technical Report No. ODL-87.10, Optimal Design Laboratory, College of Engineering, The University of Iowa, Iowa City.

Tseng, C. H. and Arora, J. S. (1988). On Implementation of Computational Algorithms for Optimal Design 1: Preliminary Investigation; 2: Extensive Numerical Investigation. *International Journal for Numerical Methods in Engineering,* Vol. 26, No. 6, pp. 1365–1402.

Vanderplaats, G. N. (1984). *Numerical Optimization Techniques for Engineering Design with Applications,* McGraw-Hill, New York.

Vanderplaats, G. N. and Yoshida, N. (1985). Efficient Calculation of Optimum Design Sensitivity. *AIAA J.,* Vol. 23, No. 11, pp. 1798–1803.

Wahl, A. M. (1963). *Mechanical Springs,* 2nd ed., McGraw-Hill, New York.

Wilson, R. B. (1963). *A Simplical Algorithm for Concave Programming,* Ph.D. Dissertation, Graduate School of Business Administration, Harvard University, Boston.

Wohl, M. and Martin, B. V. (1967). *Traffic Systems Analysis,* McGraw-Hill, New York.

Wolfe, P. (1959). The Simplex Method for Quadratic Programming. *Econometica,* Vol. 27, No. 3, pp. 382–398.

Wu, N. and Coppins, R. (1981). *Linear Programming and Extensions,* McGraw-Hill, New York.

Zoutendijk, G. (1960) *Methods of Feasible Directions,* Elsevier, Amsterdam.

ANSWERS TO SELECTED EXERCISES

CHAPTER 2
OPTIMUM DESIGN PROBLEM
FORMULATION

2.25 $x^* = (2,2)$, $f^* = 2$. **2.26** $x^* = (0,4)$, $f^* = -8$. **2.27** $x^* = (8, 10)$, $f^* = 38$. **2.28** $x^* = (4, 3.333, 2)$, $f^* = -11.33$. **2.29** $x^* = (10, 10)$, $f^* = -400$. **2.30** $x^* = (0,0)$, $f^* = 0$. **2.31** $x^* = (0,0)$, $f^* = 0$. **2.32** $x^* = (2,3)$, $f^* = -22$. **2.33** $x^* = (-2.5, 1.58)$, $f^* = -3.95$. **2.34** $x^* = (-0.5, 0.167)$, $f^* = -0.5$. **2.35** $b^* = 24.66$ cm, $d^* = 49.32$ cm, $f^* = 1216$ cm^3. **2.36** $R_o^* = 20$ cm, $R_i^* = 19.84$ cm, $f^* = 79.1$ kg. **2.37** $R^* = 53.6$ mm, $t = 5.0$ mm, $f^* = 66$ kg. **2.38** $R_o^* = 56$ mm, $R_i^* = 51$ mm, $f^* = 66$ kg. **2.39** $w^* = 93$ mm, $t^* = 5$ mm, $f^* = 70$ kg. **2.40** Infinite optimum points, $f^* = 0.812$ kg. **2.41** $A^* = 5000$, $h^* = 14$, $f^* = 13.4$ mil. dollars. **2.42** $R^* = 1.0077$ m, $t^* = 0.0168$ m, $f^* = 1.0424$ m^3. **2.43** $A_1^* = 6.1$ cm^2, $A_2^* = 2.0$ cm^2, $f^* = 5.39$ kg. **2.45** $t^* = 8.45$, $f^* = 1.91 \times 10^5$. **2.46** $R^* = 7.8$, $H^* = 15.6$, $f^* = 1.75 \times 10^6$. **2.47** Infinite optimum points; one point: $R^* = 0.4$ m, $t^* = 1.59 \times 10^{-3}$, $f^* = 15.7$ kg. **2.48** For $l = 0.5$ m, $T_0 = 10$ kN $\cdot$ m, $T_{max} = 20$ kN $\cdot$ m, $x_1^* = 103$ mm, $x_2^* = 0.955$, $f^* = 2.9$ kg. **2.49** For $l = 0.5$, $T_0 = 10$ kN $\cdot$ m, $T_{max} = 20$ kN $\cdot$ m, $d_o^* = 103$ mm, $d_i^* = 98.36$ mm, $f^* = 2.9$ kg. **2.50** $R^* = 50.3$ mm, $t^* = 2.35$ mm, $f^* = 2.9$ kg. **2.51** $R^* = 20$ cm, $H^* = 7.2$ cm, $f^* = -9000$ cm^3. **2.52** $R^* = 0.5$ cm, $N^* = 2550$, $f^* = -8000$ ($l = 10$). **2.53** (a) $R^* = 33.7$ mm, $t^* = 5.0$, $f^* = 41$ kg. (b)

$R_o^* = 36$ mm, $R_i^* = 31$ mm, $f^* = 41$ kg. **2.54** (a) $R^* = 21.5$ mm, $t^* = 5.0$ mm, $f^* = 26$ kg. (b) $R_o^* = 24.0$ mm, $R_i^* = 19.0$ mm, $f^* = 26$ kg. **2.55** (a) $R^* = 27$, $t^* = 5$ mm, $f^* = 33$ kg. (b) $R_o^* = 29.5$ mm, $R_i^* = 24.5$ mm, $f^* = 33$ kg. **2.56** $D^* = 8.0$ cm, $H^* = 8.0$ cm, $f^* = 301.6$ cm^2. **2.57** $A_1^* = 413.68$ mm, $A_2^* = 163.7$ mm, $f^* = 5.7$ kg. **2.58** Infinite optimum points; one point: $R^* = 20$ mm, $t^* = 3.3$ mm, $f^* = 8.1$ kg. **2.59** $A^* = 390$ mm^2, $h^* = 500$ mm, $f^* = 5.5$ kg. **2.60** $A^* = 410$ mm^2, $s^* = 1500$ mm, $f^* = 8$ kg. **2.61** $A_1^* = 293.7$ mm^2, $A_2^* = 65.6$ mm^2, $f^* = 7$ kg. **2.62** $R^* = 130$ cm, $t^* = 2.86$ cm, $f^* = 57\,000$ kg. **2.63** $d_o^* = 41.56$ cm, $d_i^* = 40.19$ cm, $f^* = 680$ kg. **2.64** $d_o^* = 1310$ mm, $t^* = 14.2$ mm, $f^* = 92\,500$ N. **2.65** $H^* = 50.0$ cm, $D^* = 3.42$ cm, $f^* = 6.6$ kg.

CHAPTER 3
OPTIMUM DESIGN CONCEPTS

3.2 $\cos x = 1.044 - 0.1575x - 0.353\,55x^2$ at $x = \pi/4$. **3.3** $\cos x = 1.1327 - 0.342\,43x - 0.25x^2$ at $x = \pi/3$. **3.4** $\sin x = -0.021\,99 + 1.2783x - 0.25x^2$ at $x = \pi/6$. **3.5** $\sin x = -0.066\,34 + 1.2625x - 0.353\,55x^2$ at $x = \pi/4$. **3.6** $e^x = 1 + x + 0.5x^2$ at $x = 0$. **3.7** $e^x = 7.389 - 7.389x + 3.6945x^2$ at $x = 2$. **3.8** $\bar{f}(x) = 41x_1^2 - 42x_1 - 40x_1x_2 + 20x_2 + 10x_2^2 + 15$; $\bar{f}(1.2,\ 0.8) = 7.64$, $f(1.2,\ 0.8) = 8.136$, Error $= f - \bar{f} = 0.496$. **3.9** Indefinite. **3.10** Indefinite. **3.11** Indefinite. **3.12** Positive definite. **3.13** Indefinite. **3.14** Indefinite. **3.15** Positive definite. **3.16** Indefinite. **3.18** $\mathbf{x} = (0,0)$ – local minimum, $f = 7$. **3.19** $\mathbf{x} = (0,0)$ – inflection point. **3.20** $\mathbf{x}^{*1} = (-3.332,\ 0.0395)$ – local maximum, $f = 18.58$; $\mathbf{x}^{*2} = (-0.398,\ 0.5404)$ – inflection point. **3.21** $\mathbf{x}^{*1} = (4, 8)$ – inflection point; $\mathbf{x}^{*2} = (-4, -8)$ – inflection point. **3.22** $x^* = (2n + 1)\pi,\ n = 0,\ \pm 1,\ \pm 2, \ldots$ local minima, $f = -1$; $x^* = 2n\pi,\ n = 0,\ \pm 1,\ \pm 2, \ldots$ local maxima, $f = 1$. **3.23** $\mathbf{x}^* = (0, 0)$ – local minimum, $f = 0$. **3.24** $x^* = 0$ – local minimum, $f = 0$; $x^* = 2$ – local maximum, $f = 0.541$. **3.25** $\mathbf{x}^* = (3.684, 0.7368)$ – local minimum, $f = 11.0521$. **3.26** $\mathbf{x}^* = (1, 1)$ – local minimum, $f = 1$. **3.27** $\mathbf{x}^* = (-\frac{2}{7}, -\frac{6}{7})$ – local minimum, $f = -\frac{24}{7}$. **3.28** $\mathbf{x}^{*1} = (241.7643, 0.030\,995\,42)$ – local minimum, $U = 483\,528.6$; $\mathbf{x}^{*2} = (-241.7643, -0.030\,995\,42)$ – local maximum. **3.30** $\mathbf{x}^* = (2.166\,667, 1.833\,33)$, $v = -0.166\,667$, $f = -8.333\,33$. **3.31** $\mathbf{x}^{*1} = (1.5088, 3.272)$, $v = -17.1503$, $f = 244.528$; $\mathbf{x}^{*2} = (2.5945, -2.0198)$, $v = -1.4390$, $f = 15.291$; $\mathbf{x}^{*3} = (-3.630, -3.1754)$, $v = -23.2885$, $f = 453.154$; $\mathbf{x}^{*4} = (-3.7322, 3.0879)$, $v = -2.1222$, $f = 37.877$. **3.32** $\mathbf{x} = (2, 2)$, $v = -2$, $f = 2$. **3.33** (i) No, (ii) Solution of equalities, $\mathbf{x} = (3, 1)$, $f = 4$. **3.34** $\mathbf{x}^{*1} = (0.816, 0.75)$, $\mathbf{u} = (0, 0, 0, 0)$, $f = 2.214$; $\mathbf{x}^{*2} = (0.816, 0)$, $\mathbf{u} = (0, 0, 0, 3)$, $f = 1.0887$; $\mathbf{x}^{*3} = (0, 0.75)$, $\mathbf{u} = (0, 0, 2, 0)$, $f = 1.125$; $\mathbf{x}^{*4} = (1.5073, 1.2317)$, $\mathbf{u} = (0, 0.9632, 0, 0)$, $f = 0.251$; $\mathbf{x}^{*5} = (1.0339,\ 1.655)$, $\mathbf{u} = (1.2067, 0, 0, 0)$, $f = 0.4496$; $\mathbf{x}^{*6} = (0, 0)$, $\mathbf{u} = (0, 0, 2, 3)$, $f = 0$; $\mathbf{x}^{*7} = (2, 0)$, $\mathbf{u} = (0, 2, 0, 7)$, $f = -4$; $\mathbf{x}^{*8} = (0, 2)$, $\mathbf{u} = (\frac{5}{3}, 0, \frac{11}{3}, 0)$, $f = -2$; $\mathbf{x}^{*9} = (1.386, 1.538)$, $\mathbf{u} = (0.633,\ 0.626, 0, 0)$, $f = -0.007\,388$. **3.35** $\mathbf{x}^* = (\frac{48}{23}, \frac{40}{23})$, $u = 0$, $f(\mathbf{x}^*) = -\frac{192}{23}$. **3.36** $\mathbf{x}^* = (2.5, 1.5)$, $u = 1$, $f = 1.5$. **3.37** $\mathbf{x}^* = (6.3, 1.733)$, $\mathbf{u} = (0, 0.8, 0, 0)$, $f = -56.901$. **3.38** $\mathbf{x}^* = (1, 1)$, $u = 0$, $f = 0$. **3.39** $\mathbf{x}^* = (1, 1)$, $\mathbf{u} = (0, 0)$, $f = 0$. **3.40** $\mathbf{x}^* = (2, 1)$, $\mathbf{u} =$

$(0, 2)$, $f = 1$. **3.41** $\mathbf{x}^{*1} = (2.5945, 2.0198)$, $u_1 = 1.439$, $f = 15.291$; $\mathbf{x}^{*2} = (-3.63, 3.1754)$, $u_1 = 23.2885$, $f = 453.154$; $\mathbf{x}^{*3} = (1.5088, -3.2720)$, $u_1 = 17.1503$, $f = 244.53$; $\mathbf{x}^{*4} = (-3.7322, -3.0879)$, $u_1 = 2.1222$, $f = 37.877$. **3.42** $\mathbf{x}^* = (3.25, 0.75)$, $v = -1.25$, $u = 0.75$, $f = 5.125$. **3.43** $\mathbf{x}^{*1} = (\frac{4}{\sqrt{3}}, \frac{1}{3})$, $u = 0$, $f = -24.3$; $\mathbf{x}^{*2} = (-\frac{4}{\sqrt{3}}, \frac{1}{3})$, $u = 0$, $f = 24.967$; $\mathbf{x}^{*3} = (0, 3)$, $u = 16$, $f = -21$; $\mathbf{x}^{*4} = (2, 1)$, $u = 4$, $f = -25$. **3.44** $\mathbf{x}^* = (-\frac{2}{7}, -\frac{6}{7})$, $u = 0$, $f = -\frac{24}{7}$ **3.45** $D = 7.98$, $H = 8$, $\mathbf{u} = (0.5, 0, 0, 0.063, 0)$, $f = 300.6 \, \text{cm}^2$. **3.46** $R = 7.871\,686 \times 10^{-2}$, $t = 1.574\,337 \times 10^{-3}$, $u_2 = 15.28$, $u_3 = 192.97$, $u_1 = u_4 = u_5 = 0$; $f = 30.56 \, \text{kg}$. **3.47** $R_o = 7.950204 \times 10^{-2}$, $R_i = 7.792\,774 \times 10^{-2}$, $u_2 = 15.28$, $u_3 = 97.04$, $u_1 = u_4 = u_5 = 0$; $f = 30.56 \, \text{kg}$. **3.48** $x_1 = 60.506\,34$, $x_2 = 1.008\,439$, $u_1 = 19\,918$, $u_2 = 23\,186$, $u_3 = u_4 = 0$, $f = 23\,186.4$ **3.49** $h = 14$, $A = 5000$, $u_1 = 5.9 \times 10^{-4}$, $u_2 = 6.8 \times 10^{-4}$, $u_3 = u_4 = u_5 = 0$, $f = 13.4$. **3.50** $A = 20,000$, $B = 10\,000$, $u_1 = 35$, $u_3 = 27$ (or, $u_1 = 8$, $u_2 = 108$), $f = -1\,240\,000$. **3.51** $R = 20$, $H = 7.161\,973$, $u_1 = 10$, $u_3 = 450$, $f = -9000$. **3.52** $R = 0.5$, $N = 2546.5$, $u_1 = 16\,000$, $u_2 = =4$, $f = 8000$. **3.53** $W = 70.7107$, $D = 141.4214$, $u_3 = 1.41421$, $u_4 = 0$, $f = 28\,284.28$. **3.54** $A = 70$, $B = 76$, $u_1 = 0.4$, $u_4 = 16$, $f = -1308$. **3.55** $B = 0$, $M = 2.5$, $u_1 = 0.5$, $u_3 = 1.5$, $f = 2.5$. **3.56** $x_1 = 316.667$, $x_2 = 483.33$, $u_1 = \frac{2}{3}$, $u_2 = \frac{10}{3}$, $f = -1283.333$. **3.57** $r = 4.570\,78$, $h = 9.14156$, $v_1 = -0.364\,365$, $u_1 = 43.7562$, $f = 328.17$. **3.58** $b = 10$, $h = 18$, $u_1 = 0.042\,67$, $u_2 = 0.006\,58$, $f = 0.545\,185$. **3.59** $D = 5.758\,823$, $H = 5.758\,826$, $v_1 = -277.834$, $f = 62\,512.75$. **3.60** $P_1 = 30.4$, $P_2 = 29.6$, $u_1 = 59.8$, $f = 1789.68$. **3.63** (i) $\pi \leqq x \leqq 2\pi$ (ii) $\pi/32 \leqq x \leqq 3\pi/2$. **3.64** Convex everywhere. **3.65** Not convex. **3.66** $S_1 = \{\mathbf{x} \mid x_1 \geqq -\frac{1}{6}, \quad |x_2| \leqq (\frac{1}{2})\sqrt{((x_1 + \frac{11}{12})^2 - (\frac{3}{4})^2)}\}$; $S_2 = \{\mathbf{x} \mid x_1 \leqq -\frac{5}{3}, \quad |x_2| \leqq (\frac{1}{2})\sqrt{(x_1 + \frac{11}{12})^2 - (\frac{3}{4})^2)}\}$. **3.67** Not convex. **3.68** Convex everywhere. **3.69** Fails convexity check. **3.70** Convex if $C \geqq 0$. **3.71** Fails convexity check. **3.72** Fails convexity check. **3.73** Fails convexity check. **3.74** Fails convexity check. **3.75** Fails convexity check. **3.76** Convex. **3.77** Fails convexity check. **3.78** Fails convexity check. **3.79** Convex. **3.80** $18.43° \leqq \theta \leqq 71.57°$. **3.81** $\theta \geqq 71.57°$. **3.82** No solution. **3.83** $\theta \leqq 18.43°$. **3.85** $x_1^* = 2.1667$, $x_2^* = 1.8333$, $v^* = -0.1667$; isolated minimum. **3.86** $(1.5088, 3.2720)$, $v^* = -17.15$; not a minimum point; $(2.5945, -2.0198)$, $v^* = -1.439$; isolated local minimum; $(-3.6300, -3.1754)$, $v^* = -23.288$; not a minimum point; $(-3.7322, 3.0879)$, $v^* = -2.122$; isolated local minimum. **3.87** $(0.816, 0.75)$, $\mathbf{u}^* = (0, 0, 0, 0)$; not a minimum point; $(0.816, 0)$, $\mathbf{u}^* = (0, 0, 0, 3)$; not a minimum point; $(0, 0.75)$, $\mathbf{u}^* = (0, 0, 2, 0)$; not a minimum point; $(1.5073, 1.2317)$, $\mathbf{u}^* = (0, 0.9632, 0, 0)$; not a minimum point; $(1.0339, 1.6550)$, $\mathbf{u}^* = (1.2067, 0, 0, 0)$; not a minimum point; $(0, 0)$, $\mathbf{u}^* = (0, 0, 2, 3)$; isolated local minimum; $(2, 0)$, $\mathbf{u}^* = (0, 2, 0, 7)$; isolated local minimum; $(0, 2)$, $\mathbf{u}^* = (1.667, 0, 3.667, 0)$; isolated local minimum; $(1.386, 1.538)$, $\mathbf{u}^* = (0.633, 0.626, 0, 0)$; isolated local minimum. **3.88** $(2.0870, 1.7391)$, $u^* = 0$; isolated global minimum. **3.89** $(2.5945, 2.0198)$, $u^* = 1.4390$; isolated local minimum; $(-3.6300, 3.1754)$, $u^* = 23.288$; not a minimum; $(1.5088, -3.2720)$, $u^* = 17.150$; not a minimum; $(-3.7322, -3.0879)$, $u^* = 2.122$; isolated local minimum. **3.90** $(3.25, 0.75)$, $u^* = 0.75$, $v^* = -1.25$; isolated global minimum. **3.91** $(2.3094, 0.3333)$, $u^* = 0$; not a minimum; $(-2.3094, 0.3333)$,

$u^* = 0$; not a minimum; $(0, 3)$, $u^* = 16$; not a minimum; $(2, 1)$, $u^* = 4$; isolated local minimum. **3.92** $(-0.2857, -0.8571)$, $u^* = 0$; isolated local minimum. **3.93** $R_o^* = 20$ cm, $R_i^* = 19.84$ cm, $f^* = 79.1$ kg. **3.94** Multiple optima between $(31.83, 1.0)$ and $(25.23, 1.26)$ mm, $f^* = 45.9$ kg. **3.95** $R^* = 1.0$ m, $t^* = 0.0167$ m, $f^* = 7850$ kg. **3.96** $R^* = 0.0787$ m, $t^* = 0.00157$ m, $f^* = 30.5$ kg. **3.97** $R_o^* = 0.0795$ m, $R_i^* = 0.0779$ m, $f^* = 30.5$ kg. **3.98** $H^* = 8$ cm, $D^* = 7.98$ cm, $f^* = 300.6$ cm^2. **3.99** $A^* = 5000$ m^2, $h^* = 14$ m, $f^* = 13.4$ million dollars. **3.100** $x_1^* = 102.98$ mm, $x_2^* = 0.9546$, $f^* = 2.9$ kg. **3.101** $d_o^* = 103$ mm, $d_i^* = 98.36$ mm, $f^* = 2.9$ kg. **3.102** $R^* = 50.3$ mm, $t^* = 2.34$, $f^* = 2.9$ kg. **3.103** $H^* = 50$ cm, $D^* = 3.2$ cm, $f^* = 6.6$ kg. **3.113** Not a convex programming problem; $D^* = 10$ m, $H^* = 10$ m, $f^* = 60,000\pi$ m^3; $\Delta f = 800\pi$ m^3. **3.114** Convex; $A_1^* = 2.937 \times 10^{-4}$ m, $A_2^* = 6.556 \times 10^{-5}$ m, $f^* = 7.0$ kg.

CHAPTER 4
LINEAR PROGRAMMING METHODS
FOR OPTIMUM DESIGN

4.21 $(0, 4, -3, -5)$; $(2, 0, 3, 1)$; $(1, 2, 0, -2)$; $(\frac{5}{3}, \frac{2}{3}, 2, 0)$. **4.22** $(0, 0, -3, -5)$; $(0, 1, 0, -3)$; $(0, 2.5, 4.5, 0)$; $(-3, 0, 0, -11)$; $(2.5, 0, -5.5, 0)$; $(\frac{9}{8}, \frac{11}{8}, 0, 0)$. **4.23** Decompose x_2 into two variables; $(0, 0, 0, 12, -3)$; $(0, 0, -3, 0, 6)$; $(0, 0, -1, 8, 0)$; $(0, 3, 0, 0, 6)$; $(0, 1, 0, 8, 0)$; $(4, 0, 0, 0, 1)$; $(3, 0, 0, 3, 0)$; $(4.8, 0, 0.6, 0, 0)$; $(4.8, -0.6, 0, 0, 0)$. **4.24** $(0, -\frac{8}{3}, -\frac{1}{3})$; $(2, 0, 3)$; $(0.2, -2.4, 0)$. **4.25** $(0, 0, 9, 2, 3)$; $(0, 9, 0, 20, -15)$; $(0, -1, 10, 0, 5)$; $(0, 1.5, 7.5, 5, 0)$; $(4.5, 0, 0, -2.5, 16.5)$; $(2, 0, 5, 0, 9)$; $(-1, 0, 11, 3, 0)$; $(4, 1, 0, 0, 13)$; $(\frac{15}{7}, \frac{33}{7}, 0, \frac{65}{7}, 0)$; $(-2.5, -2.25, 16.25, 0, 0)$. **4.26** $(0, 4, -3, -7)$; $(4, 0, 1, 1)$; $(3, 1, 0, -1)$; $(3.5, 0.5, 0.5, 0)$. **4.27** Decompose x_2 into two variables; 15 basic solutions; basic feasible solutions are $(0, 4, 0, 0, 7, 0)$; $(0, \frac{5}{3}, 0, \frac{7}{3}, 0, 0)$; $(2, 0, 0, 0, 1, 0)$; $(\frac{5}{3}, 0, 0, \frac{2}{3}, 0, 0)$; $(\frac{7}{3}, 0, \frac{2}{3}, 0, 0, 0)$. **4.28** Ten basic solutions; basic feasible solutions are $(2.5, 0, 0, 0, 4.5)$; $(1.6, 1.8, 0, 0, 0)$. **4.29** $(0, 0, 4, -2)$; $(0, 4, 0, 6)$; $(0, 1, 3, 0)$; $(-2, 0, 0, -4)$; $(2, 0, 8, 0)$; $(-1.2, 1.6, 0, 0)$. **4.30** $(0, 0, 0, -2)$; $(0, 2, -2, 0)$; $(0, 0, 0, -2)$; $(2, 0, 2, 0)$; $(0, 0, 0, 2)$; $(1, 1, 0, 0)$. **4.31** $(0, 0, 10, 18)$; $(0, 5, 0, 8)$; $(0, 9, -8, 0)$; $(-10, 0, 0, 48)$; $(6, 0, 16, 0)$; $(2, 6, 0, 0)$. **4.32** $\mathbf{x}^* = (\frac{10}{3}, 2)$; $f^* = -\frac{13}{3}$. **4.4.33** Infinite solutions between $\mathbf{x}^* = (0, 3)$ and $\mathbf{x}^* = (2, 0)$; $f^* = 6$. **4.34** $\mathbf{x}^* = (2, 4)$; $f^* = 10$. **4.35** $\mathbf{x}^* = (6, 0)$; $z^* = 12$. **4.36** $\mathbf{x}^* = (3.667)$, 1.667; $z^* = 15$. **4.37** $\mathbf{x}^* = (0, 5)$; $f^* = -5$. **4.39** $\mathbf{x}^* = (2, 4)$; $z^* = 10$. **4.40** Unbounded. **4.41** $\mathbf{x}^* = (3.5, 0.5)$; $z^* = 5.5$. **4.42** $\mathbf{x}^* = (1.667, 0.667)$; $z^* = 4.333$. **4.43** $\mathbf{x}^* = (1.6, 1.8, 0)$; $f^* = 18$. **4.44** $\mathbf{x}^* = (0, 1.667, 2.333)$; $f^* = 4.333$. **4.45** $\mathbf{x}^* = (1.125, 1.375)$; $f^* = 36$. **4.46** $\mathbf{x}^* = (2, 0)$; $f^* = 40$. **4.47** $\mathbf{x}^* = (1.3357, 0.4406, 0, 3.2392)$; $z^* = 9.7329$. **4.48** $\mathbf{x}^* = (0.6541, 0.0756, 0.3151)$; $f^* = 9.7329$. **4.49** $\mathbf{x}^* = (0, 25)$; $z^* = 150$. **4.50** $\mathbf{x}^* = (\frac{2}{3}, \frac{5}{3})$; $z^* = \frac{16}{3}$. **4.51** $\mathbf{x}^* = (\frac{7}{3}, -\frac{2}{3})$; $z^* = -\frac{1}{3}$. **4.52** $\mathbf{x}^* = (1, 1)$; $f^* = 1$. **4.53** $\mathbf{x}^* = (2, 2)$; $f^* = 10$. **4.54** $\mathbf{x}^* = (4.8, -0.6)$; $z^* = 3.6$. **4.55** $\mathbf{x}^* = (2, 4)$; $z^* = 10$. **4.56** $\mathbf{x}^* = (0, 5)$; $z^* = 40$. **4.57** Infeasible problem. **4.58** Infinite solutions; $f^* = 0$. **4.59** $A^* = 20\,000$,

$B^* = 10\,000$, Profit $= \$1\,240\,000$. **4.60** $A^* = 70$, $B^* = 76$; Profit $= \$1308$. **4.61** Bread $= 0$, Milk $= 2.5$ kg; cost $= \$2.5$. **4.62** Bottles of wine $= 316.67$, Bottles of whiskey $= 483.33$; Profit $= \$1283.3$. **4.63** Shortening produced $= 149\,499.2$ kg, salad oil produced $= 50\,000$ kg, margarine produced $= 10\,000$ kg; profit $= \$19\,499.2$. **4.64** $A^* = 10$, $B^* = 0$, $C^* = 20$; capacity $= 477\,000$. **4.65** $x_1^* = 0$, $x_2^* = 0$, $x_3^* = 200$, $x_4^* = 100$; $f^* = 786$. **4.66** $f^* = 1\,333\,679$ tonnes. **4.67** $\mathbf{x}^* = (0, 800, 0, 500, 1500, 0)$; $f^* = 7500$; $\mathbf{x}^* = (0, 0, 4500, 4000, 3000, 0)$; $f^* = 7500$; $\mathbf{x}^* = (0, 8, 0, 5, 15, 0), f^* = 7500$. **4.68** (a) no effect (b) cost decreases by $120\,000$. **4.69** 1. no effect; 2. out of range, re-solve the problem; $A^* = 70$, $B^* = 110$; Profit $= \$1580$; 3. profit reduces by $\$4$; 4. out of range, re-solve the problem; $A^* = 41.667$, $B^* = 110$; profit $= \$1213.33$. **4.70** $y_1 = 1$, $y_2 = 0$. **4.71** $y_1 = 0.25$, $y_2 = 1.25$, $y_3 = 0$, $y_4 = 0$. **4.72** Unbounded. **4.73** $y_1 = 0$, $y_2 = 2.5$, $y_3 = -1.5$. **4.74** $y_1 = 0$, $y_2 = \frac{5}{3}$, $y_3 = -\frac{7}{3}$. **4.75** $y_1 = 4$, $y_2 = -1$. **4.76** $y_1 = -\frac{5}{3}$, $y_2 = -\frac{2}{3}$. **4.77** $y_1 = 2$, $y_2 = -6$. **4.78** $y_1 = 0$, $y_2 = 5$. **4.79** $y_1 = 0.654$, $y_2 = -0.076$, $y_3 = 0.315$. **4.80** $y_1 = -1.336$, $y_2 = -0.441$, $y_3 = 0$, $y_4 = -3.239$. **4.81** $y_1 = 0$, $y_2 = 0$, $y_3 = 0$, $y_4 = 6$. **4.82** $y_1 = -1.556$, $y_2 = 0.556$. **4.83** $y_1 = 0$, $y_2 = \frac{5}{3}$, $y_3 = -\frac{7}{3}$. **4.84** $y_1 = -0.5$, $y_2 = -2.5$. **4.85** $y_1 = -\frac{1}{3}$, $y_2 = 0$, $y_3 = \frac{5}{3}$. **4.86** $y_1 = 0.2$, $y_2 = 0.4$. **4.87** $y_1 = 0.25$, $y_2 = 1.25$, $y_3 = 0$, $y_4 = 0$. **4.88** $y_1 = 2$, $y_2 = 0$. **4.89** Infeasible problem. **4.90** $y_1 = 3$, $y_2 = 0$. **4.91** For $b_1 = 20\,000$: $-8000 \leq \Delta_1 \leq 0$; for $b_2 = 10\,000$: $0 \leq \Delta_2 \leq \infty$; for $b_3 = 20\,000$: $-20\,000 \leq \Delta_3 \leq 0$; for $b_4 = 30\,000$: $-20\,000 \leq \Delta_4 \leq \infty$. **4.92** For $b_1 = 100$: $-34 \leq \Delta_1 \leq \infty$; for $b_2 = 80$: $-38 \leq \Delta_2 \leq 17$; for $b_3 = 70$: $-28.333 \leq \Delta_3 \leq 63.333$; for $b_4 = 110$: $-34 \leq \Delta_4 \leq \infty$. **4.93** For $b_1 = 12$: $-8 \leq \Delta_1 \leq \infty$; for $b_2 = 6$: $-\infty \leq \Delta_2 \leq 12$. **4.94** For $b_1 = 10$: $-8 \leq \Delta_1 \leq 8$; for $b_2 = 6$: $-2.6667 \leq \Delta_2 \leq 8$; for $b_3 = 2$: $-4 \leq \Delta_3 \leq \infty$; for $b_4 = 6$: $-\infty \leq \Delta_4 \leq 8$. **4.95** Unbounded problem. **4.96** For $b_1 = 5$: $-0.5 \leq \Delta_1 \leq \infty$; for $b_2 = 4$: $-1 \leq \Delta_2 \leq 0.333$; for $b_3 = 3$: $-1 \leq \Delta_3 \leq 1$. **4.97** For $b_1 = 5$: $-2 \leq \Delta_1 \leq \infty$; for $b_2 = -4$: $-2 \leq \Delta_2 \leq 2$; for $b_3 = 1$: $-2 \leq \Delta_3 \leq 1$. **4.98** For $b_1 = -5$: $-\infty \leq \Delta_1 \leq 4$; for $b_2 = -2$: $-8 \leq \Delta_2 \leq 4.5$. **4.99** For $b_1 = 1$: $-5 \leq \Delta_1 \leq 7$; for $b_2 = 4$: $-3.5 \leq \Delta_2 \leq \infty$. **4.100** For $b_1 = -3$: $-4.5 \leq \Delta_1 \leq 5.5$; for $b_2 = 5$: $-3 \leq \Delta_2 \leq \infty$. **4.101** For $b_1 = 3$: $-\infty \leq \Delta_1 \leq 3$; for $b_2 = -8$: $-\infty \leq \Delta_2 \leq 4$. **4.102** For $b_1 = 8$: $-8 \leq \Delta_1 \leq \infty$; for $b_2 = 3$: $-14.307 \leq \Delta_2 \leq 4.032$; for $b_3 = 15$: $-20.16 \leq \Delta_3 \leq 101.867$. **4.103** For $b_1 = 2$: $-3.9178 \leq \Delta_1 \leq 1.1533$; for $b_2 = 5$: $-0.692 \leq \Delta_2 \leq 39.579$; for $b_3 = -4.5$: $-\infty \leq \Delta_3 \leq 7.542$; for $b_4 = 1.5$: $-2.0367 \leq \Delta_4 \leq 0.334$. **4.104** For $b_1 = 90$: $-15 \leq \Delta_1 \leq \infty$; for $b_2 = 80$: $-30 \leq \Delta_2 \leq \infty$; for $b_3 = 15$: $-\infty \leq \Delta_3 \leq 100$; for $b_4 = 25$: $-10 \leq \Delta_4 \leq 5$. **4.105** For $b_1 = 3$: $-1.2 \leq \Delta_1 \leq 15$; for $b_2 = 18$: $-15 \leq \Delta_2 \leq 12$. **4.106** For $b_1 = 5$: $-4 \leq \Delta_1 \leq \infty$; for $b_2 = 4$: $-7 \leq \Delta_2 \leq 2$; for $b_3 = 3$: $-1 \leq \Delta_3 \leq \infty$. **4.107** For $b_1 = 0$: $-2 \leq \Delta_1 \leq 2$; for $b_2 = 2$: $-2 \leq \Delta_2 \leq \infty$. **4.108** For $b_1 = 0$: $-6 \leq \Delta_1 \leq 3$; for $b_2 = 2$: $-\infty \leq \Delta_2 \leq 2$; for $b_3 = 6$: $-3 \leq \Delta_3 \leq \infty$. **4.109** For $b_1 = 12$: $-3 \leq \Delta_1 \leq \infty$; for $b_2 = 3$: $-\infty \leq \Delta_2 \leq 1$. **4.110** For $b_1 = 10$: $-8 \leq \Delta_1 \leq 8$; for $b_2 = 6$: $-2.6667 \leq \Delta_2 \leq 8$; for $b_3 = 2$: $-4 \leq \Delta_3 \leq \infty$; for $b_4 = 6$: $-\infty \leq \Delta_4 \leq 8$. **4.111** For $b_1 = 20$: $-12 \leq \Delta_1 \leq \infty$; for $b_2 = 6$: $-\infty \leq \Delta_2 \leq 9$. **4.112** Infeasible problem. **4.113** For $b_1 = 0$: $-2 \leq \Delta_1 \leq 2$; for $b_2 = 2$: $-2 \leq \Delta_2 \leq \infty$. **4.114** For $c_1 = -48$: $-\infty \leq \Delta c_1 \leq 27$; for $c_2 = -28$: $-36 \leq \Delta c_2 \leq 28$. **4.115** For $c_1 = -10$: $-\infty \leq \Delta c_1 \leq 0.4$; for $c_2 = -8$:

$-0.3333 \leqq \Delta c_2 \leqq 8$. **4.116** For $c_1 = -1$: $-2 \leqq \Delta c_1 \leqq \infty$; for $c_2 = -2$: $0 \leqq \Delta c_2 \leqq$ 1.333; for $c_3 = 2$: $0 \leqq \Delta c_3 \leqq \infty$. **4.117** For $c_1 = -1$: $-1 \leqq \Delta c_1 \leqq 1.6667$; for $c_2 = -2$: $-\infty \leqq \Delta c_2 \leqq 1$. **4.118** Unbounded problem. **4.119** For $c_1 = 1$: $-\infty \leqq \Delta c_1 \leqq 3$; for $c_2 = 4$: $-3 \leqq \Delta c_2 \leqq \infty$. **4.120** For $c_1 = 1$: $-\infty \leqq \Delta c_1 \leqq 7$; for $c_2 = 4$: $-3.5 \leqq \Delta c_2 \leqq \infty$. **4.121** For $c_1 = 9$: $-5 \leqq \Delta c_1 \leqq \infty$; for $c_2 = 2$: $-9.286 \leqq c_2 \leqq$ 2.5; for $c_3 = 3$: $-\infty \leqq \Delta c_3 \leqq 13$. **4.122** For $c_1 = 5$: $-2 \leqq \Delta c_1 \leqq \infty$; for $c_2 = 4$: $-2 \leqq \Delta c_2 \leqq 2$; for $c_3 = -1$: $0 \leqq \Delta c_3 \leqq 2$; for $c_4 = 1$: $0 \leqq \Delta c_4 \leqq \infty$. **4.123** For $c_1 = -10$: $-8 \leqq \Delta c_1 \leqq 16$; for $c_2 = -18$: $-\infty \leqq \Delta c_2 \leqq 8$. **4.124** For $c_1 = 20$: $-12 \leqq \Delta c_1 \leqq \infty$; for $c_2 = -6$: $-9 \leqq \Delta c_2 \leqq \infty$. **4.125** For $c_1 = 2$: $-3.918 \leqq \Delta c_2 \leqq$ 1.153; for $c_2 = 5$: $-0.692 \leqq \Delta c_2 \leqq 39.579$; for $c_3 = -4.5$: $-\infty \leqq \Delta c_3 \leqq 7.542$; for $c_4 = 1.5$: $-3.573 \leqq \Delta c_4 \leqq 0.334$. **4.126** For $c_1 = 8$: $-8 \leqq \Delta c_1 \leqq \infty$; for $c_2 = -3$: $-4.032 \leqq \Delta c_2 \leqq 14.307$; for $c_3 = 15$: $0 \leqq \Delta c_3 \leqq 101.8667$; for $c_4 = -15$: $0 \leqq \Delta c_4 \leqq \infty$. **4.127** For $c_1 = 10$: $-\infty \leqq \Delta c_1 \leqq 20$; for $c_2 = 6$: $-4 \leqq \Delta c_2 \leqq \infty$. **4.128** For $c_1 = -2$: $-\infty \leqq \Delta c_1 \leqq 2.8$; for $c_2 = 4$: $-5 \leqq \Delta c_2 \leqq \infty$. **4.129** For $c_1 = 1$: $-\infty \leqq \Delta c_1 \leqq 7$; for $c_2 = 4$: $-\infty \leqq \Delta c_2 \leqq 0$; for $c_3 = -4$: $-\infty \leqq \Delta c_3 \leqq 0$. **4.130** For $c_1 = 3$: $-1 \leqq \Delta c_1 \leqq \infty$; for $c_2 = 2$: $-5 \leqq \Delta c_2 \leqq 1$. **4.131** For $c_1 = 3$: $-5 \leqq \Delta c_1 \leqq 1$; for $c_2 = 2$: $-0.5 \leqq \Delta c_2 \leqq \infty$. **4.132** For $c_1 = 1$: $-0.3333 \leqq \Delta c_1 \leqq$ 0.5; for $c_2 = 2$: $-\infty \leqq \Delta c_2 \leqq 0$; for $c_3 = -2$: $-1 \leqq \Delta c_3 \leqq 0$. **4.133** For $c_1 = 1$: $-1.667 \leqq \Delta c_1 \leqq 1$; for $c_2 = 2$: $-1 \leqq \Delta c_2 \leqq \infty$. **4.134** For $c_1 = 3$: $-\infty \leqq \Delta c_1 \leqq 3$; for $c_2 = 8$: $-4 \leqq \Delta c_2 \leqq 0$; for $c_3 = -8$: $-\infty \leqq \Delta c_3 \leqq 0$. **4.135** Infeasible problem. **4.136** For $c_1 = 3$: $0 \leqq \Delta c_1 \leqq \infty$; for $c_2 = -3$: $0 \leqq \Delta c_2 \leqq 6$. **4.137** 1. $\Delta f = 05$; 2. $\Delta f = 0.5$ (Bread $= 0$, milk $= 3$; $f^* = 3$); 3. $\Delta f = 0$. **4.138** 1. $\Delta f = 33.33$ (wine bottles $= 250$, whiskey bottles $= 500$, profit $= 1250$); 2. $\Delta f = 63.33$, 3. $\Delta f = 83.33$ (wine bottles $= 400$, whiskey bottles $= 400$, profit $= 1200$). **4.139** 1. $\Delta f = -100$; 2. $\Delta f = 0$; 3. no change. **4.140** 1. cost function increases by \$52.40; 2. no change; 3. cost function increases by \$11.25, $x_1^* = 0$, $x_2^* = 30$, $x_3^* = 200$, $x_4^* = 70$. **4.141** 1. $\Delta f = 0$; 2. no change; 3. $\Delta f = 1800$ ($A^* = 6$, $B^* = 0$, $C^* = 22$, $f^* = -475\,200$). **4.142** 1. $\Delta f = 0$; 2. $\Delta f = 2485.65$; 3. $\Delta f = 0$; 4. $\Delta f = 14\,033.59$; 5. $\Delta f = -162\,232.3$. **4.143** 1. $\Delta f = 0$; 2. $\Delta f = 400$; 3. $\Delta f = -375$. **4.144** 1. $x_1^* = 0$, $x_2^* = 3$, $f^* = -12$; 2. $y_1 = \frac{4}{5}$, $y_2 = 0$; 3. $-15 \leqq \Delta_1 \leqq 3$, $-6 \leqq \Delta_2 \leqq \infty$; 4. $f^* = -14.4$, $b_1 = 18$. **4.145** $y_1^* = \frac{1}{4}$, $y_2^* = \frac{5}{4}$, $y_3^* = 0$, $y_4^* = 0$, $f_d^* = 10$. **4.146** Dual problem is infeasible. **4.147** $y_1^* = 0$, $y_2^* = 2.5$, $y_3^* = 1.5$, $f_d^* = 5.5$. **4.148** $y_1^* = 0$, $y_2^* = 1.6667$, $y_3^* = 2.3333$, $f_d^* = 4.3333$. **4.149** $y_1^* = 4$, $y_2^* = 1$, $f_d^* = -18$. **4.150** $y_1^* = 1.6667$, $y_2^* = 0.6667$, $f_d^* = -4.3333$. **4.151** $y_1^* = 2$, $y_2^* = 6$, $f_d^* = -36$. **4.152** $y_1^* = 0$, $y_2^* = 5$, $f_d^* = -40$. **4.153** $y_1^* = 0.65411$, $y_2^* = 0.075\,612$, $f_d^* = 9.732\,867$. **4.154** $y_1^* = 1.335\,66$, $y_2^* = 0.44056$, $y_3^* = 0$, $y_4^* = 3.2392$, $f_d^* = -9.732\,867$. **4.155** $y_1^* = 8$, $y_2^* = 108$, $y_3^* = 0$, $y_4^* = 0$, $f_d^* = 1\,240\,000$. **4.156** $y_1^* = 0$, $y_2^* = 16$, $y_3^* = 0.4$, $y_4^* = 0$, $f_d^* = 1308$.

CHAPTER 5
NUMERICAL METHODS FOR
UNCONSTRAINED OPTIMUM DESIGN

5.2 yes. **5.3** no. **5.4** yes. **5.5** no. **5.6** no. **5.7** no. **5.8** no. **5.9** yes. **5.10** no. **5.11** no. **5.12** no. **5.13** no. **5.14** no. **5.16** $\alpha^* = 1.42850$, $f^* =$

7.713 29. **5.17** $\alpha^* = 1.427\,58$, $f^* = 7.714\,29$. **5.18** $\alpha^* = 1.386\,29$, $f^* = 0.454\,823$. **5.19 d** is descent direction; slope $= -4048$; $\alpha^* = 0.158\,72$. **5.20** $\alpha^* = 0$. **5.21** $f(\alpha) = 4.1\alpha^2 - 5\alpha - 6.5$. **5.22** $f(\alpha) = 52\alpha^2 - 52\alpha + 13$. **5.23** $f(\alpha) = (6.887\,47 \times 10^9)\alpha^4 - (3.6\,111\,744 \times 10^8)\alpha^3 + (5.809\,444 \times 10^6)\alpha^2 - 27\,844\alpha + 41$. **5.24** $f(\alpha) = 8\alpha^2 - 8\alpha + 2$. **5.25** $f(\alpha) = 18.5\alpha^2 - 85\alpha - 13.5$. **5.26** $f(\alpha) = 288\alpha^2 - 96\alpha + 8$. **5.27** $f(\alpha) = 24\alpha^2 - 24\alpha + 6$. **5.28** $f(\alpha) = 137\alpha^2 - 110\alpha + 25$. **5.29** $f(\alpha) = 8\alpha^2 - 8\alpha$. **5.30** $f(\alpha) = 16\alpha^2 - 16\alpha + 4$. **5.31** $\alpha^* = 0.61$. **5.32** $\alpha^* = 0.5$. **5.33** $\alpha^* = 3.35E-03$. **5.34** $\alpha^* = 0.5$. **5.35** $\alpha^* = 2.2973$. **5.36** $\alpha^* = 0.16\,665$. **5.37** $\alpha^* = 0.5$. **5.38** $\alpha^* = 0.40\,145$. **5.39** $\alpha^* = 0.5$. **5.40** $\alpha^* = 0.5$. **5.41** $\alpha^* = 0.6097$. **5.42** $\alpha^* = 0.4999$. **5.43** $\alpha^* = 3.45\,492E-03$. **5.44** $\alpha^* = 0.4999$. **5.45** $\alpha^* = 2.2974$. **5.46** $\alpha^* = 0.1667$. **5.47** $\alpha^* = 0.4999$. **5.48** $\alpha^* = 0.4016$. **5.49** $\alpha^* = 0.4999$. **5.50** $\alpha^* = 0.4999$. **5.51** $\alpha^* = 1.42857$, $f^* = 7.71429$. **5.52** $\alpha^* = \frac{10}{7}$, $f^* = 7.71\,429$, one iteration. **5.54** 1. $\alpha^* = \frac{13}{4}$ 2. $\alpha = 1.81\,386$ or $4.68\,614$. **5.56** $\mathbf{x}^{(2)} = (\frac{5}{2}, \frac{3}{2})$. **5.57** $\mathbf{x}^{(2)} = (0.1231, 0.0775)$. **5.58** $\mathbf{x}^{(2)} = (0.222, 0.0778)$. **5.59** $\mathbf{x}^{(2)} = (0.0230, 0.0688)$. **5.60** $\mathbf{x}^{(2)} = (0.0490, 0.0280)$. **5.61** $\mathbf{x}^{(2)} = (0.259, -0.225, 0.145)$. **5.62** $\mathbf{x}^{(2)} = (4.2680, 0.2244)$. **5.63** $\mathbf{x}^{(2)} = (3.8415, 0.48087)$. **5.64** $\mathbf{x}^{(2)} = (-1.590, 2.592)$. **5.65** $\mathbf{x}^{(2)} = (2.93529, 0.339\,76, 1.42\,879, 2.29\,679)$. **5.66** (5.56) $\mathbf{x}^* = (3.996\,096, 1.997\,073)$, $f^* = -7.99\,999$; (5.57) $\mathbf{x}^* = (0.071\,659, 0.023\,233)$, $f^* = -0.073\,633$; (5.58) $\mathbf{x}^* = (0.071\,844, -0.000\,147)$, $f^* = -0.035\,801$; (5.59) $\mathbf{x}^* = (0.000\,011, 0.023\,273)$, $f^* = -0.011\,626$; (5.60) $\mathbf{x}^* = (0.040\,028, 0.02\,501)$, $f^* = -0.0525$; (5.61) $\mathbf{x}^* = (0.006\,044, -0.005\,348, 0.002\,467)$, $f^* = 0.000\,015$; (5.62) $\mathbf{x}^* = (4.1453, 0.361\,605)$, $f^* = -1616.183\,529$; (5.63) $\mathbf{x}^* = (3.733\,563, 0.341\,142)$, $f^* = -1526.556\,493$; (5.64) $\mathbf{x}^* = (0.9\,087\,422, 0.8\,256\,927)$, $f^* = 0.008\,348$, 1000 iterations; (5.65) $\mathbf{x}^* = (0.13\,189, 0.013\,188, 0.070\,738, 0.072\,022)$, $f^* = 0.000\,409$, 1000 iterations. **5.67** $\mathbf{x}^* = (0.000\,023, 0.000\,023, 0.000\,045)$, $f_1^* = 0$, 1 iteration; $\mathbf{x}^* = (0.002\,353, 0.0, 0.000\,007)$, $f_2^* = 0.000\,006$, 99 iterations; $\mathbf{x}^* = (0.000\,003, 0.0, 0.023\,598)$, $f_3^* = (0.000\,056$, 135 iterations. **5.68** Exact gradients are: 1. $\nabla f = (119.2, 258.0)$, 2. $\nabla f = (-202, 100)$, 3. $\nabla f = (6, 16, 16)$. **5.69** $\mathbf{u} = \mathbf{c}/(2v)$, $v =$ Lagrange multiplier for the equality constraint. **5.71** $\mathbf{x}^{(2)} = (4, 2)$. **5.72** $\mathbf{x}^{(2)} = (0.07\,175, 0.02\,318)$. **5.73** $\mathbf{x}^{(2)} = (0.072, 0.0)$. **5.74** $\mathbf{x}^{(2)} = (0.0, 0.0233)$. **5.75** $\mathbf{x}^{(2)} = (0.040, 0.025)$. **5.76** $\mathbf{x}^{(2)} = (0.257, -0.229, 0.143)$. **5.77** $\mathbf{x}^{(2)} = (4.3682, 0.1742)$. **5.78** $\mathbf{x}^{(2)} = (3.7365, 0.2865)$. **5.79** $\mathbf{x}^{(2)} = (-1.592, 2.592)$. **5.80** $\mathbf{x}^{(2)} = (3.1134, 0.32\,224, 1.34\,991, 2.12\,286)$. **5.83** $\mathbf{x}^{(1)} = (4, 2)$. **5.84** $\mathbf{x}^{(1)} = (0.071\,598, 0.023\,251)$. **5.85** $\mathbf{x}^{(1)} = (0.071\,604, 0.0)$. **5.86** $\mathbf{x}^{(1)} = (0.0, 0.0232\,515)$. **5.87** $\mathbf{x}^{(1)} = (0.04, 0.025)$. **5.88** $\mathbf{x}^{(1)} = (0, 0, 0)$. **5.89** $\mathbf{x}^{(1)} = (-2.7068, 0.88\,168)$. **5.90** $\mathbf{x}^{(1)} = (3.771\,567, 0.335\,589)$. **5.91** $\mathbf{x}^{(1)} = (4.99\,913, 24.99\,085)$. **5.92** $\mathbf{x}^{(1)} = (-1.26\,859, -0.75\,973, 0.73\,141, 0.39\,833)$. **5.95** $\mathbf{x}^{(2)} = (4, 2)$. **5.96** $\mathbf{x}^{(2)} = (0.0716, 0.02\,325)$. **5.97** $\mathbf{x}^{(2)} = (0.0716, 0.0)$. **5.98** $\mathbf{x}^{(2)} = (0.0, 0.02\,325)$. **5.99** $\mathbf{x}^{(2)} = (0.04, 0.025)$. **5.100** DFP: $\mathbf{x}^{(2)} = (0.2571, -0.2286, 0.1428)$; BFGS: $\mathbf{x}^{(2)} = (0.2571, -0.2286, 0.1429)$. **5.101** DFP: $\mathbf{x}^{(2)} = (4.37\,045, 0.173\,575)$; BFGS: $\mathbf{x}^{(2)} = (4.37\,046, 0.173\,574)$. **5.102** $\mathbf{x}^{(2)} = (3.73\,707, 0.28\,550)$. **5.103** $\mathbf{x}^{(2)} = (-1.9103, -1.9078)$. **5.104** DFP: $\mathbf{x}^{(2)} = (3.11\,339, 0.32\,226, 1.34\,991, 2.12\,286)$; BFGS: $\mathbf{x}^{(2)} = (3.11\,339, 0.32\,224, 1.34\,991, 2.12\,286)$. **5.107** $x_1 = 3.7754$ mm,

$x_2 = 2.2835$.　　**5.108** $x_1 = 2.2213$ mm,　$x_2 = 1.8978$.　　**5.109** $x^* = 0.619\,084$.
5.110 $x^* = 9.424\,753$.　　　　**5.111** $x^* = 1.570\,807$.　　　　**5.112** $x^* = 1.496\,045$.
5.113 $\mathbf{x}^* = (3.667\,328,\ 0.739\,571)$.　**5.114** $\mathbf{x}^* = (4.000\,142,\ 7.999\,771)$.

CHAPTER 6
NUMERICAL METHODS FOR
CONSTRAINED OPTIMUM DESIGN

6.29 $\mathbf{x}^* = \left(\frac{5}{2}, \frac{5}{2}\right)$,　　$u^* = 1$,　　$F^* = 0.5$.　　　**6.30** $\mathbf{x}^* = (1, 1)$,　　$u^* = 0, f^* = 0$.
6.31 $\mathbf{x}^* = \left(\frac{4}{5}, \frac{3}{5}\right)$,　　$u^* = \frac{2}{5}$,　$f^* = \frac{1}{5}$.　　　**6.32** $\mathbf{x}^* = (2, 1)$,　　$u^* = 0$,　　$f^* = -3$.
6.33 $\mathbf{x}^* = (1, 2)$,　　$u^* = 0, f^* = -1$.　　　**6.34** $\mathbf{x}^* = \left(\frac{13}{6}, \frac{11}{6}\right)$,　　$v^* = -\frac{1}{6}$,　　$f^* = $
$-\frac{25}{3}$.　　**6.35** $\mathbf{x}^* = (3, 1)$,　　$v_1 = -2$,　　$v_2^* = -2, f^* = 2$.　　　**6.36** $\mathbf{x}^* = \left(\frac{48}{23}, \frac{40}{23}\right)$,
$u^* = 0,\ f^* = -\frac{192}{23}$.　**6.37** $\mathbf{x}^* = \left(\frac{5}{2}, \frac{5}{2}\right)$, $u^* = 1,\ f^* = -\frac{9}{2}$.　**6.38** $\mathbf{x}^* = \left(\frac{63}{10}, \frac{26}{15}\right)$, $u_1^* = $
$0,\ u_2^* = \frac{4}{5},\ f^* = -\frac{3547}{50}$.　**6.39** $\mathbf{x}^* = (2, 1)$, $u_1^* = 0$, $u_2^* = 2$, $f^* = -1$.　**6.40** $\mathbf{x}^* = $
$(0.241\,507,\ 0.184\,076,\ 0.574\,317)$; $\mathbf{u}^* = (0, 0, 0, 0)$, $v_1^* = -0.7599$, $f^* = 0.3799$.

CHAPTER 7
INTERACTIVE DESIGN OPTIMIZATION

7.1 Several local minima: $(0, 0)$, $(2, 0)$, $(0, 2)$, $(1.386, 1.538)$.　　**7.2** $\mathbf{d} =$
$(0, 0)$.　　　**7.3** $\mathbf{d} = \left(-\frac{21}{13}, -\frac{19}{13}\right)$,　$f(\mathbf{x}^{(1)}) = -0.00\,364$.　　　**7.4** Subproblem is
infeasible.　　**7.5** $\mathbf{d} = (-0.01\,347,\ -0.7663)$.　　**7.6** $\mathbf{x}^* = (0.76\,471,\ 1.05\,882)$,
$f(\mathbf{x}^*) = -4.05\,882$.　**7.7** $\mathbf{d} = (1.5294, -0.38\,235)$; $f(\mathbf{x}^{(1)}) = -3.73\,362$.　**7.8** $\mathbf{d} =$
$(-1.2308, -1.8462)$; $f(\mathbf{x}^{(1)}) = 1.53\,846$.　　**7.9** $\mathbf{d} = (-0.2353, -0.9412)$; $f(\mathbf{x}^{(1)}) =$
$-4.05\,822$.　**7.10** $\mathbf{d} = (-0.29\,412, -1.1765)$; $f(\mathbf{x}^{(1)}) = -3.11\,765$.

CHAPTER 8
PRACTICAL DESIGN OPTIMIZATION

8.1 For $l = 500$ mm, $d_o^* = 102.985$ mm, $d_o^*/d_i^* = 0.954\,614$, $f^* = 2.900\,453$ kg;
active constraints: shear stress and critical torque.　**8.2** For $l = 500$ mm,
$d_o^* = 102.974$ mm, $d_i^* = 98.2999$ mm, $f^* = 2.90\,017$ kg; active constraints: shear
stress and critical torque.　**8.3** For $l = 500$ mm, $R^* = 50.3202$ mm, $t^* =$
$2.33\,723$ mm, $f^* = 2.90\,044$ kg; active constraints: shear stress and critical
torque.　**8.4** $R^* = 129.184$ cm, $t^* = 2.83\,921$ cm, $f^* = 56\,380.61$ kg; active constraints: combined stress and diameter/thickness ratio.　**8.5** $d_o^* = 41.5442$ cm,
$d_i^* = 40.1821$ cm,　$f^* = 681.957$ kg;　active　constraints:　deflection　and
diameter/thickness　ratio.　　**8.6** $d_o^* = 1308.36$ mm,　$t^* = 14.2213$ mm,　$f^* =$
$92\,510.7$ N; active constraints: diameter/thickness ratio and deflection.
8.7 $H^* = 50$ cm, $D^* = 3.4228$ cm, $f^* = 6.603\,738$ kg; active constraints: buckling load and minimum height.　**8.8** $b^* = 0.5$, $h^* = 0.28\,107$, $f^* = 0.140\,536$;
active constraints: fundamental vibration frequency and lower limit on b.
8.9 $b^* = 50.4437$, $h^* = 15.0$, $t_1^* = 1.0$, $t_2^* = 0.5218$ cm, $f^* = 16\,307.2$ cm^3; active
constraints: axial stress, shear stress, upper limit on t_1 and upper limit on h.

8.10 $A_1^* = 1.4187$, $A_2^* = 2.0458$, $A_3^* = 2.9271$ in^2, $x_1^* = -4.6716$, $x_2^* = 8.9181$, $x_3^* = 4.6716$ in, $f^* = 75.3782$ in^3; active stress constraints: member 1-loading condition 3, member 2-loading condition 1, member 3-loading conditions 1 and 3. **8.11** For $\phi = \sqrt{2}$: $x_1^* = 2.4138$, $x_2^* = 3.4138$, $x_3^* = 3.4141$, $f^* = 1.2877 \times 10^{-7}$; For $\phi = 2$: $x_1^* = 2.2606$, $x_2^* = 2.8481$, $x_3^* = 2.8472$, $f^* = 8.03 \times 10^{-7}$. **8.13** d_o^* at base $= 48.6727$, d_o^* at top $= 16.7117$, $t^* = 0.797\,914$ cm, $f^* = 623.611$ kg. **8.13** d_o^* at base $= 1419$, d_o^* at top $= 956.5$, $t^* = 15.42$ mm, $f^* = 90\,894$ kg. **8.14** Outer dimension at base $= 42.6407$, outer dimension at top $= 14.6403$, $t^* = 0.699\,028$ cm, $f^* = 609.396$ kg. **8.15** Outer dimension at base $= 1243.2$, outer dimension at top $= 837.97$, $t^* = 13.513$ mm, $f^* = 88\,822.2$ kg. **8.16** $u_a = 25$: $f_1 = 1.07\,301E\text{-}06$, $f_2 = 1.83\,359E\text{-}02$, $f_3 = 24.9977$; $u_a = 35$: $f_1 = 6.88\,503E\text{-}07$, $f_2 = 1.55\,413E\text{-}02$, $f_3 = 37.8253$. **8.17** $u_a = 25$: $f_1 = 2.31\,697E\text{-}06$, $f_2 = 2.74\,712E\text{-}02$, $f_3 = 7.54\,602$; $u_a = 35$: $f_1 = 2.31\,097E\text{-}06$, $f_2 = 2.72\,567E\text{-}02$, $f_3 = 7.48359$. **8.18** $u_a = 25$: $f_1 = 1.117\,707E\text{-}06$, $f_2 = 1.52\,134E\text{-}02$, $f_3 = 19.815$, $f_4 = 3.3052E\text{-}02$; $u_a = 35$: $f_1 = 6.90\,972E\text{-}07$, $f_2 = 1.36\,872E\text{-}02$, $f_3 = 31.479$, $f_4 = 2.3974E\text{-}02$. **8.19** $f_1 = 1.12\,618E\text{-}06$, $f_2 = 1.798E\text{-}02$, $f_3 = 33.5871$, $f_4 = 0.10$. **8.20** $f_1 = 2.34\,615E\text{-}06$, $f_2 = 2.60\,131E\text{-}02$, $f_3 = 10.6663$, $f_4 = 0.10$. **8.21** $f_1 = 1.15097E\text{-}06$, $f_2 = 1.56\,229E\text{-}02$, $f_3 = 28.7509$, $f_4 = 3.2547E\text{-}02$. **8.22** $f_1 = 8.53\,536E\text{-}07$, $f_2 = 1.68\,835E\text{-}02$, $f_3 = 31.7081$, $f_4 = 0.10$. **8.23** $f_1 = 2.32\,229E\text{-}06$, $f_2 = 2.73\,706E\text{-}02$, $f_3 = 7.48\,085$, $f_4 = 0.10$. **8.24** $f_1 = 8.65\,157E\text{-}07$, $f_2 = 1.4556E\text{-}02$, $f_3 = 25.9761$, $f_4 = 2.9336E\text{-}02$. **8.25** $f_1 = 8.27\,815E\text{-}07$, $f_2 = 1.65\,336E\text{-}02$, $f_3 = 28.2732$, $f_4 = 0.10$. **8.26** $f_1 = 2.313E\text{-}06$, $f_2 = 2.723E\text{-}02$, $f_3 = 6.86\,705$, $f_4 = 0.10$. **8.27** $f_1 = 8.39\,032E\text{-}07$, $f_2 = 1.43\,298E\text{-}02$, $f_3 = 25.5695$, $f_4 = 2.9073E\text{-}02$. **8.28** $k^* = 2084.08$, $c^* = 300$ (upper limit), $f^* = 1.64\,153$.

APPENDIX A
ECONOMIC ANALYSIS

A.1 (a) $S_{216} = \$24\,004.31$, (b) $R = \$586.02$. **A.2** (a) $R = \$156.89$, (b) $P = \$2\,293.37$. **A.3** (a) $R = \$843.53$, (b) $P = \$67\,512.98$. **A.4** (a) $S_{24} = \$2\,392.83$, (b) $S_{730} = \$2\,394.38$ **A.5** $i = 0.02\,075$ (24.9% annual). **A.6** $\$855.01$. **A.7** $PW_A = \$598\,935.18$, $PW_B = \$518\,807.4$; $AC_A = \$60\,408.07$, $AC_B = \$52\,326.46$. **A.8** $PW_A = \$24 \times 10^6$, $PW_B = \$27.5 \times 10^6$. **A.9** $AC_{80} = \$53\,134.42$, $AC_{50} = \$53\,507.54$; $PW_{80} = \$343\,468.81$, $PW_{50} = \$345\,880.69$. **A.10** $PW_A = \$54\,668.58$, $PW_B = \$54\,889.53$. **A.11** $PW_A = \$101\,600.00$, $PW_B = \$102\,546.86$; $AC_A = \$14\,187.96$ ($MC_A = \$1\,118.70$), $AC_B = \$14\,409.08$ ($MC_B = \$1\,136.14$). **A.12** (a) $AC_A = -\$4\,475.76$, $AC_B = -\$5\,983.84$, $AC_C = -\$10\,459.63$; $PW_A = -\$38\,104.76$, $PW_B = -\$50\,943.86$, $PW_C = -\$89\,048.62$: (b) $A = \$213\,296.47$, $B = \$171\,977.88$, $C = \$295\,933.65$. **A.13** $PW_A = \$106\,567.11$, $PW_B = \$123\,283.56$. **A.14** $AC_A = \$21\,013.62$, $AC_B = \$24\,000.00$; $PW_A = \$104\,996.60$, $PW_B = \$119\,918.35$. **A.15** $PW_A = \$1\,186\,767.70$, $PW_B = \$1\,159\,216.60$; $AC_A = \$118\,676.77$, $AC_B = \$115\,921.66$. **A.16** $PW_A = \$76\,665.15$, $PW_B = \$100\,000.00$, $PW_C = \$54\,884.60$. **A.17** $PW_A = \$11\,928.68$, $PW_B = \$12\,708.79$, $PW_C = \$12\,392.64$.

APPENDIX B
VECTOR AND MATRIX ALGEBRA

B.1 $|A| = 1$. **B.2** $|A| = 14$. **B.3** $|A| = -20$. **B.4** $\lambda_1 = (5 - \sqrt{5})/2$, $\lambda_2 = 2$, $\lambda_3 = (5 + \sqrt{5})/2$. **B.5** $\lambda_1 = (2 - \sqrt{2})$, $\lambda_2 = 2$, $\lambda_3 = (2 + \sqrt{2})$. **B.6** $r = 4$. **B.7** $r = 4$. **B.8** $r = 4$. **B.9** $x_1 = 1$, $x_2 = 1$, $x_3 = 1$. **B.10** $x_1 = 1$, $x_2 = 1$, $x_3 = 1$. **B.11** $x_1 = 1$, $x_2 = 2$, $x_3 = 3$. **B.12** $x_1 = 1$, $x_2 = 1$, $x_3 = 1$, $x_4 = 1$. **B.13** $x_1 = 1$, $x_2 = 2$, $x_3 = 3$. **B.14** $x_1 = 2$, $x_2 = 1$, $x_3 = 1$. **B.15** $x_1 = 1$, $x_2 = 2$, $x_3 = 3$. **B.16** $x_1 = 1$, $x_2 = 1$, $x_3 = 1$. **B.17** $x_1 = 2$, $x_2 = 1$, $x_3 = 1$, $x_4 = -2$. **B.18** $x_1 = 6$, $x_2 = -15$, $x_3 = -1$, $x_4 = 9$. **B.19** $x_1 = (3 - 7x_3 - 2x_4)/4$, $x_2 = (-1 + x_3 - 2x_4)/4$. **B.20** $x_1 = (4 - x_3)$, $x_2 = 2$, $x_4 = 4$. **B.21** $x_1 = (-3 - 4x_4)$, $x_2 = (-2 - 3x_4)$, $x_3 = x_4$. **B.22** $x_1 = -x_4$, $x_2 = (2 + x_4)$, $x_3 = (-2 - x_4)$. **B.23** $x_1 = 4 + (2x_2 - 8x_3 - 5x_4 + 2x_5)/3$; $x_6 = (9 - x_2 - 2x_3 - x_4 - 3x_5)/2$; $x_7 = 14 - (13x_2 + 14x_3 + 8x_4 + 4x_5)/3$; $x_8 = (9 + x_2 - 16x_3 - 7x_4 + 19x_5)/6$. **B.24** Linearly dependent. **B.25** Linearly independent. **B.26** $\lambda_1 = (3 - 2\sqrt{2})$, $\lambda_2 = (3 + 2\sqrt{2})$. **B.27** $\lambda_1 = (3 - \sqrt{5})$, $\lambda_2 = (3 + \sqrt{5})$. **B.28** $\lambda_1 = (5 - \sqrt{13})/2$, $\lambda_2 = (5 + \sqrt{13})/2$, $\lambda_3 = 5$. **B.29** $\lambda_1 = (1 - \sqrt{2})$, $\lambda_2 = 1$, $\lambda_3 = (1 + \sqrt{2})$. **B.30** $\lambda_1 = 0$, $\lambda_2 = (3 - \sqrt{5})$, $\lambda_3 = (3 + \sqrt{5})$.

APPENDIX C
A NUMERICAL METHOD FOR
SOLUTION OF NONLINEAR EQUATIONS

C.1 $x^{(3)} = 0.6190$. **C.2** $x^{(2)} = 9.4249$. **C.3** $x^{(2)} = 1.5708$. **C.4** $x^{(3)} = -1.4958$. **C.5** $x^{(2)} = (3.788\ 42,\ 0.6545)$. **C.6** $x^{(2)} = (4.10321,\ 8.14637)$. **C.7** $x^{(2)} = (-3.3738,\ 0.0360)$. **C.8** $x = (3.6840,\ 0.7368)$. **C.9** $x = (-4,\ -8)$; $x = (4,\ 8)$. **C.10** $x = (-3.3315,\ 0.03949)$; $x = (-0.3980,\ 0.5404)$.

INDEX